ECOLOGY

PAUL COLINVAUX
The Ohio State University

JOHN WILEY & SONS
New York Chichester Brisbane Toronto Singapore

Production Supervisor Pamela A. Pelton
Cover design Sheila Granda
Cover photo Jacques Jangoux
Photo Editor Linda Guttierrez
Photo Researcher Dana Dolan
Manuscript Editor John Thomas
under the supervision of Bruce Safford

Library of Congress Cataloging in Publication Data:

Colinvaux, Paul A., 1930–
Ecology.

Includes indexes.
1. Ecology I. Title.
QH541.C627 1986 574.5 85-6441
ISBN 0-471-16502-6

Printed in the United States of America

10 9 8 7 6 5 4 3 2 1

PREFACE

"Ecology" is designed to be a teaching tool.

My earlier book, *Introduction to Ecology,* was written from 1968 to 1971 to "introduce" the new ideas of evolutionary ecology that occupied us in the sixties but that had not yet permeated textbooks (Orians, 1973). Since then ecology has come to near-maturity. When moral suasion from my publisher eventually put me to textbook writing once again, I decided to abandon that "introduction," compounded as it was of thoughts from the 1960s, a historical approach, and enthusiastic revisionism. Writing from 1981 to 1984, I set out to review all the basic parts of modern ecology in a way suitable for teaching. My object has been both to offer readable reviews and at the same time to give students a manual of ecological information that may add to material covered in lecture courses.

"Ecology" goes from the individual to the community—the quasi levels of integration approach. This arrangement requires less repetition than the reductionist technique of starting with communities and thus working from the big to the small. I now usually give my lectures in integration order also, though sometimes I choose to begin with ecosystems and work backwards for variety. This book can be used either way by altering the order in which chapters are assigned.

I begin the book with a review of ecological concepts of which modern university students certainly will have heard before starting the course. The concepts of niche, ecological pyramids, succession, energetics, and limiting factors are almost part of the vernacular, although possibly not correctly used. Research with my own propagules, for example, shows that something called the "pyramid of life" enters the school system as early as the third grade. Accordingly, I review these ideas at the start, providing definitions and discussions that might otherwise have been deferred until later in the text.

The beginning chapter also includes as succinct a statement of the principle of natural selection, with its companion concept of fitness, as I can manage. After this beginning I found that most of the subdisciplines of ecology fell into a natural order: individual—population and species—community.

Ecosystem processes, however, do not necessarily follow communities as a fourth level of integration. In fact, physical processes in habitats must be studied before the integration of species into communities can be understood. I recognize this truth by discussing ecosystem processes in Part Three of the book, before going to community synthesis in Part Four. Thus my levels of integration approach puts ecosystems before communities.

The chapters of the book can easily be rearranged for use in a course that begins with ecosystem function as follows:

Chapters 1 & 2	Overview and Ecosystem Energetics
14–21	Ecosystem Processes
11	Species Strategies
12	Social Systems
22–25	Community Building
26 (pp. 679–683)	Ecosystem Stability
6–10 & 13	Population Ecology
3–5	Individual Adaptation
26 (pp. 650–674)	Species Diversity

Table I offers reading suggestions for shorter courses. When time limits the material that can be covered it seems to me that the coverage of the text should be as extensive as possible, so that the inquiring student can put the shortened material in a proper context. Instructors design short courses around their own perceptions of what is most useful. But Table I lists subjects most ecologists will think fundamental.

Table I

Short Course: Levels of Integration Approach	
Chapters 1 & 2	Introduction and Overview
3 (pp. 46–53)	Efficiency of Photosynthesis
4 (pp. 74–82)	Consumer Efficiency
6 & 7	Competition and Speciation
10 (in part)	Review of Predation
11	Species Strategies
12	Social Systems
14	Biomes
16 (pp. 392–406)	Production Ecology
17	Review of Ecosystem Process
23	Ecological Succession
24–26	Community Building, Diversity, and Stability

Short Course: Reduction from Ecosystem Approach	
Chapters 1 & 2	Overview and Ecosystem Energetics
14	Biomes
3 (pp. 46–53)	Efficiency of Photosynthesis
16 (pp. 392–406)	Production Ecology
17	Review of Ecosystem Process
11	Species Strategies
12	Social Systems
23	Ecological Succession
24–25	Community Building
26 (pp. 679–683)	Ecosystem Stability
6 & 7	Competition and Speciation
10 (in part)	Review of Predation
5	Individual Adaptation
26 (pp. 650–674)	Ecological Diversity

I have avoided chapters or passages explicitly about environmental issues. Acid precipitation, for instance, is not reviewed, nor is there a statement on population problems. I am not untouched by these matters having, in fact, written a book on human population and history (Colinvaux, 1980). But I decided that a thorough survey of the basics of ecology was of value in its own right and should be treated as such in a university course. The essential background data contributed by ecology to environmental debate lie in these basics and will be found throughout the book. An extensive discussion of the atmospheric carbon cycle and the possibility of enriching the atmosphere with anthropogenic carbon dioxide will be found in Chapter 19 (Maintenance of the Air). Eutrophication is discussed at length in Chapter 21 (Limnology). The effects of clear-cutting are discussed in the contexts of the Hubbard Brook watershed study in Chapter 17. Productive limits of the earth are assessed in Chapters 3 and 16. A review of what is meant by ecosystem stability is given in Chapter 26, and so on. The reading schedule given in Table II could accompany a course for which environmental issues were an organizing theme.

Some subjects are given more extended treatment than in other texts. Biogeochemistry is one of these, particularly the maintenance of the ocean as a solution of sodium chloride and of the air as an oxygen–nitrogen mixture (Chapters 18 and 19). Another subject worthy of ecological notice is modern progress in soil classification and in the understanding of soil genesis (Chapter 20). In a one-quarter course these subjects have to be treated only briefly in lecture, but should be accessible to students.

Paleoecology, however, deserves a more central part in ecological teaching. Some of the more important hypothesis testing in ecology can be by appeal to the fossil record, particularly hypotheses about community development. In Chapter 22 I have concentrated on the use of Quaternary records by pollen analysis and paleolimnology, the record of the ice-age earth, and the reconstruction of community changes in the Holocene that led to present patterns of distribution and abundance.

Table II
Readings Organized Around an Environmental Theme

Chapter	Topic
Chapter 14	Climate and biomes.
19	Maintenance of the air. The carbon dioxide enrichment problem.
17	Habitat steady states. Watershed studies. Nutrients and fertility of terrestrial ecosystems, clear-cutting.
21 (pp. 506–516)	Water pollution and eutrophication.
2	Limiting factors, niche, ecosystem energetics.
16	Productivity. Food limits to the earth.
20 (pp. 490–499)	Soils of temperate and tropic regions.
5	Individual adaptations and optimal foraging.
6–13	Population ecology.
23–25	Succession and community building.
26	Diversity and stability of ecosystems.

My long Chapter 21 on limnology perhaps needs defense beyond the statement that it is one of my research areas. Life in aquatic systems often is starkly different from life on land. Community structure in water is most strongly dependent on predation, as predators hunt prey through lighted spaces, and plants are the smallest prey of all. Problems of adaptation or dispersal are quite different in water, and we must stretch our understanding to realize conditions for microscopic life when Reynolds numbers are low. The limnology chapter may seem long, but even this treatment is highly condensed compared with the treatment of terrestrial habitats in this and other textbooks. Far from being defensive about a whole long chapter describing lakes as ecosystems, I feel remorse at having not written a companion piece on oceans.

The book is deliberately not strong on statistical techniques. Elaboration of ordination methods, and of multivariate analyses, does not to my mind make a book or a lecture course more sophisticated; merely uninteresting to the average student. Likewise I have introduced equations only when they are vital to the argument. Words are still our best medium of communication.

In attempting a complete survey of ecology, it has been necessary to include subjects for which I have little enthusiasm, or which are coming under increasing criticism. An example is the Shannon–Wiener information statistic as a measure of diversity. I have disliked and suspected this measure since my first contact with it in the late fifties. My 1973 book expressed my doubts that it told us anything about the stability of ecosystems. One of the first formal criticisms of the measure in stability studies was done in my laboratory (Goodman, 1974). But the measure remains important to ecology for the way it was used; the old controversy about complexity and stability would not have come about without the use of this measure. Thus it needs to be covered in general texts.

Concepts like character displacement, or *r*- and *K*-selection, have their detractors, myself among them on occasion, but they need to be learned before they can be criticized. Thorough historical accounts of these concepts are essential to an understanding of modern ecology. Some of the older ideas of succession theory may seem archaic, but knowledge of them is essential to understanding of present-day attitudes. For these subjects some history is needed, even though history is always, in a sense, archaic.

Ecological succession is a subject over which consensus has changed completely in the last decade. When I wrote "Succession Revisited" in the 1973 book, I felt I was a revolutionary. Contemporary wisdom had it that succession was the ecosystem process *par excellence,* and various writers experimented with ideas of succession maximizing information or order. To submit that the essential process was no more than the inevitable replacement of opportunists with equilibrium species as I did seemed daring in a textbook. When Drury and Nisbet (1973) independently published this same view of succession they noted in their acknowledgments, "We thank three anonymous reviewers whose comments made clear to us that the traditional view of succession is alive and well among our peers." But now the strategic view of succession has become the new conventional wisdom, and the ecosystem view takes second place. Yet ecosystem changes in succession can be directional and are of interest. Even Clements' old remarks about "superorganisms" need to be understood in order to understand the latest group selection arguments of D. S. Wilson (1982) who actually uses the term "superorganism." I have tried, within the limits of space, to review these different contributions to succession theory in Chapter 23, not just to expound the latest consensus.

Proponents of group selection may call me conservative. The possibilities of structured demes are reviewed in pages 643–644, and of superpredators in pages 301–302. Group evolution is an attractive explanation for curious communities like those surrounding beetles in dung or under bark. But simple natural selection can explain so much more, or many things so easily, that conservatism seems proper in a textbook.

I have curbed my temptations to literary expression, while attempting to retain clarity and readability. Orians (1973) said of my *Introduction to Ecology* that it was "full of quotable quotes," apparently intending that this should be taken as praise for the book. Others found quotable language unsuitable in a science text. In this book I have removed the more colorful products of first draft writing. An attempt to describe ecology in more literary language can be found in my *Why Big Fierce Animals Are Rare* (Colinvaux, 1978).

Columbus, Ohio **Paul Colinvaux**

ACKNOWLEDGMENTS

This book was written after twenty years of teaching ecology and a quarter of a century of researching it. Inevitably I have borrowed from the minds of more than two generations of ecologists, my own and my teachers' generation. Most important to the formation of my thinking have been D.A. Livingstone, E.S. Deevey, and G.E. Hutchinson; my gratitude to the happenstance that placed me in such intellectual descent is not easily expressed. R.H. Whittaker served as role model and counsellor when I wrote my first text. Other teachers whose thoughts became embedded in my own were P.H. Klopfer, K. Schmidt-Nielsen, H.J. Oosting, and H.G. Andrewartha. The first draft of this book had the benefit of detailed reviews by J. Webster, D.C. Coleman, and D. Johnston. Many other reviews of this manuscript and my earlier writings have helped greatly, including those by A.J. Brook, N. Stanton, J.A. MacMahon, F.B. Golley, C.A.S. Hall, H.S. Horn, and P.D. Moore. People who have contributed in various ways on the road to an ecology text include M. Acosta-Solis, T. Ager, M.G. Barbour, E.S. Barghorn, B. Barnett, M. Bergstrom, W.D. Billings, J. Blackwelder, L.C. Bliss, S. Bolotin, I. Brodniewicz, W.S. Broecker, J.L. Brooks, J. Brown, L. Brown, L.B. Brubaker, J.B. Calhoun, R.A. Carpenter, R.G. Cates, P. Chesson, P.D. Colbaugh, J.O. Connell, J.S. Creager, M. Cunningham, E.J. Cushing, M.B. Davis, D. Deneck, T. deVries, J. Doherty, J.F. Downhower, G.M. Dunnett, W.T. Edmondson, I. Eible-Eibesfeldt, M. Ewing, R.S.R. Fitter, R.E. Flint, M. Florin, D. Frey, I. Frost, A.S. Gaunt, Z.M. Gliwicz, H. Godwin, R.P. Goldthwaite, T. Goreau, C.E. Goulden, P.R. Grant, D.H. Greegor, A.T. Grove, T. Grubb, J.P. Hailman, B. Hajek, M.P. Harris, J. Hatch, C.J. Heusser, H. Higuchi, D.M. Hopkins, S.P. Hubbel, H.W. Hunt, P.L. Johnson, D.E. Johnston, W.S.L. Laughlin, E.P. Leopold, W.M. Lewis, M. Lieberman, Kambiu Liu, S. Longenbaker, O.L. Loucks, R.H. MacArthur, P.S. Martin, R.M. May, R. McIntosh, J.H. Mercer, M.C. Miller, R.D. Mitchell, W.N. Mode, H. Nichols, H.C. Noltimier, W.J. O'Brien, K. Olson, G.H. Orians, F. Ortiz, L. Parrish, R. Patrick, R. Perry, T.J. Peterle, G.M. Peterson, K.G. Porter, F.W. Preston, C. Racine, J.L. Richardson, J.E. Richey, M.R. Rutter, E.K. Schofield, K. Schmidt-Koenig, J. Sedell, J. Shackleton, W.M. Shields, N.A. Shilo, A. Sih, C. Smith, F.E. Smith, G. Sprugel, H. and T. Steinitz, J.R. Strickler, M. Stuiver, J.C.F. Tedrow, D. Tilman, M. Tsukada, F.C. Ugolini, J. Vagvolgyi, L. Van Valen, A.L. Washburn, M. Watanabe, W.A. Watts, D.S. Webb, F.H. West, G.W. Wharton, T.C.R. White, D.R. Whitehead, T.G. Whitham, D.T. Wicklow, R.G. Wiegert, I.L. Wiggins, E.O. Wilson, W.C. Wimsatt, and H.E. Wright. Most advanced training of professors is done by their doctoral students, and I have been trained in this way by P.D. Boersma, D. Goodman, M. Steinitz-Kannan, and D. Maxwell. My colleague L. Hillis-Colinvaux is my companion in ecology as in life, and her imprint is on all my work.

CONTENTS

PART ONE

ENERGY AND THE INDIVIDUAL

PREVIEW TO CHAPTER 1

Ecology is the science that seeks to understand the distribution and abundance of life on earth. It is both an environmental science and an evolutionary science, since it works to discover the ways in which environmental resources are divided among individuals of different species. In this process species are forged and kept distinct, males are separated from females, and numbers are so regulated that the common stay common and the rare stay rare. Evolutionary ecology constantly tests the hypothesis that every individual organism acts to leave behind the largest possible number of surviving offspring, and our measure of an organism's success at doing this we call "fitness." Fitness may be increased by more births, better survival of the young, or when close relatives survive to transfer copies of an individual's genes to the next generation. Patterns as different as foraging behavior or systems of sex can be shown to maximize fitness for the individuals using them. The seeming paradox that elegant designs of plants and animals result from basically random processes is resolved because selection works non-randomly on the endless variety provided by genetic recombination in sexual systems. A greater paradox of apparent design is the way communities appear to work as complex entities, but this paradox also is resolved by ecological theory. All community members are fitted by selection to the shared reality of a physical habitat, and to the presence of each other, giving a spurious impression of true community design. Individual plants in communities act selfishly to maximize fitness, but the summed effects of plants in vegetation impose structure on habitat and community alike. Replacement series of colonizing plants, called "ecological successions," proceed as soil forms and nutrients collect, but this appearance of increasing order during succession also can be understood as a consequence of individuals behaving in ways likely to maximize immediate fitness. It was the discovery of how plant communities are regulated by habitats which were themselves influenced by vegetation that was recognized when the term "ecosystem" was first coined. A basic hypothesis of community ecology is thus that natural selection coadapts individuals of different lineages to shared environmental constraints and to the presence of each other. Ecology explores this process of adaptation, finding in it the mechanism that fashions species in the first place. But this study requires that physical processes in the habitat be understood also. Physical process in the environment fashions habitats, setting the ground rules in which the game of speciation and community building is played. Ecologists study these ground rules. Necessarily this takes them all the way from studies of process in habitats to the physics and chemistry of the biosphere, as they seek to understand what regulates the composition of the air, the salinity of the oceans, or the working of the weather.

CHAPTER 1

OVERVIEW: THE ECOLOGICAL INQUIRY

The most widely used definition of ecology is THE STUDY OF ANIMALS AND PLANTS IN RELATION TO THEIR HABITS AND HABITATS (Elton, 1927). This describes ecology as the study of how the ways of life of a place depend upon the local environment and how the environment is changed by life. It is this emphasis on environment that has given ecology its popular image of an environmental science. Yet ecology is more than an environmental science.

Ecology is also a branch of evolutionary biology. It seeks to explain how many different kinds of plants and animals can live together in the same place for many generations. Animals and plants share habitat. Sometimes they can only share for so long before some locally go extinct, but there are other circumstances when many different kinds persist in a habitat indefinitely.

In the more livable parts of the earth, in what an ecologist would call A MESIC HABITAT, there are always many species living together. A prime question of evolutionary ecology is how it came about that so many species should be present to coexist. Consider the plant species of a thickly vegetated place like a meadow. Many kinds of meadow plant get their energy from the sun by photosynthesis, their carbon from the air, and their water and minerals from the soil that they share. Evolutionary biology concludes that these many different species are the product of selection of the best adapted individuals. But the meadow shows that selection has produced many different solutions to the problem of living in a meadow. It is necessary for evolutionary biology to explain why there should be so many different kinds of plant in a meadow rather than one or a few perfectly adapted species. This is an ecological question as well as an evolutionary question, since the answers must be found in the ways in which the many species share the resources of the habitat.

A related purpose of ecology is to explain how, or to what extent, populations are regulated in nature. Wild populations certainly fluctuate but usually only within perceived limits. This fundamental fact is revealed most clearly with the everyday observation that some species are common but others rare. The common stay relatively common and the rare stay relatively rare over many consecutive generations. RELATIVE ABUNDANCE, as well as total population, therefore, must be under control.

Allied to the problem of population regulation is that of species distribution. A species that may be rare in one place often turns out to be common in another. It follows that any attempt to identify those interactions between species or the environment that regulate number must also identify processes that limit distribution. For this reason ecology is sometimes defined as THE STUDY OF THE DISTRIBUTION AND ABUNDANCE OF SPECIES.

The word "ECOLOGY" is taken from the Greek word "oikos," which means "home," or "house," or "household," or something like that, and "logos," which means "knowledge." Ecology, therefore, literally means "the study of the household," it being implied that the "household of nature" is meant. Haeckel (1866), who is generally credited with coining the word ecology more than a century ago, defined it as "the domestic side of organic life." The modern subject includes all phenomena that result from the interaction of organisms with their environment and with other organisms. Among these phenomena are the dispersal of species, the speciation mechanism itself, population regulation, life in communities, modification of the habitat, soil formation, nutrient cycling, the sources of energy used by organisms, the efficiency with which organisms use resources or energy, and the maintenance of the earth as a life support system. Some of the questions that ecologists try to answer are given in Table 1.1.

NATURAL SELECTION AND FITNESS IN ECOLOGY

Modern ecology developed after the discovery of the process of evolution by natural selection. Natural selection works by destruction; it kills individuals or it stops individuals from breeding. Change comes about because the culling process of natural selection hits some varieties harder than others. It is this endless sifting out of arrays of chance or contrived variety that ensures that existing species are suited to the environment in which they live.

An inevitable consequence of the process of natural selection is that all individuals of all living species must breed to the uttermost. Success is measured by offspring thrust into the next generation through the meshes of natural selection's net. This means that the more eggs or young an individual makes, or the more effort it puts into care of young, the more chances there are of its having survivors. The outcome of natural selection depends not only on the rate at which individuals are removed (selected against) but also on the rate at which their replacements are made (reproduction). A good working definition of the process may be written NATURAL SELECTION IS THE DIFFERENTIAL REPRODUCTION AND SURVIVAL OF INDIVIDUALS CARRYING ALTERNATIVE INHERITED TRAITS.

Charles Darwin spoke of natural selection as promoting "the survival of the fittest." He did not, by "fit," mean those who did calisthenics, or those who were clever fighters or big bullies. Being "fit" in a Darwinian world simply means escaping removal by untimely death or from failing to reproduce more successfully than others do. Fitness is a measure of success at both survival and reproduction.

Both geneticists and ecologists talk of fitness but each adapts the concept to special interests. Geneticists think of fitness in terms of *gene frequencies* and give definitions like *fitness is a measure of the relative change in the frequency of an allele owing to selection* (Valentine and Campbell, 1975). But ecologists follow more closely to Darwin's usage and think of numbers of surviving offspring rather than surviving genes. Ecologists may define FITNESS as THE INDIVIDUAL'S RELATIVE CONTRIBUTION OF PROGENY TO THE POPULATION. Fitness so defined is easy to measure, being simply the number of offspring that themselves live to reproductive age. This meaning of the word "fitness" is not the meaning of everyday speech. Ecological fitness is described as a number. If the Joneses raise four children, all of whom grow up to marry and have children of their own, then Mr. and Mrs. Jones each have a fitness of 4 (or 2 if the context suggests allowing for the fact that each parent has only a half interest in each child).

Fitness is sometimes also conveniently measured as the number of copies of a gene that appears in the next generation, regardless of which organism carries them. This is INCLUSIVE FITNESS. The usefulness of the definition is that it

Table 1.1
The Major Questions of Ecology
Ecology is an analytical, question-asking science. The questions in this list include aspects of all the major problems studied by ecologists. It should be possible to answer all these questions by application of the physical sciences and the principle of evolution by natural selection.

QUESTIONS OF DIVERSITY
Why do living things exist as discrete species?
Why are there so many different species of plants and animals?
Why do so many species divide their resources between males and females?
Why are there more species in some places than others?
Why are some species common, whereas others are rare?

QUESTIONS OF EVERY SPECIES
Why are individuals shaped the way they are?
Why do individuals do the things they do?

QUESTIONS OF BALANCE
Why do the common stay common, whereas the rare stay rare?
Are natural populations normally controlled in systematic ways?
Do predators control their prey?
How often do diseases or hunger regulate populations?
Is it possible for natural populations to have self-regulating mechanisms?
What causes "population explosions" (irruptions, plagues, epidemics)?
Are complex communities of many species more stable than simple communities of few species?

QUESTIONS OF ORGANIZATION
Why are maps of vegetation, soils, and climate similar?
Can communities be identified and described as if they were Linnaean species?
Why is the destruction of mature vegetation nearly always followed by an ecological succession of plant communities, seldom by direct regeneration of the original vegetation?
Why does the pattern of change in natural communities repeat itself to the extent that the course of succession is predictable?
Is the association of many species in a natural area the result of chance, of physical circumstances, or of an organic organizing process?
Are complex communities more efficient at using energy and raw materials than simple communities?

takes into account the copies of genes that may be carried into the next generation through relatives, notably brothers and sisters, called SIBLINGS.

The concept of inclusive fitness allows the parallel concept of KIN SELECTION as an explanation of what would seem to be otherwise altruistic acts. Any action by an individual that helps the relatives of that individual survive and reproduce necessarily gives some reward in *inclusive fitness* because those relatives carry copies of that individual's genes. The closer the relationship, then the higher the probability of sharing genes. Hamilton (1964) showed that an act of helping relatives at cost to the helper (apparent altruism) can be selected for if the gain in inclusive fitness resulting from relatedness outweighs the apparent costs in fitness of the act to the altruist itself.

The concepts of kin selection and inclusive fitness satisfactorily explain social systems where close relatives help care for offspring in circum-

QUESTIONS OF LIFE FORM

Why are there no trees in the arctic?
Why are trees evergreen in equatorial and northern latitudes, but deciduous at in-between latitudes?
Why are the plants of open water microscopic?
Why do some trees have round leaves but others have needles?
Why does a grass have spear-shaped leaves?

QUESTIONS OF PHYSICAL PATTERN

How is the ionic composition of the oceans maintained?
How is the gaseous mixture of the atmosphere maintained?
What proportion of solar energy flux is used to drive the living components of ecosystems?
Why are soils of temperate regions brown, whereas those of the tropics are red?
What sets the limit to the mass of living tissue on the surface of the earth?
What sets the limit to the energy available to life?

QUESTIONS FOR STUDENTS OF THE HUMAN CONDITION

Why does the western-style agriculture of monoculture yield more food than have any attempts at manipulating the original complex wild vegetation?
Why does western agriculture work less well in the tropics?
What sets the upper limits to food production?
How much extra food could be taken from the sea and what would set the eventual limit?
Is life on earth at risk from contamination with radionuclides or novel chemicals?
Can pollution kill a lake?
What will set the limit to the energy flux that societies of the future will be able to release?
Why do human populations continue to grow?
What influence do population growth and human ecological need have on the structure of society and the events of history?
What changes may people expect in themselves from living at high densities?
Can we trigger an ice age, and does it matter?
Is the atmosphere at risk?

stances when they do not breed themselves, particularly if their chance of inclusive fitness is higher in a cooperative breeding effort than if they attempted a family of their own with small chances of success. The concepts are also used to explain sterile castes in insects, or various patterns of warning behavior in birds or mammals that seem to put the warning individual at risk.

An object of evolutionary ecology is to try to explain the grand patterns of the distribution and abundance of life on earth as the outcome of a game of reproduction and survival played on a gaming-board made up of the physical earth. Each roll of the reproductive dice yields new varieties, and natural selection is a universal obstacle that delays some individuals more than others in the reproductive race. The game proceeds at different speeds at different times and different places, depending on the local environment. Ecologists try to understand the rules of the game, asking what gives fitness in different places or to different individuals, and why these local rules have led to the patterns of distribution and abundance of the contemporary earth.

COMMUNITIES, ECOSYSTEMS, AND ENERGETICS

Ecology used to be subdivided into AUTECOLOGY, the study of single-species populations, and SYNECOLOGY, the study of communities, though these terms are less popular now. Typical of *autecology* are studies of population regulation, or the adaptation of individuals to the environment. Even these studies of individuals lead eventually to consideration of the whole community, since other species are components of the environment. *Synecology,* the deliberate study of the community as an entity, had particular importance in the history of ecology because it led to the concept of the ECOSYSTEM.

The concept of the ecosystem was a direct outcome of intensive work by more than a generation of botanists studying whole plant communities. It was concluded that the plant community could not usefully be studied except in relationship to the total environment of the place where the community was found (Chapter 15). The areal extent of a community was set by such physical factors as soil or drainage so that the plants were linked together by their common needs and their common resources. The animals that lived on or with the plants also had great importance in shaping the plant community. The proper unit for study, therefore, was one that included plants, animals, and the physical environment. This unit was termed the ecosystem (Tansley, 1935).

The ecosystem concept has been important in developing theories of COMMUNITY ENERGETICS, since energy is needed to drive a system. That life is lived in ecosystems driven by fluxes of solar energy has become a central unifying concept of ecology. The stream of sunlight in which the earth spins is the prime energy source. Plants are the energy transformers that convert light energy into forms usable by organisms. They make carbohydrate fuels in photosynthesis and then use these fuels in the work of living. Some of the carbohydrate fuel is eaten by animals and passed from mouth to mouth along FOOD CHAINS, but more still goes to DECOMPOSERS, which are energetically very important to the functioning of most ecosystems (Chapter 16). This systems view of life is now widely understood by the general public and not just by ecologists.

Essential limits to life on earth were revealed by applying the first and second laws of thermodynamics to the ecosystem concept. The FIRST LAW OF THERMODYNAMICS states that when energy is converted from one form into another no energy is either gained or lost. This first law is also called the "law of conservation of energy" and is, of course, strictly true only if matter is considered a form of energy. The SECOND LAW OF THERMODYNAMICS states that every energy transformation results in a reduction of the free energy of the system. In the language of physicists this law states that all energy transformations result in an increase in ENTROPY, requiring that only a fraction of the energy conserved under the first law is available to do useful work within the system.

Applying the two laws of thermodynamics to the ecosystem concept leads to elementary generalizations important to the development of ecology. Most ecosystems continue to function only as long as usable energy is supplied as direct sunlight. A few ecosystems, like the floors of the deep oceans or in caves, depend on continued inputs of carbohydrate fuels supplied by export from other ecosystems under the sun. Predators should be less numerous than their prey because the movement of food along food chains represents energy transformations that cannot, according to the second law, be 100% efficient. Accordingly, large cats tend to be solitary and rare (Figure 1.1).

ECOLOGICAL ENERGETICS also is a subject of autecology, as energy budgets are constructed for individuals, with calories allocated to maintenance, growth, and reproduction. An animal or plant is viewed as an energy-getting machine. Its fitness is measured as a function of energy used for successful reproduction. This approach has the powerful advantage that life

Figure 1.1 *Felis concolor*: large predators are rare. This one, a mountain lion, was apparently treed by dogs. Mountain lions also are solitary with large home ranges needed to support each lion (Chapter 12).

processes can be measured in the common currency of calories.

The ecosystem approach also leads directly back to fundamental Darwinian problems of the number of species, because ecosystems are made of numbers of moving parts that seem, at first sight, quite extraordinary. We are used to the idea of something functional having the smallest possible number of moving parts. Keep it simple and it works, make it too complex and it is bound to go wrong. And yet, the number of individual animals and plants, which are an ecosystem's moving parts, is always very large. It is true that many of the individual animals and plants are duplicates, members of single-species populations, so that they may share a single function within the system. But there are also what appear to be quite unreasonably large numbers of different species as well (Table 1.2). The systems analyst's view of the world, therefore, brings back the fundamental ecological question "Why are there so many species?"

An ecosystem scientist must ask if this multitude of species is necessary to the smooth running of the system. Does it promote homeostasis? Are all the species precious, or is there great redundancy in the workings of natural systems that would allow some of these many species to be discarded without serious loss? It has been suggested that many species promote stability, an idea used by political activists claiming to use ecology as the basis for their arguments, as in the document *Blueprint for Survival* (Goldsmith *et al.,* 1972), where it is argued that dire consequences must follow the continued replacement of wild systems by those simpler ones of farm and city.

This general hypothesis that *species complexity leads to population stability* has now been the subject of many investigations by mathematical modelling and ecosystem simulation and the balance of opinion among theoretical ecologists is that the hypothesis fails (May, 1973; Goodman, 1975; Chapter 26). A large number of species in a community might in fact make the system unstable, making the community "fragile" as some ecologists think the tropical rain forest communities to be (Farnworth and Golley, 1973).

The relationship between complexity and stability in ecological systems cannot be examined separately from questions of environmental stability. Populations often fluctuate as a result of large-amplitude changes in such physical parameters as the weather, and these physical changes may be more important to stability than the complexity of the community itself.

Studies of energy flow into ecosystems suggest that the ultimate control of stability is nonliving, because it turns out that more energy does work in ecosystems through inanimate physical process than is at the disposal of the living inhabitants. In a temperate forest, for instance, about 98% of the incident solar energy is degraded in physical processes and only some 2% is trapped by green plants in photosynthesis (Chapters 3 and 16). This calculation suggests that 98% of the energy available to perturb an ecosystem or to keep it stable is not under control of the biota. Some of the excess physical energy actually disrupts ecosystems in storms (Figure 1.2).

Table 1.2
Living Species of Plants and Animals
These are approximate numbers culled from various taxonomic sources (plants from Jensen and Salisbury, 1972; animals compiled from Villee et al., 1973, and Orr, 1971). The actual number of living species must be greater than this. Taxonomic authorities generally believe that less than 1% of bird and mammal species remain to be described, but it is likely that a million more species of arthropods actually exist. Heterotrophic Protozoa are not included in this table. Possibly free-living flagellates, ciliates, and sarcodina number about 1000 species, but parasitic Protozoa might be almost as diverse as their possible hosts. (Pennak, 1953, suggests that at least 20,000 species of Protozoa have been described.)

Plants		
Bacteria	Bacteria and Fungi	1,600
Slime molds		500
		100,000
Blue-green algae	Algae	1,500
Euglenids		450
Golden algae, diatoms		425
Yellow-green algae		5,800
Dinoflagellates		1,000
Brown algae		1,500
Red algae		4,000
Green algae		7,000
Liverworts	Bryophytes	
Mosses		23,000
Hornworts		
Psilopods	Ferns and Allies	3
Lycopods		1,200
Horsetails		40
Ferns		10,000
Cycads	Gymnosperms	100
Ginkgos		1
Conifers		520
Gnetum		44
Dicots	Angiosperms	236,500
Monocots		48,500
Total		461,283

Animals		
Porifera		10,000
Coelenterata		9,000
Ctenophora		90
Platyhelminthes		12,700
Mesozoa		50
Rhynchocoela		650

Animals (continued)		
Gnathostomulida	Invertebrates	43
Rotifera		1,500
Gastrotricha		175
Kinorhyncha		64
Nematoda		10,000
Nematomorpha		230
Acanthocephala		500
Entoprocta		60
Priapulida		8
Sipunculida		250
Mollusca		128,000
Echiurida		60
Annelida		8,700
Terdigrada		350
Onychophora		65
Arthropoda		923,000
Pentastomida		70
Phoronida		70
Bryozoa		4,000
Brachiopoda		260
Chaetognatha		50
Echinodermata		5,300
Pogonophora		80
Hemichordata		80
Chordata		39,000
Urochordata		1,600
Cephalochordata		1
Fish	Vertebrates	(22,000)–30,700
Amphibians		2,600
Reptiles		6,500
Birds		8,600
Mammals		4,060
Total		1,208,366

Figure 1.2 The eye of the hurricane.
Part of the 98% of solar energy never captured by photosynthesis wreaks actual havoc, forcing species to be equipped to survive intermittent catastrophe.

One of the remarkable outcomes of ecosystem studies, however, is the demonstration that plant communities can confer physical stability on a habitat, despite the fact that the energy available from photosynthesis is modest. Partly this effect is due to the physical structures built by plants, the aerial parts that break the force of wind and rain and the buried root systems that alter chemical gradients in the soil. But in wetter habitats the process of TRANSPIRATION may be as important. *Transpiration* serves to intercept ground water of a whole habitat and return it to the atmosphere. Solar energy is used directly for this, the water being evaporated from the leaves. The flux of heat energy involved in transpiration commonly is many times the total flux of energy transformed to biomass in photosynthesis. Certainly some ecosystem HOMEOSTASIS is due to this direct use of solar energy by plants for transpiration (Chapters 16 and 17).

THE PARADOX OF DESIGN

All life offers the paradox of apparent design. Individual organisms seem to be very well made. Different species populations live together in communities so well put together that they seem to be organized. And yet the scientific explanations of these things offered by evolutionary theory rely on random chance. There are two orders of paradox here. The first is the one familiar to evolutionary theorists of the apparent "design" of individuals. The second paradox is the apparent "organization" of these individuals into communities or ecosystems.

The idea that something as elegant as a bird or a tree can be the product of pure chance seems, quite properly, unreasonable. This is, in fact, a stock argument of those wishing to dispute evolutionary knowledge on religious grounds. The fallacy in the argument is the assertion that evolutionary theory requires that organisms appear *only* from an accumulation of random events. All organisms are designed by the process of selection: varieties appear very rapidly because of genetic recombinations and natural selection throws out the wrong and leaves the right. There is nothing random about natural selection itself.

Not chance mutation alone but the genetic mechanism itself provides the controlled variety on which natural selection works. Genes are reshuffled at almost every reproductive act, usually as a result of sexual recombination, and progeny of sexual reproduction always differ from either parent in endless permutations.

It may be true to say that the total variety produced in a population or species is a random set of unpredictable varieties, but these varieties are in fact all subtle variants of the original design. They should all (or nearly all) be quite workable organisms, as are, for instance, all the varieties of domestic animals. But natural selection does not give a chance for most of them to live. Only the best answer to local circumstance is permitted to survive.

Understanding the distribution and abundance of life on earth thus requires understanding why genetic recombination occurs. This is to question sex. Why do most organisms reproduce by sexual means, at least some of the time? The cost to a female in fitness obviously is high, because she sacrifices a half interest in all her offspring to another, the male.

A consequence of sex is the vast variety of offspring that makes evolution possible. The elegant designs of advanced plants and animals cease to be a paradox when seen to come about when natural selection is applied to the diversity produced by sex. But this does not explain why sex itself is necessary. Evolution is a result of sex, not a cause. The advantage of sex must be given to each female parent, immediately, giving her more surviving offspring herself. How this can be is explored in Chapter 12.

The first paradox of design is thus satisfactorily resolved by modern evolutionary theory. Something as complicated as a plant or an animal can be made by selection working on an immense array of alternatives. Plenty of chance varieties on which to work arise from the very complexity of organic life. Every individual is a unique variety, and the endless inexactitude of reproduction continually compounds the chance array, generation after generation.

Ecologists tend to ask questions about the *apparent* design of organisms. They ask "why" questions like "Why do desert bushes have thorns?" or "Why do lions spend so much time lying in the shade?" They ask these questions of the purely mechanistic process of selection, really saying "How have these traits given fitness to the animals and plants that have them?"

This asking of "why" questions by ecologists is not teleology. TELEOLOGY is the DOCTRINE THAT DEVELOPMENTS ARE DUE TO THE PURPOSE THAT IS SERVED BY THEM. The development of traits, in the thinking of evolutionary ecology, is due to the random appearance of varieties and the failure to survive, or poor reproduction, of all but a few of the organisms bearing those traits. A purpose is served by the adaptations, but the development of them is not due to this purpose.

The first paradox of design, therefore, is resolved comfortably by evolutionary theory without resort to teleology. The second paradox—the apparent organization of communities—requires a different, perhaps even simpler, explanation. The observation is that whole communities are arranged in familiar ways that are repeated in similar communities. Individual species seem to have places in a community, some common, others rare, some large and dominant, et cetera.

No process is known by which an ecosystem or community can be chosen, as an entity, by selection. Even the possibility of "group" selection of single-species populations is extremely difficult to demonstrate (Chapters 12 and 25). Nevertheless, some themes in the history of ecology reflect attempts to see organizing principles in community building to explain this paradox of community design. But community design can be more easily explained as resulting from common adaptation to shared circumstances of life. The animals and plants of a place fit themselves together according to their special needs, the resulting structure having a superficial semblance to design.

ECOLOGICAL SUCCESSION

Like the concept of the ecosystem, ECOLOGICAL SUCCESSION is an idea that is discussed widely outside the profession of ecology. The basic observation is that the growth of vegetation on previously bare ground proceeds with a definite sequence of communities.

A typical succession would be a first invasion of the ground by weedy herbs, many of them ANNUAL PLANTS. This PIONEER COMMUNITY of weeds is invaded by larger, longer-living plants (PERENNIALS) or woody plants. The resulting shrub or thicket land is invaded by trees and the first woodland is invaded by still other trees. The final vegetation changes only slowly and is called the CLIMAX. Ecological successions like this can be seen in progress on waste land or on abandoned farms throughout the north temperate latitudes. The successions are called SECONDARY SUCCESSIONS, because they take place on land that originally was vegetated but that had been cleared by humans (Figure 1.3).

Rather similar ecological successions take place on bare ground that has never before been vegetated. A classic example is on glacial till being deposited by retreating glaciers in Alaska (Chapter 23). Another such succession is now (1983) in its early stages on mud-flows deposited on the flanks of Mt. St. Helens following the 1980 eruption (Figure 1.4). These are called PRIMARY SUCCESSIONS.

A striking fact revealed by these observations is that plant communities in an ecological succession appear in a fairly orderly sequence: these forest successions always begin with pioneer weeds, then come perennials, then shrubs, and finally trees. Perhaps more interesting still is the way in which the physical habitat may change in pace with the succession of plant communities. Soils develop and soil nutrients increase even as the plants go from pioneers to climax. This is particularly apparent in primary successions where an infertile, raw, mineral mass left by a glacier or volcanic blast is turned into fertile soil during succession.

A number of important questions are set by the facts of succession. Why do the plants appear in a recognizable sequence? Are the pioneer communities necessary for the full development of vegetation? Is the long sequence of ecological succession necessary for the development of fertile soils and habitats? Do the early successional communities hasten their own demise by modifying the habitat so that other plants can replace them? In answering these questions, modern ecology draws on evolutionary and population studies describing species strategies, as well as on studies of the physical system itself (Chapter 23).

THE ORGANIZATION OF THE BOOK

The book begins with some of the principal concepts of ecology, and how these ideas came about. Most important are energy flow, the ecological niche, and the concept of limiting factors. After introducing these, the narrative runs from studies of individual adaptation through populations, species, ecosystems, and communities. This is the ordering found practicable in most science courses in that it introduces subjects according to levels of complexity.

It is, however, useful to remember that working ecologists sometimes think in the opposite

Figure 1.3 Old field succession.
(*a*) A corn field is abandoned and crabgrass moves in. (*b*) Perennial Compositae replace the crabgrass. (*c*) An almost pure stand of broom sedge (*Andropogon virginicus*) invades by year three. (*d*) Pine trees grow in the broom sedge. (*e*) A pine wood shades out the broom sedge. (*f*) Oak seedlings grow out of the pine litter, but no pine seedlings grow in the wood.

(e)

(f)

order, from the complex to the simple. The first questions posed were about communities, formations, biomes, and ecological succession. These questions were tackled by the reductionist approach typical of science, that of breaking down the complex into its fundamental parts. This pattern of thought thus begins with ecosystems and ends with populations and individual organisms.

Despite the continual need for analysis, however, ecology probably is more integrative than most sciences. Ecosystems result from physical, chemical, and living processes all combined. Before community organization is reviewed in the last part of the book, therefore, there is a section on ecosystems describing the physical processes that maintain soils, the atmosphere, climate, the oceans, and lakes. Ecology, a Darwinian study at heart, yet becomes most special when it integrates evolutionary thought with ideas developed by surficial geology, climatology, and terrestrial chemistry.

Figure 1.4 Mount St. Helens mud-flow.
Bare, seed-free ground made by a volcano is a natural site for a colonization sequence called a primary succession.

PREVIEW TO CHAPTER 2

The familiar concepts of food chains and food webs became prominent in ecology sixty years ago from the work of Charles Elton, who saw their significance when studying the simple system of an arctic island. Nutrients were cycled up food chains, down to the soil, and back through the plants. Each animal of the community had a function, a place in the food chains and nutrient cycles that Elton called its "niche." Predators high on food chains were both rarer and larger than the animals of the lower links. Elton called this result "the pyramid of numbers." Large size in predators is an adaptation for hunting and subduing prey, but the rarity of big predators requires explanations based on energetics. From Elton's observation of the pyramid of numbers came the concept of energy flow, introduced twenty years later in the 1940s. The food energy of the community all comes from the photosynthesis of the plant base. Some large portion of this food energy is dispersed by respiration in every trophic level of the numbers pyramid so that only a tiny flux of energy trickles to the top predators, which are correspondingly rare. Patterns of distribution are also set by energy needs, since the home range of predators must be large for them to be supported on the small energy flux per unit area remaining for their trophic level. Forest trees are huge and live packed together, side by side, because the energy flux for plants is very large. But the big cats of the forest, so much smaller than a tree, yet live kilometers apart as they scavenge what little of that energy reaches the cat trophic level. Beneath this overview of energy constraints ecologists look at the details of species distribution using the concepts of "niche" or "limiting factors," different concepts that yet converge in practice. Elton's "niche" referred to community function, as in "there goes the top predator." This niche could be filled, in theory, by ecologically equivalent animals; a lion for a tiger, for instance. But we can also define a niche as being unique to a species as in "the niche of the American robin." Then the niche is the program of living fashioned by natural selection to let individuals of that special species feed, survive hazard, and reproduce. A niche can also be defined as an environmental space in which individuals of a species function. This niche space is not just one of geometry or position, but more particularly includes such measurable parameters of the environment as food size, food density, places of shelter and numbers of competitors or predators. Limits are set to local populations of all species by their niches since there can be no more individuals of a species than there are opportunities, or niche spaces, for their way of life. Since maximum numbers are thus set by a species' niche it follows that maximum numbers are independent of the breeding effort. Selection chooses which individuals of a generation occupy niche spaces made vacant by death, letting the surplus perish. Selection thus perfects individuals to their niches in each generation, raising the question of how evolution can happen, since adaptation is by definition always perfect. The Red Queen hypothesis answers this riddle by noting that the environment changes in every generation, either physically or by the arrival of fresh organisms, so that selection must always make changes to keep pace with environmental change. The older concept of limiting factors suggests that distributions are set by critical environmental variables close to the limits of a species tolerance. The idea converges on niche theory when some particular factor of the niche hyperspace is more critical than the rest to a local population. Most distributions, however, are set by complex interactions between organisms, not by a single measurable environmental variable.

CHAPTER 2

FIRST PRINCIPLES: ENERGY FLOW; PYRAMIDS; NICHE; LIMITING FACTORS

A FOOD CHAIN is a chain of eating and being eaten that connects large and carnivorous animals to their ultimate plant food. A classical example is:

pine trees → aphids → spiders → titmice → hawks

The food chain is written thus, with the arrows pointing the way they do, to show that food passes from animal to animal up the chain from the plants to the herbivorous aphids, thence to carnivorous spiders, which are eaten by titmice, which are eaten, in turn, by hawks.

The parallel concept of FOOD WEB is necessary because many kinds of plants live side by side, and because most animals can eat more than one kind of food. Spiders, chickadees, and hawks, for instance, can all eat several kinds of prey so that all must be part of several food chains. Part of the hawk's livelihood may come from virtually every species of plant in a complex patch of forest via different food chains that radiate downwards through the various animals of the hawk's generalized diet. The result is a web of intersecting food chains (Figures 2.1 and 2.2).

These familiar concepts of *food chain* and *food web* became particularly prominent in ecology from the work of a young researcher from Oxford University more than fifty years ago, Charles Elton, who published a textbook *Animal Ecology*, in 1927, in which he set out the importance of food chains, together with the immensely important concepts of *pyramid of numbers* and *niche* that he derived from his observation of food chains. Elton did not talk of "energetics" in discussing these phenomena; instead he talked of food and nutrients. His observations, however, led directly to models of energy flow in ecology.

Elton's most important observations were made on the tundra at Bear Island, near Spitzbergen (Summerhayes and Elton, 1923). Arctic foxes are conspicuous on the tundra and feed in the summer on birds—the resident ptarmigan, migrants like sandpipers and buntings, and various sea birds (Figures 2.3 and 2.4). The foxes evidently got their living both from the tundra and from the plants of the sea via different food chains. In winter the sea was particularly important because the foxes survived by eating polar bear dung and the remains of seals that the polar bears killed (Figure 2.5). Figure 2.6 is drawn from Elton's original diagram describing the food chains and food web important to foxes on Bear Island.

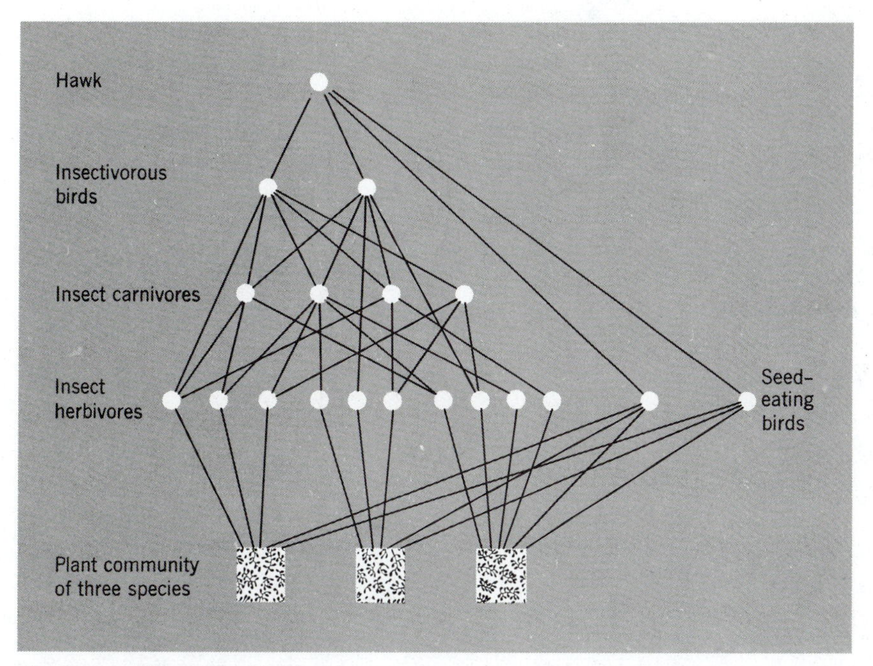

Figure 2.1 Hypothetical food web.
It is assumed that there are three species of plants, ten species of insect herbivores, four insect carnivores, two bird herbivores, two bird insectivores, and one hawk. In a real community, there would not only be many more species at each trophic level but also many animals that feed at more than one level, or that change level as they grow older. Some general conclusions emerge from even an oversimplified model like this, however. There is an initial diversity introduced by the numbers of plants. This diversity is multiplied at the plant-eating level. At each subsequent level the diversity is reduced as the food chains converge.

Elton described this transfer of food from animal to animal as the FOOD CYCLE. He did this because he observed that indestructible solid matter was passed, particularly combined nitrogen. Tundra communities tend to be poorly supplied with combined nitrogen. Elton postulated that on Bear Island much nitrogenous fertilizer came from the sea via sea birds and that this nitrogen was then cycled up the food chains and back again, hence the *food cycle*. Workers since Elton's time have been more inclined to stress the transfer of food calories through food chains and webs. The movement of calories is, of course, unidirectional, since energy is degraded in the process according to the second law of thermodynamics. Nutrients are cycled in an ecosystem but energy flows through it.

The ease with which animals could be seen on the tundra brought home another fundamental fact; that animals high on food chains were both larger and rarer than animals lower down. This result Elton called the PYRAMID OF NUMBERS, sometimes referred to as the ELTONIAN PYRAMID.

When animal size is plotted against number the result is a graph of this general form:

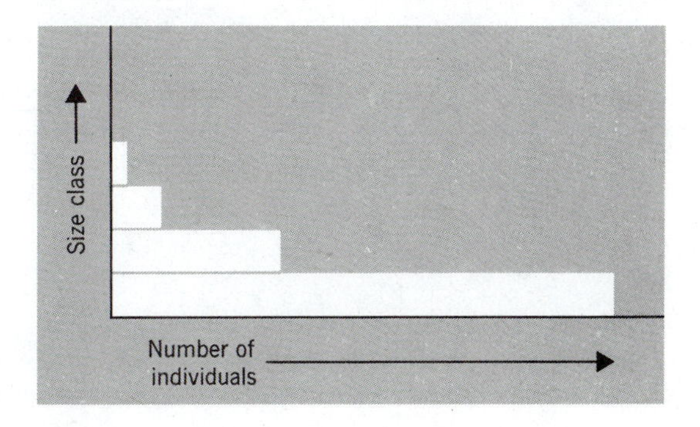

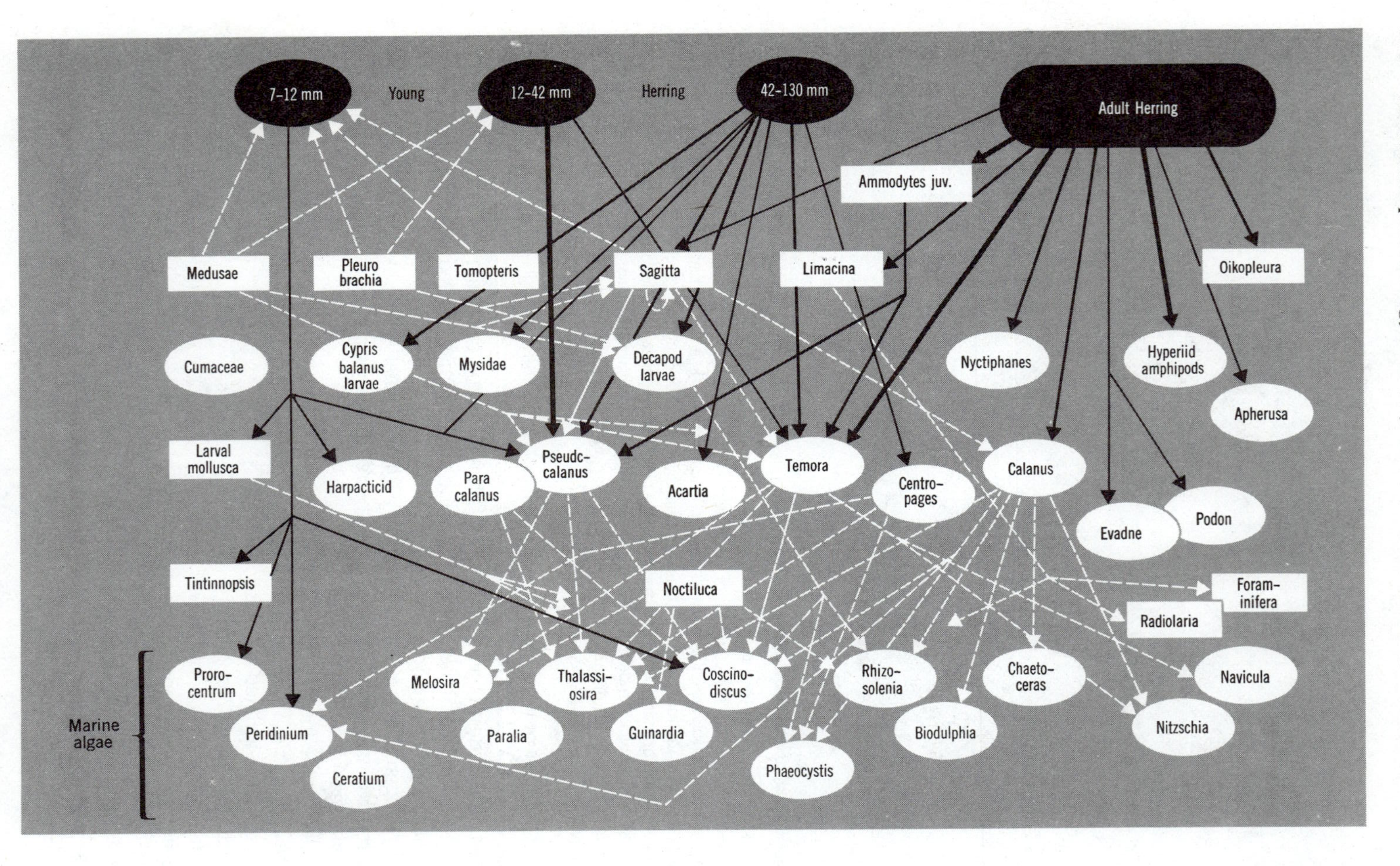

Figure 2.2 A food web for the open waters of the North Sea.
This food web was reconstructed by Hardy (1924) to describe the results of his investigations into the herring fishery of the North Sea. The herring is a carnivore that feeds several links in its food chains away from the floating algae that are its ultimate source of food. The names in the bottom lozenges are of various genera of phytoplankters. Higher up in the food chains are planktonic animals, still larger animals, and then the herrings themselves. Notice that young herring feed on smaller food than the adults, as you would expect. As the herrings age, so they feed at higher and higher trophic levels.

Figure 2.3 *Alopex lagopus*: arctic land predator. Arctic foxes are top predators on land on arctic islands, but dependent on food chains from both land and sea.

Figure 2.4 *Lagopus rupestris*: snow-adapted grouse. Ptarmigan overwinter in the arctic, connecting arctic foxes to the tundra.

Shifting the "x" axis to the middle of the figure gives the following result:

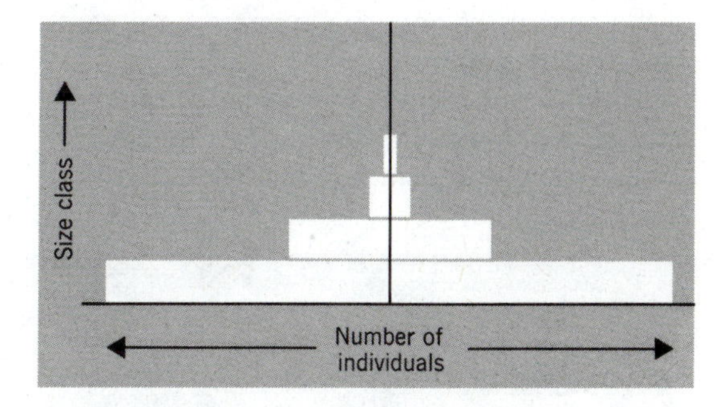

Each layer on the pyramid represents kinds of animals living at parallel levels on food chains; all HERBIVORES on one level, all PRIMARY CARNIVORES on the next level, all SECONDARY CARNIVORES on the next level, and so

Figure 2.5 *Thalarctos maritimus*: seal hunter. Polar bears are top predators of a marine food chain in the arctic, but arctic foxes scavenge seal carcasses killed by bears and eat bear dung.

on. These levels are now called TROPHIC LEVELS (after the Greek word for nursing, in the sense of a mother suckling her young, because all the animals of one level operate on a common feeding plan).

It seems that there is a *pyramid of numbers* associated with many of the earth's habitats. Figure 2.7 shows a pyramid of numbers of arthropods on a tropical forest floor, a result probably typical of most terrestrial sites. Elton (1927) described the possibilities for water, saying "In a small pond, the numbers of protozoa may run into millions, those of *Daphnia* and *Cyclops* into hundreds of thousands, while there will be far fewer beetle larvae, and only a very few small fish."

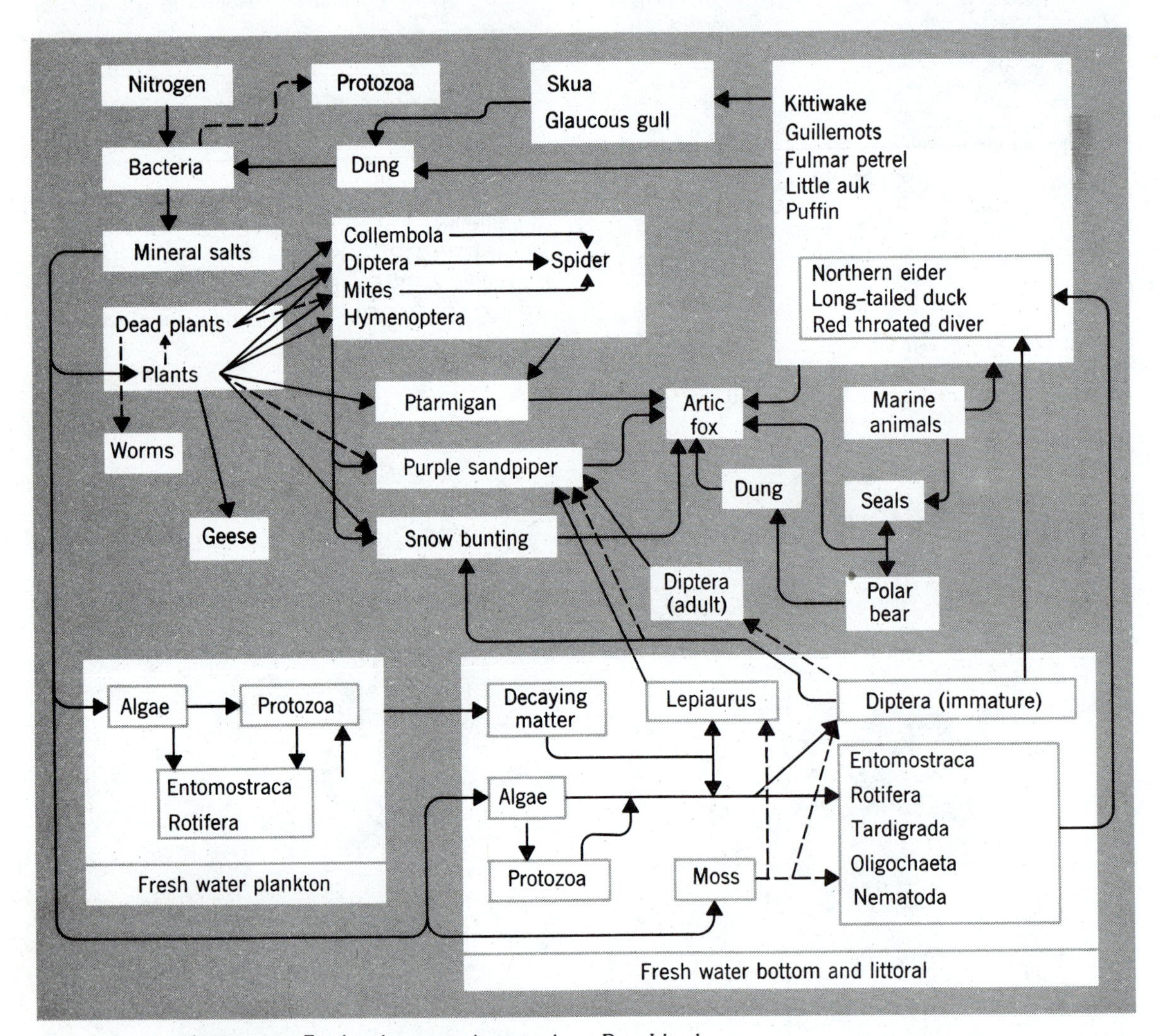

Figure 2.6 Food cycle among the animals on Bear Island.
Elton worked out the details of this food cycle by seeing in the mind's eye the movement of nitrogen with the food through the community. He was able to confirm some of the pathways by direct observation, and others (shown as dashed lines) he could only infer. He notes that the best way to read the diagram is to start with marine mammals and follow the arrows. (From Summerhayes and Elton, 1923.)

Some obvious qualifications must be made to the assertion that pyramids of numbers are everywhere; the largest animals are herbivores—elephants (which have no predators) or moose, which are preyed upon by wolves (animals smaller as well as rarer than the moose)—parasites are smaller than their prey, and so on. These exceptions have their own stories to tell. But much was to be learned by asking the question "Why are there pyramids of numbers at all, and where are they found?"

To Elton, one aspect seemed to have a clear enough explanation, and this was the separateness of the *trophic levels.* Where the food chains ran insect → titmouse → hawk, or in Elton's pond, animals came in size fractions that were remarkably distinct. There is a quantum jump in size between an insect and a bird, between a water flea (*Daphnia*) and a midge larva (*Chaoborus*), and between a midge larva and a fish. Elton explained this with his PRINCIPLE OF FOOD SIZE, saying that animals tended to be of a size that let them thrust their prey into their mouths whole. There is an advantage in an animal's being big, so that it can easily catch and eat its prey, but it must not be so big that it cannot catch enough small prey to keep itself alive. So there must be an optimum size for any animal, a size determined by the size and agility of its food. As with the *pyramid* observation it-

Figure 2.7 Eltonian pyramid of numbers on the floor of a forest in Panama.
Williams (1941) collected all the animals in small samples of the litter on the floor of the forest, counted the individuals in his catches, and sorted them into size fractions. The smallest and most numerous animals were Collembola (spring tails) and mites, both of which are herbivores or scavengers feeding in the litter. The larger, rarer animals, such as ground beetles and spiders, were carnivores.

self, there are clear exceptions to this general observation (wolves and moose again, for instance) but the explanation seems to have widespread validity wherever the pyramid structure shows up clearly. Animal size goes up by quantum jumps along the food chains of a pyramid of numbers because of successive needs to handle ever larger prey. Therefore, an *Eltonian pyramid* is properly drawn with stepped sides to reflect the change of size between trophic levels.

Instead of pyramids with stepped sides, general biology textbooks usually show some version of a figure with sloping sides as the pyramid of numbers. This may take the form of Figure 2.8, the first of its kind introduced by the limnologist Juday (1940), though the cone-shaped drawing is often embellished with pictures of big cats snarling at animals lower down, or even with

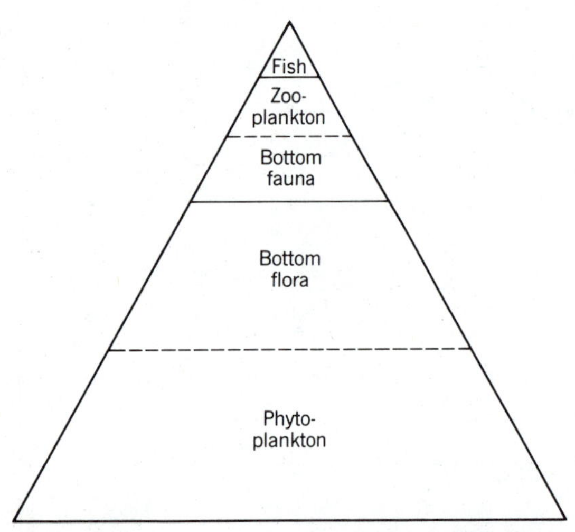

Figure 2.8 Cone-shaped version of the pyramid of numbers.
This figure, taken from Juday (1940), is probably the first pyramid figure with smooth sloping sides to be published, and it is now often imitated. Elton's original vision was of a pyramid with stepped sides, looking in profile like the stepped pyramids of Egypt. In a lake system like that studied by Juday it is probably more realistic to separate the organisms of the different trophic levels into discrete size fractions so that the steps on the pyramid show. This is because the animals of the different levels are in the main large enough to engulf their food whole.

a human on top brandishing a spear. All sizes of animals are, of course, possible but real counts are likely to reveal steps (Figure 2.7). It was these steps, when present, that led Elton to the importance of food size, itself the forerunner to the concept of *energy flow*.

An important question posed by Elton's observation of food chains and pyramids was "Why are big animals, particularly big carnivorous animals, rare?" Elton (1927) suggested that small animals reproduced more quickly than large animals, but lived shorter lives. This rapid turnover of small animals resulted in large numbers of individuals. Slowly reproducing large animals, however, could not maintain such large populations. The essence of this suggestion is that population size is a function of reproduction rate. This explanation can apply when all populations are growing but it cannot apply at equilibrium, when a high birth rate should be balanced by a high death rate. Elton's explanation, therefore, could not apply in most circumstances.

The explanation of Elton's pyramid now generally offered is based on energetics. Small animals low on the pyramid are the food of larger animals higher up so that energy, as food calories, enters the pyramid at the bottom. In maintaining life, energy is constantly being dissipated as heat. This means that the usable food energy flux received by each successive trophic level is less than that received by the trophic level below (Figure 2.9).

Living biomass is itself a store of energy and its maintenance costs are a function of its mass. It follows that less biomass can be supported by the food energy received in each higher trophic level. Even if the animals at all levels of the pyramid were of the same size, those high on food chains would of necessity be less numerous than those below. Since, because of the *principle of food size,* those high on the chains tend to be large, the available energy for body-building and maintenance is sufficient for even fewer of them.

This explanation for the pyramid of numbers has generality, being applicable to all systems in equilibrium where all populations are regulated. As long as there are food chains, pyramids of

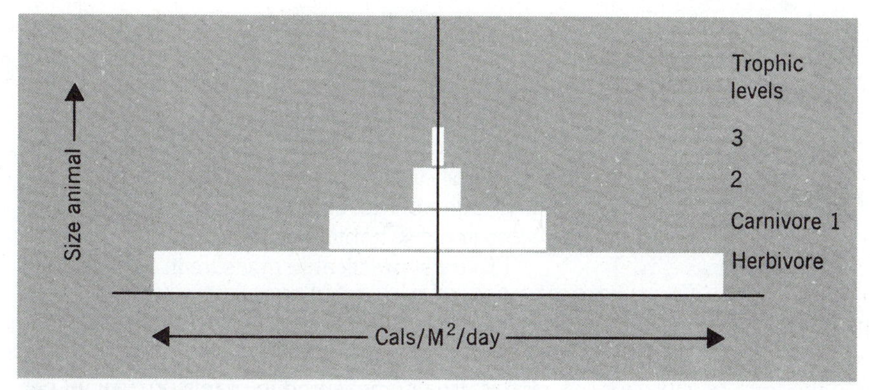

Figure 2.9 Energy pyramid.
The figure illustrates the concept of energy degradation at each transformation between trophic levels that was introduced to ecology by Lindeman and Hutchinson.

energy flow must develop. Limits are set immediately for the distribution and abundance of animals adapted to different kinds of food. Large carnivorous animals, for instance, always must be rare, whereas small herbivorous animals can be abundant.

The energy flow model, however, is very general, describing relative upper limits that may not always be reached. Many other factors are involved in fixing population number and individual size at any time or place than just position on a food chain and the principle of food size. For systems in rapid change that are not at equilibrium, reproductive and mortality rates may be definitive, as Elton suggested. In many systems the energy of plant biomass going to rot is far larger than the energy of biomass eaten by herbivores (Chapter 16), so that the distribution and abundance of local life may depend heavily upon the process of decomposition. Even more distorting is the fact that food chains may be reduced to a single link, as when large grazing herbivores feed directly on vegetation.

The energy flow model represented an important advance in the history of ecology 40 years ago. The workers who were to realize the importance of energetics in explaining this phenomenon were mainly those studying aquatic systems. Even Elton (1927) had noticed that his pyramid was nicely expressed in a pond, where the herbivorous animals tended to be small. Important to the neat expression of pyramids in ponds was the fact that most plant biomass in this system did go to true herbivores and not to rot as in a forest or prairie (Chapter 21). Aquatic scientists were active in community studies in the 1930s and confirmed Elton's observation of pyramids (Juday, 1940). Two other limnologists gave us the formal statement of the energy flow model, G. E. Hutchinson, already a senior limnologist, and Raymond Lindeman, who developed the idea in a doctoral dissertation and published the paper that was to set a new paradigm for ecology (Lindeman, 1942).

PRODUCTIVITY AND STANDING CROP

Production can be measured as biomass, mass of nutrients, mass of carbon, mass of some element important to ecosystem chemistry, or as energy. Measures of mass in ecosystem studies for various purposes are discussed in Chapters 16 to 20. But production measured as energy has particular importance in ecology. Concepts of ENERGY PRODUCTION, however, are made difficult because energy in a system cannot be measured directly but must be inferred from measures of biomass or gas exchange.

The GROSS PRODUCTION of a trophic level is THE ENERGY REPRESENTED BY THE BIOMASS PRODUCED TOGETHER WITH THE ENERGY THAT WENT INTO THE WORK OF PRODUCING IT.

$$\text{Energy (biomass)} + \text{Energy (work of manufacture)} = \text{Gross Production} \quad (2.1)$$

The energy represented by the biomass is NET PRODUCTION and the energy used in the work of manufacture is RESPIRATION. This last term seems a little strange to use for an energy flux, but ecologists have adopted the term because all energy released by organisms involves the respiration of carbon dioxide. Whether plant or animal, the doing of work depends on the oxidation of reduced carbon compounds with evolution of CO_2 in proportion to the work done. Hence activity always results in the release of CO_2 and the release of CO_2 always indicates activity. Ecologists can rewrite equation (2.1) as

$$\text{Net Production} + \text{Respiration} = \text{Gross Production} \quad (2.2)$$

where all three terms are measured in calories per unit time.

Hutchinson, in a manuscript cited by Lindeman (1942), defined the PRODUCTIVITY of any trophic level in a way that can be restated as THE TOTAL RATE OF ENERGY FLOW INTO THAT TROPHIC LEVEL. He assigned the Greek letter lambda (λ) to represent this RATE OF PRODUCTION or PRODUCTIVITY, as follows:

λ = symbol for the productivity of a trophic level

λ_n = productivity of high trophic level of a pair

λ_0 = productivity of plant (producer) trophic level.

The efficiency of energy transfer between any two trophic levels may then be written as the ratio λ/λ_{n-1} or

$$E = \frac{\lambda}{\lambda_{n-1}} \times 100 \quad (2.3)$$

where E is the efficiency of energy transfer between trophic levels. "E" also is called *ecological efficiency* or *Lindeman efficiency*.

It needs to be emphasized that the productivity λ is the *rate* of *gross production* of energy and includes both the energy of biomass and the energy being lost to respiration. It is not, therefore, possible to calculate transfer efficiencies from measures of biomass alone. This has made difficult attempts to explore the energetics of real ecosystems (Chapters 4 and 16).

The most practicable measure in field ecology is of mass present at the time of a visit, the *standing crop.* The STANDING CROP is the BIOMASS PRESENT AT UNIT TIME IN UNIT AREA and is usually measured in grams dry weight per square meter (g m^{-2}), or similar convenient units. Analysis of biomass allows calculation of standing crop in carbon or calories, or of the mass of nutrients or chemical element under investigation. Measures of standing crop, therefore, are used in many branches of ecology. When series of measurements of standing crop are made at different times, then rates of net production result. When expressed either as dry matter or as calories, these measures are the basis of PRODUCTION ECOLOGY (Chapter 16).

Standing crop must not be confused with the agricultural term "crop" from which it is derived. When ecologists talk of the standing crop of a corn field they mean not just the grain but the leaves, stems, and roots as well. Standing crops can be measured for every trophic level, and can mean the total dry weight of the herbivores of a place, of the carnivores, or of the decomposers.

Measurement of standing crop of plants and animals in many communities is a straightforward, if time-consuming, undertaking—essentially a matter of a population census coupled with weighing dried samples of the various plant and animal parts. E. P. Odum (1959) has drawn together data for standing crop in a number of habitats that can be presented in the form of pyramids of biomass (Figure 2.10). In an old field, a coral reef, a lake, and a clear-water stream, the pyramids of biomass appear similar to a pyramid of numbers, with orders of magnitude reductions in mass with each higher trophic level. This reflects the degradation of energy up the food chains as expected. In two fertile marine systems, Long Island Sound and the English

Channel, however, the pyramids of biomass are inverted (Figure 2.10).

A hypothesis to explain the inverted marine biomass pyramids is readily offered; it is that the small plants (the *phytoplankton*) that are the *primary producers* of the sea turn over very quickly because they are so short-lived. At any one time many plants are there (the pyramid of numbers is the right way up) but their tiny bodies represent very little mass, and this little mass is being constantly removed and replaced. The animals, however, include large bottom dwellers like starfishes and urchins, which are long-lived and slow to turn over. Energy is predicted to pass through the phytoplankton trophic level at a much greater rate than through the animal trophic levels, so that the energy pyramid is the right way up, like the numbers pyramid.

A test of the hypothesis requires estimating the energy flux from phytoplankton through the various consumers. The best estimate for this is the flux of dry matter calculated from growth rates of the more important organisms. In Figure 2.11 are Harvey's (1950) estimates for dry matter flux through the English Channel food chains. The daily flux of dry matter through the phytoplankton trophic level turns out to be larger than that through the animal trophic levels as predicted.

A general terminology has come into use to describe animals or plants according to their role

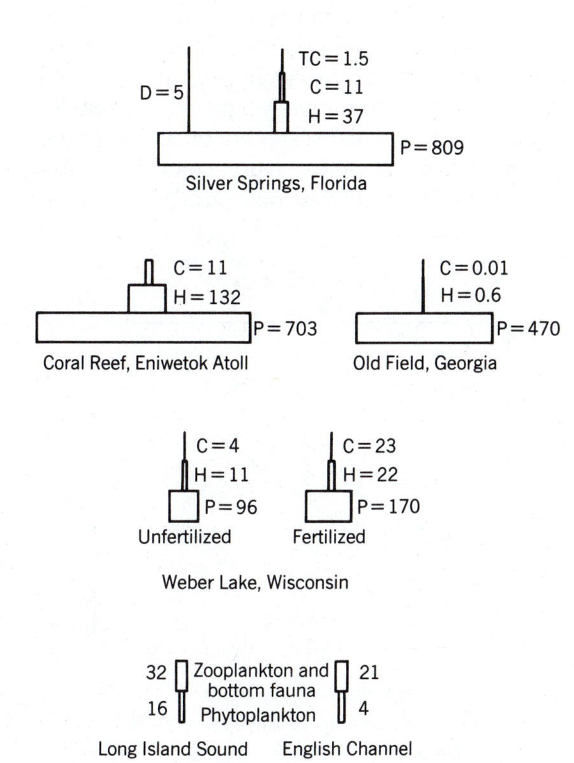

Figure 2.10 Pyramids of biomass from various communities.
In the terrestrial and fresh-water ecosystems the pyramids of biomass are the shape we would expect the pyramids of energy to be if we had sufficient data to draw them: large energy flows low on the food chains are reflected in large biomass. In the two samples of productive ocean over shallow continental shelves, however, the biomass pyramids are inverted. This has been shown to be caused by rapid turnover of biomass in the producer trophic level, which occurs because the plants of this system are microscopic and short-lived. Pyramids are drawn approximately to scale and figures represent grams biomass per square meter. P = Producers, H = herbivores, C = carnivores, TC = top carnivores, and D = decomposers. (Data from Odum (1971) from various sources.)

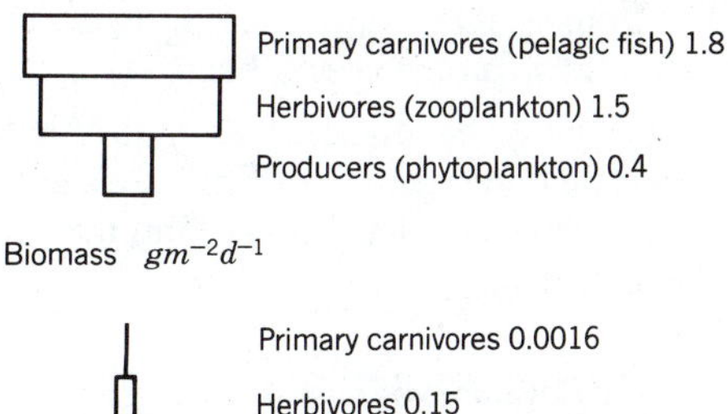

Figure 2.11 Pyramid of dry matter flow in the English Channel.
Biomass produced by the producers (phytoplankton) per day is greater than the biomass produced per day by the consumers of the English Channel ecosystem. This large flux of biomass requires a large flux of energy so these data provide a satisfactory explanation for the anomalous inverted pyramids of standing crop biomass found in shallow productive marine systems as in Figure 2.9. (Data from Harvey, 1950.)

in food chains or in the trophic level hierarchy of production.

Producers are the plants, the base trophic level that is responsible for transforming solar energy into the potential form of fixed carbon compounds.

Consumers are the representatives of all other trophic levels and include herbivores, carnivores, omnivores, and parasites. PRIMARY CONSUMERS are the herbivores (plant eaters). SECONDARY CONSUMERS are the *primary carnivores* (flesh eaters); TERTIARY CONSUMERS are the *secondary carnivores*; and so on. Animals of the highest link in any food chain are often called TOP CARNIVORES. A PARASITE CHAIN typically runs from a small animal (flea) to an even smaller animal (protozoan parasite of a flea) to an even smaller animal still (bacterial disease of the protozoan). The animals of these parasite chains may be directly assigned to appropriate trophic levels along with the other consumers.

Decomposers are organisms that feed on corpses or dead organic matter. SAPROBE or SAPROPHAGE (corpse-eating) are terms that are sometimes used instead of DECOMPOSER. A DECOMPOSER CHAIN or a SAPROBE CHAIN might run from a large organism (fungus) to smaller organisms (protozoa and bacteria).

Productivity may be qualified as PRIMARY (energy fixed by green plants) or SECONDARY (energy flowing into any level other than that of primary green plants).

HOME RANGE, ENERGETICS, AND SIZE

As animals become larger and rarer, they also have larger HOME RANGES. Elton recognized this in his original work and elaborated on the idea later when he talked of the INVERSE PYRAMID OF HABITATS (Elton, 1966). Like the original observation of the pyramid itself, this is a necessary consequence of the quest for energy. Large animals need a large energy flux and they must hunt or crop a large area to get it. Whereas plants can support themselves on the energy flowing directly onto the spot where they stand, animals that receive their energy after one or two inefficient energy transfers up food chains from the plants require the pickings of a much larger area. Rarity and being spaced out are connected states, both necessary when energy is in short supply high on food chains.

McNab (1963) compiled data for home ranges of mammals ranging in size from mice to moose and showed a strong logarithmic relationship between range and body weight (Figure 2.12). The data show that the range–weight relationship for carnivorous animals is shared by those herbivores that eat scattered, high-calorie foods like seeds. McNab said that both kinds of animals

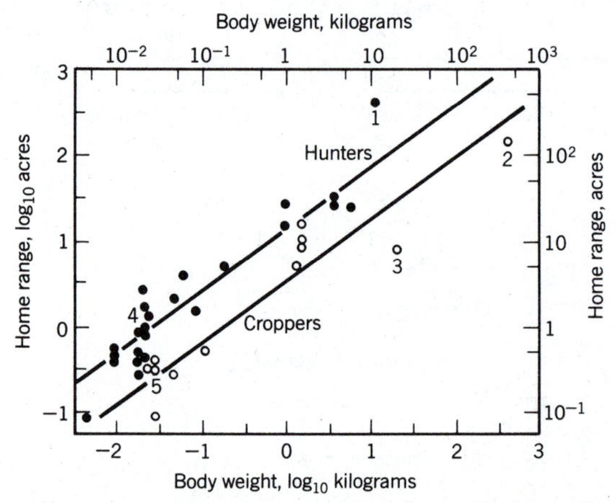

Figure 2.12 Relationship of size of home range to weight of body in mammals.
There is a significant relationship between the size of the animal and the area over which it must range to get its food. Herbivorous browsing animals like moose or deer need about four times less land than carnivores of the same size, a relationship that is easy to understand on energetic grounds. Of particular interest, however, is that herbivores or omnivores that seek out discrete particles of food like seeds need about the same area as true predaceous animals of their own size. McNab (1963) refers to both the true predators and those that seek out patchily distributed plant food with the common term "hunters." These are distinguished from the true browsing or grazing animals, which are "croppers." Open circles = croppers, solid circles = hunters, 1 = raccoon (*Procyon lotor*), 2 = moose (*Alces alces*), 3 = beaver (*Castor canadensis*), 4 = deer mouse (*Peromyscus maniculatus*). (From McNab, 1963.)

were "HUNTERS," both the true carnivorous hunters and those that "hunted" seeds or fruits. The energy flux per unit area from both kinds of food was low and animals of both habits lived well spread out. With the data available it was not possible to separate the relative size of home range of vegetarian hunters from that of flesh eaters. But much less space, about a fourth, was needed for animals that browsed or grazed on vegetation, animals that McNab called CROPPERS.

Price (1975) offers the following list of attributes that should change along a gradient from herbivory to carnivory up the food chains.

1. fewer species
2. lower population levels
3. lower reproductive rates
4. increased body size
5. increased home range
6. higher powers of dispersal
7. higher searching ability
8. higher maintenance cost
9. higher utilization efficiency of food
10. food of higher calorific value
11. reduced feeding specialization
12. more complex behavior
13. longer life expectancy

Some of these attributes are directly related to animal size, and others are consequences of a large home range and the need to hunt for food. All, however, are ultimately consequences of the energetic constraints of living high on food chains.

THE SEVERAL COMPATIBLE MEANINGS OF NICHE

Ecologists say that every species has its niche, a unique position in a food web or ecosystem, or a unique function in life, or a unique set of resources or factors needed for survival. There are several subtly different meanings behind the idea of niche, though all are linked.

1. Niche as Community Function (Class I Niche)

Each kind of animal and plant has a perceived function in an ecosystem. Elton (1927) said that, when he saw a badger as he went for an early morning stroll in an Oxford wood, he tried to think he was seeing a functionary of the local community, rather like seeing a man walk down a village street and saying to himself, "There goes the vicar" or "There goes a plumber," or "There goes a farmer." Every kind of plant or animal must have its own function in the community, its *niche*. Some niches are filled by many individuals (farmers, aphids), whereas others by few (vicar, sparrow hawk), but each holds a unique, circumscribed profession, its niche (Figure 2.13). Elton offered a formal definition saying that NICHE means THE ANIMAL'S PLACE IN THE BIOTIC ENVIRONMENT, ITS RELATION TO FOOD AND ENEMIES. Niche so defined is sometimes referred to as the ELTONIAN NICHE.

2. Niche in the Species Definition (Class II Niche)

The function of a species population in a community can also be defined from the point of view of the individual animal or plant of that population. Then a NICHE is a SPECIFIC SET OF CAPABILITIES FOR EXTRACTING RESOURCES, FOR SURVIVING HAZARD, AND FOR COMPETING, COUPLED WITH A CORRESPONDING SET OF NEEDS (Colinvaux, 1982).

Niche in this sense conveys an ecologist's view of what a species is, not a group of individuals classified by possessing a common shape and represented by type specimens in museums, but a population of individuals suited to a particular way of life for which their common shape is adapted. It is how an animal or plant wins resources that matters; and the tricks it uses to get those resources are the concerns that are really special to a species.

The niche of an American robin (*Turdus migratorius*) can serve as an example. Specific adaptations of robins include the ability to pull worms and various insect hunting habits like the

Figure 2.13 Ecological equivalents in a plankton filtering niche. Bowhead whales and basking sharks feed on small crustaceans that they are able to filter from seawater. Their functional niches in ocean communities are similar (Class I niche) and both grow larger than predators that must hunt large prey in the sea, but their ancestry and details of life histories are quite different.

turning over of dead leaves. Other adaptations include the proper response to alarm calls or to the songs of conspecifics, and the motivation and navigation skills needed for a long migratory flight twice a year. These traits add up to a highly specialized way of life that may be called the SPECIES NICHE.

Both the Class I (Eltonian) and the Class II (species) niches describe the profession of the animal, but the viewpoint is different. In the former, the American robin plays a role in the community as a puller of worms and food for hawks; in the latter, an American robin pulls worms and avoids hawks as part of a program working to thrust more robins into the next generation.

Only individuals of *Turdus migratorius* can fill the Class II niche of this species, but the community function fulfilled by American robins can in theory be filled by similar but different kinds of birds. American robins are the only common thrushes (family Turdidae) that pull worms on lawns of the Eastern United States, but in north Europe other species of thrush fill this function, in particular the European blackbird (*Turdus merula*) and the song thrush (*Turdus ericetorum*). These birds fill the worm pulling niche in Europe, though their species niches are significantly different from that of the American *Turdus migratorius*.

The species niches are, of course, equally distinct for plants. The European oak tree, *Quercus robur*, has obvious functional properties that make it distinctive. The oaks are slow growing, very long-lived trees of old forests. Their seedlings grow in the densest of shade, and can be found in the litter of needles under old pine woods where it is too dark for pine seedlings to grow (Figure 1.3*f*). Oaks can both grow in this dim light and function as adult trees in the bright light of the canopy when they inherit the habitat from the pines (Chapters 3 and 23). Other special adaptations include using environmental cues to develop leaves in the spring late enough to reduce the danger of frost but early enough to minimize insect attack, the deposition of tannins in leaves that will make them uneatable to most herbivores, and the production of fruit (acorns) suitable for transport by pigeons. Each individual of the *Q. robur* species is programmed for the same *Q. robur* niche, with of course individual variability allowed by the genetic mechanism.

In ecology, therefore, it is often convenient to think of a species in terms of *niche* rather than as a physical presence. What has been fixed by natural selection is the program of behavior, habits, and special skills that lets individuals of a species win resources. Since the shape of individuals within a species varies little, so that we can always recognize individuals for what they are, so the niche of individuals varies but little also. The *species niche* is fixed. This yields an intuitive understanding of why population sizes are so often nearly constant.

There cannot be more individuals in a species population than there are opportunities to practice its niche. The species niche, therefore, sets an upper limit to population size. The truth of this is made evident by comparing an animal (or plant) niche with a human profession. The number of people who can earn a living at professing a subject in a university, for instance, is set by the number of university professorships available. Professorships may be likened to NICHE SPACES for the niche of professing. When society reproduces more professors (by training graduate students) than there are opportunities for professing, this makes very little difference to the numbers of professors teaching in the next generation. Opportunities set numbers in a human occupation. Niche sets numbers for a species population.

If niche sets number, then it must follow that numbers are NOT set by reproductive effort. Individuals continue to reproduce rapidly, but this is because they are competing with other individuals for niche spaces in the next generation. The more offspring they have the greater are the parental chances of placing offspring in one of the limited niche spaces. But reproduction in excess of niche spaces available will not affect the size of the future population, though it does make a difference to individual fitness.

When animals or plants live in changing or harsh environments, they may never achieve a saturation of all niche spaces, as happens on the borders of deserts or in the arctic (Chapter 11). For these organisms populations fluctuate within wide limits and the rate of reproduction does influence the size of the population of the immediate future as each generation proceeds to colonize an array of niche spaces continually emptied by adversity. But for a population in a more stable environment without large periodic catastrophes *the rate of reproduction makes no difference to the size of the stable population.* This follows from the fact that species niches are fixed within narrow limits. It will be noted that this conclusion refutes Elton's (1927) original explanation of the pyramid of numbers, showing that the abundance of small animals cannot be a function of rapid breeding as he suggested.

3. Niche as a Quality of the Environment (Class III Niche)

If a niche is a property of a species it must yet be exercised in a suitable environment. Specialization for a food resource is only possible where that food resource is present. A bird with feet adapted to perching on small twigs can only live where the twigs are small, and so on. The concept of a species niche, therefore, allows the complementary concept of an environmental space in which that niche is exercised, a mirror image niche, as it were. The most useful definitions of *niche* for empirical work have proved to be those that define niche in terms of environment and resources, rather than in terms of the doings of the individual animal or plant.

MacFadyen (1957) expresses this idea with the following definition: A NICHE IS THAT SET OF ECOLOGICAL CONDITIONS UNDER WHICH A SPECIES CAN EXPLOIT A SOURCE OF ENERGY EFFECTIVELY ENOUGH TO BE ABLE TO REPRODUCE AND COLONIZE FURTHER SUCH SETS OF CONDITIONS. This definition still allows an individual niche to each species, and it still involves the primary assumption that it is the profession and doings of the species that set the niche, but it shows how important dimensions of a niche might be measured in practice.

An alternative definition is that of Hutchinson (1958): A NICHE IS A MULTI-DIMENSIONAL HYPERVOLUME OF RESOURCE AXES. Hutchinson assumes that every variable that affects a particular species can be thought of as being linear, in which event two resource axes can be expressed as *x* and *y* on a conventional 2-dimensional graph. When a third variable is added (say, perch size to food size and height above ground for a kind of small bird) then a 3-dimensional figure is needed to plot an outline of the niche that appears as a volume (Figure

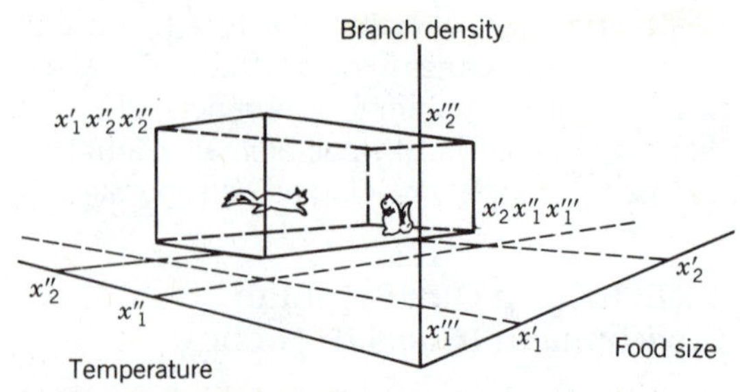

Figure 2.14 A three-dimensional orthogonal niche. Three niche axes likely to be important to a squirrel are temperature, food size, and branch density. If these are represented by x', x'', and x''', the rectangular space is a niche volume (Class III niche) of only three axes. Adding more axes to make a hypervolume can be understood intuitively and mathematically although it cannot be drawn. (From Hutchinson, 1978.)

2.14). When more dimensions than three are included the resulting theoretical space can no longer be drawn but can still be described mathematically as a hypervolume:

$$x', x'', x''', \ldots, x^n$$

where x', x'', et cetera, are niche axes.

Hutchinson's (1958) definition is effectively the same as MacFadyen's (1957) definition: the *set of ecological conditions* becomes the *multidimensions of resource axes*. Both use language that is deliberately vague because they seek to specify all the possible needs that the organism must find in the habitat. These needs include purely physical needs, like temperature, food, or nutrients, and needs that are measured by the presence of other organisms, like competitors or enemies.

But Hutchinson (1958) showed that two classes of niche axes could be separated by their probable effects on the size of the niche hypervolume. If a species population lived without competitors or other organisms that would interfere with it, then the size of the niche would be set by physical needs and food alone. Hutchinson called the resulting niche the FUNDAMENTAL NICHE of the species. But wherever the resources of the environment had to be shared with other species, as usually happens, then the niche hypervolume would be smaller. This was the REALIZED NICHE of the population.

On Niche, Habitat, and the Red Queen

"Niche" is one of those words that always needs to be considered in the context in which it is used. The Eltonian meaning of the term as a community function (Class I) continues to be used by biologists, perhaps particularly when talking about *ecological equivalents* in different communities (Chapter 7). An evolutionary biologist might talk of sunbirds in Africa filling "the same niche" (meaning they have an equivalent ecological function) as hummingbirds in tropical America. This is a hallowed and perfectly respectable usage.

Ecologists now give more attention to the other meanings. Species populations are defined as niche plus phenotype (Class II) or as the requirements that this idea invokes, the hypervolume or hyperspace (Class III). Evolutionary arguments then turn on the way members of species populations become adapted in the sense of having their niches fitted to the environment in which they must live. But this invokes the difficulty that the niche to which an organism is to be adapted by natural selection does not exist until revealed by the existence of the organism itself.

Since niches can be identified only by the organisms that occupy them, it becomes difficult to see how natural selection can adapt an organism to a niche. In a sense, organisms must always be perfectly adapted to the niche they are in and therefore cannot be adapted any further by natural selection. This apparent paradox is resolved by the Red Queen hypothesis (Van Valen, 1973). This hypothesis states that the environment of all organisms is always changing, either from climatic or habitat alterations or from the arrival or removal of other species. But if adaptation to the existing environment were already perfect, then all changes in the environment represent effective decay. Natural selection continually adapts organisms to meet these changes. The process is not one of perfecting an adaptation already effectively perfect, but of maintaining an adaptation to an environment that always changes for the worse. Evolution "runs hard" just to maintain perfection, like the Red Queen in *Through the Looking Glass*, who told Alice she ran so hard to stay where she was.

A test of the Red Queen hypothesis is provided by the fossil record. If the hypothesis is correct, then the survival time of comparable species should be roughly similar on the average. But if natural selection really worked to make species better adapted all the time, then the longer a species had been exposed to natural selection the better adapted it would be and the smaller would be its chances of going extinct. The fossil record (Figure 2.15) for a number of evolutionary lines for which good data are available shows the constant extinction predicted by the Red Queen hypothesis.

Thus it is useful in evolutionary arguments to think of a species as the phenotypic expression of a functional niche fashioned by natural selection. This niche can be defined as the environmental hyperspace occupied by individuals of the species. A large hyperspace, the *fundamental niche*, can be defined by resource and habitat parameters, which requires that the concept of niche overlaps somewhat the concept of habitat. Usually there is no confusion between the concepts of *habitat* and *niche*, though Whittaker *et al.* (1973) have suggested that it is helpful to coin a new term to combine habitat with the other axes of a niche, the ECOTOPE. The term "niche," however, continues to be preferred, and seldom leads to confusion if taken in context. Perhaps the one important cautionary note is to avoid using "niche" in the sense of "microhabitat." This usage has appeared in the older ecological literature, particularly in the sense of

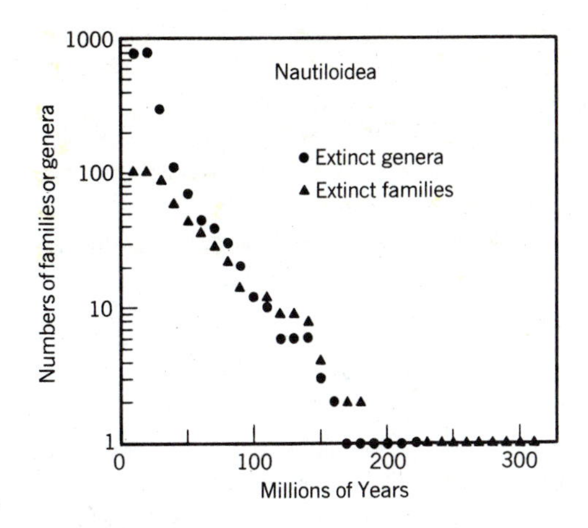

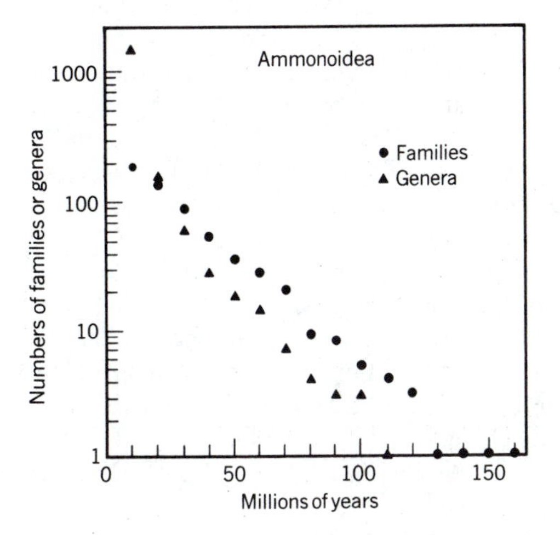

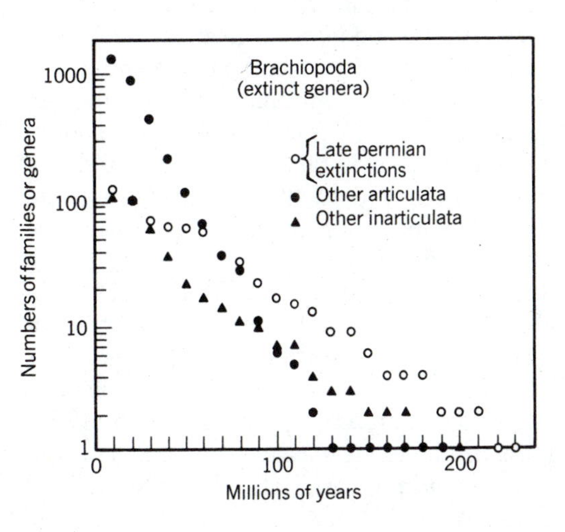

Figure 2.15 Test of the Red Queen hypothesis. The plots are survival times in two groups of molluscs and one of brachiopods. The Red Queen hypothesis predicts that survival will be independent of the age of a taxon, which the data confirm (from Van Valen, 1973.)

meaning the precise physical location of individual plants in a community or habitat. The usage is literally correct, since outside the profession of ecology a niche often means a physical place, like a slot in a wall suitable for placing a statue or a climbing boot. Ecologists now, however, always include a sense of function in their various uses of niche.

LIMITING FACTORS

Before the concept of niche was invented, the idea of environmental limits already was a major theme of ecology. It was referred to as the CONCEPT OF LIMITING FACTORS. It first became prominent in ecology from the work of the German chemist Justus von Liebig in the middle of the 19th century. Liebig discovered the importance of phosphorus to agriculture, noting that if a field were poorly supplied with phosphorus, crops would not do well on it however splendid a place the field was in other respects. Growth of a field crop, therefore, was *limited* by the *factor* phosphorus. Liebig wrote in a time when it was the fashion to express your ideas as "laws" and so Liebig called his result the LAW OF THE MINIMUM, which states GROWTH OF A PLANT DEPENDS ON THE AMOUNT OF FOODSTUFF THAT IS PRESENTED TO IT IN LIMITING QUANTITY.

The fertilizers that limit agricultural crops in this way are not the nutrients that the plants need in large amounts, like carbon dioxide and water, but those needed in comparatively minute masses. The essence of the concept is that an organism is likely to be limited by some apparently minor factor in the environment. Further studies were to show that the minor factors might actually have to be provided in combination, leading to the subsidiary principle of FACTOR INTERACTION. Moreover other factors could be present in excess, as when a soil has toxic concentrations of cobalt, or when heat or water are limiting through being present in excess. Odum (1955) summarized these results in a *combined concept of limiting factors* as THE PRESENCE AND SUCCESS OF AN ORGANISM OR GROUP OF ORGANISMS DEPENDS UPON A COMPLEX OF CONDITIONS. ANY CONDITION THAT APPROACHES OR EXCEEDS THE LIMITS OF TOLERANCE IS SAID TO BE A LIMITING CONDITION OR A LIMITING FACTOR.

This combined concept has much in common with some of the definitions of niche that seek to define an environmental hyperspace, but there are some significant differences of emphasis that lead to the concepts of limiting factors and niche being used in different ways. The concept of limiting factors:

1. Often refers to a plant process (growth) rather than to the distribution or fitness of the plants.

2. Emphasizes that a few factors alone are critical *limiting factors*, instead of all the factors (resource axes) of a hypervolume.

These properties of the concept of limiting factors are explored in the sections below.

1. Limiting Factors and Rate Response

Rates of all biological processes depend on environmental variables. Rates of processes such as growth, feeding, photosynthesis, respiration, nutrient uptake, et cetera, have important implications for fitness. These rates can be measured experimentally. Thus it is possible to use the concept of limiting factors as a guide in experimental investigations of the success or failure of organisms in nature. The approach is used, for instance, in all agricultural experimental stations that investigate what fertilizers or other factors of the environment must be supplied for maximum crop yield under local conditions.

When the response of a biological process like growth to a potentially limiting factor is plotted as a function of factor concentration up to some critical value above which there is no further response, the experiment yields a SATURATION response curve of the kind shown in Figure 2.16*a*. Biological activities that respond to concentrations of nutrients or food in this way include photosynthesis to increases in light, the kill rate of small predatory animals to increases in concen-

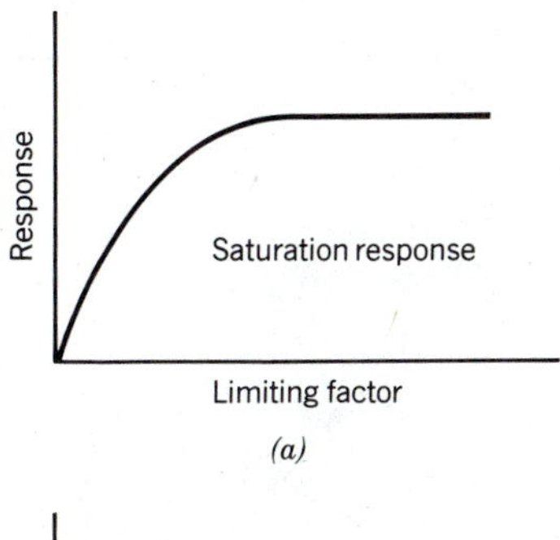

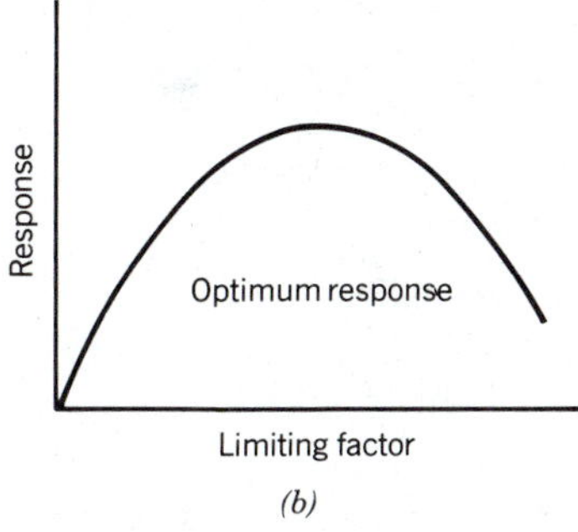

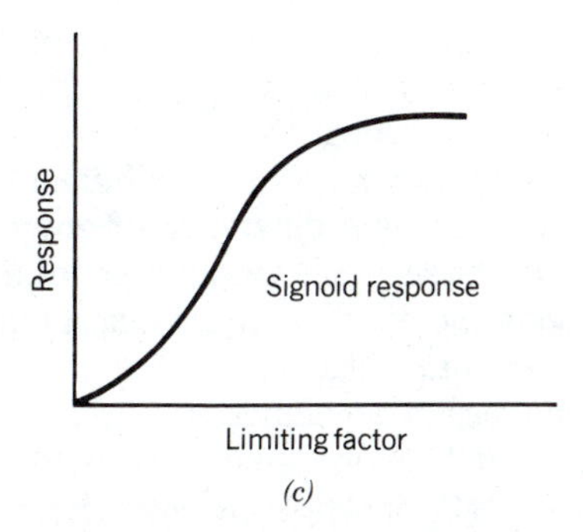

Figure 2.16 Limiting factor response curves. The saturation response in (*a*) is of the kind originally investigated by von Liebig. The optimum response curve in (*b*) suggests that distribution is set by tolerance. A sigmoid response (*c*) is unusual, probably being restricted to hunting animals that can learn behavior.

trations of their prey, and most aspects of plant growth to nutrient concentrations.

Response to the all-pervading environmental influence of temperature is different, however, because there is an optimum temperature for most biological processes at which maximum rates are achieved. Temperatures above, as well as below, this optimum temperature result in lower rates and a plot of process against temperature is likely to yield an OPTIMUM RESPONSE CURVE like that in Figure 2.16*b*. Sometimes chemical factors can yield an optimum response as well, for instance, the growth of organisms adapted to brackish water since they might be inhibited by the salinity of the open ocean in one direction and by the low concentrations of salt in fresh water in the other direction.

Most chemically based processes result in either a saturation or an optimum response but the activities of whole organisms can respond in more complicated ways. Some predatory animals with the ability to learn what to hunt can spend little time hunting a prey species in low concentrations, hunt more intensively prey at medium concentrations, or learn to concentrate on just one abundant prey item until saturated. The result is a SIGMOID RESPONSE CURVE (Figure 2.16*c*). This is a curiosity of predator–prey systems and is best discussed with other aspects of predation (Chapter 10).

2. Limiting Factors and Species Distributions

If some vital life process of a species is rate-limited in a way describable by an *optimum response curve* (Figure 2.16*b*) then it follows that the distribution and abundance of this species is likely to be set by the appropriate limiting factor of the environment. A plant may have its distribution set by temperature, water, or some necessary soil nutrient. An herbivorous animal may have its distribution set by temperature or water in a like manner, but also by the distribution of its plant food, in which event the plant food is called a BIOTIC FACTOR. The herbivore might also have its distribution set by the concentration of its competitors, parasites, or predators, which become other *biotic factors*. Similar arguments apply to predators, saprophages, and parasites.

When the whole animal or plant has its distribution set by its response to some limiting factor of the environment we can say that it is restricted according to its environmental TOLERANCES (Shelford, 1908), a statement that puts an optimum response curve into words. Organisms can

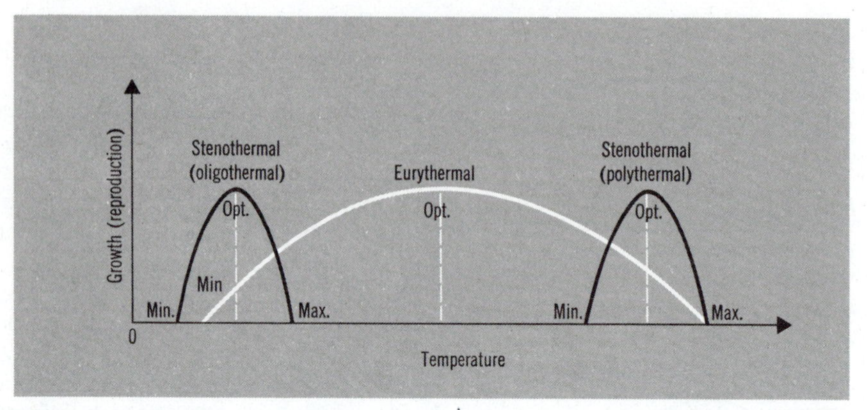

Figure 2.17 Illustration of ranges of temperature tolerance in stenothermal and eurythermal animals.
All animals are thought to be most active at some optimum condition, in this example, temperature, but some have narrow tolerances and others broad tolerances. (Redrawn from Ruttner, 1963.)

have narrow tolerances or broad tolerances, denoted by the Greek prefixes *steno-* and *eury-*; hence *stenothermal* (restricted to a narrow range of temperatures) and *eurythermal* (able to live in a wide range of temperatures) (Figure 2.17). Other common uses of the *steno-* and *eury-* prefixes are listed in Table 2.1.

Starting from the simple idea of a life-process being rate-limited by a limiting factor, we can therefore approach a concept of all species being restricted to particular places in which to live; not too hot, not too cold, of an appropriate dampness, with the right food, and with an acceptable array of natural enemies. This is a world view that has much in common with niche theory. The concept of the niche as a hypervolume of resource axes draws on the concept of limiting factors directly, saying that each axis can be treated both as a linear factor and as a factor to which an *optimum response* can be expected.

And yet there is a difference of emphasis in the niche and limiting factor approaches that has led to the asking of different ecological questions and to different results.

Limiting factors encourage ecologists to search for critical limits in the environment to explain why a species is, or is not, present. These factors tend to be properties like temperature or soil chemistry that are measured with little difficulty. These factors can be duplicated in the laboratory to check responses, or are susceptible to experimental manipulation in the field. The response of a particular plant or animal can be known with some precision. The ecologist oriented to limiting factors tends to see plants and animals as objects spread out before stresses of the environment (Figure 2.17). It is true that the limiting factor world view includes the *biotic factors* in its judgment, but it is almost inescapable that these biotic factors are of secondary importance because they are hard to measure.

Table 2.1
Terms Sometimes Used to Describe Various Kinds of Tolerance
The suffixes **-thermal** *and* **-haline** *are used quite commonly, the others less commonly.*

Terms	Refers to
stenothermal—eurythermal	refers to temperature
stenohydric —euryhydric	refers to water
stenohaline —euryhaline	refers to salinity
stenophagic —euryphagic	refers to food
stenoecious —euryecious	refers to habitat selection

Niche theorists, on the other hand, approach the issues of species distributions and abundance from the point of view of food resources and enemies as their first consideration. An animal or plant is an organism set to win resources and to reproduce. It is equipped to hunt or crop food, to escape its predators, and to stand up to its competitors. To do these things it is true that it has to be equipped to cope with the temperature, dampness, or odd chemistry of a place. But to the niche theorist these physical stresses are likely to be secondary. The unspoken assumption is that animals and plants will have been fashioned by natural selection to cope with whatever physical hardships there may be. Food and enemies are expected to provide the main rationale for a species to exist in any particular place.

Thus, although the concepts of *limiting factors* and *niche* clearly converge, in practice they have led to different results. An ecologist guided by limiting factors is inclined to measure temperature or soil chemistry first. An ecologist guided by niche theory is likely to measure food size or perch heights first. Both are valid approaches to ecological problems.

LIMITS BY TEMPERATURE OR RESOURCE: A FIELD EXAMPLE

The bottoms and sides of streams running from hot springs in Yellowstone Park are lined by algal mats. There is a temperature gradient down the stream channels, as you would expect, and there are different taxa of algae to be found in various temperature ranges. Between about 55°C and 40°C the algal mats are largely made up of filamentous blue-green algae, but these plants are rather stenothermal and will not actively grow at temperatures below 40°C, although the mats can persist at lower temperatures as long as they are not eaten or otherwise disrupted. The mats are eaten, however, by a species of fly which raises its maggots on them. The fly is very skillful at seeking out algal mats to lay large numbers of eggs on them, but the fly larvae cannot tolerate the temperatures at which the algal mats will actively grow, and hence must make do with moribund colonies in cooling water. This state of affairs can be understood from the graph shown in Figure 2.18 based on a sketch drawn for me by R. D. Mitchell, who has been investigating the phenomenon (Wiegert and Mitchell, 1972).

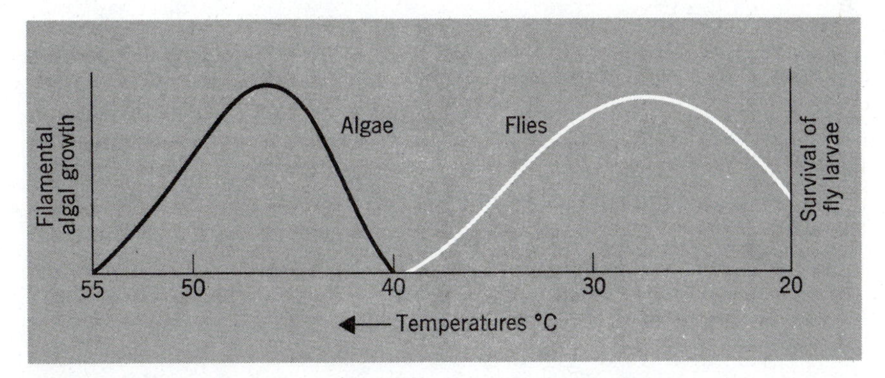

Figure 2.18 Survival and growth of fly larvae and algae they feed on in a hot spring as a function of temperature.
The flies feed on the moribund algal mat, which offers a larger standing crop of biomass at the lower temperatures to which the flies are adapted. Temperature tolerance in these flies has evolved to let them make use of the largest possible resource. (From observations of Wiegert and Mitchell, 1972.)

The tolerances of the flies are such that they are apparently prevented from eating what one might think should be their main source of food. Why should they have evolved such tolerances? Or is it that the fly engineer somehow could not make a fly which would tolerate warmer water?

The answer proposed by Wiegert and Mitchell rests on the discovery that most of the standing crop of the algae at any one time is present as moribund mat at temperatures which are near the optimum for the fly larvae. This comes about because of some peculiar properties of the spring system; the algal mats act as insulators so that the middle of an old mat becomes too cool for active growth. Furthermore, algal mats serve to divert the flow of the shallow streams, deflecting warm water, and cooling the mats still more. It thus comes about that the main food resource available to the alga-eating flies is at a temperature too cold for algal growth. The ecological strategy of the flies has been to adopt temperature tolerances that will enable them to feed on this resource. It has apparently not been worthwhile in an evolutionary sense for flies to have developed tolerance to warmer water, because the resources in that warmer water were too small to support a viable population. Knowing the temperature tolerances of those flies would be valuable to someone setting out to kill them, but is not very illuminating to someone wanting to know the flies' role in the hot spring ecosystem, and is almost useless in aiding an understanding of why that species of fly evolved.

The ecological moral of this story is that sometimes it is better to ask what an animal or plant is doing, than to ask what it can stand. For it will have evolved to tolerate what is necessary to allow it a livelihood. In some investigations it is important to ask questions about temperature. For the hot spring flies it was more illuminating to ask, "What was their resource?"

PREVIEW TO CHAPTER 3

Many of the most familiar properties of plants—their shapes, the arrangements of their leaves, and their colors—can be understood by thinking of plants as machines for the synthesis of glucose. The fuel for these machines is sunlight delivered from above. The raw materials are carbon dioxide delivered as a rare gas in the atmosphere, and a solution of nutrients in water delivered from the soil. Eighty percent of the radiant energy of the sun punches through the atmosphere and reaches the earth's surface as a characteristic pattern of wavelengths. Only half of these wavelengths, essentially visible light, have photons of energies that excite electrons in the way required by photosynthesis. Energy from absorbed photons of these wavelengths can be transformed to the energy of glucose with a maximum efficiency of 35%, but the efficiency of whole living plants is much less. Plants in experimental chambers can convert all the energy supplied to the potential energy of glucose with about 20% efficiency when light is dim, but in bright light this efficiency falls off to a maximum of about 8%. At about one-fifth of full sunlight photosynthesis becomes light-saturated. In full sunlight a leaf works at the same rate as it would at one-fifth of sunlight, thus it is very inefficient. Field measurements show that a whole crop or piece of vegetation is even less efficient, generally achieving less than 2% energy conversion. This low efficiency is partly explained as a result of light lost in warming soil or to moribund plant parts, or light degraded by the evaporation of water. It may also be that plants are built with a large array of chlorophyll pigments to let them be good scavengers of all available light in the dim hours of morning and evening. Light absorbed in these extra pigments could not be coupled to enough production sites in bright light and so would be wasted, lowering overall efficiency. But a more general explanation of the inability of a leaf to use light with an intensity of more than one-fifth of sunlight probably lies in the scarcity of the raw material needed for glucose synthesis, carbon dioxide. Synthesis can only proceed as fast as CO_2 can be extracted from the air, and this rate must be a function of the low concentration of the gas, 0.03% by volume. Supplying extra CO_2 to plants increases the rate of photosynthesis in bright light, as predicted by this hypothesis. More general support for the importance of CO_2 as a regulator of photosynthesis comes from studies of plants with specialized chemical pathways. Special enzymes and leaf structure in plants of many different genera adapt them to maintain CO_2 intakes in conditions of drought and heat. These species are the C4 and CAM plants, which pay the cost of an extra CO_2 absorption and handling mechanism by increased production in bright light. The physical designs of familiar plants can also be seen to result from the low efficiency of photosynthesis in bright light. Shapes of leaves and trees turn out to embody engineering designs that diffuse light so that as many leaves as possible receive light of optimum intensity for photosynthesis. Trees adapted to bright sunlight have leaves dispersed in depth, called a multilayer design, which serves to diffuse light at optimum intensity for photosynthesis to as much photosynthetic tissue as possible. Shade trees are monolayers designed to occlude all light at first interception. The erect leaves of grasses work as light receptors on an inclined plane, another device to increase leaf area receiving optimum light for photosynthesis. In the sea, unlike on the land, photosynthesis is more likely to be limited by light than by the carbon supply and large attached seaweeds (benthic algae) have adaptations for operating in dim light. The various pigments that make many seaweeds brown or red are additions to the pigment array employed by land plants and their function is to absorb in the parts of the spectrum that land plants reflect. Green plants in the sea approach the same goal of being "black" to the dim light of the depths with a very dense and thick array of pigments.

CHAPTER 3

THE TRANSFORMATION OF ENERGY BY PLANTS

All plants transform their energy by *photosynthesis* of reduced carbon compounds and all are equipped with the same basic chemical engineering and enzyme systems. There are, to be sure, variations on the theme of photosynthesis (like C3 and C4 pathways, or prokaryotic systems using sulfides as electron donors, see below) but these variations are minor. In general, all plants tackle the task of transforming energy with comparable chemical engineering. This engineering is probably ancient, being little altered from that operating in the first *prokaryotic* photosynthetic bacteria more than 3×10^9 years ago, with some embellishment in eukaryotic plants perhaps 0.7×10^9 years ago. It is to be expected that this long evolution has produced plants of optimum photosynthetic efficiency. This efficiency determines the energy flux available to an ecosystem.

THE ENERGY FLUX AVAILABLE TO PLANTS

A plant staked out under the sky receives only the radiant energy of the sun that penetrates the atmosphere. Not all solar energy reaching the earth does penetrate the atmosphere; nor is the flux of energy reaching a plant constant.

The flux of solar energy is measured as units (calories, kilogram calories, ergs, watts, BTUs, *et cetera*) arriving per unit area in unit time. The unit usually used in atmospheric physics is the LANGLEY (or *ly*), defined as the gram calories arriving per square centimeters per minute (g cal cm^{-2} min^{-1}), or a unit that equals a thousand langleys (kg cal cm^{-2} min^{-1}). Units of total energy like these must be used in ecology rather than light measurements like *lux* or *foot-candles*, because light measures apply only to the visible portion of the electromagnetic spectrum. Since, however, visible light is roughly proportional to total incoming radiation in most circumstances, a light-meter reading can be used to give an approximation of total radiant energy by use of a conversion factor:

$$\frac{\text{foot-candles}}{6{,}700} = \text{g cal cm}^{-2}\ \text{min}^{-1} = ly$$

(Reifsnyder and Lull, 1965).

The flux of solar energy reaching the top of the atmosphere appears to be constant within narrow limits and is referred to as the SOLAR CONSTANT. This is given an average value of

2 g cal cm^{-2} min^{-1} (2 *ly*), although there is probably an error of up to 5% in present measurements. The solar constant varies in synchrony with cycles of sunspots, but this variation is probably much less than the 5% measurement error (Lamb, 1972).

It is an open question whether the solar constant has changed more markedly in the geologic and evolutionary past but there is no evidence that requires changes in solar radiation within a time scale of interest to the ecologist. The more parsimonious theories of climatic change, for instance, manage without invoking changes in the sun's heat (Flint, 1971; Lamb, 1972).

The actual intensity of solar radiation received at the top of the earth's atmosphere, however, does change slightly over geologic and astronomical time because of rhythmical changes in the shape of the earth's orbit and angle of spin. These changes are small, though they may result in changes of average temperature of a few degrees at some latitudes, perhaps sufficient to cause and end ice ages. But their effect is unimportant to the working of contemporary plants, from whose point of view the solar constant is really constant. The energy a plant actually receives, however, is by no means constant and depends on such things as cloudiness and the season of the year.

Before solar radiation reaches a plant it may be attenuated in three different ways: by *reflection, scattering,* and being *absorbed in the atmosphere.* The largest of these three attenuations is reflection, particularly from the upper surfaces of clouds. Reflectivity has its own name in meteorology, ALBEDO. The *albedos* of different kinds of clouds are given in Table 3.1, from which it can be seen that as much as 70% of the total incident solar radiation is reflected from clouds.

Of the solar energy that is not reflected, more than 80% penetrates the atmosphere to strike the earth's surface. The portion of the missing 20% that is back-scattered is small, most of the loss being due to absorption by atmospheric gasses. This absorption is highly selective of different parts of the light spectrum with the result that the radiation reaching the earth's surface is a biased sample of the original output of the sun.

Table 3.1
Reflectivity (ALBEDO) of Clouds Compared with Land Surfaces
The most important loss of solar energy to the ground surface is from reflection on clouds, which may reflect up to 70% of incoming radiation back to space. Only snow-covered landscapes have higher albedos (up to 95%). Absorption is high on well-vegetated landscapes and in water columns. (Data from Lamb, 1972.)

Bare Crustal Surface	Albedo (%)
New-fallen snow	85 (90% has been observed in Antarctica)
Old snow	70
Thawing snow	30–65
Salt deposits from dried-up lakes	50
White chalk or lime	45
Clayey desert	29–31
River sands (quartz), wet	29
Clay (blue), dry	23
Ploughed fields, dry	12–20
Granite	12–18
Other rocks	12–15
Clay (blue), wet	16
Water, depending on angle of incidence	2–78
Vegetation	
Yellow deciduous forest in autumn	33–38
Parched grassland	16–30
Green deciduous forest	16–27
Green grass	8–27
Grain crops, depending on ripeness	10–25
Pine forest	6–19
Oak tree crowns	18
Stubble fields	15–17
Spruce tree crowns	14
Wet fields, not ploughed	5–14
Fir tree crowns	10
Clouds	
High-level clouds and cloudsheets (cirrus)	21
Middle-level cloudsheets (between about 3 and 6 km)	48
Low-level cloudsheets	69
Heap clouds (cumulus types)	70

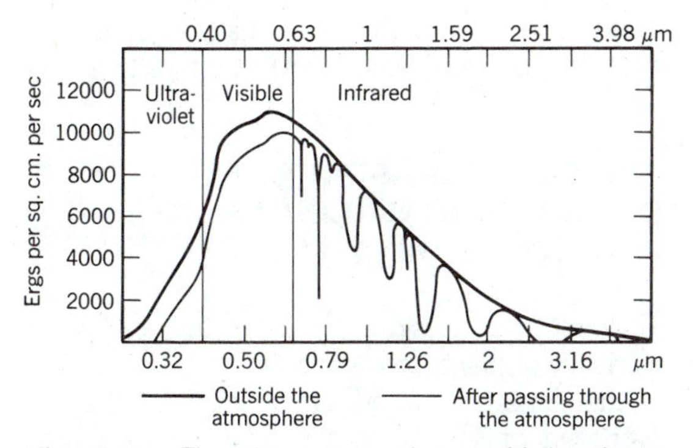

Figure 3.1 The solar spectrum above and below the atmosphere.
Although only about 20% of the incoming energy is degraded in the atmosphere, selective absorption by atmospheric gasses (particularly CO_2 and H_2O) presents plants with an irregular array of wavelengths. Photosynthesis needs energy of wavelengths roughly comparable to that of visible light. The energy flux is actually greatest (peak of the curve) in the visible wavelengths so the narrow range of wavelengths used in photosynthesis actually includes about half the total energy flux. The data are from measurements made in clear weather from a mountain observatory in Australia by C. W. Allen. (From Lamb, 1972.)

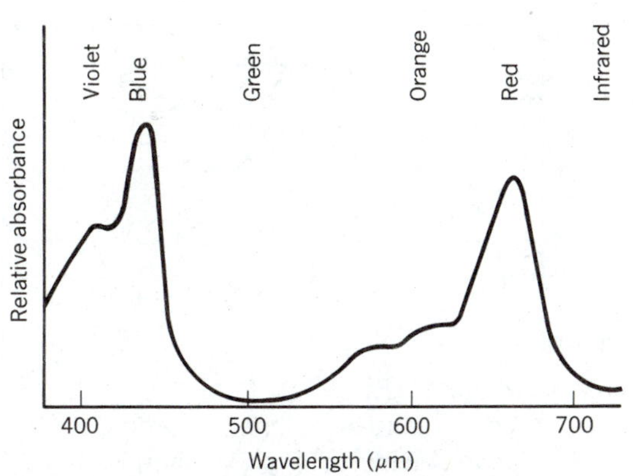

Figure 3.2 Absorption spectrum of chlorophyll *a*.
Only light with wavelengths between 0.38 and 0.78 nm have quantum energies that cause electron transitions when absorbed (Table 3.2). These are both the wavelengths of visible light and the wavelengths in which most solar energy reaches the earth's surface (Figure 3.1). Chlorophyll *a* absorbs strongly in these wavelengths. The "green gap," where chlorophyll does not absorb strongly, is plugged by secondary pigments like *carotenoids, xanthophylls,* and *phycobillins.* Electron excitations induced in these pigments by absorbed quanta are passed to chlorophyll *a* by resonance.

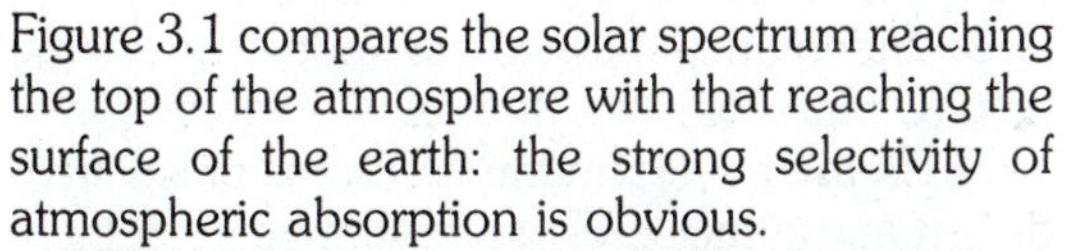

Figure 3.1 compares the solar spectrum reaching the top of the atmosphere with that reaching the surface of the earth: the strong selectivity of atmospheric absorption is obvious.

Figure 3.1 also shows that the energy of radiation is concentrated in the visible portion of the spectrum. In fact the narrow band of wavelengths that is visible to the human eye represents about half the total solar energy reaching the earth's solid surface. It would be a reasonable inference from a data set like that of Figure 3.1 alone to expect the photosynthetic mechanism to be adapted to these visible wavelengths, as it turns out to be (Figure 3.2). Photosynthesis works by using quanta of absorbed energy to displace electrons into orbitals of higher energetic states and only radiations in the visible range have quantum energies appropriate to do this (Table 3.2). Plants use the same light that we see by.

Table 3.2
Solar Energy Distribution and Photochemistry
Only radiations that result largely in electronic transitions can be used by photosynthesis. Ionizing radiations cannot be used, nor can low-intensity radiations that result largely in vibrational transitions and dissipation as heat. As a result, photosynthesis effectively uses only visible light. (From Morowitz, 1968.)

0.01– 0.2	152. 2–304.4	0.02	Ionization
0.2 – 0.38	75. 3–152.2	7.27	Electronic transitions and ionizations
0.38– 0.78	36. 7– 75.3	51.73	Electronic transitions
0.78– 3	9. 5– 36.7	38.90	Electronic and vibrational transitions
3 –30	0.95– 9.5	2.10	Rotational and vibrational transitions

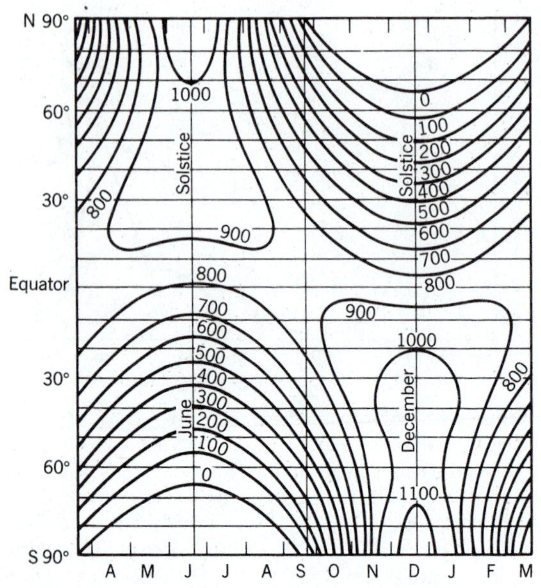

Figure 3.3 Solar radiation at different latitudes. The data are calculated values in g cal cm^{-2} day^{-1} for an earth with no atmosphere. Real values below the atmosphere will be lower (Figure 3.1), but will differ least in the wavelengths of visible light used in photosynthesis. (Data from Lamb, 1972.)

Figure 3.3 shows the average daily totals of solar radiation that would be delivered to different latitudes if there were no atmosphere. The actual radiation received at the surface of the real earth will be a reflection of this pattern, with the values being reduced by atmospheric absorption and subjected to local variations due to the patterns of clouds. Notice that the polar regions in high summer receive nearly as much energy of radiation in a day as places on the equator, because the 24-hour days that they then experience make up for the losses due to the oblique incidence of solar rays.

In summary, available light always is a skewed distribution of wavelengths, with half the energy concentrated as visible light. The overall intensity will range from almost nothing at night to the intensity of full sunlight under a clear sky. Where clouds interfere, intensity will be reduced, perhaps by as much as 70%, but the spectral composition will remain unaltered. For plants the central fact is that light is of constant quality but available in dramatically varying intensity.

THE FIRST MEASUREMENT OF ECOLOGICAL (LINDEMAN) EFFICIENCY OF PLANTS

The first and classic estimate of the ecological efficiency of plants growing in the open was made in the American Midwest in 1926 by Nelson Transeau of The Ohio State University. Transeau (1926) was interested in the very practical matter of determining if the efficiency of transformation of energy by plants was high enough for it to be reasonable to grow plants for fuel, thus freeing our society from dependence on fossil fuels like petroleum and coal. Transeau was half a century ahead of his time in this. From an ecological point of view, he set out to measure the ratio λ_0/sunlight, or the efficiency with which energy is transformed in the first trophic level of an Eltonian pyramid. He did this the year before Elton published his concepts of *food chains* and the *pyramid of numbers*, and sixteen years before Lindeman published his model of energy flow (Chapter 2).

Since there were many measures of solar radiation in the meteorological literature even then, Transeau's task was to estimate the energy transformed by plants as calories per unit time. Then, as now, it was not possible to measure the energy intake of a whole plant directly, but Transeau reasoned that all the energy transformed by a plant was used to synthesize the sugar glucose ($C_6H_{12}O_6$). By the principle of conservation of energy, the energy of all the glucose synthesized over a time interval by a plant represented the total energy transformed. Transeau's problem then became one of measuring glucose production over the lifetime of a plant.

Transeau chose corn (maize, *Zea mays*) as his plant because good data for the growth of corn plants could be culled from the agricultural literature. He then proceeded as follows: An acre of good farm land in Illinois (chosen because there were good meteorological data for Illinois) should contain 10,000 corn plants and they should take 100 days from sprouting to harvest; then . . .

Total dry weight of 10,000 corn plants (roots, stems, leaves, and fruits)	6000 kg	(dry corn)
Total ash content of 10,000 corn plants (minerals from soil left after burning)	322 kg	(ash)
Therefore, total organic content per acre	5678 kg	(dry carbohydrate)
Average organic matter contains 44.58% of carbon, therefore carbon per acre.	2675 kg	(carbon)
Converting to glucose, 10,000 corn plants produced on one acre in 100 days.	6678 kg	(glucose)

This figure of 6678 kg glucose represents the STANDING CROP of corn at harvest described as a mass of glucose. The farmer's crop, or YIELD, of grain would of course be much lower. 6678 kg of glucose was also the NET PRIMARY PRODUCTION of the field. It was now necessary to measure the glucose that had been used in respiration. Transeau had his own measures of corn-plant respiration on which to draw. He had kept corn plants in dark chambers through which he passed continuous streams of air. He collected the carbon dioxide at the outlet in an alkali solution and estimated the quantity evolved in unit time by titration. Measurements made on typical corn plants of various ages gave him an average figure for respiration of 1 percent of the mass of each plant per day, which enabled him to complete his calculation as follows:

Since the crop at the end of the season weighed 6000 kg, the average dry weight for the season was	3000 kg	(dry corn)
Average respiration was 1 percent of this which	= 30 kg	(dry corn)
Therefore, the total CO_2 released in 100 days is 30 times 100	= 3000 kg	(CO_2)
Carbon equivalent of 3000 kg CO_2	= 818 kg	(carbon)
Glucose equivalent of 818 kg carbon	= 2045 kg	(glucose)
Gross primary production of glucose equals net primary production plus respiration equals 6687 kg plus 2045 kg	= 8732 kg	(glucose)
But the energy required to produce 1 kilogram of glucose is 3760 kg cal (a figure found by bomb calorimetry).		
Therefore, total energy consumed in photosynthesis of one acre of corn in 100 days equals 8732 times 3760 equals approximately	33,000,000 kg cal	
Energy received by one acre of Illinois in 100 days equals	2,043,000,000 kg cal	
Therefore, efficiency of photosynthesis equals		

$$\frac{33 \times 10^6}{2043 \times 10^6} \times 100 = 1.6 \text{ percent}$$

This figure of 1.6% has since been shown to be in the normal range, both for agriculture and for wild vegetation (Table 3.3). These results are for plants in the aggregate, rather than for single plants, but it is a good assumption that individual plants conform to the average of their community within tolerable limits. The agricultural crops are all single-species stands, showing that the results do apply to individuals. Ecological (Lindeman) efficiency of the first trophic level, therefore, turns out to be low. It is now necessary to examine photosynthesis, and the demands of this process on a living plant, in quest of an explanation of this low efficiency.

PHOTOSYNTHESIS: ITS LIGHT AND DARK REACTIONS

Photosynthesis is a process of many stages but they may be divided into two groups collectively known as the LIGHT REACTIONS and the

Table 3.3
Ecological (Lindeman) Efficiencies of the Plant Trophic Level
Transeau's and Gaastra's data as discussed in text. Remainder rearranged from Odum (1971).

	Kcal/m²/day	Efficiency %
Transeau's cornfield	33	1.6
Gaastra's sugarbeet field	—	2.2
Sugar cane	74	1.8
Water hyacinths	20 to 40	1.5
Tropical forest plantation	28	0.7
Microscopic alga culture on pilot scale	72	3.0
Sewage pond on seven-day turnover	144	2.8
Tropical rain forest	131	3.5
Coral reefs	39 to 151	2.4
Tropical marine meadows	20 to 144	2.0
Galveston Bay, Texas (fertilized by wastes)	80 to 232	2.5
Silver Springs, Florida (vegetated bottom)	70	2.7
Subtropical blue water (open sea)	2.9	0.09
Hot deserts	0.4	0.05
Arctic tundra	1.8	0.08

DARK REACTIONS. The light reactions essentially transform quanta of light (photons) into forms of energy usable by a plant to power chemical synthesis, and the dark reactions work the synthesis itself. Both reactions continue at the same time, forming a smoothly meshed total process. But different constraints must apply to the light and dark reactions, to the energy transfer and to the chemical synthesis. The final efficiency of a plant at converting sunlight to the potential energy of glucose is necessarily a function of the efficiency of both processes.

Photosynthesis is reviewed in Figure 3.4 (see Calvin, 1976; Bonner and Varner, 1976).[1] The essentials of the light reactions are the dissociation of water into molecular oxygen, hydrogen ions, and electrons, followed by the excitation of the captured electrons by light in the boosts of photosystems II and I in sequence. The molecular oxygen is vented to the atmosphere. Both the hydrogen ions and the energy of excited electrons are used in the production of NADPH. Some energy of the excited electrons is used to make ATP from ADP. Both NADPH and ATP then serve as energy sources for synthesis of glucose in the Calvin–Benson cycles of the dark reactions.

These essentials of the LIGHT REACTIONS have in common that they can go on in a tiny, closed, transparent container. Water is needed, but then water is the commonest ingredient of living cells. Beyond water, the plant must provide the right molecular array spread out on membranes. The essential energy flux arrives in the form of radiations that penetrate cell walls and cell sap to strike the absorbing pigments spread out to receive them. But the dark reactions have to be performed in qualitatively different circumstances.

The ultimate raw material of the DARK REACTIONS is the rare gas carbon dioxide, present in the ambient air at a concentration of 0.03% by volume. This gas does not arrive at the reaction sites by itself, as light does. Carbon dioxide has to be hauled in. It is first taken up, inside the stomates, by the 5-carbon phosphorylated sugar RuBP (ribulose biphosphate, formerly known as RuDP) under the action of the enzyme RuBP carboxylase. The result is 6-carbon phosphorylated sugar that speedily breaks down into two 3-carbon compounds, phosphoglycerate (PGA). PGA is then the refined raw material for sugar synthesis in the Calvin–Benson cycle. It will be evident that the rate of the entire synthesis depends on the rate at which carbon dioxide can be dragged out of the air by RuBP carboxylase, thrust onto RuBP, broken into 3-carbon PGA, and fed into the cycle. The dark reaction, therefore, has limits set by the flux of scarce raw material (CO_2) as well as limits inherent in thermodynamics.

Measures of the rate of either the light or dark

[1] Any college biology text will review photosynthesis for the purpose of this discussion but there is a particularly clear and balanced account in D. L. Kirk, *Biology Today,* 3rd ed., Random House, New York (1980).

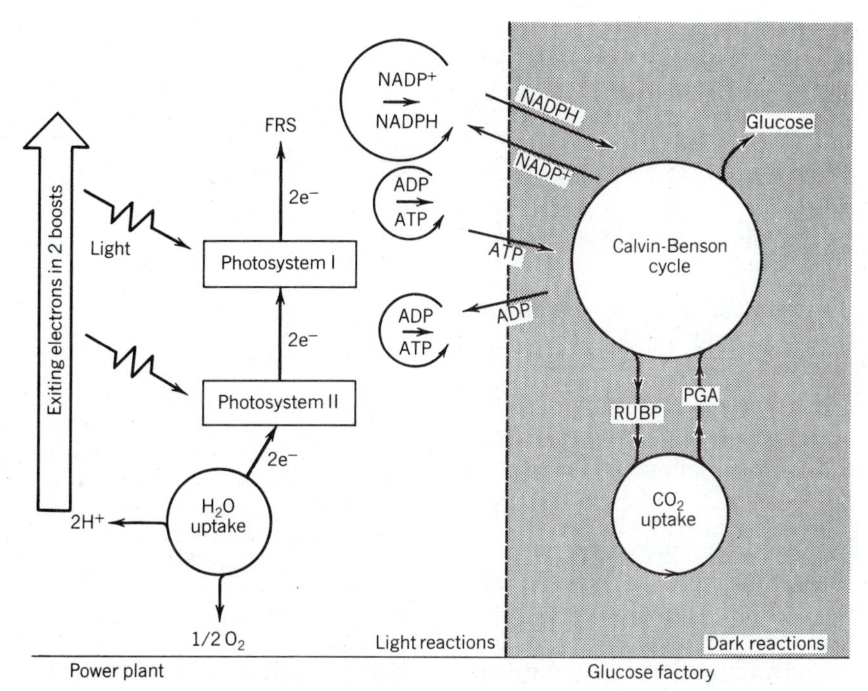

Figure 3.4 Light and dark reactions of photosynthesis.
The light reactions provide the power for the glucose synthesis performed in the dark reactions. Electrons released by splitting water are excited by light in a two-stage booster process until they have sufficient energy to reduce FRS (ferredoxin-reducing substance). FRS then drives the production of NADPH (reduced form of nicotinamide adenine dinucleotide phosphate). Some excited electrons are also used in the production of ATP. NADPH and ATP then serve as fuels for synthesis of glucose in Calvin–Benson cycles.

reactions give the rate of photosynthesis. The rate of production of oxygen monitors the light reaction and the rate of extraction of CO_2 gives the dark reaction. Production of O_2 exactly balances absorption of CO_2 because the rates of the light and dark reactions are coupled. We write the pseudo-chemical equation

$$6CO_2 + 6H_2O \xrightarrow{\text{energy}} C_6H_{12}O_6 + 6O_2 \quad (3.1)$$

This equation will certainly be familiar. It has little validity as a description of the process of photosynthesis, but the equation is a convenient expression of the gas exchange involved.

A difficulty in using gas measures to record photosynthesis is that living plants are always respiring CO_2 even as they haul the same gas into their RuDP–PGA feed for fresh glucose synthesis. In respiration they also absorb O_2, thus partly offsetting the O_2 vented from the dissociation of water. The respiration equation, less inaccurate than the photosynthesis equation, is opposite to it:

$$C_6H_{12}O_6 + 6O_2 \underset{\longrightarrow}{\overset{\uparrow}{\text{energy}}} CO_2 + 6H_2O \quad (3.2)$$

The routine way of estimating respiration is to measure gas exchange in the dark, when no photosynthesis is possible, and to amend the light measurements accordingly. This is what Transeau did when he added glucose respired to the glucose of standing crop to convert NET PRIMARY PRODUCTION into GROSS PRIMARY PRODUCTION. In general, the procedure is as follows:

1. Measure oxygen added to container in unit time in light.

2. Measure oxygen taken from container in unit time in dark.

3. Add volume O_2 *added* in (1) to volume O_2 *lost* in (2) to derive total O_2 produced by photosynthesis in unit time or

4. Add volume CO_2 *lost* in unit time in light to volume CO_2 *added* in unit time in dark to derive total CO_2 used by photosynthesis in unit time.

An obvious objection to these methods is the assumption that respiration by day and by night is the same. The volume of CO_2 respired in daylight is calculated from estimates of what the plant does in the dark. This is surely not fair. When the plant is working away at synthesis in the daylight hours it should be respiring more than at night when it is doing less obvious work. There is no satisfactory way around this difficulty (Westlake, 1963). It means that measures of rate of photosynthesis, or of production, are always probably marginally too low.

Gas exchange measurements let the rate of photosynthesis be monitored in simple laboratory systems with some precision. Cultures of green algae (*Chlorella* spp. are a favorite), seedlings, or preparations of single leaves, parts of leaves, or isolated chloroplasts can all be used. When light intensity is varied, the relationship between photosynthesis and light intensity is as shown in Figure 3.5. At low light intensities, the rate of photosynthesis is directly proportional to supplied light and appears to be light-limited. But there is always a plateau above which the rate of photosynthesis cannot be raised by increased light where the system is said to be LIGHT-SATURATED.

These data suggest that different factors limit photosynthesis at low light intensities and at high. The likelihood of this was, in fact, what prompted the original postulate that photosynthesis comprised separate light and dark reactions. F. F.

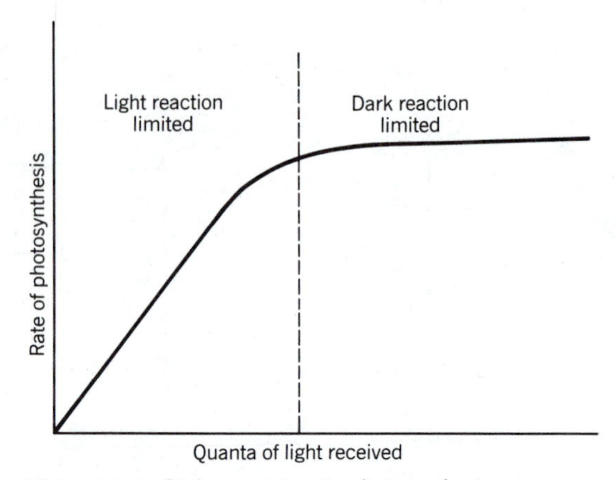

Figure 3.5 Light saturation in photosynthesis. The rate of photosynthesis cannot be increased beyond a certain level by supplying more light, showing that some other limit is applied. This observation led to the original suggestion that photosynthesis involved separate light and dark reactions.

Blackman obtained results like those of Figure 3.5 in 1905 and postulated limits independent of light (that is, dark reaction limits) when the system was light-saturated. Such dark reaction limits would obviously be applied if the flux of CO_2 provided by RuBP carboxylase achieved maximum rates before light became limiting. This hypothesis of CO_2 limitation predicts that light saturation would be delayed if extra CO_2 were provided to the system, a prediction that is confirmed by experiment (Figure 3.6). A rich supply of CO_2 increases the rate of photosynthesis in full light.

Initial reflections, therefore, suggest that photosynthesis may be limited by the flux of energy itself when light is dim but by the concentration of carbon dioxide when light is bright. This general conclusion can be both amplified and modified by closer examination of some of the systems involved.

WHAT DO WE MEAN BY THE EFFICIENCY OF PHOTOSYNTHESIS?

A biochemist's calculation of the efficiency of the whole process of photosynthesis, from the ab-

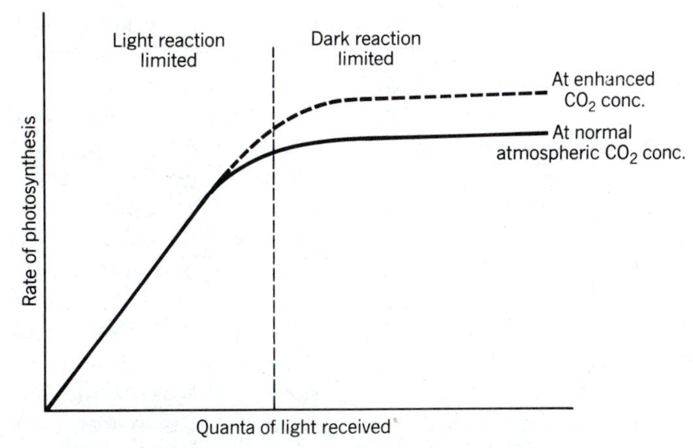

Figure 3.6 Effect of CO_2 concentration on photosynthesis.
Increasing the concentration of CO_2 in the experimental vessel raises the rate at which photosynthesis becomes light-saturated, thus suggesting that the dark reaction limit can be set by carbon dioxide concentration. Similar results have been obtained for field crops provided with enhanced CO_2 under plastic tents.

sorption of photons in photosystem II until the completion of a molecule of glucose, goes like this. The total free energy supplied in the synthesis of glucose is +686 kcal/mol

$$6CO_2 + 12H_2O \rightarrow C_6H_{12}O_6 + 6O_2 + 6H_2O \quad (3.3)$$

$$\text{Increase in free energy} = +686 \text{ kcal/mol}$$

This can be compared with the energy put into the synthesis by estimating the number of electrons that pass through the pathway in the synthesis of glucose, this number being 24. But each electron is energized twice (in photosystem II followed by photosystem I), requiring a total of 48 photons of light per electron. The longest wavelength these photons are likely to have is 700 nm (Figure 3.2), which gives a minimum of 41 kcal/mol per photon, yielding an energy input to all 24 electrons of $41 \times 48 = 1968$ kcal.

$$\text{Biochemical efficiency of photosynthesis} = 686/1968 = 35\%$$

This result is quite different from Transeau's estimate of photosynthetic efficiency for corn of 1.6%. But the different results can be understood easily when we examine what biochemists and ecologists mean when they talk of "efficiency."

The first glaring difference is that a biochemist is only concerned with the efficiency with which energy actually introduced into the system is transformed: 48 photons are used, so let us forget all the rest that might have been absorbed into a photosystem but were not. Let us even forget those that did enter the photosystem, and which excited electrons that then dropped back to lower energy states without doing useful work. Only the used energy will be counted. But calculating the *ecological* efficiency of a plant requires that all these portions of the total energy flux be included.

When photosynthesis is used by an individual plant to yield the resources it will need to gain maximum fitness, the test must be, "How much of the energy theoretically available to the plant was actually trapped and stored as glucose?" The top of the ecological efficiency equation might be the same as the biochemist's but the bottom, including as it must all the photons that got away, will be much larger.

So part of the wide gap between the biochemical efficiency of photosynthesis by a corn plant (35%) and the ecological (Lindeman) efficiency of a field of corn (1.6%) is represented by the photons that got away from the corn plants. It is easy to see where most of these escaped. In the first weeks of the corn's hundred days there was bare ground between the seedlings and most photons went to heat soil. In the last weeks of the hundred days many incoming photons struck moribund tissue of rustling yellow corn plants at the end of their lives, and these photons did no more than warm the corn. Still other photons were absorbed by water, resulting in transpiration streams through the stomates rather than in glucose.

And yet these photons that so clearly got away cannot explain all of the discrepancy between the efficiencies of biochemists and field ecologists, though they do reduce the gap greatly. The efficiency with which a vigorous, healthy, young, growing plant transforms sunlight falling directly upon it turns out to be on the order of 8% at light saturation and highest productivity.

Figure 3.7 shows data for photosynthesis measured as CO_2 absorbed by a preparation of a sugarbeet leaf at various light intensities. A curve of photosynthetic efficiency is superimposed on the curve of photosynthetic rate. Efficiency is high in dim light, being nearly 18%, but this falls to 8% at light saturation. Efficiency continues to fall at still higher light intensities, although photosynthetic rate is constant.

At very low light intensities, when light is not limiting, the efficiency of energy conversion by the sugarbeet leaf is still half the biochemist's efficiency of the process itself (18% versus 35%). Perhaps this discrepancy can be accounted for by electrons absorbed but not passed into the photosystems, because of necessary mechanical imperfections in plant design, so that the plant designer's true efficiency is closer to 20% rather than the theoretical maximum of 35% calculated from changes in free energy. Alternatively the discrepancy might result because some light was of the wrong wavelength for photosynthesis. But neither will account for efficiency falling to 8% at light saturation, and then falling further. Another explanation has to be offered for this.

Data sets like that of Figure 3.7 for sugarbeets are typical of all plants from crops to algae. Wassink (1959) got closely comparable results for a culture of *Chlorella*, for instance: 20% efficient in the dimmest light, 8% when light saturation

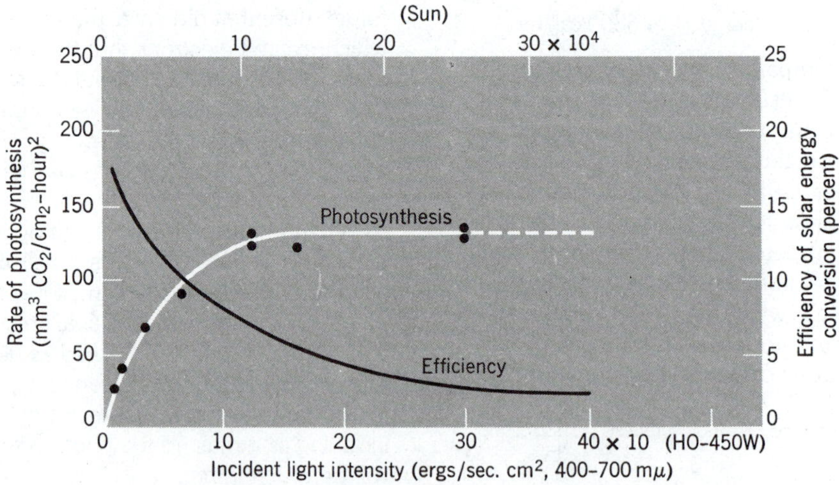

Figure 3.7 Rate and efficiency of photosynthesis as a function of light intensity. The data are for preparation of a sugarbeet leaf. The rate of photosynthesis would not be raised for light intensities much above 10 ergs per second per square centimeter, and accordingly the efficiency of the process progressively falls. It is important to note, however, that the yield of energy to the plant was highest at high light intensities even though the efficiency was low. Note also that this sugarbeet leaf in dim light was quite as efficient as a green alga culture in equally dim light. It is untrue to claim that algae are more efficient than other plants (From Gaastra, 1958.)

was reached, and even lower efficiencies at still higher light intensities. Thus algae are no more, and no less, efficient than higher plants. The limits to photosynthetic efficiency are apparently universal.

As we examine more closely alternative limits that might prevent photosynthesis in practice from reaching the high efficiencies of theory, it is necessary to remember that the actual mechanism in all plants is the same, and the working efficiency is therefore the same. Somewhere between the 20% measured in dim light and the free energy calculation of 35% is what the chloroplasts of all plants are engineered to yield. The comparatively poor actual performances of whole plants must reflect either their limited capacity to absorb energy or limits to the dark reaction, probably the supply of carbon dioxide.

THE BONNER LIGHT-ABSORPTION HYPOTHESIS

Bonner (1962) notes that all plants have what appears to be a large excess of chlorophyll molecules for the number of synthesis sites we can identify. This means that more photons are absorbed in bright light than can possibly be used for photosynthesis, and the low efficiency results. It is then necessary to explain why plants should have this wasteful excess of chlorophyll.

Bonner's answer is that plants are designed to work in dim light because dim light is what they get much of the time. Light is dim in the morning, in the evening, and in the shade of other leaves and other chloroplasts. The sun of noonday, for all its rich flood of energy, is a fleeting phenomenon set against longer periods of dimness or overcast. This means that fitness will go to plants that achieve a good trade-off between operating in gloom and operating in bright times. And this trade-off burdens a plant with surplus chloroplasts.

Operating in dim light requires that each reaction site be coupled to numerous chloroplasts to ensure that a synchronous pulse of sufficient photons can be intercepted. Unless all the photons arrive at once, enough electrons will not get their double kick through photosystems II and I for FRS to receive the energy needed to reduce $NADP^+$ to NADPH. Coupling a reaction site to a large number of absorbing chlorophyll molecules, therefore, is the only way to operate in gloom. This solution clearly works because 20% efficiency is possible in dim light. But it does mean redundant chloroplasts for the bright light operation, with a consequent decreased efficiency.

Yet, even though Bonner's explanation offers an elegant proximal cause for decreased efficiency in bright light, this still leaves the possibility of another distal cause. Part of the advantage of operating efficiently in dim light is to make the most use of the rarefied resource, carbon dioxide, for as long a time as possible. This means that surplus chlorophyll molecules are themselves a consequence of CO_2 shortage.

CARBON DIOXIDE AND THE C4 PATHWAY

Many plants give direct evidence of the importance of CO_2 uptake to the rate of photosynthesis by operating an extra CO_2-fixing system outside the RuBP system. These are the C4 plants. The carbon pump of the C4 system was discovered when the classical experimental approach of Calvin was applied to samples of sugar (Hatch and Slack, 1970). In the classical approach, leaves or chloroplasts receive isotopically labelled carbon dioxide ($^{14}CO_2$) and are then killed after short time lapses. The progress of the labelled carbon through the series of compounds leading to glucose synthesis can then be followed. In Calvin's work the labelled carbon passed rapidly through RuBP and into the 3-carbon phosphoglycerate (PGA). But when this same approach was applied to sugar cane the labelled carbon appeared first in several 4-carbon compounds instead of in PGA. Sugar cane is a "C4" plant instead of being "C3."

The essentials of the Hatch–Slack C4 system are shown in Figure 3.8. The enzyme PEP carboxylase fixes CO_2 as it enters the leaf through the stomates. CO_2 is added to the PEP (phosphoenolpyruvate), converting it from a 3-carbon molecule into a 4-carbon molecule. Oxaloacetic acid, malic acid, and aspartic acid (all 4-carbon)

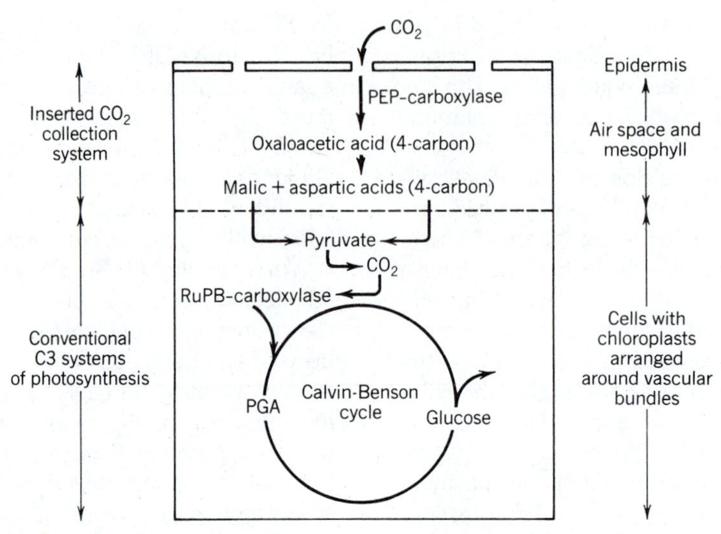

Figure 3.8 C4 photosynthetic pathway.
C4 plants have the same system for light collection and glucose synthesis as C3 plants but the chloroplasts where synthesis is performed are placed deep within the leaf around the vascular bundles. CO_2 is collected from air spaces inside the stomates by an enzyme PEP carboxylase that passes the CO_2 to a series of 4-carbon acids. These are transported across cells of the mesophyll to the cells containing chloroplasts, where the CO_2 is regenerated. In some environmental circumstances the energetic costs of this extra CO_2 transfer can be met by extra glucose production from more efficient use of light.

are then produced. These are transported across the cells of the leaf mesophyll to the interior, where cells green with chloroplasts await them. Once arrived at a production site, the 4-carbon acids are decomposed to yield their CO_2, and their residues are returned across the mesophyll for more. The regenerated CO_2 is then fed into a Calvin–Benson synthesis cycle via RuBP carboxylase in the usual way, the synthesis being powered by photons that have penetrated the transparent mesophyll to reach these production sites deep in the leaf.

Possessing a Hatch–Slack chemical system clearly incurs costs that most plants do not have to pay, because extra work is done in manipulating CO_2 and 4-carbon acids through the mesophyll. This work must detract from calories that the plant can use to win fitness. There must, therefore, be compensating advantages. These seem to be of two kinds: *better scavenging of* CO_2 and *prevention of photorespiration.* Both these advantages reflect on unsatisfactory properties of RuBP carboxylase when used, as by a C3 plant, to take CO_2 directly from the air.

The enzyme RuBP carboxylase has the unfortunate quality that, in the presence of free oxygen, it catalyzes a reaction that combines free oxygen with RuBP to split it into one molecule of PGA and some useless by-products that are oxidized back to CO_2. The enzyme thus undoes much of its own good work. On the one hand the enzyme adds carbon to RuBP but on the other it oxidizes carbon already held in RuBP and vents it back as CO_2. This curious behavior of RuBP carboxylase is called PHOTORESPIRATION (because CO_2 is respired in the presence of light). In light, and in the presence of free oxygen, this always happens. At first sight, this *photorespiration* appears to reduce very significantly the efficiency of the enzyme in fixing CO_2, though this is arguable and has been argued (Bonner, 1976). In a C4 plant this unfortunate side-effect of RuBP carboxylase is blocked, because the sites of activity of the enzyme are removed from the air and away from ambient free oxygen.

But the issue of scavenging CO_2 may be of more importance and is certainly of more direct interest to an ecologist. The rate at which CO_2 is taken from air by a C3 plant with RuBP carboxylase falls rapidly at concentrations much below the atmospheric norm of 300 ppm (*parts per million*, 300 ppm = 0.03%) and ceases entirely at about 50 ppm (Figure 3.9). PEP carboxylase of a C4 plant, however, continues to take up CO_2 at a high rate down to concentrations so close to zero that they are hard to measure. If photosynthesis in bright light is truly limited by the rate of CO_2 uptake, then C4 plants should have a decided advantage.

It can readily be shown that C4 plants can take up CO_2 from the surrounding air at faster rates than can C3 plants in a variety of experimental conditions. Such comparisons are particularly easy because it often happens that different, closely related species (*sympatric* species, see Chapter 7) occur in pairs, one using the C3 pathway and one using the C4. Both species can then be raised in a greenhouse in identical conditions and their carbon uptake measured in experimental chambers. Figure 3.9 describes the results of this procedure when applied to two species of desert saltbush (*Atriplex*), suggesting that the C4 plants were more "efficient."

Yet, to an ecologist, the idea that "efficient" plants and "nonefficient" plants could coexist is an idea so unlikely as to verge on the impossible. The "efficient" plants should win more resources, leave more offspring, and drive the "nonefficient" plants to rapid extinction. The only logical way in which this coexistence can be conceived is that the process of replacement is still in progress. There is, however, convincing evidence that no such replacement is happening. The evidence comes from studies of the spread and antiquity of the C4 pathway.

Whether a plant is C3 or C4 can be determined from dead herbarium material by using a mass spectrometer. The two enzymes RuBP carboxylase and PEP carboxylase discriminate dif-

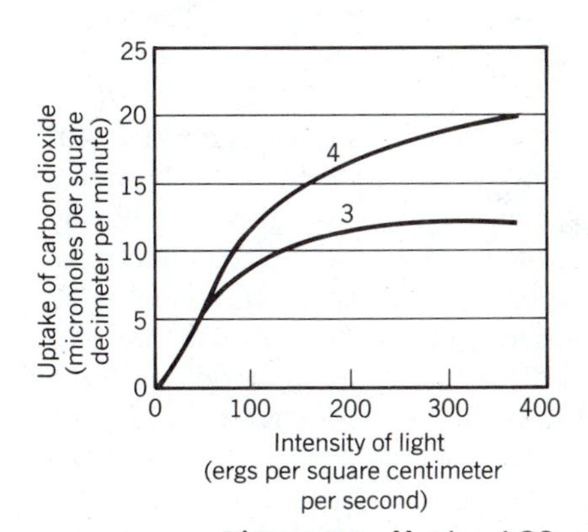

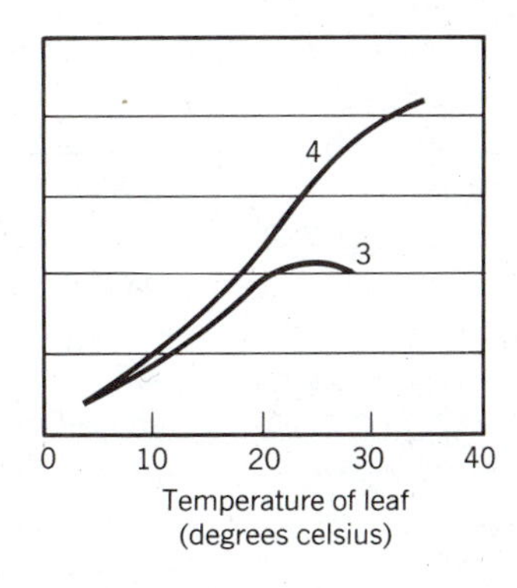

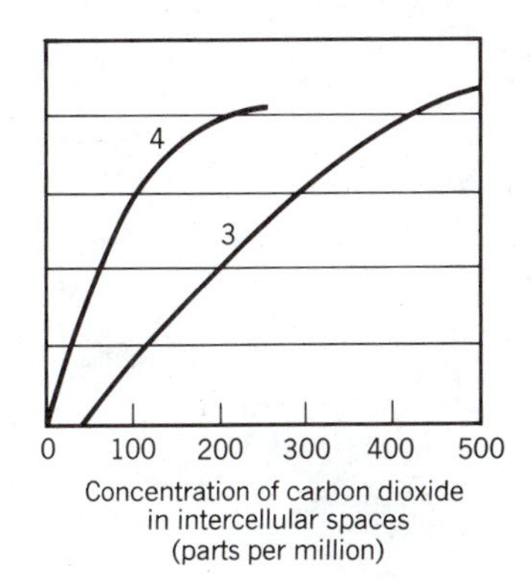

Figure 3.9 Uptake of CO_2 in C3 and C4 plants.
The data are for greenhouse-grown plants of two species of the same genus, one C3 and the other C4, grown under identical conditions. For both, the rate of CO_2 uptake depends on light energy, temperature, and CO_2 concentration, but the rate for the C4 species is always higher. The data, however, describe only the rate of CO_2 uptake of plants for the experimental runs and this is not necessarily proportional to the rate of glucose synthesis either during the run or over the life of the plant. Furthermore, a calculation of relative efficiency requires an estimate of the extra cost of the C4 operation. The plants are two species of desert saltbush *Atriplex patula* (C3) and *Atriplex rosea* (C4). (After Björkman and Berry, 1973.)

ferently against the rare stable isotope of carbon, ^{13}C, so that a mass spectrographic measure of the $^{13}C/^{12}C$ ratio on dead material reveals which pathway the living plant used. This method has shown that the C4 mechanism is possessed by more than a hundred genera in ten families of flowering plants (Burris and Black, 1976). The method has also shown that different species within the same genus possess one or the other, as in the *Atriplex* pair of Figure 3.9. When used on subfossil fragments the method has also shown that both C3 and C4 pathways were in use more than 40,000 years ago (Troughton *et al.*, 1974). Clearly the C4 pathway cannot be a new evolutionary invention that is even now changing the balance of life on earth but rather it is an alternative long available.

In fact all plants possess PEP carboxylase and the related chemistry needed to use the enzyme in this role. The C4 pathway requires more than chemical prowess—it also requires changes in leaf geometry that let PEP carboxylase be the primary fixer of carbon. This special leaf geometry, called the KRANZ SYNDROME, is illustrated in Figure 3.10. Since it occurs so widely in so many unrelated genera, only simple genetic changes can be needed to bring it about, possibly only a change at one locus. A rare intermediate condition is known (Kennedy and Laetsch, 1974). As with measurements of $^{12}C/^{13}C$ ratios, the *kranz syndrome* can be used to detect the use of the C4 pathway by fossil plants. This approach has shown the existence of C4 plants at least as early as the Pliocene (Nambudiri *et al.*, 1978).

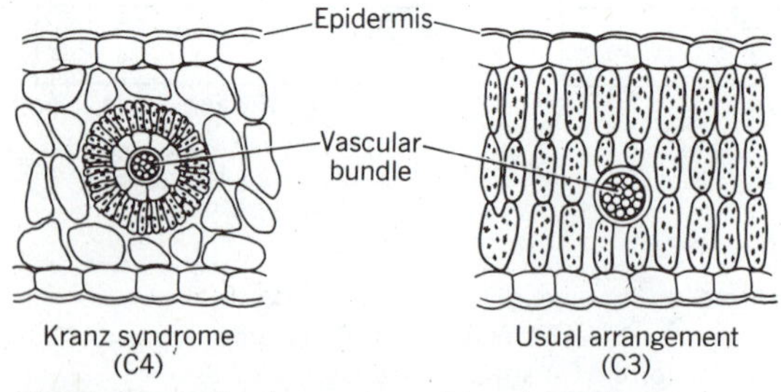

Figure 3.10 Leaf structure in C4 and C3 plants. The *kranz syndrome* of C4 plants groups cells containing chloroplasts clear of the epidermis and the ambient air. In the leaf structure of the more usual C3 plants, cells containing chloroplasts are distributed throughout the leaf.

These arguments make it clear that C4 photosynthesis is, as it were, an option always open to natural selection. Whether the individuals of a species possess a C3 system or a C4 system must depend on the suitability of one system or the other to the local circumstance. Neither can be more efficient throughout the life of an individual, or in all circumstances, because then only one system would be found. And the C3 system is the most common: it follows that in most circumstances the C3 system must be the more "efficient" despite the advantage that C4 seems to give in carbon uptake.

The reasons why C4 does not give a decisive advantage to a plant are as yet far from clear. Some reasons probably depend on the cost of operating a separate CO_2 pump, and others probably depend on the rate constants of other enzymes involved. However, the habitats of C4 plants reveal where they do have an advantage: hot, dry places, or places with special water stresses. All plants of more mesic habitats and all trees use C3. The only plant of cool temperate regions known to use C4 is the salt marsh plant *Spartina,* which may be thought to be under water stress because of the strong salt solutions in which it lives (Figure 3.11; Long *et al.*, 1975).

Plants of hot, dry places must open their stomates, exposing wet tissue to the air, when they take in CO_2. This suggests that the C4 pathway works to reduce water loss in conditions of drought. The PEP carboxylase can scavenge air within the leaf very thoroughly, allowing the stomates to be closed for longer periods, and the air spaces between the productive cells and the epidermis offered by the *kranz syndrome* can collect and store the CO_2 of respiration, thus allowing it to be recycled behind closed stomates. This water conservation aspect of the C4 pathway appears in its most refined form in CAM plants.

CAM stands for CRASSULACEAN ACID METABOLISM and is so named because it was first discovered in desert succulent plants of the family Crassulaceae, though it is used by cacti

Figure 3.11 C4 plants in a salt marsh.
A North Carolina salt marsh dominated by *Spartina alterniflora*. Use of the C4 photosynthetic pathway by this plant possibly is correlated with water stress resulting from the ionic concentration of seawater.

and other desert succulents as well (Figure 3.12). These plants operate the C4 pathway, but they work in shifts, a night shift and a day shift. At night they open their stomates, collect CO_2 on PEP carboxylase, and make the array of C4 acids. A desert plant can do all this at night with little loss of water because gas collection is powered by ATP and can be done in the dark. The next stages of production require light, but they can be put off until daylight. In the daytime, CAM plants leave their stomates shut, taking in no more CO_2. Instead they use the CO_2 stored overnight in 4-carbon acids and run their Calvin–Benson cycles in the abundant desert light but behind closed stomates. It is pleasing to think of a large cactus under the desert sun, splitting water to combine hydrogen with CO_2 in the synthesis of glucose and using the sun as power but otherwise almost entirely insulated from desert stress, a capsule journeying through the light-filled day towards the next night when it will take on supplies again.

Figure 3.12 User of CAM photosynthesis.
The thick, fleshy leaves are typical of the family Crassulaceae and hold moisture strongly defended against surrounding hot, dry air by a thick cuticle and closed stomates. CAM metabolism allows photosynthesis with the minimum exchange of gasses with the air, thus reinforcing the drought defenses.

The Hatch–Slack pathway of C4 and CAM plants is used only in hot places, or dry places, or both. Building a *kranz syndrome* leaf and meeting the costs of a PEP carboxylase pump, therefore, only offer a commensurate reward of fitness in these places. Where water is abundant, or temperature moderate, the C3 pathway is more efficient.

In different circumstance it pays
To gather gas in different ways.

Much research has recently gone into attempts at breeding C4 strains of crop plants, many of which are C3. The aim is to make them "more efficient" and thus increase yields. The breeding program will probably succeed, since the genetics changes needed to produce a *kranz syndrome* leaf seem to be few. Whether the yields will indeed be better with the C4 varieties must depend on the conditions in which the plants grow. The C3 mechanism of the wild ancestors of the crops must have been the most efficient system for the niches and habitats of those ancestors. The C4 varieties will be better only if grown where the C4 is known to be best, in relatively hot, dry places. In fact agriculture does create mini-deserts in which crops have to grow, for instance, bare fields under the scorching sun of summer. It may be, therefore, that C4 wheat will yield better than the present C3 wheat on vast people-made prairies.

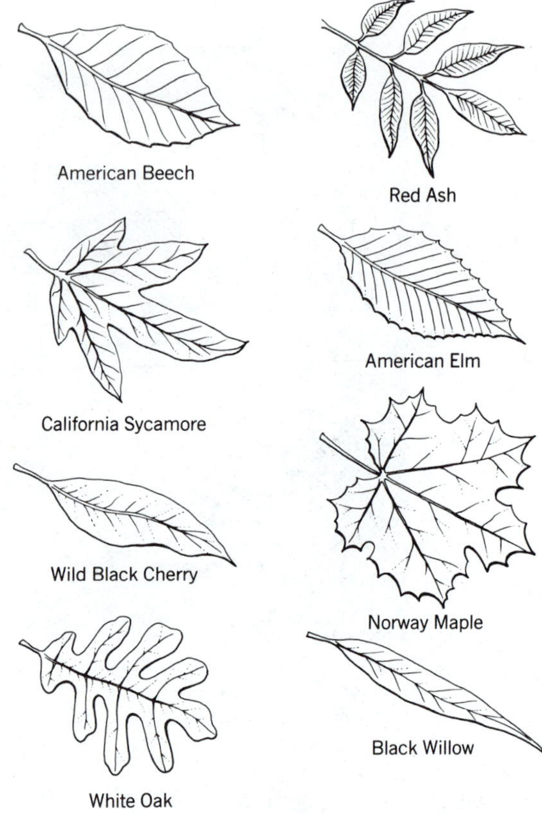

Figure 3.13 Solar panels used by trees of the temperate forests.
Trees do not make solar panels in large sheets but instead use small leaves, often of irregular design. The use of small leaves instead of large solar panels results partly from mechanical and heat budget considerations (Chapter 5), but a small leaf design also allows the diffusion of light through layers of leaves, thus increasing the effective area for CO_2 uptake.

GAS AND LIGHT COLLECTION BY TREES

A tree is an array of solar panels spread on stalks. The panels are green leaves. They are usually small, say, five centimeters or less at the widest continuous part (Figure 3.13), and this smallness is a peculiar feature of solar panels built by plants. When people build solar panels to heat a house or power a spaceship, they make them in large continuous sheets perhaps several meters across. But trees make their green panels small. Part of the explanation for this concerns adaptation to physical stress, to resisting wind and manipulating temperature (Chapter 5). But a more interesting general explanation for the smallness of leaves, and for their arrangement, comes from examining the usefulness of the small-leaf design for photosynthesis.

Although a leaf may be fairly called a solar panel, it is also a chemical factory. Each leaf therefore must have access to raw materials as well as to energy, unlike the engineer's solar panel. It turns out that an arrangement of leaves stacked in broken layers through which light streams is

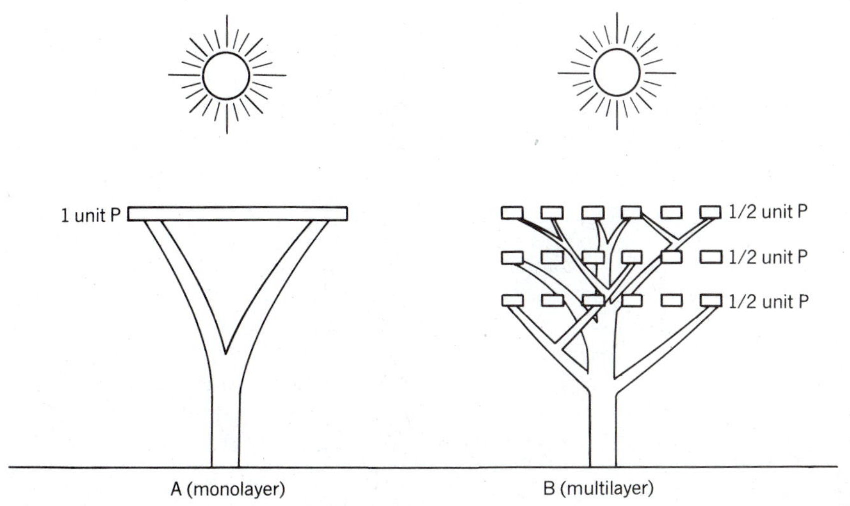

Figure 3.14 The advantage of a perforated leaf system.
Tree A has a single huge leaf that traps all incident sunlight. Photosynthesis is 1, a rate set by CO_2 uptake and light saturation. Tree B has a perforated leaf structure in layers such that each layer is half leaf and half holes. The leaf area in each layer can achieve maximum photosynthesis in the diffused light because each has a private CO_2 supply. Each layer of tree B achieves $\frac{1}{2}$ unit of photosynthesis and the whole tree B achieves one and a half times as much photosynthesis as tree A.

an arrangement to increase photosynthesis by maximizing CO_2 uptake at the production sites. What follows comes from the analysis of Horn (1971).

A large, intact sheet-like leaf spread out under the noonday sun would be light-saturated and would achieve a rate of photosynthesis set by the CO_2 supply, if no other limiting factor such as water or temperature intervened. Let the photosynthesis of a hypothetical tree equipped with a single huge leaf be one unit. If instead of an intact leaf the tree was equipped with a perforated sheet that was half holes it would achieve only $\frac{1}{2}$ unit of photosynthesis. But then it would be possible to spread out a second perforated sheet below the first to receive the diffused light (Figure 3.14). If the light on the second layer were still sufficiently intense for maximum photosynthesis then the second perforated sheet would also achieve $\frac{1}{2}$ unit of photosynthesis and the two perforated sheets would have the same total production as the one intact sheet. However, there would still be light filtering through the holes of the second sheet making it possible to put out still extra sheets below. Any photosynthesis achieved with these extra sheets would be pure gain. In bright light, therefore, an arrangement of layers of small leaves with gaps between them should be more productive than a single intact sheet-like solar collector.

An arrangement of layers like that of tree B in Figure 3.14 would work only if the lower layers of leaves could be kept clear of the shadows of those above. But shadows can be avoided provided the distance between the layers is large enough (Figure 3.15). Because of the earth–sun distance, and of the diameter of the sun, a leaf casts a shadow for about 100 leaf diameters. If a leaf is 1 cm across, therefore, the second layer would only have to be 1 m below the first to escape shadows. Leaves of conventional size (Figure 3.13) would all be suited to layer arrangements with layers up to 2 to 3 m apart—well within the design of trees.

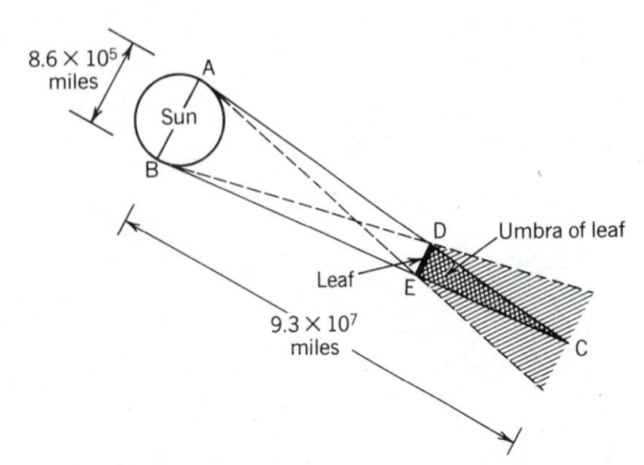

Figure 3.15 Shadow cast by a leaf.
The triangle ABC is similar to the triangle DEC. Hence (length of umbra)/(distance from tip of umbra to sun) = DE/AB. Therefore length of umbra = $(9.3 \times 10^7)(DE)(8.6 \times 10^5)$ = 108 DE. This means that a leaf casts a shadow to 108 diameters and that light farther away than 108 diameters is evenly diffused. Leaves with solid widths of 2 cm need to be in layers about 2 m apart for the lower leaves to escape the shadow of those above. (Redrawn from Horn, 1973.)

Larger leaves can be accommodated by irregular shapes that reduce their effective diameters and permit close packing of layers. Here is an explanation of the deeply notched margins of so many leaves: the notches are devices to reduce effective leaf diameter, thus turning a layer into the optimum perforated sheet.

Horn's logic leads to the conclusion that a tree under the sun, as in Figure 3.14, ought to have small leaves arranged in discrete layers as in B rather than as a single layer as in A. Suppose, however, that the tree grew in a gloomy place where there was never enough light for its top leaves to be light-saturated. None of its leaves would run short of CO_2 and there would be no photons of light wasted even by the top-most leaf. A tree in so gloomy a place might well do best with a single dense solar collector like that in Figure 3.14A, because letting light through holes would serve no purpose and might give an opportunity for some other plant to use the penetrating light instead.

Horn calls the design of Figure 3.14A a MONOLAYER and of tree B a MULTILAYER. His analysis suggests that trees adapted to life in the shade, like those of the understory of a forest, should be designed as *monolayers*, that those of open places should be *multilayers*, and that there should be a general correspondence between the number of layers a tree has and the degree of its shade tolerance.

However, real trees do not live under a sun stationary at the zenith as in Figure 3.14. They are illuminated from different directions at different hours of the day, and they also receive reflected light or light diffused through clouds. We cannot expect to find real trees built in these discrete layers but only to tend towards them.

Assuming that intact sheets of tissue for a perfect monolayer are not practicable, then monolayers must be built of smaller leaves. The Horn hypothesis then predicts that shade trees have leaves arranged close together in a non-overlapping array. This follows because any overlap of leaves in a fairly thin layer will shade out neighbors and be wasteful (leaf maintenance costs will rise without benefit in added production). Leaves in multilayers, however, can be arranged at random, or nearly so, as long as the layers are kept apart. This follows because in a pattern of leaves and holes there is little danger of serious overlap so that a random array is probably as effective as any other, granted that the direction of incoming light is not constant. Finally, random array is consistent with leaves of irregular shape, but the non-random array of a monolayer will be fostered if leaves are of regular shape without deep indentations.

Horn's hypothesis can be summarized by saying that leaves should be arranged on a tree so that the largest possible leaf area is exposed to a supply of CO_2 and to light at the optimum intensity for photosynthesis (about 20% of full sunlight). This hypothesis has the essential quality of all good hypotheses because it leads to predictions that can be tested in an attempt to falsify the hypothesis. Among the detailed predictions are the following:

1. Trees growing in the open should have their leaves arranged in depth, probably in a random array.

2. Understory trees growing in dense shade

should have their leaves arranged non-randomly and within a short vertical distance of one another.

3. A single branch of a tree adapted to growth in the open should cast less shade than a single branch of a tree adapted to grow in dim light.

4. Trees growing in the open should have small leaves of irregular shape.

5. Understory trees and shrubs should have leaves of regular shape and these should tend to be larger than leaves of trees growing in the open.

6. Total leaf area will be larger than the ground area covered in trees growing in the open but nearly equal to the ground area covered for trees growing in shade.

Predictions 4 and 5 receive qualitative support from standard observations of botanists. Many plants of the shady lower regions of tropical rain forests have very large leaves of smooth shape; the leaves of a rubber plant are a familiar example and this plant is so well adapted to shade that it survives in gloomy corridors of office buildings (Figure 3.16). Common emergent canopy trees of the tropical forests, on the other hand, are of the family Leguminoseae with their typical finely divided pinnate leaves. Familiar temperate trees that colonize open fields are ashes (*Fraxinus*) and maples (*Acer*) with their characteristic leaves of thin or irregular shape. Figure 3.17 shows how the leaves of an oak tree have only small teeth on the margin when borne by a sapling growing in the shade of larger trees, whereas they are deeply indented when exposed to the direct rays of the sun in the canopy of a mature oak tree.

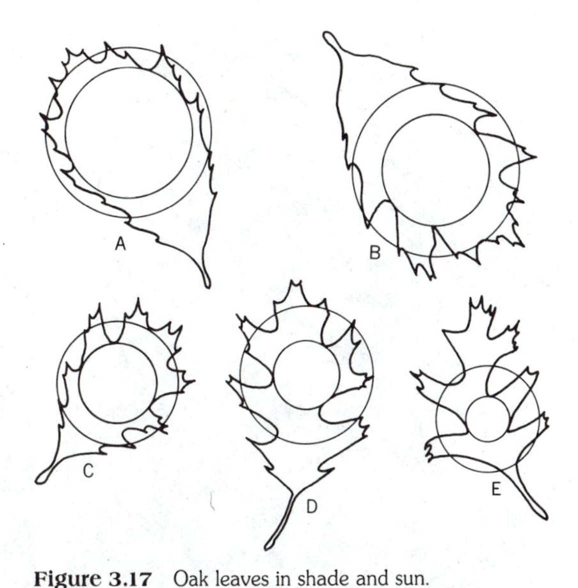

Figure 3.17 Oak leaves in shade and sun.
The leaves are all from black oak (*Quercus nigra*). (*A*) Leaf of a seedling. (*B*) Leaf from a shaded branch near the ground. (*C–E*) Leaves from progressively higher on the tree. The small circle is the largest circle that can be inscribed in each leaf. The relative sizes of the two circles show how well lobing adapts the leaf to its multilayered role of letting half the light pass through the layer of which it is a member. (From Horn, 1971.)

Figure 3.16 Germination on the floor of an Amazonian rain forest.
The seedlings germinating in the dim light appear to be of monolayer design like small umbrellas. This photograph was taken in the Oriente Province of Ecuador, at the equator and near midday, but the exposure required was 1 second at f3.5 with 200 ASA film.

These qualitative observations on leaf shape are more anecdote than hard evidence, yet they are consistent with the hypothesis. Much stronger support comes from Horn's demonstration that the relative layering of trees and relative leaf area can be measured and shown to correlate with their relative tolerance to shade (predictions 3 and 6).

Figure 3.18 Light penetration in monolayers and multilayers. Much light comes through a single branch of white pine (*Pinus strobus*) (top left) but very little through the whole tree (top right). In the eastern hemlock (*Tsuga canadensis*), however, about as little light penetrates a single branch (bottom left) as penetrates the whole tree (bottom right). This is evidence that white pine is a multilayer and hemlock a monolayer. This conclusion conforms with the known distributions of the trees, white pine being a colonizer of old fields and hemlock a tree of climax forest. (From Horn, 1971.)

To measure the relative number of layers and relative leaf area Horn worked backwards by measuring the light actually penetrating different parts of a tree. A single branch of a true monolayer, for instance, should allow virtually no light to penetrate, but a single branch of any multilayer should allow much light through (Figure 3.18).

There is usually little difficulty in deciding what is a single branch to measure, because the branching of most trees is distinct and laminar. Horn used a specially calibrated light meter to compare the light coming through a branch with the light in the open, the difference giving him a measure of relative leaf density (p) of a branch. At the same time he measured the proportion of light coming through the whole tree (u). From the three measures—light in the open, light through a branch, and light through the whole tree—he was able to calculate both relative number of layers and relative area:

$$u = (1 - p)^n$$

where n is the relative number of layers. Solving for n we get

$$n = \log(u)/\log(1 - p) \quad (3.4)$$

Table 3.4 gives an example of Horn's results. Both the relative number of layers and the ratio between leaf area and ground covered fall as succession proceeds. This data set is particularly attractive because the trees are ranked by known order of appearance in secondary succession and by known shade tolerance. Horn's hypothesis predicts that the most shade tolerant should have layers, and leaf-to-ground ratios, approaching unity and the data confirm this.

Horn's hypothesis, therefore, generates predictions that fit the facts of tree life very well. Trees in temperate forests spread their leaves in a dappled pattern against the sky as an essential strategy in the quest for energy with which to win fitness. The familiar forests of planet earth are a necessary consequence of the intensity of sunlight striking the earth and of the concentration of carbon dioxide in the terrestrial air (Figure 3.19).

Table 3.4
Tree Layering and Shade Tolerance
The tree species are arranged in the conventional order in which they appear in secondary succession. The order also represents a ranking by shade tolerance known to foresters. Horn's calculations of relative number of layers show that there is a relative shift from multilayer to monolayer designs as succession proceeds as predicted. The ratio of total leaf area to ground covered also is reduced as succession proceeds. (From Horn, 1971.)

Species	Number Measured	Percentage Light/ Branch	Percentage Light/ Tree	Number of Layers ± SE	Leaf Area/grnd. Area
EARLY SUCCESSION					
Gray birch	10	44	3.6	4.3 ± 0.4	2.4
Bigtooth aspen	6	45	6.9	3.8 ± 0.5	2.1
White pine	13	25	0.8	3.8 ± 0.4	2.9
Sassafras	3	14	0.8	2.7 ± 0.7	2.4
MID-SUCCESSION (on moist soil)					
Ash	10	26	3.0	2.7 ± 0.2	2.0
Blackgum	7	15	1.4	2.6 ± 0.5	2.2
Red maple	21	20	1.8	2.7 ± 0.2	2.2
Tuliptree	6	17	2.3	2.2 ± 0.2	1.8
MID-SUCCESSION (late on dry soil)					
Red oak	19	23	2.6	2.7 ± 0.2	2.1
Shagbark hickory	12	18	1.4	2.7 ± 0.2	2.2
Flowering dogwood	13	5	2.1	1.4 ± 0.1	1.3
LATE SUCCESSION					
Sugar maple	8	9	1.2	1.9 ± 0.1	1.7
American beech	16	6	1.5	1.5 ± 0.1	1.4
Eastern hemlock	13	8	2.1	1.6 ± 0.1	1.4

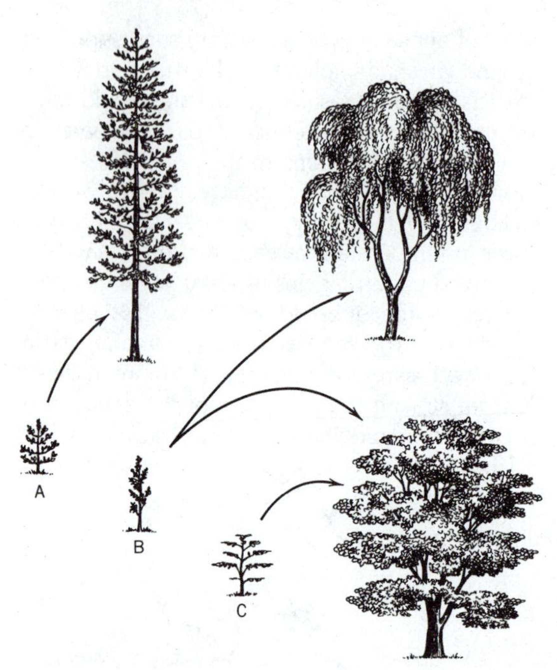

Figure 3.19 Tree shapes determined by adaptations to different light intensities.
(*A*) Early-successional multilayer. (*B*) Persistent multilayer. (*C*) Monolayer. The intermediate arrangement (*B*) is of a tree that persists after the canopy closes around it. (After Horn, 1971.)

LEAVES ON AN INCLINED PLANE

If light strikes a leaf at an oblique angle, it will be spread out and the intensity of the radiation received on unit area will be less than if the leaf is square to the sun. The light received by a horizontal leaf of length AB (Figure 3.20) would be spread over the length BC of an inclined leaf, and the relationship between the two is the familiar pythagorean one. Two effects of this relationship accrue to the inclined leaf: it absorbs less heat on unit area of surface and it spreads the light to more photosynthetic tissue at reduced intensity.

Inclined leaves are possessed by many plants, most notably grasses and cereals, and plants like these dominate many plant formations. It is a useful hypothesis to suggest that these plants have acquired fitness from their inclined leaves as multilayer trees do in Horn's model. Light intensity received by each chloroplast is reduced, though still high enough to permit photosynthesis at the maximum rate, and a larger number of active chloroplasts and CO_2-collecting stomates are brought on line. This is the solution to the excess radiant energy problem offered by plants that are too short for the geometry of multilayers to be possible.

Since heat load is reduced, it is likely that plants with long narrow leaves lose less water by transpiration than would plants with horizontal broad leaves, and this should be adaptive in dry places. The apparent adaptive advantage of the inclined leaf, therefore, offers an acceptable general explanation for the existence of grassy prairies in drier regions where trees do not grow. The shapes of the dominant prairie plants maximize production while conserving water.

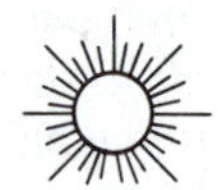

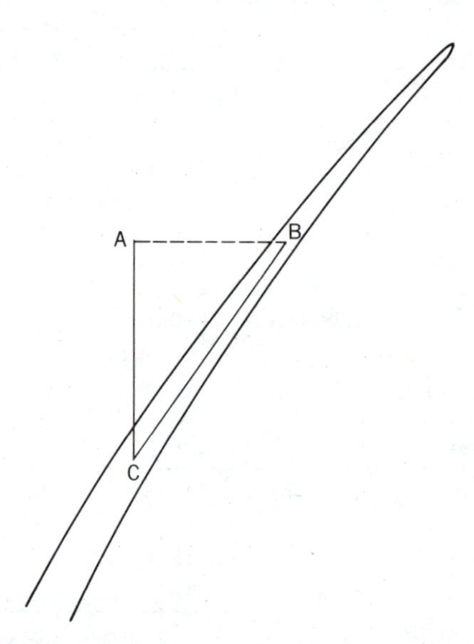

Figure 3.20 Light on an inclined leaf.
Light intensity per unit area is reduced by the oblique presentation of the leaf to the sun. This increases the number of chloroplasts and CO_2-collecting stomates that can be brought into production in bright light.

Figure 3.21 Tussock tundra: Seward Peninsula, Alaska. Dominant life form is grass-like tussocks, actually of the sedge *Eriophorum vaginatum* (cotton grass). Upper photograph shows tussock field; bottom photograph is a close-up of a tussock in flower.

Figure 3.21 shows the same life form in an arctic tundra, where the dominant plant is a sedge (*Eriophorum*). Tundra plants on permanently frozen ground plausibly may suffer water stress at some seasons. In particularly wet arctic tundra like that of the Pribilof Islands, on the other hand, the dominant plants may be not only grasses or sedges but also plants like Queen Anne's lace and lupines, whose leaves are essentially horizontal. Figure 3.22 shows Pribilof tundra of this kind. The Pribilof Islands have no permafrost but rain falls from the fog nearly every day, and the vegetation is usually soaking wet.

Many papers in the agricultural literature discuss how the attitudes and shapes of crop leaves affect net primary production (Loomis *et al.*, 1971). Rice plants, for instance, have long, narrow leaves typical of the grass family Gramineae, although some cultivated varieties hold their leaves in different attitudes. In one study it was shown that a variety with drooping leaves produced less dry matter in unit time than another variety with erect leaves. This would be expected if the leaves had to operate in bright sunlight as crop plants do.

Other well-known traits of some plants with lanceolate leaves are consistent with the hypothesis. Those that flower in the spring on woodland floors hold their leaves erect when exposed to the bright sun before the trees open their leaves, but let them droop later. Cultivated daffodils are the most familiar representatives of this strategy, the straggling leaves after the flowers have gone causing such annoyance to gardeners that they sometimes tie them in knots. The hypothesis suggests that the daffodils change their light-spreading vertical stance to the horizontal pattern when light on the woodland floor changes from bright to dim. Another example is the huge, droopy leaves of bananas that suggest the adaptation of lanceolate leaf strategy to dim light. Wild relatives of bananas grow in tropical forests beneath the canopy.

Herbs such as cereal do not have the opportunities a tree has for stacking leaves over several meters. They may, however, achieve the same effects by keeping their leaves small or narrow, because it is the width of the leaf that determines how far apart they must be if they are to avoid being in each other's shade (Figure 3.15). It has been shown (Nichiporovich, 1961) that the occlusion of skylight by leaves is given by the relationship

$$\Upsilon = 2 \tan^{-1} (w/2d)$$

where Υ is the angle of skylight occlusion, w is the width of a leaf, and d is the distance from a shaded leaf.

When $w/2d$ is large, shading of the lower leaf is complete. Shading is reduced as w is reduced. Most cereals seem to have leaf structures such that

$$d > 2w (\Upsilon < 28°)$$

Figure 3.22 Horizontal leaves in a wet tundra.
Tundra on the wet Bering Sea island of St. Paul. There is no permafrost on St. Paul and the ground is usually wet. The island also has little sun, often being covered with fog for days on end. Here the sedges and grasses of the mainland tundras share the canopy with many broad-leaved herbs.

which appears to yield an optimum spread of light.

Like a tree, therefore, a grass plant is shaped to spread light to as many chloroplasts as can be operated at the maximum rate. The erect posture helps by spreading the noonday sun over as much area of each leaf as possible. At the same time the leaves are narrow to minimize their shadows and let the diffuse light pass down to power the next narrow leaf on line.

AT THE BOTTOM OF THE SEA: PLANTS IN WATER-FILTERED LIGHT

BENTHIC marine plants (plants that grow anchored to the bottom of the sea) have photosynthetic problems quite different from those of land plants. They receive their carbon dioxide in solution as bicarbonate (HCO_3^-), and this carbon source is much more ample than the carbon source of plants relying on air. An anchored seaweed probably never encounters a carbon shortage; neither is it likely to suffer for lack of other dissolved nutrients. It is true that some nutrients are so scarce in the sea as to limit very strictly the productivity of plants of the open water (the *phytoplankton,* see Chapter 21), but water moving over a plant that is anchored brings continual fresh nutrient supplies within its reach. It is likely, therefore, that anchored seaweeds seldom encounter shortages of raw materials. Constraints on photosynthesis at the bottom of the sea are dim light, and light of selected wavelengths, caused as sunlight is absorbed by water.

Light is rapidly absorbed by water, and its energy is degraded to heat. Figure 3.23 shows how rapidly light is absorbed by even clear ocean water, the red end of the spectrum being lost with particular speed. In turbid coastal waters, absorption by suspended sediments and colloids may occlude light ten times more quickly still. Except when very near the surface, therefore, efficient photosynthesis in the sea requires spe-

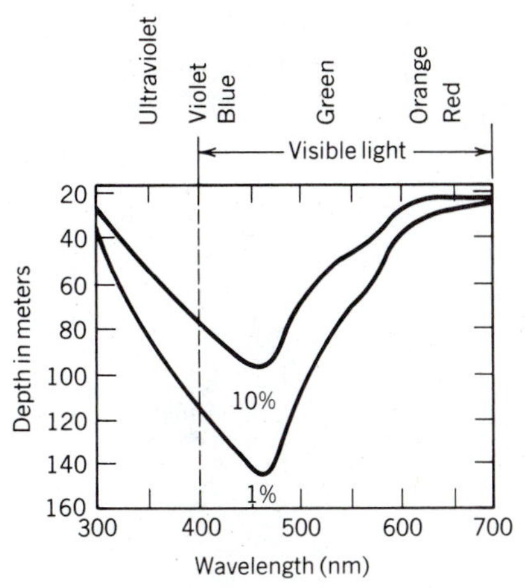

Figure 3.23 Light extinction by clear water.
Depths at which 10% and 1% of normal sunlight penetrate into the clearest seawater. Red light is absorbed rapidly but even the blue end of the visible spectrum is mostly extinguished by 140 m. The deeper a plant lives the more it is dependent on a narrow band of wavelengths. (Modified from Jerlov, 1951.)

cial attention to the absorption of light by the pigment array.

The large anchored plants of the sea are of different colors—green, red, brown, and blue-green. On the face of it, these different colors might be thought of as primary adaptations to the spectral composition of light filtered through seawater. In the shallows, light is little changed from what passes through the air, and we find green algae like the sea lettuce *Ulva* and the thin, short ribbons of green *Enteromorpha* that make rocks in estuaries slippery. But deeper in the sea the longer wavelengths are selectively absorbed and we find brown algae and red. So attractive is this line of reasoning that it was long asserted that the colors of seaweeds were correlated with depth, first the greens, then the browns, then the reds. But it is now known that this scheme is in error (Ramus *et al.,* 1976; Ramus, 1980). For instance, the shallowest algae of coral reefs are the red calcareous members of the genus *Lithothamnion,* which present their partially submerged mass both to the equatorial sun and to the fury of the waves. And green algae of the genus *Halimeda* are now known to flourish down to depths of at least 140 m (Figure 3.24).

One data set long cited to support the postulate that the colors of seaweeds were correlated with depth is the distribution of common and conspicuous plants in and below the intertidal region along northern temperate shores. Along a typical temperate coastline, like that of Massachusetts or southern England, there seems to be a rough zoning of the colors of seaweeds with depth. Near the high tide line are the green sea lettuces (*Ulva*) and *Enteromorpha.* Occupying the bulk of the intertidal region are brown seaweeds like *Fucus* and *Ascophyllum* and the kelps like *Laminaria.* Below the low watermark red seaweeds with delicate fern-like fronds become numerous. This zonation shows green algae exposed to most light, brown algae in the intertidal, and red algae only at depth. But the implied correlation with light intensity is spurious.

Direct evidence that factors other than pigments suited to different light set the depths of both brown and red algae even in the classical coastal zoning of the North Atlantic came from a study comparing distributions between Massachusetts and the Bay of Fundy in eastern Canada and Labrador: the same species of both red

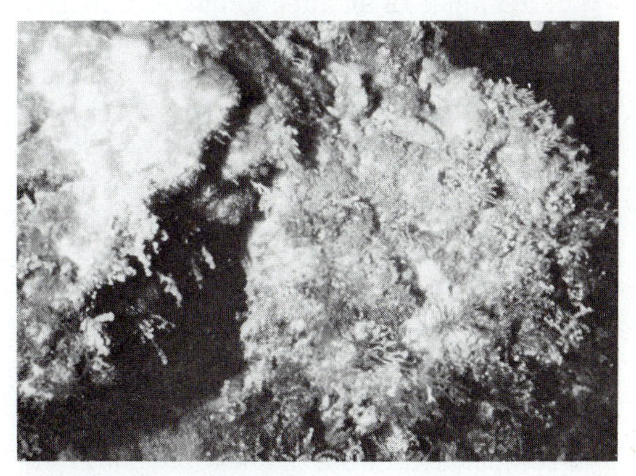

Figure 3.24 Green algae in the deep sea.
This community of *Halimeda* was photographed from a submersible at about 140 m depth on the outer face of the coral atoll of Enewetak. Light at this depth is but a trace of 1% of sunlight and yet these green algae achieve extensive cover. *Halimeda* contributes much of the carbonate mass to coral reefs.

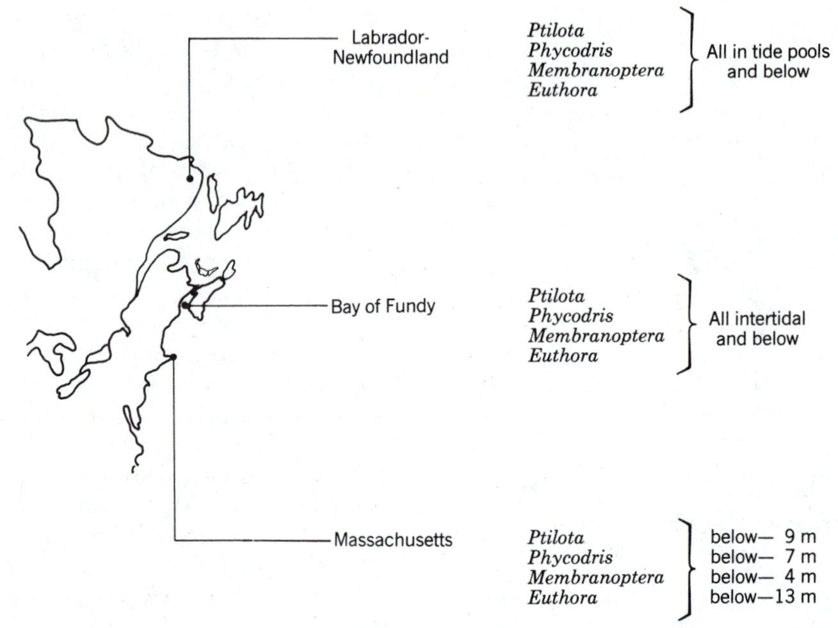

Figure 3.25 Depths of red algae in northeastern America.
The four genera of red algae are found at progressively shallower depths as one goes from south to north. It is likely that water temperature (or some correlate of temperature) controls the depth distribution of these red algae more than light does. Massachusetts data are in depths below mean tide level. (From data of Hillis-Colinvaux, 1966.)

and brown algae were found at different depths at the three sites, there being a general shallowing of the range of all three to the north (Hillis-Colinvaux, 1966). If a simple hypothesis of limiting factors explains these distributions, temperature works better than light (Figure 3.25).

The truth is that there is no simple correlation between the colors of seaweeds and their depth—browns, reds, greens, and blue-greens can be found throughout the lighted layer of the sea where photosynthesis is possible, the so-called EUPHOTIC zone. Why, then, is there such an array of colored photosynthetic pigments in sea plants, whereas almost all plants on land are of the same color, green? The answer probably lies in a combination of the dimness of light in the sea with the variety of ancient lineages of marine plants.

Green, brown, red, and blue-green algae are so different in structure that taxonomists place them in separate divisions of the plant kingdom. The zoological equivalent of the botanical division is the phylum. This means that taxonomists consider that green and red algae are as distant from each other as are worms and birds, or starfish and insects. All the green plants of the land, with the exception of the mosses, belong to a single division, the division Tracheophyta (Table 3.5). Each group of algae has inherited a pigment pattern from remote ancestors, and its color cannot be changed. An interesting question for an ecologist then becomes, "How have each of these different colors been adapted to life in the dim light of the sea?"

When light is limiting, the problem for a plant is to absorb as much of it as possible. Ideally, the plant ought to be black, though this requirement can be relaxed since some colored wavelengths are useless for photosynthesis. But "blackness" can be approached with almost any strongly absorbing pigment if this is deployed in sufficient density. Red and brown algae are dark colored—it is possible to say that they do not afford the luxury of land plants of being able to reflect in the green. Therefore redness and brownness can be thought of as adaptations to dim light at depths as in the classical depth–color hypothesis. But green algae can solve this problem with a close array of chloroplasts, each absorbing slightly in the green.

Species of the reef-building alga *Halimeda* grow from depths exposed at low tide down to at least 140 m, but their close relative *Penicillus*, the merman's shaving brush, is apparently confined to shallow water (Hillis-Colinvaux, 1980, 1984; Figures 3.24 and 3.26). The green pigment array, therefore, can be adapted to any depth and questions about actual distribution become questions of species niche and strategy, rather than questions that can be answered in terms of light as a limiting factor. The fact that sea grasses have only entered the shallows of the sea may likewise have little to do with their green pigments (Figure 3.27).

Colored algae do have elegant solutions to the absorption problem. They have accessory pigments arranged around their chlorophyll *a* molecules, and these pigments absorb many of the wavelengths that would be reflected by chlorophyll. When a plant has a battery of these pigments as well as chlorophyll *a*, it is able to absorb virtually all wavelengths that deliver photons energetic enough for photosynthesis. Figure 3.28 illustrates one such array of pigments (collectively called PHYCOBILISOMES). They form a series with overlapping absorption spectra. Energy can be passed from one to another down a chain by a process of resonance. This process

Table 3.5
Divisions of the Larger Photosynthetic Plants and Their Photosynthetic Pigments
(Adapted from Dawson, 1966.)

Division		Pigments
Chlorophyta	Green algae	Chlorophyll *a*, *b* Carotenes
Cyanophyta	Blue-green algae	Chlorophyll *a* Carotenes Phycobilins
Phaeophyta	Brown algae	Chlorophyll *a*, *c* Xanthophylls Carotenes
Rhodophyta	Red algae	Chlorophyll *a* Carotenes Phycobilins
Tracheophyta	Angiosperms Gymnosperms (flowering plants) Ferns	Chlorophyll *a* Carotenes

Figure 3.26 *Penicillus*: the merman's shaving brush. A green calcareous alga of shallow water in the tropical Atlantic Ocean. The plants grow spread out on sandy bottoms, reproducing asexually by rhizoids through the sand or by sexual episodes that result in the death of the adult plant (semelparity, Chapter 12).

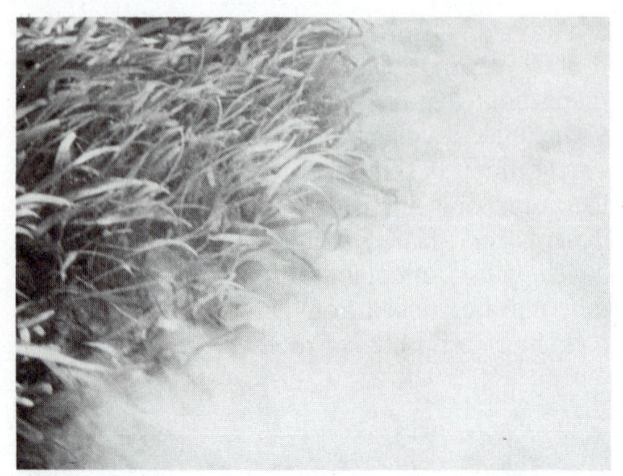

Figure 3.27 Tropical sea grasses.
Stand of sea grass (*Thalassia*) in about 1.5 m of water on a sand flat behind the fringing reef, north shore of Jamaica. Alone of the angiosperms, the sea grasses have penetrated the sea but are confined to very shallow water.

is analogous to striking the first tuning fork in a series, which in turn causes other tuning forks to resonate in sequence. At each transfer the wavelengths of induced resonance are larger and less energetic, but energy can still be passed down the chain to chlorophyll *a* with an efficiency of more than 80%.

Possession of *phycobilisomes* gives brown and red algae their colors: they reflect red and orange light when viewed in bright light. At the depths where many of them grow they may reflect very little light at all.

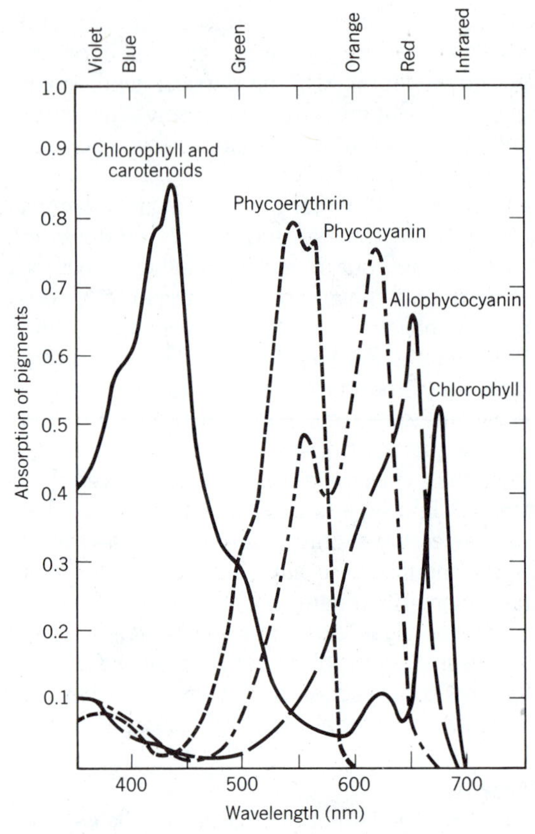

Figure 3.28 Light absorption by accessory pigments of seaweeds.
The chlorophyll *a* and carotenoid system absorbs the blue and red ends of the visible spectrum as in land plants, but brown and red algae also possess pigments that absorb light in the wavelengths reflected by chlorophyll *a*. Energy is passed to chlorophyll *a* from these pigments by resonance. (Modified from Gantt, 1975.)

COLOR CHANGES FOR DIM LIGHT ON LAND

One of the club mosses of the rain forest floor in Malaya, *Selaginella willdenovii,* has an iridescent blue color when growing in its typical forest habitat but loses the blue color when grown in full sunlight. Lee and Lowry (1975) were able to show that the iridescent blue color was caused by a thin film spread on the surface of the leaves that acted as an interference filter, rather as the coating acts on the lens of a camera. This blue-looking interference filter altered the way in which the leaf reflected light. Lee and Lowry showed this by measuring reflectance on the blue leaves and comparing it with reflectance on leaves that had turned green through exposure to direct sunlight. Their results are given in Figure 3.29. The bluing made the leaves reflect more light at the extreme blue end of the visible spectrum, at which wavelength it was least useful for photosynthesis, and increased absorption of light between 500 and 600 nm, the most useful wavelengths. The bluish cast of many tropical rain

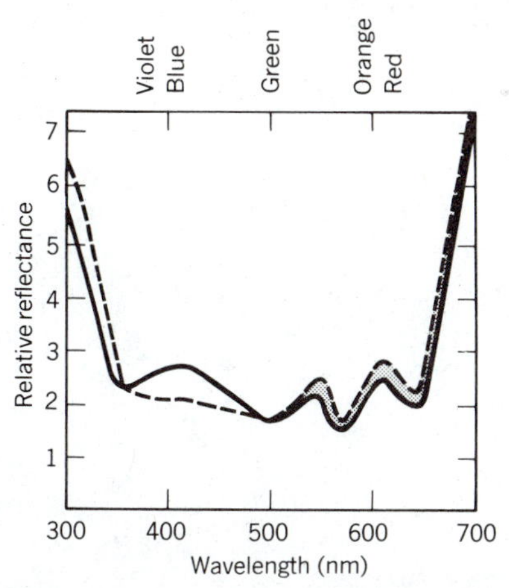

Figure 3.29 Reflectance spectra of a plant from the floor of a tropical rain forest.
Selaginella willdenovii has an iridescent blue cast to its leaves. A surface film on the leaves allows reflectance of wavelengths less useful for photosynthesis but more absorption in the energetic wavelengths. Reflectance by iridescent leaves is indicated by the solid line and that by leaves grown in bright light and without iridescence by the dashed line. The shaded region represents the extra absorption of useful wavelengths. (Modified from Lee and Lowry, 1975.)

forest leaves, therefore, may be an adaptation that maximizes the collection of light in their gloomy surroundings. The normal presence of blue-green algae on rain forest floors also may be due, in part, to the suitability of their pigment to dim light.

CONCLUDING NOTE: THE BIOSPHERE SHAPED BY LIGHT AND CARBON DIOXIDE

On land the pervading reality is that energy from light is abundant but that the raw material of carbon chemistry, carbon dioxide, is scarce. These two pervading realities are functions of the size and age of our sun, the distance from it to the earth, and the mass of the earth. They are probably unique to our planet. Because of the low concentration of carbon dioxide, land plants can afford to reflect strongly in the green. For the same reasons forest trees have small leaves instead of large sheets of light-absorbing tissue, and plants of prairies have lance-shaped leaves that are narrow in width. The proportion of the total energy flux that is trapped by the biota is low because of this same imbalance between the supply of energy and the supply of carbon as gas. Not only the familiar colors and shapes of our landscapes but also the abundance of animal life result from the peculiar supplies of light and gas available at the surface of the earth. Modest changes in these would produce life forms intensely alien to our perceptions.

PREVIEW TO CHAPTER 4

An imaginary super-herbivore ought to be able to eat a plant's sugars as soon as the plant makes them, so that none are wasted. A comparable super-carnivore ought to eat its prey as fast as its prey did its own feeding. The ecological (Lindeman) efficiencies of animals are measures of how close they come to these impossible ideals, which are ratios of gross productivities. Lindeman tried to calculate these efficiencies for the different trophic levels in lakes by reconstructing biomasses of different populations from standing crops and turnover times, then adding respiration. This method cannot work and Lindeman's results were in large error by being too big. His results gave rise to the often cited rule of thumb that Lindeman efficiencies are about 10%. Better estimates can be calculated in theory from measures of yields of corpses and respiration. Measures of efficiency of water fleas and hydra in culture by this method gave results close to the rule of thumb of 10% but measures for wolves give results of about 1%. Actual Lindeman efficiencies in nature are probably much less than 10% usually, which fully explains the shape of Eltonian pyramids. The language of production ecology is confusing because we talk of animals "producing" energy, when actually they are burning their food-calories as fuel. This semantic difficulty leads to a number of different usages in ecological and agricultural writings, which are discussed. The term "production efficiency" refers to the efficiency with which animals convert the food they actually get into net production, which is to say into actual bodies. Production efficiencies vary widely for different kinds of animals. Since the size of a population of carnivores depends on the production efficiency of its prey, these different production efficiencies have important consequences beyond the individual species. It is sometimes possible to calculate a complete energy budget for a population—work on desert rats is used as an example. That animals are in fact programmed to acquire the largest practicable energy intake is shown by studies designed to test optimal foraging theory. This theory describes how ideal animals should choose food from arrays of different sizes of items. The theory assumes that animals do maximize net production (reflecting energy intake), and that they use their time to the best advantage. The theory predicts quantitatively the mixtures of food items that animals in given circumstances should collect and these predictions have been found to describe correctly the foraging of real animals in a number of trials. Our belief that the niches of animals are such that they can maximize energy flows thus receives powerful verification.

CHAPTER 4

THE TRANSFER OF ENERGY BY ANIMALS

An individual animal may be thought of as a device programmed by a base-pair sequence of DNA to collect reduced-carbon fuel and to process as much of this fuel as possible into offspring. All animals should be highly efficient at this, since the stakes are fitness or non-fitness, survival or oblivion.

Yet the drive to energetic efficiency meets powerful constraints. Plants and prey animals have adaptations that work to prevent them from being eaten. Other organisms may compete for the food. Always food is taken against resistance. The efficiency of energy transfer, therefore, depends on fundamental relationships between animals. They are common or rare, and arranged in pyramids along food chains, according to the efficiencies with which they turn plants or other animals into food.

FUNDAMENTAL ENERGY TRANSFERS ALONG FOOD CHAINS

The efficiency of all the plant eaters combined (the whole herbivore trophic level) is a measure of how much of the potential energy of plants they convert to their own use.

$$E_h = \frac{\lambda_n \text{ (herbivores)}}{\lambda_{n-1} \text{ (plants)}} \times 100 \qquad (4.1)$$

where E_h is the ecological efficiency of herbivores and the symbols are those introduced by Lindeman (1942) described in Chapter 2.

The efficiency of the primary carnivore trophic level is

$$E_{c1} = \frac{\lambda_{n+1} \text{ (primary carnivores)}}{\lambda_n \text{ (herbivores)}} \times 100 \qquad (4.2)$$

where E_{c1} is the ecological efficiency of primary carnivores.

Equations (4.1) and (4.2) both use ratios of *gross productivity*. The ecological efficiency of the herbivore trophic level, for instance, is biomass and respiration of herbivores divided by biomass and respiration of plants. The ratios are strictly comparable to that used in calculating the ecological efficiency of plants, where the ratio was biomass and respiration of plants divided by total solar energy received (Chapter 3).

These equations describe the fact that the theoretical upper limit of energy available to a trophic level is all the energy flowing into the trophic level below. This includes even the energy that is degraded (used) in the lower trophic level and thus not in practice available for passing on. The formulation assumes that a super-herbivore would be able to get all the plant production, even before the plants respire, and a supercarnivore would be able to eat all the starch reserve of the prey without the prey using any of those reserves for running about. Only animals able to achieve these impossible ideals as food gatherers could be 100% efficient. It will be recalled that the low ecological efficiency of 2% for plants was largely accounted for by the fact that most of the solar energy theoretically available never entered active photosynthetic tissue at all. The efficiency of animal trophic levels is usually low for similar reasons—most energy is degraded in the trophic level below as respiration, or goes to decomposers, and thus can never be transferred as food.

Transferring energy from one trophic level to another involves many processes, each of which proceeds with different efficiency (Figure 4.1).

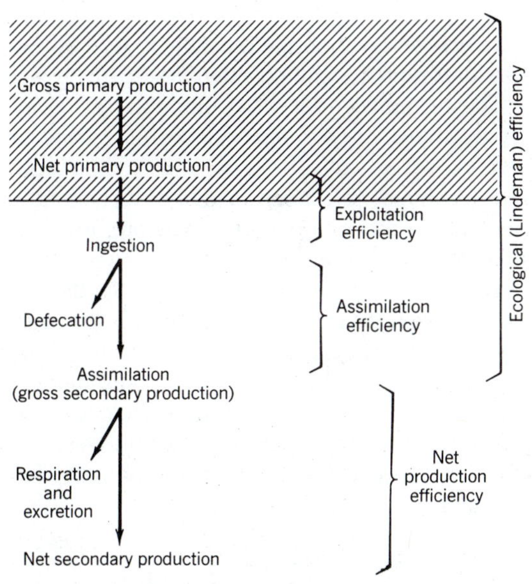

Figure 4.1 Relationships of efficiencies of energy transfer. The various efficiencies are defined in Table 4.1. All these efficiencies refer to the rates at which animals collect fuel to use to win fitness. All may be expected to be maximized by natural selection. Differences in calculated efficiencies must therefore reveal the local circumstances in which the animals live (see Table 4.4).

The efficiency with which plant or prey organisms maintain biomass per unit of respiration determines how much net production is available as food. The efficiency of cropping or hunting determines what portion of this net production is taken. The efficiency of digestion determines how much of the energy eaten goes into net production, and hence potential food for the next trophic level. Ecologists define, and seek to measure, these many efficiencies. The ecological efficiency of energy transfer between whole trophic levels depends on the cumulative effects of these other efficiencies (Table 4.1).

Lindeman (1942) made the first attempt to measure the ratios of gross productivity for whole animal trophic levels in nature, for which reason the *ecological efficiency* often is called LINDEMAN EFFICIENCY. (The term *transfer efficiency* sometimes is used instead, and plant physiologists talk of *assimilation efficiency* for the plant trophic level: the terms are synonymous.) Lindeman's results were in error by some large but undetermined amount, yet they still influence ecological judgment. The difficulties he encountered were so forbidding as to discourage many imitators. It is necessary to know that Lindeman tried to measure the efficiency we name for him, and why his method would not work, to understand some of the limits to our present knowledge.

LINDEMAN'S CALCULATIONS: THE UNSUITABILITY OF STANDING CROP DATA

Lindeman tried to calculate his efficiencies from data on standing crops in a manner analogous to that of Transeau when he calculated the efficiency of a field of corn from data on standing crop and respiration (Chapter 3). Lindeman studied Cedar Bog Lake in Minnesota for 5 years, measuring standing crops of all the more important animals and plants of each trophic level and year by year (Lindeman, 1941). The problem then was how to calculate the energy input for each trophic level over a typical year, and to do

Table 4.1
Efficiencies of Energy Transfer
The ecological (Lindeman) efficiencies of energy transfer between trophic levels are the fundamental efficiencies that set the energy flux to the next trophic levels. All other efficiencies result from processes that also affect Lindeman efficiency. All efficiencies, therefore, are related to the fundamental Lindeman efficiency.

$$\text{Ecological efficiency of plants (Lindeman and Transeau efficiencies)} = \frac{\text{Rate of assimilation (photosynthesis) by plants}}{\text{Solar flux}} \times 100$$

$$\text{Lindeman efficiency of herbivores} = \frac{\text{Rate of assimilation by herbivores}}{\text{Gross productivity of plants}} \times 100$$

$$\text{Lindeman efficiency of primary carnivores} = \frac{\text{Rate of assimilation of primary carnivores}}{\text{Rate of assimilation of herbivores}} \times 100$$

$$\text{Assimilation efficiency} = \frac{\text{Food assimilated (digestible energy)}}{\text{Food ingested}} \times 100$$

$$\text{Exploitation efficiency} = \frac{\text{Food ingested}}{\text{Available net production}} \times 100$$

$$\text{Net production efficiency} = \frac{\text{Net production (growth and reproduction)}}{\text{Food assimilation}} \times 100$$

this from the standing crop measurements alone.

All the animals and plants of the system were respiring, and this must be measured as for Transeau's corn. But all were reproducing and dying as well, some *turning over* rapidly and some slowly. Lindeman had to allow for the missing individuals as well as for the living in a system in constant flux. Lindeman tried to account for all these replacements of individuals, and for the energy that was dissipated in all this flux, by multiplying each standing crop with an appropriate *turnover time*: a week for algae, two weeks for zooplankton, a whole year for rooted aquatic plants. Then he had to add a figure for respiration to the reconstituted standing crops.

In work with plants, the measurement of respiration of an active individual is the most tricky part of an efficiency measurement because plants absorb CO_2 in the light even as they respire the gas. Plant efficiency measurements accordingly depend on unrealistic measurements of respiration made in the dark (Chapter 3). But measurements on animals are without this problem and there was a good literature of respiration rates for animals on which Lindeman could draw. But there was an insuperable difficulty in using these measurements for the calculations.

The respiration data that Lindeman used included the energy dissipated in the activity of turning over as well as that used for finding food; and turnover had been allowed for already. Lindeman's energy input was, therefore, too high by some large but unknown factor. He then compounded the error by adding estimates for animals taken by predators and animals that died other deaths and went to the decomposers. The energy represented by these bodies had already been allowed for in the turnover calculation and was now appearing in the equation twice.

Lindeman had wanted to solve the following equation:

$$\text{Productivity} = (\text{yield of dry matter in calories}) + (\text{calories respired})$$

but what he had actually done was to say:

$$\text{Productivity} = (\text{standing crop in calories} \times \text{turnover time}) + (\text{calories respired}) + (\text{calories lost to mortality})$$

This second equation has both mortality and respiration appearing twice, once disguised as "turnover" and once in their proper places (Slobodkin, 1962).

Lindeman's results, together with a more recent aquatic study using comparable logic, are given in Table 4.2. They look plausible. The highest efficiency of any animal trophic level is given as about 20%, and a conservative figure of about 10% looks to be about the norm. This became an ecologist's rule of thumb. And it explained production pyramids nicely enough: a top carnivore got 10% of 10% of the 2% of the energy that plants got from the sun.

But Lindeman's results were actually in error, as are all other estimates based on standing crops and turnover times. They err in the direction of being too high, perhaps by a large amount, because the measures of turnover and respiration overlap. At the very least, the rule of thumb of 10% efficiency had best be taken as an upper limit unless there are strong reasons for thinking otherwise. For top carnivores on land, for instance, the 10% figure may in some circumstances be nearly ten times too large (see below).

LINDEMAN EFFICIENCIES IN SINGLE-SPECIES TROPHIC LEVELS

Slobodkin (1962) showed that correct Lindeman efficiencies could be calculated by considering the fate of energy that flowed into a trophic level. This energy is either degraded within the trophic level, in which event it appears as respiration, or it is passed out of the trophic level in a dead animal body. No other fates for the energy were possible. It follows that, in a trophic level,

$$\text{Respiration} + \text{Mortality} = \text{Gross Production} \quad (4.3)$$

Thus the problem is resolved into two measurements per trophic level, but this can still be difficult. Separate measures of mortality and respiration of all the species of a trophic level in a complex ecosystem must require much work. Obvious short-cuts like measuring respiration of the whole community will not serve because the trophic levels must be separated. In practice the measures must be made species by species. Moreover the ecosystem state must not be changing. A time with few herbivores would show

Table 4.2
Lindeman Efficiencies Calculated from Standing Crops and Turnover

These estimates are in error, being too high. They are often quoted still, however, and they lead to the ecological rule of thumb that energy is usually transferred between trophic levels with an efficiency of about 10%. Real ecological (Lindeman) efficiencies may usually be significantly lower than 10%. (From Lindeman, 1942, and H. T. Odum, 1957.)

	Cedar Bog Lake (%)	Lake Mendota (%)	Silver Springs, Florida (%)
Primary consumers (herbivores)	13.3	8.7	16
Secondary consumers (1° carnivores)	22.3	5.5	11
Tertiary consumers (2° carnivores)	No data	13.0	6

low herbivore efficiency, but a time of overgrazing would show a spuriously high efficiency. No measures of Lindeman efficiency of complex ecosystems using this technique have been published, apparently because of these practical difficulties, though a few measures are available from simple systems in which the higher trophic level was occupied by a single-species population.

In the laboratory, stable ecosystems with few species can be designed in which populations are both constant and food-limited. In a pioneer demonstration of the possibilities, water fleas of the genus *Daphnia*, and *Hydra*, were used as the top trophic level (Richman, 1958; Armstrong, 1960; Slobodkin, 1962; Figure 4.2). The food in both systems was green flagellates, and the supply was manipulated until a roughly constant yield of growth and mortality was obtained. Micro-bomb calorimeters were designed to give accurate estimates of the calorific value of *Daphnia* and *Hydra* corpses. *Hydra* populations had ecological efficiencies of 7% and *Daphnia* populations of 13%. These results are close to the values found by Lindeman (Table 4.2), making it seem that his errors may not have been too large in practice. The use of the rule of thumb of 10% for transfer efficiencies was encouraged.

Colinvaux and Barnett (1979) applied Slobodkin's method to a population of wolves feeding on moose (Figures 4.3 and 4.4). The wolves and the moose live on an island in Lake Superior. The island is about 100 km long but for animals as big as moose and wolves this is a small place. In theory both wolves and moose can travel off the island by swimming or over the winter ice but in practice there is little exchange with the mainland. The experimental container called Isle Royale, therefore, is effectively sealed. Studies over many years showed how many wolves and moose lived on the island, how long they lived, and that both populations were stable (Mech, 1966; see Chapter 9). This means that

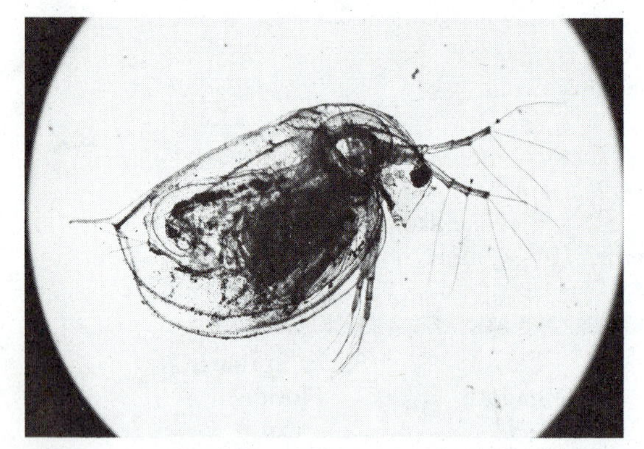

Figure 4.2 *Daphnia*: populations can be 13% efficient. Water flea populations in steady state with food supply in a laboratory system converted energy with an ecological (Lindeman) efficiency of 13%, ten times the efficiency of a population of wild wolves feeding on moose. This high efficiency of the water fleas might partly reflect the favorable conditions in the experimental chamber and partly the fact that the animals are ectotherms (Chapter 5) with low metabolic rate.

Figure 4.3 *Canis lupus*: 1.3% efficient. The ecological (Lindeman) efficiency of a pack of wolves feeding on moose is about 1.3%, or about one-tenth of that of water fleas in captivity. This low efficiency is probably typical of large carnivores of high metabolic rate that hunt prey equipped with defenses.

the prime necessities for reasonable calculations—simplicity and stable populations—were met.

Another rare circumstance favored the investigation in that it was known both what the wolves ate and the cause of all moose mortality. The wolf diet on Isle Royale is almost entirely moose, certainly 70 to 80% of the diet, anyway. And the only fate of a moose was to be eaten by wolves. The wolves killed some young moose when they could catch them, and they killed all sick or old moose. Thus there was on Isle Royale a trophic level of primary carnivores (wolves) almost entirely subsisting on a single-species trophic level of herbivores (moose). There were, of course, other carnivores on Isle Royale, like birds and small mammals feeding on rodents, but these could be expected to have little impact on the wolf–moose system.

Figure 4.4 *Alces americana*: defense of large size. Moose may have been selected for large size as a defense mechanism against big predators. They can be taken only with low ecological efficiency.

Two quantities now had to be calculated for both the moose and the wolf populations, yield of corpses (in calories) and respiration. Mortality data were available from published life tables of both moose and wolves (Chapter 9).

Respiration of wolves and moose was calculated from the established function relating body weight to respiration of homeotherms (Kleiber, 1961). About 20 wolves on Isle Royale subsisted on 600 moose. The ecological (Lindeman) efficiency of wolves in this system is 1.3% (Table 4.3).

Wolves on Isle Royale are evidently much less efficient than *Daphnia* (13%) or *Hydra* (7%) in laboratory tanks. This should not be a cause for surprise. For one thing the laboratory animals were fed simple food in circumstances designed to let them give of their best; seasons did not change in the tanks and the measurements were made over a time span when conditions for life were optimum. But the wolves had to hunt through the changing seasons of a whole year. Moreover the wolves tackled prey larger than themselves, prey which, apart from calves, they could only pull down when it was old or sick.

In conclusion, it is worth noting from Table 4.3 how critical the measurement of respiration is to the calculations. Most of the energy ingested into each trophic level reappears as respiration, not as mortality. This emphasizes the dangers of attempting calculations of the Lindeman type where measurements of turnover or activity duplicate respiration. Measures of respiration are critical.

ASSIMILATION AND ASSIMILATION EFFICIENCY

Animals take food into their guts when they eat: ecologists call this process INGESTION. But the energy of the food has not yet been absorbed by the animal because it has still to be digested

Table 4.3
Trophic Level Data for Moose and Wolves
Published data for moose and wolves on Isle Royale are used to calculate the ecological (Lindeman) efficiency of wolves. (Reworked from Colinvaux and Barnett, 1979.)

	Moose	Wolves
Population	600	20
Yield of young corpses (calves: pups) per year	52	5
Yield of adults per year	111	1.57
Total corpse yield per year	123,321,825 kcal	600,439 kcal
Annual respiration	910,349,091 kcal	12,786,382 kcal
Energy flux into trophic level	1,033,670,916 kcal	13,386,821 kcal

$$\text{Ecological (Lindeman) efficiency of wolves} = \frac{13{,}386{,}821}{1{,}033{,}670{,}916} \times 100 = 1.3\%$$

and taken through a biological membrane. Ecologists call this second process ASSIMILATION.

$$\text{Assimilation} = \text{Ingestion} - \text{Defecation}$$

For ruminant animals this definition must be modified because ruminants let ingested food ferment in the first stomach (*rumen*), a process that vents methane from their mouths. For ruminants

$$\text{Assimilation} = \text{Ingestion} - (\text{Defecation} + \text{Methane})$$

Other animals use fermentation in their guts as an aid to digestion, notably termites, so that loss of ingested energy as gaseous reduced carbon may be widespread. A more general formula for *assimilation* is

$$\text{Assimilation} = \text{Ingestion} - (\text{Defecation} + \text{gaseous reduced carbon eliminated})$$

Assimilation efficiency is a straightforward idea:

$$\text{Assimilation efficiency} = \frac{\text{Absorption}}{\text{Ingestion}}$$

where *absorption* is energy assimilated and *ingestion* is food energy eaten. *Assimilation efficiencies* tend to be high, that for the domestic cow being about 50% (Figure 4.5, Table 4.4).

The energy an animal can *assimilate* over a period of time is its GROSS PRODUCTIVITY or GROSS PRODUCTION.[1] The equivalence of

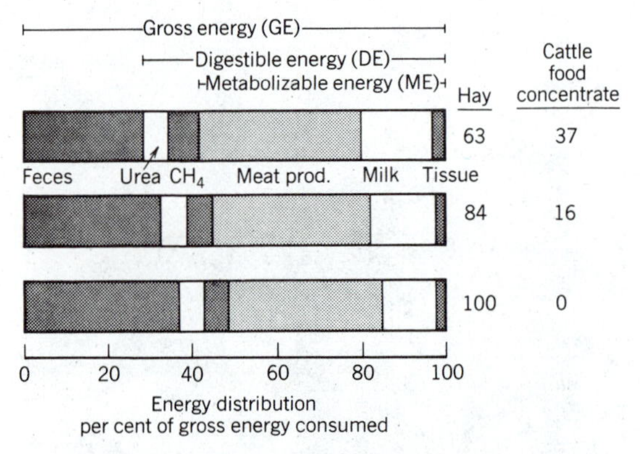

Figure 4.5 Energy assimilation by a lactating cow. The three bar diagrams show the energy budgets for a cow fed different rations of hay and cattle food concentrate. When the cow is given only hay, the assimilation rate is about 50%, reflecting both the fact that mammals have to use symbionts to digest cellulose (an extra link in the food chain) and that fermentation liberates methane. *Gross energy* in the agricultural usage is what an ecologist would call energy ingested. (Data of P.W. Moe from Church and Pond, 1974.)

[1]Odum (1955) suggests the use of *productivity* to refer to rate of production and *production* to refer to mass produced without reference to time. In ordinary English usage the word "productivity" means the "state of being productive" not a rate process at all. Ecologists seldom make Odum's fine distinction now, using the two terms almost interchangeably.

Table 4.4
Food Conversion Efficiencies Reflect Habitat and Diet
Assimilation efficiencies are functions of the kind of food eaten, being high for meat eaters and low for cellulose eaters. Net production efficiencies vary with activity and mode of temperature regulation. It is necessary to be cautious about some of the estimates that were made in difficult circumstances. The estimates for African elephants and kob, for instance, are based on very approximate data and the correct production efficiencies of these large herbivores may be close to the 11 to 12% of cattle and moose. (Data for kangaroo rat from Soholt, 1973; for moose and wolves from Colinvaux and Barnett, 1979; remainder compiled by Ricklefs, 1973, from various sources.)

Species	Habitat	Diet	Percentage Assimilation Efficiency	Percentage Production Efficiency
Aquatic poikilotherms				
Megalops cyprinoidea (fish)	Fresh water	Fish	92	35
Bleak (carp family)		Invertebrates		7
Nanavax (predatory gastropod)	Intertidal	Invertebrates	62	45
Roach (carp family)		Plants		7
Ribbed mussel	Salt marsh	Phytoplankton		30
Hyallela azteca (amphipod)	Lakes	Algae	15	15
Ferrissia rivularis (stream limpet)	Streams	Algae	8	19
Tegula funebralis (herbivorous gastropod)	Intertidal	Algae	70	24
Nematodes	Salt marsh	Detritus		25
Littorina (periwinkle)	Salt marsh	Detritus	45	14
Calospecta dives (midge)	Spring	Detritus		25
Terrestrial poikilotherms				
Harvester ant	Old field	Seeds		0.3
Saltmarsh grasshopper	Salt marsh	*Spartina*	27	37
Melanoplus species (grasshopper)	Old field	Lespedeza	37	37
Grasshoppers (3 species)	Old field	Vegetation	16	
Spittlebug	Alfalfa	Plant juices		39
Woodlouse	Forest litter	Dead leaves	33	
Mites (oribatei)	Old field	Detritus	25	22
Millipede	Forest floor	Decaying wood	15	
Terrestrial homeotherms				
Least weasel	Old field	Mice	96	2.3
Marsh wren	Salt marsh	Insects	70	0.5
Grasshopper mouse	Desert shrub	Insects	78	5.7
Kangaroo rat	Desert shrub	Seeds	87	6.0
Vole	Old field	Vegetation	70	3.0
Field mouse	Old field	Grass seed		1.8
Jackrabbit (*Lepus*)	Desert shrub	Vegetation	52	5.5
Cottontail rabbit (*Sylvilagus*)	Desert shrub	Vegetation	52	6.0
Uganda kob	Savanna	Vegetation		1.3
Cattle	Pasture	Vegetation	38	11
Moose	Forest	Vegetation		11.9
African elephant	Savanna	Vegetation	32	1.5
Wolves	Forest	Moose		4.5

assimilation with *gross production* prompts another approach to the problem of measuring Lindeman efficiencies. Measures of ingestion and defecation can be used instead of measures of respiration and mortality so that

$$\text{Ingestion} - \text{Defecation} = \text{Respiration} + \text{Mortality}$$

If measures of ingestion and defecation can be made for all the principal species populations of complex communities, then Lindeman efficiencies can be calculated for complex ecosystems.

NET PRODUCTION AND PRODUCTION EFFICIENCY

NET PRODUCTION is always visible as biomass.

$$\text{Net Production} = \text{Gross Production} - \text{Respiration}$$

and

$$\text{Net Production} = \text{Growth} + \text{Reproduction}$$

But there are some semantic traps to look out for when reading about ecological research, because "production" means different things to different people.

In many ecological writings NET PRODUCTION is simply referred to as *production* and ecologists talk of *production efficiency* when it would be formally more correct to use the term *efficiency of net production*. It is easy to see that the shorter term is more convenient and the usage does not matter when the professional parties to the discussion understand the shorthand. But it is always necessary to ask oneself exactly what is meant by the word "production" when it appears in ecological writing. The general rule seems to be, "when the word PRODUCTION is used without qualification, it is to be taken to mean NET PRODUCTION."

In ecology (but not in agriculture—see below) NET PRODUCTION EFFICIENCY (or simply *production efficiency*) is

$$\text{Net Production Efficiency} = \frac{\text{Net Production}}{\text{Assimilation}}$$

Production efficiencies of species populations or individuals have frequently been measured by ecologists because they reveal much of individual strategies. This efficiency tells what an animal does with the energy flux at its disposal, whether it uses this more to produce biomass or for activity. Table 4.4 compares the assimilation and net production efficiencies of a wide variety of animals. Some generalizations are possible from the table. Carnivores tend to have higher assimilation efficiencies than do herbivores, which is a reflection of the success of plants in evolving indigestible tissues. Homeotherms tend to have lower net production efficiencies than poikilotherms, a reflection of the high energetic cost of maintaining a constant body temperature.

Net production efficiencies can also offer their own special insight into the structure of animal communities that we seek to measure with Lindeman efficiencies. The production efficiency of the Isle Royale moose herd, for instance, is 11.9%, comparable to that of domestic cattle (Table 4.4), yet the Lindeman efficiency of the wolves was only 1.3%. In theory the wolves could have got all the net production of the moose, or close to it, by an optimum yield cropping program. If they did this the ecological efficiency of wolves would approach the production efficiency of the moose. In fact it was only about one-ninth of it (1.3% of the gross instead of 11.9%). The low Lindeman efficiency of wolves reflects not only the energy respired by moose that was impossible for wolves to get but also most of the "possible" energy they missed since their *exploitation efficiency* was low.

THE EFFICIENCY OF EXPLOITATION

After the ECOLOGICAL (LINDEMAN) EFFICIENCY, the efficiency that helps most in answering the grander ecological questions is the EXPLOITATION EFFICIENCY. This relates the food an animal can actually get to the available supply of that food. The essential difference from the Lindeman efficiency is in referring to the available net production of the food species rather than to the impossible ideal of getting at the gross production.

$$\text{Exploitation Efficiency} = \frac{\text{Energy ingested}}{\text{Net production of food species}}$$

For an individual herbivore species population this becomes

$$\text{Exploitation Efficiency} = \frac{\text{Food ingested by herbivore population}}{\text{Available net production of food plant}}$$

and for a whole herbivore trophic level this becomes

$$\text{Exploitation Efficiency} = \frac{\text{Food ingested by herbivores}}{\text{Net primary productivity}}$$

This efficiency is not only interesting in its own right but is more practical than the Lindeman efficiency since it is easier to measure (net production replaces gross production in the food trophic level). Table 4.5 lists *exploitation efficiencies* for different animals. These data reveal at a glance the relative importance of animals in different lifestyles in different ecosystems. As a cause of biospheric process zooplankters in the sea appear to be more interesting than mammals in the Arizona desert.

But there is a special word of caution about *exploitation efficiency* that concerns the state of the populations at the time of measurement. Blatant overgrazing, for instance, will result in a very

Table 4.5
Exploitation Efficiencies of Herbivores in Different Systems
Exploitation efficiencies suggest the importance of the animals to their respective ecosystems. The data need to be treated with care, however. The lemming exploitation efficiency must refer to the local population concentration of a lemming high (see Chapter 13) and lemmings may be rare at some times and in some places when their exploitation efficiency would be correspondingly low. In general, exploitation efficiencies will refer to the performance of animal populations at particular times and places. (From Chew, 1974, and Soholt, 1973.)

Consumer Category	System	Exploitation Efficiency
Mammals	Desert, Arizona	2% total NPP (above ground) 5.5% available NPP 86% seed production
Kangaroo rat	Desert, California	10.7% available NPP 95% available seed product
Mammals	Oak–pine forest, Poland	0.7% available NPP
Mammals	Taiga forest, Alaska	13.5% available NPP
Leaf feeders	Deciduous forest	3–8% of foliage
Soil-litter fauna	Deciduous forest	4–8% annual litter 1% litter
Insects, except ants	Old field, herbland	1% total NPP
Ungulates	Grassland, Africa	28–60% total NPP
Domestic animals	Rangeland	30–45% for maximum sustained yield
Lemming	Tundra monocots	93% NPP
Zooplankton	English Channel	98.5% phytoplankton
Seed predators	General	10–90% predispersal seed crop

high exploitation efficiency, but this is likely to be a short-lived phenomenon.

PRODUCTION AND RESPIRATION RATIOS

The partitioning of energy between production and respiration can be a function of the size of an animal, its trophic status, and its physiological organization. The best data to support this intuitively reasonable proposition come from plotting data for net production against respiration (Figure 4.6). This approach has been discussed for some time in ecology but had to await the accumulation of enough energy budgets to allow significant correlations (Englemann, 1966). McNeill and Lawton (1970) first showed that the P/R ratio was a function of animal organization and not of body size, finding that homeotherms could be separated from poikilotherms (warm- and cold-blooded animals, Chapter 5). Figure 4.6 is from the recent work of Humphries (1979), who repeated McNeill and Lawton's work with a larger data set than previously available, a total of 265 energy budgets.

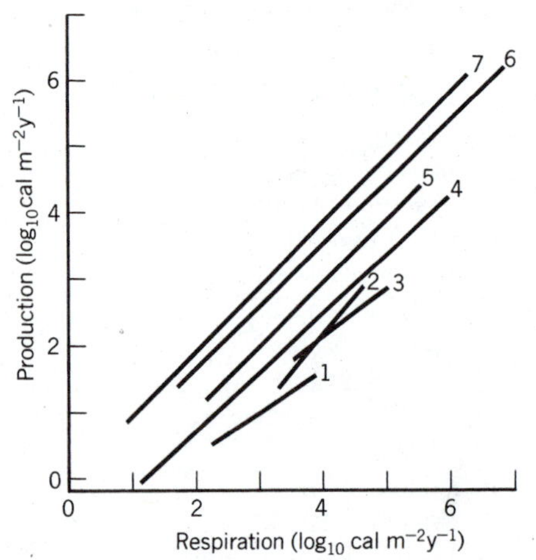

Figure 4.6 Production, respiration, and thermoregulation.
Ratio between production and respiration is a function of animal size and method of controlling heat budget. (1) Insectivores, (2) small mammals, (3) birds, (4) other mammals, (5) fish and social insects, (6) non-insect invertebrates, (7) non-social insects. (After Humphries, 1979.)

Regressions for seven kinds of animal organizations can be separated in Figure 4.6 with various degrees of confidence. All seem reasonable, though caution is necessary in placing reliance on them since the energy budgets on which they are based are subject to wide variations in method and probable accuracy.

Plots of P/R ratio have yielded a practical tool in production ecology since they can be used to estimate probable production from measures of respiration alone. This short cut has been used in a number of investigations since the paper of McNeill and Lawton (1970) and the work of Humphries (1979) encourages its validity.

SEMANTICS AND CONFUSIONS OF TERMS BETWEEN DISCIPLINES

Terms like *production, gross, net,* and *yield* are given different meanings when animal energetics are studied in the tradition of agriculture. Animal energetics overlaps the disciplines of ecology, agriculture, and animal physiology with the result that these primarily agricultural meanings find their way into literature used by ecologists. The possibilities for confusion are considerable.

For ecologists (page 103)

Energy Assimilated =
Energy Ingested − Energy Defecated

But another way of expressing this in wide use in agricultural literature is to say

Digestible Energy = Gross Energy − Fecal Energy

It will be evident that this use of "gross" does not invoke the meaning ecologists give the word. "Gross energy" to an experimenter in animal nutrition is the energy of the food given the animal to eat, and "net energy" then is this food less the indigestible portion.

Agriculturalists often refer to YIELD to mean the same things as the ecologist's *net secondary production.* But *yield* also is used in the traditional farmer's sense to mean the yield of marketable crop.

One ratio appearing in the agricultural literature has to be treated with care to avoid confusion. This relates *net production* to *ingestion:*

$$\text{Yield Efficiency} = \frac{\text{Net Production}}{\text{Energy Ingested}}$$

It has already been noted that agricultural workers often call *energy ingested* the *gross energy,* from which practice the *yield efficiency* becomes the *gross efficiency.* From similar reasoning the *net production efficiency* becomes *net efficiency* (Church and Pond, 1974). These uses of *net* and *gross* are quite different to what ecologists usually mean and yet both uses are common in the literature of animal energetics. The only advice to offer the student seems to be always to look carefully at descriptions of production ratios and efficiencies to see precisely what is meant, and do not let someone talk loosely about an animal being "efficient": ask, "Efficient at what?"

PRACTICAL MEASURES OF RATES OF ASSIMILATION

Measures of actual food intakes and actual assimilation make possible the calculation of a complete energy budget for a population, which can be a tool for answering many ecological questions. The task often requires experimental ingenuity. Results from a famous example of success with laboratory measurement were used by Lindeman for his calculations, the measurement by the Russian fisheries scientist Victor Ivlev of the assimilation rate in a mud-living worm, *Tubifex tubifex* (Ivlev, 1939).

Tubifex worms feed by ingesting mud and passing the mixture of silt and organic matter of lake bottoms through their alimentary canals and digesting what parts of it they can. Ivlev mixed some mud from the place where he found his worms with grains of platinum black so that the platinum black was evenly distributed throughout the mud. Then he put mud and worms in glass jars, the bottoms of which were made of gauze, and which were suspended in water. The worms positioned themselves so that their tails were at the bottom of the mud, projecting through the gauze into the water. The fecal pellets fell down and could be collected in a watch glass underneath (Figure 4.7). The difference in concentration of platinum black in the droppings and in the original mud gave the volume of sediment that had been ingested during the time that the worms were in the jars. Worms were weighed before and after feeding, and calorimetry of worms and mud let all masses be converted to calories. These data gave calories ingested and calories represented by growth. The difference between ingested calories and growth calories gave metabolism, and hence respiration and excretion. Ivlev thus had rate of assimilation, rate of growth, and a combined rate of respiration and excretion for common bottom-living worms.

The feeding rate of animals can be measured as the rate of food removed from an experimental system. This has proved particularly useful for studies on zooplankton (Chapter 21). Individuals or populations of the animals are introduced into containers where the concentration of their food is known, they are left for a measured time, and the concentration of the food is noted at the end of the experiment. Then

$$C_t = C_0 e^{-kt} \tag{4.4}$$

where

C_t = concentration of food at the end (time t)
C_0 = concentration of food at beginning
k = feeding rate constant.

An early demonstration of this method showed that the rate of capture of the green alga *Chlamydomonas* by marine copepods was what would be expected if the animals filtered water nonselectively (Table 4.6; Gauld, 1951). More recent work shows that copepods can mix generalized filtering with selection of food from more complex food mixtures (see Chapter 21).

The rate of ingestion also can be measured by isotopic labelling of food. This method has wide application but again can be particularly useful in aquatic systems. Kibby (1971) fed algae labelled with ^{14}C to freshwater copepods and calculated that the assimilation efficiency of *Diaptomus gracilis* varied between 45% and 83%,

Figure 4.7 Tubificid feeding experiment of Ivlev.
The inverted jar contains organic mud through which grains of platinum black have been evenly distributed. *Tubifex* worms placed in the jar orient themselves so that their tails project through the gauze that closes the bottom of the jar. Their fecal pellets can then be collected in the watch glass below, and the amount of mud they have ingested can be computed from the concentration of grains of platinum black. By measuring the calorific value of the mud it was thus possible to determine the calories ingested by the worms in unit time. (After Ivlev, 1939.)

Table 4.6
Gauld's Results for the Filtering Abilities of Copepods
The column "Mean volume swept clear per copepod" was calculated from equation (4.5). (From Gauld, 1951.)

Series	Number of Experiments in Series	Species	Number of Copepods per Vessel	Volume of Vessel (ml)	Duration (hr)	Temperature (°C.)	Mean Volume Swept Clear per Copepod (ml)
A	21	*Pseudocalanus minutus*	1	10	24	10	4.28
B	45	*Temora longicornis*	10	150	24	10	8.38
C	8	*Centropages hamatus*	10	150	24	10	12.99
D	19	*Calanus finmarchicus stage V*	1	100	24	12.5	64.36
E	77	*C. finmarchicus stage V*	1	100	18	17	71.03
F	25	*C. finmarchicus stage IV*	2	100	18	17	36.65
G	13	*C. finmarchicus stage III*	3	100	18	17	22.24

depending on the species of phytoplanktonic alga that was used as food. There are now many measures of feeding by copepods and other zooplankters using both food removal and isotope methods (Kerfoot, 1980).

The animals for which we have the best data on assimilation rates are the chickens, pigs, sheep, horses, and cows of Western agriculture, but even for these the measurements may be complicated. It is simple to monitor the food intake of a farm animal (*ingestion*) and to collect its excrement (*defecation*). But the *assimilation ratio* (ingestion/defecation) varies not only with the kind of food given the animal but also with how much food is given, how old the animal is, how healthy it is, what other food it has had recently, what other animals are with it, and so on.

Part of the excrement of the animal represents bodily excretions, like bile, voided into the intestinal tract, and the energy represented by these must be subtracted to arrive at a measure of how much of the total ingested energy of the food appears in the feces. Apparently the only way to estimate this excretion into the alimentary tract is to starve an animal and then to measure the fecal matter it produces, because the feces of a starving animal should equal alimentary excretion. This is difficult and seldom done, even for farm animals. Any such measurement carries the hazard of assuming that the performance of the body of a stressed animal (starving) is comparable to that of an unstressed animal.

Finally, some of the ingested energy of food is neither digested nor defecated but released as gas in the digestive process. This is particularly true of ruminant animals that eliminate significant masses of methane. To monitor this it is necessary to keep the animal in a closed container, to pass air through the container, and to monitor the flux of carbonaceous gas leaving it.

Apparent DIGESTIBLE ENERGY (DE), which is the difference between *ingestion* and *defecation,* of food supplied to farm animals varies between 10% or less for animals fed on milk to 60% or more for animals given what an agriculturalist calls "poor quality roughage" (Church and Pond, 1974). Perhaps 5% of DE is voided as methane by ruminants, meaning that the actual ASSIMILATION RATE or DIGESTIBLE ENERGY is less than the apparent DE by a proportional amount. Figure 4.5 gives average data for a lactating dairy cow when fed on three different rations. Collecting the data in Figure 4.5 involved estimating excretion into the gut, the flow of urine, and the gasses voided from the gut, as well as food eaten and feces dropped.

KANGAROO RATS: ECOLOGICAL LEARNING FROM AN ENERGY BUDGET

The energy assimilated by an animal goes either to maintenance or to growth and reproduction. Between reproductive episodes an animal's reproductive potential is represented by its increasing energy reserves so that reproduction can be considered a special form of growth, and we write

$$E = E_m + E_g \qquad (4.5)$$

where

E = energy flowing into the population
E_m = maintenance energy
E_g = growth energy.

This equation lets us approach the problem of measuring the energy flux without measuring either ingestion or defecation. Instead we monitor what happens to energy for individuals of the various kinds in a population (young, old, male, female, *et cetera*), calculate an energy budget for each kind, and then apply these energy budgets to the standing crop data. This approach has been used for several small rodent populations. What follows is based on work by Soholt (1973) and by Chew and Chew (1970), who used their own measurements and those of a number of other workers to produce particularly convincing energy budgets for a population of Californian kangaroo rats (*Dipodomys merriami*).

The kangaroo rat is a desert animal. It rests by day underground in a nest and comes to the surface at night for a few hours to feed. The energy budget, therefore, includes the energy needed for maintenance when the animal is resting in its burrow, the energy of maintenance when the animal is in the open at night, and the energy used to support the nighttime activity. These data have to be known for animals of various ages and for the ambient temperatures prevailing at different times of the year. All these data can be collected by measuring the carbon dioxide flux from captive animals. If growth rates are also known, either from repeated weighings of captive animals or from weights of wild animals that are repeatedly trapped, then an energy budget for the population can be calculated as follows:

$$E_m = E_r + E_a \quad (4.6)$$

where

E_r = maintenance energy of resting animal
E_a = maintenance energy of active animal

$$E_r = (T)(S_c)(E_m \cdot t_n + E_{rs}) \cdot t_s \quad (4.7)$$

where

T = time interval of measurement
S_c = standing crop
E_m = maintenance energy of animal resting in nest
E_{rs} = maintenance energy of animal resting at surface
t_n, t_s = time animal spends in rest and on surface

$$E_a = (T)(I_a)(t_s)(S_c) \quad (4.8)$$

where

I_a = estimated rate of energy expenditure for activity

$$E_g = \frac{NP_1 + NP_2 + NP_3}{GE_n} \quad (4.9)$$

where

NP = net secondary production
NP_1 = production of biomass in prenatal growth
NP_2 = production of biomass while nursing
NP_3 = production of biomass after weaning
GE_n = net efficiency of growth

$$GE_n = \frac{NP}{E_g} \quad (4.10)$$

To solve for the growth efficiency GE_n, we must know the growth energy E_g, but this cannot be found without knowing GE_n itself (equation (4.10)): hence an apparent impasse. This difficulty was avoided in a later study by Soholt (1973) that used data for growth efficiency GE_n reported in the literature for white rats, which suggested that the correct figure for kangaroo rats would be about 50%. By measuring all the other parameters needed for solving equations (4.6) through (4.10) and applying the resulting energy flux (E) to his standing crop data, Soholt calculated that 85.5 megacalories flowed through the kangaroo rat population per hectare per year.

The habitat of the kangaroo rats studied by Soholt was mostly vegetated with annual plants, although there were some perennial bushes. This meant that rough estimates of net primary productivity could be made by measuring the standing crop of plants at the end of the season. Using Soholt's data it is possible to arrive at an approximation of Lindeman efficiency as follows:

(a) net primary productivity = 1400 megacalories per hectare per year,

(b) productivity of plant species found in kangaroo rat stomachs (available productivity) = 900 megacalories per hectare per year,

(c) assume respiration is one-third of net productivity (approximation of Transeau's finding),

(d) gross available productivity = 1200 megacalories per hectare per year,

(e) approximate ecological (Lindeman) efficiency of population of kangaroo rats = $\frac{(85.5 \times 100)}{1200} = 7.13\%$.

This result for Lindeman efficiency is more doubtful than the calculations for wolves on Isle Royale (Table 4.7) because not all the assumptions of the Slobodkin approach are met. In particular there is no reason to believe that the population of rats or their food plants are at a steady state. Indeed, the very high exploitation of the

Table 4.7
Ecological (Lindeman) Efficiencies
*These results are free from the dangers inherent in using standing crops and turnover times and are offered as generally conservative figures. (*Daphnia *and* Hydra *from Slobodkin, 1962; kangaroo rat calculated from data of Chew, 1974, and Soholt, 1973; sheep from Perkins, 1978; wolf from Colinvaux and Barnett, 1979.)*

	Ecological (Lindeman) Efficiency (%)
Plankton feeders in culture	
Daphnia	13
Hydra	7
Wild herbivores	
Kangaroo rat	<7.3
Domestic herbivores	
Sheep	0.96
Wild carnivores	
Wolf	1.3

seeds on one particular plant by the rats (see below) suggests that the system was sampled at a time when rats may have been overnumerous. If the rats were overexploiting their food then the calculated efficiency is too high.

But perhaps the most striking insight coming from these studies was the immense importance of kangaroo rats to the food plant *Erodium.* Soholt (1973) used the data to show that kangaroo rats

(a) consume 6.9% of net primary production,

(b) consume 10.7% of available production,

(c) consume 95% of seeds of the principal food plant (*Erodium*),

(d) consume 90% of the total net production of the principal food plant (*Erodium*).

The impact of the rats on *Erodium* is clearly very heavy. It is this heavy impact that suggests that the rat population may have been unusually high at the time of sampling, thus yielding too high a figure for Lindeman efficiency. But it seems likely that kangaroo rats and *Erodium* are important factors in the lives of each other, and their respective niches must be adapted to allow for this.

OPTIMAL FORAGING THEORY INTRODUCED

The question "Efficient at what?" runs through all these studies of animal energetics. Always the answer depends on a point of view. Consider the difficulty of applying the word "efficient" to the performance of an automobile: a small family car run in times of gasoline shortage would be called "efficient" if it went many miles at modest speed on a gallon of gasoline, but a racing car would be called "efficient" if it achieved very high speeds for its size even though its mileage per gallon was exceedingly poor.

Many animals and plants must be in the position of the racing car when some of their decisions about using fuel are made. They are "efficient" at whatever it is that they are doing, as is required by natural selection, but this "efficiency" may have been won by a prodigal use of energy that gives them an advantage. Cheetahs running down gazelles, for instance, might have much in common with racing cars and migrating geese may observe the economics of family motoring.

However, there seems to be a common-sense way out of the "point of view" difficulty. Whatever an organism does in the quest for energy must result in the largest number of offspring reproducing in the next generation, and this suggests that it is the energy *capital* won by the individual that counts. Energy capital is measured as *net productivity.* A common-sense view of natural selection, therefore, leads to the hypothesis that all individuals should act to maximize (should be efficient at) *net production.* This postulate yields a set of testable predictions collectively known as OPTIMAL FORAGING THEORY (MacArthur and Pianka, 1966; MacArthur, 1972; Schoener, 1971; Pyke *et al.*, 1977; Krebs, 1978).

Many animals forage by taking food that typically exists in discrete packages (seeds, fruits, or prey). The food is scattered in space into patches of high density or low, and it comes in different sizes. Large food items are usually much less common than small food items, but a large food item gives a much better return for the trouble of catching it than a small food item. If an animal is to maximize net production, therefore, it must be programmed to make optimal choices about where to hunt, what to hunt, and when to switch from patch to patch or from prey to prey.

Obviously we cannot expect all classes of animal to be programmed with equal precision. A vertebrate, for instance, can learn, and so be programmed to switch from one prey to another as a result of having experienced a fuller belly in the past from a similar switch. A vertebrate, therefore, might get closer to some theoretical optimal behavior than could an insect with no powers of learning. Nevertheless, it should be possible to see if active animals do forage in ways approximating the theoretical best pattern. If they are to be efficient at maximizing net productivity, this should show.

Not all hunting animals are active pursuers of prey. Some, like spiders and lions, lie in ambush, the so-called SIT AND WAIT PREDATORS. For these to maximize net production the ambush must be well laid and well placed. But a well-arranged ambush increases the net production of these animals by increasing the kill rate, which is the number of kills per unit time. Time, therefore, is important even to patient sit and wait predators. For more active animals time is an obvious limiting resource in its quest for energy.

Maximizing the energy taken in unit time must be the goal of optimal foraging. Whether taking a particular food item is profitable or not, therefore, is determined by the energy gained per prey item as a function of time. By observing animals feeding, the time needed to catch and handle each class of prey can be measured, as can the worth of the prey in calories (or as simple weight if prey is all of the same kind). This lets actual performance be compared with theoretical performance.

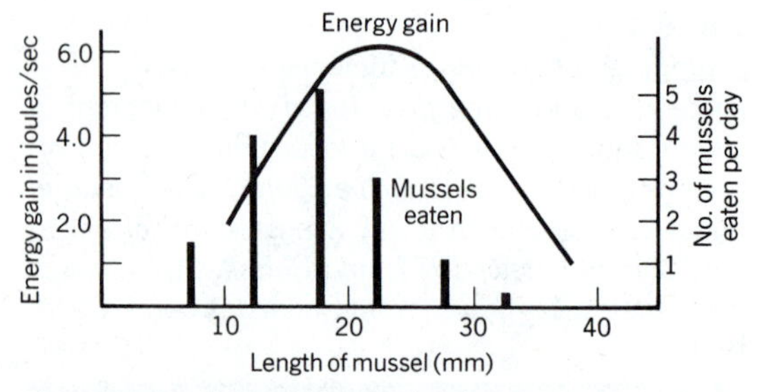

Figure 4.8 Optimal foraging by shore crabs. Large (6.0–6.5 cm) shore crabs (*Carcinus maenas*) break up mussels (*Mytilus edulis*) with their claws and eat the soft parts. When offered an unlimited choice of mussels of different sizes, the crabs concentrate on mussels that yield the largest energy return for handling time. The curve of energy per second as a function of mussel size was derived from watching crabs handle mussels of different sizes. (From Elner and Hughes, 1978.)

Figure 4.8 shows the foraging behavior of shore crabs feeding on mussels of different sizes. The curve shows the energy gained per second in taking small and large mussels. It is clear that the difficulty of breaking up a large mussel is such that the time spent yields a poor return despite the large meal that results. When given an opportunity to choose between mussels of different sizes, the crabs concentrate on intermediate sizes, where the energy return is greatest (see histograms).

Figure 4.9 shows the results of observations of pied wagtails feeding on flies of different sizes. Histograms in the top part of the diagram show that the wagtails actually concentrate their efforts on flies that are not quite the commonest size class, but that are slightly smaller. The lower part of the diagram shows that this behavior is optimal because flies of this smaller size take so much less time to subdue and eat that the energy flux received is greater.

There is now a growing body of literature that shows results comparable to these for crabs and wagtails from many kinds of birds, a number of fish, and a few insects (Krebs, 1978; Pyke *et al.*, 1977). When given simple choices, animals of many kinds are capable of making "choices" that optimize net productivity. But what if the choices are not so simple? Optimal foraging theory can

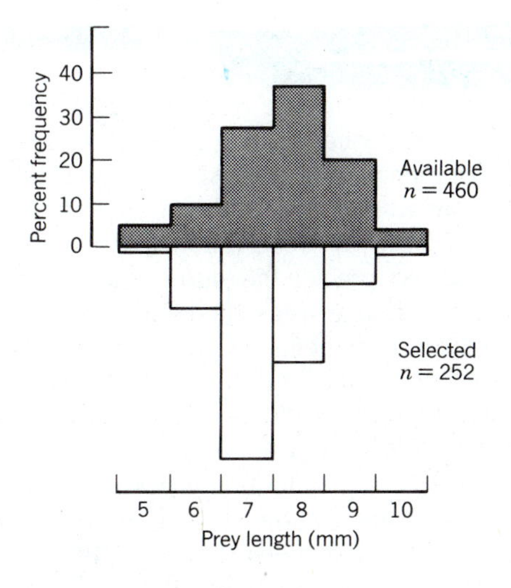

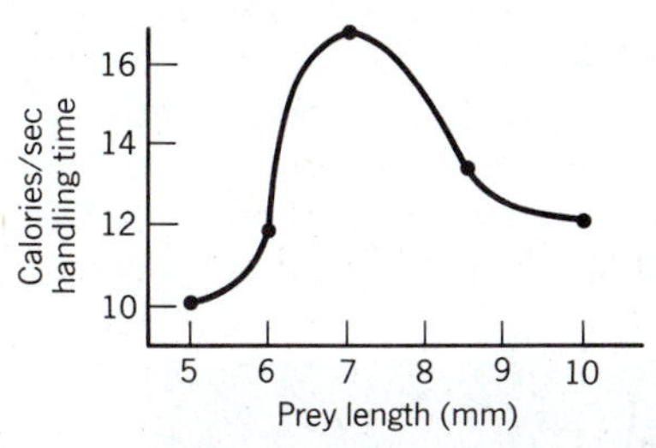

Figure 4.9 Optimal foraging by wagtails.
Wagtails fed preferentially on flies of smaller sizes, even when somewhat larger flies were the commonest (upper diagram). This behavior optimizes the energy return because large flies take proportionately longer to subdue and eat (lower diagram). (Data of Davies, 1977, from Krebs, 1978.)

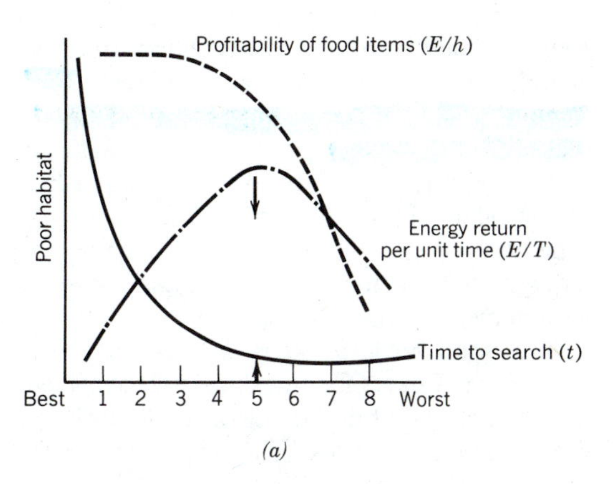

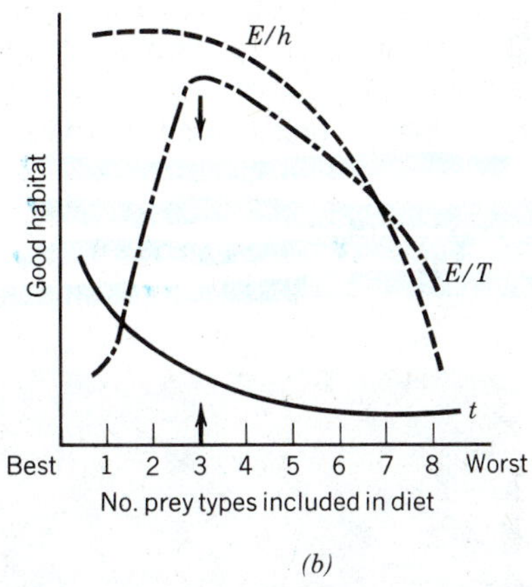

Figure 4.10 Optimal foraging in good and bad habitats. Time for search is assumed to be inversely proportional to density of prey and profitability of prey types is normally distributed. The energy return (net productivity) is a function of the time taken in collecting a number of food types and the profitability of the capture. This graphical presentation of the theory shows that where food is abundant the optimal diet of a forager should shift towards high-quality food items and a narrow range of items. (Modified from MacArthur and Pianka, 1966.)

be extended to predict successfully more complex patterns of behavior and "choice."

Consider first a rather poor habitat in which a number of different kinds of food are thinly dispersed. If a forager eats only the most desirable kind, it would spend a very long time hunting, but if it ate all kinds of possible prey, whether rewarding or not, then it would spend little time hunting. It is reasonable to assume that the time spent will in fact be inversely proportional to the choosiness of the forager, which assumption leads to the plot of curve *t* (time to search) in Figure 4.10*a*. Another reasonable assumption is that the profitability of various combinations of food items is normally distributed, an assumption that is supported by studies like those of the crabs and wagtails described above. A profitability curve (*E*/*h*—energy per unit handling time), therefore, shows high profitability for concentrating on the very best food items and low profitability for indiscriminately taking whatever is nearest. Optimal diet is now defined as that which yields the highest net productivity (energy flux per unit time = E/T) and $E/T = (E/t) + h$. The curve E/T (the resultant of the time and profitability curves) is plotted in Figure 4.10*a*. Thus an optimal diet requires that the forager includes the best five food items in its diet.

In a good habitat (Figure 4.10*b*) the profitability of the various food mixtures will, of course, remain the same, but the time taken to seek out the best items will be less. The curve t for searching time will accordingly be flatter and the resulting net productivity curve will be skewed to the left. In this hypothetical example, an optimal forager should concentrate on the three best food items to achieve an optimal diet.

Two general qualitative predictions follow from this analysis. In better habitats where food is relatively abundant foragers should choose a smaller number of food types, which is to say that in good times foragers should specialize. And in better habitats foragers should concentrate on prey of the highest quality. These are perfectly definite, and therefore testable, qualitative predictions. Moreover the argument also allows the development of quantitative predictions. For any particular system it is possible to measure values for t and E/h by observation and experiment, and thus to predict the numerical diet preferences of a forager presented with different concentrations or varieties of prey.

Werner and Hall (1974) applied optimal foraging theory to the diets of bluegill sunfish hunting three sizes of the water flea *Daphnia* in aquaria (Figure 4.11). They found that the handling time of a fish for the *Daphnia* was the same whatever the size of the prey—a gulp and the *Daphnia* was gone. This meant that the profitability of each size class of *Daphnia* was directly related only to its size, this being a crude measure of its energy content. Measuring the time taken to hunt each size class was trickier, but Werner and Hall were able to draw on a considerable literature showing that fish that hunted by sight had their

Figure 4.11 Bluegills: the laboratory optimal forager.
Bluegill sunfish have proved to be useful animals for experimental analysis of foraging behavior. One advantage is that handling time for water flea prey of different sizes is the same so that energy reward to fish per attack can be assumed to be a simple function of prey size.

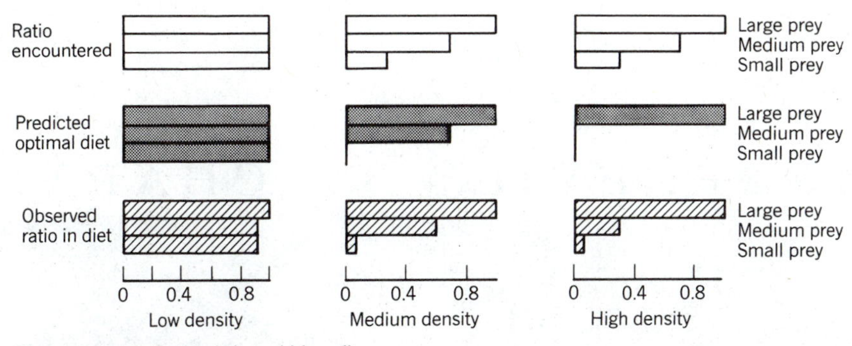

Figure 4.12 Optimal diet of bluegills.
Data are results of feeding experiments with bluegill sunfish (*Lepomis macrochirus*) feeding on three size classes of *Daphnia* in aquaria. Predicted diets were calculated from profitability, measured as size of prey, and search time, calculated from prey diameter and dispersion. Actual diets were measured from stomach contents after short trials. (Redrawn from Krebs, 1978, from data of Werner and Hall, 1974.)

hunting efficiency set by the visual size of the prey as a target. Time taken to find and attack prey was thus a function of the diameter of prey and its dispersion, which were two parameters easily measured. From these the relative time taken to find prey of each size class in different densities could be calculated. This meant that both parameters (profitability and time) needed for a predictive model were measured.

Werner and Hall proceeded by loosing ten fish into aquaria with different arrays of their three size classes of *Daphnia* and let the fish hunt for a while. They then killed the fish and found out what had been eaten by direct examination of stomach contents. They were then in a position to compare the predicted diet of the fish with their actual diet in variously stocked aquaria (Figure 4.12). The fit between expected and actual performance was very close.

There are other experimental data like those for the bluegills, notably for European great tits (Krebs *et al.,* 1977), crabs (Elner and Hughes, 1978), and water boatmen (notonectids) (Cook and Cockrell, 1978). These data suggest strongly that real animals of very different kinds are indeed programmed to forage in an optimal way, making "choices" of food that give the best return in energy for time spent. There is also an increasing body of field observation that supports this conclusion. Animals as varied as starfish, predatory birds, and waders all seem to choose from among available prey in ways compatible with a naturalist's idea of what is optimal (Krebs, 1978).

Optimal foraging theory thus gives powerful support to the postulate that animals maximize net productivity. The support is of the kind most convincing to the scientist: experimental verification of predictions that must hold if a hypothesis is correct. We may reasonably conclude, therefore, that the prime efficiency set by natural selection concerns net production.

CONCLUDING NOTE

Animals maximize their energy intakes. This can be taken as a good working principle derived from a natural selection argument, supported by observation, and confirmed by optimal foraging theory. Yet this does not mean that very high efficiencies of energy transfers result. Each animal is programmed to collect energy efficiently as that energy is available; not to forgo energy *now* in favor of energy *later,* because then others might get it. Moreover, every animal is programmed to make the best use of energy for its own purposes and not to pass energy on to others. Animals may be efficiency machines, but they are selfish machines also. A result of this is that transfer efficiencies up food chains, which we call ecological or Lindeman efficiencies, may be low.

PREVIEW TO CHAPTER 5

Plant shapes serve to regulate temperature and water loss. In hot places leaves can heat rapidly as they absorb sunlight for photosynthesis and so must be cooled. Leaves in moist forests are water-cooled, a mechanism that can regulate operating temperature within a few degrees of an optimum biochemical temperature. In deserts water cannot be spared for cooling, requiring that plants be designed to minimize heat-loading. An organ pipe cactus, for instance, exposes the minimum surface to the noonday sun and does away with leaves all together. The needle leaves of conifers let the flow of air past them be nearly laminar as a result of which leaf temperature tracks air temperature closely. Needles thus can be air-cooled in a semidesert environment where many pines live. More importantly coniferous trees can be air-warmed on cold, cloudless nights, which is probably the reason that evergreen conifers grow in northern forests. In cold arctic or alpine places plants must be designed to raise leaf temperatures as close to a biochemical optimum as possible. Trees are at a disadvantage for this in very cold places because they reach up into cold, moving air, whereas ground-hugging plants live in a thin layer of ground air warmed by radiations from the ground. This is a prime explanation for the absence of trees from the arctic. The short growing seasons of arctic and alpine places also constrain structures that can be grown before the ensuing winter, making it particularly difficult for frost-proof needles to be readied for the low air temperatures to come. These various energetic constraints on plant design explain the shapes of the great plant formations and their dependence on climate. Energetic constraints also lie behind the broad generalizations of animal distributions known as biogeographic rules—Allen's rule, Bergman's rule, and the rest. More exciting for animal ecologists are the alternative strategies used by different animal groups for regulating temperature. Cold-blooded or poikilothermic animals like reptiles or fish are best called ectothermic animals. They regulate temperature when they are active to within five degrees celsius of the temperature of homeotherms, but they use external heat sources to do so. This lets them conserve metabolic energy and go for long periods without food—and it makes them admirable ambush predators. A prime drawback is that bouts of strenuous activity must be short, since their metabolism rapidly causes oxygen debts in the muscles. Warm-blooded homeothermic animals are best called endotherms, because they regulate temperature by using metabolic heat. These have the big disadvantage of heavy energy costs for metabolism, but they are able to be active at night and in cold places. They can maintain high rates of activity for prolonged periods, letting them run down ectothermic prey or escape from ectothermic predators. They can also migrate long distances, letting them tap seasonal fluxes of resources denied to ectotherms. Insects get the best of both systems. Insects are ectotherms, with all the advantages of energy conservation and ability to exist for long periods without food that this implies, but they also are capable of prolonged exertion like endotherms. Insects avoid the oxygen debts built up in the muscles of reptiles by piping air directly to muscles in the system of tracheae. As a result insects can migrate on a scale comparable to that of endotherms. Their admirable system, however, limits size to a mass that can be served by air pumped through microtubules. The different methods of locomotion on land, whether running or hopping, are energetically roughly comparable and are determined by different evolutionary histories. Animals make use of countercurrent systems to direct or impede energy exchanges and to manipulate water concentrations. Whales and dolphins reduce temperature losses from extremities in this way and desert mammals reduce water losses. Predatory sharks are able to raise muscle temperatures to allow more rapid activity than if at the temperature of the surrounding water, a novel adaptation for an ectotherm.

CHAPTER 5

ENERGETICS AND ADAPTATION

A general conclusion from studies of energetic efficiency was that more than 98% of incident solar energy is dissipated in physical systems of the biosphere. Plants and animals have the remaining 2% to support life and gain fitness. Individual living things, therefore, cannot control their environment but must adapt to physical realities of habitats made wet or dry, hot or cold, windy or calm, by the power of sunlight wielded through physical cycles.

PHYSIOLOGICAL ECOLOGY has grown as a discipline seeking to answer questions about how plants and animals do adapt to physical reality. Often these studies are directed to understanding the unusual: deep diving by whales, survival in deserts, life at the tree line, the prodigious winter sleeps of hibernating bears. We see how basic patterns of design—warm blood, cold blood; needle leaves, broad leaves—are stretched to let their possessors tap the resources of unusual places. Often the work involves ECOLOGICAL ENERGETICS when investigating how form and function are suited to balancing heat budgets or conserving energy—Is it more efficient to run or to hop, to be warm-blooded or cold-blooded, to be evergreen or deciduous?

Many of the more obvious patterns of biogeography can be explored by *ecological energetics* and *physiological ecology*; shapes of plants in the great *formations* (biomes), for instance, or the restriction of hummingbirds or sunbirds to the tropics. But the answers are often complicated and the dictates of energetics may be only one of several, possibly conflicting, selecting pressures that lead to an observed form or function. A number of biogeographic *rules* have been inherited by modern ecology and these have been supplied in the past with simple answers that may be less than the whole truth. The following is a list of some of the better known of these "ecogeography rules."

Allen's Rule *In many species of animal, body extremities like ears, limbs, and tails appear to be smaller in varieties or close relatives living in cold regions than they are in individuals living in warm regions.* Having small ears and feet in the cold sounds sensible and it is often asserted that the rule, if true, is adaptive in that small appendages reduce heat loss. But is the rule true? A nice generalized critique is that of Scholander (1955, 1956), who notes that "cold climates do not produce a fauna tending towards large-sized globular forms with small protruding parts." It is prudent to expect that the heat-loss problem of large extremities would be solved by insulation if these extremities serve the animal well.

Bergman's Rule *In warm-blooded animals, individuals living in cold places tend to be larger than individuals of close relatives living in warm places.* It is possible to rank species of deer, foxes, gophers, bears, and rabbits by latitude, and conclude that they all get bigger in the poleward direction. When such ranking does indeed demonstrate Bergman's rule it is then possible to claim that the result is, like the observations of Allen's rule, brought about by the need of northern animals to conserve heat (large animals have smaller surface to volume ratios than small animals). This is to ignore all other pressures giving selective advantage to any particular size, such as lengths of food chains, size of food, what predators have to be avoided, for how long must food reserves stored as fat last, and so on.

Gloger's Rule *Individuals of a species living in dry climates tend to be lighter in color than individuals of the same species living in humid climates.* The standard explanation of this rule (again if it is really true) is that the different shades match background and thus serve as camouflage: light on dry, dark on wet. Other explanations, however, must be possible and one candidate is that the different colors help in balancing heat budgets in direct sun (dry habitat) or in the shade of clouds (wet habitat).

Jordan's Rule *Fishes living in cold water tend to have fewer vertebrae than fish of the same species living in warmer water.* This rule is usually offered without explanation, since there is no apparent adaptive advantage to possessing different numbers of vertebrae that can look like a "common-sense" explanation. It may be that the number of vertebrae is adaptively neutral, being a side-effect of growth in cold or warm water, or there may be adaptive advantages to differential flexibility of a spine that we do not understand.

The Mollusk Size Rule *The body size of marine mollusks tends to be larger in places of higher than normal salinity.* This rule is based on observations in northern Europe on six species of mollusk. It may, or may not, be a general phenomenon among all marine mollusks. No standard explanation is given. It must be possible to develop arguments depending on longevity and survival in fluctuating environments as well as those equating size with physiological tolerance of saline.

Polyploidy Rule *Northern, particularly tundra, and alpine plant species are more likely to be polyploid than other species of more southerly range.* This has been noted generally by taxonomists working on arctic floras (Johnson *et al.*, 1965). No simple physiological explanation is possible since it is the genetic mechanism of the plant as a whole that is altered. A reason for arctic polyploidy probably is to be found in the rapid production of ecotypes, in rapid speciation, in suitability of accepting asexual reproduction in the arctic, or perhaps in permitting phenotypic plasticity. Polyploidy is probably best looked at as a device to allow manipulation of life history, rather than as permitting appropriate physiological responses.

With the possible exception of the polyploidy rule, none of these "rules" has been shown to be generally true—Bergman's rule and Allen's rule, for instance, can only be supported by selection of data (Scholander *op. cit.*). But more serious is the way in which simple "explanations" equate correlation with cause. That the size of extremities correlates with decreasing heat, for instance, is not evidence that the changing size is an adaptation to temperature. If the "rules" are sometimes true, then they become observations that must be explained by the process of hypothesis and test like other observations in science. And when this process is applied to the "rules" it is likely that several conflicting ecological processes are involved. Alternative explanations, probably all with validity, emerge as various patterns of adaptation to environmental stress are examined.

ORGANISMS HAVE A HEAT BUDGET PROBLEM

All organisms must balance their heat budgets. They produce heat as they release energy to do the work of living and they gain heat from the sun. It follows, therefore, that they must lose heat at an equal rate or they would soon heat up to the point where their proteins were precipitated, causing death. This is true even of a plant; perhaps especially for a plant that is designed as a solar trap. Plants, therefore, must "thermoregulate" even as all animals do. No plants or animals, whether we call them "cold-blooded," "warm-blooded," "poikilotherm," or "homeotherm," can avoid the need to manipulate a heat budget.

An animal standing under the sun (Figure 5.1) receives energy from many sources: directly from the sun, considered as a point source; by the radiation of skylight, considered as a hemispherical bowl above the animal; by reflection from its surroundings; and by infrared radiations from the ground, surrounding objects, and the clouds. If the air is hotter than the animal, it may also receive heat by convection, though convection is usually a source of heat loss in real environments. And an animal receives much energy from

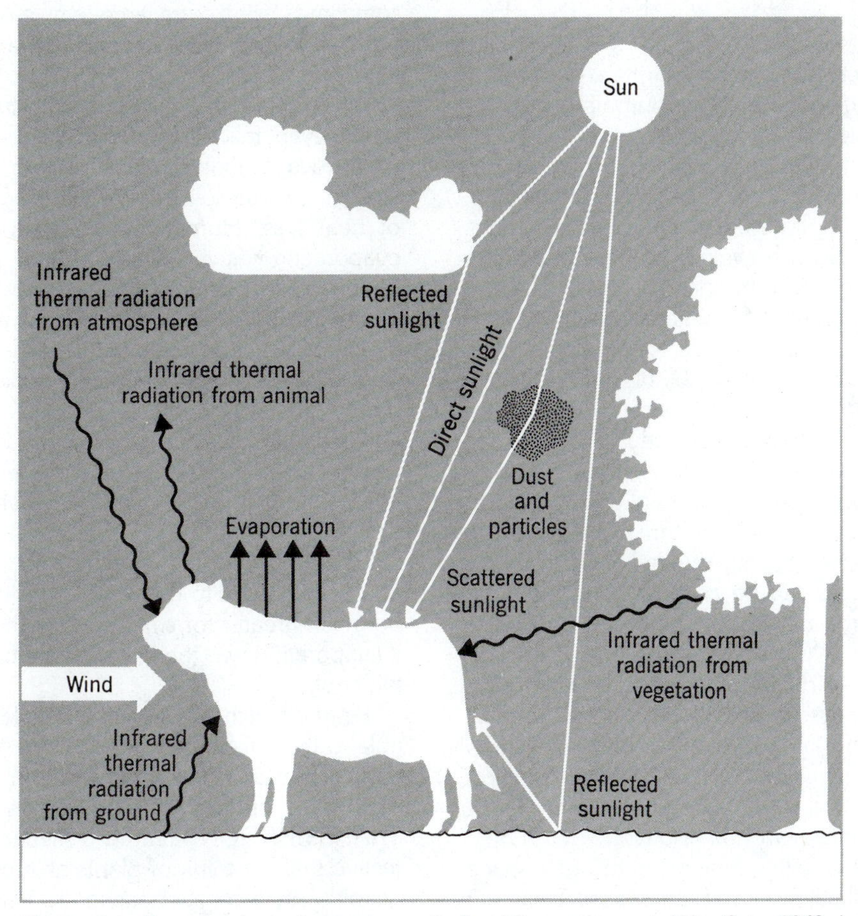

Figure 5.1 Energy exchange between an animal and the environment. (After Gates 1968a.)

its food, called metabolic energy. On the other side of the heat budget, the animal loses energy by black body radiation, by convection, perhaps a little by conduction, and by evaporation of water at the respiratory or skin surfaces. A heat budget for an animal may be drawn as follows:

$$\left(\begin{array}{c}\text{Metabolic}\\ \text{energy}\end{array}\right) + \left(\begin{array}{c}\text{energy absorbed}\\ \text{from direct}\\ \text{sunlight}\end{array}\right) + \left(\begin{array}{c}\text{energy absorbed}\\ \text{from}\\ \text{skylight}\end{array}\right)$$

$$+ \left(\begin{array}{c}\text{energy absorbed}\\ \text{from radiations}\\ \text{coming from ground}\end{array}\right) + \left(\begin{array}{c}\text{energy absorbed}\\ \text{from radiations}\\ \text{coming from clouds}\end{array}\right)$$

$$= \left(\begin{array}{c}\text{energy lost}\\ \text{by black body}\\ \text{radiation}\end{array}\right) + \left(\begin{array}{c}\text{energy lost}\\ \text{by}\\ \text{convection}\end{array}\right)$$

$$+ \left(\begin{array}{c}\text{energy lost}\\ \text{by}\\ \text{evaporation}\end{array}\right) + \left(\begin{array}{c}\text{energy lost}\\ \text{by}\\ \text{conduction}\end{array}\right)$$

Animals can act to alter their heat budgets by manipulating both inputs and outputs. Inputs are strongly influenced by the environments in which they live, yet animals can regulate them by behavior, by moving to sun or shade, for instance, or by spending time underground. Heat inputs from metabolism, however, are less under the control of the organism since they are a function of the resources that the animal uses.

Energy outputs may be manipulated in a number of ways. Heat loss by black body radiation is proportional to the fourth power of the absolute temperature of the body surface, so the animal can reduce this loss by insulating itself with fat or fur, a stratagem that keeps the body surface at a lower temperature than the body core. Alternatively both the radiation loss and convection loss can be enhanced by arranging to have a large surface area heated to nearly the temperature of the body core. This is done by animals of deserts and other hot places by having intensely vascular tissues that can be used as radiators. The large ears of desert lagomorphs (rabbits and hares) and of elephants can function in this way, and it is likely that the spiny fin of the mesozoic *Dimetrodon* also worked as a heat regulator by radiating heat away to outer space when the animal faced the sun or absorbing heat when the animal turned at right angles to the sun (Figure 5.2). It has been shown that nesting seagulls manipulate heat budgets by twisting to the sun in this way. Energy lost by convection also depends on the temperature of the surface and is manipulated at the same time as the radiation loss.

Notice that these first thoughts about animal convectors and radiators allow a fresh interpretation of some of the phenomena of Allen's rule: extremities may be large in hot places as a special adaptation to environmental heat. It is not so much that Arctic rodents and elephants (woolly mammoths) have small ears to conserve heat, as Allen argued, but more that ears were convenient structures to convert into radiators when living in a hot desert. It may be as profitable to think of the ears of animals in warm places as sometimes being extra large to take on a radiator function as to think of animals of the cold having extra small ears.

All heat budgets may be powerfully affected by the evaporation of water. Since the latent heat of the evaporation of water is high (about 590 kcal g^{-1}) water loss represents a potent source of heat loss. Humans function in deserts by evaporating water of sweat from all parts of the body surface, a practice that makes us splendidly adapted to sunbelt living as long as we have unlimited water to drink, but less adapted otherwise. Dogs and other animals cope with temporary heat overloads by panting. Yet evaporating water to balance heat budgets in hot places is not something that is equally open to all animals, the utility of the stratagem being critically dependent on body size. Figure 5.3 shows how the water demanded to balance heat budgets of animals exposed to hot desert conditions would be much greater for small species than for large if evaporation was the principal method of keeping cool.

Plants, which are essentially solar traps unable to hide from the noonday sun, often make particular use of evaporating water to balance their heat budgets. They do this with their TRANSPIRATION streams, a mechanism so effective that we think of plants as typically being "cool." Yet water shortage in dry heat or in cold wind may drive plants to other expedients that lead to the appearance of deserts or tree lines.

(a)

(b) (c)

Figure 5.2 Animal heat loss by induced radiation.
Maintaining a large surface at close to core temperature will result in a net loss of heat by radiation if surrounding bodies are at lower temperatures. The best surrounding "body" for this purpose is outer space, at which a desert jack rabbit (*a*) directs its large ears while sitting at the mouth of a burrow. It is likely that the back "fin" of *Dimetrodon* likewise allowed loss of body heat by radiation (*b*). *Dimetrodon* could also have gained heat rapidly by turning its "fin" at right angles to the sun (*c*).

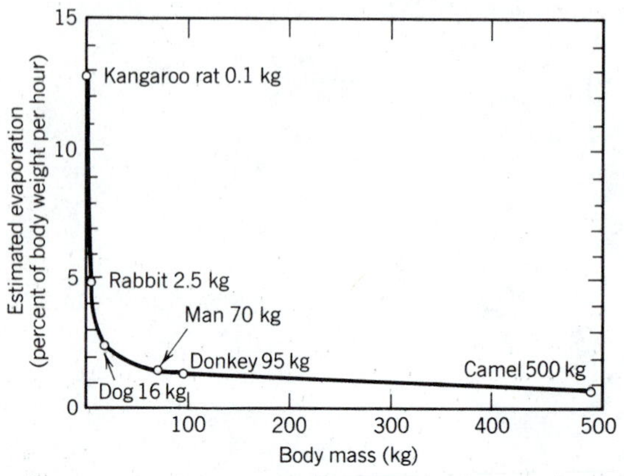

Figure 5.3 Evaporation requirements for animals exposed to hot desert conditions.
The curve is calculated on the assumption that heat load is proportional to body surface. Small animals have a body surface that is exponentially greater than that of larger animals. Animals like kangaroo rats of deserts must avoid the desert sun, since they are without the necessary large flux of water. (From Schmidt-Nielsen, 1964.)

It will be evident that any one kind of animal or plant has only limited ways to manipulate either its surface temperature or its water balance to balance its heat budget at a temperature satisfactory for life processes. Animals with good insulation of fur or feathers, and an ability to maintain the core temperature by the release of metabolic heat (HOMEOTHERMIC animals), can be well suited to life in cold places. Being large helps in the cold if you are a homeotherm, which is the observation that leads to Bergman's rule. Yet there are many small rodents in the Arctic (lemmings, for instance) who avoid the cold by living in burrows under the snow, and big grizzly bears spend much time in torpor, in a den with slightly lowered body temperature, thus minimizing heat loss. Relatively large animals that do not maintain core temperatures by metabolic heat (POIKILOTHERMS), like lizards, fail to inhabit the Arctic, presumably because they cannot survive the inevitable freezing. Yet insects of many kinds do very well in the Arctic, possibly because they seldom maintain an adult existence through the winter.

In hot places the various stratagems succeed differently. Reptiles do well in the heat as they manipulate their temperature by moving in and out of the sun, while many mammals are forced to be nocturnal. Thus, since every species is a variant on some grand engineering design (reptile, insect, bird), and each has a subdesign suited to its specific niche, any kind of animal or plant can be thought of as being restricted to a habitat where the temperature and water supply are such that it can balance its heat budget. Thinking on these lines leads to explanations for several well-known patterns of distribution, particularly to the spread of the great plant *formations*, suggesting why conifers predominate in the boreal forest, why trees are deciduous in middle latitudes, and why deserts and tundras have their special life forms.

HEAT BUDGETS AND LIFE FORMS

To make a heat budget for an animal we can start with the simplifying assumption that animals are cylindrical and with negligible appendages; not really too desperate an oversimplification considering the shape of a deer (Gates, 1968a and 1968b). The cylindrical animal then consists of a body core, an insulating sleeve of fat bound by skin, and perhaps a sleeve of fur or feathers separating the skin from the radiating surface of the animal, as shown in Figure 5.4.

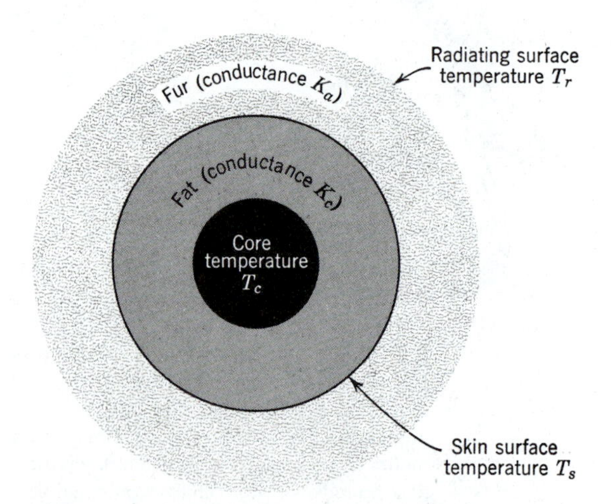

Figure 5.4 Heat balance cross section of idealized animal.

The following symbols are used in the equations below.

M = metabolic energy as cal cm^{-2} min^{-1} generated at skin surface

E_1 = energy lost by sweating cal cm^{-2} min^{-1}

E_2 = energy lost by evaporation at respiratory surface cal cm^{-2} min^{-1}

K_a = conductance of insulation (fur or feathers)

K_c = conductance of body fat

T_c = temperature of body core

T_s = temperature of body surface

Q = radiation absorbed by body surface

h_c is the convection coefficient

T_a = air temperature

ϵ = emissivity of the skin surface

δ is the Stefan–Boltzmann constant for radiation

T_r = temperature of effective radiating surface at outside of insulator

The energy conducted from the body core to the skin surface is

$$\begin{pmatrix}\text{Metabolic energy}\\ \text{generated at}\\ \text{skin surface}\end{pmatrix} = \begin{pmatrix}\text{energy passed}\\ \text{by simple}\\ \text{conduction}\end{pmatrix} + \begin{pmatrix}\text{energy lost}\\ \text{at respiratory}\\ \text{surface}\end{pmatrix}$$

$$M = K_c(T_c - T_s) + E_2 \tag{5.1}$$

If the animal has no fur or feathers the energy budget of the external surface is

$$\begin{pmatrix}\text{Energy passed}\\ \text{by simple}\\ \text{conduction}\end{pmatrix} + \begin{pmatrix}\text{radiations}\\ \text{absorbed from}\\ \text{environment}\end{pmatrix}$$

$$K_c(T_c - T_s) + Q$$

$$= \begin{pmatrix}\text{radiant energy}\\ \text{lost as black}\\ \text{body}\end{pmatrix} + \begin{pmatrix}\text{convective}\\ \text{heat}\\ \text{loss}\end{pmatrix} + \begin{pmatrix}\text{energy lost}\\ \text{from skin}\\ \text{evaporation}\end{pmatrix}$$

$$= \epsilon\delta T_s^4 + h_c(T_s - T_a) + E_1 \tag{5.2}$$

Eliminating $K_c(T_c - T_s)$ from equation (5.2) we get

$$M + Q = \epsilon\delta T_s^4 + h_c(T_s - T_a) + E_1 + E_2 \tag{5.3}$$

If the animal has fur or feathers, the energy budget at the radiating surface is given by

$$\begin{pmatrix}\text{Energy reaching}\\ \text{skin surface}\\ \text{by conduction}\end{pmatrix} = \begin{pmatrix}\text{energy lost}\\ \text{from skin}\\ \text{evaporation}\end{pmatrix} + \begin{pmatrix}\text{energy conducted}\\ \text{across insulating}\\ \text{layer}\end{pmatrix}$$

$$K_c(T_c - T_s) = E_1 + K_a(T_s - T_r) \tag{5.4}$$

and the energy budget for the radiating surface is given by

$$\begin{pmatrix}\text{Energy conducted}\\ \text{across}\\ \text{insulating layer}\end{pmatrix} + \begin{pmatrix}\text{radiations}\\ \text{absorbed from}\\ \text{environment}\end{pmatrix}$$

$$K_a(T_s - T_r) + Q$$

$$= \begin{pmatrix}\text{radiant energy}\\ \text{lost as black body}\end{pmatrix} + \begin{pmatrix}\text{convective}\\ \text{heat loss}\end{pmatrix}$$

$$= \epsilon\delta T_r^4 + h_c(T_r - T_a) \tag{5.5}$$

This last equation describes how the environmental energy flow may affect the heat budget of the animal, but it is not easy to put numbers against various terms in the equation. Consider the difficulty of measuring K_a or ϵ for a real animal, for instance. But this difficulty may be avoided as follows:

By eliminating $K_c(T_c - T_s)$ from equations (5.1) and (5.4) we get

$$M - E_2 - E_1 = K_a(T_s - T_r) \tag{5.6}$$

Solving for T_s, and substituting back into equation (5.1), we get

$$T_c - T_r = \frac{M - E_2 - E_1}{K_a} + \frac{M - E_2}{K_c} \tag{5.7}$$

Now

$$\frac{M - E_2}{K_c}$$

is the temperature difference between the core and the skin surface and

$$\frac{M - E_2 - E_1}{K_a}$$

is the temperature difference between the skin and the radiant surface. Equation (5.7) therefore shows that temperature differences between the various layers of an animal will always be proportional to the temperature difference between the core and the radiant surface. This means that *measures of temperature of the core and the surface are all that are needed to infer the ability of an animal to regulate temperature.*

This apparently rather self-evident result of Gates (1968a and 1968b) in fact gives formal understanding to some well-known properties of animals. Mammals and birds commonly maintain larger differences between core temperatures and surface temperatures than do reptiles and fish. It is not so much that one group is warm-blooded and the other cold-blooded that matters as that the former can maintain a larger temperature difference between core and surface. Warm-blooded animals can withstand greater variations in the energy flowing through their environments. This gives fresh insight into the relative successes of the two engineering designs: the cold-blooded poikilothermous designs of fish and reptiles are suited to oceans and equatorial regions because these places are where variance in environment energy flux is least.

But it is in the study of plant design that the heat budget approach has been most revealing. All plants are *poikilotherms* yet they live in habitats of every form of variance of energy flux. They must cope with these different degrees of variance almost entirely by manipulating their shapes or the flow of water.

Figure 5.5 illustrates the energy exchange between a plant and its environment. A plant receives all its energy directly from radiation or convection from the contemporary environment, except when it is metabolizing the reserves of previous years, as when a potato sprouts from a tuber. A small fraction of the incoming solar radiation is transformed into reduced carbon compounds (Chapter 3) and the rest is converted to heat. The plant must balance this income of heat by loss—through radiation, convection, and transpiration—of an equal amount.

The active parts of plants are the leaves, and these are so thin that the exterior temperature must approximate the core temperature. Insulation, therefore, is of little importance to a plant's heat budget. Leaf behavior also matters little, though there are possibilities, as when heated leaves droop through wilting or turgor changes, events that move the flat surface so that it no longer faces the sun. The adaptations of plants to different temperature regimens are, however, mostly restricted to adopting appropriate shape and to controlling the transpiration flow by manipulating the stomates.

Gates (1965) studied leaf and ambient temperatures of monkey flowers (genus *Mimulus*) in Nevada. He found that *Mimulus lewisii* populations growing at 10,600 feet in the Nevada Range on a day of still air at 19°C had leaf temperatures of between 25 and 28°C. At 1,300 feet on the same mountain the air temperature was as high as 37°C but the leaf temperature of the local monkey flowers (*Mimulus cardinalis*) was between 30 and 35°C, or very close to the temperature of those in cool air near the mountain top. The plants in cool air manipulated their heat budget so that they were warmer than their surroundings; the plants in hot air manipulated theirs so that they were cooler. The two species are not radically different in design (Figure 5.6) thus requiring that their feats of temperature regulation be done by manipulating the flow of water from the leaves by opening or closing stomates.

It seems likely that *Mimulus cardinalis* of the hotter elevations in Nevada is limited to habitats with an adequate supply of water, since these organisms, like *Homo sapiens*, solve the problem of a too hot habitat by evaporating water. The niches of *Mimulus* and *Homo* include the use of water and water becomes a *limiting factor* of their environment of such importance that *isohyets*[1] on maps might well be found to run parallel to the borders of their species distributions.

Mimulus plants can get a living by photosynthesis in hot places using broad, thin leaves, if the local species is adapted to pump large quantities of water through the leaves. But if water is not available, the flat thin leaf cannot be kept

[1]An ISOHYET is a line on a map connecting places of equal precipitation.

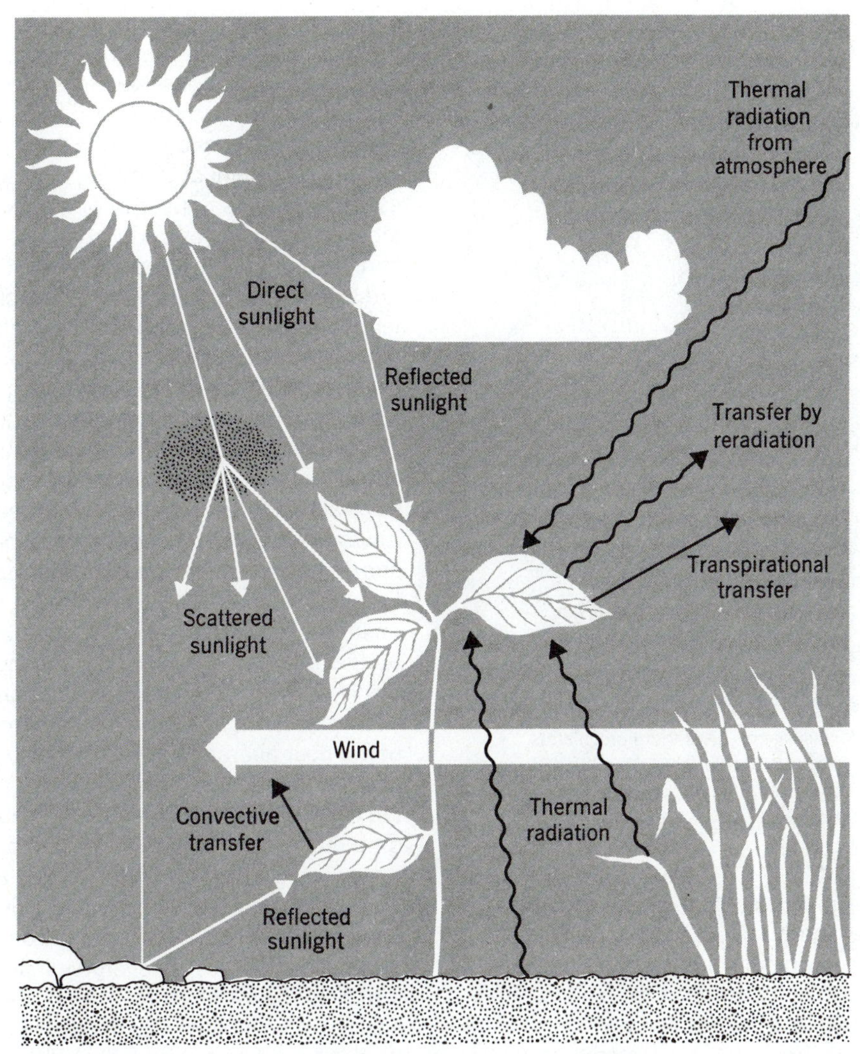

Figure 5.5 Energy exchange between a plant and its environment.
For an energy budget in the short term a small part of absorbed solar radiation is sequestered as the energy of reduced carbon, but the remainder must be balanced by radiative and connective losses. (After Gates, 1968b.)

cool and plants with this structure cannot live. Plants of hot places that are also very dry are made without conventional leaves, resulting in the barrel or organ pipe designs of the family Cactaceae (the true cacti) or their old world analogues in the Euphorbiaceae. (Figure 5.7).

HEAT BUDGETS AND LIFE FORMS OF PLANTS

The design of a cactus is known to be suited to conservation of water: the surface to volume ratio is small, the surface is covered with a thick

Figure 5.6 *Mimulus* spp.: floral thermoregulators. Monkey flowers at over 10,000 feet in the Nevada Range were found to be at nearly the same temperature as those 9000 feet lower down.

cuticle that is impermeable to water, and the stomates are few and tend to be sunk in pits. Furthermore, these plants use the C4 system of photosynthesis, or even resort to CAM metabolism so that they can reduce carbon stored overnight, working behind closed stomates when supplied with the energy of the sun (Chapter 3). These adaptations obviously make good sense for life in a dry place. But they leave the plant with a large heat budget problem that cannot be solved by the usual expedient of evaporating water. A solution in structural engineering is used instead.

Figure 5.8 offers an analysis of heat budget control in an organ pipe cactus of the type of saguaro (*Carnegiea gigantea*) of the Sonoran desert of Arizona. The tall, narrow shape presents the smallest area to the noonday sun, and the largest area to the oblique sun of early morning and late evening. It seems a good working hypothesis, therefore, that the shape of a cactus not only reflects conservation of water but also is suited to minimize heat load under the hottest sun. Cactus shapes are among the more extreme examples of plant engineering in deserts, but the essential themes appear in other taxa also—notably reduced leaves, reduced horizontal shapes, and green, photosynthetic stems. These structural designs are approximated in plants of different ancestry (Figure 5.8). They reflect the fact that in hot deserts heat budgets must be balanced without resort to evaporating water. The result is the range of life forms found in hot deserts.

In other hot places, however, water is abundant and the typical life form is that found in the tropical rain forest. Trees are evergreen and their leaves are flat sheets. The geometry of these trees and their leaves can be explained largely from the point of view of carbon dioxide uptake and optimum light distribution (Chapter 3). The heat budget problem is solved by pumping water and has little effect on engineering design.

In places that are hot in only some seasons of the year alternative structural designs exist. In middle latitudes the basic leaf and tree plan of the tropical forest is maintained, but the trees drop their leaves in winter. In the more northerly region of the boreal forest leaves are needle-shaped and are kept year-round. Some hint of an explanation of the transition from deciduous broad-leaved trees to evergreen needle-leaved forms in the north comes from Gates' (1965) analysis of heat balance in the two kinds of leaves, for the two shapes have quite different convective properties.

Heat gain or loss by convection depends on relative temperatures of the leaves and the surrounding air, on the speed of air flow, and, most importantly, on the intimacy of contact between the flowing air and the leaf surface. As air flows over a surface there is always a thin film of almost stationary air at the junction, called the BOUNDARY LAYER. This *boundary layer* turns out to be much thinner for a needle-shaped leaf than for a flat leaf, implying that convection would produce more rapid temperature changes in needle-shaped leaves. Gates was able to test this

Figure 5.7 Shapes to balance heat budgets in deserts.
Reduced leaves and vertical stance minimize heat loading from the noonday sun. The cacti of the new world (upper photo) and the Euphorbiaceae of the old world are ecological equivalents given the same shapes by natural selection to minimize heat stress in hot dry places. The extended ears of the animals in the lower photograph are working as heat-shedding radiators.

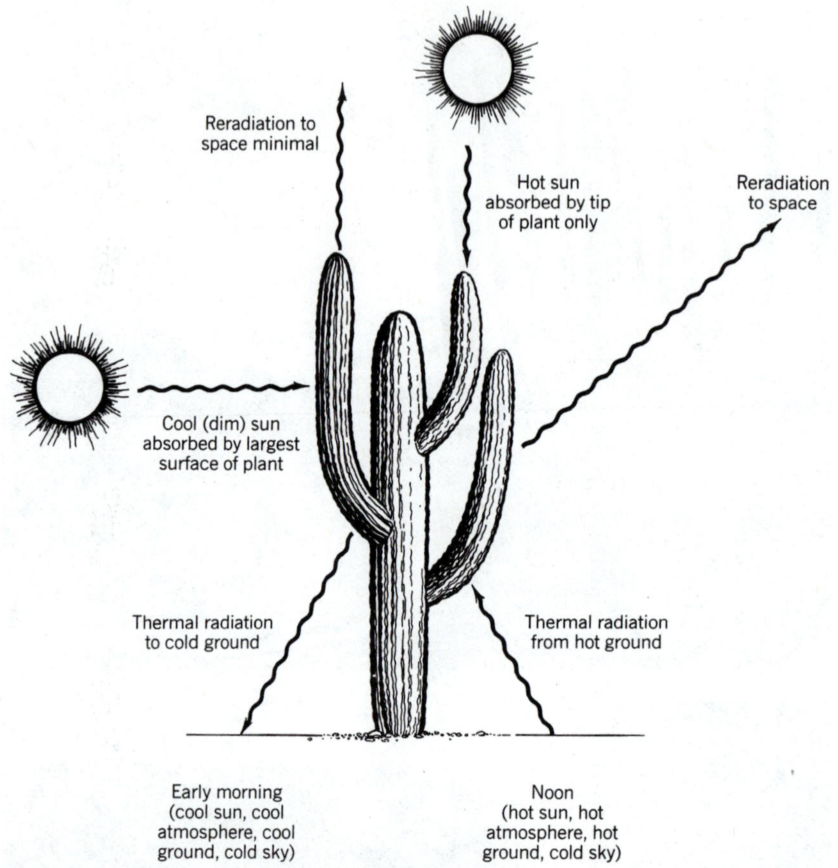

Figure 5.8 Heat budget control by a saguaro cactus.
At high noon only the point of the plant faces the sun and the largest possible surface faces the sky (which is cold since it is without clouds). In early morning and late evening the sides of the cactus face the sun to allow maximum absorption, permitting maximum photosynthesis and keeping the plant warm enough for efficient operation.

with wind tunnel experiments, using silver replicas of leaves. Electrodes fastened to the silver leaves recorded rates of heat gain or loss when the leaves were exposed to warm or cold air moving past them. A pine needle gained or lost heat many times more rapidly than a poplar leaf. A practical consequence of this is that needle shapes of leaves tend to force a passive balancing of the heat budget towards the temperature of the surrounding air.

In the boreal forest, then, trees are designed in a way that keeps their operating parts close to air temperature without further manipulation by the plant. In cool summers of boreal latitudes, possession of needles keeps plants cool with a minimum of transpiration. But the real significance of this design is probably found by night and in winter because then the leaves will be kept relatively warm at air temperature despite the tendency for them to cool much below this by radiation to the cold black body of space. Even in a northern winter, the surface of a spruce tree is warm compared to the black cloudless sky above it, and so must be losing heat by radiation

(a)

(b)

(c)

Figure 5.9 Needle-leaved trees: when cold or dry. Needle leaves lose or gain heat by convection many times faster than flat leaves. Radiation in bundles of needles is more to each other than to the sky. These combined properties result in higher leaf temperatures in cold times than would be possible for flat leaves, suggesting an advantage in the north. Needles also need less cooling by transpiration under a hot sun, suggesting that they are adaptive in hot, dry places or in dry northern summers. (*a*) A pine in a hot, semidesert place. (*b*) A pine in an old field succession exposed to relative heat and drought. (*c*) Trees of the boreal forest.

according to the fourth power of its absolute temperature. Rapid convection from the surrounding air balances this heat loss, perhaps keeping the leaf several degrees warmer than possible for a flat leaf.

Since needle leaves tend to adopt, by convection, the temperature of the air around them, the design is suited to warm deserts as well as to the dry cold of the north temperate belt. This is likely to explain why many pines exist in semiarid regions and others invade old fields in the early years of secondary succession, where conditions are often locally semiarid (bare fields, direct sun) (Figure 5.9).

Understanding the implications of having needle-shaped leaves, therefore, allows plausible hypotheses to explain their distribution. Plants with this design strategy live in places that may be seasonally hot and dry, because high heat loss by convection reduces the demand for transpired water. More interestingly, the great latitude spread of the boreal forest reflects the suitability of the design both for preventing leaf temperature falling below air temperature in cold times and for conserving water in dry times. Presumably the needle-leaved design may pay a cost in photosynthetic efficiency, since flat leaves are the rule whenever there are no constraints of temperature or water.

In the broad belt of deciduous forest, between the evergreen rain forest of the wet tropics and the evergreen needle-leaved forest of the cold temperate north, a different strategy apparently is optimum. Summers in this belt have much in common with the climate of wet tropics—they are warm, moist, and cloudy. Trees using flat leaves produce very well in these summers, but their leaves would be colder in winter than needles would. Apparently it pays in mid-latitudes to behave like a tropical tree in summer and simply close down production for the winter months. Further north the summers are so constrained (shorter, or drier, or both) that a strategy of making the most of all seasons with needle leaves may well be optimum. Forest formations, therefore, like desert plant forms, can be understood as the result of mechanical adaptation to make the most of local heat and water.

Figure 5.10 Tree line in Idaho.
Many continental tree lines look like this as trees finger their way up valleys. This Idaho scene can be duplicated in the arctic (see Chapter 14).

TREE LINE AND TUNDRA: THE ROLES OF HEAT BUDGETS AND WATER

The disjunction at a tree line is one of the most striking of natural phenomena: to one side the forest, to the other an open space, the transition abrupt enough for easy mapping (Figure 5.10). On high mountains the tree line can be sharp indeed, a progression from trees to no trees within a span of a few meters. In the arctic the tree line is a more gradual affair and follows a progressive shortening of trees across tens of kilometers. The line is broken by long fingers of trees stretching up river valleys and depends on the direction to which slopes face. Yet the change is clear enough to pose teasing questions: Why should there be an edge to the forest? Why should there be no trees in the arctic?

The puzzle of the absence of trees in the arctic is not simple. Clearly cold weather has something to do with the phenomenon but it has never been enough to make statements like "the arctic climate is too harsh for trees" because the arctic ground is often completely carpeted with herbs that survive the "harshness" well enough and woody plants of low stature live in the arctic. There is some rough correlation with the distribution of trees and of permanently frozen ground (*permafrost*), but the synchrony of tree line and permafrost is far from complete. And yet an analysis of the heat budget problem of a hypothetical arctic tree yields a plausible explanation.

A tree in the far north in the few weeks of summer must balance its heat budget when the heat input is from low-intensity sunlight supplemented by radiations from the ground and neighbors. Set against these inputs are the unavoidable losses by radiation to the sky, by transpiration (some of which will always be necessary to provide the leaves with a flux of moisture), and by convection to cold air. The arctic air is cold most of the time, and usually in motion, so it follows that the leaves of an arctic tree will always be relatively cold. This means low productivity, a serious matter for a plant with only a short growing season and the metabolic expenses of a large body to be met.

This poor heat budget outlook for an arctic tree may be compared with the prospects of an arctic herb or low shrub like a prostrate willow. Their leaves are close to the ground, very often within the layer of air that is almost stationary as winds pass over it. Still air between ground-level leaves will be warmed by radiation and conduction from the plant leaves as they themselves are warmed by the sun. A dwarf plant, therefore, will be able to maintain higher leaf temperatures in the arctic than is possible for a tree. Dwarfs also have smaller maintenance costs because their bodies are smaller. A low life form, therefore, can be expected to yield a positive photosynthetic balance in cold northern places where a tree cannot. Thus, one simple answer to the question "Why are there no trees in the arctic?" is "because a heat budget cannot be balanced at a satisfactory temperature for a tall plant in cold polar air." A more complicated answer to the tree line problem must take into account the effects of cold on the water supply, together with such matters as length of growing season on the energy budget. These complications have been studied most thoroughly for alpine tree lines.

A common pattern at alpine tree lines is a solid block of trees approaching to a definite line beyond which there are essentially no trees (Figure 5.11). Inside the block of forest, the environment of each tree must be modified by the presence of its neighbors. Outside the block, or at its edge in the front rank, trees do not have the shelter of neighbors. In a recent review, Tranquillini (1979) considered the stresses endured by trees exposed beyond the tree line. Although direct correlates of air temperature (like length of growing season) can all be shown to be correlated with the ability of seedlings to show a net gain in weight, the outstanding stress experienced by plants exposed near tree lines comes from damage by frost after drying.

Figure 5.11 Natural tree line in the Alps. The trees are pines (*Pinus cembra*). Although a few trees stand out at the edge, the forest approaches its limit as a solid block.

Conifer needles can withstand frost as long as they do not dry out, but if they become desiccated they are easily killed. In a strong frost the ground itself is likely to be frozen, particularly if the alpine snow cover is light. Trees are then denied access to water, yet the needles, and even twigs, must remain turgid. Resistance to intense cold by coniferous trees, therefore, is a function of the ability of their needles to retain water when the ground is so frozen that fresh supplies of water cannot be taken up. A well-formed needle, with a thick cuticle right to the tip, retains water well and survives long periods of freezing temperatures, but a needle with a thin cuticle does not. The ability of needles to survive frost is thus a function of the previous growing season, the length and warmth of which determine how well-formed the needles are at the onset of winter (Table 5.1).

Table 5.1
The Liability of Immature Spruce Needles to Winter Desiccation.
The needles of Picea engelmannii *were sampled at tree line. By February the water content of needles that started the winter without being matured had seriously decreased. Winter desiccation is a crucial factor in setting tree lines. (From Wardle, 1968.)*

	Mature Needles	Immature Needles
Shoot length (cm)	0.07–2.7	0.4–1.2
Needle length (mm)	12–16	4–11
Needle spatial density (No. per cm shoot)	29–44	50–72
Water content (% d.w.)		
17 Nov.	134	152
22 Dec.	125	130
3 Feb.	126	51

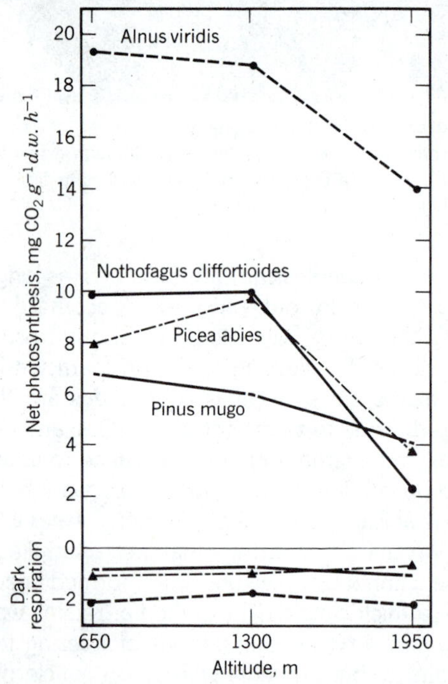

Figure 5.12 Effect of altitude on net photosynthesis and dark respiration of seedlings.
The four species were grown in pots at the three altitudes and net photosynthesis and respiration were measured as gas exchanged under controlled conditions. (From Benecke, 1972.)

This work on alpine trees shows that physiologically induced drought in winter can be a cause of tree death, but that this death is linked to the length of the preceding summer. Net production during the summer, however, is partly determined by the temperature at which a tree balances its heat budget (as discussed above), and this is a function of summer temperature. Low-growing plants high in the mountains or in polar regions produce more in summer and make leaves better able to resist desiccation in the physiological drought of a cold winter. This combined explanation of why trees do not grow in cold seasonal places applies alike to the tops of high mountains and to the arctic (Figure 5.12).

Heat budget analysis can be extended to the prevailing life form in many of the familiar vegetation types of the world (the plant *formations*, see Chapter 14): no trees in the arctic, needle-leaved trees in the less cold north, broad-leaved evergreen trees in the wet tropics, *et cetera* (Figure 5.13).

ECTOTHERMY OR ENDOTHERMY: WHAT ARE THE ADVANTAGES?

Animals have long been divided into "warm-blooded" (birds and mammals) and "cold-blooded" (the rest, but reptiles are particularly called to mind). The terms HOMEOTHERMIC (keeps the same temperature) and POIKILOTHERMIC (the body temperature varies) commonly replace "warm" and "cold" bloodedness. This makes the point that the important difference is not temperature alone but the way in which one group can maintain a constant temperature independent of the environment whereas the other cannot. Fish (poikilotherms) must let their body temperature be close to that of the water around them, but whales stay warm even in cold water.

Talking of homeothermy or poikilothermy in land animals can be misleading, however, because active animals of either design regulate their body temperatures within fairly narrow limits, to within a few degrees of 38°C. Poikilothermic

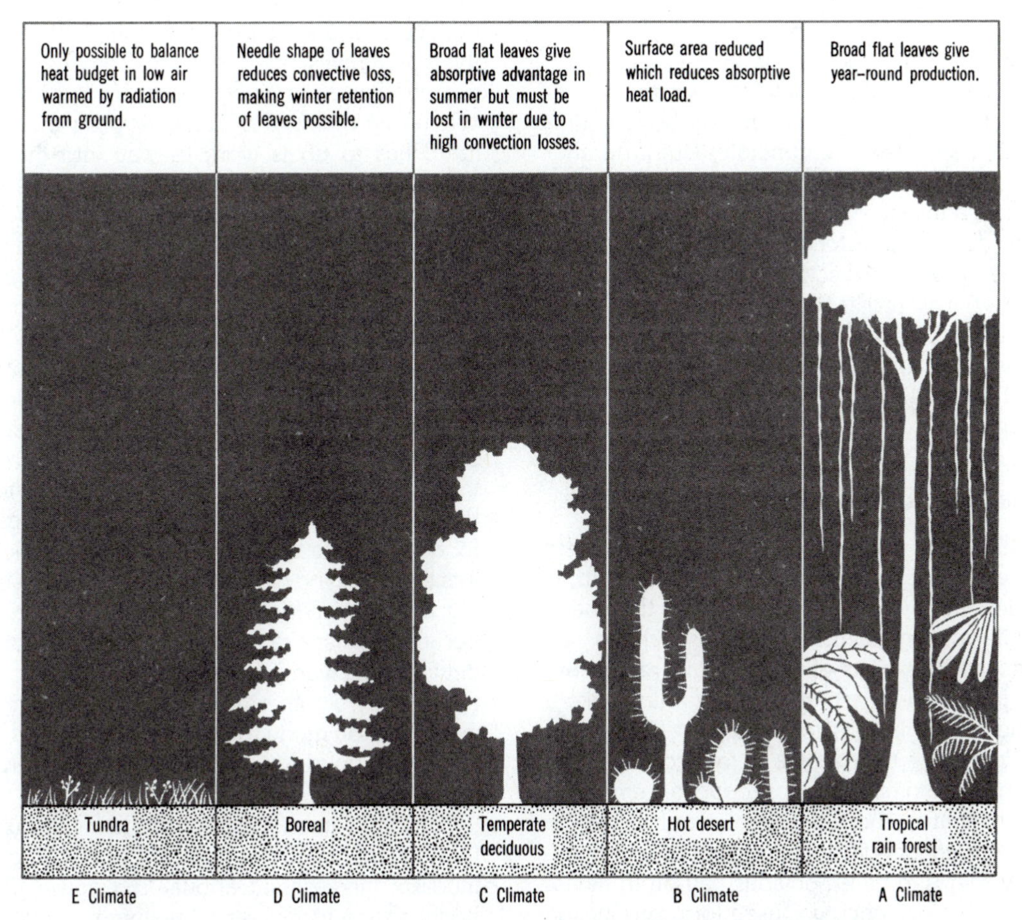

Figure 5.13 The heat budget explanation for the life forms of plant formations. Where water is abundant, heat overloads can be shed by transpiration. In deserts and in cold places the absence of water imposes design constraints and introduces other complications. For details see text. Climate types A–E are from the Köppen classification (Chapter 14).

reptiles do let their body temperature change when they are active or when they rest, so that at night, for instance, their temperature may be close to that of the air. But hibernating mammals let their temperature fall also. Hummingbirds of the high Andes Mountains let their tiny bodies cool at night, thus slipping into a state of torpor, a stratagem that saves energy for them just as it does for the reptiles called poikilotherms (Calder and Booster, 1973). Clearly the terms "homeothermic" and "poikilothermic" do not offer an unambiguous classification.

The terms ENDOTHERMIC and ECTOTHERMIC are more revealing for ecologists, since they direct attention to the energy source used to reach the temperatures at which the animal is active. Endothermic animals use energy ingested as food to heat their bodies and ectothermic animals tend to use solar power for their heating.

Ectothermy is the more ancient plan, and it is still used by most living species. In one sense, therefore, ectothermy can be called the most "successful" plan, which is a good observation with which to start a discussion. It is endothermy (in the guise of homeothermy) that is often thought to be the "better" or "more successful." Endotherms regulate their temperatures precisely, with advantages in running body chemistry that are easy to enumerate. But if number of species or individuals among the followers of a plan is a measure of "success" then there can be no doubt that ectothermy wins. What, then, are the properties of ectothermy that lead to this success? It is profitable to consider first ectothermy in the lives of tetrapod vertebrates, the reptiles and amphibians, with which tetrapod vertebrate endotherms (mammals and birds) can be compared directly.

Benefits and Costs of Ectothermy

Ectothermy is one way of manipulating a heat budget and regulating body temperature. In its simplest form, the hot animal seeks the shade and the cold animal the sun. This is a cheap way of meeting the costs of temperature regulation. Few food calories are used by an animal like a reptile to regulate its temperature. The resting metabolic rate of a reptile is low compared to that of a mammal and its *production efficiency* is high (that is, the ratio of calories ingested to calories used for growth and reproduction is high). There seems to be an obvious pay-off in fitness for this because energy is used for reproduction rather than for keeping warm.

Pough (1980) examined production efficiencies for 16 species of small mammals and birds and 8 species of reptiles and amphibians of comparable size and found that the efficiency of the endotherms was 1.4% whereas that of the ectotherms had an average of 43.6%. Ectotherms, then, seen to be remarkable machines for making much biomass or many offspring from the ingested energy available. Pough (1982) calls ectotherms *low-energy systems*.

When food is scarce, an ectotherm can relapse into torpor. It then uses very little energy. This ability to switch off as an energy consumer for long periods lets ectotherms exploit episodic or periodic resources in ways that are very different for endotherms. The lizard *Sauromalus obesus*, for instance, relies on desert plants for both calories and water. When the plants dry up, the lizard crawls into a rock crevice (where it can stay cool) and remains there for up to 8 months. When rain makes the plants grow again, all the lizard has to do is work its way into the sun, warm up to a nice operating temperature near 38°C, and start feeding again (Nagy, 1972). A similar ability to be active and grow when food suddenly becomes available is shown by many snakes that can do well eating bird-eggs though these are available for only short periods.

An even more striking result of the property of eating when there is food and relapsing to torpor when there is none is the ability to switch from food to food as animals grow. To ectotherms this opens the possibility of life histories that include larvae or small young that feed on food quite different to that of adults. Table 5.2 lists the changing diet of a snake with age, showing a series of food preferences possible only because the snake can make do with nothing between the various crops of small frogs. This property of ectothermy is used to even greater advantage by many fish and invertebrates where dispersing young feed on different food in different places at different times.

Yet other possibilities of the low-energy system are small size and long shape. Resting metabolism increases at about 0.75 power of the body mass, which means that energy at rest goes

Table 5.2
Changes in Diet of Growing Garter Snakes (*Thamnophis sirtalis*)
The food of the snake changes as it grows. Snakes can endure long times between supplies of food of different sizes because the energy cost of their maintenance in a state of torpor is low. (Data from Fitch, 1965.)

Length (cm snout to vent)	Prey	Mass of prey as % Mass of Snake
20–29	Earthworms	3.8
30–39	Small frogs	5.6
40–49	Large frogs	94
≥50	Mice	50

up very steeply as animals get smaller. This is not very serious for ectotherms because resting metabolism is always low, since at rest the animals are cold. But for endotherms to be small involves a heavy energy cost (Figure 5.14). Mice, shrews, and chickadees are the smallest endotherms, and yet they are huge by the standards of most ectotherms. Even most lizards and amphibians are much smaller than chickadees and may be only one-tenth the mass. Ectothermy, therefore, opens all the advantages of small size (ability to hide, short growth time to reproduction, possibly larger food resources). It is one of the keys to the success of insects.

Ectothermy also allows the long, thin shape of a snake that would impose a ruinously expensive energy cost for an endotherm because of the large surface area over which heat would be lost. The peculiar hunting methods of snakes must be successful by any test of what "success" means because there has been a massive radiation of snakes since they first appeared in the Cretaceous period and they have occupied all habitats except the very coldest. The serpentine shape must be intimately linked with these hunting methods and this success, yet this life form and life style are forbidden to endotherms because of the energy cost. Even the somewhat elongated bodies of weasels apparently require twice the maintenance costs of a normal mammal of comparable size (Brown and Lasiewski, 1972).

Finally, ectothermy allows short bursts of violent activity (flight or fight) at minimum cost. It does this by using anaerobic metabolism to support the activity (Figure 5.15). Anaerobic metabolism is itself an expensive way of releasing energy, because the lactate molecules stored must be resynthesized to glycogen after the period of intense metabolism is over (Figure 5.16). But the cheaper way of releasing energy for a violent burst of activity (at least in large animals: insects have solved the problem) requires the high basic metabolism of an endotherm, a cost that is quite out of proportion to the small expense of synthesizing a little glycogen from lactate.

One little noticed pattern among reptiles and amphibians is that they occupy the size classes between insects and birds or mammals, letting them extend into niches denied to endotherms by energy costs and denied to insects by such factors as gas exchange and growth by *ecdysis*

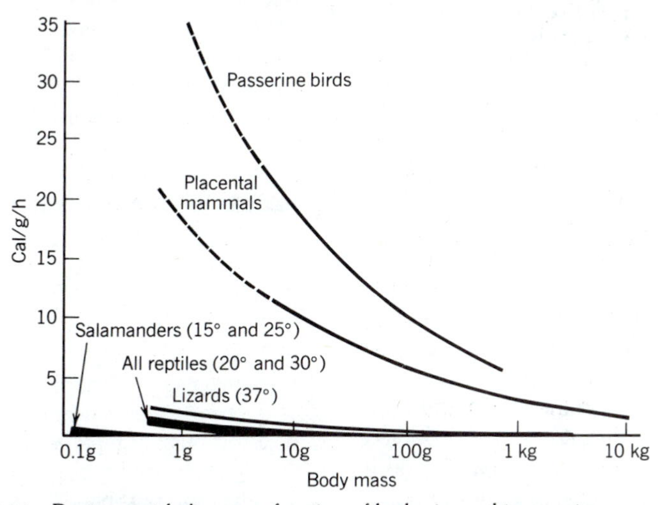

Figure 5.14 Resting metabolism as a function of body size and temperature. Ectothermic animals can be small because their energy consumption for maintenance is low. A bird weighing 1 gram would use about 30 times the energy of a reptile of the same weight. Not surprisingly, 1-gram reptiles exist but 1-gram birds do not. (From Pough, 1982.)

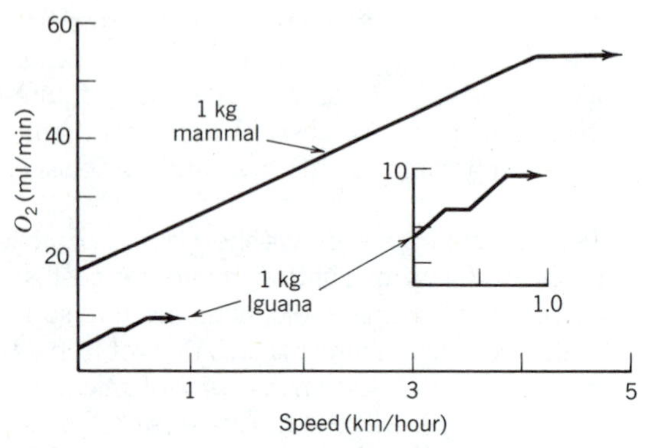

Figure 5.15 Oxygen consumption and speed in endotherms and ectotherms.
The data for a green iguana are from measurements on captive animals. The curve for a mammal of the same size is calculated from generalized mammalian data. Mammals attain five times the oxygen consumption of reptiles. (From Bennett and Ruben, 1979.)

Figure 5.16 Komodo dragon: low-energy system as predator.
Ectotherms can deliver short bursts of very high activity, a pattern suited to large ambush predators in places with sufficient sun to warm their bodies for the spurt. Crocodiles are the classic example on the contemporary earth, but the Komodo dragon shows what was once a common life style.

(moulting the rigid exoskeleton). Some useful properties of ectothermy are summarized in Table 5.3.

Benefits and Costs of Endothermy

Endotherms can do things that ectotherms cannot. Their burst of energy, for instance, can last longer. An endotherm can use body chemistry that releases energy from ATP by continual oxidation, a property attested to by the fact that the oxygen consumption of a mammal under hard exercise may be five times that of a lizard being similarly driven. The lizard quickly builds an oxygen debt and has to reduce activity for a long while for lactate to be resynthesized to glycogen, but a mammal can keep going with only a modest and manageable oxygen debt. Hard-running time for a lizard is never more than a few minutes, but mammals can produce a marathon runner.

Endotherms can penetrate the shade and hunt by night, undertakings that are virtually denied to large ectotherms in colder habitats (Figure 5.17). The advantage of this to endotherms is more time to forage. The endothermic animal uses fuel rapidly to stay warm and keep going, but it has the compensation of having more time

Table 5.3
Special Properties of Ectothermy
Ectothermy represents a strategy for heating the body into an activity temperature range suited to foraging as cheaply as possible. Extra advantages accrue because the animal easily relapses into torpor at other times. Table compiled from the analysis of Pough (1982).

1. Resting metabolic rate low
2. Production efficiency high
3. Adverse periods easily passed in torpor
4. Maximum advantage taken of episodic gluts of food
5. Remaining motionless while cryptically colored; a cheap defense against predators
6. Short burst of rapid activity (flight or fight) possible by anaerobic means
7. Small size possible
8. Growth from small to large facilitated
9. Serpentine or elongated shape possible

Figure 5.17 Endothermic hunter of the night. The high-energy system of endothermy is inefficient in its use of fuel energy but makes possible activity when ectotherms are sluggish.

in which to collect its fuel. It can properly be called a *high-energy system.*

Clearly a high-energy system does not let an animal be an efficient producer of growth or young if "efficiency" is taken to mean the ratio of energy absorbed to energy put into growth and reproduction (*production efficiency*). Like racing cars, endothermic animals do not use fuel "efficiently" but use large quantities of energy for speed and other activity, and yet it is also clear that many animals gain fitness by this "inefficient" behavior.

Ectothermy and endothermy are two quite different strategies for gaining fitness. Each has properties that the other lacks. Both systems allow regulation of the body temperature to within narrow limits during the period of activity, and tetrapods using both systems operate at near 38°C, a temperature apparently required by their common body chemistry. Endothermic tetrapods keep their temperature within about a degree of this optimum but even lizards are stenothermic to the point of not allowing their body temperature to fluctuate by more than 5°C when active. Endotherms can forage more easily in the shade, the cold, and the dark; they can also indulge in protracted chases, either as pursuer or as fugitive. But ectotherms can feed in good times, pass the bad times in torpor, and spread through habitats and food webs as they grow by switching from food to food, and they can escape, or mount an ambush, as a torpid, semianimate, camouflaged object hidden in the background. The difference between the two strategies is based in energetics, so they are well described by Pough's (1982) terms "low- and high-energy system." They represent quite different possibilities for obtaining a living. When the small endothermic tetrapods, which would one day give rise to mammals and birds, first arose, the number of species possible on the earth was multiplied by all the possibilities of foraging at night and of the more relentless and persevering kinds of hunting.

Some Natural History of Endothermy and Ectothermy

That the inefficiency of squandering energy as heat and movement can bring remarkable returns in fitness is shown nicely by the natural history of hummingbirds. A hummingbird is tiny, with a large surface area, and is an active forager so that it loses heat rapidly. It must meet its high energy costs from continued high-energy intakes, and yet there are many kinds of hummingbird, each with large populations, throughout the warmer parts of the Americas. This appears to be because the hummingbird life style can be used by plants to give them fitness by spreading pollen. Plants with nectaries well supplied with nectar to meet the high energy demands of hummingbirds are at a selective advantage over plants that fail to meet the energy extortions of hummingbirds.

A parallel process in Africa, starting with different bird stocks, has led to another array of related species of high-energy systems called sunbirds. They are somewhat larger than hummingbirds and drink their nectar while perching as much as possible, but they are in all other respects very much like hummingbirds (Figure 5.18; Wolf and Hainsworth, 1982). Sunbirds and hummingbirds succeed because other living things

Figure 5.18 Nectar feeders: the ultimate high-energy systems.
Both the hummingbirds of the new world and the sunbirds of the old world support high activity in tiny bodies on copious supplies of high-energy food (nectar). Plants support this profligate energy addiction of the birds by using them as "flying penises" for the plants' own reproduction (see Chapter 25).

can gain fitness by supplying them with energy for their profligate habits.

A life of fishing in a tropical river is apparently well suited to a low-energy approach like that of crocodilians. The river bank is handy as a place to raise temperature under the sun, the ambush and dash of a reptilian carnivore appears to be a good fishing technique, and the river provides food in many sizes, most of it small, of which the changing appetites of a growing ectotherm can take good advantage. Possibly the one necessary adaptation for a tetrapod ectotherm to thrive as a riverine fisher is a defense against endotherm predators hunting the banks for animals recovering from torpor in the sun (Figure 5.19). Large crocodiles clearly have the defenses needed to support their low-energy foraging system. They have remained unchanged as the perfection of fishers of tropical rivers for a hundred million years.

In theory, the crocodilian ectotherm, low-energy system must often have advantages for foraging in shallows close to ocean shores also, but the strategy does not seem common outside tropical rivers. Reflections on the Galapagos marine iguana may suggest why this is. Marine iguanas (*Amblyrhynchus cristatus*) crop seaweeds just below high tide line, penetrating water of cold upwellings to do so. The animals are black and well adapted to rapid absorption of the equatorial sun when lying on shore (Figure 5.20), but they are known only from the Galapagos Islands. No other herbivorous ectothermic tetrapod forages at sea anywhere in the world. A working hypothesis to explain the success of the strategy in the Galapagos and nowhere else is that there are no mammalian predators on the Galapagos Islands, and no avian predators able to attack adults. If there were, an iguana still torpid from its cold swim or huddled with its fellows

Figure 5.19 Low-energy ambush in armor.
For a large predator in tropical rivers ectothermy has not been bettered: safe on the bank behind armor and weapons while heating up, and fast in the water.

on a cold, cloudless night would be easy prey. But most of the warm coasts of the earth are in reach of endothermic predators. Thus, as admirably efficient, and even elegant, as the life style of the marine iguana is, there is no fitness to be gained by using its energetic logic on most coastal environments of the contemporary earth.

Oxygen Transport and the Success of Insects

The most numerous practitioners of ectothermy are insects and arachnids (spiders and allies), which compress size far more than is possible for vertebrates, and they do so without paying the cost of limiting the duration of intense activity. The secret of this success appears to be their unique system of gas exchange with the air.

Insects deliver oxygen to the muscles directly through a system of branching tubes and tubelets called TRACHEAE and TRACHEOLES. They do not have to carry oxygen in solution at all, nor do they have to return carbon dioxide in solution. Air flows, partly passively and partly through pumping motions of body movements, through the short tubes directly to where it is needed. In a sense, each tissue does its own breathing. This makes possible tiny ectotherms

Figure 5.20 The only marine lizard.
The Galapagos marine iguana is able to operate its low-energy system to feed on seaweeds in cold water because the Galapagos shores are without large endothermic predators (mammals). Or at least they were until people introduced dogs to the islands.

Figure 5.21 Migrating ectotherms.
Monarch butterflies overwintering in California. They have flown south from all over the northwest having accomplished journeys comparable to those of migrating birds. Like the sustained flight of bees, this is possible because of the air supplied directly to their muscles in the system of tracheae.

that can sustain high activity, essentially until their energy reserves run out (Figure 5.21). Butterflies and dragonflies can fly non-stop for hundreds of kilometers, and heavy bees hum and hover at speed all day long under the sun (Figure 5.22). Perhaps more remarkably still moths fly on cold nights, and the hours of darkness are fit for the activity of insects though it may be a time of torpor for most ectothermic tetrapods.

Summary

Ectothermy and endothermy allow a variety of different strategies for winning fitness. In animals of large size, endothermy permits the rapid movement of oxygen in solution and hence the maintenance of activity almost independently of sunlight. The cost in calories is high, but endotherms can occupy niches denied to large ectotherms and so pay this cost. Large ectotherms can tap low or intermittent fluxes of calories and convert them efficiently into offspring, thus occupying niches denied to endotherms. Ectotherms can also be much smaller than endotherms, a property that always makes ectothermic animals more numerous than endothermic animals. The most efficient designs of ectothermy are those that do not need to transport oxygen in solution but can transport air directly to the tissues. This extremely effective process of gas exchange, however, is usable only over very short distances.

Most of the smallest metazoan land animals are arthropod ectotherms using tracheal air transport. The next size class is occupied almost entirely by tetrapod ectothermic vertebrates, except in wetter places where molluscan and annelid systems primarily suited to water can operate. As the largest animal sizes are reached, homeothermy becomes increasingly important

Figure 5.22 Ectothermy with oxygen pump.
A bee flies on when an ectothermic vertebrate collapses because air pumped directly to muscles in systems of tiny tubes allows muscle metabolism to continue without the long periods of rest needed by vertebrate ectotherms.

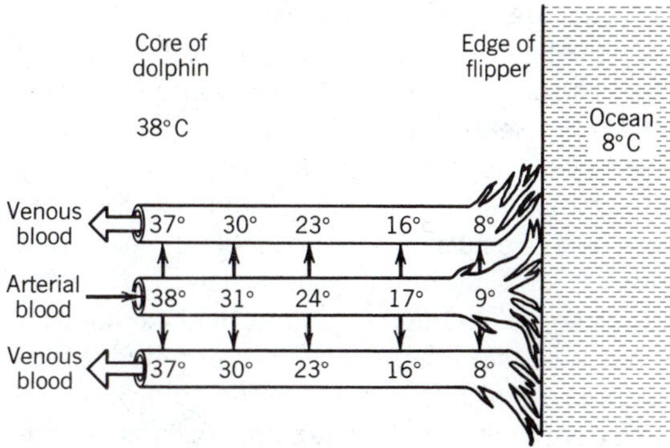

Figure 5.23 A countercurrent flow heat exchanger.
In the flipper of a porpoise, each artery that goes towards the skin is wrapped in a bundle of veins carrying blood away from the skin. This arrangement means that the outgoing arterial blood is always bathed in venous blood of slightly lower temperature. Arterial blood is progressively cooled as it approaches the cool surface of the fin and venous blood is progressively warmed as it approaches the region of the core temperature. Small arrows from artery to veins in the figure show that a constant temperature gradient is maintained. This heat exchanger can cool arterial blood, or warm venous blood, to within 1°C of the ambient temperature. The result is that there need be very little heat loss from flippers despite the absence of insulation in their thin cross section.

except in special circumstances (crocodiles in rivers) where low energy requirements give ectotherms a continued advantage.

COUNTERCURRENTS IN THE SERVICE OF ENERGETICS

The number of habitats that can be penetrated by both endothermic and ectothermic systems are often increased by the presence of *countercurrent* flows in organ systems. By use of COUNTERCURRENTS animals manipulate the temperature of their surfaces, thus increasing heat loss in warm places and reducing heat loss in cold places. Countercurrents also serve to regulate the gain and loss of water, letting large animals exploit both deserts and the salt oceans.

The principle of COUNTERCURRENT FLOW is used by marine mammals to reduce the loss of heat from flippers, exposed as they are to the rush of water thirty or more degrees celsius colder than the temperature at which the blood is regulated. And yet the tissues must be supplied by blood. A countercurrent flow is used as a heat exchanger, cooling the outgoing blood with the return blood so that the surface temperature of the fin is kept close to the temperature of seawater, a condition in which there would be no loss of heat. This arrangement (Figure 5.23) will serve to illustrate the principle of countercurrents wherever they are used.

When hot and cold liquids flow in opposite directions down narrow pipes (capillaries) laid side by side there will always be a small temperature difference between the adjacent flows (Figure 5.24). Arteries and veins in a porpoise flipper are arranged such that bundles made up of veins are wrapped around arteries. The outgoing arterial blood is progressively cooled and the returning venous blood is warmed. Instead of losing heat to the environment the heat is "recycled" as much as is thermodynamically possible within the flipper. In practice the heat loss of an operating porpoise flipper in cold water may be small (Schmidt-Nielsen, 1975). This manipulation of countercurrents as heat exchangers is an essential adaptation that permitted endothermic animals to operate in the cold oceans.

Countercurrents opened the oceans to endotherms in another way also, by letting them remove excess salt from seawater. The oceans are a solution of NaCl at a concentration of about 35 parts chlorine per thousand (35 ppt), for reasons examined in Chapter 18. For a tetrapod endotherm, like a porpoise or a bird, to operate there, some sort of still (distillation device) must be provided to manipulate solute concentrations. Countercurrents probably make possible the necessary stills.

Marine mammals have kidneys that can excrete salt from seawater. Mammalian kidneys work by enhancing the effect of active transport of ions across biological membranes with a COUNTERCURRENT MULTIPLIER. The basic plan is shown in the familiar Henle's loop, first described from the human kidney (Figure 5.24). The essential quality of a *countercurrent multiplier*, as opposed to the heat exchanger discussed above, is that active transport across cell membranes maintains the gradients between opposing flows. The countercurrents are used to multiply changes in solute concentrations achieved by simple membrane systems. Countercurrent flow is arranged by doubling a tubule back on itself, which allows fluid in a semipermeable tube to pass back through the medium already traversed in the first part of a loop.

Figure 5.25 illustrates the mechanism of Henle's loop. The thick arrows pointing out of the ascending tube of the loop describe active transport of sodium ions. This results in an ambient sodium concentration greater than that in the adjacent portion of the descending loop, which drives passive transport of sodium into the descending fluid (thin arrows). The active transport

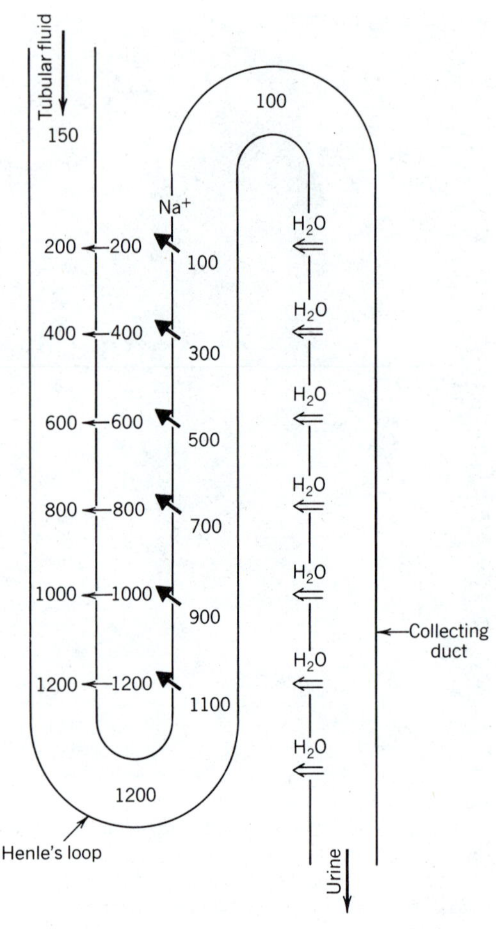

Figure 5.24 Henle's loop: a countercurrent multiplier. For details see text. (From Schmidt-Nielsen, 1975.)

mechanism maintains a constant small concentration difference throughout the system but the countercurrent arrangement multiplies this many times, finally achieving a far greater concentration of solutes than the active transport systems can produce across a simple membrane.

Marine mammals have so refined the system that they can remove the toxic salt from seawater. Countercurrents let them drink seawater; countercurrents let them use an endothermic strategy although immersed in a cold fluid of high specific heat. By these two devices a new swarm of niches was found in the world oceans (Figure

Figure 5.25 Triumph of countercurrent systems.
Dolphins and other cetaceans conserve heat and distill salt water using countercurrent flow devices. By these physiological adaptations the sea was opened to air-breathing endotherms.

5.25). Whales, both filter feeders and Eltonian predators like sperm whales and porpoises, eat the same kinds of foods as many sharks of comparable size. But their basic endothermy, coupled to access to the flux of atmospheric oxygen, lets them exploit different patterns of these resources. Diversity of life in the oceans was thus increased.

An adaptation probably involving countercurrent multipliers is also necessary to marine reptiles and birds. For a bird to live far out to sea it must sometimes ingest salt water or salty food, but the avian kidney has not been boosted to cope with salt concentrations of the ocean any more than the human kidney has. Instead marine birds have salt glands. The salt glands are usually in pairs in the seabird's head, and they secrete strong solutions of sodium chloride into the nasal passages where the salty solution is lost as drops. Similar pairs of glands are present in the heads of birds as diverse as penguins, gulls, petrels, and oceanic ducks, showing that the structures have appeared several times in evolutionary history. Many studies have illustrated the salt-concentrating power of bird salt glands (Schmidt-Nielsen, 1975). Probably this performance relies on countercurrent multipliers as do conventional kidneys.

With the provision of salt glands many novel opportunities were opened for bird niches. Some are obvious, like taking to a steady diet of fish as penguins or cormorants do, but others are more subtle. Many seabirds, like pomarine jaegers (*Stercorarius pomarinus*), for instance, use the oceans to overcome one of the main disadvantages of endothermy, the poor ability to take advantage of highly intermittent or unpredictable food supplies. The jaegers spend most of the year, and most of their lives, wandering the oceans far from land. They are able to meet their metabolic needs from scavenging the surface of the ocean. But the jaegers reproduce by using resources of the land—lemmings or other small rodents of the arctic (see Chapter 13). The supply of lemmings is only available in the spring, and it may not be available at all. The response of an ectotherm like a snake to this dilemma would be to relapse into torpor for most of its life, only boosting its temperature and activity when lemmings were about. Jaegers have adapted the endotherm design to this problem by meeting the maintenance costs of endothermy with calories taken from the world oceans.

Fish, too, make good use of countercurrents. There is a countercurrent arrangement in every fish gill, where water flows across tissue supplied with blood flowing in the opposite direction. Water full of oxygen encounters blood whose dissolved oxygen is only a little less. When the water is more depleted of oxygen from its passage through the gills it meets blood at the lowest oxygen tension so that the oxygen gradient between water and blood tends to be maintained. A similar countercurrent arrangement is found in the *rete mirabili* tissue of teleost swim bladders and is responsible for maintaining partial pressures of gas in the bladders.

Some fast-moving fish use heat exchange countercurrents at the periphery as is done in the flippers of porpoises. Fast-moving sharks and tuna do this (Schmidt-Nielsen, 1972). The system allows the animals to maintain temperatures in the swimming muscles well above that of the surrounding water, providing them with a modified endothermy. Warm muscles give these predatory fish the ability to maintain rapid movement in pursuit of food, thus letting them function as do endothermic predators on land. Sharks, therefore, have some of the energetic adaptations of mammalian predators. Countercurrent heat exchangers are particularly necessary for these fish because they are without significant insulation and would not be able to maintain high core temperatures without a system for cooling blood before it reaches the periphery.

In hot deserts countercurrent systems are of use for the conservation of water. The simplest arrangement is the countercurrent multiplier in the kidneys of desert rodents, which let them conserve water by excreting highly concentrated urine. But more elegant are countercurrent heat exchangers that lower the temperature of exhaled air, a device used by a number of desert animals to restrict the loss of water from the necessity of breathing. Breath tends to be wet because exhaled air is warmed to close to the core temperature of the body and because it has been in contact with the wet surface of the lungs. Water lost by breathing, therefore, may be considerable, particularly when the inhaled air is dry as in deserts. For a desert rodent, so constrained for water that it can live on the water held in dried seeds in its diet without the need for drinking, this loss from the lungs could be serious and should be reduced. Comparison of the water in exhaled air of desert and nondesert animals shows that the breathing loss is in fact controlled in desert animals (Table 5.4).

Exhaled air of desert rodents is dried by being cooled before it is released from the nostrils. A countercurrent heat exchanger is arranged in the convoluted nasal passages so that incoming air cools the outgoing air, an arrangement comparable to the opposed blood vessels in the flippers of porpoises. As the outgoing air cools it loses water, which condenses in the nose or is absorbed by the incoming air as this gains temperature. Evaporation within the nasal passages can actually cool the exhaled air below the temperature of the outside air, again without loss of water to the animal. An easily detectable consequence of this mechanism is that the animal

Table 5.4
Evaporation in Exhaled Air
Animals adapted to desert life lose much less water in exhaled air than do other animals. This is a consequence of an efficient countercurrent heat exchange that cools, and hence dries, the outgoing air. Evaporation is expressed relative to oxygen consumption to permit comparisons of different-sized animals. (From Schmidt-Nielsen, 1972.)

Desert species	Evaporation mg H_2O $ml^{-1}O_2$ cons.
Kangaroo rat	0.54 ± 0.04
Pocket mouse	0.50 ± 0.07
Golden hamster	0.59 ± 0.05
Mesic species	
White rat	0.94 ± 0.09
White mouse	0.85 ± 0.07
Human (lungs only)	0.84

has a cool nose. But the important consequence detected by natural selection is that the animal does not lose water from its cold nose while breathing.

This adaptation is also one of the many that equip camels, the most celebrated of desert animals, to exploit the carbon resources of desert lands. They cool and dry their exhaled air, they manipulate the temperature at the periphery with fur insulation, they concentrate urine, they tolerate desiccation, and they store energy in a fatty hump; the whole making a package of adaptations that stretches endothermic energetics to survival through the leaner times of desert living (Figure 5.26).

ENERGY, MOVEMENT, AND OPPORTUNITY

Apart from maintenance, the major cost to be met by the energy budget of an animal is movement. The animal must move to where the food is, to shelter, or for purposes of flight or fight.

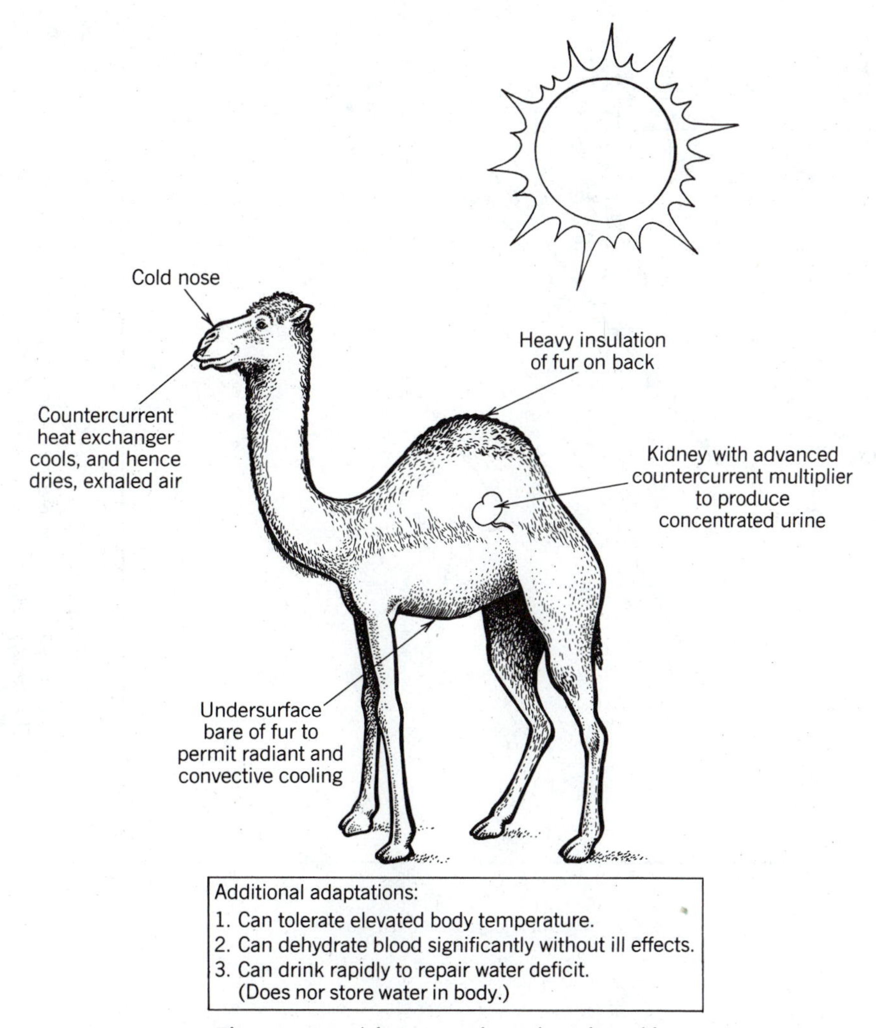

Figure 5.26 Adaptations of camels to desert life.

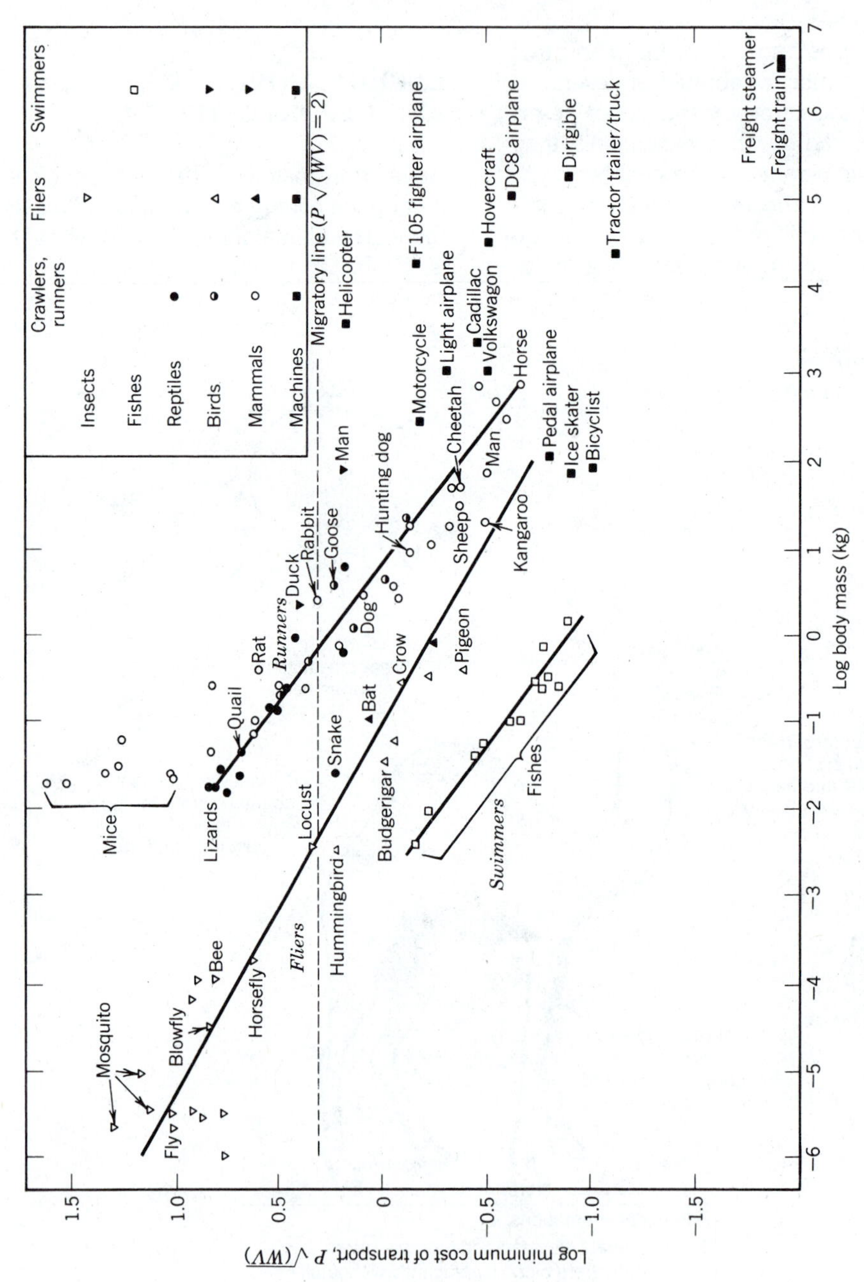

Figure 5.27 Energy costs of swimming, flying, and running.
The data suggest that mass to be moved is the decisive constraint once the organism is committed to a particular gait. (From Tucker, 1975.)

Natural selection arguments require that movement be efficient, and yet many kinds of locomotion are in use. There are different patterns of flight, different gaits on land, running or hopping, two legs, four legs, or six.

In Figure 5.27 the costs of swimming, running, and flying are compared for animals of different sizes and for various machines. The animal data are derived from measurements of metabolic rates of experimental animals in motion as follows:

Pi = metabolic rate = power output

V = velocity

W = weight = mass × gravity
= the force to be overcome.

Then

$$W \times V = \text{Power}$$

and

$$\text{Cost of transport} = Pi/WV \qquad (5.8)$$

This function has the advantage of allowing the costs of animals of different sizes to be compared. In Figure 5.28 these transport costs are plotted against body masses on logarithmic scales. Swimming and flying use less energy than running, but otherwise the mass to be moved seems the determinant function (Tucker, 1975; Schmidt-Nielsen, 1972).

These data reveal some apparent puzzles. Why are flying and swimming less costly than running? A flying animal has to support its weight in a diffuse fluid medium by the expenditure of energy and a swimmer has to force its way through a dense viscid medium (Figure 5.28). By contrast, a runner encounters negligible friction from the air and an amount so small from contact with the ground that it is within the error of our measurements of metabolic rate. Why, then, is running so costly in energy? The answer must lie in the energy spent in extending muscles and levers, and in absorbing changes in angular momentum as heavy limbs continually change their direction of motion. Birds must pay these costs too, but their greater speed spreads the cost. Greater speed in proportion to muscular movement is also the reason why cycling is more

Figure 5.28 Energy efficiency in transport: geese in transit.
Bird flight seems to be energetically more efficient than running only because of the greater speed attained. A principal energetic cost still seems to be that of returning heavy limbs to the starting position for the next power stroke, just as in running. The vee formations may allow geese to save work by riding the wave of turbulence of the goose ahead and to one side, a physical mechanism constraining social behavior (see Chapter 12).

efficient than running. For a swimmer designed on proper hydrodynamic lines, these costs of muscle flexure and angular momentum are much reduced and a correspondingly greater part of the cost goes to overcoming friction. Swimmers that work by thrashing limbs about, like ducks and people (Figure 5.28), face both kinds of cost and are inefficient.

Common sense suggests that there will be different levels of efficiency within any class of locomotion depending on such things as numbers of limbs and gait, yet these turn out to be small. For instance, there has been speculation in the literature that the hominid line met special energetic costs when the change from four-legged locomotion to bipedalism was first made. This has been tested by Taylor and Rowntree (1973), who trained chimpanzees to work treadmills both with two legs and with four. No difference could be detected in the metabolic rate of the chimpanzees whichever way they worked the treadmill.

Of greater interest is the evidence of kangaroos. In Australia large kangaroos fill the niches (in the Eltonian, Class I niche sense) that are filled by ungulate grazers on all other continents. Kangaroos hop. It is not certain if hopping gives kangaroos an energetic advantage or disadvantage. In Figure 5.28 they are shown as being slightly more efficient than quadruped ungulates, but the difference is small when the difficulty of measuring metabolic rate in a running kangaroo is considered. But why should the principal grazing endothermic tetrapod consumer on one continent hop while its equivalents on all other continents run on all fours? The answer seems to have nothing to do with energetics.

Marshall (1974) traces the evolution of the kangaroo foot and compares it with the feet of *artiodactyles* (two-toed ungulates like cows) and *perissodactyles* (one-toed ungulates like horses). From the original 5-toed plan, cows run on toes III and IV, whereas horses run on toe III alone (Figure 5.30). But the ancestors of kangaroos had toe III fused to toe II, so that it was off center and not available to serve as an ungulate hoof. This left toe IV to be the principal prop of kangaroo motion when its ancestors took to a grazing life on the plains. The attachment of toe IV to the complex of bones in the ankle is profoundly different from the attachment of the other toes and results in a different positioning of stress and balance (Figure 5.29). Marshall argues that this anchoring of toe IV allowed the ancestral animal to gather speed by a process of spring and ricochet but not by the weight transfers required of ordinary running. Kangaroos hop, therefore, not because hopping is more or less efficient than running but because this was the best solution to the speed problem. Grazing herbivores need speed if they are to achieve fitness in a world stocked with predators. Making a fast herbivore from the Australian marsupial line required that the animal be a hopper.

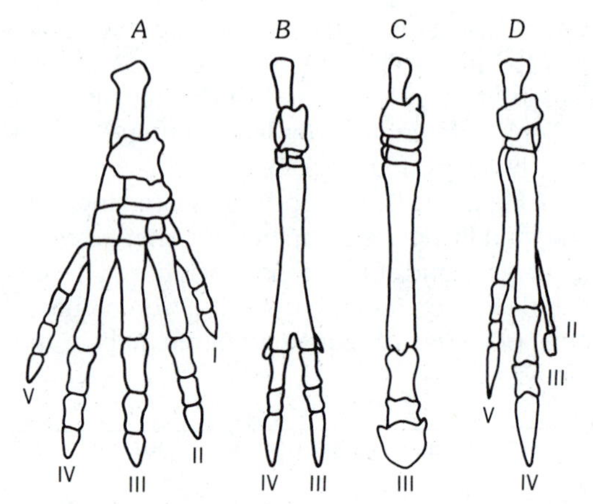

Figure 5.29 Ungulate and kangaroo feet. Artiodactyles run on toes III and IV and perissodactyles on toe III alone. But kangaroos run on toe IV alone, since toe III was fused to toe II in the ancestral stock. (*A*) Hypothetical placental ancestor; (*B*) artiodactyle; (*C*) perissodactyle; (*D*) kangaroo. (From Marshall, 1974.)

Wondering why a kangaroo hops is like wondering why a spider has eight legs but an insect only six. These arrangements resulted from selection in distant ancestral lines to adapt animals to the niches then open to them. Subsequent adapting of these arrangements to other niches was always done starting with the engineering plan at hand. Natural selection cannot go back and start over. But what is interesting to an ecologist is that the way of life dictates the tasks that the structure must perform. A tasty grazer had best run fast: it matters not whether it hops or sprints. Intense activity requires that the body be held near 38°C; it matters not whether this regulation is achieved by ectothermy or endothermy. Really prolonged activity requires aerobic metabolism, whether achieved through endothermy and an efficient blood-vascular system or by the direct piping of air.

PART TWO

POPULATION AND SPECIES

PREVIEW TO CHAPTER 6

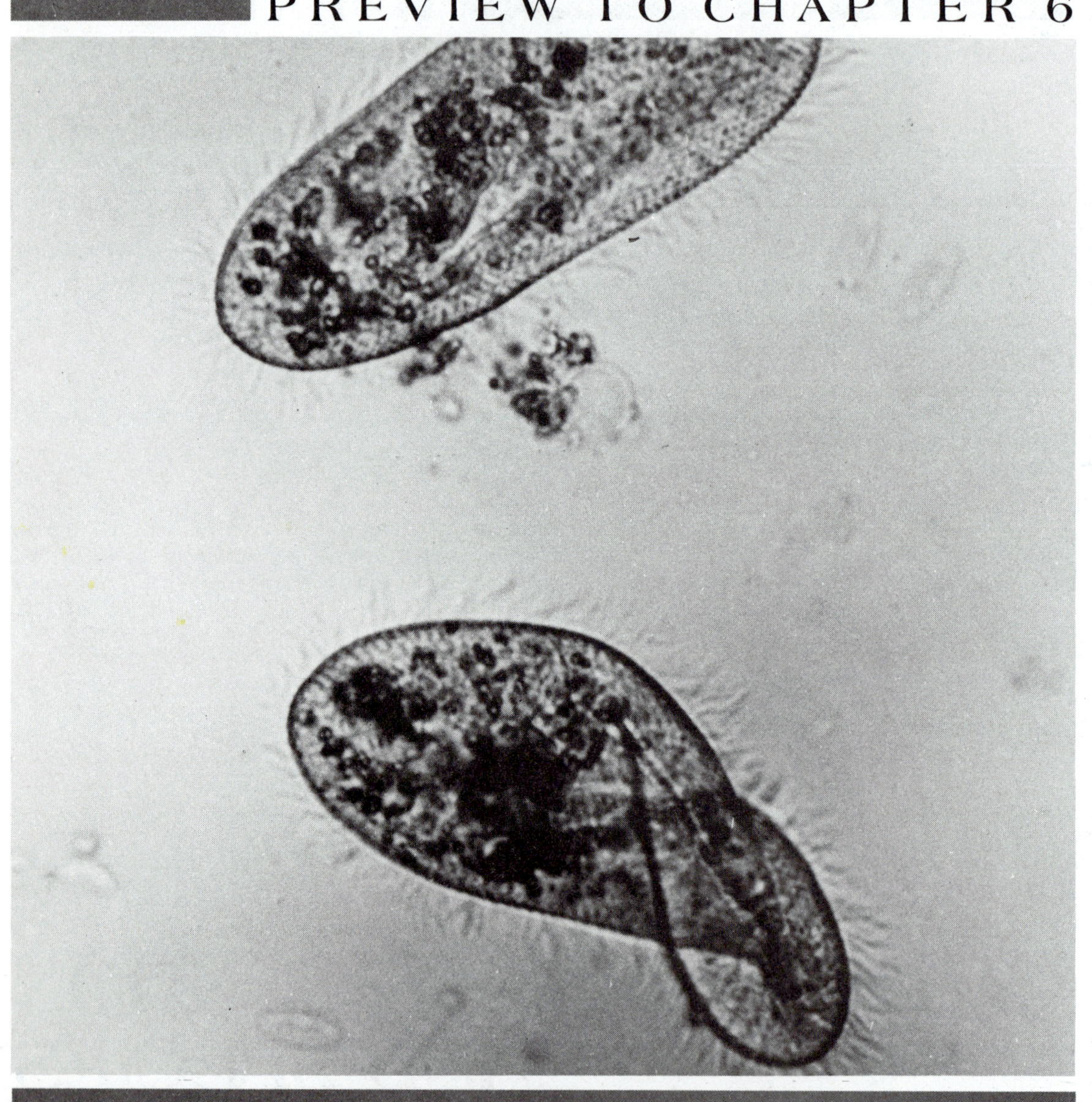

The ideas of a population equilibrium and of separate species being preserved by natural selection both invoke the concept of competition. Competition is defined in ecological terms, both within species and between species. S-Shaped (sigmoid) population histories are observed in the laboratory, and they may be explained by a hypothesis that states that competition increases as populations are crowded. This hypothesis may be stated in mathematical form with the logistic equation. The hypothesis can be investigated in laboratory cultures with single-species populations of simple animals like paramecia, fruit flies, and flour beetles. Sometimes the postulated competition for food or space can be demonstrated directly, and at other times crowded animals may be shown to interfere with each other by cannibalism or by fouling the habitat. These processes are all density dependent and control by them is allowed in the logistic hypothesis along with simple competition. But the true test of the logistic hypothesis comes when it is used to predict the outcome of competition between different species in confinement. This development, provided by the Lotka–Volterra equations of competition, predicts that competition must be muted or it will lead to the total extinction of one species. The prediction has been tested in numerous laboratory experiments and shown to be true, hence this result can be stated as the principle of competitive exclusion, known in ecological shorthand as "one species: one niche." Two ideas of paramount importance in ecology are established by the experimental validation of the logistic hypothesis. One is that populations in nature may often be under density-dependent control. The other is that species populations are kept distinct by the selective removal of individuals that tend to compete with individuals of other species.

CHAPTER 6

THE EQUILIBRIUM MODEL AND THE LOGISTIC HYPOTHESIS

A fundamental question of ecology has always been, "What are the natural checks on number?" In trying to answer this question, ecologists found themselves feeling for answers to a more fundamental question still, the reason for the distinctness of species.

A *population* is A GROUP OF ORGANISMS COEXISTING AT THE SAME TIME AND PLACE AND CAPABLE FOR THE MOST PART OF INTERBREEDING. But a *species* is A POPULATION OR SERIES OF POPULATIONS THE INDIVIDUALS OF WHICH ARE CAPABLE OF INTERBREEDING FREELY WITH EACH OTHER BUT NOT WITH MEMBERS OF OTHER SPECIES. The regulation of numbers within a population inevitably is linked to the processes that keep separate the species themselves—because a species is essentially a population kept separate from other populations. An outcome of the study of population regulation has been a theory to explain the existence of distinct species.

Questions about checks on numbers and the origin of species are, of course, questions that Darwin asked. He thought of a world in which every kind of animal or plant could leave far more offspring than were needed to replace the parents and that this must mean a *struggle for existence* in which the surplus perished. Selection operated during this struggle, and a species changed even as populations were kept down. Darwin was suggesting that a full understanding of what regulated populations must also show why species are distinct. We have found this to be true.

Essential to the struggle-for-existence view of life is the idea of COMPETITION. Gaining fitness may be viewed as struggling for the opportunity of filling one of the niches vacated by death in the next generation (Chapter 1). This struggle at once suggests the idea of competition within the species population: INTRASPECIFIC COMPETITION. But there must also be the possibility of struggle for resources with neighboring populations of other species: INTERSPECIFIC COMPETITION.[1]

The study of both *intraspecific* and *interspecific competition*, therefore, has been central to attempts to understand both the regulation of populations and the uniqueness of species.

[1]The student not familiar with Latin should note that *intra* means "within," as "on the inside," and *inter* means "between." They sound similar to Anglo-Saxon ears but their meanings are opposites.

COMPETITION OCCURS WHENEVER A VALUABLE OR NECESSARY RESOURCE IS SOUGHT TOGETHER BY A NUMBER OF ANIMALS OR PLANTS (OF THE SAME KIND OR OF DIFFERENT KINDS) WHEN THAT RESOURCE IS IN SHORT SUPPLY; OR IF THE RESOURCE IS NOT IN SHORT SUPPLY, COMPETITION OCCURS WHEN THE ANIMALS OR PLANTS SEEKING THAT RESOURCE NEVERTHELESS HARM ONE ANOTHER IN THE PROCESS (Andrewartha and Birch, 1954).

"Competition" is a word with a clear meaning, valid and hallowed in English usage. There is competition whenever two or more individuals or groups "strive together" (the literal meaning of the Latin roots) for something in short supply. People compete for prizes, and only one of them, or one group of those competing, can win the principal prize. Yet competition between individuals for niches must be subtly different from competition between long-distance runners for a trophy and the other usual examples provided by Webster and the Oxford English Dictionary. In ecological competition whole life styles must be pitted one against another. Real success is measured a generation away in the success of offspring at reproducing themselves. Meanwhile, there must have been many interactions with neighbors and other animals and plants of the community in which it lives; perhaps competition for food energy, success at rapid dispersal to occupy an empty habitat or nesting place, better success than others at escaping predators, or the opportunity to find refuge in frost or drought.

To go out to look for and demonstrate competition in nature, therefore, can be a difficult undertaking. Yet the concept of competition has provided field ecologists with many of the tools they use, tools forged not in the field but in the laboratory.

SIGMOID GROWTH IN LABORATORY POPULATIONS

Every beginning biology textbook has a graph of population growth that is S-shaped:

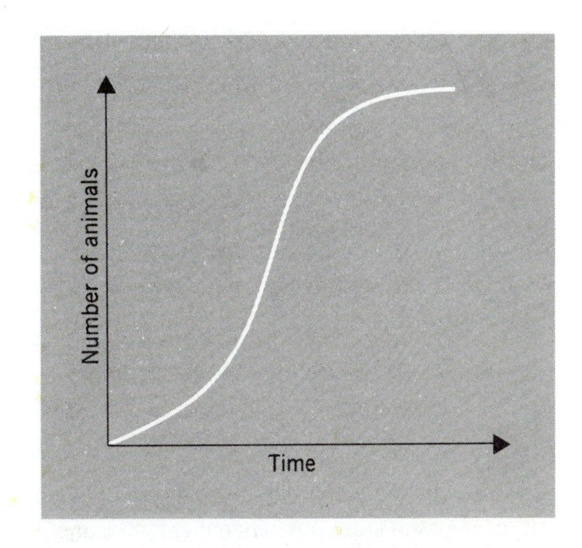

Various captions might come with the figure, but the impression is often given that this is a generalized curve for the growth of a typical population. This is misleading. What these S-shaped (usually called SIGMOID) curves actually describe is population growth as observed in a number of experimental systems where populations of single species are kept under strict confinement. No such complete histories have been seen in the wild, though two histories of bird invasions are known that come close (Chapter 7). The real significance of *sigmoid* population histories to ecology is that, being simple, they let us make plausible hypotheses to explain them. Moreover, having been observed in experimental systems, tests for the hypotheses are possible. A typical sigmoid population history in the laboratory is illustrated with the cultures of *Paramecium* kept by the Russian ecologist G. F. Gause[2] (1934).

Paramecia are protozoans that feed on bacteria and other small particles. Gause found that he could make a standard medium of oatmeal which served as excellent food for paramecia,

[2]The final *e* in "Gause" is pronounced, making the word Gauzer. This is not only phonetically correct but avoids confusion with the mathematician "Gauss," whose work is also often cited by ecologists.

and one which he could add to their water in measured amounts. In a typical experiment, he placed 40 individuals of *Paramecium caudatum* in small tubes containing 10 cc of water and a few drops of his oatmeal medium. Each day thereafter he would take out a subsample with a pipette to count the paramecia present, which gave him an estimate of the total population. He would also spin the tubes in a centrifuge, which drove the animals to the bottom and allowed him to pour off the old water with its unused suspended food in order to replace it with fresh. There was thus a daily constant but limited input of food to the system. Under these conditions, the paramecia reproduced quickly at first, their numbers increasing exponentially until the water was cloudy with them. But when there were about 4000 animals in each tube, the rate of growth leveled off, and the population did not grow any more. The top of the sigmoid trace had now been reached. The population remained at the same level indefinitely, as long as the daily food ration was provided.

Similar sigmoid population histories result in laboratory cultures of yeasts, molds, bacteria, various protozoa, and even insects, such as fruit flies in banana mash and *Tribolium* beetles in flour. All that is required is an animal of simple wants, a way of confining those animals, and an energy source. Attempts to contrive sigmoid population histories in the laboratory for more complicated animals with more complicated wants do not work. Mice, for instance, reproduce well enough in large pens until they are crowded, after which they suffer mass death or failure to breed, even though given plenty of food, water, and bedding (Calhoun, 1962). This history is far from being sigmoid.

There have been two approaches towards a formal theory to explain sigmoid population histories when they do occur. Most useful has been the hypothesis that the history reflects changing intensity of competition as the animals crowd, and this is the basis of modern understanding. But an alternative approach has given us the term *environmental resistance*, which is still used in ecology. This approach will be considered first.

Chapman (1928) thought it profitable to look at the restraints imposed on populations as they levelled off at the top of the "S" in terms of the resistance of the environment to the population pressure. He was attracted by the analogy of electric current flowing in a wire, as described by Ohm's law, to that of a growing population of animals. Ohm had shown that the current flowing at any point in a wire was equal to the potential divided by the resistance of the wire (a constant now measured in *ohms*). The resistance was given by the following equation:

$$\text{Resistance (ohms)} = \frac{\text{Potential (volts)}}{\text{Current (amperes)}} \qquad (6.1)$$

A population of animals might be said to have a potential in its innate ability of its members to reproduce, called the *intrinsic rate of increase* and referred to by the symbol *r*. For Chapman, *r* was the animal equivalent of voltage—the pressure for driving units of animal into the container.

It is important to distinguish between *r* and simple physiological reproductive power or *fecundity*. FECUNDITY is THE RATE AT WHICH AN INDIVIDUAL PRODUCES OFFSPRING (fecundity is usually only expressed for females). The INTRINSIC RATE OF INCREASE is a population characteristic. It takes into account the fact that all individuals do not have equal fecundity and it refers to their reproductive performance in the actual population in which they live. *r* also takes into account death rates; *r*, therefore, is a statistic rather than a parameter and is defined as the EXPONENTIAL GROWTH RATE OF A POPULATION WITH A STABLE AGE DISTRIBUTION.

If *r* is the voltage of the system, then the number of individuals present at any one time is the "current." Chapman wrote, in imitation of Ohm's law,

$$\text{Environmental resistance} = \frac{\text{Intrinsic rate of increase}}{\text{Population at equilibrium}} \qquad (6.2)$$

Since the intrinsic rate of increase and the population could be measured for real animals, it should be possible to solve for environmental resistance and to describe this as a number. There would then be possible the rather heady prospect of describing all the environments of animals in terms of simple numbers.

Unfortunately, as was soon realized, Chapman's analogy was unsound. An electric current flows through a wire, but there is no wire down which a population flows. To liken an environment, the total experience of an animal in which it finds its niche, to a wire is absurd. Nothing is fixed or permanent about an environment, and such key parts of it as the food supply are used up as the population "flows." Stating a numerical environmental resistance was in practice merely a cumbersome way of stating census data.

The term *environmental resistance* is a legacy of this abortive attempt to measure the controlling influence of environments. The term has lingered, perhaps, because it has a catchy sound. Yet, as described in a later section, Chapman's use of flour beetles to measure environmental resistance led him to study properties of crowded populations in quantitative ways that, in the service of the logistic hypothesis, led to many modern ideas on competition theory.

THE LOGISTIC HYPOTHESIS

It is natural to suggest that sigmoid growth histories reflect changing intensities of competition. A few animals in a container with plenty of food should not suffer any competition: *fecundity* would be high, the *generation time* would be short, and the *intrinsic rate of increase* would be high. Soon there would be many animals in the container, all reproducing rapidly. Population growth would be exponential or geometric. But then the animals would become crowded and *competition* should be important to life in the container. Fecundity, or survival, or both would drop; the death rate would rise. Population growth would slow until it ceased altogether with that degree of crowding where births exactly balance deaths. This hypothesis is valuable because it can be set down in formal mathematical terms:

$$R = rN\,(1 - N/K) \qquad (6.3)$$

or

$$\frac{dN}{dt} = rN\,(1 - N/K) \qquad (6.4)$$

where

R is the population growth rate (best defined as the differential dN/dt)

r is the intrinsic rate of increase

N is the number of animals present at time t

K is the number of animals able to live in the container at population equilibrium.

Equation (6.4) is called the LOGISTIC EQUATION. It has been of interest to biologists since its properties were first pointed out by P. F. Verhulst in the middle of the last century (Hutchinson, 1978). It neatly describes the hypothesis that population growth in a closed container is first damped, then regulated, by intraspecific competition.

Crucial to the logistic statement is the concept of CARRYING CAPACITY (K). K is a constant, a property of the container, and is expressed as a number—the maximum number of individuals that can persist under the conditions specified. The logistic hypothesis states that as the population number, N, approaches the saturation number, K, then rN becomes zero and population growth ceases.

The logistic equation, therefore, is merely a mathematical description of a hypothesis. The logistic equation, in its differential form, does NOT describe a sigmoid growth history. A plot of changing values of dN/dt against t results not in a sigmoid graph but in a parabola:

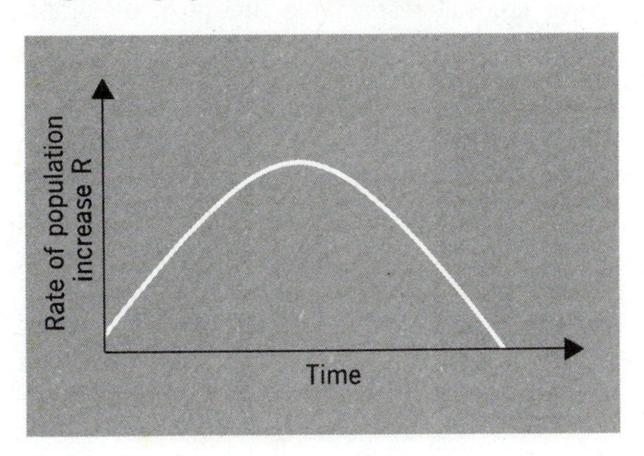

The integral form of the equation describes the sigmoid history of numbers postulated by the hypothesis:

$$N = \frac{K}{1 + e^{-rt}}$$

The differential form of the equation, however, is more useful, being a statement of the hypothesis put forward to account for the sigmoid change in number. The rate of population growth, dN/dt, is postulated to be parabolic, first rising and then falling as the container becomes crowded.

Saying that competition controls experimental populations when they are crowded and writing a logistic equation are two ways of stating the hypothesis. There is a third way—population appears to be brought under control by DENSITY-DEPENDENT FACTORS. Competition is *density dependent;* the more crowding, the more competition. But many other processes that might not automatically be called competition also act in a density-dependent way. Cannibalism might occur in crowds, or crowded animals might spoil food even though they did not actually compete for it, or the chance of finding a refuge might be less in a crowd. So the logistic hypothesis is also a hypothesis of control by density-dependent factors.

The idea that animal populations in nature are commonly under density-dependent control, sometimes by competition, sometimes by disease or predation, is widespread. Most statements about the *balance of nature* necessarily rely on accepting the hypothesis of density dependence as probably reflecting reality. That these views are still respectable among professional ecologists is a consequence of the fact that we have generally failed to falsify the logistic hypothesis, even after much work.

There have been two main approaches used to test the logistic hypothesis. The first is to examine experimental populations for direct evidence of the postulated competition or density-dependent control. The second, and more powerful, has been to manipulate the logistic equation to predict the resulting populations at equilibrium when populations of different animals were kept in the same containers.

DENSITY DEPENDENCE IN EXPERIMENTAL POPULATIONS

The detailed working of competition and density dependence in single-species populations has received extensive study in the laboratory, particularly using cultures of the fruit flies or flour beetles. The geneticist's fruit fly, *Drosophila melanogaster*, is especially suitable, since so much is known of the care and management of *Drosophila* cultures. The generation time is only 9 days (1 day as an egg, 4 as larvae, 4 as pupae), and a colony started with two pairs of flies leads to a maximum population of over 200 individuals in three weeks. The food supply in a culture bottle is fixed so all flies eventually die of starvation at about the 50th day. Figure 6.1 shows a typical population history from inoculum to population collapse. Reasons for depression of population growth in all stages of the life history can be studied by manipulating the density of crowds artificially, for instance, by putting a full crowd of 200 adult flies into a fresh culture bottle with unexhausted food and measuring the rate of egg laying. In these experiments it has been shown that females lay fewer eggs when crowded and that this reduced egg laying is largely a result of starvation because the space each female has in which to feed is reduced in crowded bottles. Increased death rates of larvae and pupae in crowded cultures could be due to direct starvation or they may result from interference or even cannibalism, but that they are density-dependent deaths is certain (Pearl, 1932; Pearl and Parker, 1922; Bodenheimer, 1938; Robertson and Sang, 1944 reviewed in Colinvaux, 1973).

Studies with flour beetles, such as *Tribolium confusum*, reveal how crowded animals may interfere with each other apart from direct competition for food. The beetles are kept in dishes of flour through which they burrow, a condition that ensures an abundance of food at all times. Nevertheless, population histories are sigmoid, reaching equilibrium with a density of about 4.4 beetles per gram of flour (Figure 6.2). Much experimental work has revealed at least three density-dependent pressures working against egg laying (fecundity) and hatching of eggs. Adult

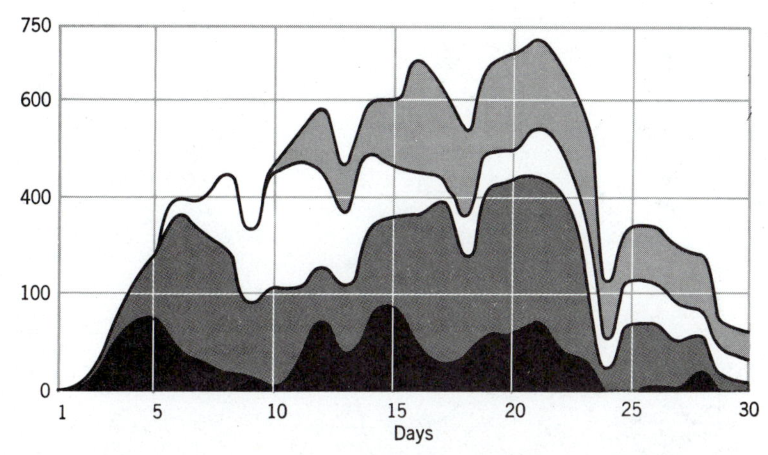

Figure 6.1 Population growth in *Drosophila* bottles. Reconstructed actual history of a population of flies in a bottle containing culture medium from the initial inoculation until the starvation that ensues when all the food is gone. About five generations are spanned. Notice that the survival of all stages in the life history declines as the population becomes dense. (After Bodenheimer, 1938).

beetles are egg cannibals, eating every egg they encounter in their passage through the flour (Figure 6.3). Second, egg laying is directly inhibited by secretions of beetles that collect in old flour, notably the chemical ethylquinone secreted by male beetles when the mating act is disturbed by the arrival of a third beetle. But, third, additional social pressures reduced egg laying in crowds because in experiments where both egg cannibalism and the accumulation of ethylquinone were prevented fecundity still dropped at high density (Figures 6.4 and 6.5). In addition to these constraints on egg laying or survival it can also be shown that mortality of larvae and adult beetles increases in crowds, particularly when they bite one another (Chapman, 1928; Allee *et al.*, 1949; Park *et al.*, 1965; Boyce, 1946; Rich, 1956).

The work with these various experimental animals, with *Paramecium*, fruit flies, and flour beetles, as well as with many others, has shown convincingly that populations of real animals can be controlled by density-dependent factors in the manner required by the logistic model. They may be controlled by starvation, which is to say by simple competition for dwindling food, or they may be controlled by other density-dependent mechanisms. It is very clear, however, that the logistic model can only be a very simplified analogy of what actually happens. In the equation

$$\frac{dN}{dt} = rN\left(1 - \frac{N}{K}\right)$$

the carrying capacity of the environment "K" is set by whatever limits the density of the animals, and may usually be expected to have many components. In some simple systems, as perhaps in *Paramecium* cultures or the egg laying of fruit flies, K may be set by direct competition for food; but more usually a complex of many factors must set K, as the *Tribolium* studies so clearly show. The matter is made more complex still by the fact that different mechanisms may be at work in different stages of the lives of the animals. In the holometabolous insects so widely used for laboratory studies this difficulty is readily apparent. Eggs, larvae, pupae, and adults are, in a sense, different organisms, depending for their survival on different mechanisms. Any long-term population balance for the whole species

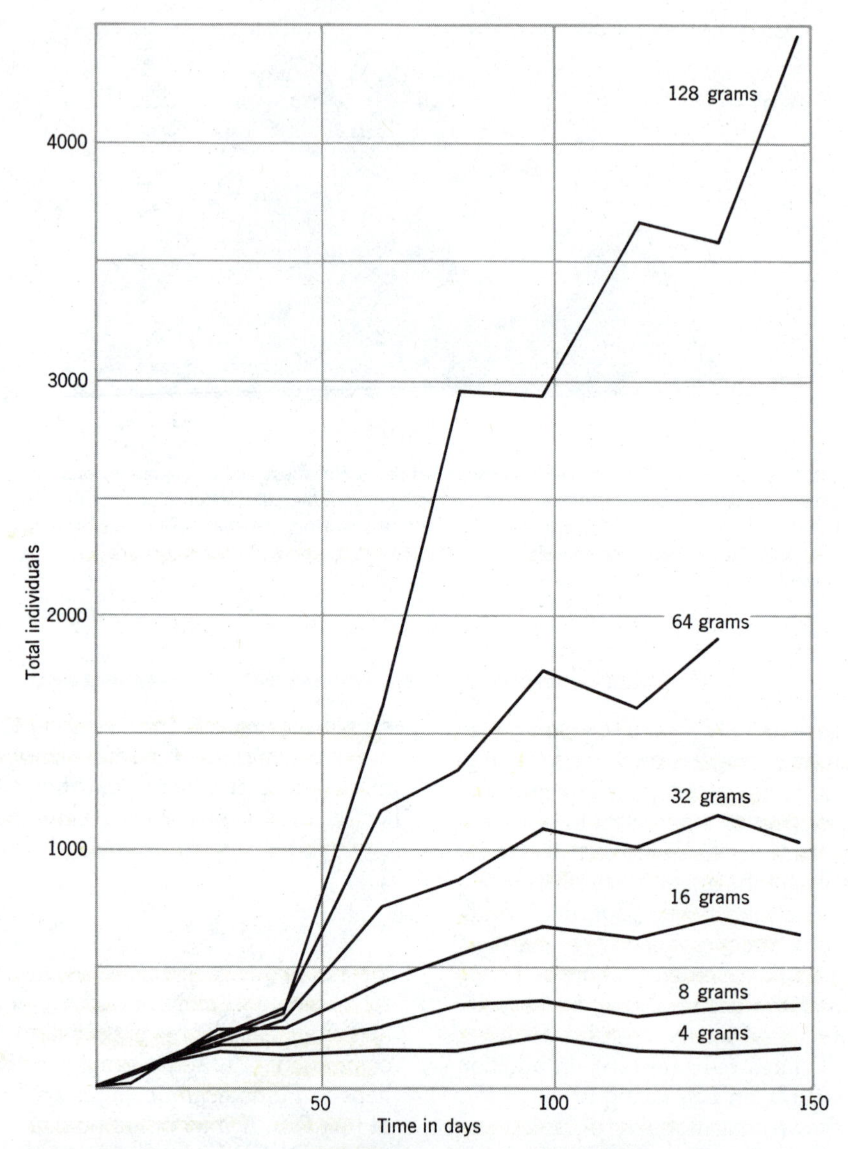

Figure 6.2 Chapman's results for the growth of flour beetle populations in different volumes of flour.

In these experiments the flour was always changed at each count, so that in none of the cultures were the animals ever short of food. Yet population growth always stopped at about the same density, 4.4 beetles per gram of flour. Chapman concluded that egg cannibalism was the main density-dependent factor leading to population equilibrium. Later work showed that reduced fecundity was more important than cannibalism and that many other factors were also involved.

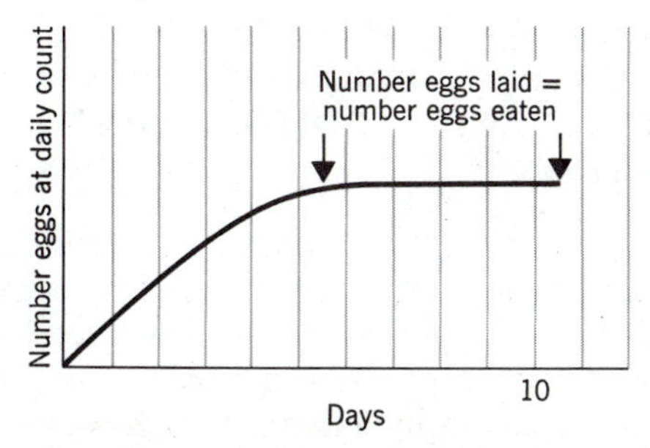

Figure 6.3 Egg cannibalism in *Tribolium*.
Population history of a system in which eggs are changed every five days to prevent hatching. Adult numbers are held constant. The only population events possible are egg laying and egg cannibalism.

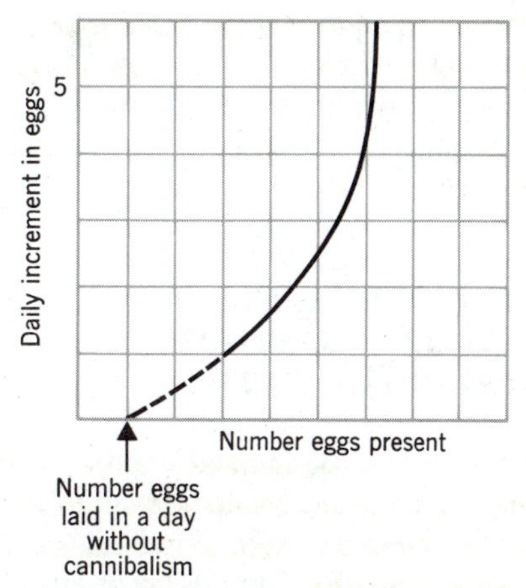

Figure 6.4 Total fecundity in the absence of cannibalism.
Using a series of systems with only beetles and eggs, the adults being at different densities, it is possible to measure the effect of crowding on egg laying.

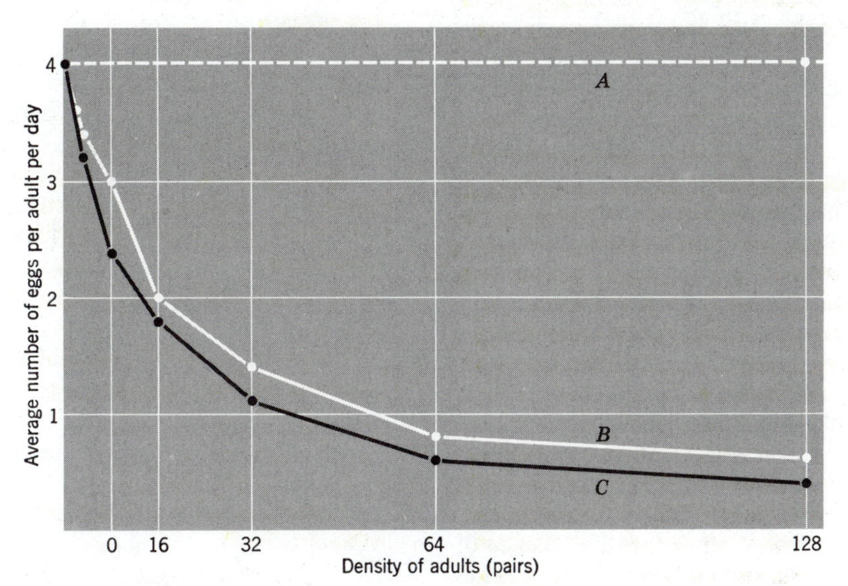

Figure 6.5 Daily egg production of *Tribolium confusum*.
A, average number of eggs per day that would appear at different densities of adults if there were no cannibalism and no reduction in fecundity. *B*, average real fecundity per adult per day. *C*, average apparent fecundity per adult per day. *A* minus *B* at each density will give an approximation of the reduction in real fecundity for each density. *B* minus *C* at each density will give an approximation of the egg reduction due to cannibalism for each density. (From Boyce, 1946.)

must be an integral function of the production and survival of each stage in the life history, so a realistic mathematical model of the growth of such a population must be very complicated.

REVIEW OF RESULTS OF EXPERIMENTAL STUDY OF SIGMOID GROWTH

A frustration of experimental work comes from the way it is vulnerable to the argument that results may be manipulated, artificial, or not relevant to wild nature. The design of any laboratory experiment in which animals are kept alive automatically ensures that the conditions of the logistic model are met, for the space assigned to the animals must be limited and they must be fed. If we make animals compete for food, it can be argued that we should not draw too much conclusion from the fact that they then do compete for food.

G. E. Hutchinson expresses this dilemma in an elegant way by telling his students that those who design a population experiment with laboratory animals are really making a simple analog computer. The behavior responses of the animals are like the pulse inputs of computers using an electric input analog. The animal experiment can thus be designed to yield any desired result, and the growth curves coming out of a population experiment are dependent on the "data" that are fed to the system.

Experiments with single-species populations, therefore, show that *niche theory, competition,* and *density dependence* are concepts that plausibly apply to real animals but the experiments do not, of themselves, show that the balance of populations in wild nature is set in comparable ways. The work paves the way for an understanding, therefore, but is not conclusive.

The real power of the logistic hypothesis lies in the way it can be used to predict the outcome of competition between different species. Use of the hypothesis in this way led to much of our understanding of why species are unique and how they are formed.

THE LOGISTIC HYPOTHESIS PREDICTS EXTINCTION WHEN INTERSPECIFIC COMPETITION IS STRONG

If populations of two species compete for a resource it must follow that the carrying capacity of container or habitat for each is reduced from what it would have been if one species was in sole occupancy. Intuitively, it might seem obvious that making different species populations compete would force the numbers of each down at equilibrium. But the logistic hypothesis allows us to develop a mathematical model to predict the relative sizes of two competing populations in a formal way divorced from intuition. This model turns out to predict extinction of one of the competing pairs, a result not at all in keeping with intuition. It is this prediction, and its test by experiment and observation, that lies behind the modern ecological view of the significance of distinct species.

The necessary manipulation of the logistic equation was performed by two mathematicians working independently in the early part of this century, Vittorio Volterra in Italy (Chapman, 1931) and Alfred Lotka (1925) at Johns Hopkins University. The resulting equations are known as the LOTKA–VOLTERRA EQUATIONS OF COMPETITION, and the term *Lotka–Volterra equations* is one of the stock phrases of ecology. The math is as follows:

For two competing species 1 and 2, the population growth curve of species 1 living alone is

$$\frac{dN_1}{dt} = r_1 N_1 \left(1 - \frac{N_1}{K_1}\right) \tag{6.5}$$

and the population growth curve of species 2 living alone is

$$\frac{dN_2}{dt} = r_2 N_2 \left(1 - \frac{N_2}{K_2}\right) \tag{6.6}$$

When two populations are grown together, the growth rate of neither population will be influenced by the presence of a few individuals of the other species at first because there is still lots of space and food. But when the animals are numerous, such that N_1 approaches K_1 or N_2

approaches K_2, the effects of competition should be significant. Competition must be dependent on numbers and the strength with which each individual is able to compete. The growth curve of a population of species 1 in the presence of species 2 is, therefore, given by

$$\frac{dN_1}{dt} = r_1N_1\left(1 - \frac{N_1}{K_1} - \frac{\alpha N_2}{K_1}\right) \tag{6.7}$$

where α is the coefficient of competition of population 2. If population 2 does not compete, contrary to our expectations, then α will be zero, αN_2 will also be zero, and the history of population 1 will be quite unaffected. But if there is competition, then αN_2 will be positive, will combine with N_1, and will reduce the term

$$\left(1 - \frac{N_1}{K_1} - \frac{\alpha N_2}{K_1}\right)$$

to zero before the saturation value of population 1 is reached. Assigning the coefficient β to represent the competition of the other species we may write a pair of equations as follows:

$$\frac{dN_1}{dt} = r_1N_1\left(1 - \frac{N_1}{K_1} - \frac{\alpha N_2}{K_1}\right)$$

$$\frac{dN_2}{dt} = r_2N_2\left(1 - \frac{N_2}{K_2} - \frac{\beta N_1}{K_2}\right)$$

The outcome of the competition between the species would be revealed when the system finally came to equilibrium. Then

$$\frac{dN_1}{dt} = \frac{dN_2}{dt} = 0$$

and

$$r_1N_1\left(1 - \frac{N_1}{K_1} - \frac{\alpha N_2}{K_1}\right) = r_2N_2\left(1 - \frac{N_2}{K_2} - \frac{\beta N_1}{K_2}\right) = 0$$

and

$$\begin{cases} \dfrac{dN_1}{dt} = r_1N_1\left(1 - \dfrac{N_1}{K_1} - \dfrac{\alpha N_2}{K_1}\right) \\ \dfrac{dN_2}{dt} = r_2N_2\left(1 - \dfrac{N_2}{K_2} - \dfrac{\beta N_1}{K_2}\right) \end{cases} \tag{6.8}$$

Although these equations cannot be solved simultaneously by analytical techniques, it can be shown that they predict that coexistence between strongly competing pairs of species, when under the constraints of the model, is impossible. This conclusion suggests why species are distinct in nature. Existing species populations are descendants of individuals that have survived competitions in which less distinctive individuals were the losers. Speciation is, in part, a process of avoiding competition by promoting difference.

Graphical Approach to a Solution

The competition described in equation (6.8) can occur in four ways:

Model 1 Species 1 may compete much more strongly than species 2 (β is large compared with α).

Model 2 Species 2 may compete much more strongly than species 1 (α is large compared with β).

Model 3 Both species compete strongly (α and β are both large).

Model 4 Both species compete weakly (α and β are both small).

Or, stated more formally:

Model 1 $\alpha > K_2/K_1$, $\beta < K_1/K_2$.
Model 2 $\alpha < K_2/K_1$, $\beta > K_1/K_2$.
Model 3 $\alpha > K_2/K_1$, $\beta > K_1/K_2$.
Model 4 $\alpha < K_2/K_1$, $\beta < K_1/K_2$.

The simplest model to consider is one that allows only one species to compete strongly. Let us first suppose that species 1 is the strong competitor, making β large and α small (Model 1). The inhibitory effect of any individual of species 1 on its own population is $1/K_1$ and on the population of species 2 is β/K_2. If an individual of species 1 treats an individual of species 2 as one of its own kind $\beta = K_2/K_1 = 1$ and it is clear that for species 1 to compete more strongly with species 2, $\beta > K_2/K_1$. If the reciprocal competition is weak, we have

$$\alpha < \frac{K_1}{K_2} \quad \text{and} \quad \beta > \frac{K_2}{K_1}$$

To understand the necessary outcome of such competition, it is convenient to express the possibilities in a diagram. This can be done by plotting all possible combinations of the two species as number of species 1 against number of species 2. If alone, species 1 can have no more than K_1 individuals. If it lives with K_2 individuals of species 2, there must be K_1/α individuals. The line K_1, K_1/α therefore represents all possible numbers of species 1 for which the derivative is zero. Conversely, the line K_2, K_2/β represents all possible numbers of species 2 for which the derivative is zero. When species 1 is the stronger competitor, as above, we have the graph shown in Figure 6.6. In any mixture of the two populations whose plot falls in the white hatched area, species 1 can still produce more individuals while species 2, already above its saturation value, must decline.

In the second of the possible models, species 2 is the stronger competitor such that $\alpha > K_1/K_2$ and $\beta < K_2/K_1$. The outcome of this competition is illustrated in Figure 6.7. Now the area between the lines of maximum abundance is shaded black to show that species 2 may increase in it whereas species 1 must decline. The competition ends at K_2 with only population 2 surviving, and there is an equilibrium population of K_2 individuals of species 2 having the habitat to themselves.

In these models of competition between the strong and the weak there can be no surprise that the strong is predicted to achieve an annihilating victory. That victory should be just as complete when the strong meets the strong is not, however, so intuitively obvious. When the strong meets the strong $\alpha > K_1/K_2$ and $\beta > K_2/K_1$ then

$$\frac{K_1}{\alpha} < K_2 \quad \text{and} \quad \frac{K_2}{\beta} < K_1$$

so that the lines describing possible populations of each cross as illustrated in Figure 6.8. In the black triangle, species 2 can expand but not species 1, and a plot of the relative populations will move to the left with time until only species 2 remains. But in the white triangle, species 1 is favored, and the competition proceeds until K_1 individuals of species 1 have the habitat to themselves. The outcome of the competition between mutually strong contenders thus depends on the initial concentrations of the two populations, but

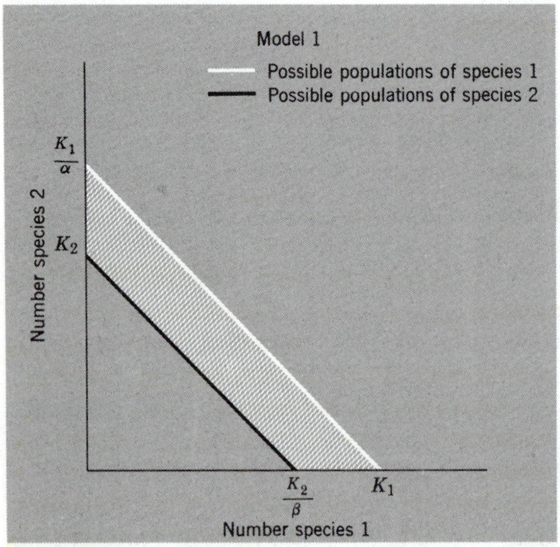

Figure 6.6

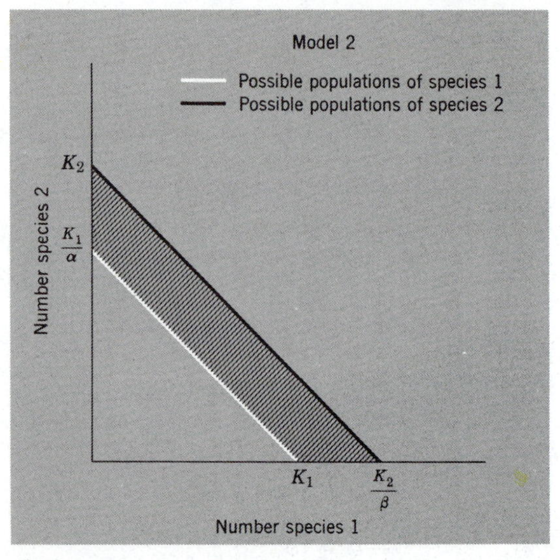

Figure 6.7

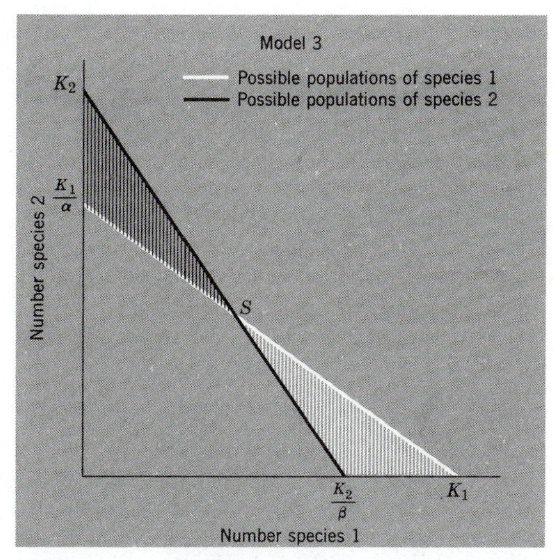

Figure 6.8

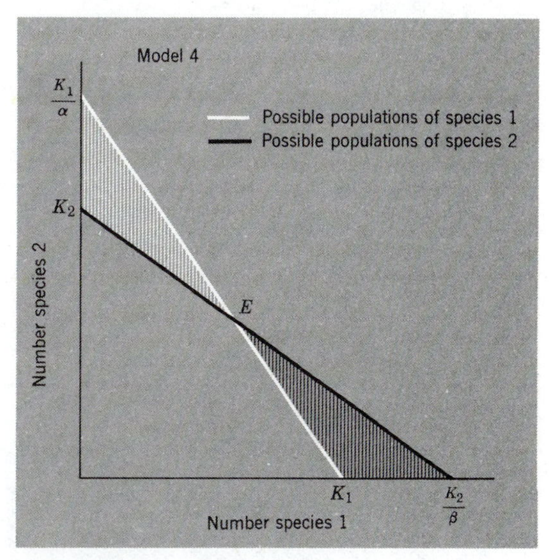

Figure 6.9

the final victory is always absolute. There is also a theoretical possibility of equilibrium at point *S*, where the two populations are beautifully balanced so that both persist. This equilibrium, however, has numbers so nicely balanced around a point of stability that any perturbation such as would be expected in a real environment would upset it.

There remains the fourth model, that of populations that compete only weakly. When the weak meets the weak $\alpha < K_1/K_2$ and $\beta < K_2/K_1$ so that $K_1/\alpha > K_2$ and $K_2/\beta > K_1$ when a plot of competing populations is as shown in Figure 6.9. The white triangle in which species 1 has the advantage is now at the left of the diagram, and the plot of a population mixture in this region must move to the right with time. But when point *E* is passed, the plot enters a black region where species 2 should be favored. The tendency would be for the relative proportions to move back to the left. Thus, there should be a stable equilibrium at point *E* so that both populations persist indefinitely. For weakly competing populations, then, the Lotka–Volterra equations predict that both should persist indefinitely, their populations fluctuating only gently about equilibrium levels.

EXPERIMENTAL TESTS OF THE LOGISTIC HYPOTHESIS

The logistic hypothesis predicts that:

(i) strongly competing species cannot coexist indefinitely and

(ii) where populations of different species do share a resource their competition is muted and weak.

These predictions can be tested by laboratory experiment. The classical work of the Russian biologist G. F. Gause provides the basic tests of the hypothesis. Gause (1934) explored the Lotka–Volterra competition equations and their implications with a series of experiments with populations of simple organisms, notably yeast plants and protozoans. It can be argued that his work was not designed primarily as a formal test of the logistic hypothesis, since Gause was engaged in a general experimental investigation into as many aspects as possible of struggles for existence in experimental populations. His work, however, together with that of other experimentalists, does provide formal tests of the hypothesis.

Gause (1934) grew various species of *Paramecium* on his oatmeal medium in centrifuge tubes as described earlier. Many species of *Paramecium* did thrive on this medium and attained population equilibria after a sigmoid growth history. The centrifuge tube habitats were structurally simple and provided only one kind of food so that animals as similar as different species of *Paramecium* should be expected to have to compete if introduced into the same tube. When an inoculum of both *Paramecium caudatum* and *P. aurelia* was introduced into the same tube population histories resulted as shown in Figure 6.10. *P. caudatum* is a relatively large and slow-growing species, whereas *P. aurelia* is smaller and reproduces more quickly. In combined culture, the population of the rapidly reproducing *P. aurelia* rose more quickly, and continued to grow even as the population of *P. caudatum* first levelled off and then declined to extinction. The prediction of the Lotka–Volterra competition equations is fulfilled by this experiment, the inevitable victory by the stronger suggesting Model 1 or 2.

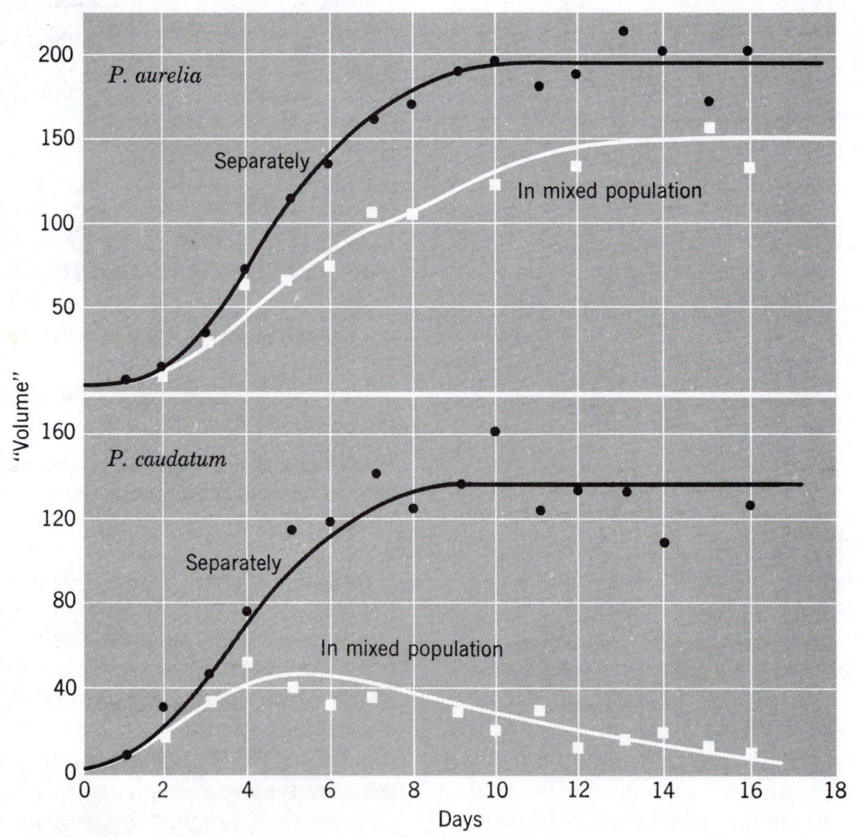

Figure 6.10 Competition between species of *Paramecium*.
Example of data used by Gause to confirm the Lotka–Volterra prediction that closely related species could not coexist when forced to share the same niche. Under the conditions of this experiment (daily changes of water and constant inputs of food) *Paramecium aurelia* always persisted, while *Paramecium caudatum* died out. Gause was able to alter conditions so that *P. caudatum* could inevitably win the competition instead. Notice that the size of each population is expressed in volumes, a device used by Gause to eliminate the effect of different sizes of the two species from his graphs. (After Gause, 1934.)

The niche requirements of *P. aurelia* appear to have been more closely met in centrifuge tubes of oatmeal medium changed daily than were those of the vanquished *P. caudatum*. There should, however, be different circumstances in which the outcome of the experiment could be reversed. Gause tried a simple change in experimental procedure. Instead of daily changing the medium in which the paramecia lived he left the old medium in place and added a daily dose of food concentrate. When this was done, the outcome of the competition was reversed. It is thought that the new conditions favored *P. caudatum* because the changed experimental conditions allowed chemicals secreted by the paramecia to accumulate in the water. Many protozoa do secrete substances that inhibit the growth of other protozoa (ECTOCRINE SUBSTANCES). These substances might serve a species of *Paramecium* as a competitive mechanism. It is at least a good working hypothesis that the ability of the slow-growing *P. caudatum* to outcompete the fast-growing *P. aurelia* in this second series of experiments was due to its relative immunity to the ectocrines that collected in the centrifuge tubes. The important aspect of this second series of experiments, however, is the continued demonstration that only one of a pair of strongly competing species continued to exist.

Gause made many other experimental tests of the Lotka–Volterra predictions, using yeasts as well as protozoa. His systems were always simple: small closed containers with food energy supplied in one simple form. The possibilities of different ways of life in these containers were deliberately few so that the animals should compete. There was little chance of the animals avoiding the competition by living different lives, as by occupying different niches. And then one of Gause's *Paramecium* experiments had an unexpected outcome; both populations persisted.

When *Paramecium aurelia* and *P. bursaria* were grown together, neither became extinct, but the population of each levelled off at about half the number it would attain when alone (Figure 6.11). Gause (1936) soon saw why this was so. When the two species were coexisting they were separated in space, *P. aurelia* being in pure culture at the top of the tubes and *P. bursaria* concentrated at the very bottom of the tubes, there being a narrow zone of overlap in the middle where both species could be found. Furthermore, Gause discovered adaptations of the two species that gave an understanding of the different life styles they were able to adopt in their unnatural habitat of centrifuge tubes, one at the top and one at the bottom. Food was concentrated at the bottom of the tubes as bacteria in the culture medium settled out, suggesting that the bottom of the tube would be a good place to live for a bacteria-eating protozoan. However, this high density of bacteria drew heavily on the reserve of dissolved oxygen in the water so that the bottom of the tube tended to become anoxic; food was available but lack of oxygen tended to deny this food to an aerobic animal like most paramecia. *P. aurelia* suffocated in the bottom water and was denied the food resource.

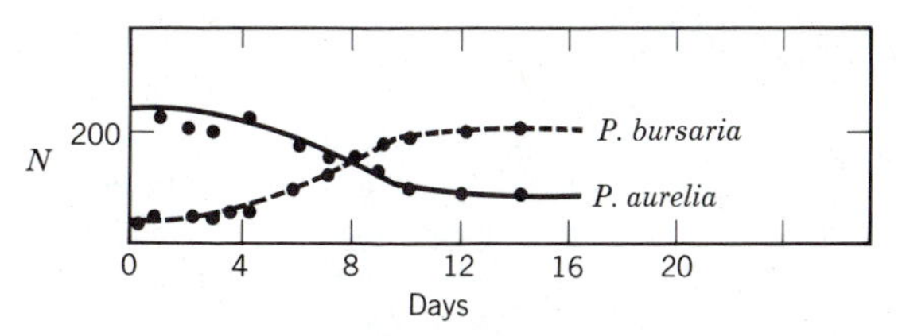

Figure 6.11 Competition between *Paramecium aurelia* and *P. bursaria*.
An inoculum of *P. bursaria* is introduced into a *P. aurelia* culture at the start of the experiment.

But *P. bursaria* had a private oxygen supply. It is a "green" *Paramecium* possessing symbiotic algae called *Zoochlorella*. In the light, these *Zoochlorella* performed photosynthesis, produced oxygen, and let their carriers, members of the *P. bursaria* population, forage in the anoxic bottom water (Figure 6.12). Without this adaptation, the *P. aurelia* population was forced to rely on the bacterial food in the upper water, where it apparently had a competitive advantage. The *P. aurelia* and *P. bursaria* system, therefore, was an example of Model 4 of the Lotka–Volterra formulation, that which allows coexistence if competition is substantially avoided.

Gause would now conclude that the Lotka–Volterra predictions were verified, so their

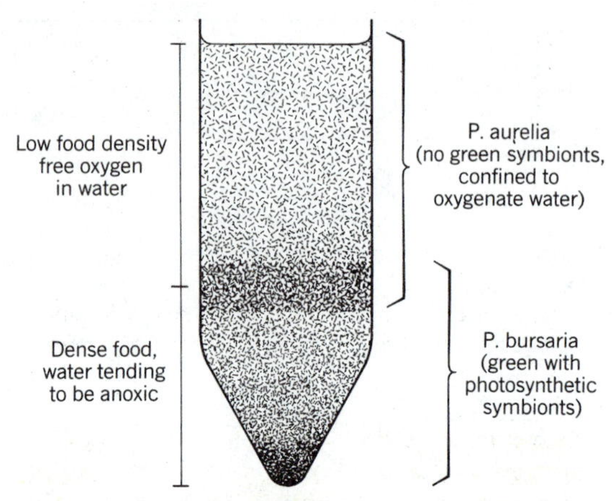

Figure 6.12 Coexistence of *Paramecium aurelia* and *P. bursaria*.
The green *P. bursaria* can survive in anoxic water where food is concentrated but *P. aurelia* cannot. The two species coexist with minor competition in the zone of overlap.

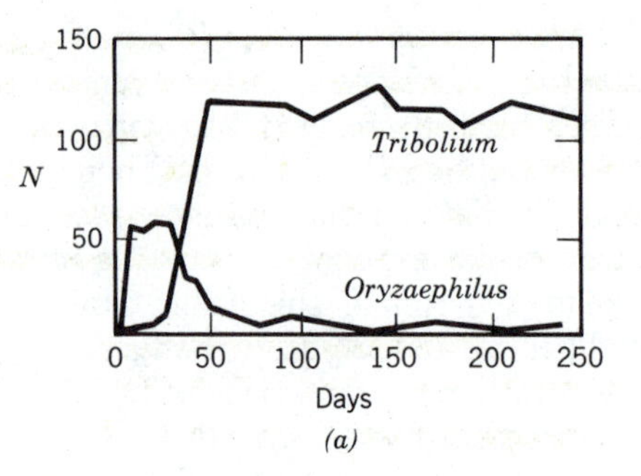

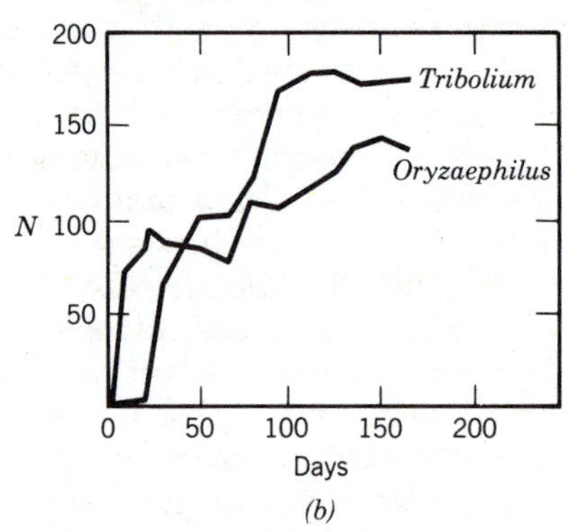

Figure 6.13 Competition between flour beetles of different genera.
When cultured in dishes containing nothing but flour, *Oryzaephilus* was eliminated always (*a*). When short lengths of capillary tubing gave *Oryzaephilus* safe sites for pupation, the populations coexisted (*b*). (From Crombie, 1946.)

predictions could be stated as a working principle. It is now known as the GAUSE PRINCIPLE, or, using the language of Hardin (1960), the PRINCIPLE OF COMPETITIVE EXCLUSION, and may be stated as:

> *"Stable populations of two or more species cannot continuously occupy the same niche."*

Or, more simply, as:

> *"One species: one niche."*

Others have followed Gause's example, choosing their favorite animals and matching species against species. Provided the systems were simple enough, the results have always been as expected; one population of strongly competing pairs has always died out. Figure 6.13 shows the results of competition between flour beetles in systems very like those used by Park described earlier for single-species populations, where cannibalism, particularly of eggs, was shown to be the most restrictive density-dependent pressure regulating population size. Crombie's (1946) work showed that flour beetles of different genera could be launched against each other in dishes of flour to the utter elimination of one kind by the other (Figure 6.13*a*). Crombie was able to show that *Tribolium* was the more voracious of the two, having a larger appetite for eggs and being particularly likely to eat the pupae of *Oryzaephilus*. *Oryzaephilus*, however, could not eat *Tribolium* pupae. This meant that *Tribolium* did more damage to *Oryzaephilus* as it blundered through that unnatural medium of flour than its population received from *Oryzaephilus* in return. *Tribolium* triumphed in the competition and eventually occupied the flour dishes alone, in pure culture.

This competition between the flour beetles shows the usefulness of the definition of competition supplied by Andrewartha and Birch given at the beginning of the chapter, which stressed that competition may sometimes involve physical harm. Food is not in short supply for the beetles, and neither is living space. One species eliminates another merely because one accidentally harms the other as it carries on its normal feeding process in unusual circumstances. There is no competition for a single prize, yet the way the animals interact means that one population leaves surviving offspring whereas the other does not. Natural selection recognizes this competition even though the meaning has been strained somewhat from classical English usage.

Understanding why *Tribolium* could eliminate *Oryzaephilus* from a dish of flour through an appetite that took in pupae and eggs as readily as its normal herbivorous diet let Crombie design systems in which predation by *Tribolium* would be less possible. He mixed short lengths of capillary tubing in the flour that could be used by *Oryzaephilus* as hiding places. *Oryzaephilus* then pupated in the tubes where the tunneling *Tribolium* did not come across them and, in these flour and tubing mixtures, both species coexisted (Figure 6.13*b*).

A feature common to all these experiments is that there are likely to be surprises both as to which potential competitor succeeds and to what curious circumstance may lead to the actual outcome. The Lotka–Volterra–Gause model of the struggle for existence, therefore, tells us only that direct competition is always fatal for one of the protagonists but gives us no insight into stratagems that might avoid such competition and thus be preserved by natural selection.

GENERAL IMPLICATIONS OF THE LOGISTIC HYPOTHESIS

The most far-reaching of the implications of this work is a view of species kept distinct by competition. Because individuals vary in every population, the chances of being eliminated by interspecific competition vary also. Individuals most different from an invading population will compete less, will have a greater chance of survival, and will gain more fitness. Interspecific competition, therefore, must work to eliminate individuals of deviant habit. The result, over time, is populations cast in the narrow mould of the parental populations. It is this pattern that we note when we describe life on earth as being made up of distinct species.

A second major implication of the work is that it gives strong support to the view that populations in nature may be under density-dependent control. Control by density dependence, particularly competition, is the assumption on which the logistic hypothesis is built. This hypothesis predicts that interspecific competition must be minimized at equilibrium. The prediction is confirmed by experimental test. Moreover the model of species as populations isolated to avoid competition offers a plausible explanation for production of distinct species by natural selection. Thus the underlying assumption that populations are usually under density-dependent control has strong attraction.

Some ecologists, persuaded by these arguments, have worked hard to discover if density-dependent control can be demonstrated in wild populations, but others have noticed that many if not most wild populations fluctuate so widely as to suggest a strong random element in their histories. The development of population ecology over the last two decades has seen prolonged debate between those whose overview of nature is that density dependence and competition are ubiquitous and those whose view is stochastic, seeing nature as holding much randomness with rare interludes of equilibrium. In this debate a prime argument of the density dependence school has been the apparent success of the logistic hypothesis.

PREVIEW TO CHAPTER 7

A species population consists of individuals that compete more with each other than with members of populations of other species. This view is derived directly from the idea of competitive exclusion, but it was foreshadowed by the use of the word "niche" and similar terms in the early development of ecology. Naturalists had talked about a species having its own niche or "mores" and so being able to coexist in communities, a custom that left ecologists ready to accept the principle of competitive exclusion. Formal statement of this principle leads to the prediction that wild populations of animals must share resources in ways that minimize competition. Such assemblages of species that share and partition common resources are called "guilds." When members of a guild are closely related it is inherently likely that individuals of different component species should compete so that the results of selection to avoid competing might be particularly evident where closely related animals live in sympatry. Many studies of sympatric pairs and guilds reveal mechanisms that result in lessened competition, particularly good data being from cormorants, cone shells, warblers of spruce trees, New Guinea pigeons, and two guilds of lizards. Selection that results in structures or habits that prevent competition is said to cause "character displacement." In this process individuals of sympatric species populations that tend to compete strongly with each other (strong interspecific competition) are at selective disadvantage. Individuals that compete almost entirely with their own kind (intraspecific competition) leave more surviving offspring, with the result that the sympatric species populations diverge. The idea of character displacement is powerful in explaining many facts about the distinctness of species but it has proven to be very difficult to demonstrate that the process actually works in nature. Divergence of character can be the consequence of selection working in several parallel ways at the same time as individuals are selected to concentrate on different kinds of food (food-tracking) or to escape predators even as selection also works to lessen competition. It is likely, however, that reduction of competition by character displacement is of widespread importance in establishing a species population as one collection of individuals genetically isolated from other species populations.

CHAPTER 7

THE ECOLOGY OF SPECIATION

The concept of COMPETITIVE EXCLUSION provides an alternative version of what is meant by Linnaean species. Traditional definitions stress that individuals of a species can mate with their own kind but not with individuals of other species. The alternative view stresses competition, noting that individuals that suffer too much interspecific competition have a reduced chance of leaving viable offspring. Individuals of a species population must be both alike each other and unlike individuals of other species populations with which they must live. Thus, A SPECIES IS A NUMBER OF RELATED POPULATIONS THE MEMBERS OF WHICH COMPETE MORE WITH THEIR OWN KIND THAN WITH MEMBERS OF OTHER SPECIES. This definition was inherent in ecological thinking from quite early on.

The California naturalist J. Grinnell, who was one of the first to use the word niche in ecology (Hutchinson, 1978), was talking of species in an ecological sense near the turn of the century when he said "Two species of approximately the same food habits are not likely to remain long evenly balanced in numbers in the same region. One will crowd out the other" (Grinnell, 1904). This is essentially a statement of Gause's results made thirty years in advance. It was also made within nine years of the publication of the first textbook of ecology (Warming, 1895; English edition 1909).

Shelford (1911) argued in a like manner that each species must have its exclusive set of habits and he coined his own term to do duty for the modern *niche*, a species' MORES. Shelford studied the fish in a series of ponds left behind old sand dunes that marked abandoned shorelines of Lake Michigan near Chicago. They were long narrow ponds, running parallel with the modern shore and separated from each other by ridges of sand (Figure 7.1). It was clear that they were arranged in order of age, with the youngest close to the lake and the oldest farther away, some thousands of years probably being represented. Until the builders of Chicago ran roads and railroads across them, there were water connections between the various ponds so that they could all be reached by fish and other organisms from the lake. Young ponds near the lake had little rooted vegetation and sandy bottoms; older ponds were progressively more choked with plant growth, and with accumulated debris on the bottoms. Collections from a series of the ponds revealed a succession of fish communities, ranging from those that liked the bare-bottom, clear-water, lakeside ponds that were still sometimes scoured by storm waves of the main lake sweeping over the dividing spit, to those living in the oldest, weed-choked, and stagnant ponds farthest from the lake (Table 7.1). Shelford first tried to explain the sequence of fishes in terms of *succession theory*, at that time becoming a dominant

Figure 7.1 Shelford's parallel ponds beside Lake Michigan.
The ponds were all connected to each other and the lake by narrow ditches through which fish could pass. They represent an age sequence: youngest near the lake, oldest farther away. Young bare ponds near the retreating lake held different fish communities from the old weed-choked ponds farther away. Shelford argued that each fish species positioned itself in a pond suited to its *mores*, a term that anticipated our modern use of niche. (From Shelford, 1911.)

Table 7.1
Fish Species in Ponds of Different Ages beside Lake Michigan (From Shelford, 1911)

Common Name	Scientific Name	Ponds Young ⟵⟶ Old							
Large-mouthed black bass	*Micropterus salmoides*	*							
Bluegill	*Lepomis pallidus*	*							
Blue-spotted sunfish	*Lepomis cyanellus*	*						*	
Pumpkin-seed	*Eupomotis gibbosus*	*							
Warmouth bass	*Chaenobryttus gulosus*	*							
Yellow perch	*Perca flavescens*	*	*						
Chub-sucker	*Erimyzon sucetta*	*	*	*					
Spotted bullhead	*Ameiurus nebulosus*	*	*	*					
Pickerel	*Esox vermiculatus*	*	*	*	*	*		*	
Mud minnow	*Umbra limi*	*	*	*	*	*		*	
Golden shiner	*Abramis crysoleucas*		*	*	*	*			
Yellow bullhead	*Ameiurus natalis*			*					
Black bullhead	*Ameiurus melas*			*	*	*	*	*?	
Dogfish	*Amia calva* (juvenile)								*

idea in ecology (Chapter 23). As the ponds aged they were physically changed by the activities of rooted plants, perhaps aided by the fish. But he made clear that fish, which could swim from pond to pond throughout the system, survived only where their individual habits, physiology, behavior, and mode of life were appropriate. The MORES of the fish must be right. It is a pity Shelford did not go on to say explicitly "one species: one mores," thus anticipating Gause directly. However, the word "niche" is easier to use; it is clearer and more compelling than "mores."

The work of Lotka, Volterra, and Gause, therefore, became available to working ecologists who already were well prepared by their own field studies to accept it. Gause's experiments made clearer the implications of niche and led to a formal statement of the exclusion principle. It was this formal statement that was so useful, based as it was on the classical scientific method of model-making and experiment. It at once suggested that formal tests of the principle with field data were necessary, and what was once the subject of musings by gifted naturalists alone became a working body of hypotheses to guide the field work of a young discipline. The first general test of the exclusion principle was already at hand—it lay in the experience of classical taxonomy.

Species are classified by shapes of their bodies, but these shapes reflect function so closely that you can usually deduce much about the niches animals occupy merely by looking at a corpse; talons mean a carnivore, hooves mean fleetness of foot, opposing thumbs mean climbing trees, and so on. The deduction of niche from shape is what palaeontologists do every time they reconstruct the life of an extinct animal. To a remarkable extent, therefore, museum taxonomy had already tested the exclusion principle; indeed, taxonomists may be said to have found it on their own, for there are groups of organisms whose shapes are so similar that they have been classified by function. Pathogenic bacteria may be classified by testing them against a host, and parasitic nematodes by the plant hosts in which they are found. This experience of classical taxonomy explains why the exclusion principle could gain ready acceptance as a working tool. A species could be thought of as a morphological expression of the animal's way of life, of its niche. Animal and plant species were unique; they reflected unique niches; one species, one niche. But more formal tests of the model came from studies of closely related species living together—studies designed as deliberate tests of the exclusion principle.

NICHE SEPARATION BY CLOSELY RELATED SYMPATRIC SPECIES

The implications of the competitive exclusion principle are particularly interesting for pairs of closely related species that live close together, species that are called SYMPATRIC (literally "same country"), because these species might be expected to have similar niches (because closely related) and, therefore, be in danger of competing. How might pairs of sympatric species avoid competition? In the 1930s, David Lack set himself the task of testing the exclusion principle against all the pairs of closely related sympatric bird species in the British list. He could not find enough data on feeding habits to be sure about many, but for all those for which there were good data he was able to show that they fed differently (Lack, 1944, 1945).

One of the nicest sets of data Lack could find was for the two species of British cormorant, the common cormorant *Phalacrocorax carbo* and the shag *Phalacrocorax aristotelis*. These two birds are strikingly similar (Figure 7.2). They live on the same stretches of shore, they both feed by swimming underwater after fish, and they both nest on sea cliffs. Lack used data on stomach contents collected by government agents investigating complaints against cormorants by fishermen to determine their diets.

Shags, by far the more abundant of the two species, ate mostly sand eels and sprats. The common cormorants ate various things, particularly shrimps, as well as a few small flatfish but no sand eels or sprats. The food of the two species was obviously quite different, so that they

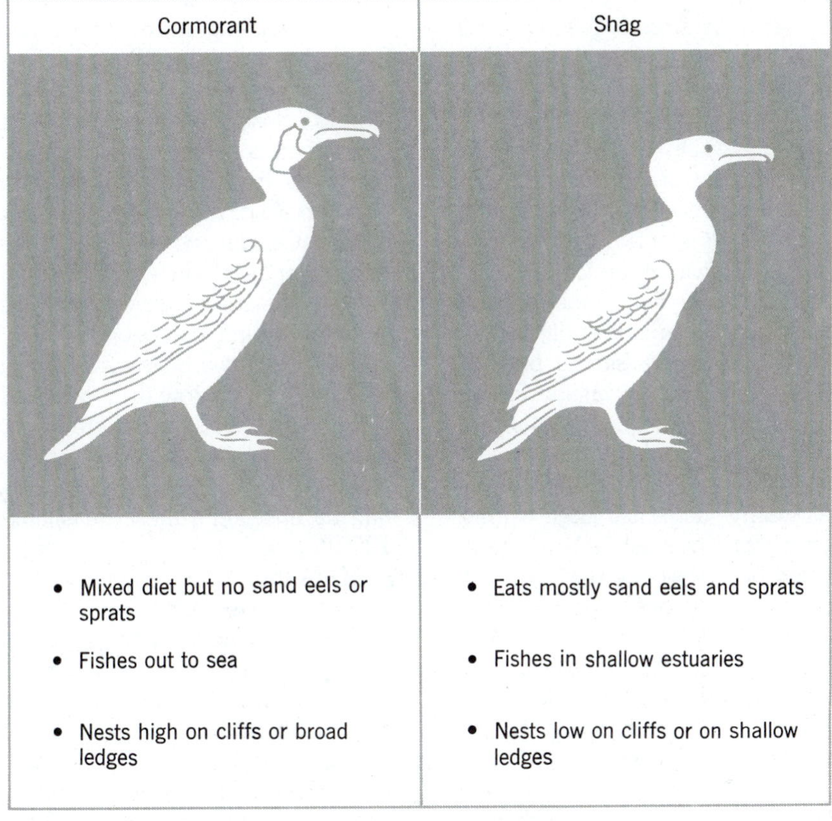

Figure 7.2 Closely related sympatric cormorants in Britain.
These close relatives, so similar to look at, get their livings in quite different ways and so do not compete. We infer that their ancestors were separated by character displacement and that the species were then preserved by competitive exclusion as genetic isolating mechanisms evolved.

avoided competition and the exclusion principle was upheld. The fisheries study had shown further how the catching of different fish was ensured, because the shags did their fishing in shallow estuaries while the common cormorants went farther out to sea. Lack was also able to show that the nesting requirements of the two birds were different even though they did nest on the same cliffs. The shags nested low among boulders or on narrow ledges, whereas the common cormorant nested on the high tops or on broad ledges. In short, these closely related birds, so similar to look at, had niches that were quite distinct. In their normal lives they were unlikely ever to come into serious competition.

This kind of study has now been repeated many times, for the discovery of a pair of related animals living together is sure to set an ecologist to finding out how. The exclusion principle is one of the few firm anchors in this diffuse science, and we can use it as a physicist uses the general principle of the conservation of mass, as a base from which to make an assault on complexity.

Three yellow weaverbirds of the genus *Ploceus* breed side by side in one colony stretching nearly 200 yards along the shore of Lake Mweru in Central Africa, and the man who found them promptly shot a few to see what they were eating (White, 1951). The stomachs of one species had hard black seeds in them, those of the second soft green seeds, whereas those of the third held nothing but insects.

Rapacious gastropods of the genus *Conus*, the pretty cone shells of the collector, live with many species on Hawaiian reefs; but careful mapping of their distribution shows that they have divided the littoral zone into six or more narrow strips so that their ranges scarcely overlap (Figure 7.3). They have divided up the feeding grounds in much the way that Gause's *Paramecium aurelia* and *P. bursaria* had divided up his culture tubes; there was a simple physical division of the available space. But the *Conus* species also specialized in food, as Kohn (1959) was able to show by keeping them in aquaria or by cutting open wild *Conus* that he saw feeding. Some ate small worms, some ate large worms, some ate different sorts of snail, and the deep water species used their poisoned darts to catch small fish. The whole series of *Conus*, so apparently alike, apart from differences in size and pretty markings, had found a set of specialized ways in which they could each carry on the trade of poisoned dart-wielding carnivorous snails on the same stretch of coast without coming into competition.

Herbivorous animals are provided with easy opportunities for food specialization by the great variety of plants making up vegetation, the effects of such food specialization being particularly obvious among herbivorous insects. Any amateur lepidopterist knows that you must get the right food plant if you are to rear caterpillars successfully implying that the closely related butterflies that flitter together in clouds over a meadow are avoiding competition by growing as larvae on different kinds of meadow plant. The plants may be said to force this specialization with specialized chemistry that makes them inedible to all but the specialist herbivore (Chapter 25).

Migratory warblers are mostly so similar, even in coloring, that learning to tell all except breeding males apart is one of the trials of a bird watcher. In the eastern United States there appears each spring a particularly frustrating assortment of them flying north from their tropical winters along common flightways to their common breeding grounds in the woods of New England and eastern Canada. Five species, in particular, nest in Maine and Vermont. The five birds are closely related. The vegetation in which they breed is spruce forest without obvious variety. The beaks of the birds are all the same size, and alike, suggesting they can eat the same food. Investigations of stomach contents have shown that their food is, indeed, roughly the same. The food of closely related sympatric cormorants and weaverbirds had been found to be different, but these little warblers even had tastes for food that were alike. How then can they occupy different niches? How can there be more than one species of them? One of the foremost ecological theorists, Robert MacArthur, earned his doctorate by answering these questions (MacArthur, 1957).

MacArthur spent many long hours in the springs of several years watching warblers. Each time he saw one he noted exactly where it was; on top of a tree, at the side of a tree, on the ground, flying about; and he started a stop watch so that he could measure in seconds just how long the warbler spent where it was. This was a tedious and time-consuming undertaking, for the little birds are hard to see in the dense spruce forest, and they never remain in sight for more than a few seconds. But eventually MacArthur had watched for so long that he could be certain of where each kind of warbler spent most of its time. It was clear that the warblers worked in substantially different parts of the trees (Figure 7.4). The birds did trespass on each other's space somewhat, but MacArthur was able to show that other behavioral traits stopped them from poaching many of each other's caterpillars even then. He timed the motions of each kind of warbler, noting how long each spent hovering, running along branches, or slowly plodding, and was able to show that each species had a characteristic pattern of doing things. One was more active than others; another was more deliberate. There seemed little doubt that these different activities reflected different hunting methods. One kind of warbler got caterpillars on tops of needles, another kind got caterpillars hidden under needles,

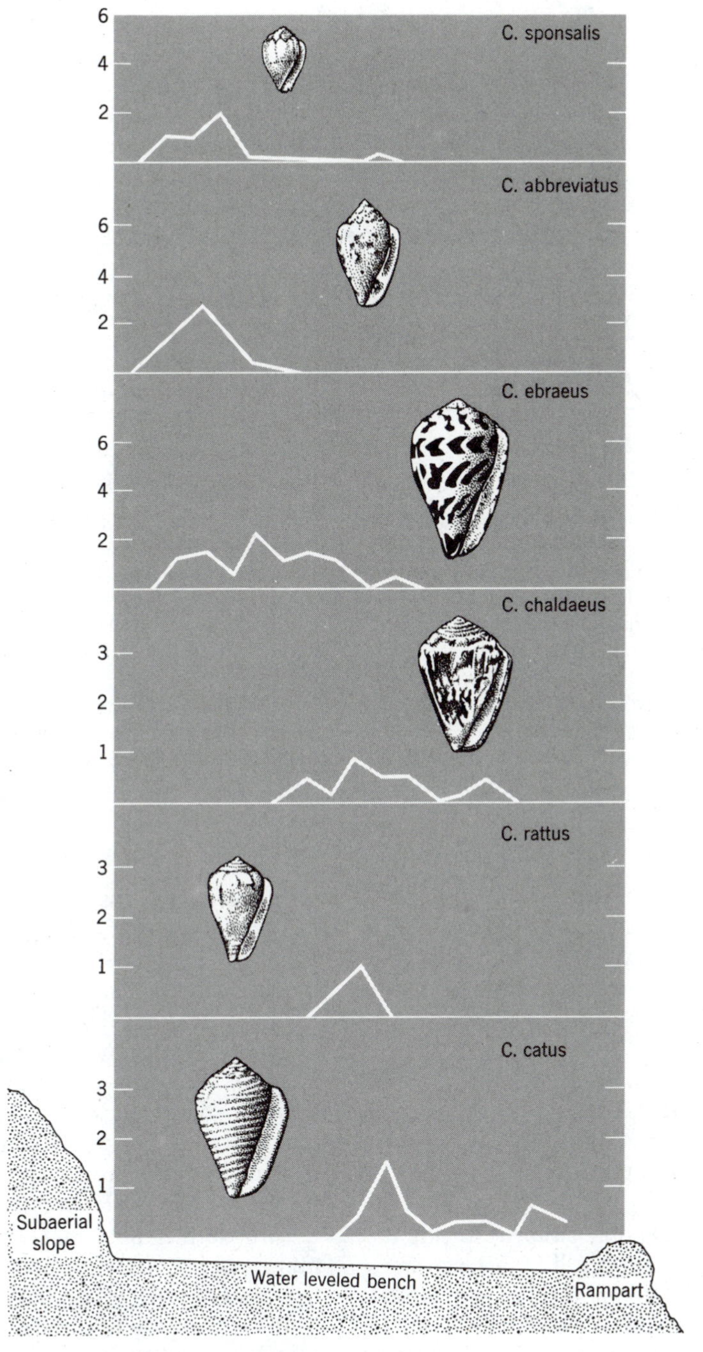

Figure 7.3 Closely related sympatric cone shells in Hawaii.
Animals of the genus *Conus* are carnivorous snails that hunt with poisoned darts. These sympatric Hawaiian species live in parallel strips along the shore where each catches a unique array of prey animals. (From Hutchinson, 1965.)

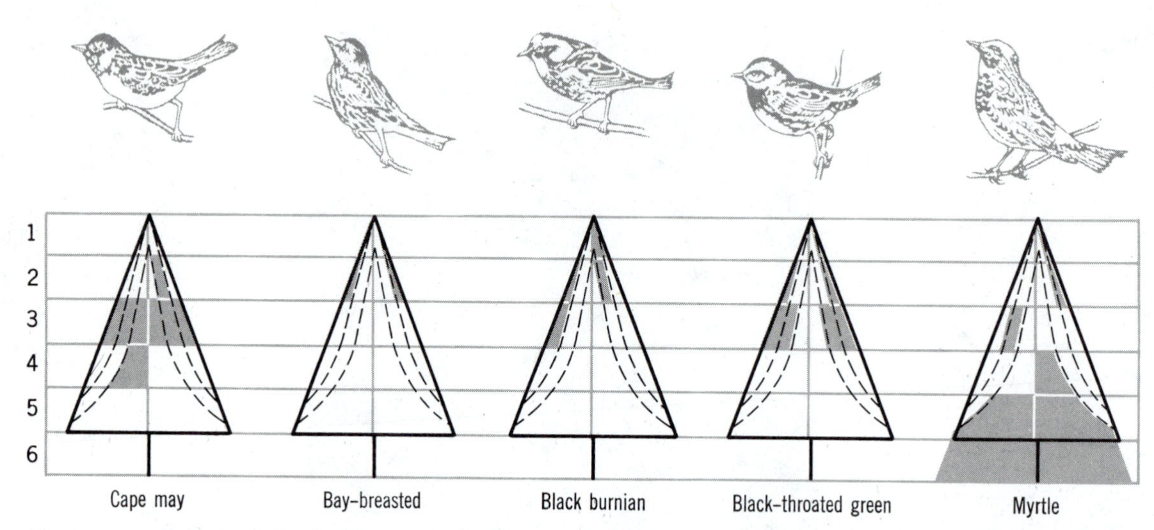

Figure 7.4 Separation of niches by warblers eating the same kinds of caterpillars.
The five kinds of warbler feed on bud worms on the same spruce trees. These diagrams illustrate MacArthur's study, which showed that the birds hunt in different parts of the trees, so that each kind has a private crop. MacArthur divided the trees into five layers, and then considered each branch as having a top, a middle, and a bottom, which let him divide each tree into fifteen compartments. A sixteenth compartment was provided by the ground underneath. MacArthur then noted in which compartment each warbler occurred, timing some of them to see how they divided their time between compartments. The stippled areas show where they spent more than 50% of their time or where they were on more than 50% of his observations. The results of timing the birds are on the left of each tree; the results of other observations are on the right. It is evident that the birds go some way to avoiding competition by hunting in different parts of the trees. Different methods of hunting further separate them. (From MacArthur, 1958, and Hutchinson, 1965.)

and so on. Even though a warbler might poach another warbler's space, and hunt the same kind of caterpillar there, the two might still minimize competition because the hunting method of each caught a different portion of the total crop.

THE CONCEPT OF GUILD: NUMERICAL ESTIMATES OF RESOURCE PARTITIONING

When many species living in sympatry divide a resource they are called a GUILD, a term first used by Root (1967). In any guild of sympatric species like the cone shells or New England warblers there must be, if the concept of competitive exclusion is valid, a set of muted competitions between guild members. Each individual species avoids head-on conflict with other species in the guild, yet there must be some overlap between niches causing some competition for resource, however muted. The Lotka–Volterra–Gause model of competition, therefore, offers a chance to measure and describe aspects of GUILD COMPETITION and NICHE-OVERLAP, and does so, moreover, in ways that generate predictions about species relationships testable by further observation. MacArthur and Levins (1968) developed this approach using MacArthur's data for the New England warblers (Figure 7.4).

The warbler data make it possible to make numerical estimates for coefficients of competition between species on the assumption that the warblers compete whenever they stray into each other's part of the spruce trees. In this guild, the assumption is reasonable because all the warblers take the same caterpillar food: time trespassing = time competing = a numerical estimate of a coefficient of competition. Thus the relative frequency of trespass and the relative frequencies P_{ih} and P_{jh} of the *i*th and *j*th species in the *h*th microhabitat measures competition relative to P_{ih}^2 and

$$\alpha_{ij} = \sum P_{ih} \cdot P_{jh} \Big/ \sum P_{ih}^2 \qquad (7.1)$$

where α is the coefficient of competition. For each of four of the five species of warbler in Figure 7.4 MacArthur knew the mean number of pairs (N_i) in the 5-acre woodlot he studied and he could calculate the *competition coefficients* from his stop watch data and equation (7.1). At equilibrium the Lotka–Volterra statement of relative populations is

$$\frac{dN_i}{dt} = N_{i}r_i (K_i - N_i - \alpha_{ij} N_j) = 0 \qquad (7.2)$$

Evaluating K_i from the set of values for N_i and α_{ij} now lets equation (7.2) be solved for each of the four species and the results can be expressed in matrix notation as

$$K = A\,N \qquad (7.3)$$

where K and N are the columns giving the saturation values K_i and the actual populations N_i and

$$A = \begin{matrix} \alpha_{11} & \alpha_{12} & \alpha_{13} & \alpha_{14} \\ \alpha_{21} & \alpha_{22} & \alpha_{23} & \alpha_{24} \\ \alpha_{31} & \alpha_{32} & \alpha_{33} & \alpha_{34} \\ \alpha_{41} & \alpha_{42} & \alpha_{43} & \alpha_{44} \end{matrix}$$

Results for the four warblers; myrtle, black-throated green, blackburnian, and bay-breasted; were

$$A = \begin{matrix} 1 & 0.490 & 0.480 & 0.420 \\ 0.519 & 1 & 0.959 & 0.695 \\ 0.344 & 0.654 & 1 & 0.363 \\ 0.545 & 0.854 & 0.654 & 1 \end{matrix}$$

The resulting matrix (A) is best called a GUILD MATRIX, a term that is not only consistent with the ecological usage of *guild* but that avoids the frequently used *community matrix*, which would give a meaning to *community* quite unlike that generally meant in ecology (Hutchinson, 1978).

From this analysis it is possible to make statements such as:

"Blackburnians have a smaller competitive effect on bay-breasteds than on myrtles."

"Black-throated greens are strong competitors of bay-breasteds."

"The presence of blackburnians actually increases the chances of bay-breasteds being able to live in the habitat."

and so on (Hutchinson, 1978). Experiments or observations can then be sought that will test the accuracy of these statements.

Investigating guilds of coexisting resource-sharing species with numerical measurements following MacArthur's example has been an active field of ecology (Schoener, 1974).

A guild of lizards living together in the Australian desert has been investigated by Pianka (1969). The lizards are of distinctly different sizes, though otherwise similar and classified in the same genus (Figure 7.5). Pianka found that the lizards took insects of different sizes so that most of the food supply of each was a private preserve, little poached by the other species (Figure 7.6). Notice that there was niche-overlap since all three could eat some of each other's food, demonstrating some competition, though muted. If only one of the three species had arrived at a suitable habitat by migration we could expect it to exploit

Figure 7.5 *Ctenotus leonhardii*, nearly life size. A guild of three species of *Ctenotus* coexists in the Australian desert.

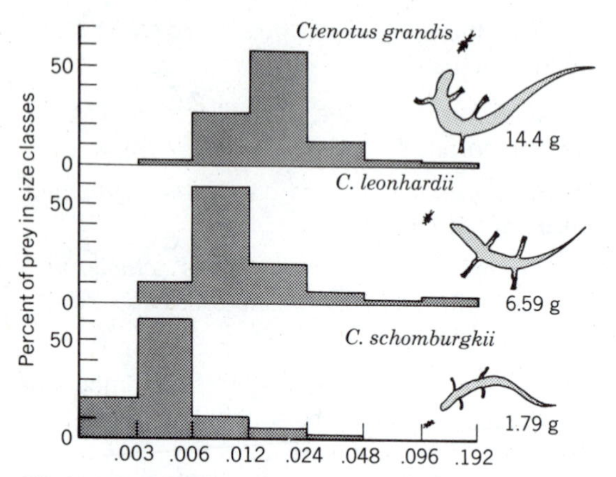

Figure 7.6 Simple resource partitioning in a guild of lizards.
The three *Ctenotus* lizards are sympatric in Australia. The three differ sharply in size and are dependent on characteristic sizes of prey. Prey size is given in a doubling scale; weight of lizards in grams is given at the right. (Pianka, 1969.)

food in all size ranges, perhaps resulting in a larger population. If the other two species then arrived, they would take up their specialist shares. We say there would be RESOURCE PARTITIONING, and the *realized niche* of the first species would become smaller.

A more subtle separation of lizard niches has been found in Jamaica (Schoener and Schoener, 1970; Schoener, 1974). *Anolis* lizards of the lowlands live in bushes, where several species hunt insects through the foliage together, rather as MacArthur's warblers hunted caterpillars on the same spruce trees. Like the Australian guild of *Ctenotus* lizards, the *Anolis* differ in size so that different species tend to take different food. This niche separation by size seems to be taken further in *Anolis*, separating even the sexes into big and little and thus giving each sex a private food supply. But the bushes cause a further complication.

Different *Anolis* species and sexes are found in particular portions of each bush, again like the warblers on spruce trees. This means that some species are on thin twigs whereas others are on thick twigs. If simple mechanics decided where lizards should be, the small lizards would be on thin twigs and the large lizards on the thick ones, but the spread of lizards through a bush does not correlate little lizards with little twigs and we see the peculiarity of the largest lizards perched precariously on the thinnest twigs (Figure 7.7). The best explanation for this is that if all small lizards congregated on small twigs they would compete for food of the same size. When large lizards live on small twigs they avoid competing with other large lizards that live on the large twigs. Natural selection has equipped some large lizards with special adaptations for living on small twigs in the form of a long, narrow body and short-femured limbs.

A spectacular guild of closely related sympatric pigeons has been found by Diamond (1973) in the rain forests of New Guinea (Figure 7.8). The largest pigeons weigh nearly 20 times the smallest. All fourteen species in the guild eat fruit, high in the branches of the trees, but the fruit comes in roughly four sizes only. Each size of fruit is divided up by a subguild of pigeons according to position on the tree, with the small pigeons foraging on the outer and thinner twigs. These herbivorous pigeons in New Guinea divide a tree by twig size, unlike the carnivorous lizards in Jamaica studied by Schoener. The pigeons can do this because there is only a single species of each size class and no complication of different sizes by sex or youth. In one sense, the lizard guilds are more complicated because their ectothermy lets them grow through different size classes (Chapter 5).

The above case histories can be multiplied many times from the experience of field ecologists, though they are not always documented with such thoroughness. Sympatric species can always be shown to partition essential resources in ways that minimize competition. There appear to be no exceptions. These observations lead to identifying a simple mechanism for the evolution of species by natural selection, the concept of *character displacement*.

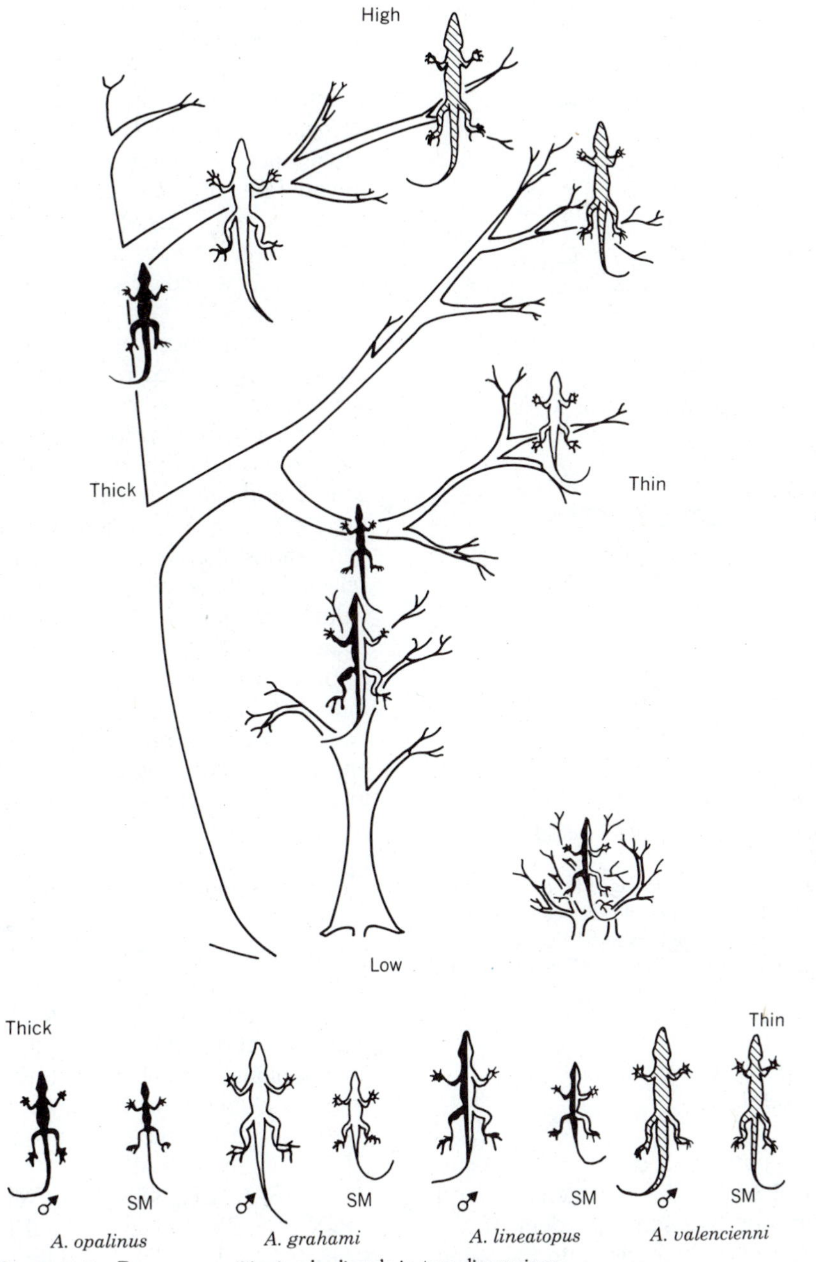

Figure 7.7 Resource partitioning by lizards in two dimensions.
Four species of *Anolis* are sympatric in bushes. Size differences, of sexes as well as of species, allow specialization on different sizes of food. Physical separation then keeps individuals of similar size apart, even if this means making a large lizard live on thin twigs. (Schoener and Schoener, 1970.)

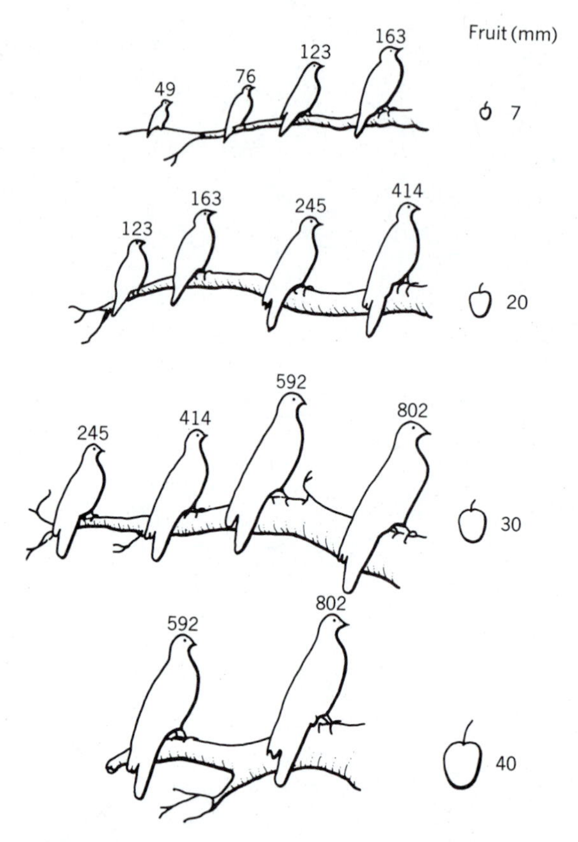

Figure 7.8 Resource partitioning in a guild of tropical pigeons.
Fourteen rain forest pigeons are arranged in four subguilds by food preference and each subguild divides its resource by perch size and position in the tree. Pigeon weights are in grams, and fruit diameters are in millimeters. (Diamond, 1973.)

CHARACTER DISPLACEMENT IN SPECIATION

Consider two hypothetical, closely related species A and B of similar but not identical food habits, and suppose that some accident of history has introduced both species into the same habitat so that the two must try to partition their resources as in Figure 7.9. Suppose further, which would certainly be true of wild populations, that the food preferences of different individuals of each population varied considerably, some eating larger food than the rest and some eating smaller. The result would be that some individuals of species A would tend to compete with some individuals of species B, but that other individuals would escape interspecific competition altogether. In Figure 7.9*a* only individuals of each species that concentrated on the resources in zone 2 would be subject to competition. Concentrating on the resources of zone 2, therefore, means that time and energy have to be spent in competition that would be saved for individuals feeding in zone 1 or zone 3. This fact must have a direct consequence for reproductive success because energy spent on competition is energy denied the business of making young. Individuals of A feeding in resource zone 1 should leave more surviving offspring than those in resource zone 2, and individuals of species B should leave most offspring by feeding in zone 3. In a few generations, therefore, we should expect natural selection to have produced populations of individuals programmed to concentrate on resources outside the zone of competition and the resource use of the local populations might look more like Figure 7.9*b*.

The process of natural selection must have as one of its effects this tendency for competition to be lessened as the generations succeed each other. Competition is inherently likely to reduce the reproductive effort, and there must always be selective advantage in being equipped to avoid competition. Thus we are able to predict, by a natural selection argument, that the niches of similar species should be different, the same result predicted by the Lotka–Volterra–Gause equations and that is so strongly upheld by field investigations. Natural selection will work to make *characters diverge*, as Darwin put it in the *Origin of Species*, or, as ecologists now say, by CHARACTER DISPLACEMENT.

CHARACTER DISPLACEMENT IS THE PROCESS BY WHICH A MORPHOLOGICAL CHARACTER STATE OF A SPECIES CHANGES UNDER NATURAL SELECTION ARISING FROM THE PRESENCE, IN THE SAME ENVIRONMENT, OF ONE OR MORE SPECIES SIMILAR TO IT ECOLOGICALLY OR REPRODUCTIVELY (Grant, 1975). This statement defines a process that must, in theory, hap-

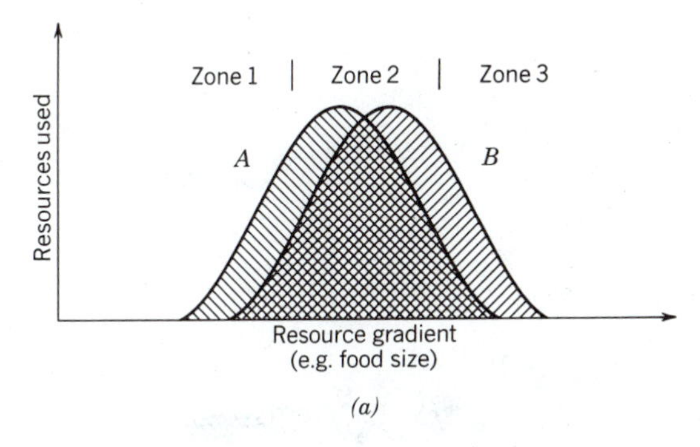

(a)

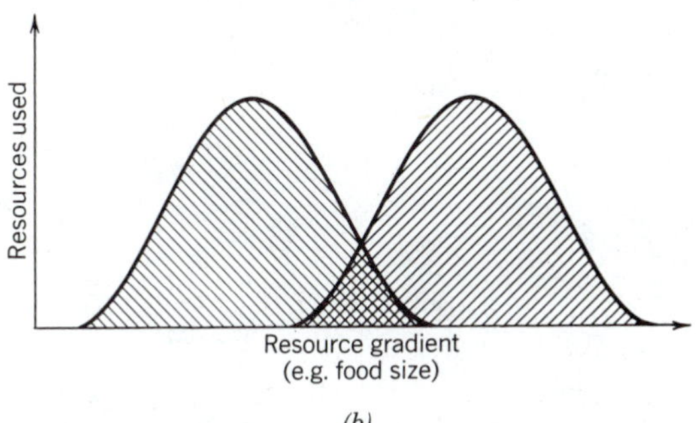

(b)

Figure 7.9 The concept of character displacement. Individuals of species A and B whose niche preferences bring them into competition will leave fewer surviving offspring than individuals who avoid competition. In time niche-overlap will be reduced and the spread of resource use in sympatric populations will change from a pattern like that in (*a*) to a pattern like that in (*b*).

pen. *Character displacement* provides both the mechanism and the theory to explain how real populations of real species are always adapted in ways that mute competition. Yet the process may not come into play very frequently since it is only necessary in the rare circumstance that diverging and genetically different populations come together by chance. The story of the rock nuthatches of Iran is instructive.

Nuthatches (family Sittidae) are small birds with long woodpecker-like bills that search for food by climbing up and down trees or rocks hunting in cracks and crannies. The physical adaptations for this way of life (long bills, short stubby tails, and big feet) give nuthatches a characteristic appearance, making them all look much alike. Strikingly similar are the common species of Greece and Turkey, *Sitta neumayer*, and of central Asia, *Sitta tephronota* (Figure 7.10). The nuthatches of these sites 1000 miles apart, so closely related and so similar, should be occupying closely similar niches. What differences exist between them may be explained in terms of the geographical isolation of their breeding populations. Where their ranges come into contact, as they do south of the Caspian Sea in Iran, we should expect the species to come into competition. What we in fact find is that throughout Iran both species coexist, but that they are here so strikingly different that there is never any difficulty of telling them apart (Vaurie, 1951). Their bills are of different sizes and one bird has a thick black stripe from eye to shoulder, whereas the other has almost lost its eye stripe. The different bill lengths suggest different feeding habits, and the different markings suggest distinctive patterns that could be used in recognition of suitable mates.

An explanation of the nuthatch distributions and their differences in sympatry that is very satisfying to ecological thought is that differences between the two species have been emphasized by natural selection in the zone of overlap. Competition was avoided by individuals that were most different and these left more surviving offspring. At one time there was a common ancestor to both species, but then chance isolated populations in Greece and in Central Asia and the characters of the two populations, so far away from each other, slowly diverged. Then another accident, perhaps the ending of the last ice age, let the two populations spread and merge in Iran. Selection then favored avoidance of competition and brought out the differences in the populations that we see. This is a good explanation, one that has *generality* because it shows us how all guilds of sympatric but scarcely competing species came about. The concept was accepted very readily by ecologists when it was first put forward by Brown and Wilson (1956) in a seminal paper in which the term *character displacement* entered the ecological literature.

Brown and Wilson began their analysis by discussing Vaurie's nuthatches, and then went

Figure 7.10 Character displacement in nuthatches.
Where the two Asiatic nuthatches are sympatric, in Iran, they may be easily distinguished by bill size and eye stripes, but individuals from far parts of their range (allopatry) are very similar. Evidently selection has favored individuals who are most different in the region of overlap.

on to give other examples of similar patterns of species of similar form in *allopatry* (literally living in other countries) but with marked differences in sympatry, and they cited data from various birds, fishes, frogs, beetles, ants, and crabs. Gause had shown us that long-continued competition was impossible and, when it was forced upon sympatric populations, it led to *competitive exclusion* of one of them. But Brown and Wilson showed us that in the real world, where populations were large and various, and where habitats and resources came in many shades of difference, *character displacement* was more likely than competitive exclusion. Coexistence was made possible because *fit* individuals (those that survived to breed) were the different ones, the ones that avoided competition with its perils of exclusion.

Yet it turns out that the actual history of the nuthatch populations must have been more complicated than the simple explanation of character displacement allows. Grant (1975) made an exhaustive reinvestigation of all the nuthatch data preserved in the world's museum collections, examining a total of 514 specimens from localities all over the range of both species (Figure 7.11). He also went into the field in the region of overlap to look at the living birds to seek direct evidence of their food preferences and behavior and to see if they really lived in the same habitats in Iran. An essential part of Grant's findings was that there are no dramatic changes in either population in the Iranian zone of sympatry. Except in the matter of the eye stripes, the Iranian populations could be fitted to a gradient of types that continued into the surrounding regions where a species lived alone, the regions of *allopatry*.

Figure 7.12 shows how the known facts about the nuthatch sizes, bill lengths, and other niche characteristics could result from the overlapping of diverging clines, even in the absence of selection for differences in the zone of sympatry. Grant showed that gradients of size ran east to west for both species, and that they would probably each be their present size in Iran even if the other species was absent. Furthermore bill length is a function of body size (a result of *allometric* growth) so that big or little bills may have nothing to do with food preference *per se* but may result from difference in size alone. Grant's data do not suggest that there has been no selection for size or bill length in Iran, merely that the data *do not require character displacement* for them to be explained. Whatever caused the clines in size to be produced by natural selection probably explains the differences in Iran, not necessarily the presence there of the other species.

Grant does show, however, that eye stripes are not clinal and that their distribution is like that shown in Figure 7.12*b*. Moreover the individuals that have small eye stripes in Iran probably pay a cost for the loss of pigment around the eye. Pigment gives a very decided advantage in regions of strong sunlight as a device for reducing glare and improving clearness of vision. Grant showed that eye stripes were important signals in helping the nuthatches choose mates of the right species. He tested wild nuthatches in the breeding season with models and stuffed skins to see what the response would be. Sexual displays, aggression, or indifference showed clearly whether the model was recognized as being of the same species or not. Grant found that the length of the eye stripe was all-important and concluded that reproductive selection for large or small eye stripe caused the difference in the zone of sympatry.

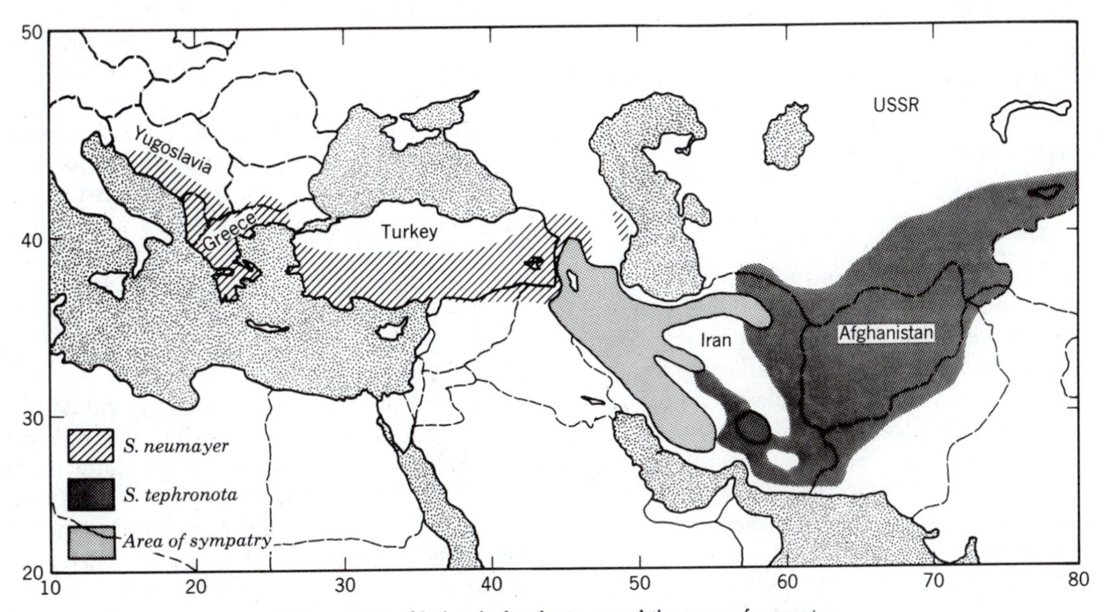

Figure 7.11 Nuthatch distributions and the zone of sympatry.

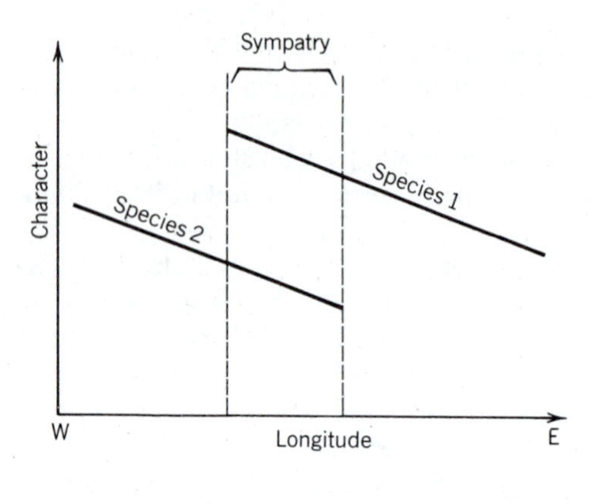

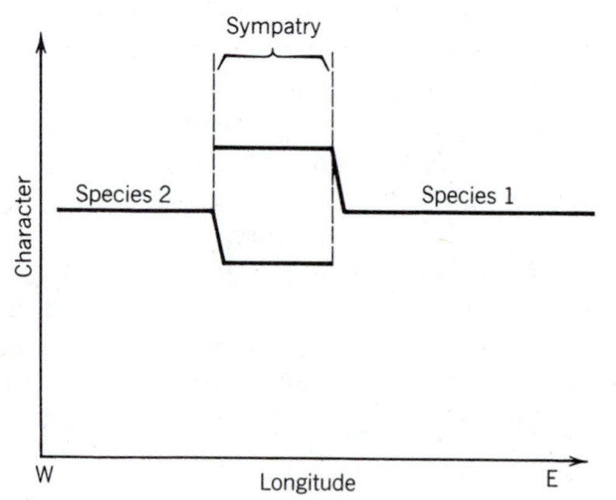

Figure 7.12 Alternative explanations of differences in sympatry.
If two species have characters like size graded in space (a *cline*) then they may automatically have different characteristics in the zone of sympatry (*a*). If differences at sympatry are solely the result of character displacement, characters should be distributed in space as in (*b*).

Thus Grant's conclusions amount to this: the obvious differences in size and bill that seem to let the nuthatches avoid competing *could* be just the result of history, a result of adapting to geographic spread starting from different places, but selection has displaced the characters needed for species recognition. Selection, therefore, has certainly worked to emphasize living in the separate niches and has helped to isolate the species where the ranges overlap.

Another much-quoted observation that fits some but not all of the requirements of character displacement concerns two species of Darwin's finch. The Galapagos finches of the genus *Geospiza* occupy many niches not occupied by finches in other parts of the world, no doubt because other passeriform families are absent in the Galapagos. The finches eat a great variety of seeds, insects, and foliage. Their beaks are of many different sizes. Lack (1947) found that where the two species *G. fulginosa* and *G. fortis* lived together, as they do on two of the larger Galapagos Islands, their beak sizes are quite separate with no overlap. But on the tiny island of Daphne there was only *G. fortis*, and these birds had beaks covering the range of sizes of both species on the larger islands. On the equally tiny Crossman Island there was only *G. fulginosa*, and these birds also have beaks of the wider size range (Figure 7.13).

A perfectly plausible explanation for these finch data is that beak size on the larger island where both species live has been separated by character displacement. The finches on the small islands, having no competitors, have been able to feed on a wider range of food sizes free from competition; the resource did not need to be partitioned and individuals with median-sized beaks fed best and left most surviving offspring. Neat though this story is, however, it cannot be completely true, because the food resources on the smaller islands are not the same as those on the large island.

Beaks have to be suited to the size of seed that is available to be eaten. The beak data themselves suggest very strongly that Daphne and Crossman Islands have seeds of only median size suited to beaks of median size (Figure 7.13). Furthermore, selection must be working on the plants that provide the seeds as well as on the finches that eat them. Selection for seed size in the depauperate vegetation of a tiny island like Daphne, where seeds are hunted by only one kind of finch, must be quite different from selection for seed size in a very large island like Santa Cruz (Figure 7.14). The beak sizes of Figure 7.12 have resulted from these many different selection pressures and not just by discrimination against competitors on the large islands.

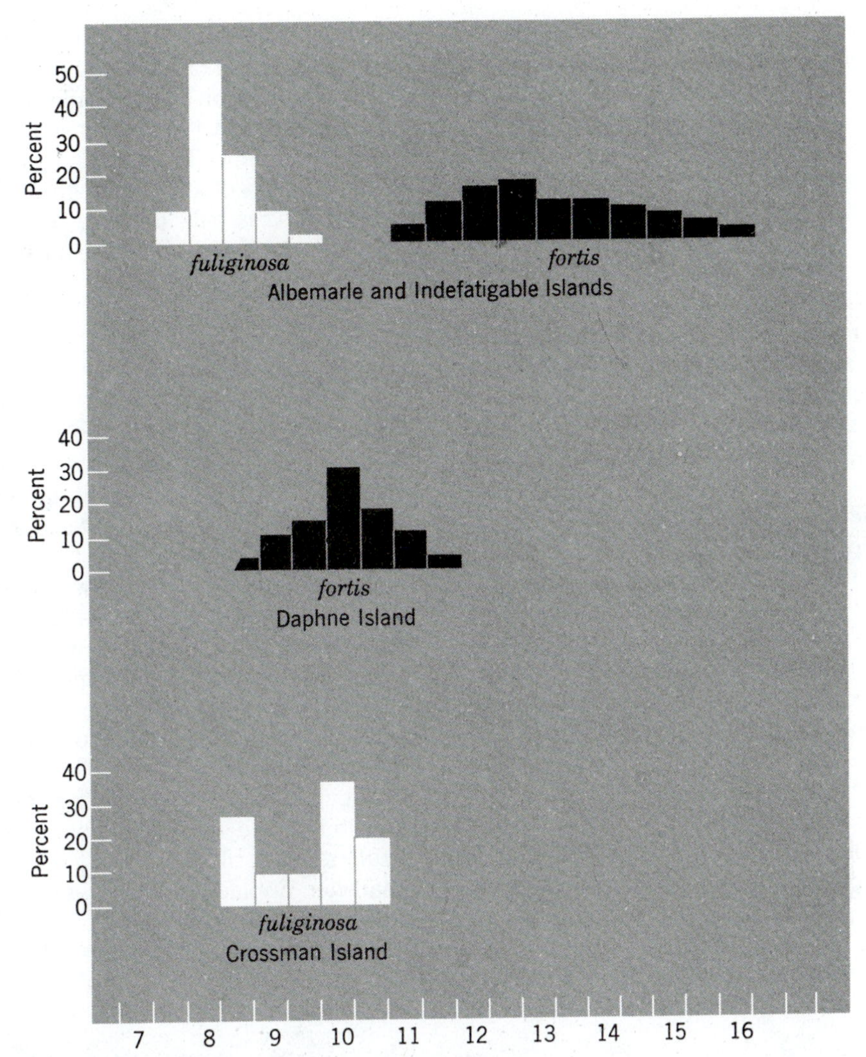

Figure 7.13 Character displacement in Darwin's finches.
Histograms of beak depth in *Geospiza* species (Darwin's finches). Measurements in millimeters are placed horizontally and the percentage of specimens of each size are shown vertically. If the beak depths on Daphne and Crossman Islands are indicative of optima in the absence of close competitors, the displacement on islands with competitors is most easily interpreted as character displacement. An alternative explanation is that the sizes of available seeds vary on large and small islands, and that natural selection tracks seed size with bill size. (Lack, 1947.)

A HISTORICAL RECORD OF SPECIES INTERACTIONS

Water ferns of the genus *Azolla* on the Galapagos Islands reveal a different history, suggesting how ranges may be kept separate in nature by competition when differences of species in allopatry are not sufficient to permit a tenuous first coexistence of immigrants. *Azolla* grows floating on small ponds like duckweed, completely covering the water (Figure 7.15). There are few ponds suitable for *Azolla* in the whole Galapagos Archipelago, only perhaps a dozen at most (Colinvaux, 1968). These now support populations of only one species, *Azolla microphylla*. The plants are apparently spread from pond to pond on the feet of birds, and it seems that the original *Azolla* colonials must have been brought in this way 1200 kilometers from the South American mainland to a suitable pond. Once established, the local Galapagos ducks should quickly spread the ferns to all other Galapagos ponds, after which there should be little chance of a second species ever getting established following another chance transoceanic flight, since the niche of *Azolla* floating on a pond must be about as restrictive and simple as those that Gause contrived for his *Paramecium*. It is, therefore, not surprising that only one species of living *Azolla* has been found on the islands.

The record of *Azolla* spores in drill cores of mud from a Galapagos lake shows the history of the genus on the islands (Schofield and Colinvaux, 1969). *Azolla microphylla* lives there now and the spores, which are released into the water, are abundant in surface mud. But in mud deep down in the sediments, shown by radiocarbon dating to be more than 48,000 years old, we find only spores of another species, *Azolla filiculoides*, a plant known from South America but not now growing anywhere in the Galapagos. How had such a changeover occurred? The ancient mud is separated from the mud of the last 10,000 years by red clay deposits, suggesting that the lake had been dry for a long period. This suggests that drought had exterminated that ancient population of *A. filiculoides*. When the lake again filled with water many thousands of years later, chance brought a migrating bird with a piece of the other species on its foot, and a second species held the lake against invasion. This, at least, seems to be the most likely explanation of the fossil history.

A CAUTIONARY NOTE

The character displacement model of speciation is attractive to ecologists as a satisfying generalization and it has comfortable theoretical roots in the logistic hypothesis. But ecologists are becoming cautious that it may be one of several mechanisms that work in parallel. Selection should work to remove deviant individuals in many circumstances other than when they compete too strongly with other species, perhaps because they are predator or accident prone. These other selection pressures for species conformity have the theoretical advantage that they are less dependent on populations being crowded near equilibrium and so will work for populations with large random fluctuations. But even though character displacement now seems less overwhelmingly important than it once did, the model continues to be important to evolutionary thinking.

Figure 7.14 (*a*) Isla Santa Cruz, (*b*) Isla Daphne.
The tiny island of Daphne is only an islet in a bay of the much larger Santa Cruz. *Geospiza fortis* populations have characteristically different beak sizes on each island.

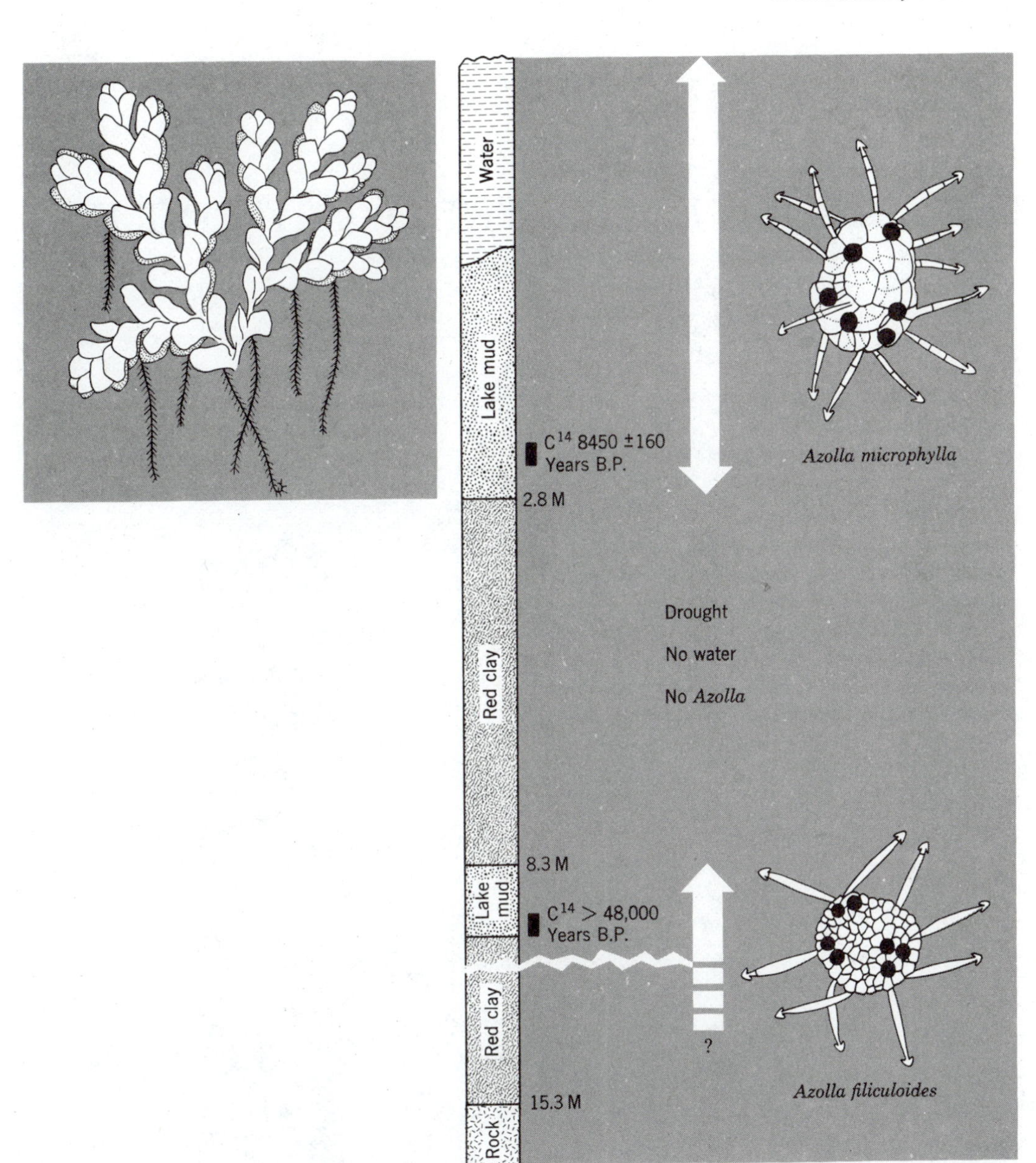

Figure 7.15 Water ferns: competitive exclusion.
Water ferns of the genus *Azolla* loose their spores into the water encased in structures armed with hooks and called massulae. The shapes of these structures are different in the different species. The lake dried up for many thousands of years in its middle history, after which the new lake was occupied by a different species of *Azolla*, and the only one now known from the Galapagos Islands. This history probably illustrates competitive exclusion, the triumphant species each time being that which reached the Galapagos first by chance transport from mainland South America in each of the wet epochs.

PREVIEW TO CHAPTER 8

Ecologists have long argued about the real importance of competition and density dependence in wild populations. For some, the success of the logistic hypothesis in predicting competitive exclusion and the distinctness of species was a compelling reason to expect competition and density dependence to be ubiquitous. This view received further strong support from census data of populations of breeding birds in the spring of temperate latitudes, showing numbers to be so constant from year to year that density-dependent control seems inescapable. It proved difficult, however, to show how density-dependent mortality actually occurred in overwintering birds and it later came to be realized that constancy at the spring census was a consequence of territorial behavior rather than winter death. For ecologists working with insects the hypothesis of density dependence tended to seem less satisfactory. Insect numbers usually fluctuate widely, especially in seasonal or marginal habitats. Populations of grasshoppers and thrips in southern Australia were studied intensively and were found to be typical for insects living in similar habitats. These insect populations rose rapidly in favorable seasons, then suffered mass death when the weather changed. The species avoid extinction in the lean times only by special adaptations that tide a few individuals over or by dispersal from other habitats. Competition and density are usually unimportant to these insects and their populations may be said to be under density-independent control by the weather. After much debate, it now seems certain that different populations are controlled in different ways, depending on local circumstance and life history. A side issue of this debate was the question of whether herbivores, particularly insects, ever eat down their food so that they are food-limited. Proponents of both density-dependent and -independent control of herbivore numbers argued that the herbivores could not in fact be food-limited because the earth was always green with vegetation. This *green earth argument* is examined and it is found that local greenness of the earth is due to the efficiency of plant defenses against herbivore attack, and to the fact that plants tap very large energy supplies with which to make good their losses to herbivores. The greenness is not necessarily a consequence of herbivores being restrained by either weather or predators. Since nearly all life is lived in fluctuating environments, it is necessary to consider the effects of random or stochastic death on interspecific competition. The classical Lotka–Volterra model assumed constant environmental states when predicting the outcome of competition at equilibrium. Adding terms to the equations to allow for chance death gives different results when the equations are solved for growth rates for populations close to K and for numbers that are very low. Coexistence is inhibited even further in crowded populations. Coexistence may be possible when chance keeps numbers low, even though coexistence is excluded when numbers are high.

CHAPTER 8

DENSITY-DEPENDENT AND DENSITY-INDEPENDENT FACTORS IN POPULATION CONTROL

Populations of animals in the wild can be kept in check by several different mechanisms. Checks to population growth may be applied evermore strongly as individuals are crowded together, which is called DENSITY DEPENDENCE, or by factors like drought or fire that kill whether individuals are crowded or not, so-called DENSITY INDEPENDENCE. Another possibility is predation by efficient predators that keep local prey populations low. These different mechanisms apply in different circumstances that are now generally understood.

An ecologist of the 1980s talks of the kind of life history that leads to one kind of population control or another, or of the physical environment or habitat in which a particular kind of control is expected, or both. But the acceptance of these alternative patterns of population control has come about only after lengthy argument. For at least ten years in the 1950s and early 1960s many ecologists debated which were the more important mechanisms of population control in nature, seeking to show that density dependence, or independence, or perhaps effective predation, was the more universal mechanism. Out of these debates came the modern understanding.

A key year was 1954, when books were published setting out rival views of the matter, "the natural regulation of animal numbers" (Lack, 1954) and "the distribution and abundance of animals" (Andrewartha and Birch, 1954). The Lack book, drawing its evidence from studies of vertebrates, notably birds, set out the view that most populations most of the time were under density-dependent control. The Andrewartha and Birch book, using insect data and the experience of entomologists in strongly seasonal climates, argued that numbers in the wild rarely reached a density-dependent equilibrium since populations were usually kept at low levels by accidents of weather or by devastating predation. The present consensus evolved slowly through controversy as these rival views, with their impressive marshalling of supporting data, were resolved.

The argument for density dependence perhaps had the most beguiling intellectual appeal, for this was based on the experiments and theory developed from the logistic hypothesis (Chapter 7). Using the logistic equation to describe population growth makes the assumption that populations will be controlled by competition or other density-dependent processes. Development of the logistic then predicts competitive exclusion and character displacement, finally resulting in that satisfying ecological explanation of why species are so distinct and separate (Figure 8.1). If the existence of all species is indeed explained by this reasoning, then it follows that the original assumption of density dependence in the control of all populations is valid. This was the argument developed by Lack (1954) and others who thought like him. When they marshalled evidence to try to demonstrate that natural populations actually were under density-dependent control, this was only to bolster a position that they considered to be intellectually almost impregnable from the start. They said that numbers in nature usually are constant from year to year because populations are in equilibrium such that[1]

$$\frac{dN}{dt} = rN\left(\frac{1-N}{K}\right) = 0 \qquad (8.1)$$

Andrewartha and Birch (1954) established their alternative view of population control from empirical data that seemed to show conclusively that particular insect pests lived in populations the size of which fluctuated from rarity to superabundance in ways that could not easily be associated with ideas of a density-dependent equilibrium. They went from these data to argue the position that, since some animals were not under density-dependent control, then perhaps almost no animals were, and they began to look for weaknesses in the by then conventional theories of competition, including the ideas of competitive exclusion and the consensus on speciation that had evolved from it.

[1]For symbols see Chapter 6.

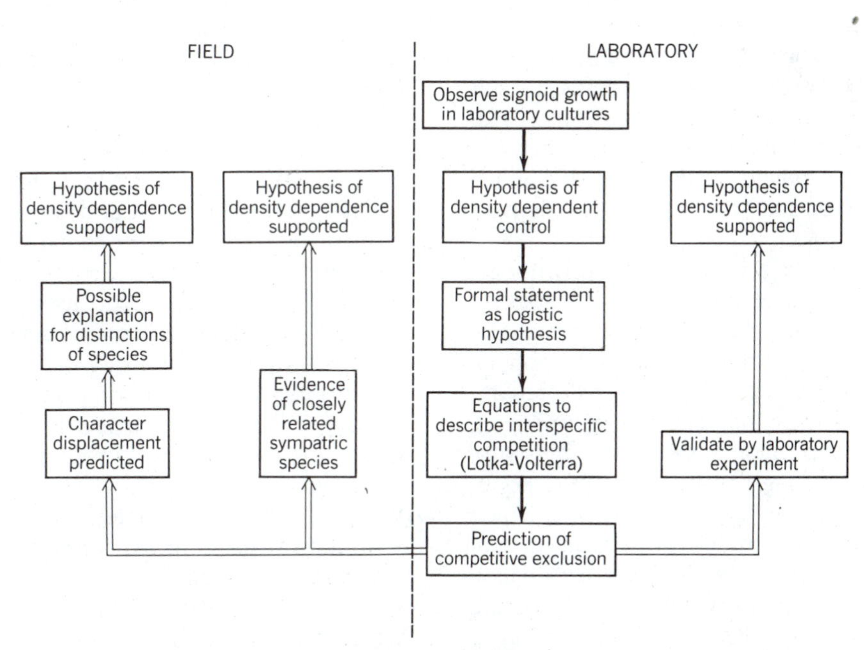

Figure 8.1 Initial arguments for density dependence.
The diagram summarizes the arguments discussed in Chapter 6.

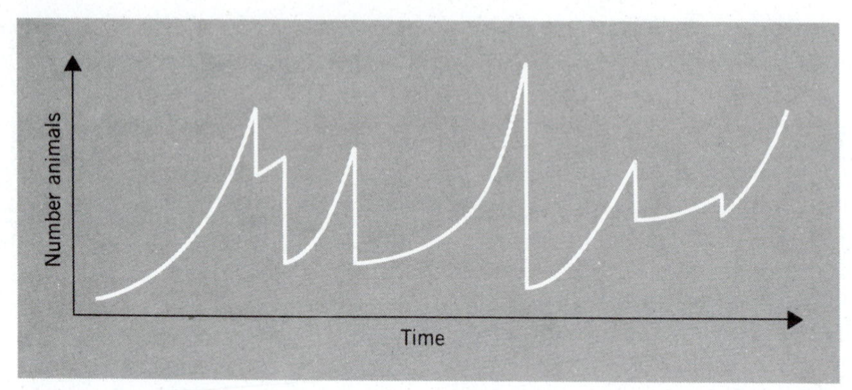

Figure 8.2 Hypothetical population history where a density-dependent population equilibrium is not attained.

Control without density dependence is control by DENSITY-INDEPENDENT FACTORS. The sort of thing implied is illustrated by the wholly imaginary population plot of Figure 8.2. This population is not held at an equilibrium, but neither is it allowed to grow without check. It looks to be under control in some sort of *density-independent* way. In any habitat with a seasonal climate this process might seem as plausible as control by density dependence. The population follows the weather or vicissitudes of changing seasons, rising rapidly in favorable times and suffering mass death or failure to breed when seasons change: not immune to all possibility of competition, certainly, but rarely becoming sufficiently crowded to experience density-dependent checks.

Predation also is a process that need not act in density-dependent ways in all circumstances. If a predator is very efficient, the prey might be reduced to very low numbers with the population hardly ever rising. These scattered prey animals would be little influenced by competition, and the low population is set by the *density of predators* not by the density of the prey population itself. Furthermore the prey population should experience density-independent fluctuations, even in its low state. Students of predation have concluded that many populations in nature can in fact be controlled by a combination of efficient predators and seasonal changes in this way (Chapter 10).

The books by Lack and Andrewartha and Birch reflected widespread interest in mechanisms of population control and were written as numerous ecologists prepared reviews on the subject, often espousing competition, or fluctuating environments, or predators as principal agents in population control (Nicholson, 1954; Hairston *et al.*, 1956). Attempts to reconcile these opposing viewpoints led to the realization that all were correct in special circumstances. A second important outcome was that they produced current theory on species strategies.

Ecologists learned to separate animals into groups according to their life history phenomena, noting whether species were adapted as colonists, being highly fecund and vagrant, with a high probability of suffering density-independent death (so-called opportunist or *r*-strategists), or animals adapted to live at a population equilibrium under conditions of density dependence (so-called *K*-strategists; see Chapter 11). Since these ideas are so important in modern ecology it is useful to review the terms of the debate between the rival proponents of density dependence and density independence, which is the subject of this chapter.

In recent years the old controversy has been taken up in connection with testing mathematical models (Lewin, 1983). Quite different models result from the deterministic ideas of competition theory than from the probabilistic or stochastic assumptions of density-independent vagaries of weather. Since the natural world allows both mechanisms to exist in different circumstances, aspects of the debate can never be resolved.

EVIDENCE THAT BIRD POPULATIONS ARE NORMALLY AT EQUILIBRIUM

The common European heron (*Ardea cinerea*) is both conspicuous itself and builds its large nests high in the trees of breeding colonies. Some heronries in England are known to have existed for centuries, so that we know that the birds have maintained themselves for long periods. Since 1928, amateur naturalist groups have kept count of the number of occupied nests in many heronries, providing a remarkably complete annual heron census for large parts of England. Lack (1954) collected the census data from the two best studied districts to produce heron population histories (Figure 8.3). Several things about these heron histories are revealing. The numbers do fluctuate from year to year but, except after the winter of 1947, not by very much. The record is, in fact, just the sort of thing that should be expected if the populations were in balance and controlled as a function of density. That there should be a marked fall after the winter of 1947 was also striking, for the English winter of 1947 was notoriously bad. Usual English winters are so mild that lakes and ponds are free of ice for much of the time, and rivers and estuaries nearly all of the time, but in 1947 there was a long freeze of the kind common in continental countries. This was particularly hard on herons, and dead herons were found by frozen lakes that winter. In the spring of 1948 there were many

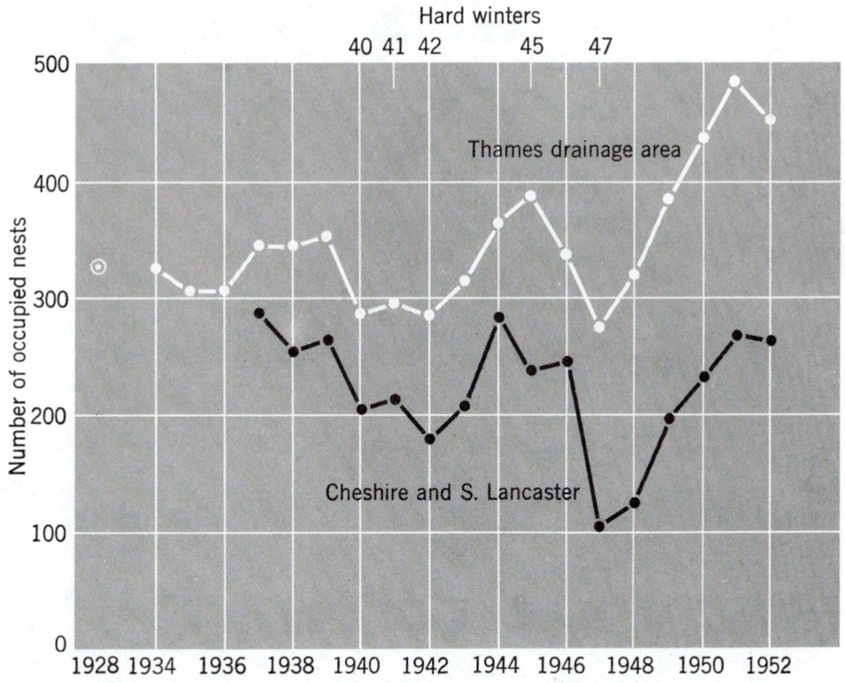

Figure 8.3 History of breeding populations of herons in England.
The herons were censused by counting occupied nests in ancestral heronries each spring. The population is remarkably constant from year to year, strongly suggesting density-dependent control. The particularly hard winter of 1947 reduced the numbers of herons breeding the following spring, but the population quickly regained its old level in subsequent years. (From Lack, 1954.)

Figure 8.4 Stork nest on a village roof.
Storks that build nests on the roofs of houses are easy to census during the breeding season (Figure 8.5).

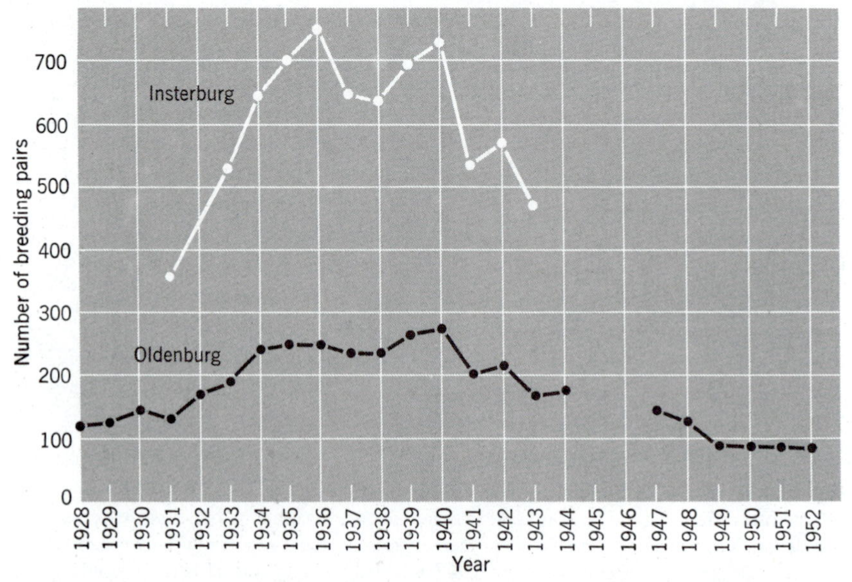

Figure 8.5 History of breeding populations of the white stork (*Ciconia ciconia*) in parts of Germany. (From Lack, 1954.)

empty nests in the ancestral heronries, but the populations were quickly built up in subsequent years. A catastrophic event had severely reduced population, but in doing so the pressures of crowding on the survivors appeared to have been released so that heron numbers could expand back to their old levels.

Lack found that data similar to the heron census existed for the white stork (*Ciconia ciconia*) in Germany. Storks build huge nests on the roofs of houses in German villages and are, like the herons, very faithful to the homes of their ancestors. No elaborate sampling system was needed to take the census of these birds, and the counts could be relied on. Once again the results suggested a population in balance (Figures 8.4 and 8.5).

Census data of a different kind are available for the great titmouse *Parus major* in a Dutch wood. The bird naturally nests in holes in trees, but it prefers nest boxes. All that need be done to count the breeding titmice in a wood is to place many more nest boxes than there are breeding pairs there, and to count how many are occupied. This novel method of census has provided a long history of a local Dutch population of great tits going back to 1812 (Kluijver, 1951). The results (Figure 8.6) again show fluctuation about a mean, although a rather wide one in which the densest population is four times the sparsest. But there is about this history nothing of that steady rise followed by unexpected fall that should denote the fluctuations caused by random catastrophes, nor is there a rhythm that might reflect recurrent seasonal crises. The tits lay six or more eggs at a time, showing that fluctuations could be very wide if there was no density-dependent control; for a population might multiply 36 times in two years, if not checked.

Bird census data are, like those for herons and titmice, nearly all for breeding pairs in the spring, because counts at other times of the year are not reliable. Census in midsummer would, of course, reveal much larger populations, since

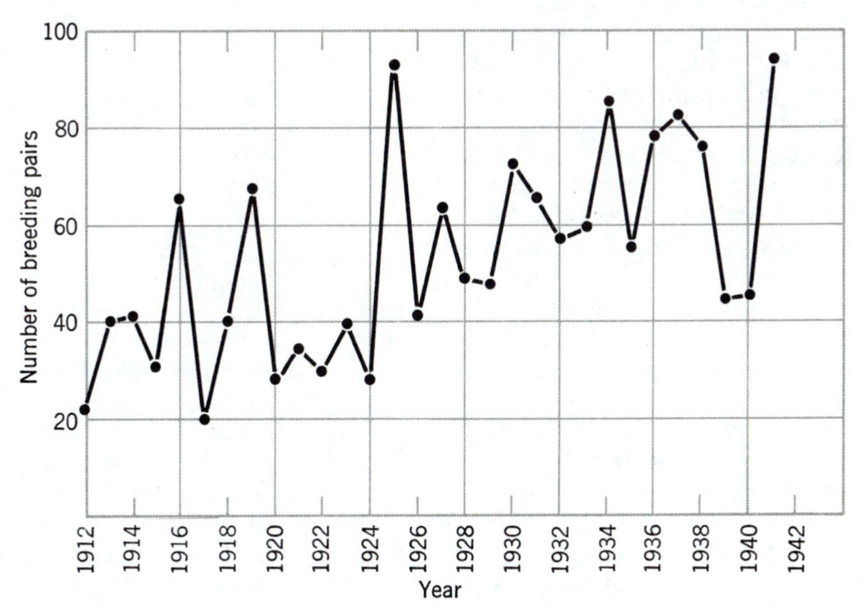

Figure 8.6 History of the breeding population of the great titmouse (*Parus major*) in a wood in Holland.
The population of the tits fluctuated more than that of the herons and storks, perhaps reflecting clutch sizes of up to six at a time, but the population still seems to be regulated within comparatively narrow limits. (From Lack, 1954.)

the year's fledglings would be included. If the records of spring census do really require density-dependent control, therefore, this must be applied between the end of summer and the time for the next counting in the following spring. The most likely season for this control is the winter and considerable effort has gone into seeking evidence for this, as discussed in a later section.

THE EVIDENCE OF IRRUPTIONS AND INVASIONS

Some bird populations fluctuate dramatically, sometimes rising to several times their usual level. Lack (1954) called these sudden rises IRRUPTIONS and showed that their histories were completely compatible with the general hypothesis of density dependence.

All bird watchers know that there are seasons when birds normally thought to be rare are suddenly abundant. In England, crossbills (*Loxia curvirostra*) suddenly appear in flocks in suburban gardens one winter, though no one has seen a crossbill in the county for more than a generation (Figure 8.7). In New England snowy owls (*Nyctea scandiaca*) may appear one winter. And there are many other well-known examples of such *irruptions*, typically featuring northern birds like grosbeaks, waxwings, and the like that arrive in great numbers in the settled temperate belt one winter to everyone's surprise. The following year they are absent.

Figure 8.7 Genesis of an irruption: *Loxia curvirostra* on nest.
Crossbills are birds of the European continental boreal forests. After unusual breeding success in some springs large numbers of crossbills migrate in the following winter to places where they are usually never seen. Such a population event is called an irruption.

Lack's explanation of bird irruptions is that they result from a combination of chances that gives the birds quite unusual success at breeding in a single spring. Some chance combination of weather and food supply means that all bird couples raise large families at the same time. By the time the young are fledged, therefore, the parental population has been multiplied several times, much more than usual. But as soon as winter comes the result is crowding and intense intraspecific competition with strong density-dependent pressures on individuals. One response of the crowded young to this is more migration than usual, hence the sudden descent on the civilized southlands. The crowd then disappears so that only the usual number return to the breeding sites next spring. Lack argued that, far from being an exception to the principle that bird populations are under density-dependent control, irruptions provided strong evidence in support of that rule. An accident of weather brings on a glut of birds but density-dependent culling quickly removes the surplus, bringing the population back to the usual equilibrium number in a few months before the next breeding season.

Crucial to understanding irruptions in bird populations is the concept that the birds always breed to the limits set by the food supply and their natural fecundity. This means too many birds for the normal equilibrium every year. In this sense, every year is an irruption year of sorts.

But when spring food and weather are favorable, the irruption is larger than usual and becomes noticeable.

A cardinal fact about irruptions is that, after they are over, the numbers of the birds drop back to roughly the numbers that lived before the irruption. A population controlled entirely by the vagaries of weather would, of course, also suffer irruptions, indeed they should be common events, but the numbers of survivors after the irruptions should be quite unpredictable. The adversity that would reduce this kind of irruption should act in a haphazard way, leaving a band of survivors whose numbers should bear no relationship to the numbers living before the irruption. That bird populations seem to be stabilized at their old levels following irruptions is thus additional evidence that their populations are normally under density-dependent control.

Bird Invasions

Two records of bird invasions are consistent with the density-dependent model because they suggest sigmoid growth histories in the wild. One is the record by Einarsen (1945) of numbers of pheasants on a small island off the U.S. West Coast following introduction (Figure 8.8). Population growth was clearly exponential but the pheasants were shot when the military occupied the island before it was known if the population would level off to complete a sigmoid history. Even if the sigmoid history had been completed it could, of course, be argued that pheasants on a small island were like paramecia in tubes or beetles in flour—experimental animals in a closed container that would be constrained by contrived circumstance toward intraspecific competition and sigmoid growth.

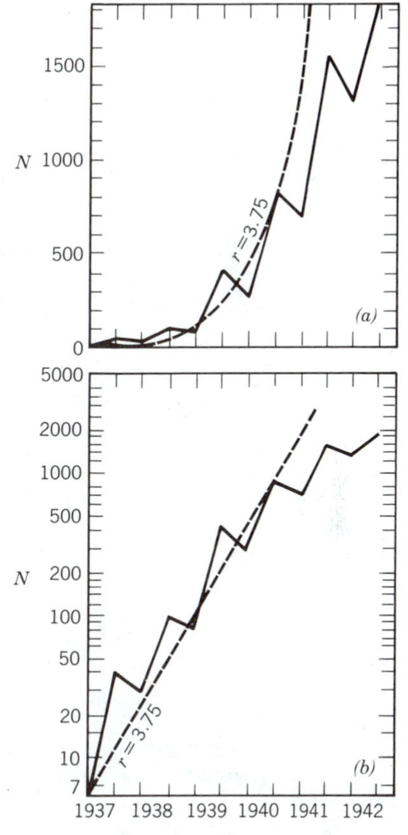

Figure 8.8 Exponential growth of a population of pheasants (*Phasianus colchicus*).
The pheasants were introduced to a small island, after which the population expanded exponentially. (*a*) Arithmetic plot, (*b*) Log plot. Broken lines are hypothetical exponential growth. (From Hutchinson, 1978.)

No such criticism can be offered on the population history of turtledoves in Britain, an event that happened after Lack (1954) set out the thesis of density dependence. The data are given in Figure 8.9. Anecdotal evidence suggests that turtledoves are now at a constant or equilibrium population in Britain. Unfortunately the regular census was abandoned in the late sixties, just when density-dependent constraints might have been demonstrated (Hutchinson, 1978). This sole example, however, does suggest that wild birds can enter new territory, expand their numbers rapidly to fill all niche spaces, and come under density-dependent control.

DENSITY-DEPENDENT WINTER MORTALITY

The discussion so far has not identified the density-dependent factors claimed to regulate bird numbers. All that has been done is to show that

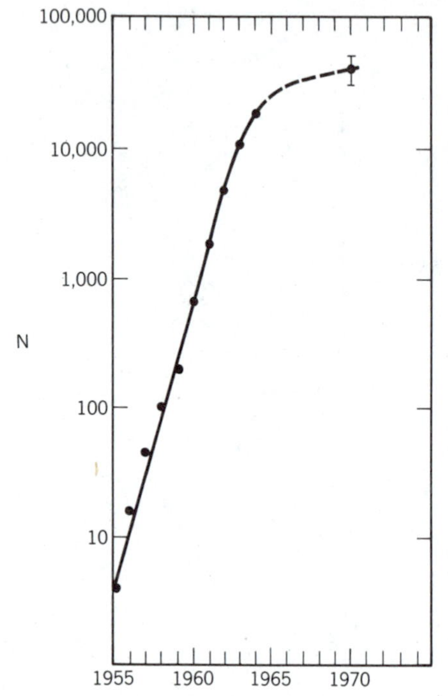

Figure 8.9 Exponential growth of a population of turtledoves (*Streptopelia decaocto*).
Turtledoves colonized Britain in 1955 and seem to have reached an equilibrium population in the early 1970s. (From Hutchinson, 1978.)

any density-dependent constraints must operate between the fledging of young in the summer and the hatching of next year's young the following spring.

In his original summary, Lack (1954) suggested that density-dependent starvation and predation in winter were the controlling factors. It has, however, been difficult to demonstrate that the required density-dependent death in winter actually operates. Intuitively it does seem likely that predation would work, because bird predators might search ever more thoroughly as the winter advances and the food supplies are not renewed—perhaps the hunt gets so difficult as the prey become rare that a population of typical size survives at the end of each winter. Unfortunately it has not been possible to demonstrate effective winter predation in field studies. It may happen, but we see so little of what goes on between wintering predators and prey that we cannot say it does.

At first sight winter starvation does not seem very satisfactory in practice either. It can be argued that real food shortage in a winter is likely to produce mass death rather than a density-dependent winnowing that would let a few healthy individuals through. A winter is a long time to last when food is really short and small homeotherms like birds need a large daily flux of energy to stay alive when temperatures are low. Thus prolonged food shortage might well kill them all, or at least kill somewhat independently of density.

Yet what may be important is not the food available all winter long but just the food of the worst or coldest day. This suggestion seems first to have been made by Fretwell (1972). In the worst days birds can probably forage over only a very small part of their range to avoid exposure or because food was frozen, covered by snow or otherwise unavailable. At the same time a homeotherm's need for food on such a day is very strong and failure to get food might be fatal. On the very coldest days of winter, therefore, birds congregate on the few most favorable feeding grounds where competition must be strong. Survivors of this competition in the bad days would be able to find adequate food in the less extreme days that follow.

Fretwell (1972) developed this argument to predict that birds would be selected to be of the right size for optimal foraging on the coldest days. This offers the chance of examining species populations to see if individuals are indeed adapted to be efficient competitors for the food that is available on the worst days, rather than for food available on average winter days or on spring or summer days. Adequate data to test these predictions do not seem to be available as yet.

The behavior of birds on cold days as predicted by Fretwell, however, has been demonstrated for some overwintering birds in Ohio by Grubb (1975, 1978). Grubb observed flocks of wintering birds throughout the winter in a region of varied topography, partly wooded and supplied with valleys. The winters in Ohio are seldom extreme, but there is much day to day variability in temperature. On most days birds can forage in open woods or high in trees where ample food can be found. But on very cold days birds are found only in defiles or in trees. On the coldest days of all, the rare days when the Ohio winter temperature falls to less than −10°C, birds can be found only in the shelter of bramble or honeysuckle thickets where they are well out of the wind. Competition for the food in these restricted sites is expected to be intense.

An alternative to density-dependent winter death is regulation as a consequence of bird territorial behavior in the spring, a possibility examined in Chapter 12. Probably the size of many breeding populations of birds actually is a by-product of territorial behavior and not of winter mortality at all. When this is true, the important competition is for the resources represented by territory. The setting of an equilibrium population in this way is still consistent with the logistic hypothesis.

THE ARGUMENT FOR WEATHER AS AN AGENT OF CONTROL INDEPENDENT OF DENSITY

Population histories of two insect pests of southern Australia provided Andrewartha and Birch (1954) with examples that seemed to argue strongly that density dependence was not important for insects.[2] These were a swarming grasshopper and a thrips, the population data for which are discussed in the following sections.

The Grasshopper *Austroicetes cruciata*

Much of southern Australia has a Mediterranean climate, with moist winters and dry summers. Winter and spring rains flush a short-lived green pasture, and allow a crop of wheat to be grown. The winters are so mild that water rarely freezes, but the summers are hot and dry. And on top of this pattern of sharply alternating seasons there is much uncertainty of how much rain the spring will bring, since the wettest year of a 20-year period may have more than three times the rainfall of the driest. This is an environment of much uncertainty for its natural animals, one in which the vagaries of weather might be expected to bear heavily on the fortunes of populations.

The grasshopper *Austroicetes cruciata* is a pest of agriculture of this region. From 1935 to 1939 grasshopper swarms were present every spring, closely watched by entomologists who then could do little more than watch. But in 1940 the swarms went away, fortunately for ecological theory with the entomologists still closely watching. They saw the animals die.

Austroicetes cruciata is well adapted to the sharply alternating season of southern Australia (Figure 8.10). There is never more than one generation in a year. The animal passes the long succession of inclement months of the dry summer and the cool winter as dormant eggs, eggs in that quiescent, durable state of some insect stages called diapause. The rising temperature of spring breaks diapause, and the young nymphs hatch ready to feed on the green herbage of spring. Within 50 days the grasshoppers are mature, but the drought of summer is coming. The females must find enough food to stock their developing ovaries with fertile eggs, and the success of reproduction becomes a race against time. If the spring is wet enough, the grass stays green a long time and many batches of eggs are laid safely in the soil for the long sleep of diapause. But if the rains fail, the adults die before they can lay their eggs. The result is a low or absent population the following year.

Birch and Andrewartha (1941) watched a mass death of grasshoppers in the spring of 1940, after first noting the ominous low rains of the previous winter and early spring, rain only a third of that usually expected. They saw the young hatch in great numbers from the eggs as the coming warmth broke diapause, and then saw the animals vanish as they starved to death. Very few were left when the mature swarms should have been around, although there were pockets of survival, particularly near irrigated wheat fields. Here, however, flocks of birds harried them, and

[2]But see Nicholson (1954) for a defense of the importance of density by an entomologist.

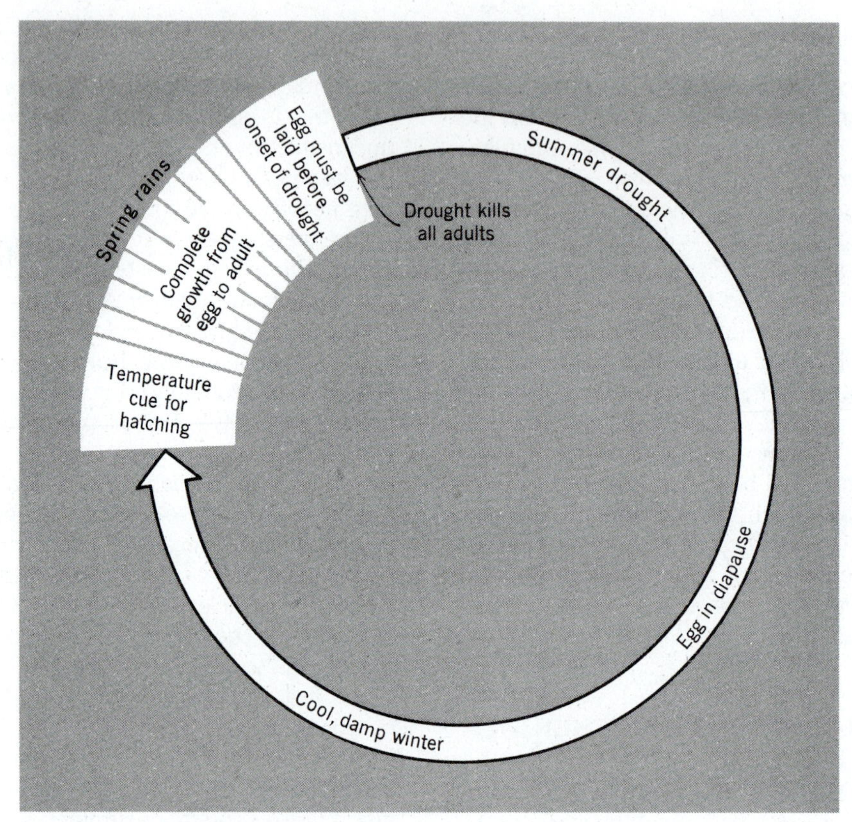

Figure 8.10 Life cycle of *Austroicetes cruciata.*

cleaned up nearly every one. It seemed not unreasonable to suppose that the bird population of large areas was feeling the insect shortage of the drought and, being able to search country, sought out the last populations of grasshoppers and reduced even these to low numbers.

There can be no doubt that the grasshopper population was cut down by the weather in a catastrophic way. Theirs was not the sort of mass death suffered by overwintering birds following one of Lack's irruptions. After an irruption the pressure of crowding increases mortality only until a normal breeding population is left, but from the grasshopper swarms there were practically no survivors, the food supply having failed completely. The species itself could continue to exist only because over the vast extent to its range there were likely to be a few isolated, favored places where an adult could lay a purse of eggs before hunger swept it away. Replacing the swarm with another swarm must take years of good fortune in breeding.

Austroicetes cruciata has obviously been able to last as a species despite the catastrophes to which its homeland is prone, for the sort of mass death of grasshoppers that Andrewartha and Birch witnessed must be a frequent event. The grasshoppers are beautifully adapted to exploit an ephemeral season, passing most of the year in the security of diapaused eggs from which they hatch only on the temperature cue of the rising warmth of spring. The species has found a strategy for letting it live where few other herbivores survive, so that it has the fresh annual growth almost to itself. But it must meet the occasional catastrophe, and this it apparently does by dispersal. The adults can fly long distances, and do so, spreading their kind over hundreds of square miles. When disaster strikes, there is a high probability that some grasshoppers somewhere will survive.

In their semidesert range it seems fair to argue that the grasshoppers suffer virtually no depression from intraspecific competition, for their numbers are always too low. Interspecific competition is even less likely. And the available data show that predation, although present, does not act in a density-dependent way. The grasshoppers are opportunists, living exposed to catastrophe but free from density-dependent constraints.

At the desert's edge all this seems true, but what of the population of grasshoppers farther away from the desert? What limits the animals at the southern edge of the range where rains are more predictable? Davidson and Andrewartha (1948) collected data to show that distribution as well as abundance was set by weather, suggesting that the animals never lived outside the control of weather.

The area in which swarms of the grasshoppers had been common formed a belt running parallel with the coast but some distance inland (Figure 8.11). The country on the seaward side of this line was comparatively moist, but that to the north was dry grassland bordering the central Australian desert. The grasshoppers could probably not swarm farther inland than they did because there every summer would be a disaster summer in which grasshoppers would usually starve without laying eggs. Immigrant adults do often start colonies to the north, but they always die out. To test this hypothesis, Davidson and Andrewartha (1948) mapped the regions with ratios between precipitation and evaporation of 0.25, an index known to describe roughly the edge of desert vegetation, and the resulting isopleth convincingly defines the northern limit of the grasshopper swarms (Figure 8.11). Beautifully adapted though the grasshoppers were to the vicissitudes of southern Australian seasons, they were not able to thrive beyond this line because it was too dry there.

To the south, however, it was not too dry, and yet the grasshoppers never managed to swarm. Davidson and Andrewartha thought it must be too wet, and were able to match another precipitation/evaporation isopleth, one suggesting the edge of wetter times, to the southern boundary (Figure 8.11). But this does not mean that wetness *per se* sets a limit to the spread of the grasshoppers. Dryness sets a northern limit to the drought, as we know because Andrewartha and his colleagues actually observed herbs and grasshoppers dying in the drought, of the drought, but the moister climate of the coast is not so moist that grasshoppers are drowned or their eggs flooded. What the comparative moistness probably does do is to allow other animals inimical to the grasshoppers to thrive there: predators, competitors, or pathogens. These agents must certainly be dependent on less dry weather, but they should act in density-dependent ways when they do act. In moister climes, then, *Austroicetes cruciata* is likely to be limited to populations well below swarm numbers by density-dependent effects of the environment, but in the dryer climate of the north it is freed from those restraints.

The Thrips *Thrips imaginis*

Thrips imaginis, like most thrips, lives in flowers, where it feeds on pollen and soft tissue. The thrips are tiny black slivers of insects, whose general appearance is familiar to anyone who has sniffed a rose. One or two of them in a flower do no damage, since their food supply is then far in excess of their needs so that their feeding is scarcely noticed. But 40 or more thrips in one apple flower can so damage the ovary that no fruit is set, and *Thrips imaginis* has sometimes descended on the orchards of southern Australia in such densities so that the apple crop failed. When this happens, the fruit growers of whole regions of southern Australia suffer together, showing that whatever causes the thrips outbreaks must be operating over a very wide area. From the start of the investigations into the cause and cure of thrips epidemics, it seemed that the controlling agent must be the weather.

Thrips imaginis, like the grasshopper *Austroicetes cruciata*, is a native to southern Australia, but it copes with changing seasons and unpredictable weather in quite different ways. At no

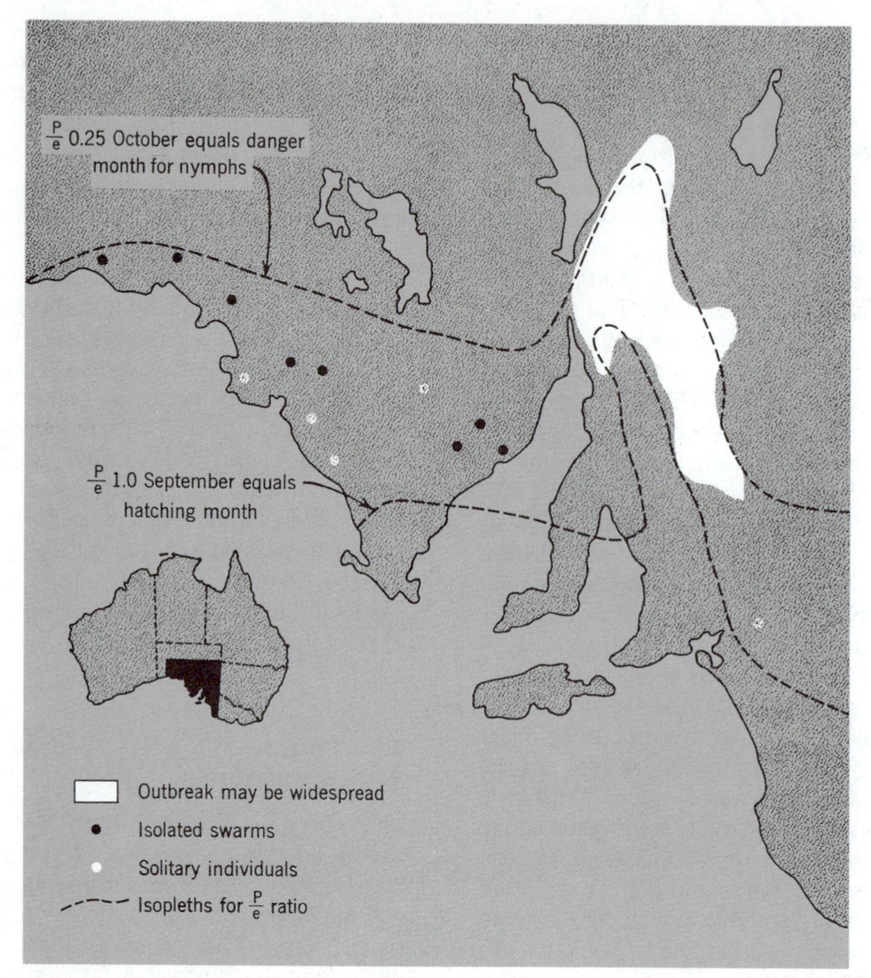

Figure 8.11 The area of South Australia where *Austroicetes cruciata* may, during a run of favorable years, maintain a dense population.
The isopleth (precipitation/evaporation ratio) lines show that the distribution parallels a climatic distribution. The northern limit may be set by the shortness of spring, but the southern limit is probably set by other animals (see text). (After Andrewartha and Birch, 1954.)

stage in the thrips life history is there a diapause like that which safely brings the eggs of *Austroicetes* through the drought of summer, so that the animals must be active in all seasons. Most of the lives of the thrips are spent in flowers, which young adults find by wandering flight. The eggs are laid in the flowers, and the young grow to their full size in the same flowers. But then they must walk down from the plant, bury themselves in the ground, and pupate; after which each young adult must fly away from its perch on the ground in quest of a flower in which to pass the rest of its days (Figure 8.12). The search for a flower appears to be the one desperate enterprise in the life of this animal. If flowers are scarce the chances of finding one must be poor, and then the thrips will die without laying eggs. But, if it finds a flower, then its chance to reproduce is virtually assured.

In the cool winter enough flowers are left to tide the animals over. Summer is the really bad time, because then flowers are scarce and a pop-

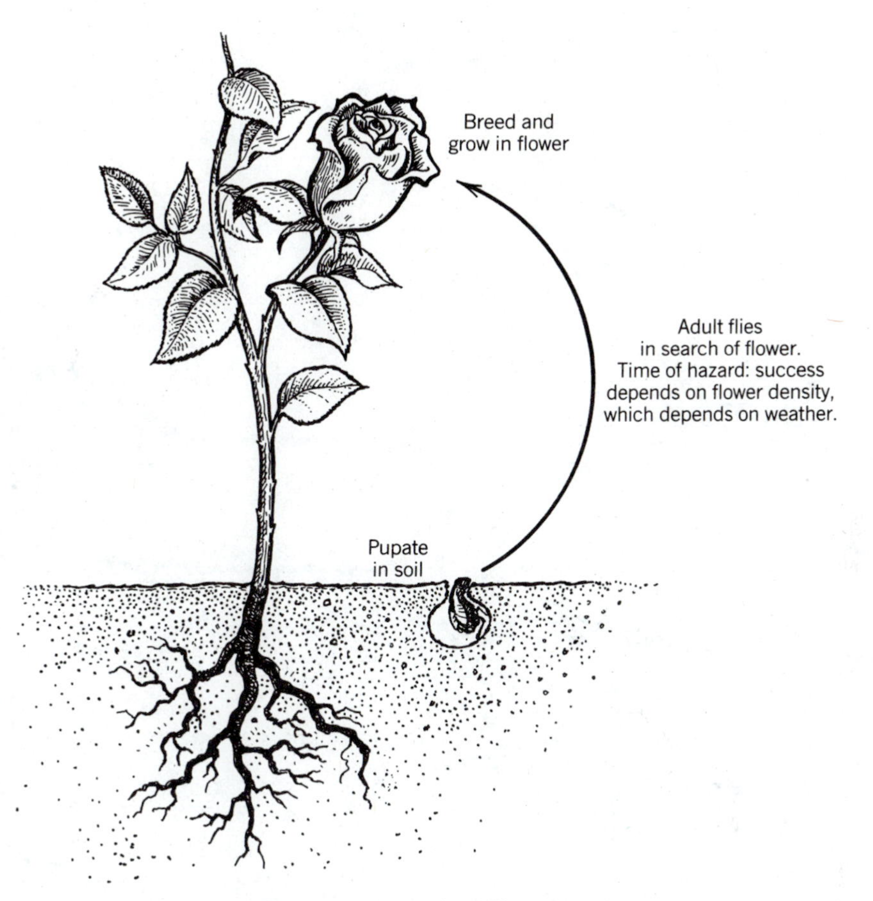

Figure 8.12 Life cycle of *Thrips imaginis.*

ulation persists only if a few individuals find surviving flowers in sheltered places; but the strategy of the species accepts that every summer is a disaster summer, a time when most of the thrips that were reared in the spring must die because they fail to find any food. The few that do find flowers, however, have all the food they need, and so leave descendants to start the population booming again when next the rains provide many flowers for wandering thrips to find. Every year the population must rise in the spring, crash in the summer, and hold its own from scattered bastions of flowers throughout the rest of the year. The monthly population of thrips is thus determined by the number of flowers, which is in turn determined by the weather.

Davidson and Andrewartha (1948) monitored the population size of *Thrips imaginis* over the changing seasons as a function of the weather. Their object was to test the hypothesis that the population size of thrips was directly dependent on the weather, and they then had the hope of using information about the weather to predict thrips outbreaks. They ran a census of thrips numbers by daily counts over fourteen years, their method being to count the thrips in a single rose taken from the same experimental gardens every day. Figure 8.13 gives their results for seven consecutive years. The annual spring outbreak (November to December in Australia) varied in timing and magnitude, though it was always present. Davidson and Andrewartha used these

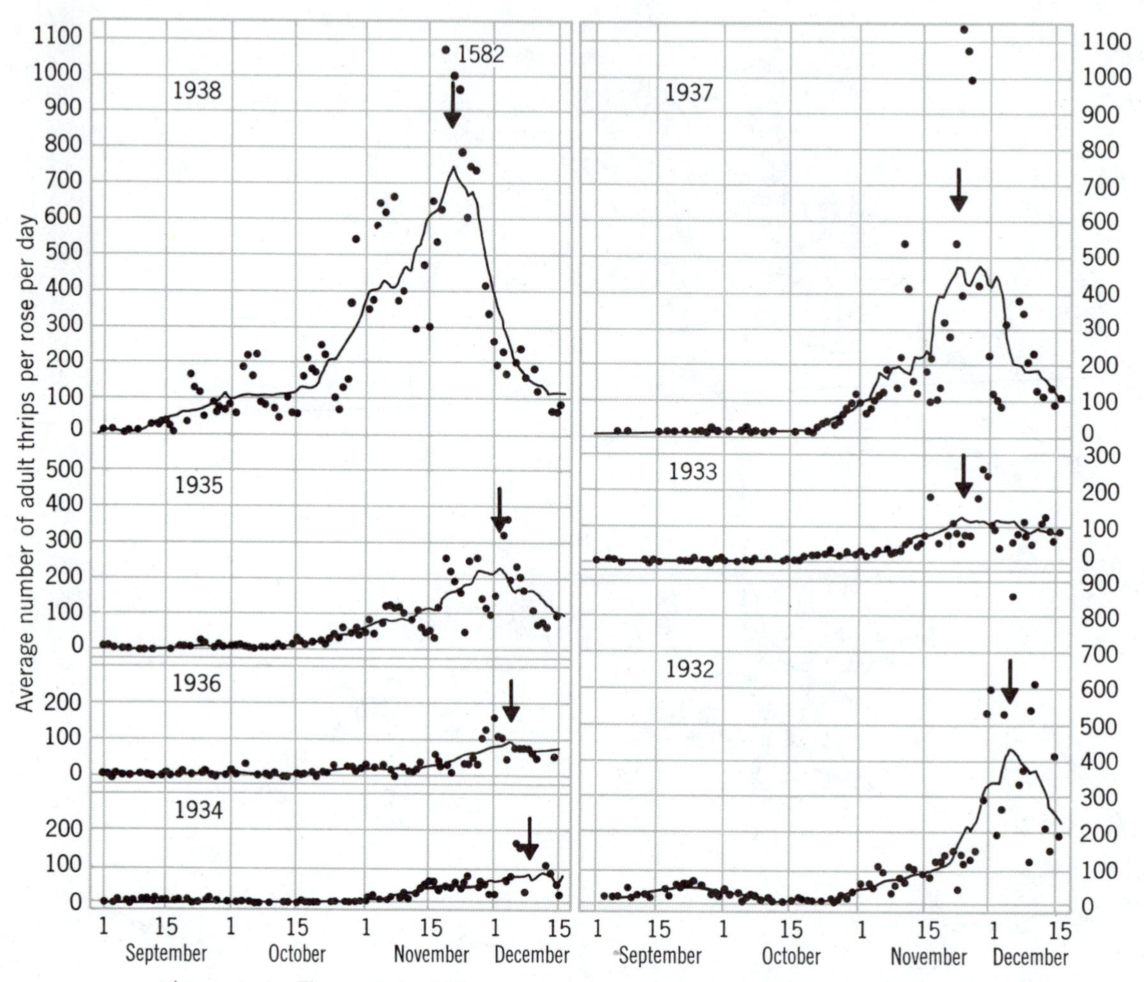

Figure 8.13 The numbers of *Thrips imaginis* during spring each year for seven consecutive years.
The numbers in late winter were the same each year, suggesting control at this low plateau by density-dependent factors. But the blooming of many flowers in the spring always caused an irruption of thrips. The size of the irruption was dependent on the length of the flowering season, so that the maximum size of the irrupted population was independent of density. (Notice that the Australian spring comes in December). (From Andrewartha and Birch, 1954.)

data to calculate correlations between the onset and size of thrips outbreaks with local climatic data, then applied the resulting regression equations to the weather of succeeding years to predict the expected size of thrips outbreaks. Their results are given in Figure 8.14, where it is shown that they were able to predict the number of thrips to be found in a rose for the years 1932 to 1944 with considerable precision.

Davidson and Andrewartha had by this exercise demonstrated that the numbers of thrips every spring were controlled by weather in a catastrophic and density-dependent way, suggesting that *Thrips imaginis* has a population history comparable with that postulated in Figure 8.2. Like the Australian grasshoppers, the thrips of peak populations were killed in ways that were neither caused by, nor triggered by, the density

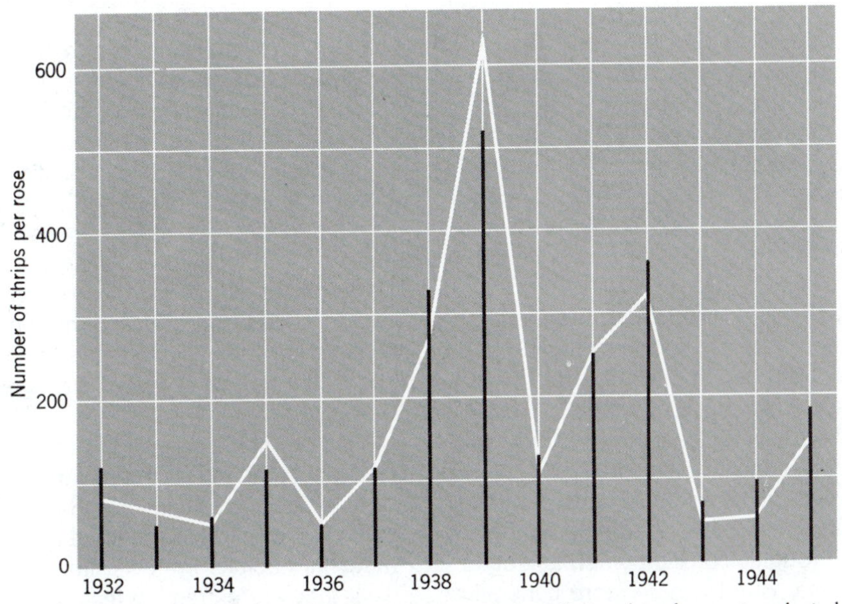

Figure 8.14 Correlation of size of thrips' irruptions and predictions based on meteorological data.
Four parameters of weather were used to predict the relative size of thrips' outbreaks in successive springs (continuous line). The bars describe the actual size of outbreaks as recorded in the roses of the garden of the experiment station at Adelaide. (From Davidson and Andrewartha, 1948.)

of their own populations. This is not density-dependent death. It can be argued that there is a density-dependent quality about the chances of any *individual* thrips dying, because the choice of survivors who find refuge must be affected by the number seeking refuge, but this is not density dependence as understood in the logistic model. The actual number of survivors is set by the number of refuges available, and this is not a property of the population.

Reconciling Hypotheses of Density and Weather

There was, and is, no doubt that grasshopper and thrips populations are always still growing in the spring or summer "highs" before the weather changes and the numbers fall. These climbing populations never reach equilibria set by density-dependent checks and negative feedbacks of the kind described by the logistic hypothesis. Andrewartha and Birch said that these populations were "controlled" by weather in a density-independent way.

There is good reason to believe that very many insects have population histories comparable to those of the grasshoppers and thrips. Populations of these insects rise in the summer but probably do NOT come under density-dependent control; the population history probably is NOT sigmoid; populations are probably cut down by the arrival of bad weather in fall or winter while the numbers are still rising exponentially.

And so it is possible to argue that insect "highs" often do not come under density-dependent control as required by the logistic hypothesis. This seems to be a disquieting conclusion because the ecological view of species is based on arguments that assume density dependence and that require that there be competitive equilibria established. How can the idea of an insect species fashioned by character displacement between pairs of populations at equilibrium be reconciled with the conclusion that population highs of many insects never reach equilibrium? One answer is that the insects *lows* may very well be at equilibrium.

In Figure 8.13 can be seen the suggestion that thrips populations are at equilibrium in winter—the animals irrupt from roughly similar winter populations in the spring of every year. Andrewartha and Birch (1954) denied that there was density dependence in this, saying that the low populations every winter were a function of the number of flowers serving as refuges—flower numbers were roughly constant when averaged over a large area. But it is now more generally believed that density dependence can be demonstrated in the thrips winter data (Varley *et al.*, 1973).

It must be remembered that there are many generations of thrips in every calendar year. The *seasons* come upon thrips as chance good or bad *years* come on a bird or mammal. Probably the best way of looking at the thrips story is to accept the evidence for density dependence in winter. These winter populations are the "normal" equilibrium populations of thrips. Then there are irruptions when spring comes along, with each irruption eventually cut down by a change of weather, after which the original species equilibrium is re-established.

In the more extreme versions of a weather-dependent life history, true population equilibria might be attained only rarely, even in the population lows. *Austroicetes cruciata* fits this model, at least over part of its range. These grasshoppers have only one generation a year. When grasshoppers are all but wiped out by failure of the spring rains, the future of the species in the desert regions seems to depend on chance survival of a few individuals somewhere over a range of hundreds of kilometers, and this has nothing to do with crowding or density.

But perhaps the grasshoppers do not rely entirely on so chancy a way of recovering from disaster. It seems likely that grasshoppers will always persist at the southern edge of their range where there is always moisture (Figure 8.11). In this region, local populations may well be at equilibrium, suffering intraspecific competition, interspecific competition, and predation, all acting in density-dependent ways. From these crowded local populations winged migrants could be the founders of fresh northern populations after each disaster. An extreme variant on this grasshopper theme is provided by locusts, whose lives, and plagues, are but a magnified version of this history.

The profitable way to look at the population histories of the many kinds of insect that peak in summer and crash later is that they have life strategies nicely programmed to win fitness in marginal habitats in good times. They disperse well, they are extremely fecund, and they have short generation times—these three traits win much fitness for the individuals that have them. But the life strategies also include adaptations for life in the lean times for the generations that must face those. There can even be selection for different genotypes in the generations of boom and bust. The animals, and the weed plants that live that way, are OPPORTUNISTS with life history phenomena adapted to exploit seasonal or transient gluts of resources (Chapter 11).

But weather does cause the populations of many short-lived animals or plants to fluctuate very widely above a baseline set by density-dependent factors or a finite supply of refuges.

WHY THE EARTH IS GREEN

When the argument over the possibility of control by weather was at its strongest in the early 1960s an argument came to be used on both sides of the debate that turned on the observation, or claim, that the earth was green.

The *green earth argument* goes like this: The earth is carpeted green with plants, which means that there should always be plenty of plant food available at the surface of the earth. But most animals are herbivores. If these herbivores were limited by food energy, and if they competed for food, then we should expect the food supply of herbivores to be constantly used and restricted, in which event the earth should not be green but eaten piebald by the herbivores. Therefore, since the earth is green, herbivore numbers are kept down in some way.

Andrewartha and Birch (1954) postulated that what kept the herbivore numbers down was weather, just as it kept down the numbers of the two herbivores they themselves had studied, the Australian thrips and grasshoppers. They realized that some insect plagues do overeat their plant food supplies, and that large mammals such as goats do sometimes destroy their range, but they claimed the supposed greenness of the earth as evidence that this is not usual. Thus did their theory of density-independent control by weather gain generality. Some insects are controlled by weather; most animals are herbivores; the herbivores eat so little that the earth is green; therefore, most animals are probably controlled by weather.

But an alternative explanation for the green earth is that herbivore numbers are kept down by predators (Hairston *et al.*, 1960). In this model herbivores are harried by predators so that competition for food is not important in the life of an herbivore. The competition that matters is between predator and predator over the bodies of herbivores. Plants are the real gainers as predator stalks prey. This argument gains power when it is considered that most herbivores are insects, because insects can be shown to suffer very heavy predation in some circumstances (Chapter 10).

Debate over whether herbivores were generally kept down by predators or by the weather was often heard among English-speaking ecologists in the 1960s. It has died away for several reasons, but principally because it was seen to be uninteresting. Sometimes it could be shown that individual insect herbivores were kept below their food limits by weather. At other times predators did the trick. But there were also many examples when it was possible to show that some herbivores were indeed food-limited, that they really did exist in equilibrium populations set by the food supply.

And yet there remains the statement that the earth is green, with its suggestion that plants, as a whole trophic level, are not seriously reduced by herbivore attack. Why is this? One way of answering is simply to refute the claim that the earth is green—for it is not, the typical color of the land being the brown of desert, rock, or dead vegetation. This can be seen in photographs from satellites. In many of the less well watered parts of the earth plants exist in tenuous equilibrium with both herbivores and each other, competing for space and water, having only short-lived flushes of activity when they turn green.

More important probably is the way plants are found to resist herbivore attack. Plants turn out to have many effective defenses against herbivores, from being uneatable or poisonous, to growing tall or dispersing out of the reach of specialized herbivores (Chapter 25). To the extent that plants are successful in frustrating herbivores, the earth is green (where it really is green) because herbivore numbers are kept down neither by weather nor predators but by the plant's own defenses.

Of even more generality is an argument based on energetics. Plants are not energy-limited, in the sense that they are usually exposed to more solar insolation than can be used in photosynthesis. But plants are space-limited, and any unit of bare space in a watered place may be treated as representing an unused unit of photosynthesis. This means that whenever a plant is destroyed or cut back by an herbivore, a space is cleared for another plant, or part of a plant, to grow.

An array of many species of plants under herbivore attack is in the position of a defending army provided with unlimited reserves. As fast as a gap is torn in the vegetation, so that gap can be filled by another plant, perhaps of a different species. The energy available to the herbivore trophic level is necessarily a small fraction of the energy available to the plants (Chapter 4). So the green parts of the earth stay that color because herbivores, *as a whole trophic level*, cannot command an energy flux sufficient to keep land free of plants. This is why a well-grazed meadow remains green.

THE LOGISTIC HYPOTHESIS FOR FLUCTUATING ENVIRONMENTS

The classical Lotka–Volterra equations of competition assume constant environments. But real world competition takes place in environments that fluctuate, particularly with the seasons. Recently attempts have been made to investigate the effects of random perturbations on Lotka–Volterra models of competing species to simulate the added effect of seasonal or density-independent mortality. This work suggests that the outcome of competition in the presence of outside sources of mortality depends on whether this mortality occurs when populations are low and growing or when they are large and beginning to compete.

In this approach the Lotka–Volterra equations are extended to include terms for random or stochastic changes in number brought on by forces outside those described in the equations themselves. We then have pairs of equations describing two populations, N_1 and N_2, whose numbers fluctuate randomly even as they proceed in their systematic approaches to K_1 and K_2.

There are two extreme conditions: random destruction is applied either at high densities, as the populations approach equilibrium and exclusion, or at low densities.

Mathematical manipulation of pairs of logistic equations incorporating stochastic terms requires unusual mathematical skill. Moreover, different mathematical techniques are needed for the examination of growth rates at high and low densities (*a* and *b* of Figure 8.15). But the separate solutions for the two cases have now been found, first for the condition near equilibrium (May, 1973), and then for low densities (Chesson and Warner, 1981).

If stochastic or random death is applied to both populations when near equilibrium, the outcome of the competition is not changed, but the process of exclusion is accelerated (May, 1973, 1975). This suggests that the outcome of competitive interactions when populations are near

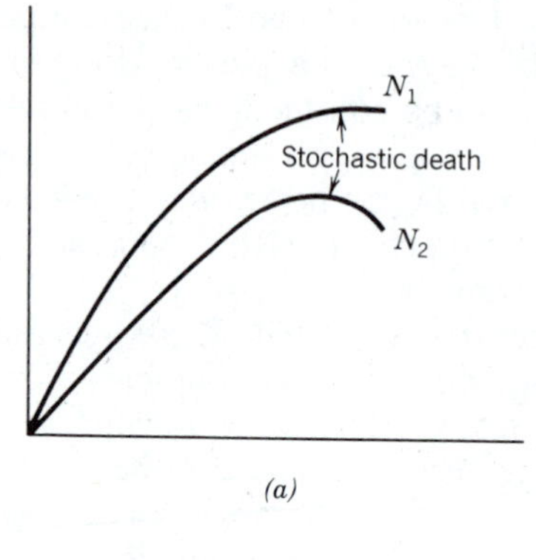

(*a*)

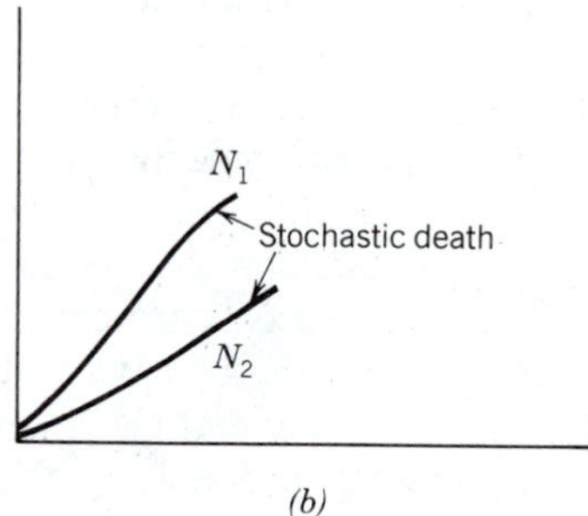

(*b*)

Figure 8.15 Insertion of random death as a stochastic process on interspecific competition.
Random death due to environmental fluctuation can be applied as N_1 approaches K and N_2 has a negative growth rate (*a*) or at low populations when both N_1 and N_2 have positive growth rates (*b*).

K is even more clear-cut in a real world of random hostilities of the weather than in simple laboratory systems. This result can be used to argue that in places of equable climate like the wet tropics the little random fluctuation that is present strengthens the effects of competition. An equable climate allows populations to grow close to K but what few changes are present will serve to knock out the weaker competitor rapidly. This argument suggests that communities of the wet tropics are unstable although complex.

But if the stochastic death is applied to populations at low densities the effect may be to boost the future growth rate of the weaker competitor disproportionately (Chesson and Warner, 1981). The likelihood of this result can be gauged intuitively. At low population densities extinction from random death is always likely so that individuals have been selected for adaptations that help them survive disaster. They will not have been so strongly selected for competitive ability. The poorer competitor of a pair may, therefore, be the better equipped to escape disaster. As a result, the weaker competitor attains the more rapid growth rate in the face of environmental change.

These recent refinements of the Lotka–Volterra model suggest, therefore, that fluctuating environments increase the chance of competitive exclusion and local extinction when populations are dense or near equilibrium, but decrease the effects of competition when populations are low.

PREVIEW TO CHAPTER 9

Data from the census of populations are arranged in a standardized manner, the result being called a life table. The conventions were first established by actuaries of the insurance industry, and the work of human demographers is used by ecologists as the basis for studies in animal demography. Problems of census are different for human and animal populations, so life tables must be constructed from different data. Animal numbers are counted by subsampling or by mark and recapture, but age at death may be hard to measure. A characteristic of life tables is that all columns can be calculated from any other. Generally the most useful column for ecologists is that describing survivorship, and this can be expressed conveniently as a graph. Human populations with a high life expectancy tend towards one extreme form of survivorship curve, showing a plunge with the accelerated mortality of old age. The other extreme is the intense mortality of young, followed by long survival, of animals like barnacles and starfish, whose swarms of pelagic larvae suffer mass death. Life tables can be a powerful tool for management of wild populations, as when they show what age classes can be harvested without serious population consequences. Some of the best data come from the Pribilof fur seal herd, life tables for which suggest that maintenance of seal numbers requires protecting females in the first three years of life as a prime consideration. A prime disadvantage of life table data for ecologists, however, is that the intrinsic rate of increase (r) cannot be calculated from the information included. Thus the numbers needed to predict changes in population of the kind modelled by logistic equations cannot be derived directly from life tables. But insight is given into the natural hazards encountered by different species, and into the species strategies that have evolved to counter these hazards. For instance, animals like mountain sheep and moose under wolf attack suffer mortality when very young and not again until very old, unless they are sick. Gray squirrels in an agricultural countryside, on the other hand, suffer modest but constant mortality throughout life.

CHAPTER 9

LIFE TABLES

For the study of population dynamics in the wild it is necessary to manipulate data from census to reveal the way numbers are changing through time. A convenient tradition for doing this is the device called a LIFE TABLE. Ecologists learned of this approach from the life insurance industry.

Typically an insurance actuary seeks to predict the expectation of life of people of varying ages. This is done from census data of a population for a single year, which shows the number of people in the population and the age of each. Also known are the number dying in the year, and the age at which each died. For convenience these data are converted to proportions of a standard-size population, usually 100,000, which is called a *cohort*. The people of a cohort can be separated into so many of each age (called an *age class*) and the number dying in each age class can be tabulated. From these data it is merely a matter of arithmetic to calculate, and tabulate, the death rate (or *mortality*) as a proportion of those surviving to each age, the deviation of age of each age group from the mean age of the population, and the expectation of life of each age group. The resulting table is called a *life table* (Tables 9.1 and 9.2). It is intended to predict what will happen to the members of the population from what has happened, since that is the desire of the insurers, but it will not do this very accurately for it assumes that the future will repeat the past, that the mortality will not change with time, and that there will be neither immigration nor emigration. But, as long as human populations change so fast, these inaccuracies cannot be avoided.

Ecologists want the sorts of data that appear in an actuary's life table, but would like to use the observation of a long time span to describe the actual population rather than to base the life table on the conditions at one census. Ecologists are under no compulsion to use the latest records of ever-changing conditions of life to speculate about future trends, and in this they have the advantage over an actuary. But they also must work with animals that do not fill out census forms with their dates of birth, nor whose passing is noted by a death certificate.

Various sampling and census techniques have been developed to estimate the number of inconspicuous or elusive animals (and most animals are one or the other), although the best always leave some uncertainty of the true number. Very numerous small organisms, such as insects or plankton, may often be sifted out of representative samples of their habitats, killed, counted, and the total population calculated by extrapolation. Less numerous or more motile organisms must be counted alive. They must be caught, marked, and let go again, and the number is estimated from the success of trapping.

One way of using trapping data is the *mark and recapture* method (or *Lincoln index*).[1] After setting many traps, and marking and releasing

[1]A practical guide to mark and recapture, with other sampling methods and statistics, is given in Tanner (1978).

Table 9.1
Human Life Table
Data are for the whole population of the United States in 1977. (From National Center for Health Studies, 1980.)

Age	${}_nL_x$	l_x	${}_nd_x$	${}_nq_x$	$\mathring{e}_x$
0–1	98,751	100,000	1,421	0.0142	73.2
1–5	393,693	98,579	268	0.0027	73.2
5–10	491,106	98,311	167	0.0017	69.4
10–15	490,355	98,144	173	0.0018	64.5
15–20	488,723	97,971	499	0.0051	59.6
20–25	485,756	97,472	650	0.0067	54.9
25–30	482,517	96,822	637	0.0066	50.3
30–35	479,306	96,185	677	0.0070	45.6
35–40	475,369	95,508	928	0.0097	40.9
40–45	469,565	94,580	1,428	0.0151	36.3
45–50	460,552	93,152	2,222	0.0239	31.8
50–55	446,727	90,930	3,379	0.0372	27.5
55–60	426,258	87,551	4,861	0.0555	23.5
60–65	396,531	82,690	7,095	0.0858	19.7
65–70	356,669	75,595	8,868	0.1173	16.3
70–75	305,147	66,727	11,768	0.1764	13.1
75–80	238,929	54,959	14,550	0.2647	10.4
80–85	164,964	40,409	14,596	0.3612	8.2
85+	165,352	25,813	25,813	1.0000	6.4

Table 9.2
Terms Used in Life Table Studies

FECUNDITY	is the potential reproduction of offspring, the maximum possible.
FERTILITY	is the actual reproductive performance of a population, measured by a reproductive rate. (Synonyms are BIRTH RATE and NATALITY RATE.)
NATALITY	is the number of offspring produced per unit population per unit time.
MORTALITY	is the number of individuals per unit population dying in unit time.

The following symbols are used as headings to life tables.

x	Age at the beginning of time interval; number of whole intervals lived.
l_x	SURVIVORSHIP: number of survivors of the cohort at exact age x.
L_x	AVERAGE SURVIVORSHIP: average number of survivors between ages x and $x + 1$.
d_x	MORTALITY: number of individuals dying between ages x and $x + 1$.
q_x	AGE-SPECIFIC MORTALITY RATE: probability of the L_x individuals dying before the age $x + 1$.
e_x	EXPECTATION OF LIFE at exact age x.

the catch, there will be a definite number of marked animals mixed in the wild population. If traps are again set, some of the animals caught the second time will have been caught before, and hence will be marked. The proportion of marked to unmarked in the trapped sample will be the same as the proportion of marked to unmarked in the whole population, if the traps take a truly random proportion of the population. Since the total number of marked animals is known, an estimate of the total wild population can then be calculated from the following simple equation:

$$N = M\,(R/C) \tag{9.1}$$

where

N = number in the population to be estimated
M = known number of marked animals in the population
C = total number captured in sample
R = the number of marked animals in the sample (that is, the number recaptured).

Unfortunately the assumptions behind this simple math rarely hold true. The difficult assumptions to meet are as follows:

1. The recaptured sample must be random so that a marked animal has the same chance of being caught as an unmarked animal. This is a hard condition to meet. Animals do not mix at random, nor is one individual as trap-prone as another. Much ingenuity in designing trapping programs and sampling statistics is required to minimize this difficulty (Overton, 1971).

2. The marks must be permanent and unambiguous.

3. Marked animals must have the same rates of death or emigration as unmarked animals (in particular, having been trapped must not frighten them away).

4. There must be no unrecognized recruitment (births or immigrants) between release and recapture.

Aging animals is usually more difficult than counting them, and age is a vital datum for constructing a life table. Some animals obligingly carry indices of age with them, such as wear on teeth and rings on scales, but most do not. Trapping and marking can be used as a guide sometimes, for at least it is known how long a recaptured animal has lived since it was first marked. This method is attractive for work with birds because they can be banded as young in their nests, but bird studies suffer the difficulty that usually few recaptures are made, most of the banded birds simply disappearing.

With sessile animals it is sometimes possible to mark a whole population and record the subsequent population history. Two attempts to do this have been important in the early development of life table techniques in ecology. One was the marking of a population of the sessile rotifer *Floscularia* by introducing particles of carmine into the water in which they lived. The animals build tubes from bits of substrate and continually enlarge them so that the population was dated by particles of carmine in the tubes. Growth of individuals, mortality, and replacement could then be monitored (Edmondson, 1945).

Many sessile marine animals have pelagic larvae that settle on hard surfaces, where they fasten and remain for the rest of their lives. If a rock surface is cleaned, or if a fresh surface like a sheet of glass is provided, population histories can be recorded by mapping the distribution of individuals on subsequent visits. Early work by Moore (1934) developed this technique and it has been used more recently not only for life table studies but for observations on competition (Connell, 1961). An obvious disadvantage of this technique is that no record is left of the mortality of the pelagic larvae, doubtless the greatest mortality suffered by the population.

The first life table for large feral animals was constructed for Dall mountain sheep (Figure 9.1) in the Mt. McKinley National Park in Alaska by Deevey (1947) using the published data of Murie (1944). The data set was an estimate of the age at death of a sample of 608 individual sheep as deduced from the annual rings on the horns

Figure 9.1 *Ovis dalli*: their bones yielded the first life table.
All dead Dall sheep are eaten by wolves, but their skulls lie on their barren habitat and can be aged. The sheep are killed when young or old but the middle years are safe.

of sheep skulls found on the tundra. Taking a thousand individuals as his cohort, Deevey constructed the life table given as Table 9.3. The table shows that mortality was high for young and old sheep but that there was a period of middle age when the animals were relatively immune from death. Since Murie (1944) had shown already that the principal cause of mortality for the sheep was attack by wolves, Deevey concluded that fit, healthy, adult sheep must be able to escape from wolves either by running or by the group action of herding, and that only the feeble young or the feeble old could be cut out or run down by the wolves.

Ecological conclusions of considerable interest thus emerged from the Dall sheep life table. Wolf predation can serve as a population-regulating mechanism for the sheep mainly through culling the very young before they have reached reproductive age. But the extent of this predation must depend on the size of the wolf population and this must be in part a function of the energy available to wolves from their catch of old sheep. If the wolves killed two many young sheep, there would be fewer old sheep for the wolves to eat some years later, with consequent repercussions for their own population. There must therefore be a rather complex relationship between the number of wolves and the number of sheep, an equation of energy balance whose solution yields a death rate for sheep.

Deevey (1947) used the possibilities of his sheep life table, together with a discussion of the early work with bird-banding, barnacles, and mark

Table 9.3
Life Table for Dall Mountain Sheep
A small number of skulls without horns, but judged by their osteology to belong to sheep nine years old or older, have been apportioned **pro rata** *among the older age classes. This table was constructed solely from examination of a collection of sheep skulls from Mount McKinley. It reveals that most mortality was suffered by the very young and the very old, which was as expected because wolves were the main source of mortality and they were known to hunt the weaker animals. (From Deevey, 1947.)*

x	x'	d_x	l_x	$1000\ q_x$	e_x
Age (*years*)	Age as Percent Deviation from Mean Length of Life	Number Dying in Age Interval out of 1000 Born	Number Surviving at Beginning of Age Interval out of 1000 Born	Mortality Rate per Thousand Alive at Beginning of Age Interval	Expectation of Life, or Mean Lifetime Remaining to Those Attaining Age Interval (*years*)
0–0.5	−100	54	1000	54.0	7.06
0.5–1	−93.0	145	946	153.0	—
1–2	−85.9	12	801	15.0	7.7
2–3	−71.8	13	789	16.5	6.8
3–4	−57.7	12	776	15.5	5.9
4–5	−43.5	30	764	39.3	5.0
5–6	−29.5	46	734	62.6	4.2
6–7	−15.4	48	688	69.9	3.4
7–8	− 1.1	69	640	108.0	2.6
8–9	+13.0	132	571	231.0	1.9
9–10	+27.0	187	439	426.0	1.3
10–11	+41.0	156	252	619.0	0.9
11–12	+55.0	90	96	937.0	0.6
12–13	+69.0	3	6	500.0	1.2
13–14	+84.0	3	3	10,000	0.7

and recapture, to illustrate the possibilities for life tables in ecology. Since then the constructing of a life table has become an important ecological tool.

A history of mortality quite different from that of mountain sheep is revealed when a life table is constructed for the gray squirrel (*Sciurus carolinensis*) in a Virginia woodlot (Figure 9.2). Mosby (1969) monitored a squirrel population by mark and recapture for six continuous years. Evidence of longevity provided by the marked animals was supplemented by estimating age from the size, weight, development of genitals, and the state of fur of the animals as they were caught. Mortality was recorded by the failure of animals to return to traps. In the resulting life table (Table 9.4) it is apparent that mortality was constant throughout life once the animals were six months old and hence likely to appear in traps. This is quite unlike the survival pattern of the Dall sheep, since there was no middle age of low mortality. Apparently this population of gray squirrels was exposed to forms of death against which there was no special defense in being prime and fit. These squirrels lived in a woodlot of an agricultural countryside where some of their original predators might have been lacking, a circumstance that invites the hypothesis that their life table records this predator-free existence. Other explanations must be possible, but the illustration serves to show how life table data invite ecological hypotheses for further testing.

Figure 9.2 *Sciurus carolinensis*: life table of constant death rate.
Mark and recapture data for gray squirrels in a woodlot show that death rates are constant throughout life, possibly a consequence of removal of predators by humans.

SURVIVORSHIP (l_x) EXPRESSED GRAPHICALLY

Different survivorships can be read from life tables simply enough, but they can be expressed even more clearly as graphs.

Figure 9.3 shows Deevey's (1947) computed survivorship curve for the Dall mountain sheep compared with survivorships of Edmondson's carmine-labelled rotifers and of Canadian herring gulls (*Larus argentatus*). Strangely enough, life expectancy of a tube-living rotifer seems to parallel that of sheep hunted by wolves, with long expectancy of life once the early years are safely behind it. The comparison is not really fair, however, because the rotifer life is taken, for the purposes of the analysis, to start when a tube is first formed, when in fact there is a highly vulnerable larval stage to be passed before the animal settles to form its tube. The herring gull data, like that from many other birds, suggest a pattern more like that of Mosby's squirrels, namely, a steady hazard throughout life. It is hard, on the basis of this, to think that birds are commonly hunted by predators who must seek the weak and infirm in the way that wolves do. Herring

Table 9.4
Life Table for American Gray Squirrels in a Woodlot
The data for this table were collected by mark and recapture techniques over six years. Unlike Dall sheep, the squirrels suffer roughly constant mortality throughout life. (From Mosby, 1969.)

x	l'_x	d_x	l_x	$1000q_x$	L_x	e_x
Age	Number Surviving in Age Class x	Number Dying in Age Interval per 1000	Number Surviving Beginning of Age Class per 1000	Mortality Rate per 1000 at Beginning of Age Interval	Average Number Living between Two Age Intervals	Mean Expectation of Life Remaining
0.5 to 1	93	204	1000	204	898	1.86
1 to 2	74	452	796	568	570	1.21
2 to 3	32	193	344	561	248	1.13
3 to 4	14	97	151	642	103	0.94
4 to 5	5	43	54	796	33	0.72
5 to 6	1	11	11	1000	6	0.50

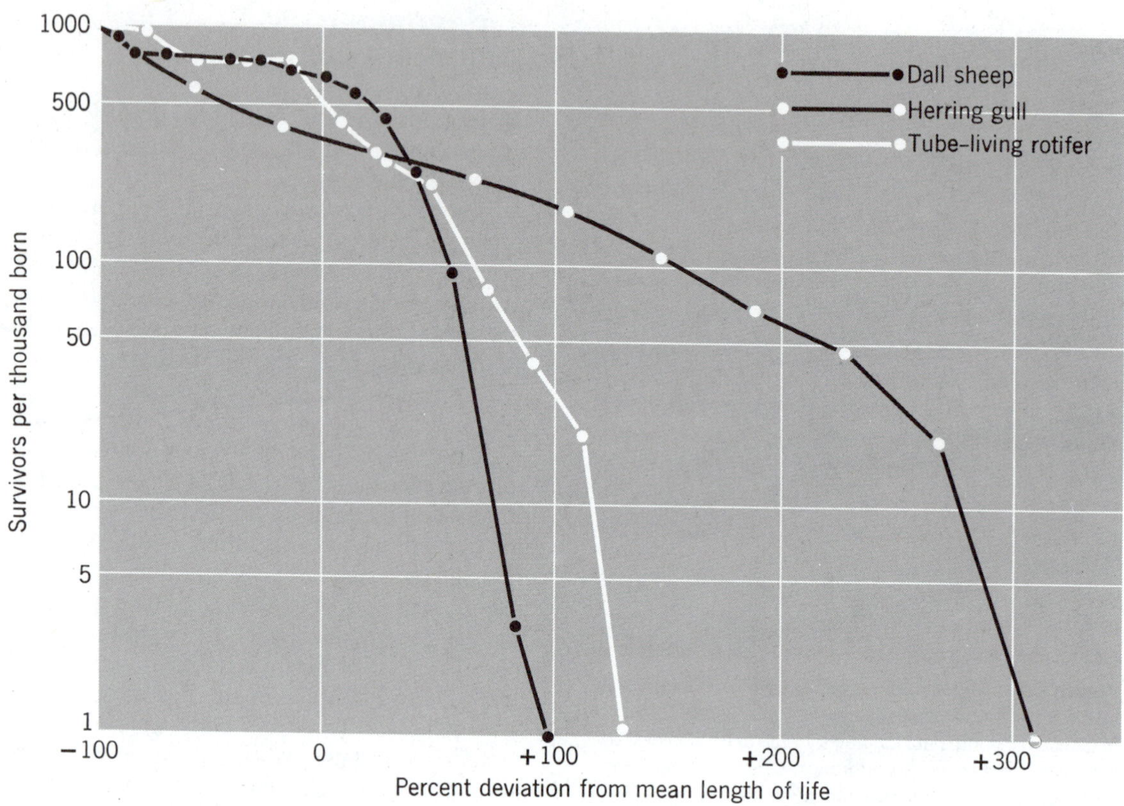

Figure 9.3 Survivorship curves of different animals drawn to a common scale. The tube-living rotifer has a secure middle age, like the Dall sheep, but herring gulls suffer constant mortality like the woodlot squirrels. (Redrawn from Deevey, 1947.)

gulls are probably little afflicted by predators anyway, and small birds may be nearly as conveniently taken by a hawk when they are healthy as when they are in poor condition.

In an early work on life tables in ecology, Pearl (1928) described the extreme forms that survivorship curves could take (Figure 9.4). The type I curve describes a population in which there is little early death as nearly all individuals live until they die of old age. This is the type of survivorship taken as a goal by modern medicine, but it must be extremely rare in nature—no obvious examples other than humans come to mind. Perhaps a majority of populations of wild animals approach the type III condition, where there is mass death of the very young, but secure life until old age for the few survivors. Many marine animals with pelagic larvae, like barnacles, echinoderms, and mollusks, have life histories that tend to the type III pattern. The type II curve is the intermediate state, a constant chance of death curve like that of woodlot squirrels or herring gulls discussed earlier.

Survivorship curves describe the fates of cohorts, but different cohorts can have different fates. Figure 9.5 shows the different survivorships of citizens of the Roman Empire, in Rome itself and in Roman North Africa. The base data are not of true cohorts, since the data were collected from epitaphs spanning some four centuries (Macdonnell, 1913). The data represent only Roman citizens with sufficient affluence to

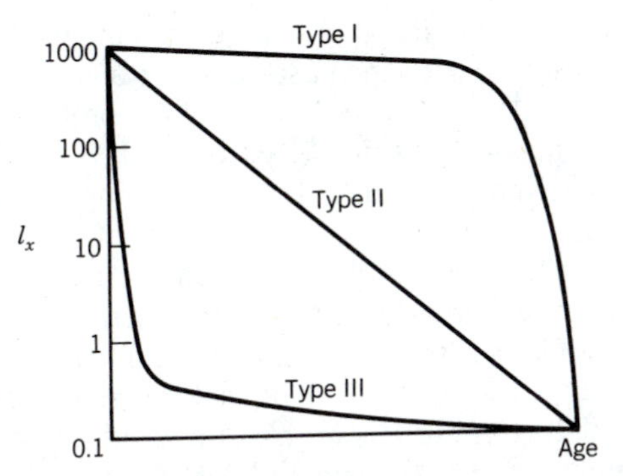

Figure 9.4 Hypothetical range of survivorship curves. Type I is approached by modern human populations in affluent societies. Type II represents constant mortality. Type III is heavy juvenile mortality and a relatively secure old age, a pattern found in many marine animals with pelagic larvae. (After Pearl, 1928.)

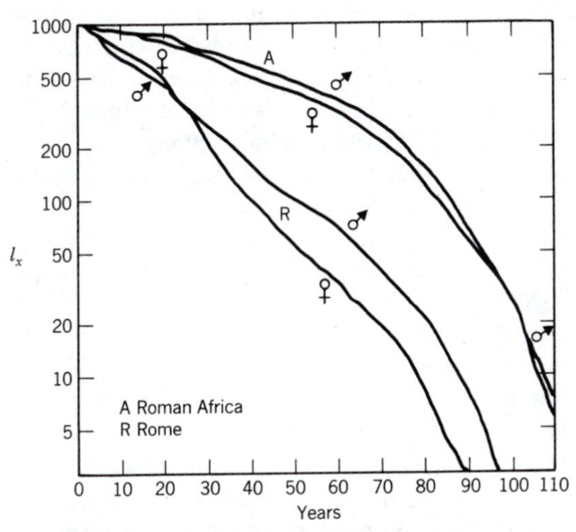

Figure 9.5 Survivorship in the Roman Empire. The data are for Romans sufficiently affluent to have earned an epitaph on stone. Even this class survived better in North Africa than in the city of Rome itself. Rome was apparently an unhealthy place throughout life. (Data from Macdonnell, 1913, and Hutchinson, 1978.)

earn an epitaph, yet there is a clear difference in the fates of even these people. The city of Rome itself was apparently a poor place to live compared with the provinces.

Figure 9.6 describes survivorship of two different cohorts of the tubed rotifers studied by

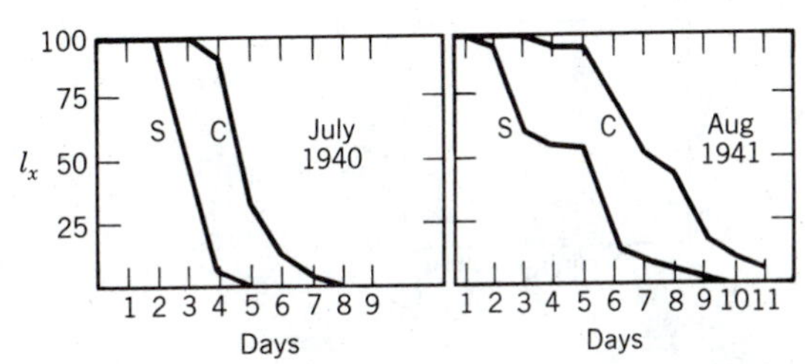

Figure 9.6 Survivorship of solitary and colonial rotifers. The data are for the *Floscularia conifera* populations studied by Edmondson (1945) and described on page 268. Animals living in colonies survived better than animals living alone. (From Hutchinson, 1978.)

Edmondson (1945) on two consecutive years. Some of the rotifers lived solitary lives anchored directly to plants, but others were colonial, living attached to each other in stalked arrays. The colonial life style improved survivorship markedly.

A survivorship curve is a good guide to the life history phenomena of a population or species. A type III survivorship curve (Figure 9.4), for instance, suggests strongly that the species has a breeding strategy of allocating resources to a large number of tiny offspring. If predation is important at any particular stage in a life history, a survivorship curve may hint at it. Data on survivorship, therefore, makes possible hypotheses about the lives of individual animals that can be tested in the field. Alternatively, survivorship can be predicted, at least in general terms, from the knowledge of the life history of a species. Models of how individuals behave lead to predictions about survivorship that can be tested against reality. This becomes a powerful tool in investigating species strategies (Chapter 11).

LIFE TABLES AS A MANAGEMENT TOOL

Many census data collected in commercial harvests, particularly of fish or whales, can be used to construct life tables. The life table then becomes the basic data set on which management decisions are made. One simple, possibly useful, hint to what is happening to a population is to construct AGE PYRAMIDS for the population.

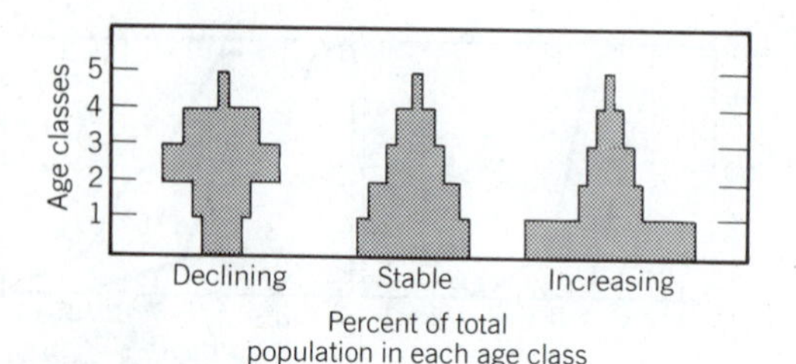

Figure 9.7 Idealized age pyramids.
The conclusions that the populations are declining, stable, or increasing come from considering the relative number of young animals. The conclusions only hold, however, if survivorship is not markedly different between the age classes.

In Figure 9.7 three hypothetical age pyramids are shown for declining, stable, and increasing populations. But care must be taken in deducing the actual dynamics of a population from such a data set, for much will depend on things such as the survivorship of each age class.

But the dynamics of a population can, in principle, be discovered from life table data. These dynamics are described by Lotka's equation thus:

$$\int_0^\infty e^{-rx} l_x m_x d_x = 1 \tag{9.2}$$

where r is, as usual, the intrinsic rate of increase and the other symbols are the standard life table terms.

From (9.2) it can be shown that the stable age distribution associated with a life table is

$$C_x = \frac{e^{-rx} l_x}{\int_0^\infty e^{-ry} l_y d_y} \tag{9.3}$$

where c_x is the fraction of the population aged x to $x + dx$.

In principle, Lotka's equation can be manipulated to model the effects of changing survivorship on growth rates and age distribution. Put in other words, we can predict the effect of changing catch rates on future yields. But there are two sorts of difficulty encountered in practice. One is the difficulty of assuming relationships between fecundity, growth, and population density when knowledge of species life histories may be incomplete. The other is the mathematical intractability of the expression. It is not possible, for instance, to write an explicit solution for r as a function of N in this system of equations. And yet r (the rate of increase) is what a manager needs to know and N (the numbers present) is what can be measured (Goodman, 1980; Beddington and May, 1977; Beddington and Taylor, 1973).[2]

Some of the useful information obtainable for planning a harvest is illustrated by work with the Pribilof fur seals *Callorhinus ursinus* (Figure 9.8). These pelagic animals haul ashore once a year on islands in the Bering Sea to breed, most of them on the two main Pribilof Islands, where they are easily rounded up and killed. Since people will pay good money for their fur, the ease with which the animals can be collected resulted in mass destruction of the herds in the late nineteenth century. Enlightened self-interest by the governments of Russia and the United States, who owned the islands, and Japan, in whose waters the seals fish in winter, led to an embargo on killing seals that was policed by warships. The herds recovered in a few decades and have been managed by international agreement ever since, the objective being to obtain the largest possible yield of furs by planned annual culling. Life table data have been gathered along the way that can be used to suggest ways of improving management.

Fur seal natural history is unusual. The seals are polygamous, with mature males guarding harems of females on the breeding beaches (Chapter 12). Successful bulls may guard, and mate, from 20 to 80 cows. The sex ratios, however, are roughly even—a fact which meant that the simplest possible management strategy of killing the surplus males and leaving the females alone worked to preserve the herds at a safe level, even when hunting was resumed. Each mature cow gives birth to a single pup each year, on the beach, guarded (or held captive) by the harem master bull. Shortly after giving birth the cow ovulates and is mated by the bull of her harem. The pup from this mating will be born on the beaches one year later, after the cow has

[2]An exposition of the mathematics of demography can be found in Keyfitz (1968) and the literature is reviewed in Goodman (1979).

Figure 9.8 *Callorhinus ursinus*: beach counts yield female life tables.
The fur seals are polygynous (Chapter 12) with large bulls maintaining harems on the breeding beaches. Management is to kill surplus males for skins but data for females from half a century of counting give guidance to management.

spent the intervening months fishing in the North Pacific. This is possible because implantation of the embryo into the fur seal uterus is delayed long enough for the next birth to occur after twelve months.

This seal life history suggests that the effects of survival on population growth can be modelled from data on females alone. Is the established policy of killing only bachelor males the best possible? Or can both the sizes of the herds and the yields of furs be increased by manipulating the age structures of females?

For an all-female life table fecundity must be expressed as the number of females of each age class bearing a female pup. The curve suggests that the ten years of middle life are the most reproductive (Figure 9.9); however, a measure of the reproductive success of the whole female population, the *net maternity function*, shows that most of the reproduction comes from the efforts of females in the first five of these ten age classes (Figure 9.10). The reason for this can be

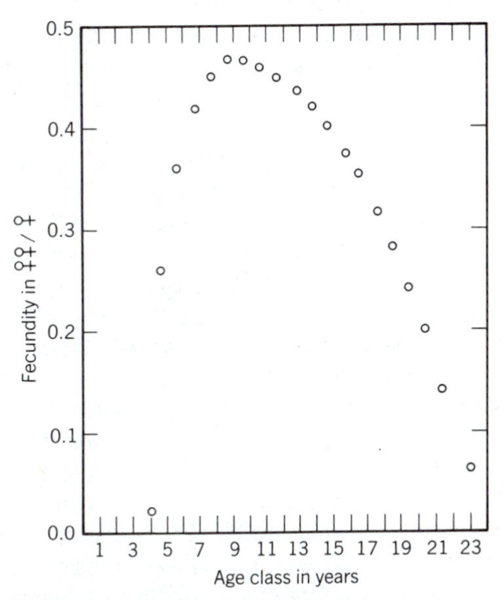

Figure 9.9 Fecundity of Pribilof fur seals.
The data in this figure, and in Figures 9.10 to 9.12 also, are for female offspring only and are all taken from Goodman (1980).

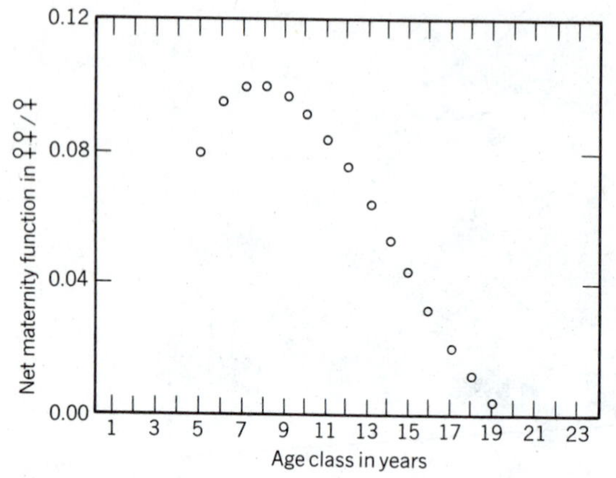

Figure 9.10 Reproductive success of Pribilof fur seals as a population.

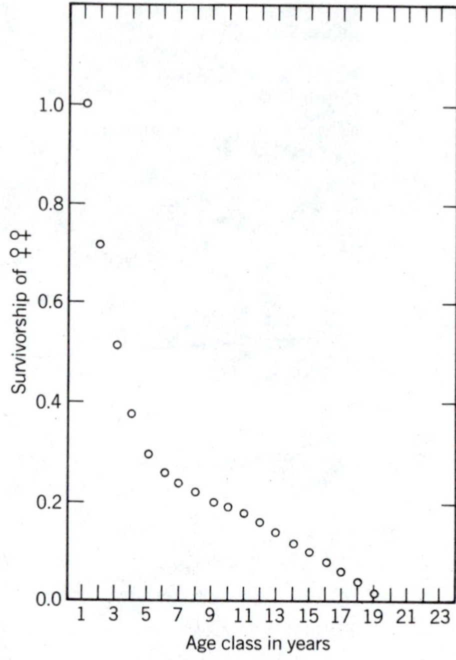

Figure 9.11 Survivorship of female Pribilof fur seals.

seen by consulting the survivorship curve for females (Figure 9.11). It turns out that most of the actively reproducing females are in the earlier reproductive age classes because not many survive into the later period. Individual cow fur seals can have a long reproductive life but the population of those actively reproducing is young (Goodman, 1980).

It is apparent from Figures 9.9 to 9.11 that population growth of fur seals is sensitive to both *age-specific fecundity* and *survivorship*. Proper management requires that these effects be understood. But here we come up against the prime difficulty of life table mathematics, that r cannot be expressed as a function of life table data for N, as described earlier. Goodman (1980) offers an ingenious solution to this mathematical problem by solving for the *sensitivity of r* rather than for r itself. In Figure 9.12 the sensitivity of the increase parameter r to age-specific fecundity is plotted as a function of age class. Inspection of this figure makes it clear that population growth is far more sensitive to changes in fecundity and survival of young animals than of older ones. Moreover, in these critical young classes survival matters more than fecundity. The message to the herd managers seems clear enough: protect cows through the first three years of life as a first objective.

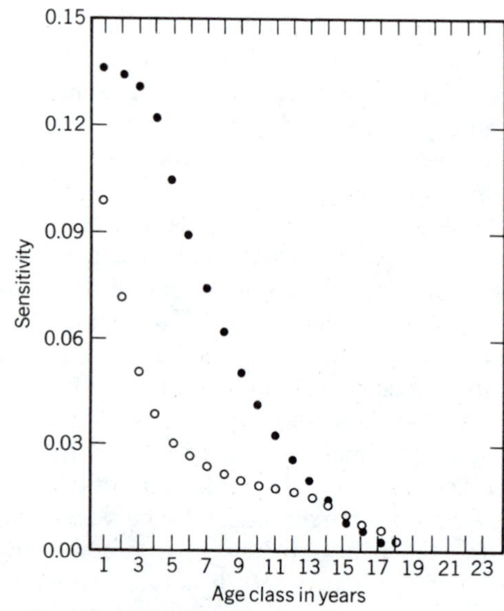

Figure 9.12 Sensitivity of population growth of Pribilof fur seals to changes in female fecundity (open circles) and survivorship (solid circles).

A LIFE TABLE FOR MOOSE

Life table data are available for a population of moose and the wolves that prey on them at Isle Royale, where a fortunate combination of circumstances makes census of both moose and wolf numbers less difficult than elsewhere. Most important is the isolation of the animals on the island, which restricts immigration and emigration, and the fact that the island is a reserve of the U.S. National Park Service where hunting is prohibited. In the heavy snows of winter the animals can be seen and counted from the air. Mech (1966, 1970) found that low-altitude flying on regular transects, when the noise of the aircraft set moose running, made direct counts of the moose possible. In winter the wolves hunted moose as a pack that could be followed from the aircraft, allowing a count of wolf numbers. Male moose could be distinguished from females from the air by their antlers, and the proportion of cows running with calves beside them allowed a calculation of the birth rate. Data for moose mortality came from examining moose remains, thus providing the numbers needed for the calculation of a life table for moose (Table 9.5).

The moose life table shows that the survivorship curve is sigmoid, or type II (Figure 9.1), inviting the hypothesis that moose mortality is primarily caused by wolf attack on the young or the old as it was for Dall mountain sheep. Mech (1966) was able to test this hypothesis directly from his observations. Examination of moose remains confirmed that all had been eaten by wolves. More important were observations of wolf hunting itself. In an aircraft, Mech followed the wolf pack as they hunted no less than 69 times and nine times he circled close overhead all the way from the find to the kill. Twice his pilot landed him quickly enough in the vicinity of a kill for him to run to the victim, chase off the hungry wolves as they fed, and examine the prey before it was much eaten away. Both these animals were old. These direct observations showed that the wolves did not kill moose of the active middle years. Fit moose were chased but turned at bay with such preparations for self-defense that wolves abandoned the attack (Figure 9.13). Apparently

Table 9.5
Life Table for Moose on Isle Royale
These moose are preyed upon by wolves, as are the Dall mountain sheep, and, like the sheep, suffer most mortality when young and old. (From Mech, 1966.) The data have not been converted to a standard cohort, and wear classes are used instead of years, since the correlation between the two is not precise.

Wear Class	Number of Moose Remains Found	Percentage of Total Mortality	Mortality	Population
I	1	2.56	2.17	85.00
II	2	5.12	4.34	82.83
III	1	2.56	2.17	78.49
IV	4	10.24	8.68	76.32
V	8	20.48	17.36	67.64
VI	8	20.48	17.36	50.28
VII	6	15.36	13.02	32.92
VIII	5	12.80	10.85	19.90
IX	4	10.24	8.68	9.05
IXA				.37
Total	39	99.84	84.83	492.80

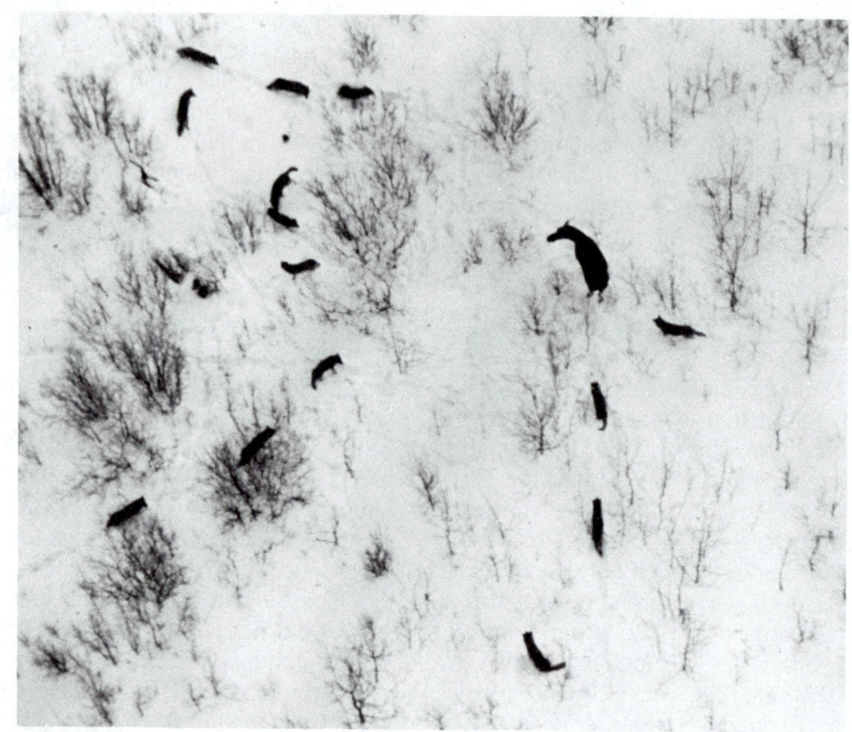

Figure 9.13 *Alces americana*: stands his ground under wolf attack.
The wolves left the moose shortly after this photograph was taken. The moose was in the prime of life and too dangerous a proposition for the wolves. Immunity of prime moose to wolf attack is reflected in their life table.

a moose in the prime of life is so formidable an opponent that wolves will avoid it if safer victims are available.

The moose life table data on Isle Royale, therefore, allow valid inferences about the behavior of moose predators over the time interval sampled. The relationship between moose defense against predation and the pack-hunting ability of wolves resulted in the low ecological efficiency of about 1.3% described in Chapter 4.

PREVIEW TO CHAPTER 10

Predation can be studied from different points of view. It is a powerful selective force that molds species strategies and the activities of predators often impose structure and order on communities. But in this chapter we concentrate on predation as it affects populations of the prey, asking if predators are important in setting the balance of nature. Since many ideas of the importance of predators at controlling their prey have been influenced by erroneous conclusions about the effects of predators on deer these errors are first examined. There is no valid evidence that high populations of deer in postsettlement America were a consequence of shooting wolves and mountain lions. A predator–prey system is amenable to simplifying assumptions and mathematical modelling. The simplest differential equations predict coupled oscillations between populations of predators and prey. Oscillating populations can only be contrived in the laboratory, however, if the prey are given the opportunity to hide or disperse away from the concentration of predators during predator population highs; otherwise the prey are exterminated. More complex difference equations that provide for lag effects predict local extinctions, not coupled oscillations. No coupled oscillations have been demonstrated in nature, not even among arctic fur-bearing animals. Dispersal of prey away from concentrations of predators would be favored by a patchy environment. This can be mimicked in laboratory systems, where both predators and prey can be shown to exist in low numbers as they perform a sort of hide-and-seek around the patches. In two celebrated examples of biological control of pests (ladybugs on scale insects, caterpillars on cactus) the final result was also low populations of predator and prey chasing each other around environmental patches. It is argued that large predators like wolves or big cats seldom exercise comparable control on their prey because their killing power is inadequate to take any but the sick, the old, and the young. The killing power of many small predators like wasps or spiders may be so great in some circumstances that they can act like ideal predators in simple mathematical models, a circumstance that may be behind some of the success stories of biological control. But there is also reason to believe that many small herbivores have evolved successful defense mechanisms and that these animals are taken by predators only when sick or old. The population effects of predation will not only result from their changes in number through breeding but also from changed behavior when prey is abundant. Ecologists have adopted the terms *numerical response* and *functional response* for these two properties of predators. Functional responses of simple animals without powers of learning (most predators) are limited and do not result in density-dependent control because their killing powers become saturated when prey is abundant. But vertebrate predators capable of learning are shown to be able to concentrate on abundant prey to the exclusion of rare items from the diet. This functional response of vertebrates can result in significant density-dependent control up to a limiting prey density.

CHAPTER 10

PREDATION: THE POPULATION CONSEQUENCES

This chapter examines the effects of predation on population sizes of both the prey and the predator itself. Predation can also be studied from other points of view. It is useful, for instance, to examine the possible tactics of a predator, asking questions about the alternative ways in which a predator might hunt safely and about the possible responses of a species under attack (Chapter 11). Then again the act of predation gives structure to communities so that much of the work of studying the process of community building becomes a study of the relationship between predator and prey (Chapters 21 and 25). But predation is so obviously a possible population regulator that this role is worth special notice.

THE PREDATOR DEFINED

A PREDATOR IS AN ORGANISM THAT USES OTHER LIVE ORGANISMS AS AN ENERGY SOURCE AND, IN DOING SO, REMOVES THE PREY INDIVIDUALS FROM A POPULATION. This definition allows the concept of predation to be extended to include herbivory as well as carnivory. Working ecologists now often talk of "predation" when describing sheep hunting grass or squirrels searching for nuts. Also included in the definition are PARASITOIDS, which are organisms that, although parasitic, kill the host. The important *parasitoids* are animals like hymenopterans whose larvae feed on the body of the host, which does not die until the *parasitoid's* development is complete. True parasites are not included in the definition, since the body of the host is not removed from the population, and the same argument excludes herbivores when the damage they do to the plant is minimal. These parasites and herbivores, however, do have effects on the population of their prey, and can be considered alongside true predators.

CLEARING THE AIR: THE KAIBAB FICTION EXPOSED

The development of studies of predator and prey has had to contend with anecdotal evidence, some of it quite unreliable. Among the anecdotes is the story of the mule deer on the Kaibab Plateau, whose numbers were claimed to have risen drastically after the native predators were shot out. A whole generation of ecological thinking was influenced by this tale and it is still repeated in many introductory biology textbooks. The anecdote is now known to be without value as evidence for the role of predation, however, and it makes an instructive case history of what to avoid.

The Kaibab Plateau is a wild area bordering the Grand Canyon, and it was declared a national park in the conservationist era of Teddy Roosevelt at the turn of the century. All the big predators of the park were then shot out by government hunters while, at the same time, a complete embargo was placed on the hunting of deer. Then, in the 1920s and 1930s came reports of too many deer in the park. Unfortunately, we have no good census figures for deer present in the park at any time in its early history, only various visitors' and wardens' "estimates." These are perhaps good enough to suggest that the herd tripled or even quadrupled its size in 20 years, although they by no means prove that it did. Such an increase need not be surprising for any place as disturbed as the Kaibab had been, for the frontier had passed through the Kaibab 20 years before it had been declared a park. Ranchers had herded 200,000 sheep, 20,000 cattle, and many horses there, and they had doubtless set their usual fires (Rasmussen, 1941). No doubt they had shot many deer too. That the removal of such exploitation as this should result in quadrupling the deer herd need surprise no one and there is no need to invoke "shooting predators" to explain it. Such are the facts of the matter as well as they can be discerned (Caughley, 1970).

This true story became embroidered with political arguments about management practices on the Kaibab Plateau. Various visitors to the park remarked on the "great numbers of deer," and some said the herd should be culled before the range was damaged by overgrazing. These matters reached the press and led to a fateful progression of assertions about what was happening, as shown in Figure 10.1. Diagram *A* gives the data as reviewed by Rasmussen (1941). The solid line gives the record of the deer population as estimated by forest rangers whose job it was to manage the park. Their estimates do not represent true census, since none was attempted, but they are the most reliable data available. More striking are the far higher estimates of a series of visitors to the park given as open circles. None of these estimates is based on census either; their value as evidence may be likened to the value of hearsay evidence in a court of law. But by accepting the lowest and highest of them it is possible to draw the dashed curve of diagram *A*, which presents the exciting prospect of a herd rising from 4000 animals to 100,000 before a precipitate decline.

With this postulated history of the Kaibab deer herd in print in an ecological journal, the anecdotal evidence became ammunition in a conservation debate. The supposed history was used to support the notions first that shooting predators brought on a plague of deer, second that the deer then destroyed their food supply by overgrazing, and third that there was then mass death, which returned the herd to low numbers. In the ensuing political argument the supposed history of the dashed curve was cleaned up and redrawn in the form of diagram *B* (Leopold, 1943). It will be noticed that this diagram has acquired a beautiful sigmoid curve (for which there were absolutely no data despite the little crosses drawn on the line) and that there has been selection of the forest ranger estimates when the population was supposed to be low but of the most extreme visitor estimates when the population was claimed to be high. Some telling remarks and claims were added to the diagram to illustrate points for the politicians for whom it was intended. Unfortunately this attractive diagram of an intriguing anecdote was reprinted in a standard ecology text, the authors of which did not realize that the data were spurious (Allee *et al.*, 1949). It was because of this that the history of the Kaibab deer herd became the example of the supposed controlling influence of predators quoted in so many beginning biology textbooks.

But the history is not true. There is no evidence that numbers of deer on the Kaibab grew exponentially, nor that there was sigmoid growth. We do not know if there really was a population explosion of deer on the Kaibab at all, nor if the

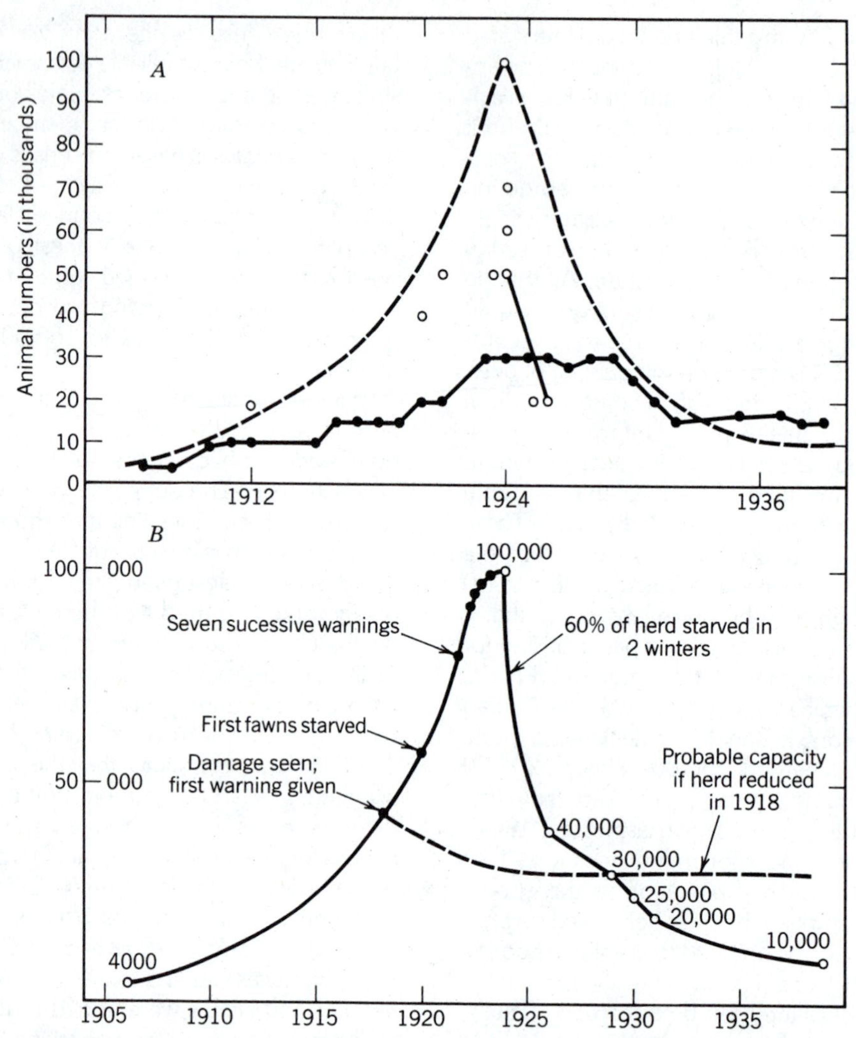

Figure 10.1 The Kaibab deer herd fiction; a history of embroidered data. (*A*) Population estimate of the Kaibab deer herd, copied from Rasmussen (1941). Linked solid circles are the forest supervisor's estimates; circles give estimates of other persons, and the dashed line is Rasmussen's own estimate of trend. (*B*) A copy of Leopold's (1943) interpretation of trend (after Allee *et al.* (1949) and Caughley, 1970.)

population "crashed." All we have is evidence of a not unreasonable fluctuation of numbers following disturbance. No data at all connect this fluctuation in deer numbers with predators.

There appear to be no properly documented histories of large herbivore populations that can be examined in lieu of the Kaibab fable. Even if there were, they probably would tell us little. Suppose the census of the Kaibab deer herd had been well and properly done from, say, 1870 to 1940, so that we had a real population history to talk about. Suppose further that the flights of imagination making the fiction actually were true—that there really had been logistic growth up to 100,000 deer and then a crash—this would tell us nothing at all about the effects of predators. Our hypothetical deer herd could have been responding to changes in forage, or to climate, or to disease just as easily as to changes in numbers of predators. Mule deer and the like often do well on land managed for range, forestry, or farm, being browsers well suited to systems disturbed by human activity.

It will be evident that understanding the effect of populations of wild predators on wild prey requires not only a properly censused population history of the prey but analyses of causes of death—we must show directly that predators have an effect. Alternatively we require control populations, some under predation and some without. This is very difficult to achieve with animals like deer and wolves or mountain lions.

MODELS OF PREDATION

The relationship between predator and prey is an attractive subject for numerical modelling, since both populations can be described by equations showing how numbers of individuals change over time. Unfortunately it is difficult to make numerical models that are realistic, the principal difficulty being the excessive simplifying assumptions that are necessary. The simplest algebraic models tend to predict extinction after a few generations. Simple calculus models predict coupled oscillations that are rarely, if ever, found in nature. The algebraic models assume that the kill rates achieved by predators are direct functions of the predator population, so that the kill rate changes only as the numbers of predator change. More complex models allow predators different killing abilities according to the density of the prey. Finally, simulation models seek to represent all the important interactions of both predator and prey. Constructing and testing these various models have proceeded alongside the natural history of predator–prey systems.

The Simplest Algebraic Models

An ideal predator is one whose every hunt is both successful and results in unit increase of fitness. For the ideal prey there is fitness granted in proportion to the time elapsed before it is caught. This ideal system is purely *numerical* since all responses are changes in numbers of both predator and prey. A perfect hymenopteran PARASITOID hunting caterpillars that face no other significant cause of death might meet these twin ideals. The perfect hymenopteran ends every hunt by stinging a caterpillar and laying its egg under the caterpillar's skin, being assured thereby of one unit of fitness after its larva has destroyed the caterpillar. For the prey caterpillars, success at hiding from wasps leads to emergence as a moth and fitness. Fecundity of the predator in this ideal system sets the kill rate, since every wasp egg is implanted in a caterpillar. Fecundity of the prey species sets the availability of prey. The outcome of the clash between populations of these ideal predators and prey becomes a function of their initial population densities alone.

If the fecundity of predator and prey are equal, there will always come a time t_g when the last generation of prey is eliminated and both populations perish. This is given by

$$t_g = n/p \tag{10.1}$$

where n is the initial density of the host and p is the initial density of the parasite (De Bach, 1964).

If the host can outreproduce the parasite, its population can escape the disastrous effects of being attacked provided that

$$\frac{p}{n} > \frac{(h - s)}{s} \tag{10.2}$$

where h is the fecundity of the host and s is the fecundity of the parasite.

This very simple algebraic model effectively allows only one ending to the contest: extermination. If the prey is not quickly killed off as a consequence of not being able to outbreed its attacker, its population will grow until some non-predation curb is applied, and then the predators will quickly catch up, attain the critical density necessary for them to exterminate their prey, and proceed to do so.

The obvious difficulty with this numerical model is that the predator is too efficient. Predators that kill on contact are inherently likely to exterminate prey. Realistic models must let some prey escape, must allow the prey to outbreed the predators in some circumstances, and must introduce lags into the population response of the predators such that the time never comes when too many highly efficient predators are hunting too few helpless prey. To improve the realism of models, ecologists turned to calculus.

The Coupled Oscillation Hypothesis

Calculus makes possible simplifying assumptions that are less unreal than those required by algebraic models. Rate of growth of predator populations is a function of the rate of successful encounters with prey and the rate of predator deaths from all sources. Rate of growth of the prey population can be assumed to be a function of the natural rate of increase when without predators together with the rate of fatal encounters between prey and predators. Then, for the prey population, species 1,

$$\frac{dN_1}{dt} = r_1N_1 - \gamma_1N_1N_2 \qquad (10.3)$$

and, for the predator population, species 2,

$$\frac{dN_2}{dt} = \gamma_2N_1N_2 - d_2N_2 \qquad (10.4)$$

where

r_1 = intrinsic rate of increase of prey
N_1 = number of prey, N_2 = number of predators
γ_1 = fraction of contacts that prove fatal to prey
d_2 = rate of death of predators in absence of prey
γ_2 = rate of growth allowed predator per unit contact with prey.

Equation (10.3) relates change in the prey population to its intrinsic rate of increase, as in the logistic equation. The limit to growth, however, is seen to be set not by carrying capacity, K, but by the rate of fatal encounters with the predators.

Equation (10.4) ignores the intrinsic rate of increase of the predator population as being irrelevant. Real predators, says the equation, have their rates of growth set by the food supply, which is in turn set by the rate of fatal encounters with prey. Furthermore, no carrying capacity for the predator population is assumed because limits to predator numbers are assigned to a death rate from unknown causes, possibly old age.

Equations (10.3) and (10.4) are known as the Lotka–Volterra equations of predation. Like the competition equations based on the logistic (Chapter 6), they were independently derived by the two workers as the logical application of calculus to predator–prey systems (Lotka, 1925; Chapman, 1931). Like the Lotka–Volterra equations for competition, the predation equations make simple assumptions that can tempt experiment with simple laboratory systems designed to meet those assumptions.

Formal solution of the equations yields the following:

$$\left(\frac{N_1y_2}{d_2 \exp \frac{N_1y_2}{d_2}}\right)^{d_2}\left(\frac{N_2y_1}{r_1 \exp \frac{N_2y_1}{r_1}}\right)^{r_1} = C \qquad (10.5)$$

where C is a constant of integration. There are two simple ways of expressing this result graphically: as plots of numbers of predators against numbers of prey, or as plots of both populations against time (Figures 10.2 and 10.3). *The solution predicts that the numbers of both predator and prey should oscillate and that the oscillations should be coupled.* This is the basis of THE COUPLED OSCILLATION HYPOTHESIS.

Figure 10.2 is the numbers against time plot. When predators are scarce but prey numerous, the predators should be able to build their population quickly, inevitably reducing the population of their prey. Eventually the predators should be numerous and the prey scarce, at which time the predators should compete with each other so vigorously for the prey food that they would suffer enormous mortality from starvation. The predation pressure on the prey would then be relaxed, and the prey population should expand, thus completing the cycle. There should always be "too many" predators or "too many" prey. Figure 10.3 is the numbers of predator against prey plot. The numbers circle a singular point, there being no stable number of either.

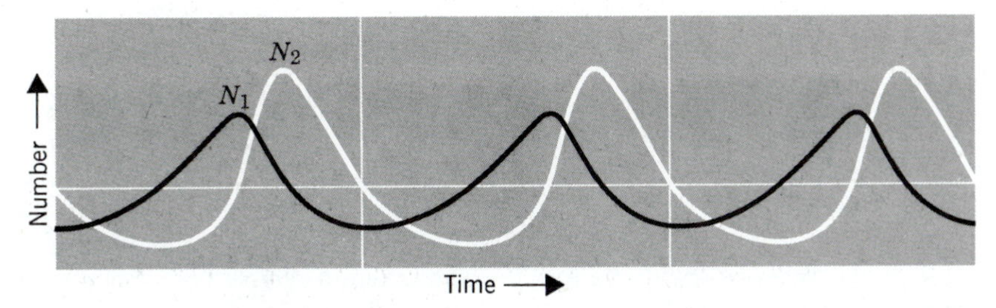

Figure 10.2 Coupled oscillations predicted by the Lotka–Volterra equations for predation. (After Volterra, from Chapman, 1931.)

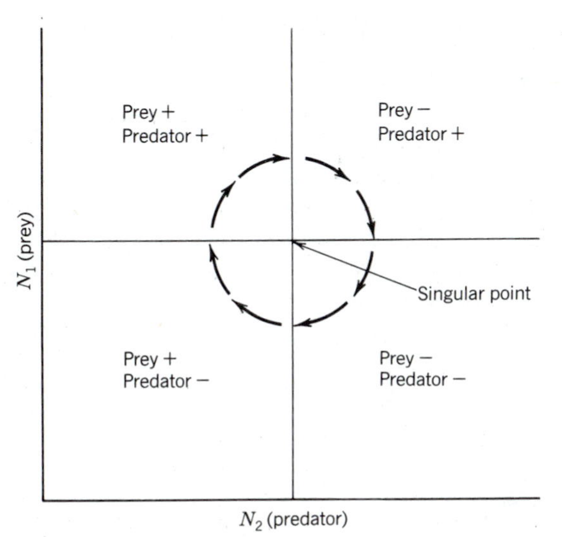

Figure 10.3 Population trajectories in a perfect coupled oscillation.
Numbers of both predators and prey circle a singular stable point in perpetually balanced imbalance. No such system has been demonstrated in nature.

The idea that oscillations in related processes should be coupled is a general one with many examples in studies far removed from ecology. Economists are familiar with the idea as the theory of supply and demand, and a form of it was well and bitterly known in the American Midwest of the depression years as the "corn–hog" cycle. When corn sold well, farmers planned next year to keep more corn for sale even if it meant rearing fewer hogs. This resulted in a corn glut and falling prices. Many farmers responded by keeping their corn back next year to fatten hogs instead, which resulted in a glut of hogs and a shortage of corn. Recurrently there would be a glut of hogs or a glut of corn, each glut being attended by a crop of farmer bankruptcies and by a shortage of one of the coupled commodities. In engineering, such coupled relationships form the foundation of servomechanism theory, and they are discussed in physics texts under the heading "the theory of coupled oscillations."

The simplest possible dynamic model of a predator–prey system, therefore, predicts *coupled oscillations.* Numbers of predator and prey should oscillate, first the predators gaining the upper hand, then the prey population growing rapidly after the predators had declined. This model does not do violence to general natural historical experience for numbers of many common animals do fluctuate somewhat from year to year. Where systems are known to be simple, as in arctic regions with few species, quasi-rhythmic fluctuations in number are a familiar property of life; the lemming hordes that come and go, fluctuating hare numbers, and good years and bad for the arctic fur trade (Chapter 13). This simplest model has, therefore, an inherent plausibility. Although modern ecologists no longer consider the model realistic, and have generally given up the quest for coupled oscillations in nature, the initial plausibility of the model generated much interest and debate in ecology and has shaped modern thinking. Particularly interesting were the laboratory studies developed to test the model.

Laboratory Tests of Coupled Oscillation Theory

A variety of laboratory systems have been used for experiments on the generation of coupled oscillations. Results show that oscillating numbers of both predator and prey can be generated, but only if special provisions are made for the escape of the prey when the predator populations are dense. If prey cannot find refuge, the prey becomes locally extinct and the predators starve. Well-known examples of these experiments are the work of Gause (1934) with protozoa and Utida (1950, 1957) with insects.

Gause (1934) used for his prey populations cultures of *Paramecium* like those used in his competition experiments (Chapter 6) and for his predator another protozoan *Didinium*. This animal is about the same size as a *Paramecium*, and it seizes its quarry one at a time to suck out their contents with a tubular proboscis (Figure 10.4). Both animals reproduce by binary fission. Experiments were begun by introducing a few individuals of *Didinium* into a flourishing *Paramecium* culture; the *Didinium* quickly reproduced at the expense of its prey until every *Paramecium* was killed and eaten. Every *Didinium* then died of starvation. This result is given in Figure 10.5 as a prey against predator plot. This shows why a coupled oscillation did not develop. When the relative numbers reached the quadrant where prey should be increasing as predators declined according to the model, the real life *Paramecium* prey fails to increase. The experiment shows that the assumption of the model that prey could escape their predators and replenish their population after the predators begin to starve could not be met in the experimental system.

Gause then altered his experimental design to simulate survival in the prey population. He reinoculated his cultures with a few fresh individuals of *Paramecium* after the original population had perished and the *Didinium* were beginning to

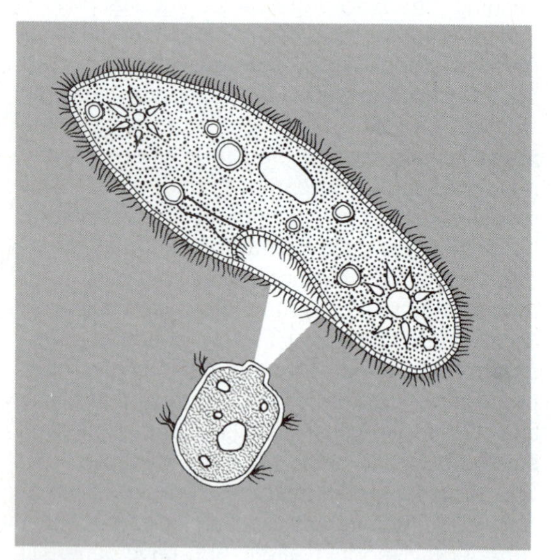

Figure 10.4 *Didinium nasutum* devouring *Paramecium caudatum*.
Didinium is so effective a predator of *Paramecium* that introducing a few into a *Paramecium* culture results in the complete extermination of the paramecia. A *Didinium-Paramecium* system can be made to oscillate by adding "immigrant" paramecia at the time when the *Didinium* population is reduced through starvation.

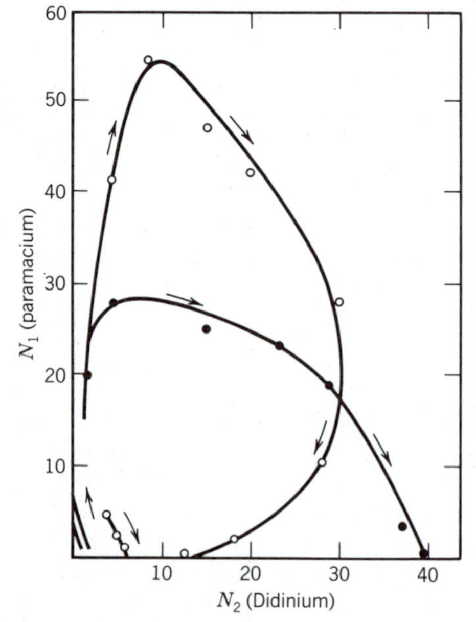

Figure 10.5 Trajectories of populations of *Didinium* and *Paramecium*.
Compare with Figure 10.3 for perfect coupled oscillations. (From Gause, 1934.)

starve. These fresh paramecia he called "immigrants." If this infusion of immigrants was timed correctly, and repeated with every population low, the populations did oscillate in the way required by the model (Figure 10.6).

Systems of bean weevils and parasitic wasps devised by Utida (1950, 1957) generate the required coupled oscillations without the necessity for the periodic arrival of immigrants from outside the system. Cultures of the weevil, *Callosobruchus chinensis*, can be maintained in petri dishes 1.8 centimeters by 8.5 centimeters in which 10 grams (about 50) of beans, replaced every three weeks, are sufficient food to maintain a population of several hundred weevils. The air space in a small petri dish is also quite large enough to maintain a population of the tiny parasitic wasp *Neocatolaccus mamezophagus*. These wasps hunt the fat, final instar larvae by feeling for them in their burrows within the beans with their long ovipositors and laying eggs beneath their skin. The experimental design allowed a population of weevils started with eight pairs to develop for ten weeks, when the population size was between three and four hundred. Four pairs of wasps were then introduced into each dish. Census was taken every three weeks, using ether to quiet both populations. The animals were gently disentangled from the remnants of old beans for counting, after which the beans were replaced by fresh ones. Figure 10.7

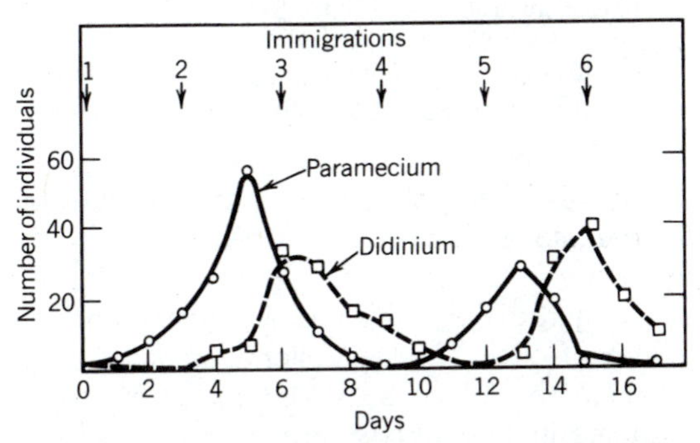

Figure 10.6 *Didinium–Paramecium* system supplied with "immigrant" prey after each *Didinium* high.
Coupled oscillations have been achieved by artificially meeting the requirements of Lotka–Volterra equations that prey can avoid extinction by hiding or dispersing. (From Gause, 1934.)

gives an example of the results. Irregular, but undoubtedly coupled, oscillations have persisted in these systems for more than 110 generations.

The bean weevil and wasp system in a petri dish apparently provides for prey escape without the need for "immigrants." It may be that the physical habit of burrowing into a bean gives the weevil larvae a reasonable chance of avoiding attack, although this may be doubtful because of the skill of wasps for feeling down tunnels with their long ovipositors. A more likely cause of escape is that the generations come to overlap slightly so that a few weevil larvae are too young to be attacked while the wasp plague is at its height. This is inherently likely because the mean wasp generation is a few days less than the mean weevil generation, and yet the wasp must stay synchronized to the weevils by its necessity of attacking only last instar larvae.

Another mechanism allowing some weevils to escape derives from the random element in wasp hunting, which means that some wasps must attack larvae that other wasps have already attacked. Wasp parasitism, although completely lethal, does not kill at once so it is possible for the same prey to be attacked by a number of different predators, each of which delivers a potentially fatal blow. When many wasps hunt a few larvae by a process of random search, some larvae will be hit only once, but others will be hit many times. A few may not be hit at all, and these may provide the adult weevils that can rebuild the prey population after the wasps have died.

Coupled oscillations, therefore, can be achieved in specialized laboratory systems. The vital provisions seem to be that only the two species be present and that the prey be provided with a refugium that remains inviolate however dense the predators might be. If not a strictly literal refugium, then some chance function in the encounters must be present to provide a high probability of a few prey surviving the lean periods. The chance of these constraints being met in the wild may not be large.

Some population histories in the wild have attracted interest for fifty years because they look superficially like records of coupled oscillations.

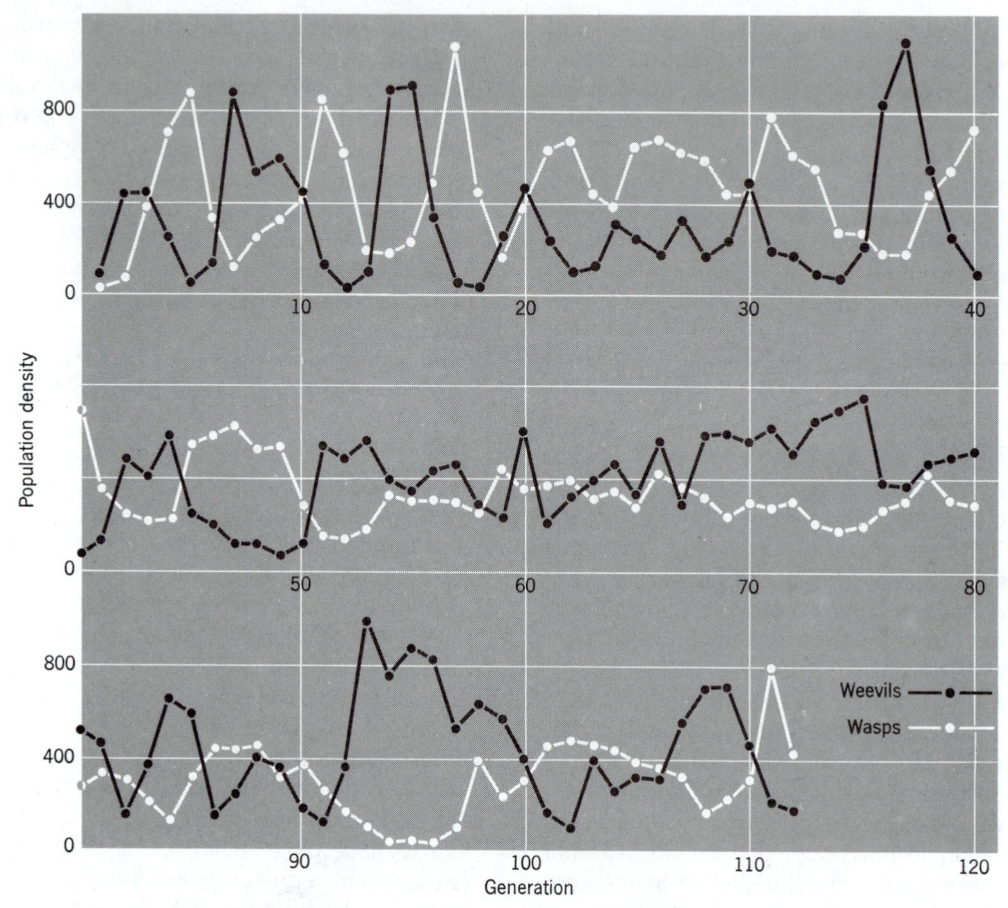

Figure 10.7 Coupled oscillations in Utida's wasp–weevil system.
The system consists of beans (renewed at intervals), weevils, and a parasitic hymenopteran in closed petri dishes. The populations oscillate without further manipulation, apparently because some weevils escape during dense predator populations because the wasp life cycle is not perfectly synchronized with the weevils or because some larvae are overlooked by the wasps, which sting others repeatedly instead of spreading their attacks evenly. (From Utida, 1957.)

These are the histories of fur-bearing animals and their prey in arctic and subarctic regions. But the great preponderance of the evidence is against these histories reflecting true coupled oscillations.

Figure 10.8 shows a data set of the relationship between hare numbers and lynx numbers over a number of years in Canada and is a figure that appears in many basic biology texts with the hint that this describes coupled oscillations between predators and prey. Doubt that this is so can be quickly gleaned by a close inspection of the figure, for this shows that peak densities of the predator are almost synchronized with those of the prey. Compare this figure, for instance, with Figure 10.2, which is a direct graphical treatment of the Lotka–Volterra predator–prey equations as drawn by Volterra. The predator numbers in Figure 10.8 are essentially superimposed on the prey numbers and do not alternate with them as in Volterra's diagram.

The lynx predator in Figure 10.8 seems to be "coupled" to the hare prey only in that the population highs and lows are nearly synchronous

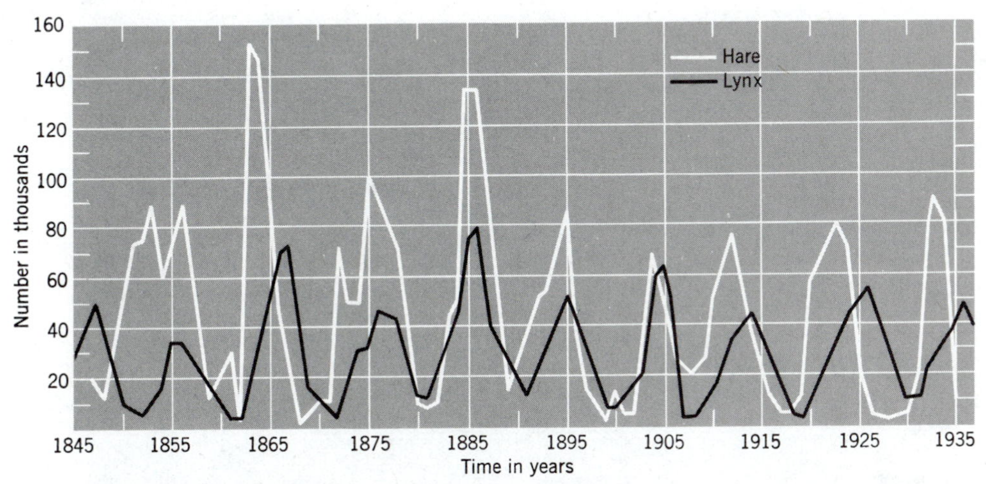

Figure 10.8 Population histories of lynx and snowshoe hare.
These histories are compiled from the records of the Hudson's Bay Company trading posts and represent skins traded. Since the trapping effort was the same each year, the fluctuations must mean that numbers of animals fluctuated by a proportionate amount. These are not coupled oscillations in the sense of the Lotka–Volterra models for the peaks of predator and prey numbers are synchronized. Predator populations are helplessly tied to prey populations. (Redrawn from MacLulich, 1937.)

(Finerty, 1980). In general, modern mathematical treatment shows that histories of natural fluctuations like these do not fit the model of coupled oscillations (May, 1975). No true coupled oscillations have been demonstrated outside the laboratory. The matter of fluctuating arctic populations is examined more fully in Chapter 13.

DISPERSAL, PATCHINESS, AND LOCAL EXTINCTIONS

The unreality of the Lotka–Volterra equations in describing the activity of real predators and real prey is obvious. A most glaring lack is the failure to provide lag times in the responses of both predator and prey to change in the numbers of the other. Early work incorporating lag times by Nicholson and Bailey (1935), however, suggested that introducing lag times would cause the system to oscillate towards rapid extinction (Figure 10.9). This model of Nicholson and Bailey, therefore, predicts just the sort of results found by Gause when he loosed *Didinium* on *Paramecium*, namely, efficient predators achieve local extinction.

The Nicholson and Bailey results can be reconciled readily enough with the consequence of providing "immigrants" to replace vanished prey as Gause did, or results like those of Utida where the animals had somewhere to hide, by including the effects of patchy environments or chances of dispersal into the system. A patchy environment promotes dispersal and reinvasion, since it suggests survival elsewhere, or a population high elsewhere, whence a patch devastated by predators can be reinvaded. A celebrated experimental demonstration of how this might work in practice is provided by the work of Huffaker.

Huffaker (1958) loosed a carnivorous mite, *Typhlodromus occidentalis*, on an herbivorous mite, *Eotetranychus sexmaculatus*. The herbivore ate oranges and could massively infest their surfaces. When the predator was loosed on such infesting mites, it killed them voraciously, expanding its population until its descendants totally consumed the prey, after which the predators died of starvation. This was Gause's *Didinium*

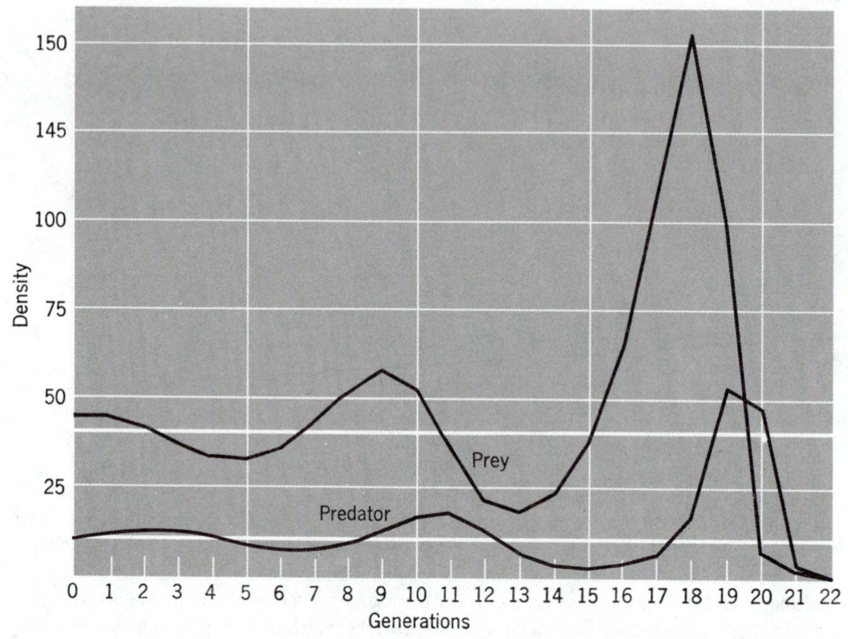

Figure 10.9 Nicholson and Bailey model of predator–prey system. This model is a development of the Lotka–Volterra model, but one that allows for various lag responses. The prediction is that the populations oscillate into extinction, the same conclusion suggested by simple algebraic models. Patchy distributions of insect pests may reflect the operation of such processes in nature. (From Nicholson and Bailey, 1935.)

story all over again, an encounter from which some of the prey could survive only if they could manage to get away. Huffaker then devised a universe in which it was possible for some of the persecuted prey to flee to a place of safety to build up a new population before wandering predators found them again. Oranges were dispersed among rubber balls of the same size on trays and were separated by barriers of vaseline or oil that were more or less difficult for the mites to cross (Figure 10.10). The proper mix of rubber balls and oranges let the total population persist indefinitely: predators exterminated prey on an orange in one part of the system, but prey were meanwhile building a population on a distant orange until the first predators dispersed across the vaseline gap. The only management needed for the system to be perpetual was the replacement of oranges as they went bad (Figure 10.11).

Huffaker's environment was patchy on a scale viewed by a mite. In each patch there was annihilation, but over the whole habitat both predator and prey persisted. Two principles seem to emerge from this experiment: that predators can, at least in some circumstances, keep down the overall numbers of their prey and that, when they do, we should expect patchy distributions of both predator and prey. These principles seem intuitively reasonable and are useful in that they can be examined against the experience of naturalists.

Entomologists have commonly noted that infestations of pests are often distributed in local patches. One tree in an orchard, or one branch, may be heavily infested with mites or caterpillars or bugs, whereas other trees or other branches have almost none. At other times, census of the same orchard may show that different trees and branches carry the infestations. The total population of pests in the orchard may well be the same, but they are differently distributed in dense

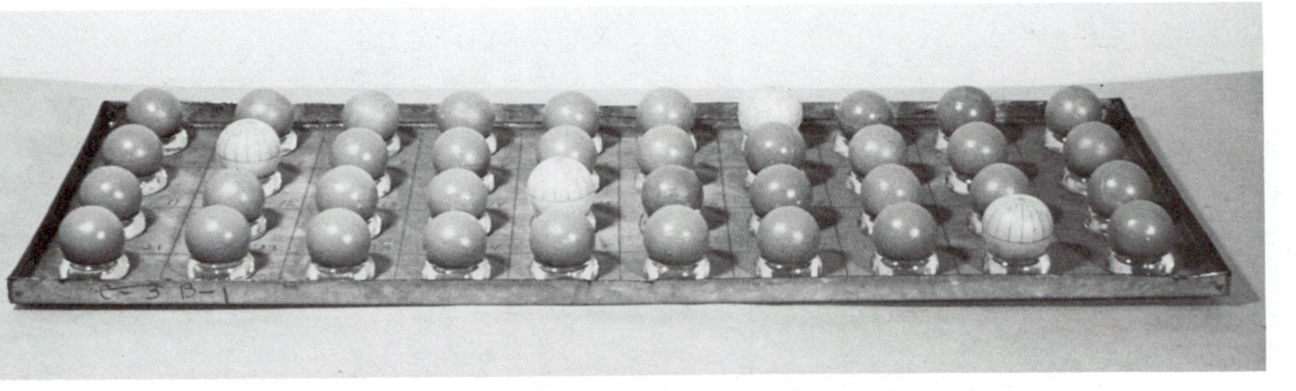

Figure 10.10 The Huffaker universe of oranges and rubber balls across which predator mites hunt herbivore mites.
Herbivorous mites infest oranges. When a predator mite finds a colony, the predator population increases until the herbivores are exterminated. This universe, in which travel from orange to orange is impeded by rubber balls and moats of grease, provides just enough opportunity for some herbivore mites to disperse away from the focus of predation and to build a new population on another orange before the predators again find them. The result is a perpetual game of hide-and-seek among the oranges in which neither population becomes extinct within the universe.

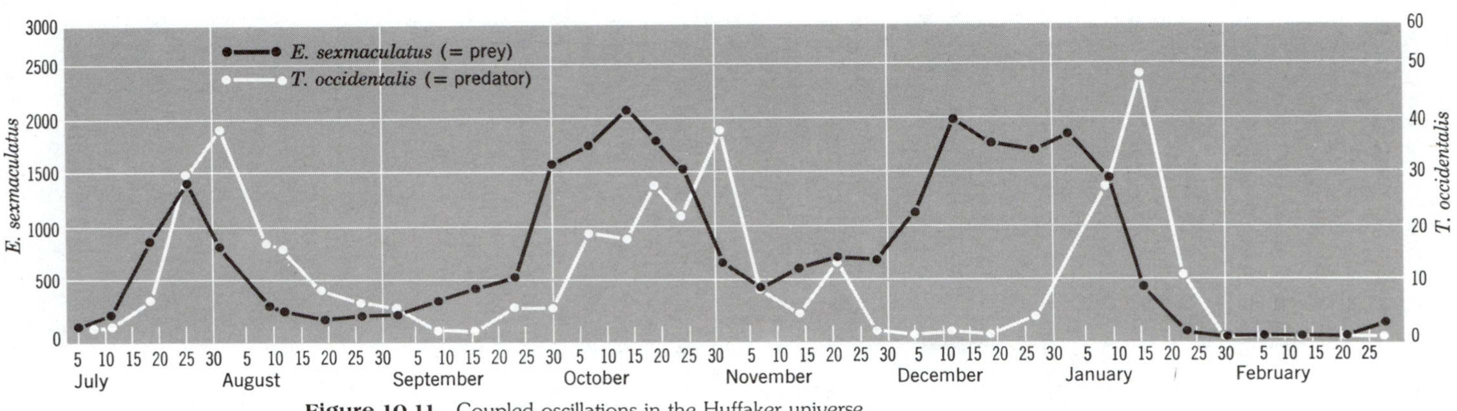

Figure 10.11 Coupled oscillations in the Huffaker universe.
These oscillations are apparently the result of the game of hide-and-seek among the oranges and rubber balls; see caption to Figure 10.10 and text. (From Huffaker, 1958.)

patches. These patterns are consistent with predator–prey relationships, provided the predator involved is so effective that local annihilation of the prey is possible.

Models of Predator Functional Responses

The mathematical models described so far assign only one response of predators to increased density of prey: an increased population of predators through reproduction. This is called a NUMERICAL RESPONSE. But it is inherently likely that predators might change their attack or feeding behavior according to the density of prey, perhaps from the lonely stalk to the fox-in-a-hen-run mode of behavior. Changed habits of the predator would be called a FUNCTIONAL RESPONSE (Solomon, 1949).

The simplest imaginable functional response of a predator to changes in numbers of its prey is no response at all. The predators kill a constant proportion of the prey. They do this by eating without regard to such mundane things as full bellies or running out of time in which to catch or chew prey, managing the while to reproduce themselves (*numerical response*). It is obvious that real predators could not destroy prey regardless of satiation, ease of catching, and shortage of time; yet the predators in Lotka–Volterra and Nicholson–Bailey type models are tacitly assumed to act in this way. The FUNCTIONAL RESPONSE CURVES of these impossible predators to different densities of prey are given in Figure 10.12.

A formal analysis of alternative functional responses has been provided by Holling (1959, 1965, 1967). Holling and his colleagues in Canada undertook this analysis so that components of behavior could be quantified and the habits of different kinds of predators simulated by computer. The goal was predictive models of predator–prey interactions that could be used in planning programs of biological control.

The simplest behavior by a real predator that ate as fast as possible until satiated, or until its powers of catching and engulfing prey were stretched to the limit, Holling (1965) called a TYPE 1 FUNCTIONAL RESPONSE (Figure 10.13). A filter-feeding animal that took without discrimination all the food collecting on its filters would show this response. Filter feeders that have been examined (Chapter 21) are now known to discriminate somewhat, suggesting that a true type 1 response may be rare.

An interesting property of the type 1 functional response is that it does not provide a density-dependent effect of predator on prey. It will be recalled that the classical model of *density dependence* allows that predation might be one of the density-dependent factors of the environment that tend to curb populations in a gradual way leading to balance (Chapters 6 and 8). Yet it is evident that predators with a type 1 functional response will have less and less relative effect on prey as the highest densities are reached.

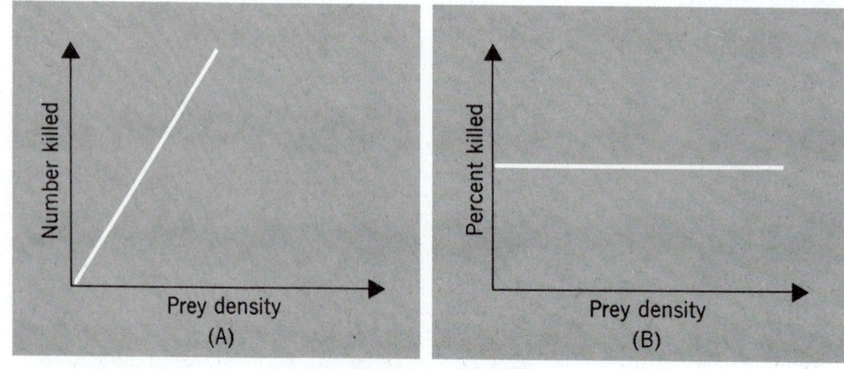

Figure 10.12 Functional response curves for a predator with unlimited appetite.

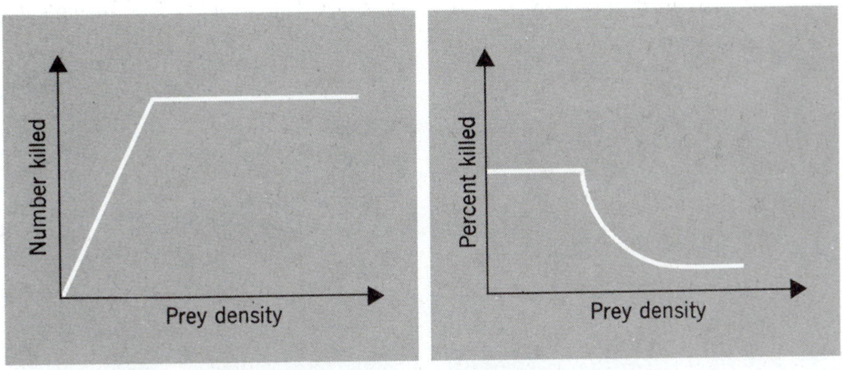

Figure 10.13 Type 1 functional response curves.

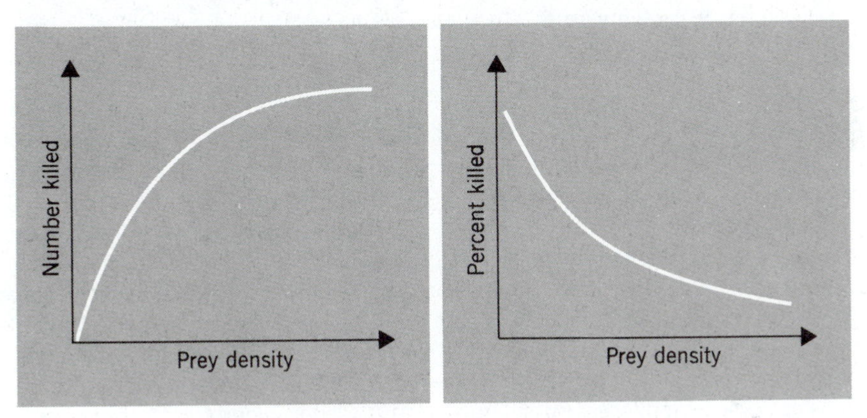

Figure 10.14 Type 2 functional response curves.

For the total effect of the predator population to be density dependent their numerical response must be relied on. The more widespread TYPE 2 FUNCTIONAL RESPONSE (Figure 10.14) also fails to have a density-dependent effect. Predators that behave in this way approach their saturation intake of prey more gradually as the density of prey increases than do those with a type 1 response, but when saturation is reached their impact on the prey population falls also. Holling found that water insects like coroxid bugs and dytiscid beetles functioned in a type 2 way when exposed to various concentrations of prey in the forms of mosquito larvae and tadpoles. Similar responses have been noted for fish and for wild parasitoid wasps hunting sawfly cocoons (Burnett, 1958).

For simple mechanistic sorts of animals either a type 1 or type 2 function response is all that one might expect. Neither can help to exercise density-dependent control except at low densities when it does not matter. For the kill rate to go up as density goes up, which is what one requires of a good controlling agent, the predator must increase its kill rate as the opportunity for killing increases. It turns out that vertebrate predators are capable of this density-dependent killing. They learn to hunt that which is most common, devoting more and more of their time to abundant prey as its very abundance makes it conspicuous and easy to catch. Holling called this widespread vertebrate behavior a TYPE 3 FUNCTIONAL RESPONSE (Figure 10.15).

Holling had a case study of vertebrate predation on hand in his laboratory that could be used to test the proposed type 3 functional responses of a set of vertebrate predators. The predators were two species of shrew and a mouse.

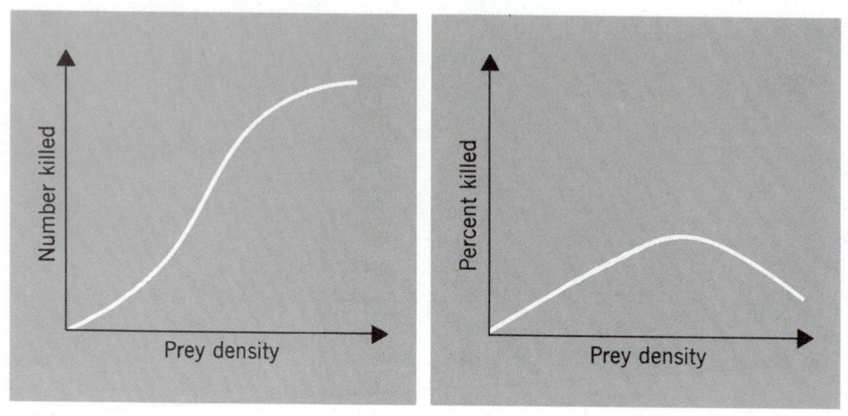

Figure 10.15 Type 3 functional response curves.

The prey were cocoons of the sawfly *Neodiprion sertifer*, the larvae of which are a major pest of Canadian pine plantations. An agricultural research station had tried controlling the sawfly populations in experimental plots with sprays of virus and this program had resulted in different plots with different concentrations of sawfly cocoons among the litter. Holling had available plots with cocoon densities varying from a few to up to 39,000 cocoons per acre.

Cocoon densities were monitored by sifting standard soil samples. The kill rate of each of the three vertebrate predators could also be monitored in this system because they left the empty cocoons behind when they had eaten the pupa inside. Furthermore each of the three predators opened a cocoon in a characteristic way so that it was possible to say how many cocoons had been taken by each predator. The predator populations were monitored by standard mark and recapture techniques (Chapter 9). This provided all the data needed to plot the functional response of each species to changing densities of the common prey (Figure 10.16).

Numerical response curves for the predators were also available from mark and recapture measurements over a number of years (Figure 10.17). The total effect of the three predators on the prey population results from the numerical and functional responses combined (Figure 10.18). The total percent killed curve for all three predators combined is also plotted in Figure 10.18. Up to considerable densities of prey the total percentage kill rate goes up sharply, showing that combinations of predators that switch to concentrate on abundant prey might be expected to have a powerful population effect.

The data of Figure 10.18 can be used to develop a model to describe circumstances in which a prey population actually is contained by a combination of predators with type 3 functional responses. Population equilibrium for the prey is established when deaths from predation exactly balance births. If x is the replacement rate of the prey, there must be x percent predation to control its numbers. In Figure 10.19 this kill rate, x, is plotted on a generalized curve based on Figure 10.18. The response curve is shown as a band

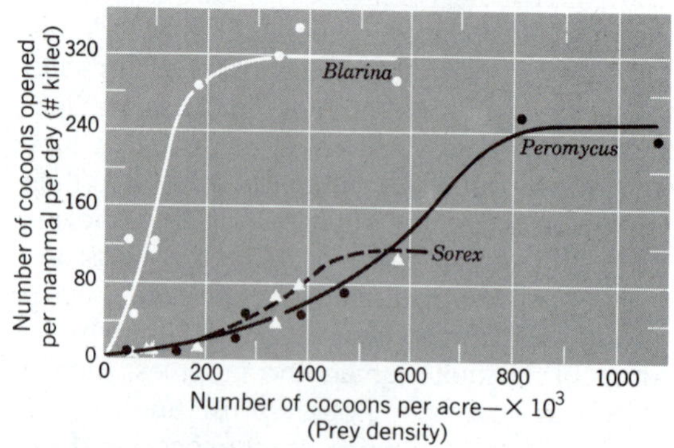

Figure 10.16 Type 3 functional response curves of mammalian predators.
These shrews and a mouse were hunting the same prey, sawfly cocoons, on the floor of a wood. (After Holling, 1959.)

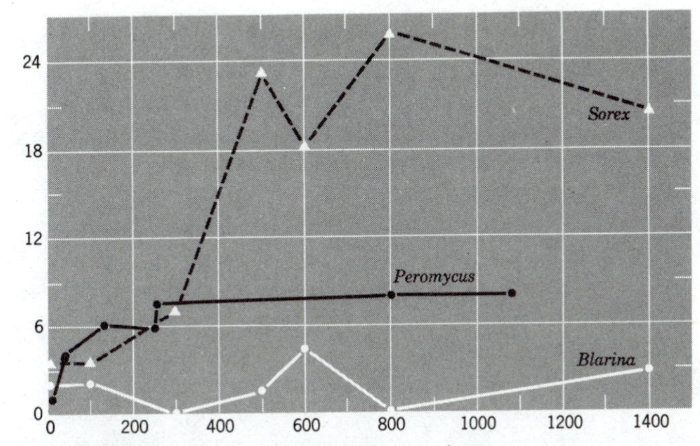

Figure 10.17 Numerical response of mammalian predators.
The curves describe population changes in the shrews and mice in the same system as in Figure 10.24. Only the shrew *Sorex* showed a positive correlation with the density of prey. (After Holling, 1959.)

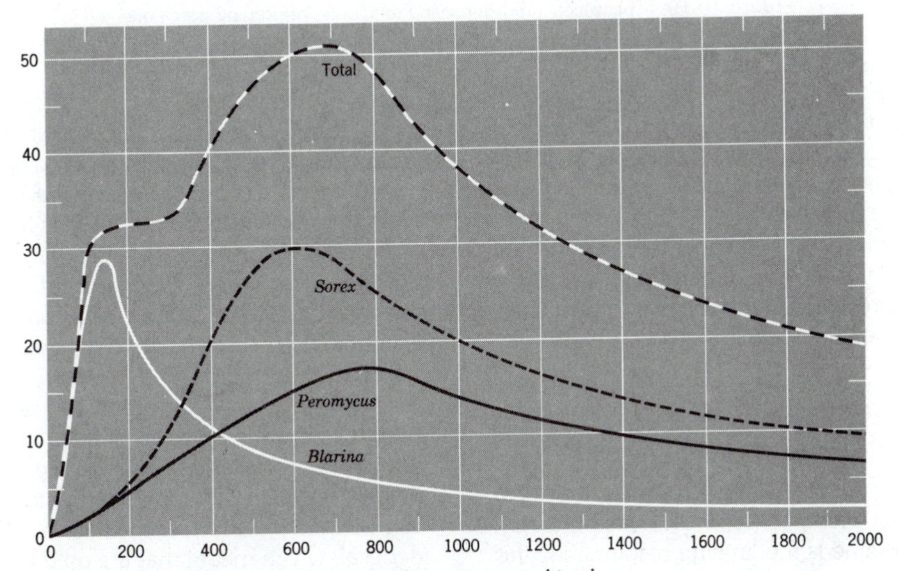

Figure 10.18 Numerical and functional responses combined.
The two effects are additive, and their overall shape suggests that control by the predators is likely to be effective up to a critical prey density. But above this density, the ability of the prey to reproduce should exceed the rate of predation and control would be lost.

to represent variation in the effectiveness of predation.

The death rate will equal the birth rate for all prey densities between *A* and *B*, and between *C* and *D* (Figure 10.19). If the prey density is between *B* and *C*, the death rate will exceed the birth rate and the prey will be suppressed until it again comes under control at *AB*. Only if prey density is beyond *D* will prey numbers be out of control so that the prey population continues to expand. For this to happen the prey population must climb across the *A*–*D* gap without check.

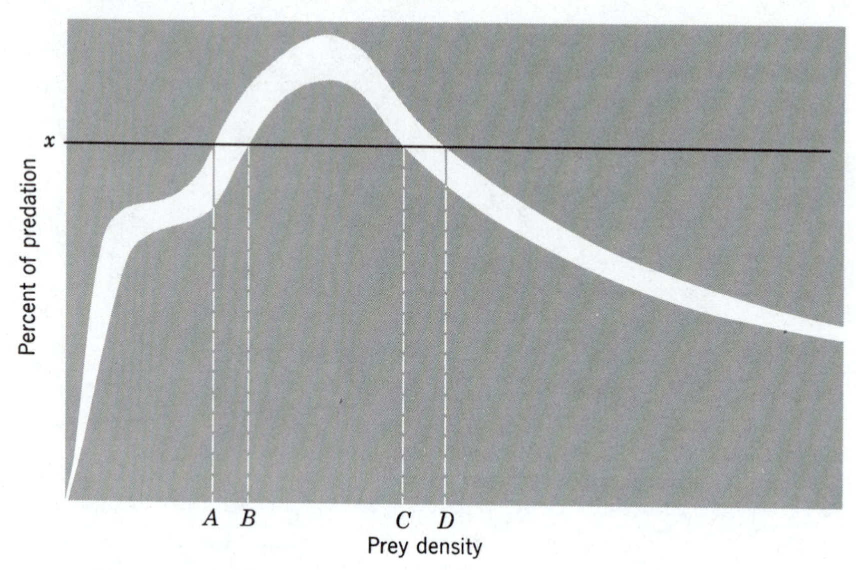

Figure 10.19 Theoretical model showing regulation of prey by predators.

This is likely only if *A* and *D* are close together, and this in turn is possible only if the number of kinds of predators are few. Even in this three-predator system, the chance of a prey species breeding so rapidly as to grow despite the combined numerical and functional responses of the predators is not large. A system with still more vertebrate predators would be even more likely to control prey numbers.

The Time Constraint (Sandpaper Disk) Model

A curvilinear functional response like that of a typical insect predator (type 2) invites the suggestion that time is the limiting resource for this kind of predator. This is the premise used in optimal foraging theory (Chapter 4). The time available to an active predator must be allocated between hunting, searching, attacking, and eating. As more time is spent eating abundant prey that is easy to catch, so there is less time available for hunting.

Holling (1959) tested the time-limiting hypothesis by simulating predation in the laboratory. The experimental predator was a blindfolded volunteer who hunted disks of sandpaper scattered over a table top by dabbing at them with a finger. When the volunteer felt sandpaper, she picked up the disk, thus removing a prey individual from the population and introducing a handling time into the model. Her hunting and killing of disks scattered at different population densities were timed. The result of this simulation was a curvilinear type 2 functional response as predicted by the time-limiting hypothesis (Figure 10.20).

In this simple model, the identification of the quarry (feeling the sandpaper under a finger) was virtually instantaneous, once the finger had completed its search. The time taken to pick up a disk was always the same, since the disks were identical, which meant that the only variable was the time spent searching. It seemed inescapable that it was, indeed, the compression of this searching time that caused the proportional decline in hunting success, as the hypothesis claimed. As a further test, Holling altered his model to introduce another time-consuming activity; he asked his blindfolded assistant to search for the disks by tapping with the blunt end of a pencil instead of her finger. She now hunted by sound, which introduced a slight time-consuming uncertainty into her attack procedure. Her re-

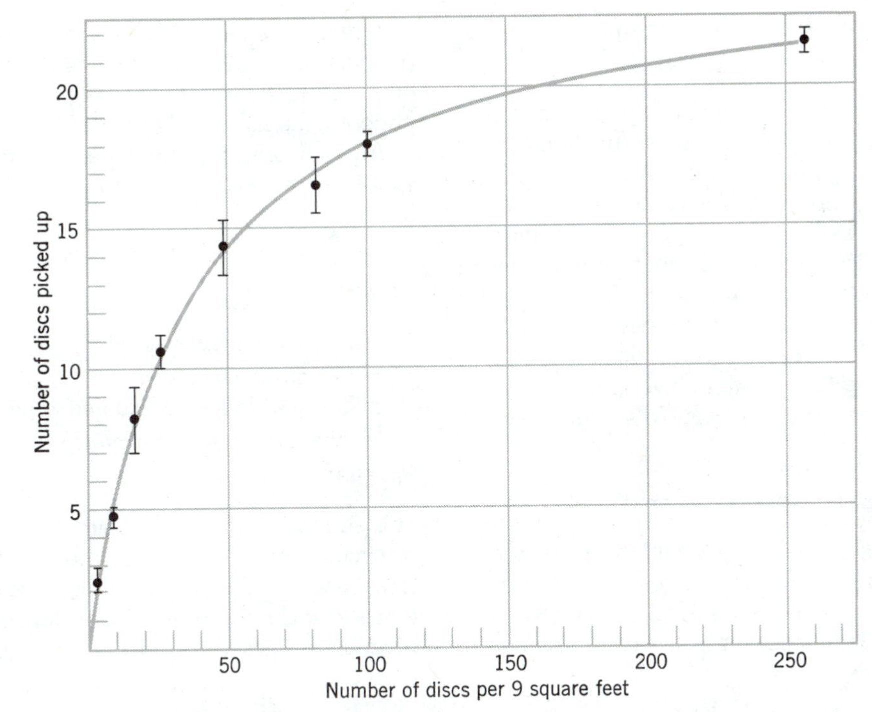

Figure 10.20 Functional response of a blindfolded human hunting sandpaper disks with a finger.
The number of disks that could be found in a set time increased with density but at a decreasing rate. This result suggests that the time available for each hunt was being compressed as the number of kills went up. Insect predators hunting insect prey were found to respond to density in a like manner. (After Holling, 1959.)

sponse was still curvilinear, but the slope of her "number killed" curve was lowered a little as predicted.

It should be noted that predators can be limited by satiation as well as by available time. A satiated animal would hunt no more, even though prey was abundant. Mammalian carnivores certainly act in this way and it is possible that other predators do on special occasions. A curvilinear type 2 functional response, however, is more parsimoniously explained by the time-limiting hypothesis.

Computer Simulation of Simple Predation

Leopold (1936), in his classic work on game management, split predation into five components, as follows:

1. The density of the game population.
2. The density of the predator population.
3. The predilection of the predator, that is, its natural food preferences.
4. The physical condition of the game and the escape facilities available to it.
5. The abundance of "buffers" or alternative foods for the predator.

These five components give the basis for models of predator–prey interactions that are potentially realistic. Using the terminology of Solomon (1949), the five components become:

1. Prey density.
2. Predator density.
3. The functional response of the predator.

4. The functional response of the prey.
5. Availability of alternate foods.

It will be noticed that Leopold's "predilection of the predator" is measured as its functional response and has thus been quantified. Describing the various escape possibilities for the prey as a "prey functional response" potentially introduces a difficult process to measure but this problem can be avoided if the model is applied to immobile prey like sawfly pupae in cocoons. For a complete quantitative model of populations changing through time, two extra components must be added to Leopold's original five:

6. Numerical response of the prey.
7. Numerical response of the predator.

This list of components identifies those functions of real animals in predator–prey systems that must be quantified if realistic simulations are to be made. In practice these seven components may readily be subdivided into still smaller components, as experience with real animals suggests, a proceeding called "experimental component analysis" (Holling, 1965).

An equation for the type 2 functional response of an insect predator is of the general form

$$N_a = a(T_t - t_h N_a)N_0$$

where

N_a = number of prey attacked
a = rate of successful search
T_t = time the predator is exposed to prey
T_h = time spent in handling prey
N_0 = prey density.

This equation describes the simplest possible set of circumstances describing the number of attacks made by this predator. Obvious additional terms to include are a satiation function, terms describing digestion, and so on. Figure 10.21

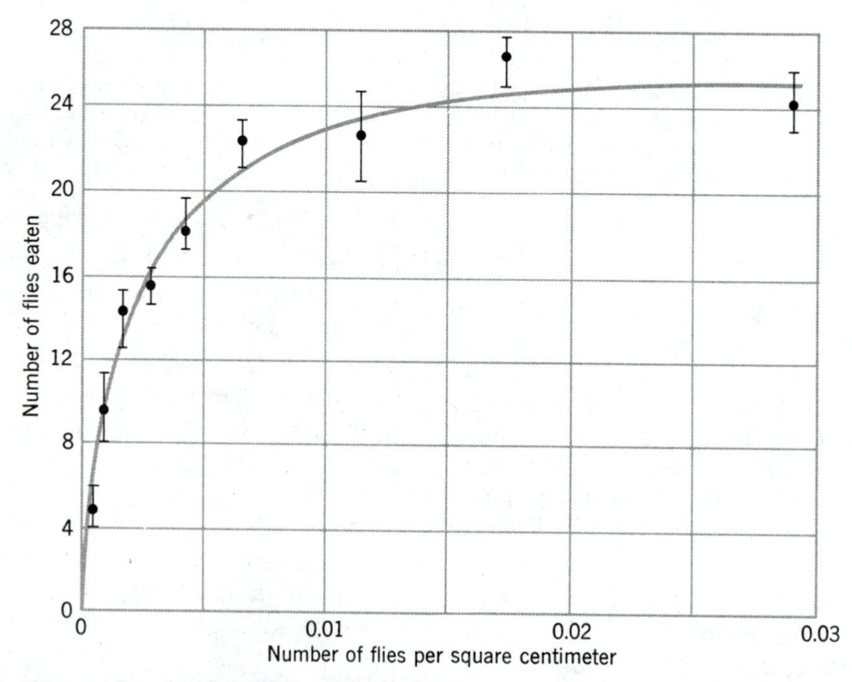

Figure 10.21 Functional responses of individual preying mantids (*Hierodula crassa*). The mantids were fed adult houseflies. Their catching prowess increased with density but at a decreasing rate, a result similar to that of the blindfolded human experiment. (After Holling, 1965.)

Figure 10.23 Rangeland in Australia before and after the moth *Cactoblastis cactorum* was introduced.

Larvae of *Cactoblastis cactorum* feed only on species of *Opuntia.* Introduction of the moth to rangeland dominated by *Opuntia* led to rapid population growth of the moth whose larvae almost totally destroyed the cacti in every part of the range. Once the wave of caterpillars passed, both the cactus and the moth became rare, with isolated patches of cactus springing up until a wandering moth finds them and lays a clutch of eggs.

is that of the wet tropics, with minimal change in precipitation across seasons, and the virgin vegetation is usually rain forest. Insect life is extremely varied, and among this array of insects are many herbivores that can eat rubber trees and oil palms. A farmer, reflecting on his great troubles with pests in temperate orchards, might well be expected to quail before the prospect of planting trees in such an insect-ridden place; and yet tropical planters have generally had less trouble with plagues of pests than have orchard growers in north-temperate climes. One hypothesis to explain this freedom from pests is that wild arthropod predators do the job of pest control for the farmers. In a place without winters, the resident spiders, carnivorous beetles, and parasitic hymenopterans might maintain high populations year-round and thus hunt the pests to rarity as do the agents of deliberate biological control. Some misadventures with spray programs in Malaysia suggest that this is the truth of the matter.

In the late 1950s many estates growing oil palms in Malaysia suffered plagues of defoliating caterpillars, and the estate managers called in entomologists to advise them. One of the entomologists was able to work out a detailed history of the insect plagues and to explain how they came about (Wood, 1971). The start of the story seems to have been a small alarm, the appearance of sufficient cockchafers on one plantation to worry the local manager. The cockchafers were not doing much damage, but the manager thought he ought to spray against them as an insurance, so he sprayed the plantation with DDT and noted the fact in his monthly report. Shortly afterward there were outbreaks of leaf-eating caterpillars on that part of the estate, so the manager sprayed again, and also sprayed the neighboring areas as well "to contain the pests." The next thing that happened was that caterpillars became a serious trouble over the entire sprayed area. Other planters learned of it, became convinced that they were faced with a general outbreak of insect pests, and began spraying too.

Wood (1971) found that the only change in the environment or management of the estates that was synchronous with the pest outbreaks was the spraying with insecticide itself. He postulated that the pesticide had killed natural predators of the system, thus allowing the herbivorous pest insects to increase in number. This hypothesis received immediate encouragement from the fact that the worst of the defoliators was a bagworm, a species of moth whose caterpillars lived in shared cocoons. It seemed likely that the communal cocoon would protect caterpillars from the spray. Hymenopteran parasitoids of these caterpillars (about 20 species of these were present) would be more exposed to the sprays than their prey and thus would be selectively killed. The test of this hypothesis was, of course, to stop spraying and see what happened, which Wood recommended. To a large extent his stratagem worked.

But Wood's remedy of stopping spraying was not always completely successful. Plenty of the hymenopteran parasites soon reappeared, but the bagworm populations sometimes remained at a disturbingly high level. Wood concluded that this was because the regular application of sprays had synchronized the life cycles of pests so that now they were all of a common age. The life cycles of predators would also have been synchronized. There might be periods, for instance, when there were large caterpillars everywhere but no flying adult wasps to hunt them, so that whole generations were able partially to escape predation. Only when the many local populations once more had generations out of phase would control be as complete as it was before spraying began. Wood's practical answer was to attack the secure generations of caterpillars with stomach poisons, that is, with sprays that had to be eaten to kill and that were thus only deadly to herbivores.

This and other similar case histories extend data on population control by predators from artificial or simple systems with one predator and one prey to natural complexes of many species of each living in their natural physical environments. But even the complex array of predators and prey in the Malaysian plantations operated in a habitat of artificial regularity over which the games of hide and seek and search and destroy were played, for oil palms and rubber trees are

Figure 10.24 Low pest plantation: oil palms.
In plantations in Malaya, like this in Ecuador, pest outbreaks are rare. The regularity of the environment, plus the large array of insect carnivores, apparently serve to keep pest populations low and patchy as in Huffaker's experimental design. (Compare with Figure 10.10.)

spaced out like orange trees or Huffaker's oranges (Figure 10.24).

Conclusions Suggested by Experience with Biological Control

Some predators can be very effective at controlling the numbers of their prey in nature if special conditions are met. The predators must be able to disperse well and to be very effective at killing prey when they are found. Conversely the prey must be susceptible to predator attack, with only modest means of defense. Essential to effective control by the predator is a pattern of dispersal by the prey that allows a few prey always to survive within the normal searching distance of adult predators. This pattern of dispersal may well be encouraged in agricultural systems by the regular patterns in which crops are planted.

Apparently these several requirements for successful control are met only infrequently even in agricultural systems, because most attempts at biological control fail. When biological control does work, circumstances always are comparable to the experimental conditions needed to perpetuate predator–prey systems in the laboratory. These examples show that some predators can control some prey, but they do not provide convincing evidence of how in general predators affect the populations of their prey in nature.

DO LARGE PREDATORS CONTROL THEIR PREY?

Predation would not be effective if prey species had evolved an effective defense that made them largely immune from attack over part of their life cycle. One of the more basic defense mechanisms is large size, suggesting that the predators of large animals may be less important in population regulation.

It is a commonplace of natural history that much of the food of wolves and big cats consists of the old and the sick. That this is true for wolves attacking moose is shown by the data from Isle Royale described in Chapters 4 and 9. Even when wolves prey on the less formidable Dall mountain sheep, they still do not kill animals in the prime of life (Chapter 9). In these systems the predators are unlikely to achieve local extermination of their prey. Predation will have a population effect only to the extent that young prey animals are killed.

These predators of large prey act principally as a check on fecundity. When the prey animals are mammals bearing only one or two young at a reproductive episode any reduction in fecundity from predation must be important. The danger to individual fitness of loss of these few offspring is shown by the tenacity with which many mammalian mothers defend their young. But the loss to the intrinsic rate of increase of the whole prey population will be important only if the predator population is relatively large, and the number of predators will depend upon the supply of old and sick prey animals for them to eat. There should thus be a complicated relationship between the populations of large predators and large prey: predator numbers are determined by the supply of old and sick animals (essentially by natural mortality), but the prey population, including the supply of the old and the sick, is influenced by predation on the young. The population consequence of this predation cannot be as significant as the consequence for small or helpless prey of being attacked by effective predators that can strike at individuals in the prime of life (Figure 10.25).

Circumstances can arise when wolves will be less discriminating about the prey they choose to attack. Apparently they attack fitter animals when food is in short supply, a circumstance that can happen if the predator population is locally too dense or if there is an unusual shortage of prey. Probably the commonest way for this to come about is for the prey population to suffer loss independent of predator attack, perhaps as the result of epidemic or pandemic disease. With the prey population made unusually low, the hungry predators could then make the survival of the remaining prey difficult so that predation by large predators may well act to increase fluctuations in prey numbers caused by other factors.

Figure 10.25 *Canis lupus*: super scavenger.
Many of the prey animals of wolves and other large mammalian predators are old and sick animals. Wolf and big cat predation has the most population consequence when young animals are killed or when the prey population is depressed for some reason other than predation, when too many predators begin to kill prime animals.

It is not so obvious that solitary hunters who kill by stealth, like the large cats, should take only the sick. Schaller (1967) has suggested that tigers may take the fit as commonly as the unfit, but his accounts of how tigers killed tethered buffalos does not suggest that killing is safe or easy even for a tiger. The tigers ran at their prey, half climbed on their backs, wrestled them to the ground, then dodged the flailing hooves to seize the buffalo by the neck. The buffalos took some 10 minutes to die, suggesting that killing sick animals should be a safer undertaking. It is not improbable that a lurking tiger might normally let the more formidable animals pass unmo-

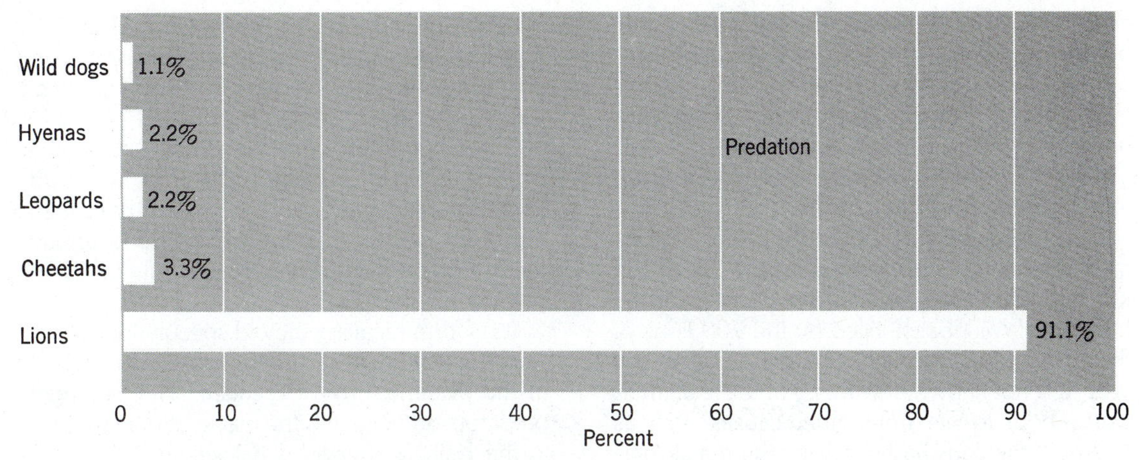

Figure 10.26 Predation on western white-bearded wildebeests.
Lions were the only serious predators of wildebeests in the Serengeti, but it only took a relatively small fraction of the wildebeests dying every year to support the known lion population. More important causes of death in wildebeests were disease and young animals getting detached from their mothers in crowded herds. (After Talbot and Talbot, 1963.)

lested. A recent study of the puma (*Felis concolor*) in Idaho by Hornocker (1969) (Chapter 12) suggests that much of puma behavior can be explained as directing the animals to achieve quick kills without getting hurt. Nevertheless Schaller's view needs to be given due weight, and it may be that big cats are more likely to kill healthy prey than are canids. As with wolves, this killing of the healthy is more likely when the prey are scarce and should have the consequence of increasing loss of population initially caused by disease or starvation.

Lions and Wildebeest of the Serengeti

Hard data for the limited effect of large predators on their prey come from studies of herds of game animals on the East African plains. Many kinds of ungulate live together on the plains and they are hunted by many kinds of large predator. Best known are the herds of Serengeti reserve where the commonest ungulates are the wildebeests (*Gorgon taurinus*), of which there are more than 200,000. Seven hundred lions hunt the Serengeti, feeding almost entirely on wildebeest, and there are also leopards, cheetahs, hyenas, and hunting dogs taking their separate tolls in different ways, though these are less important than lions (Figure 10.26). It is known that the lions live almost entirely on wildebeest.

Talbot and Talbot (1963) collected data on both wildebeest and lions with a view to studying population regulation of both (Figure 10.27). A

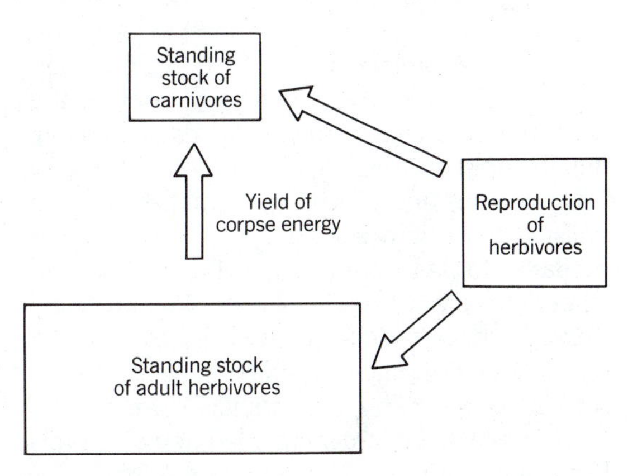

Figure 10.27 Model of population consequence of predation in large animals.
The population of carnivores is a function of the yield of old and sick individuals of prey. But recruitment to prey is a function of the take of young animals by the predators.

crude census of the wildebeest herd was made by driving a series of transects across the plains by overland vehicle, and from aircraft. Counts from several years suggested that wildebeest numbers were roughly stable, thus strongly suggesting that there was density-dependent death in the Serengeti wildebeest herd.

Predation by lions was easy to calculate since many observations suggested that each adult lion kills one wildebeest a week. A safe range for the number of wildebeest killed by the 700 lions of the Serengeti is between 12,000 and 18,000 a year. It is not known how many of these animals were about to die from other causes such as disease if the lions had not killed them first (perhaps most of them) but even if they were all healthy animals the population consequence of this killing is not very important to a herd of 200,000 wildebeest. Lion predation, therefore, does not control the numbers of the principal lion prey, wildebeest.

The Talbots went on to produce a plausible explanation of how the wildebeest are controlled, despite the low incidence of predation. Death from disease seemed to be more important than death from lions, but even so, the death rate of adults from all causes that could be identified was too low to stabilize the population. The factor that did work was density-dependent death of young that had lost their mothers.

Wildebeest cows always seemed to give birth to live calves successfully, but the calves then had to follow very closely to their mothers as the herds moved. In large crowded herds, the young, quite simply and literally, often got lost. Many fell to hyenas and jackals, no doubt, but it was unlikely that there should be enough of such predators to clean them all up in the short season when lost calves were available. Most probably starved and were picked up by vultures. The loss of calves was density dependent since the babies seemed to be safe by their mothers' sides in small herds. Only in the larger crowded herds was this loss significant.

These data are not sufficient to indicate the importance of lion predation of young wildebeest on the intrinsic rate of increase of the wildebeest herd. All the predators combined probably do have a powerful effect by removing young. It is true that the young killed by jackals or eaten by vultures may have been vulnerable because parted from their mothers in the crowd, yet this parting need not have been serious were the predators not present. Individual fecundity, therefore, probably was greatly influenced by predation.

It is worth recalling the old speculations about predators and deer on the Kaibab plateau in light of the evidence from Serengeti. The Serengeti predators do not act in the way assigned to them in the Kaibab anecdotes. Closer to the Kaibab theoretical idea are the predator–prey interactions of successful biological control in which predators reduce the prey to scattered populations that are far less than the prey food supply would warrant. Removal of predators like these would produce the sort of population explosion of prey imagined at the Kaibab. But large predators hunting large prey cannot achieve this sort of control essentially because the prey defenses are too good.

POPULATION CONSEQUENCES OF PREY DEFENSE

The evolution of an effective defense against predators may be simplest for large animals. Large size itself is an effective deterrent, as we have seen, because it makes mechanical requirements for predators that are not easily met. The largest herbivorous mammals may be almost predation free as a result; for instance, it must seem doubtful that even a sabre-toothed tiger could kill an adult elephant like a mammoth with safety. The population dynamics of very large herbivorous animals, therefore, may have fecundity lowered by predation but the actual arbiters of their population size must be found elsewhere. When prey is small the opportunities for defense are fewer and the chance of a predator evolving adequate killing power is greater (Figure 10.28).

Figure 10.28 The advantage is with the attack. Web-building spiders have a suite of weaponry adequate for the safe killing of prime specimens of prey as large as themselves. Comparable advantages for the attacker are less easily obtained when large predators attack large prey. Selection, however, provides many alternative defenses for small prey from dispersal and camouflage to noxious taste, mimicry, or armor (see Chapter 25).

Large predators hunting small prey, such as birds attacking caterpillars or shrews attacking sawfly cocoons, are quite different from wolves attacking moose. Even small predators like spiders or wasps can have quite adequate killing abilities to destroy prey without danger to themselves. These possibilities of the killing power of small predators makes possible the prey reduction of successful biological control.

Another case study of control that illustrates the possibilities of superior killing power is the devastation of rabbits in Britain by the disease myxamatosis, a mosquito-born virus that almost eliminated the huge rabbit population of the island in two to three years. Against this deadly attack even the relatively large prey initially had no defense.

Yet 20 years after the myxamatosis attack large numbers of rabbits again live in Britain and these have immunity against the disease. There has been very strong selection for individuals with a natural immune system to the disease or with the ability to acquire active immunity to it. The attack resulted in selection for an appropriate defense. We should look, therefore, for signs of effective defense against even the most formidable of invertebrate predators. If such defense is possible it is likely that it has been found by natural selection.

It is easy to list common defenses used by invertebrates. The secretion of noxious substances by arthropod herbivores is widespread: millipedes, aphids, beetles, caterpillars, ants: noxious secretions can be found in almost all groups, and they often work. T.C.R. White reports seeing an aphid escape from a New Zealand black widow spider after using a secretion that had the spider occupied, in apparent distress, with trying to scrape the secretion off its leg. Many caterpillars that are attacked by ichneumonid wasps thrash about so wildly that the wasp is frustrated. Many pupae rapidly develop hard exoskeletons through which the sting of the wasp cannot penetrate. And there are always the great varieties of ways of hiding, including spiders that vibrate on webs and consequently blur their image sufficiently to deceive spider-hunting wasps. White (1978) argues that these various adaptations might mean that even most arthropod herbivores are victims to predation only when they are sick, young, weak, or old, just like moose or wildebeest. If this is so, then dramatic population consequences of predation are no more likely in small animals than in large animals.

There is obvious conflict between the views of White (1978) and like thinkers and those who think that herbivore numbers are kept down by predators, and who cite examples of biological control as evidence that small predators have deadly effects on their prey. This takes us back to the green earth argument again (Chapter 8). It is prudent to allow that it is possible that both mechanisms operate. In some circumstances small predators can be deadly and in these circumstances the low populations of prey are a result of predation. But in other circumstances a successful defense, whether overt (spraying noxious chemicals, fighting back) or covert (hiding, being cryptically colored), denies predation an important impact on the prey population.

ERRINGTON'S STUDY OF MUSKRATS IN IOWA

A study of middle-sized predators and prey, mink (*Mustela vison*) hunting muskrats (*Ondatra zibethica*), suggests that in this size range too predation may have only a peripheral effect on population (Errington, 1946, 1963).

Muskrats are particularly hard to observe, passing most of the daylight hours under water or hidden in burrows, but Errington learned to discover their presence and activities from "reading of sign," which he describes as "studying the meaning of tracks and trails, of diggings and cuttings and heapings, of food debris and droppings, or miscellaneous traces, of blood, fur, wounds, and carcasses." These qualitative bits of evidence became quantitative when Errington used them to identify the presence of individuals and to map the members of populations.

Errington trapped muskrats on a large scale, marked them, and released them. He was given sample carcasses by the local fur trappers for autopsy, particularly for estimating the numbers of offspring that had been born to females from the number of scars of old placentas in their ovaries. He studied the possible predators of muskrats, notably minks, in the same way, and was able to detect when they were feeding on muskrats from the remains in their droppings. He came to the conclusion that nearly all the muskrats eaten by predators in his part of Iowa were individuals that were doomed to die from some other cause, such as disease, if the predators had not eaten them first.

The lives of Iowa muskrats were strongly influenced by the marked seasonal changes of the local climate, by the alternation of warm continental summers and bitter winters, by the chances of floods, and the chances of drought. Muskrats established in fine summer weather in their home ranges seemed almost immune to predation, probably because they had somewhere to run to when approached by a mink or because they faced the potential aggressor so confidently that they made the risks of combat unacceptable to it. But if they were flooded out, or left exposed by drought, or suffered epidemic disease, the minks killed many of them. Sometimes the onset of some calamity attracted predators to such an extent that whole populations of muskrats were almost wiped out.

SUMMARY: SOME FACTORS THAT INFLUENCE THE POPULATION CONSEQUENCES OF PREDATION

The importance of predation as a population regulator depends on the following properties of the prey:

1. Ability of prey to defend itself.

2. Ability of prey to escape detection.

3. Prey densities obtainable in the absence of predators (potential of prey as food).

4. Prey dispersal in the absence of predators.

Natural selection must constantly be directing prey structure and function in the direction of better defense and escape (1 and 2). The extent to which these abilities are achieved probably is largely a function of how the necessary adaptations are compatible with other requirements of the species niche. Food-getting can only be compromised so far in the cause of a better defense and some ways of life are inherently more dangerous than others. Being a planktonic larva in the upper reaches of the sea, or a small fish, for instance, seems to be potentially more dangerous than the life of a tubeworm.

That defense must sometimes be compromised also is suggested by the gaudy colors of some vertebrates in the breeding season—the fitness won by attracting a mate with spectacular colors is apparently greater than the fitness lost, on the average, by being a conspicuous mark for predators. But the general conclusion has to be that natural selection will have worked to increase the success of defenses as much as other necessary traits allow. The defense of large size may be almost completely successful so that predation on large herbivores is only important in fine-tuning a population in that the young are victims.

When prey are of low density for reasons not connected with predation, or are widely and irregularly dispersed, predation will be unimportant (3 and 4). In these circumstances the time spent searching by predators will be high per unit energy won and the predator population will be relatively low. The predators cannot achieve a high numerical response. Any factor setting low prey populations even in the absence of predators, therefore, will mute the pressure of predation.

Two factors probably operate extremely widely on herbivores in this way—weather and plant defenses. In marginal or seasonal habitats where opportunist animals live (Chapter 8) herbivore populations will usually be low because of destruction by weather. Plant chemical defenses are often so good that most arthropod herbivores survive poorly as juveniles in most years because of protein malnutrition (White, 1978). Whenever plant defenses do achieve major success, then predation on herbivores will be relatively unimportant as a population process.

The importance of predation in population control also depends on the following properties of the predators:

5. Attack properties of the predator.

6. Numerical response of the predator (a function of fecundity).

7. Functional response of the predator.

Both the killing power of the predator and the fecundity will be under strong natural selection to accommodate predator to prey. However, natural selection may not have available suitable varieties of attack to meet even simple defenses. The defense of large size, for instance, is easily overcome by human weapons, beginning with a piece of pressure-flaked chert tied to a long stick (Figure 10.29), but random mutation does not provide such weapons for other animals. As a result, large size may be nearly completely effective as a defense against all but human predators.

Figure 10.29 Super predator.
No active defense found by natural selection is of use against sharpened sticks and powers of entrapment. Small size, dispersal, and high fecundity are then the only defenses.

Even defenses like secreting mucous slime (as done most effectively by slugs) may be very hard to meet by selecting from random traits. So it may be that selection for a good defense is frequently more effective than selection for a good offense.

The functional response of a predator is determined by its ancestry and mode of life. Predators of generalized taste and powers of learning, essentially vertebrates, are capable of concentrating on a prey that becomes common and thus of having a strong density-dependent effect on the prey population.

We can expect predation to have important population consequences when the following apply:

8. The prey niche is such that dense populations would be possible in the absence of predators.

9. Prey is evenly dispersed.

10. The prey niche is such that the possibility for effective defense is low.

Dense, evenly dispersed prey (8 and 9) is the pattern against which the successful programs of biological control were launched. It is the pattern in experimental systems where the predator eventually causes extinction, or else persistence in a pattern of hide and seek, and is also the pattern in which natural predators control pests of crops unless reduced by pesticides as in Wood's palm and rubber plantations.

It is when the prey niche precludes effective defense (10) that predators turn out to have the most profound effects on the structures of communities (Chapters 21 and 25). An obvious example is an open-water floating life that requires relatively small size. The presence of fish in a lake, for instance, may prevent larger species or individuals of zooplankton from living there at all (Brooks and Dodson, 1965; Chapter 21).

PREVIEW TO CHAPTER 11

Different life histories can be thought of as different strategies for gaining fitness. Life histories suited to roles as colonists include provision for rapid multiplication and rapid dispersal, the life histories, for instance, of plants we call weeds. These are *opportunist* species, also called *r-strategists.* Colonists are also sometimes thought of as *fugitive* species, since they are not well adapted to enduring competition and may respond to pressures of competition by emigration. The opposite strategy to that of an opportunist or fugitive is the life history of an *equilibrium species,* also called a *K-strategist.* Opportunists should have populations controlled by density-independent means, whereas equilibrium species should be under density-dependent control. A continuum of possible strategies between the opportunist and equilibrium extremes is predicted, a concept that led, among other things, to a hypothesis of the causes of secondary succession. The terms *r-selection* and *K-selection* were introduced to suggest the ways selection should work within populations colonizing islands. The original opportunistic colonists of an island would eventually become crowded. Selection should then work to preserve individuals who could live at high density, which is to say individuals suited to life when numbers equal *K*, the carrying capacity. The colonists were *r*-selected and their descendants living at equilibrium come under *K*-selection. *K*-selection within populations can be demonstrated in the laboratory, and there are some suggestive examples from the field. Alternative strategies also apply to reproduction, those that rely on numerous small eggs and those that rely on a few large young. *Small-egg strategies* are suited to dispersal of many propagules and allow high fecundity, but the costs of a small-egg strategy are the likely death of most of the young. This cost is avoided, at least in part, when a large-young strategy is followed, since the death rate of young is kept to a minimum. The return in fitness on calories invested by an animal that raises a few young at a time may actually be better than the return gained by a highly fecund animal, most of whose offspring die. When animals with a large-young strategy raise a family or clutch, this should be regulated to an optimum size, since attempting to raise one more large youngster than resources will allow imperils the survival of the whole clutch. Good field data are offered to show that clutch sizes of birds are regulated as expected. The special cases of conservative breeders like seabirds and condors are discussed, showing that their low reproductive effort is an adaptation to scarce or unpredictable food resources. This low reproductive effort requires the parallel adaptation of a long safe life for the adult. The long juvenile period and distinctive coloring of juveniles of familiar birds like seagulls are adaptations that serve breeding strategies when extra food needed for breeding is scarce or unreliable. The opposed advantages of *semelparity* (breeding once in a lifetime) and *iteroparity* (repeated reproductive episodes) are discussed. Optimal foraging theory is reviewed and it is noted that time to search and handle food may not be the limiting variables for some animals. Insect herbivores tend to be food specialists, as the chemistry of food species becomes important both to the quality of the food and to the ability of the adult to detect the appropriate host plant. Ideal optimal foraging powers may also be restricted by sensory limitations that may program the possible choices of an animal to something less than is theoretically optimal. A study of notonectids (backswimmers) is used to illustrate that some animals can optimize their behavior in the face of two conflicting variables, food intake and danger of predation.

CHAPTER 11

STRATEGIES OF SPECIES POPULATIONS

Ecologists now make much use of the word "strategy." They have borrowed it from the military, though they sometimes use it in ways more appropriate to the contrasting term "tactics" than to strategy proper. But whether an instructor at a staff college would approve our catholic meaning for "strategy" or not, the word is very useful. By STRATEGY we mean ANY PATTERN OF LIFE HISTORY OR BEHAVIOR OF AN INDIVIDUAL, OR OF ALL MEMBERS OF A POPULATION, THAT IS ADAPTED TO GAINING FITNESS BY THE EFFICIENT COLLECTION OR USE OF RESOURCES. We talk of strategies for breeding, strategies for foraging, strategies for avoiding predators, dispersal strategies, or the r and K strategies of opportunist or equilibrium species.

Consciousness of the importance of different strategies to the study of populations grew out of the debate over the importance of density-dependent and density-independent control of populations (Chapter 8). The debate made clear that there were two extreme types of life history common in nature, the one suited to marginal habitats with fluctuating weather and the other for more equable places where a population equilibrium seemed more likely. The two kinds of species were at first known as *opportunist* or *equilibrium* species, respectively, though the names r and K, loosely used, are now more commonly given to them. Parallel work, particularly with birds and insects, showed how adaptations for breeding were suited to these diverse strategies and there developed what perhaps may be likened to a new paradigm in ecology in which species were assigned a place on a continuum of possible strategies, with far-reaching consequences for our understanding of how species coexist in communities.

OPPORTUNIST SPECIES

In marginal habitats many life histories are clearly adapted to rapid population growth during short favorable seasons, with other adaptations letting individuals survive the hostile times that inevitably come. These are the life histories of OPPORTUNIST SPECIES. Necessary to the opportunist life style is the ability to disperse rapidly so that remote habitats can be colonized, or recolonized, quickly from the places of refuge. Essential to the opportunist strategy, therefore, is acceptance of a high chance of death in the lean seasons. Individuals compete to win fitness with a survival and dispersal package for their offspring where the best results come from meeting adversity with a large number of survival units widely scattered.

Archetypal of the opportunist strategy was Andrewartha's Australian grasshopper, *Austroicetes cruciata* (Chapter 8), but comparable strategies are found in all short-lived animals of highly seasonal environments. The strategy also is adapted to exploiting ephemeral habitats of equable places and is the strategy followed by many of the familiar plants we call "weeds."

These weeds are adapted to placing a propagule, usually a seed, in a chance patch of bare earth, to grow rapidly in that earth, and to set seed before the benefits of a private plot of bare ground are lost as other plants crowd in. The trick of colonization may be worked by saturation of a landscape with airborne seeds, like those of dandelions (*Taraxacum*), which ride the wind beneath a pappus of fine hairs, or by leaving very resistant seeds in the soil as time capsules ready to germinate when the land is bare again. Either way, the chances for survival of any seed are slim but the chance for fitness of the parent plant is good because each surviving seed grows in a highly favorable place freed from interference by other plants. Growth of a plant from one seed can be followed by a large seed-set of its own in the next generation.

We conceive, therefore, of extreme opportunist plants or animals as being well adapted to exploit chance good times, whether these occur in time or space. Rapid reproduction, high fecundity, excellent powers of dispersal, short life, and the ability to sit out hard times in a refuge are characteristic of this life style. Such animals and plants can attain only local or short-lived population equilibria and their population histories reflect strong influences from density-independent factors. Extinction of local populations is common and population numbers fluctuate widely.

FUGITIVE SPECIES

Another way of looking at dwellers in marginal habitats is to say that they are fugitives from more desirable places, for one who journeys to the desert must have gone away from a more fertile place (an ecologist might say a more *mesic* place). Yet what could be the advantage of fleeing from the desirable or the permanent for the harsh or ephemeral? The answer has to be that the flight is from the other inhabitants of permanent or desirable places. This logic leads to a view that opportunist species are not so much those that take advantage of scarce and fleeting habitat as they are species on the run from competition. They make the best of what is left for them.

Hutchinson (1951) first made clear that an *opportunist species* should usually also be a FUGITIVE SPECIES. He used data of copepod distributions, in particular observations of Elton published nearly a quarter of a century earlier.

Elton (1927) noticed that ponds associated with the Oxford municipal sewage works usually supported populations of just one species of copepod, *Eurytemora lacinulata,* but that farm or village ponds in the region had other copepods of the genus *Diaptomus.* To Elton the important difference between the sewage-works ponds and the others was difference in age. Ponds in the sewage works were part of the filter system and they were regularly drained and cleaned out. *Eurytemora* always appeared in these ponds of perpetual youth but *Diaptomus* never did, even though *Diaptomus* was the commonest copepod of local bodies of water. Elton suggested that *Diaptomus* was somehow inimical to *Eurytemora* so that *Eurytemora* could live only where *Diaptomus* could not get.

A further hint of what might be happening among the copepods was given by the pattern of distribution in a river estuary near Liverpool, 200 miles to the north. The brackish waters of the estuary, like the Oxford filter beds, supported a population of *Eurytemora lacinulata* but no other copepods (Figure 11.1). Elton argued that this distribution had something to do with salinity,

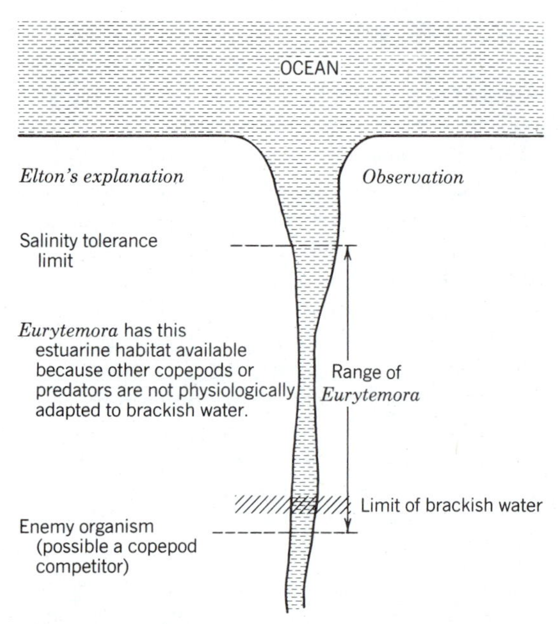

Figure 11.1 Copepods in an estuary.
Elton's (1927) observation of the distribution of *Eurytemora lacinulata* in an estuary in northern England.

since one could correlate the abundance of *Eurytemora* with the reach of brackish water. But the riddle of distribution could not be solved by a hypothesis of salt toleration alone because elsewhere *Eurytemora* thrived in fresh water. Elton reasoned that this copepod must be prevented from penetrating to the fresher reaches of the river by some hostile animal, not *Diaptomus* this time because the river was without *Diaptomus.*

Elton thought of *Eurytemora lacinulata* as an animal that normally had its range set by other animals with whom it could not persist. In Oxford village ponds this other animal was *Diaptomus,* and in the fresh-water reaches of a Liverpool river some animal unknown. It was clear that *Eurytemora* could live in places where these possible enemies could not live; in fresh filter beds to which others could not disperse, and in brackish water that others could not tolerate.

Hutchinson (1951), in his essay "Copepodology for Ornithologists," restated Elton's old conclusions with more definition, having at his disposal the results of Gause's work on competitive exclusion and the studies of ornithologists on density dependence (Chapters 6 to 8). The hostile animals in old ponds or fresh river water might well be formal competitors and *Eurytemora* might well be removed from those places by competitive exclusion. The necessary adaptations included rapid powers of dispersal and tolerance of unusual physical conditions. The result was a true *opportunist species* though Hutchinson called it a FUGITIVE SPECIES.

This concept of a *fugitive species* can be applied conveniently to Andrewartha's Australian grasshoppers also. It was shown (Chapter 8, Figure 8.13) that the range of the grasshopper to the south was bounded not by desert but by the wetter, fertile regions of the coast. It seems likely that there were no physical properties of this mesic coastal strip that would prevent grasshoppers from living there, and indeed many other species of grasshopper probably did live there. In this milder region it seems likely that crowding, density dependence, and equilibrium populations are all possible. *Austroicetes cruciata* may well suffer competitive exclusion in these regions. At the edge of the desert to the north *Austroicetes* is both a fugitive and an opportunist at the same time.

One extra point of value comes from this consideration of opportunists as fugitives in that the logic shows how false conclusions may come from using physical limiting factors to understand distributions. *Eurytemora* in an estuary is found only where there is brackish water, and *Austroicetes* is found only north of a certain precipitation line, but this does not mean that you can explain these distributions by measuring salinity or rainfall. The distributions most probably result from the activities of other animals and it is these activities that we must discover if we are to understand the causes of distributions.

EQUILIBRIUM SPECIES AND THE STRATEGIC CONTINUUM

An EQUILIBRIUM SPECIES represents the alternative strategy to that of opportunist or fugitive species. This is the strong competitor that sends the opportunist away to the desert, or to ephemeral habitats, as a fugitive. Essential to life at a species equilibrium are adaptations that allow persistence. Each individual of an equilibrium species should compete strongly with its own kind and with others, and be programmed for life in a crowded habitat. Dispersal is less important than persistence; recovery from adversity less important than perseverance; high fecundity of less importance than survival of young in a crowd. The result is a life history distinctly different from that of opportunist or fugitive species.

The adaptations that make for an equilibrium species are essentially opposite to those of an opportunist. This is most clearly seen by considering the calorie budget of the animal. If the animal, throughout its life history, is made to invest calories in mechanisms for defense or endurance, then these calories are not available for extra reproduction. An animal can use the strategy of putting calories into structures that lead to persistence or the strategy of putting all possible calories into reproduction so that genes persist even though individuals perish. Different strategies represent direct trade-offs.

In many circumstances we can expect equilibrium species to replace opportunists. This would be expected, for instance, in the Oxford filter beds studied by Elton (*op. cit.*) if they were neglected and became permanent ponds. The Elton–Hutchinson hypothesis then predicts that the *Eurytemora* populations would become extinct in the ponds and only the more slowly dispersing *Diaptomus* would be found.

But it is also possible to see how opportunist and equilibrium strategies give two different answers to the same difficulties presented by a marginal habitat. This is nicely shown by plants of the Sonoran desert of Arizona or of the coastal regions of the Galapagos Islands (Figure 11.2). Both places are difficult for plants not only because water is scarce but because the supply of

Figure 11.2 Arid land for two strategies.
A semiarid habitat on Galapagos James Island. *Bursera graveolens* trees have a long-lived equilibrium strategy, putting out leaves during rare periods of rain but being dormant through most of the year as in this photograph. Other plants are adapted to this same pattern of semiaridity with rare periods of rain by an opportunist strategy of short life, semelparity, and survival of the drought by seeds. In a wet season these opportunists would carpet the ground green.

water is highly irregular. Many months with no rainfall at all may be followed by short seasons of ample, or even excessive, rain. Both opportunist and equilibrium strategies offer solutions to living in these places. The opportunist strategy is simply that of the annual weed that scatters seeds that lie on the desert soil until made to germinate by the coming of rain. The young plants grow very rapidly in the bright desert sun following rain, with plenty of space in which to grow. But they start flowering and setting seed almost immediately. They remain small, making no elaborate stem or root system packed with reserves for the future. All calories gained that are surplus to the immediate needs of metabolism are put into flowering and seed production. The result is the desert made colorful by a carpet of flowers following rain.

Yet Arizona and the Galapagos are well supplied with large plants in their deserts: cactus, thorn bushes, dry bushes of sage (*Artemisia*) and other composite shrubs with leathery or pubescent leaves, and even small trees. These meet the problems of desert living with long lives that span many of the scarce seasons of rain. They survive the lean times, in the main, not as seeds but as stems or roots, sometimes even with leaves, all heavily protected against water loss or grazing animals. They too grow rapidly in the rains but very many of their surplus calories go into making stems or thorns, laying down insulation, and waterproofing. There may be comparatively little energy left for reproduction so the seed-set may be less than that of an annual plant, or the plant may flower and set seeds only at very wide intervals, like the celebrated century plant.

The equilibrium strategy in desert plants is expected to result in a population equilibrium with numbers being set by density-dependent factors. Field observations are consistent with this, most notably the strikingly even spacing of many desert shrubs. Figure 11.2 shows a stand of the small tree *Bursera graveolens* on the Galapagos island of Santiago (James). The trees give the impression of a commercial orchard, so evenly spaced are they. The *Bursera* trees in the figure are without leaves, which is their normal aspect. Their primary adaptation to an equilibrium strategy in the Galapagos desert is to put out leaves when it rains and to drop those leaves again when the ground dries up. They are deciduous, but not as trees of temperate regions are deciduous, because there are no seasonal rhythms to their growing and dropping of leaves. If you want to see a *Bursera* tree with leaves you have to time your arrival at the Galapagos with the coming of rain and this may be difficult. Field parties from my laboratory made repeated visits to the Galapagos for several years before we found the *Bursera* trees with leaves.

Since equilibrium and opportunist strategies are opposite extremes, it should follow that compromise strategies are possible. Indeed, the very concept that these strategies represent trade-offs between spending calories on extra fecundity or extra powers of persistence suggests compromise. We should expect most living things to have a strategy that is a compromise between the extremes, letting us talk of a possible *continuum of strategies between extreme opportunists and extreme equilibrium species*. Such a strategic continuum seems to be illustrated by the plants of a classical old field succession.

Secondary succession in old fields (Chapters 1 and 23) develops as annual weeds, perennial herbs, shrubs, small trees, and larger trees occupy the site in succession. Annual weeds are classical opportunists and mature trees of the climax may be thought to be the ultimate equilibrium species with an enormous investment in structures of persistence. The intermediate species of the succession can be seen to invest ever more strongly in persistence: perennating organs, stout woody stems, increasing size. Since allocating calories is always a zero-sum undertaking, with calories spent on one thing being unavailable for another, then it seems a safe argument that fecundity must fall as strategies approach those of the climax species. Plants of the secondary succession, therefore, illustrate several stages along a strategic continuum. Indeed, the argument can be taken one step further, so that the secondary succession can be said to be a "consequence of having species with a variety of strategies living in the same country" (Colinvaux, 1973; Chapter 23).

r- AND *K*-SELECTION

Imagine a young volcanic island, recently raised from the sea and without terrestrial animals or plants. This island will be colonized rapidly by migrants arriving from across the sea. There will be plants coming as wind-blown seeds, birds and insects coming on their own wings, seeds and small animals clinging to the birds, and many things from small mammals and lizards to living plants drifting out on floating flotsam. But all these immigrants will have in common an ability to disperse. The overwhelming number of arrivals, perhaps all, will be opportunist or fugitive species well provided with mechanisms for dispersal.

Once settled on the island, the opportunist immigrants, adapted as they are to high fecundity, should achieve rapid population growth. Soon the island will be crowded with their descendants and these ought to establish population equilibria. But these animals and plants are not adapted to crowded lives at equilibrium; they are usually fugitives from other species that do better in crowds. On their island, weather may not strike them down and the better competing equilibrium species cannot reach them. Natural selection should then work to find the most competitive varieties from the island's opportunists. This is the process now known as "*K*-selection."

The original migrants were highly dispersing, high fecundity individuals. This means that the *intrinsic rate of increase r* of their populations was large. But they now have to live in populations holding the saturation number, *K*. The symbols "*r*" and "*K*" come, of course, from the conventional writing of a logistic equation (Chapter 6). Life at *K* means that resources have to be diverted to structures and habits that serve persistence or competition, taking away from the reproductive effort. Using the island argument, MacArthur and Wilson (1967) said that the migrants were descended from populations that were originally *r*-selected but that the attainment of an unaccustomed species equilibrium on the island meant that they would be subjected to strong *K*-selection.

Notice that the terms *r-selection* and *K-selection* are not themselves synonymous with the terms *opportunist* or *fugitive* and *equilibrium* species, though they do invite comparison. The new language speaks to selection within existing species populations, not to comparison between species. However, it is easy to modify the terms to make them equivalent. Opportunists and fugitives are then called *r-strategists* and equilibrium species are called *K-strategists*. These terms have nearly eclipsed the earlier words in contemporary ecology. Instead of the continuum of strategies between extreme opportunist and equilibrium species, we talk of the *r* and *K* continuum.

The concept of *r*- and *K*-SELECTION (as opposed to *r* and *K* strategies) is powerful since it leads to testable hypotheses. The postulated selection ought to occur rapidly. Furthermore *K*-selection of an originally *r*-selected population will take place in a rapidly reproducing population with short generations and this offers hope for experiments to show results reasonably quickly. A particularly attractive experimental design was worked out by Ayala (1968).

Ayala worked with populations of *Drosophila* in typical laboratory culture. As had been well established (Chapter 6) *Drosophila* populations achieved population equilibria in standard laboratory bottles if the culture medium was maintained. Wild *Drosophila* populations have traits that are consistent with an opportunist strategy and *r*-selection; they have high fecundity and must disperse to find discrete parcels of food. But when crowded in laboratory bottles they have no need for dispersal traits but need instead to compete for a fixed food supply. *K*-selection should be in progress. But this might be slow because the amount of genetic diversity present in small laboratory populations should not offer selection a large substrate on which to work. Ayala decided to increase the genetic diversity of his flies by irradiating them. If powerful *K*-selection then worked on the irradiated flies, the survivors should be more skilled at life in crowded bottles than control populations of unirradiated flies. This should mean that descendants of the survivors of the irradiation episode would be able to live at higher densities than flies in the control populations. Figure 11.3 gives Ayala's results.

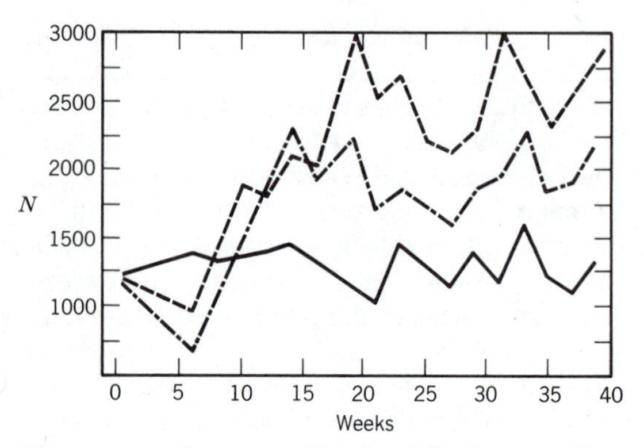

Figure 11.3 Experimentally induced *K*-selection in *Drosophila birchii*.
Two irradiated populations (broken lines) eventually reach higher equilibrium numbers than a control population (solid line). Notice that the first effect of irradiation is a population decrease, presumably due to irradiation damage. Selection then finds individuals able to live at high densities. (From Ayala, 1968.)

The new species equilibrium attained after irradiation allowed, as predicted, more flies to coexist. *K*-selection, therefore, was demonstrated.

Finding evidence for the *r*- and *K*-selection process at work in the field is more difficult. We can find plenty of examples of opportunist or equilibrium species whose ancestors we imagine to have been *r*- or *K*-selected, but finding the process itself at work is another matter. An elegant study on dandelions, however, seems convincing. The common dandelion (*Taraxacum officinale*) is an opportunist weed, presumably *r*-selected, but there are sites where dandelions persist at high density for long periods, sites like old grazed meadows, for instance. In these we would expect *K*-selection. Studying selection in dandelions is helped because the plants set seeds asexually, that is, they are *apomictic*. Apart from mutations, therefore, there should be no genetic change from generation to generation. Yet there are a number of different dandelion genotypes, each reproduced asexually. They are called BIOTYPES, and they can be recognized by isozyme patterns revealed by electrophoresis. Selection in dandelions, therefore, has only a few biotypes from which to choose and we can recognize those biotypes with a little routine laboratory work.

Solbrig and Simpson (1974) found three populations of dandelions near the University of Michigan growing on sites that were differently disturbed, but within 500 m of each other. One site was in the path of student traffic and was much trampled with bare ground showing, one site had less traffic, and the third was in the corner of a meadow, mowed annually but scarcely visited. It was reasonable to expect that there was stronger competition between dandelions in the scarcely disturbed meadow than on the other sites and thus more *K*-selection. Solbrig and Simpson were able easily to show that there was a different mix of biotypes at the three sites (Table 11.1). The "D" biotype did well in the undisturbed meadow, for instance, though poorly in the walked-over sites. Selection, therefore, was demonstrated[1]: but was this *K*-selection or selection for some unknown physical tolerance? This could only be answered by experiment. Solbrig and Simpson took a hundred plants from each site and grew them together in a garden. They found that the "D" biotype in the garden grew to be bigger, and to set less seed, than the "A" biotype, and that the data for the other biotypes were consistent. Therefore the biotypes did differ in the ways expected if they represented stages along an *r* and *K* continuum. Selection of biotype for the three sites was *r*- and *K*-selection and not just selection for physical tolerances.

Suggestive results also come from studies of fecundity in separate populations of fish. A study of the lake whitefish (*Coregonus clupeaformis*) is an example (Bidgood, 1974). In two Alberta lakes whitefish live at different densities. In Buck Lake the population is low, and this seems to be due to an array of predators and large competing fish. The whitefish of Buck Lake, therefore, are not likely to suffer much intraspecific competition. In nearby Pigeon Lake, however, there is a dense population of whitefish. These whitefish have slow growth rates, suggesting that they are indeed suffering strong intraspecific competition.

[1]The presence of different biotypes at the three sites could conceivably be due to *founder effects,* that is, the sites were by chance colonized by different biotypes in the first place and these subsequently bred true. The subsequent demonstration of phenotypic traits suited to the sites, however, makes this explanation implausible.

Table 11.1
Percentage of Each of Four Biotypes in Three Michigan Populations of Dandelions
(From Solbrig and Simpson, 1974.)

Habitat	Number in Sample	Biotypes A	B	C	D
1. Dry, full sun, highly disturbed	94	73	13	14	0
2. Dry, shade, medium disturbed	96	53	32	14	1
3. Wet, semishade, undisturbed	94	17	8	11	64

The high population density in Pigeon Lake can be explained because there are fewer predators and competitors than in Buck Lake; moreover records spanning 18 years show that whitefish numbers have progressively risen to the present high densities as numbers of predators have declined (perhaps they were fished out). We might expect that *K*-selection has been getting progressively more severe over 18 years in Pigeon Lake. Figure 11.4 gives Bidgood's (1974) data for fecundity in the two whitefish populations. Lower fecundity at Pigeon Lake suggests that the population was indeed more *K*-selected. This example is less satisfactory than the dandelion study, however, because interspecific competition is present to reinforce intraspecific competition in the low population of Buck Lake. We cannot say with certainty that reduced fecundity is not merely the usual result of interspecific competition (Chapters 6 and 12).

STRATEGIC THINKING ABOUT REPRODUCTION

Differential reproduction is essential to the concept of *r*- and *K*-selection: the more calories spent to exist at *K*, the less calories are available for reproduction. But there are other ways in which reproductive strategies influence fecundity and these are not necessarily dependent on *r*- or *K*-selection. Animals subject to both kinds of selection can lay large eggs, or small eggs, or they can bear their young alive. The young can fend for themselves from the day they are hatched, or they can be looked after by their parents. Young can be defended by parents or become the prey of cannibals. Even plant seeds can be large or small, protected from seed predators or provided with lures so that animals may seek them out, produced all at once or scattered through the seasons.

The most obviously different of extreme strategies are that of laying large numbers of tiny eggs and that of raising a few large young. Hutchinson (1978) has described these two strategies as PROFLIGATE or PRUDENTIAL reproduction: Perhaps the terms SMALL-EGG STRATEGY and LARGE-YOUNG STRATEGY convey the two alternatives more clearly. These opposing strategies must be explained as working to give fitness to individuals of the species that use them, though obviously in widely different ways.

Profligate animals with a small-egg strategy make small eggs, each capable of developing into a sexually mature adult. The cost of each egg is small so that many can be made for a modest investment in calories. Relative fitness of an individual becomes a function of the number of eggs laid so that natural selection will program all individuals to make as many eggs as possible. This, surely, is the simplest solution to the fitness problem: design tiny young and make many of them.

But real fitness is not simply a measure of fecundity; it is the number that survive to reproduce. Few of the tiny young hatched from small eggs in fact live long enough to breed because they die by accident, by predators, or even by

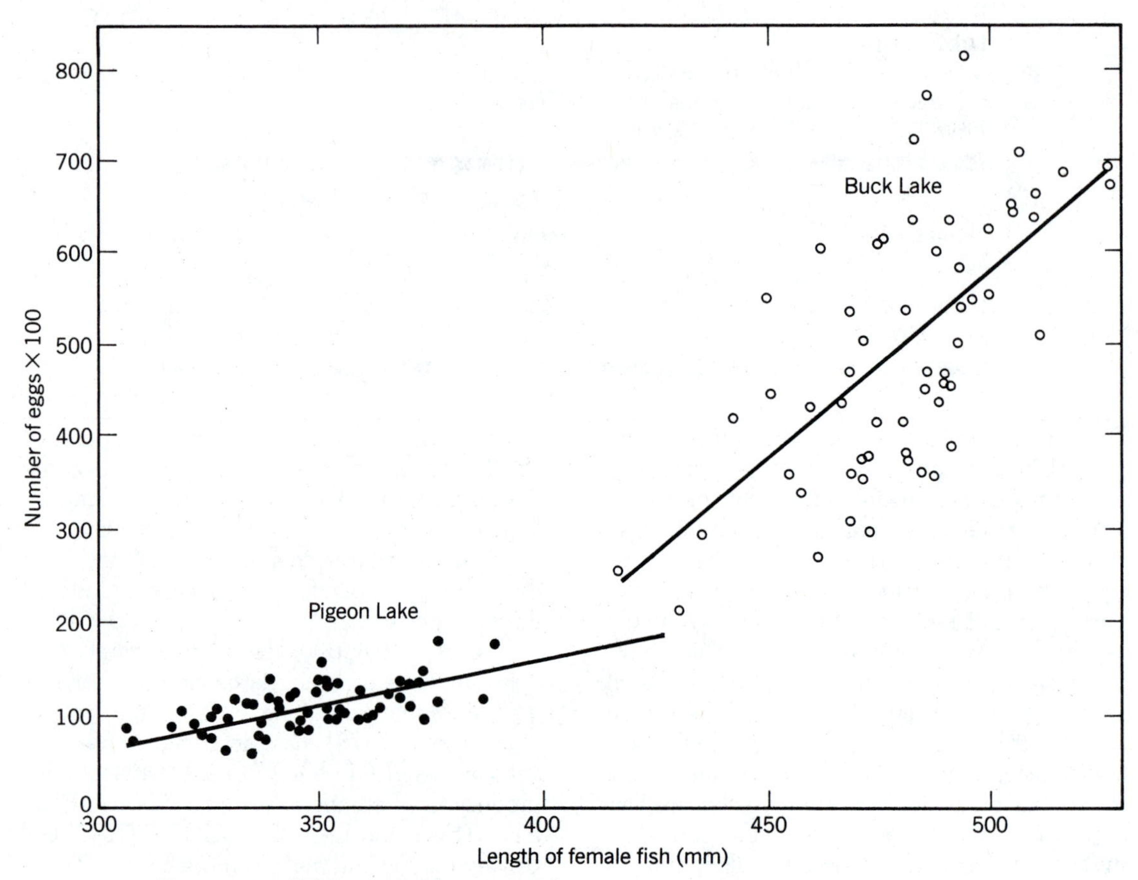

Figure 11.4 *K*-Selection in fish.
The data are for two populations of whitefish (*Coregonus clupeaformis*) in Alberta lakes. In Pigeon Lake the whitefish live at higher density. The data suggest that not only are the crowded Pigeon Lake fish smaller but that they lay proportionately fewer eggs. (From Bidgood, 1974.)

cannibalistic adults who are their parents' neighbors. Calories invested in these many young that die are wasted. Although, therefore, at first sight the strategy of many small eggs seems a good way of flooding the future with one's descendants, from a cost effective point of view it can be a very poor way. A female salmon, for instance, spends a year or two foraging so that she can lay many eggs that will be eaten by other animals.

Prudential reproduction of the large-young strategy is a quite different outlay of resources that in many circumstances can bring a larger return on investment. The mother makes large young (a reptile's egg, for instance). The mother's capital of calories will not let her make so many of these large eggs, but the young will each have more chance of survival than, say, a fish fry. Making a few large eggs may mean more surviving offspring that themselves live to reproduce than investing the same calories in many more small eggs. It can be, therefore, that restrained fecundity can result in greater fitness.

The logical extension of the large-young strategy is to reserve some calories for defense and nurture of the young. Many birds are programmed to lay few eggs, then to stop laying even though they have food resources to lay more. They divert these extra resources to feeding and defending their young. Animals that bear their young alive, whether mammals, viviparous reptiles and fish, or even viviparous insects, carry this process one step further. Essentially they are making their young even bigger at the time of birth by letting them start development within the mother. All these large-young strategies should be seen as ways of maximizing fitness by safeguarding the original investment. When done really well there may be no wasted calories at all and every calorie spent in reproduction goes to a young individual that eventually will itself reproduce. Contrast this idea with the certainty of mass death and mass waste that must result from broadcasting tiny eggs.

Precocial and Altricial Breeding Strategies

Obvious variants of the large-young strategy are shown by birds that have naked, helpless offspring (*nidicolous* or *altricial* birds) and those whose feathered young can forage for themselves near their mother (*nidifugous* or *precocial*).[2] The nidifugous or precocial life style is possible only where food is of a kind that can be found by baby birds. Game birds like pheasant and partridge, and many water birds like ducks and shore birds, are precocial and nidifugous. The food of the chicks in these species can be found by run, swim, and peck behavior. On the other hand, typical nesting songbirds, or hawks and owls, are altricial and nidicolous because insects, seeds on bushes, and animal prey can be collected only by the adult behavior of flight. And so the food supply provides the selection pressure that drives towards either the nidifugous or the nidicolous condition. It is not surprising to find that eggs of nidifugous and precocial birds are generally larger: parental investment has gone into a large egg from which develops a more advanced young, instead of hatching a less developed chick and then feeding it.

A first thing to be stressed about the opposed reproductive strategies of profligate and prudential reproduction is that they are alternative ways of pursuing fitness; NOT alternative approaches to population problems. A small family of large young does not represent restraint on breeding when compared with some insects and fish engaged in profligate reproduction. The real difference is not in reproductive effort but in the way that resources are disbursed to promote reproduction. It can be argued, for instance, that a strategy of a few large young will actually tend to increase population faster than would a strategy of many small eggs because the large young give a better return of survivors for every investment calorie.

Clutch Size Regulation

Large-young strategies are of two kinds: one-at-a-time or a clutch-at-a-time. The typical ungulate mammal like a cow bears one young at a time and then raises it to be self-supporting before the next reproductive episode. Fitness of the mother is a simple function of how long she can stay alive and active as a reproductive machine. But when a number of young are born synchronously, as with most mammalian carnivores and most birds, then there is the added complication of regulating the size of clutch.

If a bird laying a clutch of eggs is programmed to lay one egg *fewer* than its neighbors, and if the neighbors are able successfully to fledge a chick from their extra egg, then the bird with a small clutch will be at a selective disadvantage and its genes will disappear from the population. So much is obvious. But a bird might be at even greater disadvantage if it were programmed to lay one *more* egg than its neighbors because the resulting extra mouth to feed might endanger the

[2] These terms are all from Latin roots. PRECOCIAL comes from the direct transfer of the Latin word *praecox* meaning "ripe before its time," and itself derived from *prae-* (before) and *coquere* (to cook). ALTRICIAL is derived from the Latin *altrix* (nourishment) and thus implies young that need feeding as opposed to PRECOCIAL young that can feed themselves. The Latin *nidus* means a nest and *fugere* is to flee, so that NIDIFUGOUS birds have young that leave the nest as soon as they hatch, which implies that they probably are also precocial. The Latin *colere* means to cultivate, so a NIDICOLOUS bird is one that feeds (cultivates) its young in the nest, requiring that the young are certainly altricial.

survival of the whole brood. Raising a clutch of young, therefore, requires that *clutch size be regulated to an optimum.*

Regulating clutch size in birds requires that the female be programmed to stop laying when the optimal number of eggs has accumulated. This could be arranged in two ways: the number of eggs could be fixed by a genetic mechanism or the bird could be equipped with a sensing device that regulated the number of eggs according to the resources available in the local habitat (Figure 11.5). Both patterns are known to exist in different species of birds.

The best-known data for fluctuating clutch size are those for the European robin given by Lack (1954). Robins (*Erithacus rubecula*) are found all across Europe from North Africa to Finland, everywhere as year-round residents (Figure 11.6). The average clutch size corresponds closely with latitude, there being more eggs in the typical nest in the northern parts of the range. This observation certainly shows that clutch size is regulated to suit environmental circumstance. One environmental condition that might favor the larger clutches of the north is the longer day in which the parents may forage, thus providing more food. It may also be that the northern spring and early summer have insects in greater abundance than are found in lower latitudes, or that there are fewer competitors for them. These arguments suggest that clutch is being adapted to food supply available to the parents. Another factor that may be involved is predation. If the warmer latitudes were home to more predators then the parents should minimize the number of visits they make to their nest each day, a requirement that suggests a smaller clutch (Skutch, 1949; Cody, 1966). Apparently female robins are able to assess such things as available food or danger from predators and this assessment regulates egg laying. This is no great feat because a simple hormone response to maternal fat, or feeding rate, or number of alarms is all that is needed.

More spectacular regulating of clutch size is shown by snowy owls (*Nyctea scandiaca*) of the arctic. Snowy owls rear their young on the meat of lemmings and other arctic rodents whose numbers fluctuate widely from year to year (Chapter 13). In a high lemming year on the arctic coast of Alaska each owl nest may hold ten or so eggs, and the owls raise and fledge all the hatchlings, but in a low lemming year there may be only one or two eggs, if the owls breed at all. Snowy owl physiology apparently can de-

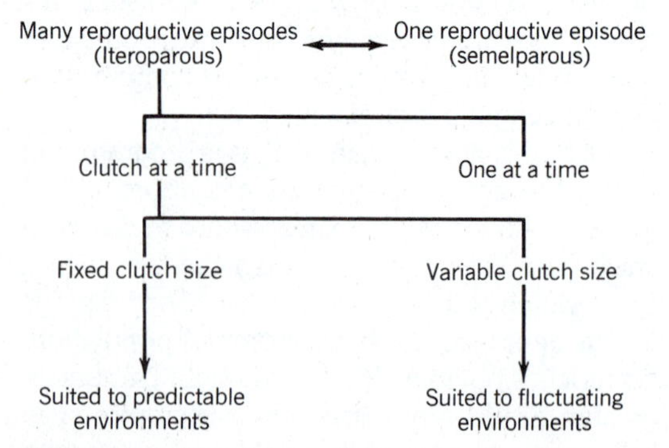

Figure 11.5 Variations on the strategy of producing and nurturing large young.

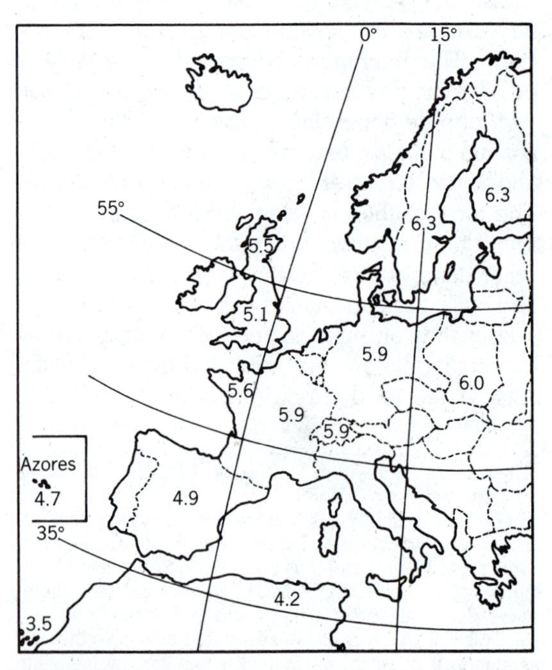

Figure 11.6 Average clutch size of the European Robin (*Erithacus rubecula*) along the north–south axis of its range. (From Lack, 1954.)

tect the density of lemmings: perhaps excitement at seeing how much succulent food runs about does the trick.

There is now a large literature showing that clutch size in birds may be correlated with the food supply. Lack (1954, 1968) first produced persuasive summaries testifying to this and there have been numerous studies since. We take this literature as providing sufficient test of the hypothesis that clutch should be regulated to an optimum in animals that lay large eggs or nurture their young as to raise this concept to a working principle.

REPRODUCTIVE STRATEGY AND THE *r* AND *K* CONTINUUM

Profligate reproduction with a small-egg strategy seems well suited to a species that must disperse rapidly and well. Classical weed plants with small, wind-blown seeds are certainly profligate reproducers: Andrewartha's Australian grasshoppers (Chapter 8), mosquitoes whose larvae live in temporary puddles, and flies that raise young in isolated bits of rotting fruit (*Drosophila*) or meat (bluebottles, genus *Lucilla*). All these use a small-egg strategy and all are opportunistic or fugitive *r*-strategists. There seems, therefore, to be a good argument that *r*-selection should lead to profligate, small-egg reproductive strategies.

Yet the broadcasting of small propagules is also done by plants of climax forests and by long-lived animals (many fish) that might well live near equilibrium numbers. Trees of the climax forest that use wind as a dispersal agent have small, though not necessarily tiny, seeds. Sugar maples (*Acer saccharum*), co-dominants of the beech–maple forests of the American Midwest, for instance, make a large number of smallish seeds, each equipped with a wing-like structure. The sugar maples are true equilibrium plants with massive trunks, long-lived, and with the habit of setting aside energy reserves for the future. But their numerous wind-dispersed seeds give them some capabilities as colonists so that it is not impossible that they might sometimes colonize bare ground like true opportunists, thus preempting the herbs of the secondary succession. Perhaps the opposite extreme for seed production in plants is shown by coconut trees, which have truly huge seeds and yet may be one of the most widely dispersed trees in the world, able to cross great oceans.

If we think of a small-egg strategy as being well suited to opportunist and *r*-selected populations, then the converse ought to be true also: a large-young strategy should be suited to equilibrium or *K*-selected populations.

What seems true is that a small-egg strategy has particular value to many life styles that require wide dispersal, because dispersal is always a gamble in which many of the would-be migrants must perish. Making these migrants small is then actually prudent because the inevitable losses are minimized. The gambling analogy is particularly appropriate here. Some chips (dispersed seeds) do find a good site, effectively hit the jackpot, and produce a large return in fitness in the next generation. The strategy works in a patchy environment when the cost of dead offspring is the price of finding the right patch with the lucky ones. But this does not mean that the strategy is without worth to other life styles (like maple trees).

Animals with a large-young strategy include both opportunist and equilibrium species. This was early shown for birds by MacArthur (1958) in the course of his classic study of sympatric warblers feeding on spruce budworms (Chapter 7). Spruce budworms are available in quite different amounts in different years, presenting the birds with a fluctuating resource, yet it turns out that three of the five kinds of warbler that feed their young on budworms always lay the same number of eggs (Figure 11.7). Only two of the five warblers tracked budworm abundance with clutches of different size and MacArthur suggested that these might fairly be called opportunistic or fugitive species (this was before he and Wilson introduced the items *r* and *K*). How a fixed clutch adapts warblers to a food source that fluctuates from year to year is not so obvious, but perhaps the adaptation reflects past conditions before human fire-control and other

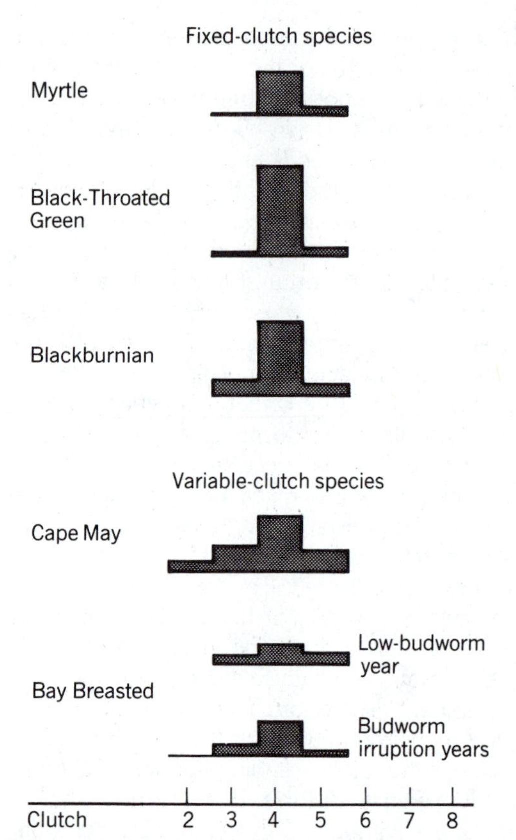

Figure 11.7 Clutch size in five warblers feeding young on spruce budworms.
Only Cape May and bay-breasted warblers respond to food density by changing clutch size. These may be thought to be more opportunistic or *r*-selected than the other three species. (Data from MacArthur, 1958: see also Figure 7.4.)

Figure 11.8 *Spheniscus mendiculus*: parent that abandons young.
The Galapagos penguin lives in an unpredictable environment, using inshore food supplies of intermittent upwellings to feed young. When upwellings cease during breeding, which happens more often than not, the parents abandon the young. This breeding behavior works to produce fitness because the adults are long-lived and feed safely out at sea.

forestry activities promoted outbreaks of spruce budworms.

It might be expected that a fixed clutch size would be well suited to predictable conditions and life at a species equilibrium. Most birds do in fact have fixed clutch sizes, which, given the evidence that birds tend to be under density-dependent control (Chapter 8), seems consistent with this expectation. But a fixed clutch can be the answer to unpredictable environments also in some circumstances. The fixed clutch size (two eggs) of the Galapagos penguin turns out to be an effective response to a highly unpredictable food supply in this way. Galapagos penguins (*Spheniscus mendiculus;* Figure 11.8) wander the seas throughout the 1000 square kilometers or so of water that surround the Galapagos Archipelago, but they nest only on the western coasts of the westernmost islands (Figure 11.9). This distribution can be understood from a knowledge of ocean currents of the region because the coasts where the penguins breed are those that intercept, from time to time, the submerged Cromwell current coming from the west. When this cold current strikes a Galapagos island, cold, nutrient-rich water upwells along the shore causing a bloom of algae and thence of fish (Maxwell, 1974). This productive water provides the food on which penguins rear their young. But the arrival of an upwelling is highly unpredictable; it may happen in any month of the year and the

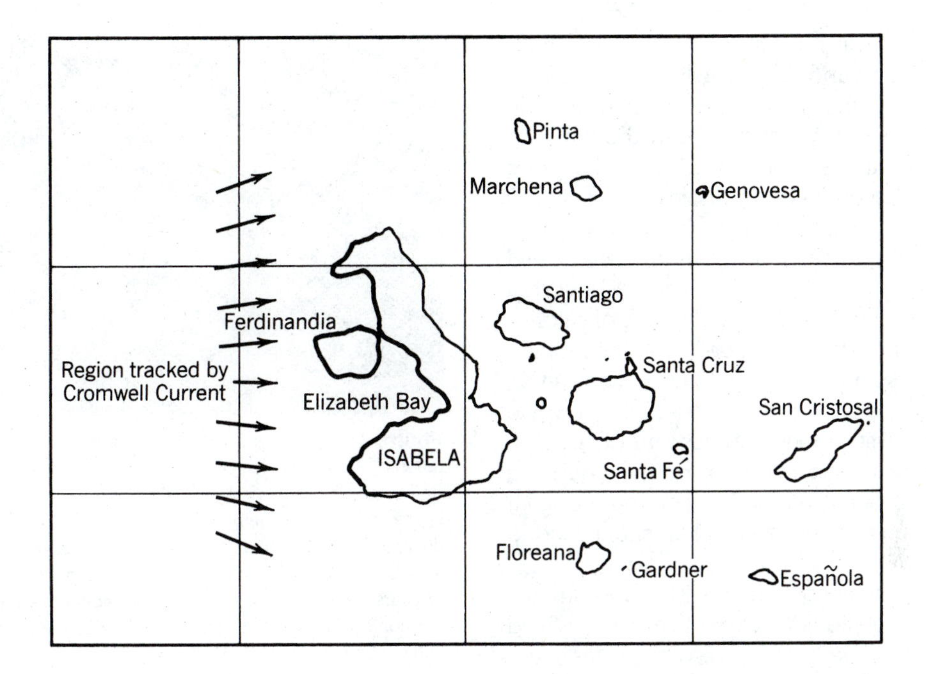

Figure 11.9 Breeding sites of Galapagos penguins (*Spheniscus mendiculus*). The penguins breed only on the westernmost coasts in the archipelago where waters of the Cromwell Current (Equatorial Undercurrent) upwell from time to time. Adult penguins can be found throughout the waters of the archipelago but they do not come ashore on other islands. (From data of Maxwell, 1974, and Boersma, 1977.)

upwelling may last for days only, or for weeks. The penguins are equipped with a breeding strategy, involving a fixed clutch of two eggs, that lets them cope with the problem of an unpredictable food source. The main elements of this strategy, as worked out by Boersma (1977), are as follows:

1. Moult as soon as an upwelling is detected (probably using water temperature as the cue).

2. Copulate as soon as moulting is complete.

3. Lay two eggs at once and start incubating.

4. Abandon the weaker chick when finding sufficient food is difficult.

5. Abandon nest and both chicks at once if the upwelling fails.

6. Live many years.

The striking thing about this breeding strategy is the readiness of the penguin mothers to abandon chicks in times of adversity. Galapagos penguins do not have human ideas of motherhood. Indeed, Boersma (1977) found that the usual outcome of a breeding effort was abandoning *both* chicks, and it was rare that both were raised. A successful breeding effort, in penguin terms, is to raise one chick, and the most usual result is total failure to rear any.

The proximal reason for a Galapagos penguin to abandon chicks is obvious, because it is done when the upwelling fails. "No upwelling" means little food within swimming range of the nest. If the penguin mother tried to sit out the failure of an upwelling, not only would there be no food for chicks but she herself might starve. The proper response is for her to swim away to the open sea where she can survive herself and come back to breed during another upwelling.

Two eggs rather than one also make sense in the penguin's circumstances, even though Galapagos penguins seldom raise more than one chick. The second egg is an insurance policy against loss by accident—incubated, or hatched, and ready to go if needed. The premium paid on the insurance policy is the calories invested in one egg, roughly the same number of calories needed for three days' metabolism of the mother. So two eggs can be the optimum clutch even when only one chick is to be raised. This is quite a common tactic among seabirds and others who face uncertain food supplies for the young, a tactic that ensures a chick is coming along at minimum cost, ready to exploit chance good times.

All five of the penguin habits listed above make sense as adaptations to exploit very short-lived and unpredictable gluts of resources. Moulting before breeding (special to Galapagos penguins), for instance, ensures the mother is in her best shape to go fishing in the cold water of the upwelling. After that, the habits are simply suited to rushing one chick through as fast as possible, but with the best chance for cutting losses and getting out if this proves impossible. Yet this behavior confers fitness only if the mother lives to breed again another day. A long, safe life for the mother is vital to the habit of abandoning chicks in lean times.

The Galapagos penguin clearly has a cautious, conservative attitude to reproduction that is an adaptation suited to exploiting an intermittent, yet ample, flux of food. Many other seabirds behave in a similar way, or they may show the behavior in more extreme form. Some, like most albatrosses, lay only one egg at a time and seem to be under more strict limits from food even than the penguins. The red-footed booby (*Sula sula;* Figure 11.10) of the Galapagos is of this kind (Nelson, 1968). These boobies do not breed at all until they are five years old, then lay one small (compared with body weight) egg at intervals of 14 months or so. Adults are long-lived, perhaps living ten, twenty, or more years.

Figure 11.10 *Sula sula*: conservative breeding on short rations.
Red-footed boobies nest in bushes on Galapagos Isla Genovesa, but wander the ocean at other times. Food for reproduction is always scarce. Like many other seabirds their adaptation for breeding is a long juvenile period of learning (5 years), long life, one egg at a time, slow growth of young, and long adult life. Half the young die in the nests, apparently from lack of food.

50% of the chicks die or are abandoned, so the reproductive rate of this species is extremely low. What seems to have happened for these birds is that food is always in short supply so that the rapid feeding of young is impossible; likewise the chance of a chick's starving must always be high. Since food is scarce, the chicks grow slowly; and this increases the cost of a chick because their metabolic needs must be met for a longer time. As chicks become more difficult to feed, and more expensive in the long run, so there is a greater value in keeping the parent alive for later reproductive episodes.

Many animals with a large-young strategy can be forced into this pattern of life history. Scarcity of resource first decrees that clutch size be small and growth rates slow. After this the mathematics of a life table show that survival of parents is an important component of fitness. Figure 11.11 shows a plot from Goodman's (1974) life table calculations of the extra young needed to compensate in fitness for the death of a parent, assuming there is only one reproductive event in a lifetime (*semelparity*). For breeding in the first year the parent can be replaced by a single egg

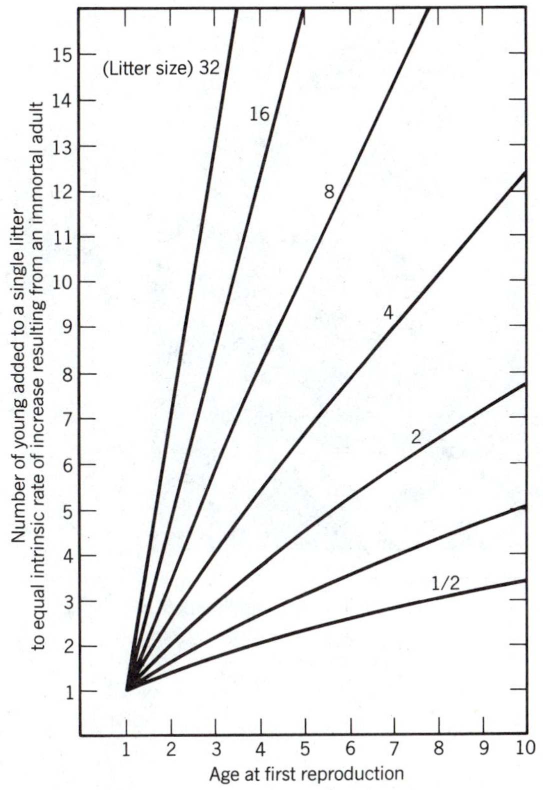

Figure 11.11 Increased clutch needed to compensate for adult death.
Fitness represented by survival of a parent goes up with the age at first reproduction. Data are plotted for a number of litter sizes, showing the extra eggs needed by a species that breeds only once (semelparous) to compensate for the death of the parent. (From Goodman, 1974.)

but as the breeding parent is older its replacement cost goes up until a parent with a litter size of one egg who is five years old is worth three additional eggs, and so on. This plot reveals the hard facts behind the life history of typical seabirds. Food is hard to get, which forces clutch size to be small. Furthermore, the difficulty of finding food requires that birds be experienced before breeding, so they cannot breed in their first year. Living longer than a year before reproduction increases the value of the adult as a unit of fitness, which increases the selection pressure for adult survival still further. The result is a strategy of long life and low reproductive effort.

This logic explains some common facts of natural history. Familiar seabirds of the north like herring gulls (*Larus argentatus;* Figure 11.12) live as immature birds for several years and these immatures are marked so that they are easily distinguished from breeding adults. The immature birds are still learning to forage under all circumstances until they have sufficient skills to be able to become parents with minimum risk to themselves. Their distinctive markings probably free them from wasting energy and time on sex encounters while they learn.

More important are the population consequences of long life and conservative breeding. As Mertz (1971) was first to point out, a species with this strategy is threatened with extinction as soon as any significant mortality of adults occurs. This is why efforts to save the California condor from extinction show so little success. Condor breeding strategy is like that of the red-footed boobies, perhaps actually an extreme version since the adults can live fifty years. They breed very sparingly and are likely to abandon their young in adversity in the expectation that they will live to breed another day. Continued increased mortality from firearms cannot be offset by reproduction in birds with this breeding strategy.

Condor, booby, and penguin life histories make them poor colonists. They cannot exploit new environments by rapid reproduction. Therefore, it is safe to say they are not opportunist species or *r*-strategists. On the face of it, they certainly seem to fit a general description of an equilibrium species or a *K*-strategist. Calories are diverted from reproduction to persistence and there is a stable population in the absence of strong environmental change. In fact the population will be stable for any environmental change less than that which imposes excess mortality on adults. And yet there must be a little discomfort in calling these birds equilibrium species, or in saying that they are *K*-selected. Their life histories are perhaps better understood as a result of selection to cope with a particular pattern of resources than as a response to crowding or competition.

Figure 11.12 *Larus argentatus*: juvenile status proclaimed by color.
In herring gulls, as in many long-lived seabirds, the non-breeding juveniles are clearly marked. Juvenile years of learning skills at foraging apparently are needed before care of offspring can be attempted and the dark coloring may serve to spare the juveniles unrewarding social encounters.

VARIETY AND CONSEQUENCE OF LIFE HISTORY CHOICES

In both animals and plants some species breed but once in a lifetime, SEMELPAROUS species, and others breed repeatedly, ITEROPAROUS species. We saw above (Figure 11.11) that when resources for reproduction are scarce, iteroparity gives an advantage in fitness. But this clearly is not generally true. Many species are semelparous in nature: annual weeds with a single flowering episode before death of the parent and the Pacific salmon, which return to the rivers of their birth but once, lay eggs, and die, are the most spectacular examples of semelparous species.

When an organism matures, reproduces once, and dies there is nothing wrong with the logic that says the death of the parent can be compensated by a single addition to the clutch: from the point of view of fitness one female equals one egg. This was first formally shown using the mathematics of life tables by Cole (1954) and the conclusion is now referred to as "Cole's result." This, however, leads to disturbing conclusions like saying that a ten-kilogram female salmon represents no more chance to the fitness of its genotype than a tiny egg. The resolution of this paradox is that fitness is secured only by survival to the next reproductive episode, and whether survival to another reproductive episode is more likely for egg or fish may not be clear-cut. Whether individuals of a species are *semelparous* or *iteroparous* must depend on resources available for reproduction and the chance of survival to the next reproductive period. We have seen that birds like boobies and condors, which use large-young strategies and face scarce resources, are forced into iteroparity. Probably iteroparity is nearly always necessary for animals with a large-young strategy because of the high cost of the young and the high survival power of adults.

But for a small-egg strategy the issue is not

clear-cut. For a salmon, for instance, the choice is not actually between one fish and one egg, but between one fish and very many eggs. The advantage then seems to swing towards the eggs. But some salmon are semelparous and others (the common Atlantic salmon, *Salmo salar,* for instance) are iteroparous. Whether one strategy or the other gives the advantage must depend on factors like the local changes of seasons, or local predators, which influence the survival of young salmon.

In plants the change from semelparity to iteroparity can be seen to represent a continuum of strategies, essentially parallel to the opportunist–equilibrium, *r–K* continuum itself. Extreme annual weeds are semelparous, perennials are iteroparous, and iteroparity seems to be prolonged through a variety of strategies roughly mirrored in the progress of a secondary plant succession. Prolonged iteroparity in plants may be accompanied by delays in age of first reproduction, just as it is in long-lived seabirds. Many trees, for instance, do not set seed until they are several years, or even decades, old. Plants with life spans of considerable length, too, may be semelparous like a Pacific salmon. Species of the green alga *Halimeda* are a case in point: the green plant turns white overnight as all its tissue is mobilized to form a cloud of zoospores that, when released, leave behind only the skeleton of white carbonate (Figure 11.13).

These variations on a theme of semelparity and iteroparity provide a very large number of possible ways in which life histories may be fitted to patterns of environment and resource. They present not so much a continuum of possibilities as endless permutations. And each variation must have consequences both for the history of the population and for the structure of the community in which the individual lives. Large young, small young; semelparous, iteroparous; early reproduction, delayed reproduction; helpless young (*nidicolous*), precocial young (*nidifugous*); all are strategies that work in some circumstances, all are selected to confer fitness on individuals, but all have consequences for populations and communities.

Figure 11.13 *Halimeda simulans*: semelparous ocean plant. Green algae of the genus *Halimeda* are important benthic plants of tropical coasts. The photograph is of a plant in culture about to release gametes. The entire organic content of the thallus has been mobilized to produce the gametangia, seen as dark, grape-like clusters on the tips of segments. Only a dead, white, calcareous skeleton remains after gamete release.

THE *r* AND *K* CONTINUUM: REVIEW STATEMENT

The concept of *r*- and *K*-selection was described by MacArthur and Wilson (1967) when the old arguments about the prevalence of density dependence or independence (Chapter 8) may have been getting a little stale. There was, in the middle 1960s, a general consensus emerging that a variety of different population histories and ways of control were possible and that they were very much functions of different kinds of life history. The idea of a continuum between opportunist and equilibrium strategies was talked about by some ecologists, and they argued that these opposed strategies led to either density-independent or density-dependent population control. A main contribution of "*r* and *K*" was to let this consensus emerge. Many ideas dropped into place when the language of *r* and *K* was used instead of talking of opportunists, fugitives, or population equilibria.

Pianka (1970) gave us a summary statement of how *r* and *K* could be used to organize so many ideas in contemporary ecology. Table 11.2 is from Pianka's summary, essentially as he published it. Some parts of the table, like the reference to MacArthur's broken stick model (Chapter 23), speak to debates now mostly past. But the properties listed in Table 11.2 generally still serve as a statement of likely consequences of a species having an opportunist or equilibrium life strategy in extreme form.

THE WAYS OF A FORAGER

What has come to be known as *optimal foraging theory* was introduced in Chapter 4. The theory argues from premises based on natural selection that individuals should act to maximize their energy intakes. As far as is practicable, each individual should adopt a feeding mode or strategy from several available to it that would yield the best return in food energy and hence in fitness. The PROFITABILITY of prey is

$$P = \frac{\text{Energy gained } (E)}{\text{Handling time } (h)} = \text{cals sec}^{-1}$$

Table 11.2
Pianka's Correlates of "*r*- and *K*-selection"
*The columns could be renamed **r**-strategists and **K**-strategists, or opportunist strategy and equilibrium strategy. (From Pianka, 1970.)*

	r-Selection	*K*-Selection
Climate	Variable and/or unpredictable: uncertain	Fairly constant and/or predictable: more certain
Mortality	Often catastrophic, nondirected, density independent	More directed, density dependent
Survivorship	Often type III (Chapter 9)	Usually types I and II (Chapter 9)
Population size	Variable in time, nonequilibrium; usually well below carrying capacity of environment; unsaturated communities or portions thereof; ecologic vacuums; recolonization each year	Fairly constant in time, equilibrium; at or near carrying capacity of the environment; saturated communities; no recolonization necessary
Intra- and interspecific competition	Variable, often lax	Usually keen
Relative abundance	Often does not fit MacArthur's broken-stick model (Chapter 25)	Frequently fits the MacArthur model (Chapter 25)
Selection favors	1. Rapid development 2. High r_{max} 3. Early reproduction 4. Small body size 5. Semelparity: single reproduction	1. Slower development, greater competitive ability 2. Lower resource thresholds 3. Delayed reproduction 4. Larger body size 5. Iteroparity: repeated reproductions
Length of life	Short, usually less than 1 year	Longer, usually more than 1 year
Leads to	Productivity	Efficiency

Energy gained per unit time (E/T) is given by

$$E/T = E/t + h \qquad (11.1)$$

This theory reduces the variables likely to influence a forager to the simplest three possible: calorific value of prey, time taken to handle prey, and time taken to hunt prey. If these three were all that an animal needed to take into consideration the following quote of MacArthur (1972) applies:

> *An animal should elect to pursue an item if, and only if, during the time pursuit (and consumption) would take, it could not expect both to locate and catch a better item.*

There we have the strategy of a forager in the simplest possible world.

Some examples were given in Chapter 4 of circumstances in which real animals actually do conduct their affairs as if guided only by the assumptions of this simplified world. The best known is the system in which bluegill sunfish are shown to take the size and density classes of *Daphnia* prey predicted by the model (Chapter 4; Werner and Hall, 1974). But a system of bluegills hunting water fleas in glass tanks, and those other examples given in Chapter 4, were chosen for study precisely because they were simple. The animals were given choices of food of the kind envisaged in the model; also it was likely that each animal was capable of the sorts of choices expected. In the real world choice may not be so simple, there may be other consequences to take into account, or the properties of the animals themselves may restrict choice.

Many animals feed in ways that make assumptions of optimal foraging theory uninteresting, notably the herbivorous larvae of holometabolous insects (insects with a complete life cycle of egg, larva, pupa, and adult). Caterpillars of Lepidoptera (butterflies and moths) forage where they hatch—the "choice" of forage is made for them by their mother when she laid her eggs. And the choice of these Lepidoptera, and other holometabolous herbivorous insects, is very restricted, sometimes being restricted to plants of a single species and very often to plants of a single genus (Table 11.3). There are probably two families of constraint that lead to this specializing in forage: the nutritive or chemical properties of the food and the sensory powers of the adults to distinguish between plants (Mitchell, 1981).

When caterpillars are on the wrong food plant many of them die. This is because of the complex chemistry of plants, itself enhanced by natural selection because this reduces the attack of foragers. A caterpillar can eat the food plant of its species because it can detoxify the chemicals in that plant. The feeding strategy of a lepidopteran species, therefore, is to use a highly specialized digestive process on the leaves of a particular target plant. Effective foraging for such a species is dependent on the ability of adults to recognize the right kind of plant and safely to lay eggs there.

That so many insect herbivores can eat only a small range of plant food (Table 11.3) illustrates a second general point about foraging strategies—that calories are not always a good measure of the worth of a food item. Food must not be poisonous. More than this, food must contain protein, nitrogenous bases, and other essential items of diet; it must be nutritious. It is thus easy to imagine animals refusing to take the mixture of food particles prescribed for them by optimal foraging theory if some other diet gave better general nutrition.

It must also be likely that many animals are programmed to forage in a satisfactory manner in the most common circumstances so that they will only "optimal forage" in the circumstances for which they are best suited. This may explain the fact that honeybees learn to persist with visits to blue flowers much more easily than they learn to persist when visiting white flowers. Bees can be trained to visit flowers of either color, but they are quick to switch from white to blue, though slow to switch from blue to white—even when the reward of sugar is clearly in favor of going to white flowers. Honeybees seem to have been programmed to "expect" good returns from blue flowers, presumably because blue flowers usually give a better reward in their natural habitat (Waddington and Holden, 1979).

These various properties of insect life histories

Table 11.3
Plant Species Preferences in Herbivorous Insects
Surveys of species (number in parentheses) in the insect families listed that feed over large areas on plants of one, or more than one, genus. Numbers are the proportions of the total sample showing each preference. By far the commonest pattern is for insect herbivores to be restricted to eating plants of a single genus. (From Mitchell, 1981.)

Group (Number of Species)	Number of Genera of Host Plants Used			Area
	1	2–5	>6	
1. Females choose larval host				
Cynipidae (547)	0.998	0.002	—	North America
Psyllidae (78)	0.974	0.026	—	Britain
Tenthredinoidea (166)	0.801	0.181	0.018	North America
Cerambycidae (advanced) (317)	0.511	0.423	0.066	North America
Chrysomelidae (139)	0.468	0.467	0.065	North America
Macrolepidoptera (766)	0.412	0.422	0.166	North America
Cerambycidae (primitive) (117)	0.383	0.603	0.014	North America
2. Females found colonies				
Aphidoidea (640)	0.803	0.109	0.088	Britain
Aphididae (249)	0.771	0.209	0.020	Pennsylvania
Thysanoptera (80)	0.475	0.125	0.400	Britain

and sensory systems thus place constraints on their foraging behavior. The animals are not only concerned about the shortness of time in which to feed, nor only with the caloric value of food. They have even more pressing concerns. This does not mean, of course, that they do not forage for the "best" return, merely that their behavior cannot be predicted entirely from considerations of handling time and search time. It remains true, however, that many animals do feed in ways that make the use of time important and on food items that are of comparable nutritional value. Optimal foraging theory has been a powerful tool for asking questions about how these animals live (Chapter 4). Of particular interest is how the approach can be used to study the strategies used by some animals to avoid predators while pursuing the best possible feeding habits.

Notonectids (backswimmers) are predatory insects that are hemimetabolous (no larval stage, the young instars being essentially miniature versions of the adult, though without wings and sex organs). In the species *Notonecta hoffmanni* all young instars and the adults can hunt and kill the same prey, mosquito larvae or floating insects trapped on the stream surface. But the larger individuals are also cannibals that prey on their own young and this means that the youngsters move into a patch of water where the feeding is good at their peril because their big relatives will certainly be there. Young *N. hoffmanni* must sometimes be faced with the "choice" of increasing their food intake by going to where most food is, but where there is a risk of being killed, or feeding where it is safer but where their food intake is small. Sih (1980) set up experiments to find out if young *N. hoffmanni* could find optimal solutions to this problem.

In Sih's experiments the notonectids were kept in small laboratory tanks. They were fed on wingless fruitflies, which struggled, helpless, on the surface. A wooden railing cut the surface to make an inner area separated from a channel around the edge so that the fly food could be concentrated in the middle, or placed in middle and edge in different concentrations. All the notonectids could be allowed to swim freely under the wooden railing, or else the water, too, could be divided into inner and outer areas by hanging a net from the railing. Young instars could be placed with various concentrations of food in outer and inner compartments alone and their foraging successes tallied. They could then be placed in similar concentrations of food but with a few large notonectids loose in the water. It was found that the young were in fact seldom caught and eaten but that they spent so much time avoiding their big relatives that their feeding rates suffered. Then came the crucial experiments of giving the youngsters choices of peril or want. When a sufficiently dense concentration of fruit flies floated in the inner ring, three adult notonectids remained in that ring even though there was no net to stop them swimming to the outer compartment. The youngsters could then hunt in safety in the outer ring or swim in to where both food and the predators were. It turns out that the first instar (very small) of *N. hoffmanni* avoided the center and that they penetrated this region as they grew into larger instars (Figure 11.14). Sih concluded that the animals could, and did, achieve the pattern of behavior that led to the largest possible calorie intake for each mixture of food and predator. They must have been using an estimate of time between successful hunts as their guide because, when they had to spend too much time escaping predators, their food intake went down and it was more profitable to go where rare food could yet be caught more quickly.

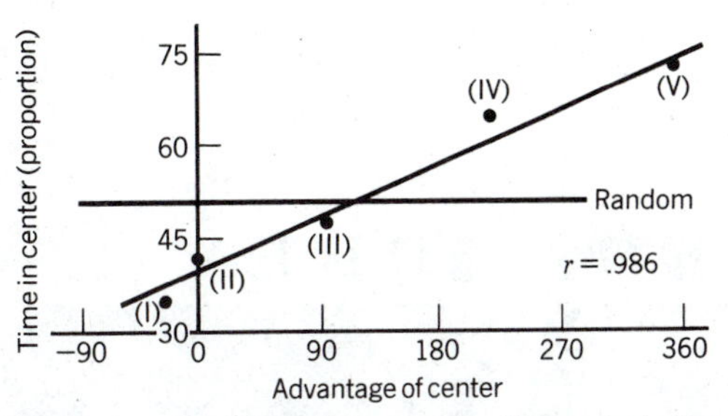

Figure 11.14 Optimal foraging by *Notonecta hoffmanni* young that includes predator avoidance.
The line is a linear regression between the relative advantage (measured as feeding rate) in a region rich in food and predators and the amount of time spent in this region when given a choice. Roman numerals are the five instars, number V being the adult predator itself. (From Sih, 1980.)

STRATEGIC THINKING FOR COMMUNITY ANALYSIS

Talking of life history strategies has, perhaps, given its best service to ecology when applied to the study of communities. We can look at a community as an aggregate of species populations put together as different life history strategies mesh with one another. Rates of colonization, powers at holding ground, powers to invade, timing of active or reproductive periods, the ways of foragers, strategies of predators, or escape from predators: all these are the stuff of which community synthesis is made. Each species has a niche, and each niche is filled by individuals maintaining themselves with a particular life history and strategy for breeding. As the niches are packed together, a community results.

PREVIEW TO CHAPTER 12

Sexual reproduction requires that each female shall surrender half of her investment in every egg or young she produces, which means that she must receive a very large compensating benefit. If a female is very fecund only a small portion of her reproductive output normally will result in surviving offspring and the few survivors will only succeed after competing successfully with their siblings and with offspring of other females. The chance for a female to have her own offspring win, therefore, will be greatest if she enters the competitive lottery with a variety of contestants, each of which is subtly different. In this way sex can be explained as a device useful to females, the benefits of which far outweigh the cost of surrendering a half share in each egg. Parthenogenetic animals are usually only modestly fecund and live where rapid population growth is possible for a few generations at a time. Asexual reproduction doubles fecundity in these episodes of rapid growth. Sex then gives no benefit and is deferred until some distant generation faces crowding or changed physical circumstances, when the parthenogenetic sequence is interrupted by a sexual episode. The reason for sex in low fecundity animals like birds and mammals is less understood, since meeting the high cost of males from a small clutch seems difficult. It may be that varied offspring are particularly needed in these species to meet intraspecific competition. A consequence of sex is that females usually invest far more heavily in each offspring than do males. This initial inequality is the cause of many of the varieties of mating and social systems that we observe. Polygyny is common because males can increase their fitness by inseminating many females. The actual size of a harem in polygyny is a function of resource availability, and of the opposed interests of males and females. Monogamy is an extreme form of polygyny in which the harem size is one; it occurs when resources are either scarce or scattered, so that these cannot easily be sequestered by one male, or when the male must raise its own investment in each offspring to be comparable to the investment of the female in order to secure his own fitness. Territoriality in monogamous songbirds is behavior that essentially results in self-spacing by the males. This behavior can grant fitness through sequestering of resources, through allowing familiarity with a neighborhood, through conserving energy in the food search, and through pair bonding. The behavior is adapted to polygyny in some birds when resources are sufficiently patchy so that territorial males can sequester particularly good patches for themselves. It may then be more to a female's advantage to join a harem on a good patch than to accept a monogamous male in a poor territory. Territorial behavior also occurs in polygynous animals where the young are not raised on the territory if there is a male against male competition for mates. Some of the extreme consequences of this kind of behavior for males are large size and gaudy colors, traits illustrated by Pribilof fur seals and the birds of leks. Many selection pressures also should lead to large, socially dominant females, but in some familiar animals with polygynous territorial males, the males are larger and thus socially dominant. Complex social mechanisms can only be maintained by natural selection if there is advantage to the loser as well as to the winner in an encounter. In every case it must be fit to quit in the sense that the loser stands a greater chance of breeding elsewhere, or in another season, thus compensating for accepting immediate humiliation and denial of a chance to breed. A curious consequence of the various forms of territorial breeding behavior is that a limit is set to the population. This is an accident of the behavior arising from the fact that space is limited. The remarkable constancy of bird numbers from year to year as shown in breeding census results from territorial behavior. Group living outside the breeding season can have numerous advantages in some circumstances, chief among which are help in finding food by foraging in flocks and defense against predators. All forms of group behavior can be fully explained as resulting because the activity gives advantage to every individual in the group. That the behavior sometimes has consequences for whole populations or groups does not imply that there is a group purpose or group selection.

CHAPTER 12

ECOLOGY OF SOCIAL SYSTEMS

Individuals live together with their own kind, often paired off in sexes, sometimes in complicated social hierarchies, sometimes in dense aggregates like flocks or schools, and sometimes spaced out so that we talk of species being "territorial." Ecologists can examine all these social structures, asking of each that prime question, "What is there in it for the individual?" We expect living in a flock, staking out a territory, or the vagaries of sexual behavior to serve the fitness of each individual participant. Our working hypothesis is that everything done by a social animal, however bizarre the behavior seems, has the result of increasing the fitness of the individual. Some ecologists, it is true, do investigate the possibilities of group selection, asking if behavior can serve the group at the expense of the individual. But before examining ideas of group advantage it is both parsimonious and prudent to see how much we can explain in terms of individual advantage. We find, in fact, that the logic of individual advantage offers convincing explanations for many strange patterns of social existence. As strange as any is the phenomenon of sex, and understanding the basis of sex provides a key to understanding much else of social behavior.

THE CHOICE BETWEEN SEXUAL OR ASEXUAL REPRODUCTION

Most living things come in two sexes, and only one of these sexes lays eggs. At first sight the advantage of this is not obvious, particularly from the point of view of the female. Each female gains fitness by reproduction; if she uses the common small-egg strategy, she lays as many eggs as possible. But sex means that she gives a half genetic interest in all her eggs to another, since half the genes passed to the next generation in her eggs come from her mate. How can it be to a female's selective advantage to act in this way? or, putting the question in another way, "Who needs males?"

Costs of sex to females are actually of three kinds. First is the loss of a half interest in the genes of her offspring, called THE 50% COST OF MEIOSIS (Williams, 1975)—the deliberate halving of the mother's contribution to her progeny by splitting her chromosome pairs in a meiotic division. Second is the cost of the recombination of genes, which is the immediate consequence of sex. This cost is the RECOMBINATION GENETIC LOAD. Perhaps biologists seldom think of genetic recombination as "bad" because we learn early in our training of all the exciting possibilities inherent for evolution in the process. But from the point of view of an individual seeking fitness by reproduction, a scrambling of genes is not so obviously a "good" thing. Scrambling genes means that the offspring will not be like the mother, but some basic ecological ideas such as character displacement and competitive exclusion suggest that in fact offspring should be very like their mothers. Indulgence in sex puts the comforting carbon copy of a mother out of reach.

A final extra cost of sex to the female is the COST OF MATING (Daly, 1978). Sex display and the other activities of mating require energy and involve danger. These costs fall most heavily on the male, but the female has her share, including energy costs of (1) sexual mechanisms, (2) mating behavior, and (3) escape from unwanted sexual attentions, together with risks from (4) predation, (5) disease transmission, and (6) injury inflicted by the males. The adverse effects of sex to a female, therefore, include the *cost of meiosis* (50%), *recombination genetic load*, and the *cost of mating*. The advantages of sex must outweigh all three combined for sexuality to exist.

It is an old idea in biology that the advantage of sex lies in producing new varieties to cope with "changed conditions." But for this hypothesis to be acceptable, the concept of coping with changed conditions has to be so phrased that the advantage can be seen to be immediate. The female who engages in sex must be expected to benefit herself so that her personal fitness is boosted more than enough to offset the 50% cost of meiosis. Notable among recent work showing that variable offspring can confer immediate advantage is that of Williams (1975), whose analysis follows.

Consider a very fecund organism like a codfish laying millions of eggs, or an elm tree scattering tens of thousands of seeds. Each fish or tree will saturate the habitat with so vast a number of her own young that the chances of survival for each are almost vanishingly small. Typically only one can replace the mother in time and the rest must perish. This is particularly clear if we forget the codfish and think only of the elm tree, because an elm can saturate the ground underneath itself with baby elms growing side by side. Only one, however, will make it to the canopy and ruthless competition will do for the rest. Yet this competition will not only be with siblings (brothers and sisters) but also with the dispersing young of other elm trees and other tree species. It is in this competition with the offspring of other trees that the advantage of variety can be seen to lie.

An asexual elm would saturate the ground beneath itself with a clone of identical copies. Williams suggests that this approach to the pending competition is like trying to win a lottery by photocopying your one lottery ticket over and over. For the uncertainties of the struggle to come it is much better to enter the game with many different tickets. With thousands of tries at its disposal, a highly fecund organism can afford to waste half its stake if this provides many chances of winning a single game.

If we enlarge the game to include all the habitats to which elm or codfish might disperse, the logic still holds. Occupying any habitat should involve competition between a host of applicants. Entering these many lists with a variety of different combinations, each subtly different, should be better than having all one's entries identical (Figure 12.1).

Sex, therefore, makes sense for the highly fecund. Many animals and plants are highly fecund. Thus the phenomenon of sex is satisfactorily explained for some organisms. But there are many species that use asexual reproduction, and many less-fecund animals like mammals and birds that retain sex. These also must be explained by a satisfactory theory.

Parthenogenesis

A common form of asexual reproduction in animals is PARTHENOGENESIS,[1] which is THE DEVELOPMENT OF AN EGG INTO AN EMBRYO WITHOUT ITS BEING FERTILIZED. This

[1]From the Greek *parthenos*, a virgin.

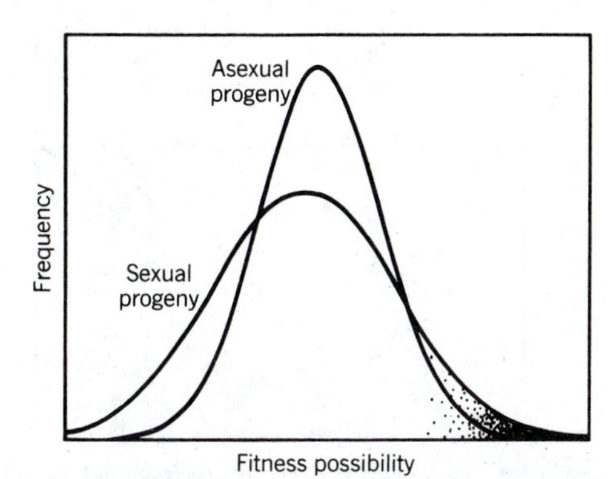

Figure 12.1 Model of advantage and disadvantage of sex. Sex yields progeny with a lower average possibility of fitness but also a low frequency of progeny with a very high possibility of fitness (black stippled area). Sex will be advantageous if there are enough progeny to give a high probability that some with a very high possibility of fitness are included, otherwise asexual reproduction gives the advantage. (From Williams, 1975.)

essentially is the avoidance of sexual recombination of genes by animals that are otherwise equipped for sex. Common examples are aphid clones and many planktonic animals of freshwater lakes, particularly rotifers and cladocerans. These animals exist for many successive generations entirely as clones of females. They lay diploid eggs that were formed by avoiding the meiotic division. The 50% cost of meiosis is not paid and the young are all carbon copies of the mother. There may, of course, be some mutation, but this is a very small source of variety.

But a characteristic of these parthenogenetic animals (with very few exceptions) is that they do engage in sexual recombinations at intervals after several asexual generations. These sexual episodes usually precede or are synchronous with environmental change. The selective advantage to individuals in these HETEROGONIC life cycles can best be grasped by noting that the animals that live through them occupy temporary habitats. The life cycle can be said to begin with a colonization event.

A colonist may be a winged aphid settling on a growing shoot or an individual water flea (*Daphnia pulex*) hatching from an egg in a puddle that has filled with rain water. The new habitat is not crowded, so food is abundant and competition is at a minimum. Good times are at hand, though they will not last indefinitely. Evidently it should be good strategy to reproduce as rapidly as possible and the simplest way to do this is asexual reproduction, which has every offspring a daughter, herself creating more daughters. Fecundity is doubled and a large array of descendants is available to contest the pond or shoot when the crowded days of competition do appear. If the habitat persists after it is crowded, and the shoot or pond were colonized originally by two or more females, then there will be competition between clones. The best clone will come to occupy the habitat all to itself and one founder female will have achieved locally that ultimate fitness of filling a universe with copies of herself.

But a plant shoot will die and a puddle will dry up. Aphids produce another winged generation and water fleas make ephippial eggs that can lie resting in dry mud until the rains return. From these a new generation of colonists must come, but conditions in the habitats of the future may not be like those of the past habitat, or the emigrants may have to compete with others who were there before them. The emigrant generation, therefore, is produced sexually to provide variety. Animals with heterogonic life cycles, many generations of which are asexual, should be considered to be primarily sexual though with sex repressed for rapid reproduction in good times.

A key part of the Williams argument is that PARTHENOGENETIC animals are only modestly fecund. They cannot overwhelm their shoot or mud puddle with propagules the way an elm tree saturates the surrounding ground with seedlings. Within their one habitat, therefore, they cannot make enough varieties by sex to pay the cost of meiosis, and they boost fecundity instead by saving resources that could have been spent on males.

Zooplanktonic animals like rotifers and cladocerans (but not copepods) are parthenogenetic in permanent lakes. For most of these an unfavorable season, typically winter, ends the run of parthenogenetic generations and is passed in dormancy following a sexual episode. The open water of next year's spring must be fought for with many varieties because next spring and summer are hardly likely to be like last year's. Interestingly, this may not be true in the arctic where one short summer is like another. Hibbert (1981) has suggested that clones of *Daphnia* in arctic lakes may be as old as the lakes they dominate, perhaps having persisted for several thousand years. These arctic *Daphnia* make their resting eggs parthenogenetically too. Hibbert's evidence is that each of several arctic lakes holds a distinct clone of *Daphnia*, the members of which can be recognized by electrophoresis.

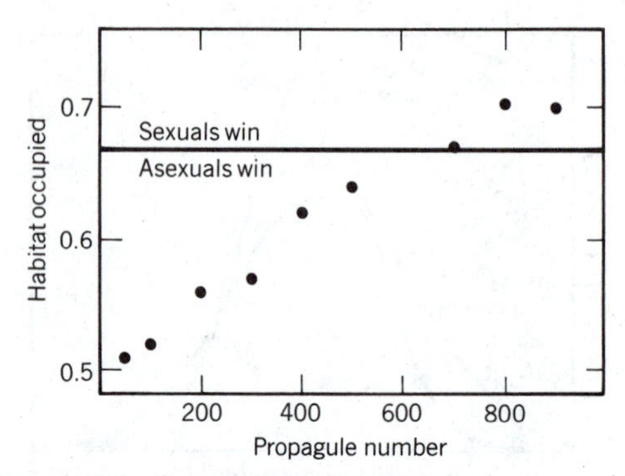

Figure 12.2 Computer simulation of elm–oyster model. As the number of propagules increases so the proportion of the habitat won by sexual progeny goes up. The horizontal represents the level above which sexual progeny have more than twice the success of asexual progeny and so compensate for the 50% cost of meiosis. (From Williams, 1975.)

Williams' Classification

Williams summarizes part of these arguments about the origin of sex with colorful names for the common variations he found, and these names are now widely quoted. They are:

The Aphid–Rotifer Model

Parthenogenetic clones compete in a confined space, but the habitat is ephemeral. New colonists are more likely to include the winning genotype of future clonal competitions if they have the variety given by sex.

The Strawberry–Coral Model

This is the version of the aphid–rotifer model applied to sessile plants and animals. These can spread locally by vegetative means, since the parent has already shown her superiority. But dispersion and colonization of other habitats comes under the same constraints as a dispersal episode by aphids or water fleas. Strawberries and corals respond with both sex and high fecundity, the strawberries with very numerous small seeds taken by birds in fruit and the coral animals with large numbers of planktonic larvae.

The Elm–Oyster Model

Highly fecund sessile animals and plants, for which competition for space is always intense, are required to use only sexual reproduction (Figure 12.2). This is a lottery so intense that a novelty is needed to win.

Unexplained by this analysis is the reason for sex in birds, mammals, and other low fecundity animals. How can an animal with a clutch of four, or two, or even one, pay the 50% cost of meiosis, the cost of genetic recombination, and the cost of mating? The logic predicts, therefore, that birds and mammals should be without sex. But the truth is that they are sexual animals without exception.

Two kinds of answer are offered to this paradox. The first is that birds and mammals have inherited sex systems from highly fecund ancestors and are now stuck with them. It certainly seems clear that all tetrapods evolved from fish stocks in which the high fecundity of small-egg strategies was usual, and it can be argued that the elaborate system of meiosis, gamete production, and fusion to form a zygote is not easily lost. Parthenogenesis, however, is possible in many animals, including such vertebrates as salamanders, making this argument of inherited sex less than satisfactory.

More encouraging is an argument based on the kinds of habitats in which low fecundity sexual vertebrates live (Glesener and Tilman, 1978). These are habitats with tolerably stable physical conditions but where the animals are exposed to strong biotic stresses, or where they live in BIOLOGICAL ACCOMMODATION (Sanders, 1968). When the young engage in intraspecific competition they do so against stresses, or a setting, largely defined by the other animals and plants of the place. The variety of biologic stresses or limits imposed in this way may be complex. Thus competitive success may turn on subtle variations that give an extra premium to the recombinant advantages of sex. Furthermore, the cost of sex to many of these females is contained because the males help with parental care or provide other services. In monogamous, territorial birds, for instance (see below), it may be that a breeding pair has an optimum clutch size twice that attainable by a single parent, thus completely offsetting the 50% cost of meiosis. The other costs of sex might easily be paid by added competitive advantage for the offspring in a world of biological accommodation.

Inbreeding

One obvious way of avoiding the cost of meiosis is for a female to fertilize herself. The system of meiosis is then intact but the offspring have only the mother's genes. Some plants do this to some extent, but those that are self-fertile usually have systems to keep "selfing" to a minimum. Not only are many flowering plants likely to fall under the logic of the elm–oyster model, and so require sex, but selfing invokes the well-known dangers of inbreeding: the likelihood of making offspring homozygous recessive for deleterious genes.

Dangers of inbreeding are taken very seriously by students of animal populations and much thought is given by ecologists to how these dangers are minimized in small populations. When young birds, for instance, return to nest near where they were raised it seems inherently likely that they would mate with close relatives. How, asks the field ecologist, is a sufficient rate of outbreeding maintained? And yet inbreeding is, in theory, a way of reducing that 50% cost of meiosis, since a female will be accepting back from her relative many genes she has already. Recently Shields (1982) has argued that so many birds are programmed to nest where they were raised that this trait must be selected for rather strongly. He suggests that biologists have been wrong to be so concerned about the possible consequences of inbreeding and that partial inbreeding is in fact desirable. In Shields' view the legendary swallows come back to Capistrano every year specifically because this results in mating with their cousins. For Shields' view to become accepted it will be necessary to show how these species can compensate for the damage caused by the appearance of deleterious recessive genes in homozygous array as a result of this inbreeding. But the argument is certainly interesting, giving another slant on the incest problem.

Another stratagem for reducing the evils of sex is that of polyploidy, a strategy followed by some plants. A POLYPLOID race or species has some multiple of the usual number of chromosomes, a proceeding that results in asexual reproduction. Polyploid plants are particularly common on arctic tundras. The speculation is perhaps allowable that fecundity is particularly low in the arctic because productivity is low, so that the 50% cost of meiosis is particularly burdensome. The possible costs of polyploidy might the more easily be met.

We must conclude that our theory of sex in low fecundity populations is still in an unsatisfactory condition. We cannot say for certain why birds and mammals have sex, though the answer may well lie in providing variety necessary for life in an environment where survival depends on coping with other organisms, both predators and competitors, rather than with physical adversity. But the fact that sex does exist has immense social consequences for animals. Females do need males. And because the interests of males and females must differ, many different social patterns result.

THE RESOURCE BASE OF MATING SYSTEMS

When eggs and sperm are of different sizes there are powerful consequences for the lives of the animals that make each. This is because the investment of a parent in each of its offspring is so different, ranging perhaps from the micrograms of organic matter in a typical sperm cell for a male to the kilograms in a mammalian baby for a female. Much in the sex and social lives of animals can be traced to this fact. Males and females use each other selfishly for the purpose of promoting their own individual fitness, but, because of the initial disproportionate investment in young, they do so from different points of view.

The initial inequality of investment brings a selective pressure for males to seek to be polygamous. A male can produce many more sperm than a female can produce eggs, so the obvious solution to his fitness problem is to spread his sperm over the egg production of many females. The need of the female to have all her eggs fertilized can be served by any virile male. From the narrow point of view of fertilizing eggs alone, therefore, the rest of a male's sex life is unimportant to a female. Thus males can win extra fitness by mating with many females, but a female needs only to be mated by one male. This is the reason why most known mating systems are POLYGYNOUS,[2] in which one male mates with several females. Both *monogamy* and *polyandry* (one female mates with several males) are much less common.

But fertilizing an egg to form a zygote is not all that is needed to gain a unit of fitness—the zygote must first grow into a reproducing adult itself. The mating behavior of both males and females is influenced by this fact. Females should choose to mate with a male only when his personal qualities, and local circumstance, are such that there is a high probability that their offspring will survive. Conversely a male should persevere with his promiscuous polygynous behavior only if this yields more surviving offspring, not just more inseminations.

[2]These names for mating systems are all from Greek words: *polus* = many, *gamos* = marrying (hence polygamy is "many-marrying"), *monos* = single, *gune* = woman, *andros* = male.

Polygamy and Monogamy in Vertebrates

The rival interests of males and females in mating systems have been explored in recent years particularly by examining size and structure of harems in polygynous mammals and birds (Armstrong, 1955; Snow, 1963; Verner, 1964; Crook, 1965; Lack, 1968; Orians, 1969; Downhower and Armitage, 1971; Williams, 1975). A male with a harem will both gain and lose by adding females. Inseminating one more female directly increases the number of his potential surviving offspring but this potential can only be realized if resources are available to raise the infant. The added female may mean more competition among females for resources, which may cost the male dearly as many of his offspring have reduced chances for survival. If the male contributes to care of the offspring, then there will be the extra penalty from spreading his services more thinly.

The interests of the male, therefore, suggest that the size of a harem should reflect the flux of resources available for raising young. With copious resources males might be little needed for parental care, nor might female competition be serious. Then the male can afford to be promiscuous and ambitious as he gathers a harem. Parallel arguments apply to females.

If resources are abundant and a harem is relatively small it may well be that a female suffers no loss of fitness from harem living. But it is not to the advantage of a female to join a harem where females already compete for resources or parental help from the male. A female offered the services of a male with overcommitted resources should reject him and find another male not so committed. Relative size of a harem, therefore, should be as much a result of female choice as it is of male choice.

Yellow-Bellied Marmots

These ideas have been tested against field data on marmots in Colorado by Armitage (1962) and Downhower and Armitage (1971) with illuminating results. Yellow-bellied marmots (*Marmota flaviventris*) make homes in burrows and their young are nursed in the burrows. The animals must forage within a short distance of their burrows and so are dependent on the resources of the immediate neighborhood for raising young. They are polygynous. Males establish themselves in an area, seek to exclude other males, and mate all females within the area. A male can maintain a harem for from one to three breeding seasons and he may have from one to several females in his harem.

Downhower and Armitage (1971) measured the number of young in the litters of every female in seven harems, finding that there was a strong negative correlation between the size of litter and the number of females in the harem (Figure 12.3). Evidently it is in the interest of a female to have a male to herself, that is, to be monogamous. Female choice, therefore, should be driving selection towards monogamy.

Downhower and Armitage also measured reproductive success as *yearlings raised*, which is a much closer approximation to fitness than litter size. Most yearlings per female resulted when a female had a male to herself and the number declined with litter size (Figure 12.4, straight line). But the number of yearlings raised per harem-master male shows a different result (Figure 12.4, curve). A male had his best success when he had a harem of three females. There was a suggestion in the data that trying to manage a harem of four or five would give poorer returns for a male than making do with one.

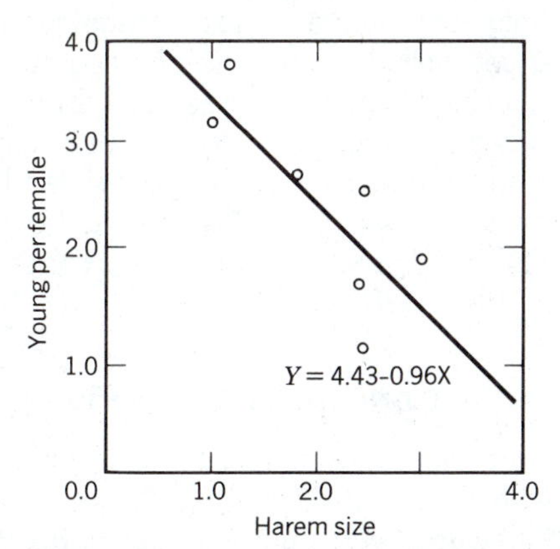

Figure 12.3 Effect of harem size on litter size in yellow-bellied marmots.
Females had largest litters when they had a male to themselves, and litter size fell progressively with increased size of harem. (From Downhower and Armitage, 1971.)

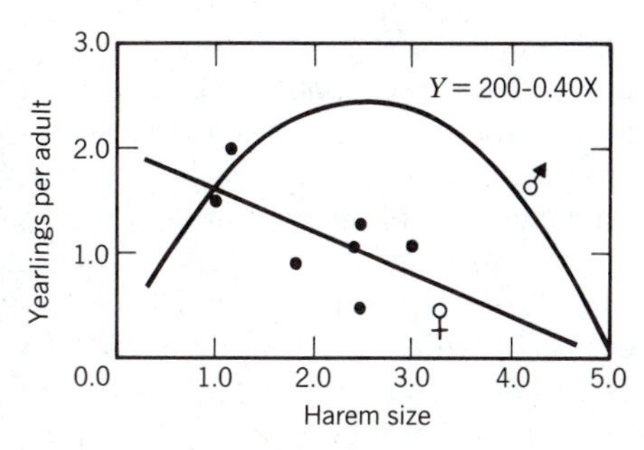

Figure 12.4 Effect of harem size on yearling production by yellow-bellied marmots.
Yearlings per female were highest for monogamy as was litter size (Figure 12.4). But yearlings per male given by the curve were highest for intermediate sizes of harem.

These data for marmots show that what is optimum for females is not optimum for males, and *vice versa*. Females should choose monogamy, and males should choose to live in a harem with three females. Both males and females have coercion at their disposal. Males establish themselves on favorable ground, so that a male with a nice piece of real estate freed from other males coerces with the promise of resources. The female coercion is her ability to choose both male and real estate with which to live. Since each sex can exert itself to try to achieve its personal optimal mating arrangement, it is not surprising that a trade-off should often result. Downhower and Armitage report that the most frequent arrangement was in fact bigamy.

This analysis of marmot behavior reveals why polygamy is so common. The best arrangement for females may very frequently be one male, or one patch of resources, to herself; but the best arrangement for males will usually be to mate a number of females. If resources are relatively easy to come by, it should nearly always be possible for a male to raise his fitness through polygamy, though unlikely that a female can. The actual polygynous arrangement that results will come about from a trade-off between selection for females best able to reduce harem size and males best able to increase it.

Monogamy

The unusual arrangement of MONOGAMY apparently represents total victory for females in what is usually a trade-off. Since total victory in any dealing is an improbable state, it seems likely that monogamy must actually result when this is in the male's best interest also. This is likely when resources are uncertain in time or space. Scarce resources could increase competition between females in a harem of two so much that the male should win more fitness if he made do with one female. Furthermore, scarce resources should put a particular premium on the contribution of the male to parental care. With the need both to reduce female competition in his own interest and to maximize his return on parental care, it may be that a male is best served with a harem size of one. Male and female choices are then identical and monogamy results.

The most abundant and conspicuous monogamous animals are birds with helpless (altricial) young. Their monogamy probably results from the demand on resources caused by their own breeding strategy. All of them require their young to grow extremely quickly, either because the flux of resources they use is short-lived, like the insects of a northern spring, or to keep as short as possible the time when nests should be at risk from predators. Not only must female competition be reduced to the minimum but males must often be required as foragers for the young. For a bird like a robin, therefore, males and females have an identical optimal harem size of one. Comparable patterns of available resources and needs of the young should be the cause of the few cases of monogamy in mammals also.

THE CONCEPT OF TERRITORY INTRODUCED

If males are to hold harems they must often do it by laying claim to space, as they certainly must discourage other males from coming near. If females are competing for food for their young, they might well do so by making other females keep their distance, which again means defense of space of some sort. Innumerable observations of natural history suggest that animals often do defend space; we say that they defend *territories*. A TERRITORY is often defined as THE DEFENDED PART OF A HOME RANGE.

An animal must have a place in which to live, and its powers of locomotion are limited. The animal can only range so far and we call the area of its wanderings its HOME RANGE. But *home range* must also be a function of resource flux, since we expect an animal to wander as far as is needed to secure resources for living. It is likely that intraspecific competition has something to do with setting the size of home range.

Many animals will offer very active aggression to others of their kind, particularly to males, who approach them near a heartland within the home range. Sometimes the way of life of an animal does not provide it with what we can easily recognize as a home range, such as, for instance, for a migrating animal almost constantly on the move, yet individuals may still be aggressive to their own kind if an intruder gets too close. What results is a perimeter of defense that an animal on the move may set out around itself and proceed to defend with vigor. We refer to the AREA WITHIN A DEFENDED PERIMETER as the animal's TERRITORY.

Howard's Concept of Territoriality

The word "territory" entered the usage of ecology with the work of an English ornithologist, Elliot Howard, who published in 1920 a book with the title *Territory in Bird Life*. Howard noted the habit most clearly in monogamous songbirds, the breeding pairs of which apparently drove all conspecifics away from the area around their nests, an undertaking involving both song and combat. Earlier workers had realized that the displays were related to breeding, but it was easy to misread the conspicuous singing and aggression of male birds with simplistic ideas about how the bird's sexual affairs were conducted. Howard began a careful analysis of the development of the behavior through a breeding season. If a bird was a winter resident, like the common English yellowhammer (a sparrow-sized finch, *Emberiza citrinella*), the first signs of the behavior were when males began to leave the mixed flocks in which they had overwintered.[3] Each male took to spending more and more time around one particular perch, from which he sang. This spread the males out across the countryside. The males became irascible, rushing at any other male that came near the favored spot. Often this resulted in one of the aggressive displays that others had interpreted as combat for a mate, but Howard noted that early in the season no females were present to watch the combat. The fighting was both ritualized and purely male to male. The consequence was that the original bird was left alone in the area near his perch.

Songs of yellowhammers and the rest did seem to attract mates, however, because each male was joined eventually by a female, and each pair nested close to the perch where the male had first experimented with song. Apparently singing did work as an effective beacon for the sexes, necessary for motile animals like birds particularly if they had arrived as migrants from far away. But the singing and irascibility to other males continued after the nests were made. Moreover, females shared in the defense activities, against both intruding females and intruding males. The complete package of behavior resulted in pairing of birds and their separation in space so that each pair nested and raised their altricial young in isolation from others of their kind.

Howard listed the possible advantages of this behavior and its spatial consequences to monogamous birds. Each pair:

[3]See page 293 for an account of the causes of flocking behavior.

1. Had a private feeding area near the nest. This would keep journey time down to a minimum, making the behavior compatible with the requirements of optimal foraging theory. In that the behavior keeps other birds out of the feeding ground, the behavior also is an effective mechanism for competition.

2. Spent most of its time on familiar ground, which should be a significant help when coping with predators. Not only would the birds learn where to hide and what were escape routes but their constant patrolling of their own neighborhood would let them keep watch for enemies approaching their nest.

3. Mixed little with others of its kind, which might prevent their contracting diseases or infecting their young.

4. Was kept together as a breeding unit for as long as it was necessary to raise the young. The habit provided a pair bond, almost serving as a marriage contract.

Selection should act to preserve the behavior on all four counts. But all four serve the breeding strategy; perimeter defense, therefore, is an adaptation for breeding. Modern ecologists tend to stress the importance of the advantage of a private food source, but Howard suggested that the most important was the forging of pair bonds. It should be impossible for territorial songbirds to raise their helpless young without some device to keep the sexes together, and here was one such device—shared hostility towards all comers. It is hard to imagine a simpler one.

Howard was encouraged in his view that forming a pair bond was the principal advantage by the fact that something very like this perimeter defense occurred in other birds that used no feeding ground near the nest. Cliff-nesting birds are the obvious example. The European razorbill (an auk, *Alca torda*), for instance, nests on ledges crowded with other razorbills (Figure 12.5). They forage out to sea where they cannot defend feeding areas, but each pair has a small area around its nest—more or less where it can reach by stretching its neck—that it defends against other razorbills just as tenaciously as yellowhammers

Figure 12.5 *Uria aalge*: territoriality on a nesting ledge. Behavior of nesting auks, like these murres on the Pribilof Island of St. George, in keeping each other at beak's length let the concept of territoriality be extrapolated from songbirds to cliff-nesting birds that lacked feeding territories.

defend their territories. Howard reasoned that the behavior was essentially the same. But the razorbills won no private foraging, nor protection from predators and disease as a result. What was left was pair-bonding.

Modern work, however, has shown that pair-bonding can be achieved quite independently of anything easily recognizable as "territoriality." Brewer's blackbirds (*Euphagus cyanocephalus*), for instance, are monogamous but nest in large colonies in the Pacific Northwest, where they show little behavior that a modern ecologist could call "territorial" (Orians, 1980). They spend their winters in Mexico, then migrate northwards in the spring to breed. But when they arrive on the breeding grounds the Brewer's blackbirds are already paired off into monogamous couples. Clearly they have managed to forge excellent pair bonds without the rituals of spacing that Howard saw in territorial birds.

Yet Howard had shown that the fetching ways of many birds in the spring, from the singing in a dawn chorus to ritual displays, could be understood as adaptations for breeding success from which both partners benefited. His analysis of the general advantages that could accrue, from increased food to social relationships, was sound. Sixty years later we know that there can be more complexities in territorial behavior (Klopfer, 1969; Davies, 1978; Orians, 1980) but the thrust of Howard's pioneering work remains. Animals space themselves out because this leads to increased fitness in a number of ways. We shall find that the manner of the spacing is a function of resources in time and space as with other strategies for living.

TERRITORIALITY IN POLYGYNOUS SPECIES

Although spacing behavior in monogamous songbirds first earned the name "territorial," polygynous species defend space too. In some, like the red-winged blackbird, the behavior clearly is similar to that in monogamous birds in that resources used for rearing young are partitioned between territorial, but polygynous, males. But in others, like the white rhinoceros and fur seals, the behavior seems almost entirely a way of rationing the chance to mate among males. These different functions of territoriality lead to quite similar patterns of male behavior and male spacing.

Territories in Which Young of Polygynous Species Are Reared: Red-Winged Blackbirds and Antelopes

Many detailed studies of territoriality in the polygynous red-winged blackbird (*Agelaius phoenicius*) are available (Orians, 1969, 1980; Horn, 1978). The birds nest in marshlands over much of North America, and may migrate or travel about in flocks in winter (Figure 12.6). In spring, males establish territories in marshes and along lake shores in the classic manner; singing, displaying, and engaging interlopers in ritual combat. Females arrive in the territory, build nests and mate; but one male may attract up to a dozen females to his territory. Some males are not so fortunate and may even be monogamous. Since it is established that the sex ratio is one to one in redwings, other males must be more unfortunate still and have no mates at all.

Figure 12.6 *Agelaius phoenicius*: territorial harem in rich habitat.
Red-winged blackbirds are both territorial and polygynous. This combination of traits is possible because they nest in marshes, like this nest in the cattails, and marshes are both of limited area and sufficiently productive of insects that one male can sequester sufficient resources for several families to be raised.

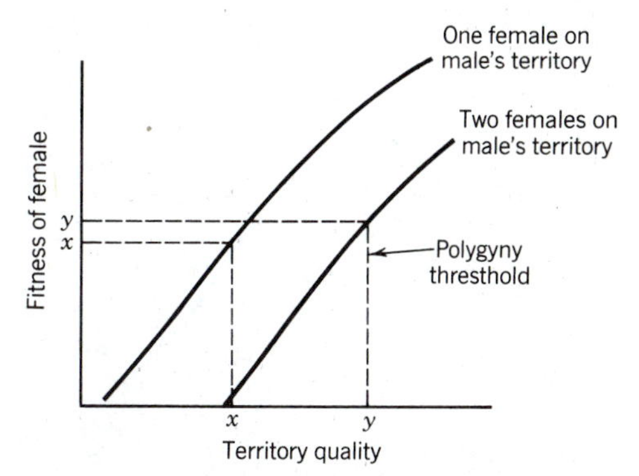

Figure 12.7 The polygyny threshold model.
The model assumes that fitness of a female is a function of the quality of a territory, of the parental contribution of the male, and of the number of other females already there. (From Orians, 1969.)

Orians (1969) suggested that this pattern resulted because marshlands were patchy environments such that some territories held much more food than others. A female ready to mate might find herself with the choice of going to a male with a poor territory and no mate or to a male with a good territory who already had several mates. Her behavior should be to go where she can breed with most hope of success. Her best choice might very well be to join a harem and share in the rich food resources available in that territory.

Orians put this explanation forward as a hypothesis, suggesting that there should be a POLYGYNY THRESHOLD over which the fitness of a female is greater if she accepts polygamy (Figure 12.7). To test this hypothesis Orians needed to show that females did achieve more fledglings when in harems and that territories were of "higher quality" where males had established large harems. Confirming data have come from many years of work (Orians, 1980). The first eggs laid are always on the territories that come to house the largest harems (Figure 12.8) and the risk of predation was markedly less for nests in the larger harems (Figure 12.9).

It was found to be very difficult to measure the quality of a territory because this would mean assessing food supply in a way that would be

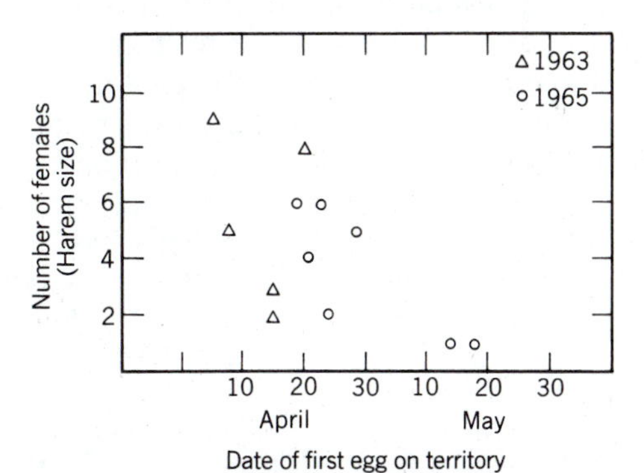

Figure 12.8 Correlation between harem size and date of first laying.
Data are for red-winged blackbirds near Seattle, Washington, in 1963 and 1965. (From Orians, 1980.)

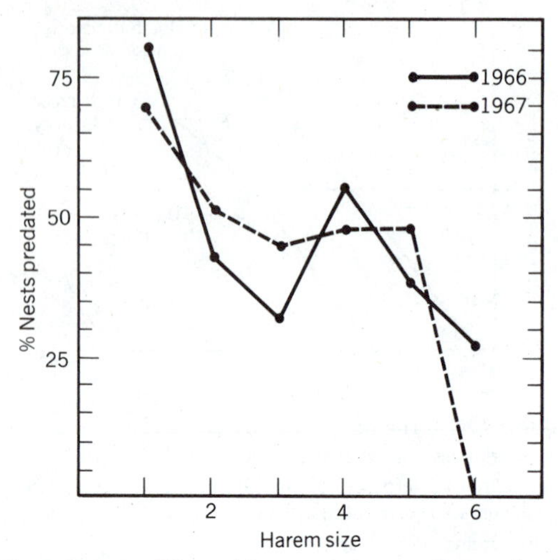

Figure 12.9 Effect of harem size on predation of red-winged blackbird nests. (Data of C. R. Holm, from Orians, 1980.)

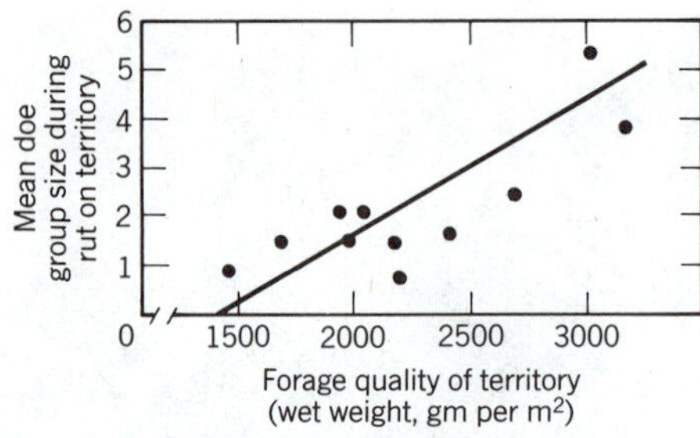

Figure 12.10 Harem size and quality of territory in the pronghorn.
Males holding territories with good forage attracted, and mated, most females. (From Kitchen, 1974.)

convincing to a red-wing. But Orians was able to show that a patch of ground holding a territory with a large harem would hold a similar territory and harem in other years. Males might change but the relative density of nests in recognizable acreage remained the same from year to year. This is strong evidence that the final arbiter of monogamy, polygamy, and harem size is in fact the quality of the habitat, principally food. In this patchy environment it is possible for one male to defend an area with food sufficient for many females. In less productive habitats, where most other kinds of songbirds live, a male cannot do this and monogamous territories result.

Evidence of a similar arrangement in antelopes is provided by a study of the pronghorn (*Antilocapra americana*), the antelope of the American West, which showed that males with good pasturage protected from other males were able to attract, and mate, most females (Figure 12.10).

Male Mating Territories: The White Rhinoceros

Males of the white rhinoceros (*Rhinoceros unicornis*) set about their arrangements for breeding in a manner reminiscent of males of monogamous songbirds. A sexually active male establishes a perimeter he will defend against other males. He fails to sing as charmingly as a songbird, but dribbles urine and establishes piles of dung around his perimeter instead. And he patrols. If another mature male enters his territory the two meet head to head and there is what Owen-Smith (1971) calls "a tense but silent confrontation" (Figure 12.11). The intruder backs away and leaves the territory.

All this follows the songbird pattern nicely, but the sequel is quite different. Young males, old males no longer sexually active, and females not in heat wander in and out of the rhinoceros territory at will. This freedom to wander is doubtless necessary because all available habitat is divided into comparatively small territories by the breeding males, whereas females with young need a much larger home range, 10 to 15 km^2, in which to wander.

Yet all mating takes place in a territory. Cows in near-oestrus interest males for two to three weeks. When a near-oestrus cow enters a territory the resident male endeavors to prevent her leaving, nudging her away from the periphery with his horn and becoming vocal to the point of making a special squealing sound. The cow can escape if she makes a determined run for it and will be pursued only 2 to 300 meters outside the territory. But usually she can be persuaded to stay until she is mated. Once mated, however,

Figure 12.11 *Rhinoceros unicornis*: a tense but silent confrontation.
Territoriality in the white rhinoceros serves males by providing mating stations. The behavior spaces out males in a way similar to the spacing of territorial songbirds but details of the selective advantage of the behavior are quite different.

she goes her own way, carrying on with her wanderings across the territories of many bulls.

This story of the white rhinoceros is particularly valuable because none of the advantages set out for territoriality in monogamous songbirds by Elliot Howard and others seem to apply. Cows and young get no food advantage from the behavior, risks from predation (if there are any for this animal) are not changed, and they have no pair bonds to worry about. Yet the pattern of establishing and maintaining a territory is remarkably similar. In the rhinoceros the behavior serves male fitness by resolving competition between males for mates with the minimum of risk to males. Benefit to females would follow because choosing to mate with a male on his territory should result in her having sons likely to have the genetic requirements for winning a territory and fitness in the next generation. Territoriality in both monogamous birds and polygynous rhinos aids the breeding effort, therefore, but in quite different ways.

Territoriality and Sexual Dimorphism: The Pribilof Fur Seal

Polygynous mating systems often result in both territorial spacing and differences between the sexes called SEXUAL DIMORPHISM. The Pribilof fur seals *Callorhinus ursinus* illustrate both (Petersen, 1968). They breed in harems on beaches with 20 to 80 cows per bull being possible (Figure 9.8). Male harem-masters weigh up to 500 kg though mature females weigh only 80 kg. Both this dimorphism and the mating system of the fur seals can readily be understood from knowledge of the resources used for reproduction: ample fish within swimming distance of shore and a beach covered with fog to shelter the cold-insulated animals from the sun. Suitable beaches are scarce, and breeding males fight for short lengths of these beaches, which they then hold as territories where they inseminate females, somewhat after the manner of white rhinoceros males on their much larger territories.

But the fur seal males face the difficulty that they cannot feed on their territories and they dare not leave them to go fishing. Accordingly fur seal bulls must be able to starve from the day they haul ashore to fight for territory until the end of the breeding season. The bulls, therefore, must be large, both to carry reserves of fat and because size is an advantage in struggles for territory in an animal that is ungainly on land. Being large also means being old, and male fur seals do not breed until they are six to nine years old, whereas females breed at age three. The growing years are accommodated by males living together on bachelor beaches during the breeding season, until they become sexually active.

The females have a clutch size of one, but infant survival is high and rapid population growth is possible (some properties of fur seal life tables are given in Chapter 9). Females increase their fitness by prolonging their reproductive life, which they do by reaching breeding age when much younger than the males and when much smaller. Evidently their smaller size also is set by minimizing metabolic demands of the mother to allow for carrying and nursing a large baby. Females give birth on the breeding beaches and are ready for mating shortly after. Implantation of the fertilized egg into the uterine wall is delayed so that the next birth will be exactly a year away.

A female swimming towards the breeding beaches with her baby to be born within a few days, and surveying the line of males along the beach, must find a good safe spot for the pup she is about to deliver. This same place must be good for her own health too, safe on land from social stress and with safe access to the sea as she comes and goes to nurse the pup. And she needs to ensure as best she can that her next egg will get good sperm, well supplied with genes that lead to survival and future fitness. She probably chooses both the properties of the beach and of the bull who guards it and cows can be seen swimming up and down offshore for some time before they land, as if surveying the beaches. We do not know how much the final choice is one of bull or beach, but the point is moot because the bulls themselves will have tried to choose the best beach. If she chooses a bull with a large harem she may be choosing sons that themselves will be successful harem-masters.

Conclusion: The Resource Base Sets the Pattern of Territoriality in Polygyny

A common denominator in the territoriality of polygynous blackbirds, rhinoceroses, and fur seals is that males identify essential resources and manage to sequester them from other males. For the blackbirds the resources are food for raising young and for the two large mammals resource is merely space in which to court and copulate. Patterns of both male and female behavior that result, and even size and life history phenomena of individuals of both sexes, then develop from these territorial habits of the males. The actual result is a function of the quality and abundance of the original resource or space for which the males of the different animals must compete.

LEKS AND SEXUAL DIMORPHISM

In birds with precocial young food can be so abundant that males may be dispensed with for the raising of those young. Like the bull fur seals discussed in the last section, the males seem to be of little use to females except as purveyors of semen. A result found in ten different families of birds is communal mating displays of polygynous males on central breeding grounds called a LEK or an ARENA (Wilson, 1975). This result may also follow if resources are hard to defend (Emlen and Orring, 1977). The consequence for males is a life devoted to competing with other males for the right to inseminate.

Perhaps the most refined lek of all is that of the sage grouse (*Centrocercus urophasianus*) of the American prairies (Wiley, 1973). In their short

Figure 12.12 *Centrocercus urophasianus*: sexual dimorphism in the arena.
All mating in the sage grouse is by a few males occupying tiny territories on a mating ground called an *arena* or a *lek*.

mating season a hectare of prairie is dotted with perhaps 400 male grouse in tiny territories and between them they will mate 400 females. But 10% of the males will in fact do 75% of the mating and a very few males indeed might do much of the work even of the favored 10%. The reason for success seems to be ownership of one of the few favored territories, usually adjacent to each other and only a few meters square (Figure 12.12). Females come only to these favored places to mate. Male success seems to depend entirely on being able to hold one of the favored territories, and they work their way, perhaps over several seasons, in towards eventual ownership of one of the choice positions, perhaps sneaking in a few matings on the way up to the top.

Success at breeding for lek males depends on their properties of display and their skills at territorial encounter. They become large, gaudy, vocal, and perhaps bizarrely beautiful. The most remarkable examples of all are males of the birds of paradise (Figure 12.13). It is not known how perilous a handicap this appearance is, but it must surely make the males a mark for predators. Females with serious family duties to face are inconspicuous (Figure 12.12). Notice, however, that gaudiness of the male is needed to impress other

Figure 12.13 *Paradisaea rubra*: danger in sexual paradise.
A male bird of paradise in flapping wing display at another male. In some polygynous mating systems male display must be so important in male to male competition that costs of increased predation can be more than paid for.

males, rather than females. It is a moot point how lek females choose with whom to mate. Possibly they actually choose males rather than real estate and the "best" males congregate at the center of the lek. Yet it is difficult to see how useful male phenotypes could be recognized behind all that show. On the other hand, it is also difficult to see how a small piece of ground occupied by a male has significance to the female. Possibly the answer is that females simply choose the middle of the lek out of social efficiency. Some selection pressure certainly will exist for females to choose males if genes for showiness increase the fitness of their own sons.

The word "lek" was originally applied only to birds but ecologists now use it to describe similar behavior in other animals. Many ungulate mammals perform in the same way, most notably the Uganda kob (*Kobus kob*, a kind of antelope; Buechner and Roth, 1974). The male kobs arrange themselves rather as the sage grouse males do, crammed into a network of tiny territories on the breeding grounds. Youthful males are not tolerated there at all, but wander at the periphery in bachelor herds. This is comparable to the habits of bachelor fur seals, also hangers-on at the periphery of the action. Males of both species are waiting for strength and learning that will let them enter the breeding zone. Females needing sperm thread their way through the maze of male territories until they let themselves be stopped and mated by a male more determined than the rest. In this respect the system echoes that of the white rhinoceros, but the territories through which the females wander are tiny. The kob system is particularly illuminating because the arenas are not on good forage but on more or less barren ground of little use to the animals otherwise. Male territoriality appears to serve male fitness only, though females use it as a sperm bank.

Some fruit-eating bats, harvesting ants, and Hawaiian fruit flies mate in leks, and there are probably many other examples from all phyla (Wilson, 1975). In all, the males must be showy and large, not primarily to attract females but to state their status for the benefit of other males.

Sexual Dimorphism and Social Dominance

Size matters in establishing social dominance; but either sex can be the larger. In lek species, males are large or gaudy, as in other familiar vertebrates, but superior size for females probably is equally common. In most fish females are the larger sex; the same probably is true for most spiders. In hawks, owls, and many game birds females are conspicuously larger than males, undoubtedly with the appropriate social consequences probably including female dominance. Understanding of these patterns of SEXUAL DIMORPHISM is in its early stages.

Studies of dimorphism have been influenced by male social supremacy in the human cultures that produced modern science. Showy, big males have been noted and treated as the norm, whereas larger, dominant females have received less attention. Most telling of these attitudes is the way that some writers have referred to largeness in females as *reversed sexual dimorphism*, the implication being of course that males are expected naturally to be big and dominant. Actually it may well be that large females are commoner than large males. What is certain, however, is that patterns of sexual dimorphism arise as selection works to fashion the size of each sex to the needs of individual self-interest.

Since the primary reproductive individual is the female, it is prudent first to look at the possible advantages of large or small size to females before taking the male point of view.

Benefits of large size to females

- maximize fecundity
- early births in seasonal environments
- maximize survival through short periods of food shortage
- maximize fat reserves for lactation
- maximize success in female to female conflict
- maximize success in female to male conflict

Benefits of small size to females

- lowered metabolic requirements giving benefits for

a. survival through prolonged periods of food shortage
b. minimizing maternal metabolism to maximize resources for infants in species that care for their young

These lists could be extended but they probably include the commonest factors leading to selection for either largeness or smallness in females.

Maximizing fecundity probably is the reason for the large females in most fish and spiders, since the number of eggs laid is a function of the mass of the female, a point that Darwin noted. Large size also allows early egg laying in seasonal environments because the large maternal body allows energy to be stored. This may be the reason for the large body size of female hawks that migrate long distances: the large female has fat reserves both for the long flight and to make eggs when she arrives on the breeding grounds. Male hawks, having no eggs to lay, can be smaller (Downhower, 1971). That large size also allows increased supplies of milk in mammals, and increased success in conflict, needs no elaboration.

Possible advantages of smallness for females seem fewer and probably all reflect the fact that a smaller body requires less food to maintain it. Small size thus should permit survival through long periods when food is in short supply. Periods of extreme food shortage, however, might promote large size for then the time of shortage can be survived on fat reserves. The food supply will thus result in selection for smallness or largeness depending on both the absolute amount available and the intermittency of its availability. For females that care for young, particularly mammals with long lactation periods, the food supply that each female can win must support both mother and infant, producing a selection pressure to minimize the mass of the mother.

Males have fewer reasons to be large since the biomass and maintenance costs of sperm are minimal. It might therefore be expected that small males should be the norm in nature, which is perhaps the truth of the matter. Males should be large only when large size permits survival in lean times or when conflict is particularly important. These pressures will lead to large males, however, only in circumstances where no offsetting strong selection for smallness is present.

In lek species, and in polygynous species where males can sequester resources, males become large because of the importance of male to male conflict. This says nothing about the expected size of females in those species. Female size will depend on selection for female self-interest. Work on the cause and consequence of size dimorphism in polygynous primates is helpful. Reproductive groups of primates, consisting of one large male and several smaller females with their young, may hold together for years, until, in fact, the male dies or is displaced by another (Struhsaker, 1977). The replacement male begins his tenure of the harem by killing the young babies sired by his predecessor (Figure 12.14). This serves his fitness well because the babies have none of his genes, whereas he can sire their replacements. It does not serve the fitness of their mothers, however, but the mothers cannot adequately defend their young because they are too small to stand up to the male. Obviously this reveals a selection pressure for the females to be large enough to fight off the male. The fact that the females are not large, therefore, demonstrates that the opposing selection pressures for smallness are the stronger. Hrdy (1981) suggests that in primates the important selection for smallness is the need to keep the mother's maintenance costs down in the interests of prolonged lactation. She provides particularly illuminating data for Hanuman langurs (*Presbytis entellus*) in India. When the new male begins his infanticide he is resisted not by the mothers but by the grandmothers. This suggests that grandmothers, being near the end of their reproductive lives, can gain no more fitness by mating with the new male and so are best served by defending existing babies carrying their genes.

Selection may also work to keep the sexes at different sizes to minimize intraspecific competition. This mechanism may be among many operating for the Galapagos marine iguana. Figure 12.15 shows basking marine iguanas (*Amblyrhynchus cristatus*) at Punta Espinosa on Isla

Figure 12.14 *Presbytis entellus*: male dominance and infanticide.
Polygyny in Hanuman langurs results in group living with one large, sexually active male to each group. When eventually he is replaced by a younger male, the newcomer kills the babies of the troop, the selective advantage of which (to the male) is that group resources go to raising his own offspring. This behavior is possible because the males are larger than females, who cannot successfully defend their babies.

Figure 12.15 *Amblyrhynchus cristatus*: male segregation.
The crowded Galapagos marine iguanas on this rock at Point Espinosa are all males. The rock fronts on deep water, suggesting that the heavy males can reach resources on the sea floor that are denied to lighter females. Assemblies of females are found close to shallower water nearby.

Fernandina (Narborough). All the animals in the photographs are males. P.D. Boersma counted the animals in this herd, finding that there were usually more than two thousand present, though they came and went. Females sun themselves in separate herds. Figure 12.16 is Boersma's sketch map of all the iguanas on the point. Also on the map is the depth of water close inshore. The large herd of male iguanas lies close to deep water and the females close to shallow. The iguanas feed on benthic algae, either at the bottom of the sea or when the algae are exposed at low tide. The male iguanas are much bigger and heavier than the females, so it may be, then, that the big, heavy males can forage in deep water more successfully than can the smaller females, perhaps because the larger males will cool more slowly in the cold water.

It is appealing thus to argue that the iguanas are equipped to feed in different places as an adaptation to increase the resources available to them, but we cannot be certain about this. For instance, why shouldn't males cheat and use their weight to elbow females aside from the shallows? One possible answer is because there is in fact better pickings in the deep water where the females cannot go. We do not have data to test this hypothesis.

All we can be sure of about the Espinosa iguanas is that the males are indeed larger, that the sexes are separated, and that the males do forage in places that might be hard for females to reach. It seems prudent to suggest that this sexual dimorphism evolved from selection for reproductive prowess within each sex, the males becoming large to overawe other males and the females remaining small to optimize the energy available for their huge eggs. Occupying different parts of Punta Espinosa may be as accidental a consequence of different size as is infanticide by male primates.

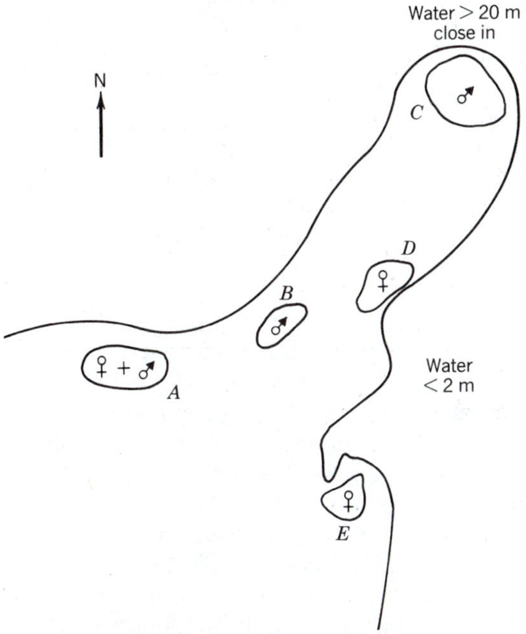

Figure 12.16 Subpopulations of male and female Galapagos marine iguanas.
The large assembly of males (Figure 12.15) is at the edge of deep water whereas many of the smaller females are close to shallow water. (Original data of P. D. Boersma from Pta. Espinosa, Fernandina.)

TERRITORY AS NON-RANDOM SPACING

Animals sometimes can be found to be spaced out when no obvious defense behavior is evident. Perhaps the spacing is achieved by mutual avoidance between these animals rather than by active defense. Animals so spaced would seem to live on territories though their behavior would not fit the definitions of territoriality given earlier. Davies (1978) has suggested that TERRITORIALITY should be redefined as ANY SPACING OF ANIMALS THAT IS MORE EVEN THAN RANDOM. An example of what is possible is given by the territoriality of mountain lions outside the breeding season.

A population of the mountain lion (*Felis concolor*; Figure 1.1) in Idaho has been studied in winter (Hornocker, 1969). Lions were tracked in the snow and individuals were identified by running them down with tracking dogs and subduing them with anesthetic darts, after which they were tagged. Hornocker handled a total of 43 different individuals, catching some repeatedly and making a particular study of 9 individuals

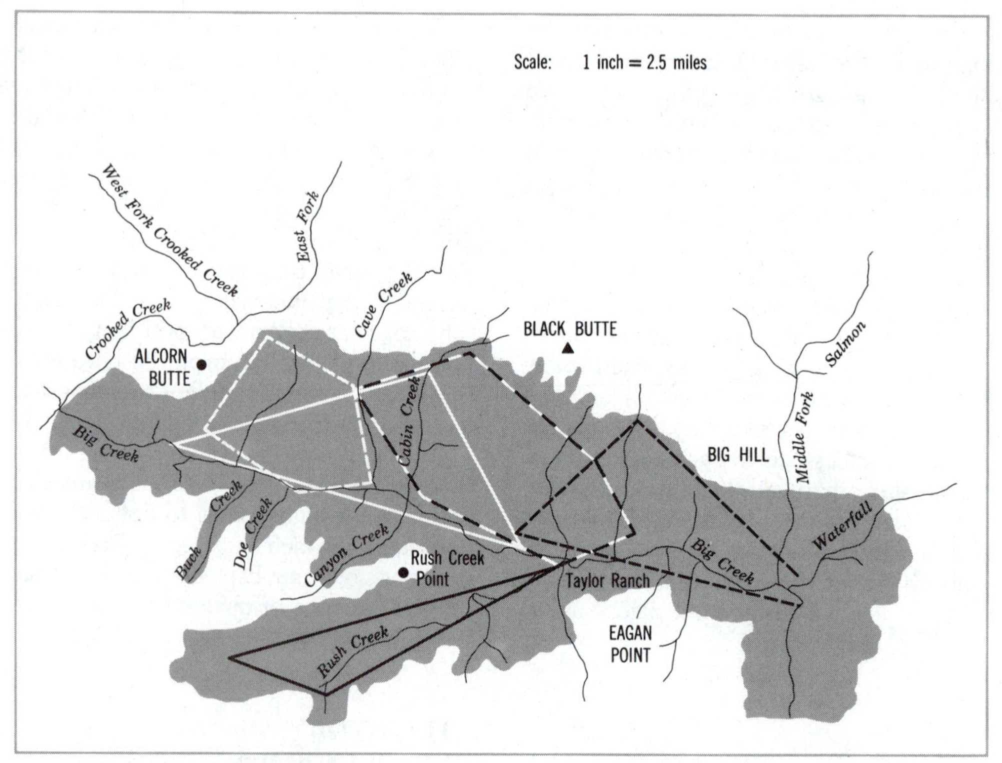

Figure 12.17 Minimum winter home ranges of mountain lions in Idaho. These home ranges were mapped by following the lions' tracks in the snow. The ranges were fairly discrete, and it seemed that the animals were seldom in the overlapping portions at the same time. One home range was used by two male lions, and another by a male and three females, but the other ranges had solitary occupants. The spacing of the lions in winter can be explained as a mechanism for ensuring that each is able to hunt an area where the deer have not been disturbed by other lions. (After Hornocker, 1969.)

caught a total of 59 times. Figure 12.17 is a map of the areas roamed by some of the best known individuals. These animals apparently lived separate lives in winter, the range of each being as separate as should result if the home ranges were in fact defended territories. And yet there was no evidence that the lions did defend their territories. Younger lions in particular wandered widely across the territories of older, more established animals in their hunting. Once Hornocker actually caught a young animal at its kill in the territory of an old established male, and found tracks of the owner first approaching, then turning away, from where the younger and weaker beast was feeding.

The lions marked their passage with scrapes and excrement, as do domestic cats. The trails revealed that all the lions, young and old, males and females with cubs, turned away from the presence of others. The territorial spacing of individuals, apparently so similar to those of songbirds in the spring, was achieved by mutual avoidance. Established animals had home ranges that were as distinct as the territories of songbirds, but they did not defend them.

Hornocker postulated that the survival value of this system of spacing by avoidance was that it let each lion hunt deer that had not recently been molested by other lions. The lion's technique of hunting by stealth requires that the target deer be unsuspecting so that lion attack has more chance of success when launched against deer that have not been disturbed recently. The winter territorial behavior in mountain lions results because the lions avoid each other as a necessary requirement of the hunt.

THE LOSER'S VIEWPOINT AND SOCIAL HIERARCHIES

Most territorial encounters are ritual combats with minimal actual damage inflicted, and yet one individual breaks off the engagement. One protagonist wins and the other loses. This means that the loser breaks off the combat without gaining his objective and without being hurt, raising the important point of the possible advantage of territorial encounters to the loser.

When an animal finds itself up against a determined defense it has the options of fight or quit. If it presses into battle the chance of victory complete enough to allow breeding must be small. This is most obvious if we think of an intruding male of a monogamous species of songbird. For the defender there may already be eggs in a nest and much investment to defend. His stakes are high and he cannot afford to quit. But the intruder stands almost no chance of an eventual gain in fitness by pressing home the attack, because continual battle, whether ritualized or serious, does not provide a satisfactory setting for reproduction. The intruder who quits, however, has a definite chance of finding a territory and a mate somewhere else; or if not, then next year.

For an intruder it is nearly always "fit to quit." Perhaps the only exceptions are for polygynous, resource-sequestering males with large harems at stake. Numerous surplus males may be driven to gamble and to risk destruction. In these species fighting may be less ritualized and involve some chance of wounding. As with fur seals, however, an aspiring young male will seek out a harem-master well past his prime if possible. The old defender himself usually quits when it is clear he is losing.

It is vital to realize that success in a territorial encounter is not genetically preordained. All individuals are equipped with the full complement of the behavior. This behavior requires that a defender of his established territory be confident, and that a challenger always retreat from a show of confidence. All learn by experience; all have the gene for flight to fight another day. This normal behavior must be maintained by natural selection since both the overaggressive and overpusillanimous will lose fitness by not knowing the proper time to quit.

Losing a territorial encounter may be likened to being low on a dominance hierarchy. Gregarious animals often spar together or fight, just as do territorial birds, but may go on living together afterward all the same. After the first combat, however, they adopt perpetual positions of victor and vanquished, with one of the pair becoming a low-status animal that always gives way to the other. A series of pair encounters may establish a complete hierarchy for the group, with one individual being socially dominant overall, and the status of the others being arranged in descending order. Such social structures in animals were first clearly recognized in domestic hens by the German biologist Schjelderup-Ebbe (1935) in the 1920s, and it is from his observations on the pecking of hens by each other that the term "peck order" has come into common use.

There seem to be real advantages to individuals in living in socially stratified societies, since many experimental and field studies show that the individuals of settled societies feed better and are more successful at rearing young than those in groups afflicted by the constant strife that accompanies the reception of a stream of outsiders (Allee *et al.*, 1949). Probably this is because the animals who live in a socially-settled state are spared the waste of time and energy that social strife would cause.

The acceptance of a lower position in a hierarchy can grant more fitness than refusal, because the strife that comes from refusal must lower fitness more than would meek acceptance of low status. Accepting low status now might very well make possible higher status later with its chance of successful reproduction. As with the loser in a territorial encounter, the socially submissive individual must not be genetically programmed to be subservient. The issue must be determined by circumstance and experience (Christian, 1970). An illustration of what is possible comes from intriguing studies on record in which a hierarchy seems to be circular so that *A* is dominant to *B* is dominant to *C* is dominant to *D* is dominant to *A* (Allee *et al.*, 1949).

THE POPULATION CONSEQUENCES OF TERRITORIAL BEHAVIOR

If survival and success at breeding are enhanced by territorial behavior, as its widespread incidence suggests they must be, the effect on population should be to increase it. Territorial behavior becomes one more tactic in the race for maximum population. But there is a possible corollary of the behavior, which Howard (1920) first pointed out in his book and which became a source of heated argument among biologists. *There cannot be more pairs of breeding animals than there are territories in which they can breed.* At once this introduces the idea of population control. Territory is defined as space, and the total space available to animals is limited. If, as Klopfer (1969) puts it, "the size of the territory cannot be reduced beyond a certain point, and if successful reproduction requires that the bird possess a territory, the regulatory function of territories becomes a function that is beyond dispute."

Populations might be expected to increase in size until they reach a limit set by the number of possible breeding territories, after which the population is kept constant from year to year. Surplus animals become territorial outcasts. Denied a home, they perish. Only the normal population, set by the available territories, persists from year to year, unchanged except by chance catastrophe. This opens the possibility of a general mechanism setting limits to the numbers of many of the larger animals, a mechanism more certain than predation, more precise than weather, triggered by density yet not proportional to density, a mechanism that might uniquely account for the remarkable constancy in the numbers of many birds and other vertebrates (Chapter 8).

At one time this hypothesis received formidable opposition from some of the most articulate of zoologists (Lack, 1966). They argued that increasing numbers should always compress territories and they suggested that for an animal to be expelled from a territory so that it could not breed should not be permitted by natural selection. But this objection has been met by the "fit to quit" arguments put forward in the last section. Restriction of the breeding population by space available for territories is perfectly compatible with individual pursuit of fitness.

The hypothesis that shortage of territorial space limits numbers, therefore, is plausible. This hypothesis makes the prediction that stable populations of territorial animals are accompanied by surplus populations of non-breeding individuals who have been denied territories. The test is to remove territorial animals and watch to see if their territories remain vacant (the null hypothesis) or if they are quickly occupied by members of a surplus population. Several such tests have now been made and they show conclusively that the predicted surplus populations do exist and that, therefore, territoriality does limit population.

An early and celebrated removal experiment on territorial songbirds in the boreal forest of Maine was undertaken with a different objective in mind, that of measuring the effect of the birds as predators of spruce budworms (Stewart and Aldrich, 1951; Hensley and Cope, 1951). The experimenters first made a census of all the breeding pairs of birds on a 40-acre section of forest, then removed all the birds they had identified by shooting. In the first year of the experiment they identified 148 pairs of various species as present in their forest, but they shot 302 adult male birds in the following two weeks, at which time territorial birds still sang all over the wood. When the work was repeated the following year the census of breeding birds gave 154 pairs, a figure remarkably close to the 148 pairs of the preceding spring. This second year 352 males were shot in the control effort. This study demonstrated conclusively that a floating population of male birds existed that was fully capable of occupying territories but was excluded from doing so by other birds.

One criticism of the Maine study is that the shooters bagged hardly any females (Brown, 1969). Of the 10 commonest species present not a single female was replaced, or, at any rate, not one female more than could be accounted for by the original census was collected. The best explanation for this seems to be that, by the time the territories were made vacant by death, surplus females were no longer in breeding condition although surplus males were.

Removal experiments have not been tried on mammals, fish, dragonflies, butterflies, and limpets as well as on birds (Davies, 1978). Figure 12.18 gives the result of one such experiment for the great tit (*Parus major*) in England. The best great tit habitat was in a patch of woodland that was completely parcelled out into territories. When six pairs were shot on their territories, space was reparcelled with four more pairs moving in. Krebs was able to show that these four pairs were previously living in nearby hedgerows. They were not true floaters but birds in suboptimal territories.

It now seems certain that a consequence of territorial behavior is to set a limit to population. "Territoriality" is the process that results in the remarkable constancy of breeding birds from year to year that is revealed by spring census (Chapter 8). We can expect population consequences whenever animals space themselves out. Territoriality, however, is NOT a population control device. The population consequences are accidental side effects of behavior fashioned by natural selection to serve the needs of individuals.

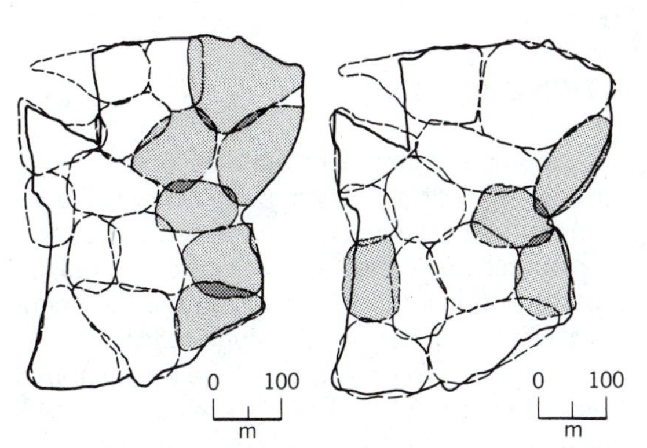

Figure 12.18 Removal experiment to test the prediction that surplus birds are denied territories.
The male great tits (*Parus major*) holding the six territories shown shaded in the left-hand diagram were shot. Four new males moved in (shaded, right) and the remaining residents expanded their territories. (From Krebs, 1971.)

THE GENERAL ADAPTATION SYNDROME AND THE HYPOTHESIS OF SOCIAL STRESS

The GENERAL ADAPTATION SYNDROME (GAS) is recognized in medicine as the response of people to stress (Selye, 1950). A number of related hormonal changes occur, including a high titer of adrenocorticotrophic hormone (ACTH) in the blood and enlarged adrenal medullas. The syndrome can be induced by a variety of stresses, from physical exhaustion to hurt, disease, worry, or fear. GAS apparently records a general mobilization of the body's resources to overcome stress, and thus is adaptive. But since GAS requires changes in hormonal balance, all functions of the individual are influenced, undoubtedly including the reproductive function. Stress, therefore, may affect breeding. Moreover it can be shown that extreme stress produces an extreme GAS reaction that, far from being adaptive, is called *stress disease.*

Stress disease can be induced in laboratory rats and mice by crowding them, as in a typical laboratory experiment on population growth. If pairs of white rats are placed in ample cages, with ample food, water, and bedding, they reproduce so that the population in the cage tends to grow exponentially for a time. But the population does not complete a sigmoid growth history (Chapter 6). Instead a whole generation of rats fails to reproduce, either having no sexual activity, or resorbing embryos, or through showing abnormal behavior like eating their young or being hyperaggressive against mates and offspring. These crowded rats exhibit physical signs of the general adaptation syndrome in extreme form, the most obvious being grossly swollen adrenal glands (Christian, 1950). The whole population has suffered stress disease with consequent total failure to breed.

This discovery suggests the hypothesis that crowded animals in the wild might sometimes suffer shock disease from the effects of crowding itself, behave abnormally, and so suffer population losses. Field biologists frequently observe population "crashes" of animals like rodents or deer when populations suffer sudden heavy losses after a period of sustained growth. Often it is thought that the food supply fails, but the *shock disease hypothesis* offers an alternative explanation. Stress of crowding and starvation combined overloads the general adaptation response and the animals die of shock.

It must first be noted that the GAS usually must be truly adaptive, being provoked to ensure life and breeding, not to end it. One telling set of data on this point comes from comparing adrenal glands of placid laboratory rats with those of wild rats trapped in sewers, when it is found that the sewer rats have the larger adrenals. Apparently stresses of sewer life provoke the truly adaptive response of the GAS, letting rats endure the vicissitudes of sewer life. But this does not exclude the possibility that gross overcrowding in the wild can produce the diseased form of the GAS with consequent population collapse (Christian, 1950).

Data from both wild hares and deer have been cited as possibly being due to hormonally induced shock. The snowshoe hare of arctic Canada has a population that cycles, reaching very high densities about every 10 years (Chapter 13). Hares of the dense populations caught in traps have been known to die suddenly and in convulsions. Autopsy of these animals reveals physiological abnormality, including low blood sugar and fatty degeneration of the liver (Green, 1930). This syndrome is consistent with the shock disease hypothesis. Crowding caused stress and overresponse by the general adaptation syndrome. The extra stress of capture could not be accommodated, adrenal exhaustion ensued, and death followed.

A direct test of the hypothesis was attempted on a herd of sika deer (*Cervus nippon*) by Christian *et al.* (1960). The deer had been introduced onto a small island in Chesapeake Bay, Maryland, and had grown to a density twenty times what is thought to bring on starvation in deer populations in comparable habitats, suggesting a population crash might be imminent. Animals were sampled and autopsied before, during, and after the crash that followed. Adrenal glands of crowded animals before the crash were twice the weight of those in the depopulate aftermath, yet there were no signs of disease detected. The most provocative observation, however, was that none of the animals that died gave evidence of starvation, all apparently having adequate fat reserves. The actual cause of death of these deer remains unknown, but outright starvation is excluded. Shock disease remains a possible cause of the deaths but is by no means demonstrated.

These field data from hares and deer certainly are suggestive. It may be that death in wild populations can come from stress, but it is exceedingly likely that the hormone syndrome (GAS) that actually kills is always acting as a proxy for the real cause of death, the ultimate stress as it were. No conclusive data show that wild animals die from the effects of GAS that were not doomed to die anyway. If it eventually can be shown that individuals of some population somewhere do die from stress when they might not have died otherwise, this effect should be seen as comparable to the population effects of territoriality, as an accidental consequence of a process that grants fitness to all the animals that possess it.

LIVING IN GROUPS

Animals do not just space out (territoriality), they also live in groups. Some group living is the direct result of sexual needs, mediated by patterns of resource as we have seen. Harem-masters impose group living on fur seals and social ungulates as their pattern of polygyny reflects the resources used. The polygyny of the lek is another form of group living. These social groups result from the reproductive needs of individuals, both for sex partners and for resources to raise the young. Once these pressing needs result in group living then social mechanisms evolve as each individual acts to improve its own fitness within the group (Alexander, 1975). Studies of how individual fitness leads to group or social phenomena are now often called SOCIOBIOLOGY (Wilson, 1975).

But there are other forms of group living that are less directly related to reproductive pairing. Perhaps the most familiar is the flocking of birds, particularly in winter. Birds live in flocks, not just one species together but sometimes as mixtures of species; we see them plunging along a winter hedgerow ahead of us as we talk, or crossing the sky to a winter roost. Then again there are huge one-species aggregations, like menhaden schools of a million fish off the North Carolina coast or the thousands of starlings that fly in formation at dusk across ancient squares in European cities.

A requirement of the theory of natural selection is that each participant in these aggregations, large or small, gets a personal advantage from the arrangement. These advantages can be many.

Individual Preference for Foraging in Flocks

Consider first the possible advantages of flocking to winter flocks of birds in northern woods and fields that can be seen feeding together (Cody, 1974; Bertram, 1978).

1. By feeding where others feed, patchily distributed food is found efficiently (information exchange).

2. Travelling with a flock means that the individual will not waste time hunting over the ground that other birds have already covered.

3. The feeding of other birds lets them act as "beaters" that expose prey. For instance, insect feeders might do well catching insects disturbed by birds ahead of them.

4. Being in a flock, both single species and mixed, can serve as a defense against predators.

If an individual is not constrained to hunt near one space, as around a nest, each of the first three in the list, those concerned with efficient foraging, should be powerful reasons for joining a group. Some ways of feeding, of course, preclude group foraging, as apparently does the hunting of mountain lions (see above). Otherwise joining a group may save considerable time and effort in the food quest, provided that competition within the group for food is not so intense that the losses from going alone are not balanced by the occasional extra benefit of a good food find. The decision to forage with the flock or alone, therefore, will be a function of food size, food abundance, food dispersion, and competition for food. In many circumstances, feeding with the flock should pay the individual in foraging efficiency.

Defense against predators is a different kind of reason for foraging in a flock. In one sense joining a flock may seem a poor thing to do because the flock is conspicuous so that a predator may find it easily. Yet there are many ways in which this danger is offset by protection within the flock. Four such advantages are given below.

Individual Safety from Joining Flocks

First, vulnerable individuals can hide in flocks. As noted in Chapter 10, many predators select the young or the weak. These individuals may be less obvious to a predator when in a group. Wildebeest, for instance, apparently hide their young from marauding hyenas by passing them to the opposite side of a herd (Kruuk, 1972). It seems likely that juvenile or weakened birds should get a similar protection from hawks in a flock.

Second, individuals get better advance warning of predators if they have many companions also on the lookout. This advantage is self-evident and is supported by field observations. An elegant study by Kenward (1978) demonstrated the effect directly by flying a trained goshawk at English wood pigeons. The goshawk's strategy was to attack on the ground as the pigeons fed, and the pigeons could escape if they got up flying speed in time. Kenward assumed, therefore, that the hawk had been spotted at the instant the pigeons took off. His data (Figure 12.19) show that hawks were spotted much more quickly by flocks than by single birds. Moreover the kill rate of the hawk was significantly higher when it was launched against single pigeons or small flocks

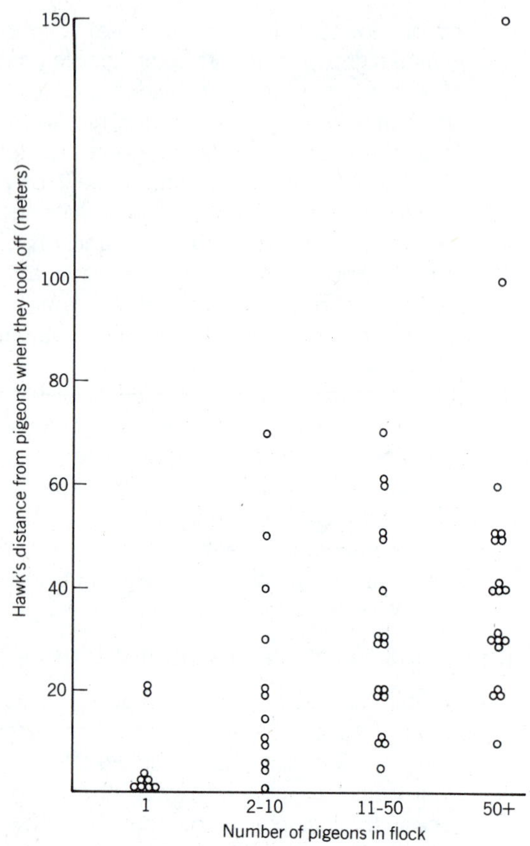

Figure 12.19 Hawk detection by pigeon flocks.
When a trained goshawk was flown at English wood pigeons on the ground the pigeons took flight earlier if they were in larger flocks, suggesting that they detected the hawk more quickly when in flocks.

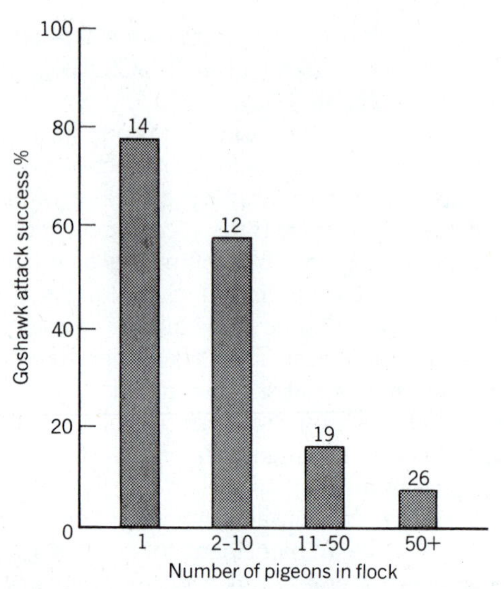

Figure 12.20 Success rate of goshawk attacking pigeons in flocks.
Attack by a trained goshawk rarely resulted in capture of a pigeon from a large flock, although most attacks on single pigeons were successful.

(Figure 12.20). This higher kill rate was probably due to the lack of early warning but it may also have been influenced by the fact that solitary birds tended to be social outcasts, either young or weak.

Third, individuals benefit because a predator may be deterred or confused when attacking members of a group, the "safety in numbers" hypothesis. This can be important in large animals, as when a ring of musk oxen face a wolf pack, but must be less important in flocking birds. However, Tinbergen (1951) long ago suggested that hawks attacking in flight could be frustrated if the prey was so bunched up that one bird could not be struck down with the hawk's talons without a real risk of colliding with a second. A midair collision at the speed with which a hawk attacks probably would mean debilitating, and hence fatal, damage to the hawk. The closed or massed formations of birds like starlings might protect individuals from hawks. But it may be that *confusing* the predator is the more important effect: a sustained chase and strike by a fast-moving predator at a dissolving cloud of numerous prey may reduce its success rate (Figure 12.21).

Fourth, the individual may benefit from the simple statistics of the encounter, on the logic that the chance of being a victim goes down with the number of alternative victims available. An animal will always be safer with others from a single attacker than if it is on its own. The habit of many animals of trying to get to the center of a crowd when predators approach is of this kind. The resulting dense huddle may be more vulnerable to predators than a loose assembly (or it may be less vulnerable) but each individual will still benefit by trying to get into the middle (Figure 12.22).

Figure 12.21 Possible confusion of a predatory fish by shoaling.
The behavior of the prey (*Stolephorus purpureus*) in scattering may so confuse the predator (*Euthynnus offinus*) that the attack is frustrated. Alternatively, the advantage to each individual may be that the chance is increased that neighbors will be the victim.

Figure 12.22 *Ovibos moschatus*: response to uncertainty.
When musk oxen huddle together this may benefit each individual by giving an appearance of menace or perhaps simply because the numerical chance of being a victim decreases in a crowd. Or the behavior may have more subtle origins and the appearance of a defensive ring is deceptive.

Large mammalian carnivores find additional advantages in group living because they hunt together in packs, enabling them both to overwhelm large prey as wolves do or so to confuse prey that victims will flee straight to a cooperating neighbor, a tactic often used by lions (Schaller, 1972).

Yet there must be compensating disadvantages in joining a group because solitary animals exist. Many advantages of solitary living lead, as we have seen, to territoriality. The increasing size of groups should limit the advantage that an individual can gain by joining so that it might be better to stay alone rather than join a crowd. Beyond this is the fact that most animals are themselves both predators and prey at the same time. Each has to allow for what is happening in the trophic levels above and below. Bertram (1978) offers the model of probabilities given in Figure 12.23 that should determine the several different responses of a predator faced with the task of attacking groups of different size.

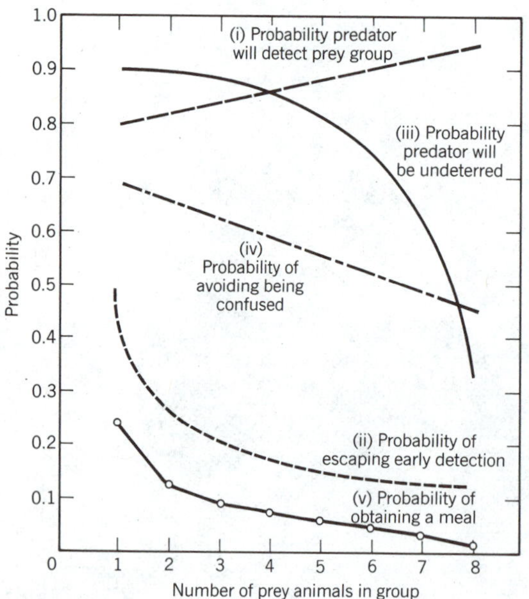

Figure 12.23 General hypothesis of the effect of group size on predation.
A solitary predator is imagined attacking groups of varying size. The probabilities shown in the models are all hypothetical. (From Bertram, 1978.)

In addition to the needs of sex, of eating, and of avoiding being eaten, group living also can serve simpler physical needs. Many animals huddle together to keep warm. A celebrated early ecological observation is that of Allee (1926), who found that woodlice huddle together to keep moist. It is possible that the "V" formations of migrating geese and others develop as individuals position themselves on patterns of turbulence behind the wing tips of those in front, and it has been argued that individual fish in schools are motivated by comparable considerations of fluid dynamics.

A FIRST HYPOTHESIS OF GROUP SELECTION

Groups can be fully and sufficiently explained as properties emerging as individuals pursue individual fitness. And yet ecologists often have been fascinated with looking beyond the individual and thinking in terms of the supposed advantage to the group as a whole. This approach leads to suggestions that, since every individual in the group benefits, *group advantage* is what matters. This is to claim that the traits are preserved by GROUP SELECTION. Most ecologists do not accept that group selection is necessary to explain these observations. Geneticists can show that except in very restricted circumstances, group selection is either not possible or is trivial (Wiens, 1966). But the idea is constantly resurrected. A most powerful statement in this cause was made by Wynne-Edwards (1962).

Wynne-Edwards was struck by the population consequences of many forms of social behavior. He was, for instance, convinced of the truth that territorial behavior set limits to population when many ecologists still doubted this. He also noted that setting a limit to population was probably a consequence of all systems of social dominance and hierarchy at some time or another. An animal low on a peck order, for instance, might be so low as to be expelled from the group altogether, in which event there were many fates waiting for it. The necessary expulsion would become ever more likely as crowding increased so that the social mechanism of expulsion should be the act that set the population size. Probably most ecologists now think that Wynne-Edwards was correct with this observation. But then there is a grand parting of the ways.

These population limits set by territoriality and various forms of social organization are most parsimoniously viewed as accidental consequences of the behavior. The behavior results, as we have seen, from the doings of individuals pursuing fitness. Everything done is in pursuit of individual fitness, even the consent to be one down in a social encounter. But Wynne-Edwards postulated that population regulation actually was the *purpose* of the behavior. When a territorial bird failed to breed, or a rodent was expelled from a society, this, in his view, was an act of self-denial done for the good of the group. And he went further still.

He went on to postulate that nearly all crowds of animals in nature had the ultimate function of population regulation for the good of the group. When starlings flew in close formation this was a way of sensing the density of the crowd so that they might adapt the breeding effort accordingly. When menhaden swam together a million strong, this told how many menhaden there were, with consequences for reproduction. Singing at the dawn chorus was a device to tell all birds their population density, as was the singing of frogs, crickets, and cicadas. Wynne-Edwards made a special point of studying leks, failing to see the purpose of all that display if the strutting males were not really showing off to count heads among themselves as a step towards family planning. He said that all these displays were not just for individual benefit but for group benefit. He called such a display EPIDEICTIC DISPLAY. The display measured density: a dense crowd should result in a lowered reproductive effort, a sparse crowd in numerous offspring.

For the general hypothesis of *population control by social mechanisms* to stand, two requirements must be met: a mechanism to translate the crowd measures of epideictic display into the appropriate fecundity, together with a theory of group selection that should give selective advantage to the behavior. In social vertebrates, at least, the mechanism is not hard to imagine, as a hormonal feedback similar to the general adaptation syndrome would do. But the second requirement of allowing birth control for group advantage has been an insuperable obstacle to the hypothesis. It is now given little credence.

CONCLUDING NOTE: SEX, INDIVIDUAL CHOICE, AND POPULATION LIMITS

Animals often live in groups because this serves individuals well and grants them fitness. Other animals live in social systems that force them to spread out, the phenomenon we call territoriality, or groups may be spread out and kept that way by a group defense resulting in group territories. Every one of these systems can be fully and sufficiently explained as resulting from individual choice. The ultimate arbiter of which kind of social system shall prevail is the pattern of resources that is being used. But superimposed on the pattern of social dispersion set by resources is the pattern set by sex. In most animals the rules of reproduction, whether sexual or asexual, are set by environment, its resources and its hazards. The important consideration is which system serves females best; to make eggs parthenogenetically or to meet the meiotic cost of males by winning a commensurate benefit. Varied offspring make sex profitable for most females so that bisexualism is the norm. In that male contribution to the progeny is less than female contribution, this produces strong pressures towards polygyny, though a crucial selection pressure within polygynous systems is female choice. An inevitable consequence of many of the forms of social behavior that result from group living and sex is the imposition of population limits. This is particularly true for territoriality but may also result from some systems of social dominance. Yet these population effects are always consequences of behavior that gives extra reproductive success to individuals.

PREVIEW TO CHAPTER 13

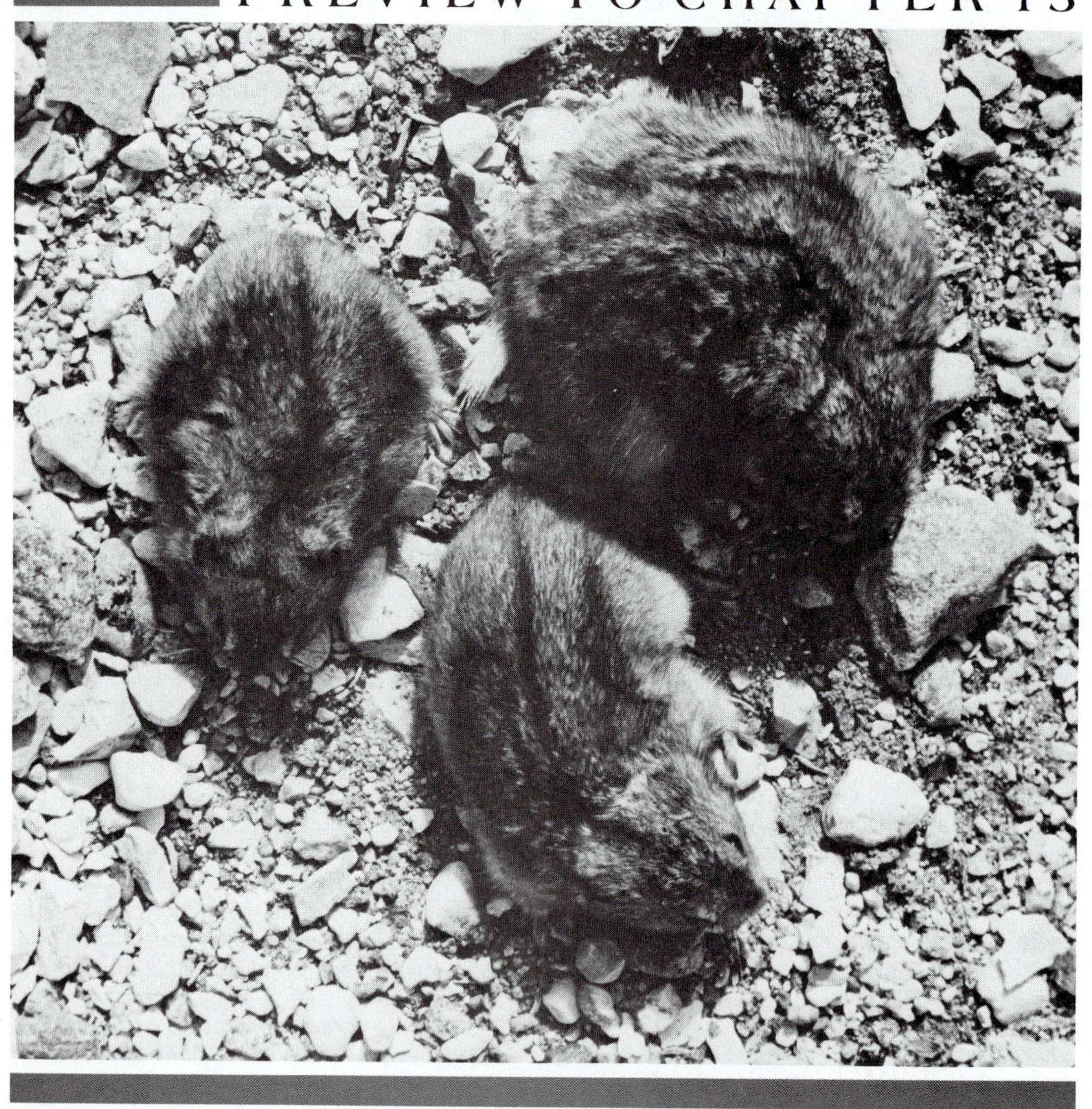

No single mechanism is responsible for the control of natural populations. Some populations sometimes can exist close to equilibrium, with birth rates balanced by death rates, but more often numbers fluctuate between wide limits as changing or repetitive environments persistently alter the balance between births and deaths. But the possible causes of death vary from species to species and from time to time. Failures to be born or to find a mate are as important as actual death, being equally effective at removing a unit of reproduction from the next generation. Birth rates vary, in some species actually being kept low by patterns of behavior that work to regulate births according to supplies of resources. This birth regulation always gives advantage to individuals by increasing the eventual number of surviving offspring over what would result from more initial births. Natural birth regulation, therefore, always works to increase the rate of population growth. Selection for group birth control is highly unlikely. Although a plausible mechanism, for the selection of groups of predators that are below maximum efficiency as hunters has been put forward, the consensus of ecologists is still that all regulation of births works to achieve optimum reproductive success and that all control of populations is by some form of environmental resistance. For the minority of species that attain lasting population equilibria, competition for vacant niche spaces is the decisive mechanism. Short-lived opportunistic species, or those suffering heavy predation, may persist with constantly fluctuating local populations. Some of the longest-lived animals and plants, particularly those of conservative breeding strategies, live in populations that are always changing very slowly and that may reach neither equilibrium nor extinction before the onset of climatic or environmental change. The ultimate arbiter of population history of any kind of animal or plant is its own life history adapted to particular patterns of resource. The history of studies into fluctuating numbers of arctic fur-bearing animals is reviewed. These animals are now known to be truly cyclic but it is also well established that these cycles do *not* result as coupled oscillations between predator and prey *nor* between herbivore and plant food. Nor are the cycles caused by weather acting as a forcing function. Population histories described by differential equations based on the logistic equation but incorporating lags in population responses do tend to fluctuate rhythmically around stable points, suggesting that some cyclicity may be inherent in population dynamics. Superimposition of predation onto fluctuating populations of prey might amplify this process, particularly if predators are subject to local extinction when prey numbers are low. An alternative family of explanations comes from ecosystem models of populations in patchy environments where local crowding causes emigration to second-class habitats serving as population sinks. Oscillations are inherent in these systems also, but the oscillations would be cyclic only if a timing device is incorporated in the models. It is suggested that the timing device, together with the mechanism of synchroneity, can be found in the way the lemming life cycle is adapted to the fluctuating arctic climate. Lemming reproduction is synchronized by weather so that populations of large areas rise or fall together. Lemmings are also short-lived, so that a poor breeding season is followed by a population crash as the old generation, born synchronously, also dies synchronously. The timing device may be a function of fecundity and life-span.

CHAPTER 13

NATURAL REGULATION OF NUMBER (WITH AN ESSAY ON RODENT CYCLES)

A struggle for existence inevitably follows from the high rate at which all organic beings tend to increase. Every being, which during its natural lifetime produces several eggs or seeds, must suffer destruction during some period of its life, and during some season or occasional year, otherwise, on the principle of geometrical increase, its numbers would quickly become so inordinately great that no country could support the product. Hence, as more individuals are produced than can possibly survive, there must in every case be a struggle for existence, either one individual with another of the same species, or with the individuals of distinct species, or with the physical conditions of life.

Numbers of all animals and plants must be under control or the earth would be quickly overrun. This is clear enough. Darwin stated it with drama in the above passage from the *Origin of Species*. A requirement of selection theory is that each individual reproduce to the uttermost; ample evidence shows that they all can and do, therefore there must be death and destruction.

Death can come in many forms, some of them not obvious. Perhaps the commonest way of "death" is not being born at all because the stresses of environmental life prevent a parent from laying an egg or carrying an infant. Perhaps almost as important is the concealed form of death that comes from failure to mate. From the population point of view an individual is just as "dead" if it is a lifelong bachelor or spinster as if it were killed at sexual maturity. The following seems to be a complete list of possible forms of death:

1. starvation (outright energy deficiency: food calories or light)
2. malnutrition (lack of nutrients, or water, or essential compounds)
3. predation
4. parasitic disease
5. accident (probably weather induced)

6. failure to be born
7. failure to find a mate.

Obviously items 6 and 7 in this list can be debited to births instead of crediting them to deaths in the ledger, and they are so treated in life tables (Chapter 9). But it may be more interesting to think of them as forms of death. A population can be controlled by invoking any or all of the seven ways of death. What is remarkable is that individual genomes pass through these hazards from parent to parent at all, not that populations are controlled (Figure 13.1).

The importance of these various ways of death varies widely with different life history strategies, producing quite different patterns of population control. But before examining these in more detail it is well to look at the possibility that some populations somewhere might be regulated not by death control but by birth control.

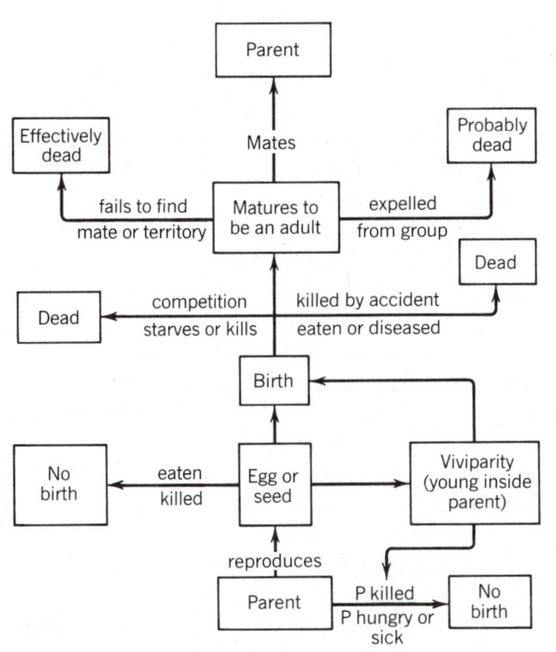

Figure 13.1 The fitness stakes.
Populations can be controlled at any or all of the obstacles in a parent to parent cycle.

GROUP SELECTION IS NOT LIKELY AS A POPULATION REGULATOR

Birth control could be used as a population regulator only if group selection prevails over individual selection. It requires that the individual sacrifice itself in the interest of the group or population, a proposition that cannot easily be reconciled with the principle of natural selection. What should be needed is that an individual wins more fitness by the self-sacrifice of curbing reproduction than can be won by simple breeding. No formal or adequate theory is available that permits this except in closely limited circumstances (see also Chapter 25 for D. S. Wilson hypothesis).

The immense difficulty of imagining a mechanism that would allow group selection to promote birth control can be illustrated by considering an intriguing recent model of how predation might work to cause the survival or extinction of whole groups—hence GROUP SELECTION (Gilpin, 1975). The proposed mechanism works to restrict the efficiency of predators.

Suppose that groups of animals, both predators and their prey, live in isolated patches, a condition in fact common in nature. Always there will be strong selection pressures for predators to become more efficient, as, of course, there are opposite pressures for the prey to achieve a better defense. But in some group, somewhere, an adaptation will appear that gives individual advantage by improving the foraging efficiency of predators that have it. This adaptation will be selected strongly within the confines of the group. Gilpin (1975) shows mathematically that this will result in rapid extinction of the predator population within the group as they reduce the prey to numbers so low that predators cannot find food to reproduce. A whole group of predators then goes extinct and the patch is recolonized by less efficient predators from elsewhere. The whole proceeding relies not only on life in patches, but on slow rates of gene exchange between groups (slow colonization or immigration) and rapid effects within the groups. A nice balance is required between the rate at which a group suffused with a fatally efficient genome goes extinct and the rate at which colonists with genomes that allow persistence arrive.

Another way of explaining the Gilpin model is to think of the system set up by Gause (Chapter 10) in which *Didinium* preys on *Paramecium*. In Gause's experiments the *Didinium* was a devastatingly efficient predator that always eliminated its prey, and hence itself, from any culture tube. Gilpin's model requires that there be other strains of *Didinium* that are less devastating, living in culture tubes (patches) elsewhere. *Didinium* individuals of this less devastating kind will survive and their groups with them. These individuals can then recolonize patches (culture tubes) where a whole group of superefficient *Didinium* has gone extinct.

The sequence of events in the Gilpin predator–prey group selection model is:

1. mutation or recombination produces an extraefficient predator

2. selection favors this efficient predator such that all predators within the group come to be of this kind

3. extinction of all individuals of the group

4. recolonization of the vacant patch with persistent, less efficient predators from other groups.

It is possible that the Gilpin model describes what has actually happened in nature at times. Certainly computer simulations show that the effects are realistic. But this is group selection against efficient predators, not against efficient breeders.

To transfer this Gilpin model to the population argument it is necessary for extra good reproduction to harm an isolated group. That a variety with improved powers of reproduction should replace all other varieties in a group is, of course, certain. Gilpin's model then requires that these reproductive powers should lead the group to rapid extinction. The same constraints of timing and rate of immigration, *et cetera*, apply as to the predator–prey model. Even within these constraints it is difficult to see how high reproduction can be a road to group extinction in the way that high efficiency in a predator might.

Super efficiency as a predator can have malignant properties; this is easy to see. But how can super reproduction be malignant? Super-reproducing individuals should have their offspring constrained in all the familiar ways, if not by accident or failure to mate, then by competition among themselves. In a very real way, super reproduction is impossible. It is true that if a way can be imagined that a group of super-breeders can breed itself into rapid extinction, then the progeny of lesser breeders in other groups might well colonize the vacated patch as in the Gilpin model. Group selection would then have resulted in birth control. Even then, just the right array of patches, immigration rates, and extinction rates would have to apply. The possibility must be taken to be highly improbable, though interesting.

Other mathematical models have been offered of circumstances in which group selection can be seen to be possible (see, for instance, Griffing, 1967). Yet to allow birth control, these models face the immense difficulty of showing that high individual reproductive effort can so escape ordinary constraints as to lead to disastrous consequences. Group selection probably will be shown to have operated in some populations for some traits, but not to have caused reduced reproduction.

More cogent is the argument that all we know of the social habits of individuals can be fully and sufficiently explained by a hypothesis of individual selection. This is therefore the parsimonious explanation for the behavior we see, including all the group or social phenomena listed, for instance, by Wynne-Edwards (Chapter 12). We do not need group selection.

The position of most modern ecologists, therefore, is essentially that stated by Darwin in the *Origin of Species*. Every individual breeds to the uttermost. Individuals may regulate births to achieve optimum reproductive success but never to influence population size. All control of populations is by environmental resistance: by preventing births or causing deaths.

THE UPPER LIMIT: A POPULATION EQUILIBRIUM

If a species equilibrium is set by intraspecific competition it follows that niche sets number. This proposition appears to be self-evident. At the species equilibrium all niches are filled. The only way of increasing the population still further is by *K*-selection for individuals able to live at higher densities, essentially in smaller niches. Population size continues to be set by the prevailing niche and the number of niches is the carrying capacity, *K*. This pattern is interesting only because it requires that *when a species equilibrium is set by intraspecific competition, the size of a population is independent of the breeding effort.*

In this special case of a single-species equilibrium, then, we come to the conclusion that reproductive rate is without population consequence. High fecundity or low, it makes no difference to the population, only to the individuals as they compete with their neighbors for a chance to occupy one of the finite number of niches.

But in the real world, species do not live in total isolation but in systems of diminished competition with other species around them. For animals and plants at a species equilibrium set by interspecific competition the reproductive effort might matter more. This is because the strength of the competition that a species population offers to another is in part a function of its reproductive rate (Chesson and Warner, 1981). It will be recalled that the outcome of some of Gause's competition experiments between species of *Paramecium* were determined by the relative reproductive rate (Chapter 6). Competition between species should always be influenced in this way by reproductive effort. On the other hand, reproductive effort is itself a function of the success at competition, because competitive struggle yields the resources used for reproduction. Even where interspecific competition sets an equilibrium, therefore, it is probably still true to say that niche size sets number, and that reproductive effort is of little consequence.

Where there is a true equilibrium, then, numbers are set by competition for niches and for the resources needed to sustain life in a niche. This competition can work by arranging death at any of the stages of a life cycle shown in Figure 13.1. Possibly the most perfect constancy of number results when the crucial "deaths" are failures of adults to find mates. This is apparently the pattern in territorial birds and results in that remarkable constancy of the spring census. All niches are full every spring with surplus applicants always ready to restore the breeding population after any loss. Deaths at all other stages in a life cycle are likely to be density dependent, though this is not required as long as *some* deaths are density dependent.

WHEN EQUILIBRIUM IS SELDOM ATTAINED

Some populations rarely attain equilibrium. The most obvious examples are opportunist weeds and insects, but it may also be true for some of the longest-lived organisms, conservatively breeding animals, and very long-lived plants.

Opportunists with short lives, high fecundity, and high powers of dispersal are prone to density-independent control. Their most massive source of death is probably destruction of seeds, eggs, or new-hatched young. This mortality is decreed by their small-young breeding strategy that sends such vulnerable offspring out to fend for themselves. The mortality may sometimes be a function of density, particularly if their numbers attract learning predators with a type 3 functional response (Chapter 10), but is more likely to be density independent as the dispersing young reach hostile environments or hostile weather.

Occasionally it must happen that circumstance preserves a large percentage of opportunists' young so that many hatch and many grow to maturity. The resulting very rapid population increase can lead to crowding and equilibrium conditions. *K*-selection is then likely with a tendency for density-dependent pressures to appear at all stages in the life cycle.

But the lives of opportunists are also fugitive lives adapted to temporary, unstable, or seasonal habitats. When the habitat changes there can be very widespread and density-independent death. It may be particularly likely that this catastrophic loss is more by failure to mate or failure for eggs to be laid than by actual destruction. Death of the individuals might well be by old age, perhaps accelerated by starving. But the effect is reduction, sometimes massive reduction, of the population.

It is a little unsatisfactory to call this process whereby numbers of opportunists never reach equilibrium "control." Numbers are not so much under restraining control or regulation as they are in a continual state of change. The habitat varies, the resource base varies, and the reproductive success varies accordingly. The birth rate/death rate ratio keeps changing. During the low population period of the hostile season numbers can sometimes be kept from falling still lower by density-dependent survival, perhaps by escape from predators when numbers are very low. But in other circumstances the continued existence of a species may depend on the probability that there will always be a few survivors from the very large initial populations when the birth rate/death rate ratio changes back again. These populations have a definite probability also of going extinct as random death brings the survivors to zero.

Very different from opportunists are long-lived birds with conservative breeding strategies, birds like penguins, boobies, and California condors (Chapter 12). The numbers of these change only exceedingly slowly. The adult population is not affected by short-term fluctuations of weather or resource, provided these are not so severe that they kill adults. Populations can change extremely gradually for very long periods, so long in fact that neither equilibrium nor extinction might be reached before there is long-term climatic change. Major changes in climate occur on time scales of only a few thousand years or less; for instance, the last ice age ended just 10,000 years ago (Chapter 21). Far smaller changes than this might make a world quite different to live in for boobies or condors, perhaps changing vegetation cover from one formation to another or altering the productivity of an ocean. Some of the more conspicuous of long-lived animals, therefore, look to have remarkably constant populations that are not so much under control as suffering continual slow change.

This effect may be even more pronounced for forest trees with lives and generation times an order of magnitude longer even than condors. If, for instance, the Amazon basin was a seasonal place without tropical rain forest in the last ice age as is possible (Prance, 1982; Colinvaux, 1979) then the forest is only 10,000 years old, perhaps not long enough for tree populations to have reached equilibrium numbers in a place where so many species might compete. Many histories of temperate forests by pollen analysis (Chapter 21) show that there have been progressive changes throughout most of the 10,000 years of the Holocene. Figure 13.2 shows the slow invasion of the American Midwest by hemlock (*Tsuga*) over the last 6000 years. It does look as if hemlock range and numbers were stable since about 2000 years ago, suggesting that a species equilibrium with density-dependent control was established eventually. This equilibrium was preceded by 4000 years of colonization. The process is so slow that often in the history of a tree the total population will be slowly changing in response to changing climate.

SMALL MAMMAL CYCLES INTRODUCED

Throughout recorded history people have at times been beset by plagues of mice. The mouse plagues always appeared suddenly, worked their mischief, and went. People have always tried to defend themselves. When ancient Philistia was afflicted by mice, the Philistine priests were induced to return the Ark of the Covenant and they placed in it golden images of "the mice that mar the land" (1 Samuel 6:5) as their hope for respite. More recently people have tried traps, poison, or rodent diseases. As Deevey (1959) points out, what these methods have in common is that they all work. This reveals one of the more remarkable aspects of rodent plagues: an end

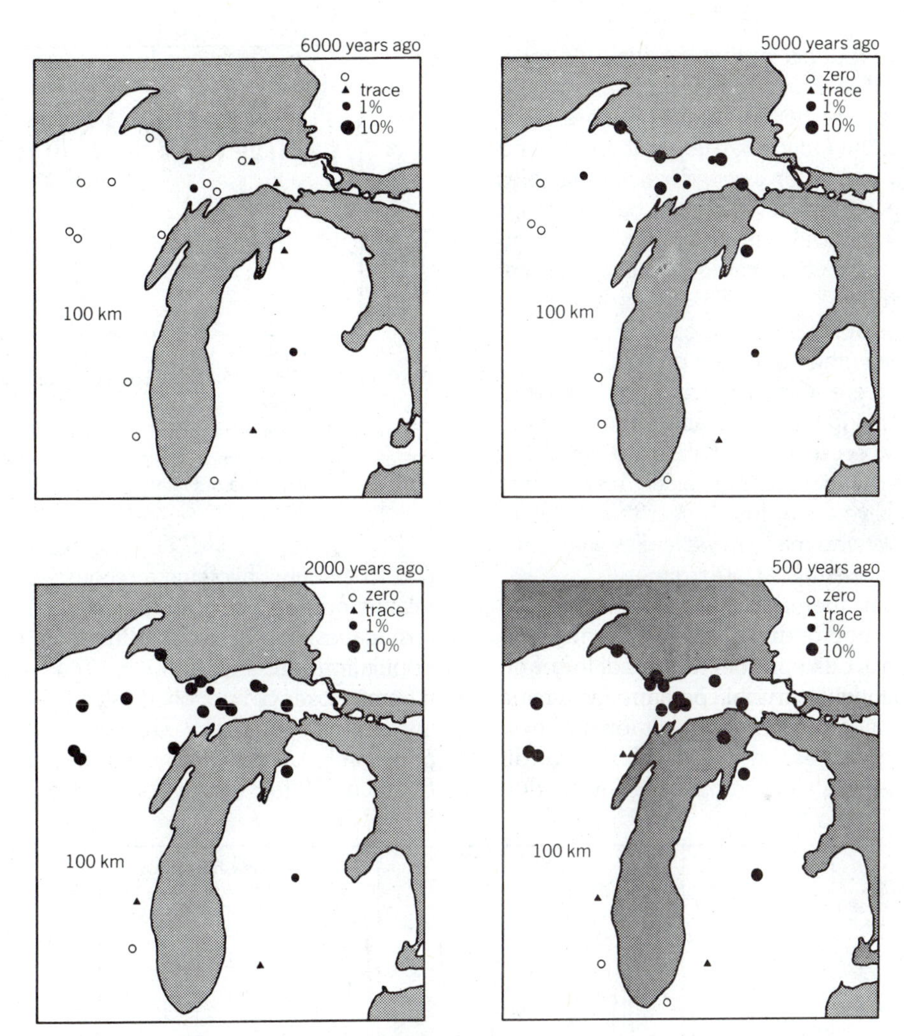

Figure 13.2 Migration of hemlock in the American Midwest.
The data are percentage hemlock (*Tsuga canadensis*) pollen in lake sediments dated by radiocarbon. The process of invading the region west of Lake Michigan took 4000 years to complete. (From Davis, 1982.)

that comes as suddenly and completely as the beginning. The history of a mouse plague tells of a sudden spectacular increase in numbers, often over a huge extent of country. For a season or two the mice rampage, destroying crops and running about in the open. Then suddenly, quietly, most of them vanish.

The most spectacular rodent plagues of all are those of lemmings, animals that not only sometimes appear in great multitudes but that also appear to come and go with almost regular periods. After a number of early notices of this phenomenon in naturalist literature the matter was formally documented by Elton (1942). The best evidence for cyclic numbers of arctic mammals come from the fur trade, particularly from Canada where missions of the Moravian church and the Hudson's Bay Company had been trading with natives for more than 100 years. The main trade items were skins of predators, but the predators ate rodents so that the fur yield must give some indication of rodent populations.

Two staples of the fur trade of those parts were skins of the arctic fox (*Alopex lagopus*) and the Canada lynx (*Lynx canadensis*). The sales records of these skins revealed that bumper crops of fox skins came about every four years, and about every ten years came bumper crops of lynx skins (Figure 13.3). The fur industry was one of boom and bust, with the booms and busts following each other in regular rhythm ever since the records were started more than 100 years ago. The trapping effort probably varied, but not enough to obscure the fact that the supply of animals was changing (Winterhalder, 1980). More foxes were alive every fourth year than at other times, and likewise many more lynxes were alive every tenth year than at other times.

The trappers themselves knew the immediate cause of their booms and busts—fluctuating supplies of food for the fur-bearing predatory animals. It was long known that lynx preyed almost exclusively on arctic hares and trappers knew that a good hare year was a good lynx year. In places arctic hares were trapped for fur in addition to the lynx, allowing a reconstruction of the relative numbers of both (Figure 13.4). This history leaves little doubt that the fates of the two populations are linked.

Arctic foxes did not often catch hares, but lived instead on mouse-like animals, on voles, and particularly on the various species of lemming. Every good fox winter followed a summer in

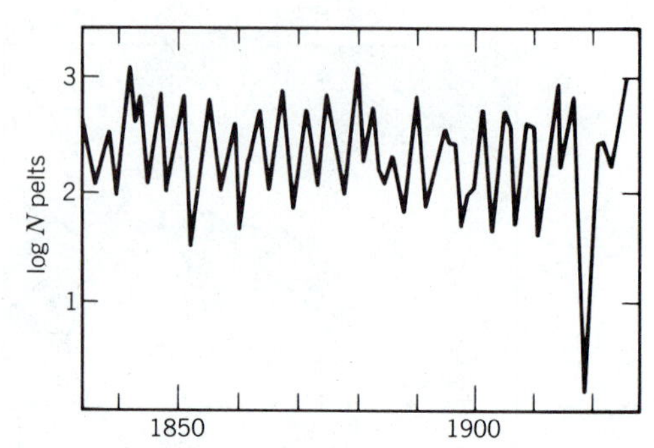

Figure 13.3 Fox (*Vulpes fulva*) skins sold to Moravian missions in Labrador.
The fluctuating numbers record population events in both the predators (foxes) and their prey (arctic rodents, particularly lemmings). (Data of Finerty, from Hutchinson, 1978.)

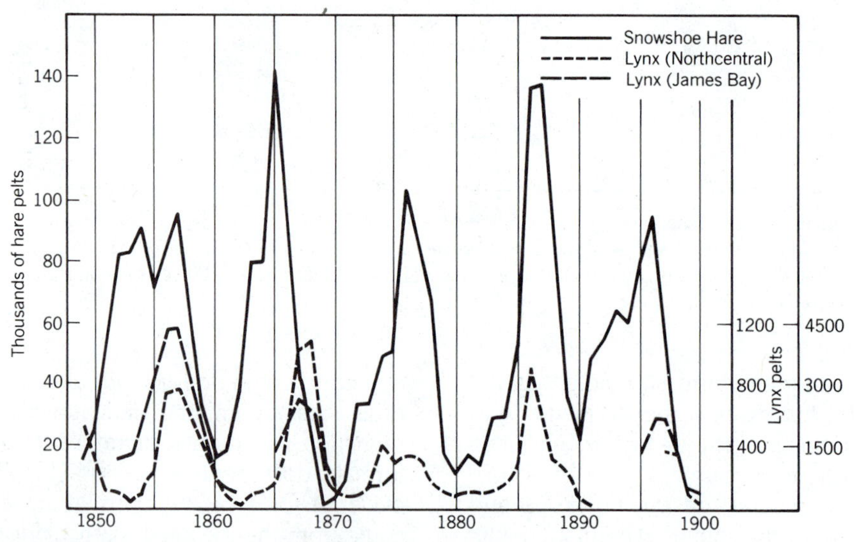

Figure 13.4 Relationship of hare and lynx populations in the Canadian arctic.
Data are from sales records of traders dealing in both hare (*Lepus americanus*) and lynx (*Lynx canadensis*) skins. Lynx records are from two nearby trapping regions. The data show not only that the populations are linked but that numbers of the predator follow numbers of the prey. (From Finerty, 1980.)

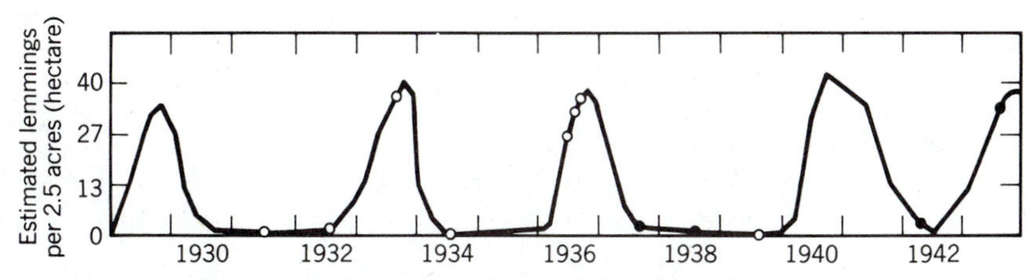

Figure 13.5 Lemming cycles in northern Manitoba. Data are from field censuses by Shelford. (From Finerty, 1980.)

which trappers' diaries recorded that the tundra was alive with little brown mice. The trappers noted the presence of the lemmings only casually, but modern studies show that fox numbers are indeed linked to lemming numbers (Finerty, 1980). About every fourth year on the tundra lemming populations are many times as big as in the lean years in between (Figure 13.5). For a trader, every summer of a lemming high means a winter following in which a rich harvest of fox skins is brought for sale.

Lemming "highs" are roughly four years apart, but hare "highs" are about ten years apart. It now turns out that two kinds of lemming and some true mice (Muridae) undergo a four-year cycle and that muskrats (*Ondatra zibethica*) undergo a ten-year cycle like the hares. Four- and ten-year cycles also appear in a large array of predators in every instance when it is clear that a close relationship exists between numbers of predator and one of the prime kinds of rodent prey (Table 13.1).

These are true cycles, not just progressions of random events that seem to imitate cycles.

Table 13.1
Arctic Animals for Which 4-Year and 10-Year Cycles Have Been Demonstrated
For each cycle an array of predators specialize on the cycling numbers of the primary rodent prey. (From Finerty, 1980.)

	Lagomorpha	Rodentia	Carnivora		
Cycle	**Leporidae**	**Muridae**	**Canidae**	**Mustelidae**	**Felidae**
4-year	*Lemmus* *Dicrostonyx* spp.	*Alopex lagopus* *Vulpes* spp.	*Martes americana* (?)		
10-year	*Lepus americana*	*Ondatra zibethica*	*Alopex lagopus* *Vulpes* spp. *Canis latrans*	*Martes americana* *Martes pennanti* *Mustela vison* *Gulo luscus* *Mephitis mephitis*	*Lynx canadensis*

Sometimes it does happen that a series of unconnected ups and downs can look cyclic, a happening for which ecologists have been wary ever since the mathematical possibilities inherent in random numbers were pointed out to them by Cole (1954). But the cycles of arctic populations are real.

A satisfactory demonstration of the reality of these cycles is provided by Finerty (1980) using autocorrelation and spectral analysis of the data. Autocorrelation is a measure of correlation between an observed series of numbers and that same series if initiated after a time lag. Autocorrelation between peaks in a sine curve displaced by one-half wavelength, for instance, gives an autocorrelation of $r = 1$. The technique used with small mammal cycles is to convert data to logarithms and use a computer to generate autocorrelograms. Figure 13.6 shows the results of this technique applied to Maklulick's (1937) data of lynx skins traded in the Mackenzie region of Canada. Correlation coefficients are high, showing that the peaks and troughs are not just periodic but also truly cyclic. Finerty (1980) applies this technique to data for most of the animals in Table 13.1, showing that they all experience true population cycles (Figure 13.6).

It is, therefore, clear that lemming and other arctic small mammal populations are not only periodic but cyclic. There is rhythm to their fluctuating numbers, whether an internal rhythm or one imposed by the environment. We thus have a number of definite questions to answer:

1. Why are these populations periodic?

2. Why are the fluctuations cyclic?

3. Why are the cycles synchronized over very large areas?

4. How are population highs generated?

5. How do the individuals in the dense populations of a high die or are otherwise removed?

We still argue about these questions, which means we do not yet have all the answers.

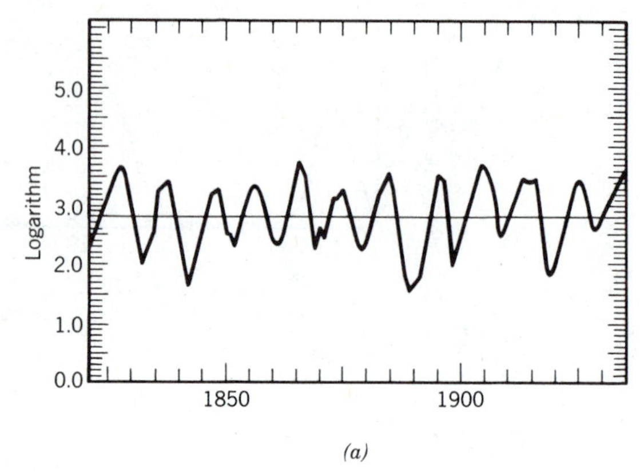

(a)

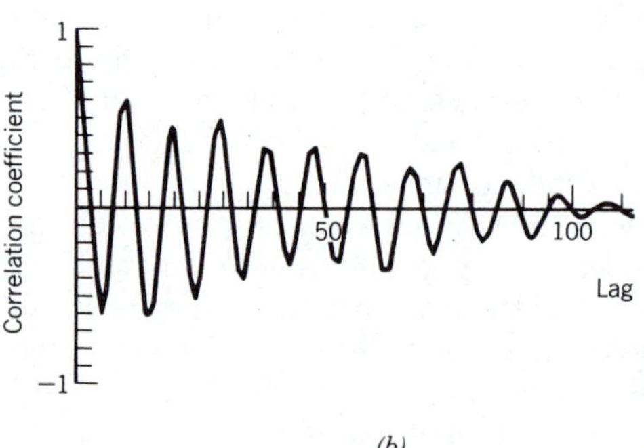

(b)

Figure 13.6 Cyclicity of lynx populations.
(*a*) Logarithms of numbers of lynx skins traded in the Mackenzie region. These data show strong periodicity that is shown by the autocorrelogram (*b*) to be truly cyclic. That the peaks do not follow the decades precisely in (*b*) shows that the cycle is not precisely 10 years but actually about 9.6 years. (From Finerty, 1980.)

Coupled Oscillation Hypotheses

Once the coupled oscillation hypothesis of the Lotka–Volterra predator–prey system became part of general ecological knowledge in the 1930s, these arctic animals seemed to provide the obvious example. The first explanation of the cycles was, therefore, that they were caused as predators first overate their prey and then starved. But evidence has accumulated suggesting that this explanation will not do. Many places can be found where rodent numbers oscillate in the absence of any local population of predators at all, so

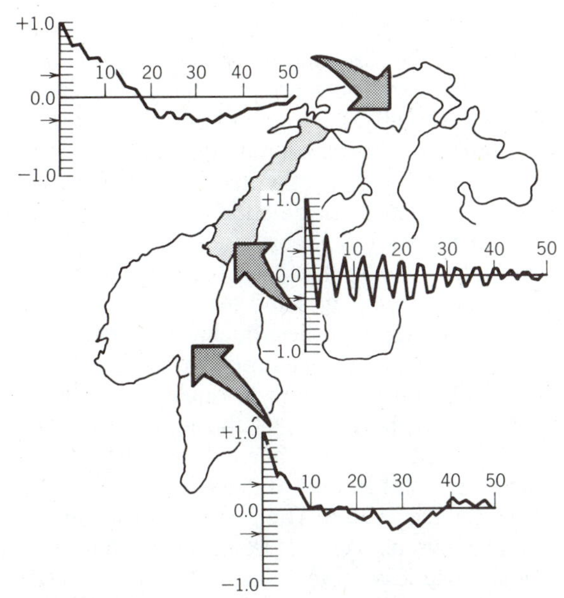

Figure 13.7 Cyclicity of foxes in Norway. Autocorrelograms of fox numbers in three regions of Norway from 1880 to 1926. Cycles are revealed in the central section but not in the other areas. These are the cycles of approximately four years that reflect cycles in the lemming prey. (From Finerty, 1980.)

that predators could not be responsible for these fluctuations (Finerty, 1980). In Norway the killing of predators for bounty over about a hundred years kept predator populations low, but the populations of their prey cycled just the same (Figure 13.7). In addition the numbers of predators killed for bounty went up every four years when lemming highs caused rare and harried predators to be a little less rare than in other years (Elton, 1942). These data suggest that predator numbers respond to cycles of prey numbers but that prey numbers cycle for some reason outside the predator–prey system.

Perhaps the most persuasive evidence is in the population records themselves. Highs of predator and prey do not appear to alternate as required by the Lotka–Volterra formulation, but are more nearly synchronous (compare Figure 13.4 with Figure 10.2). The populations are very nearly in phase, as would happen if predator numbers helplessly followed numbers of prey, victims of the cycles rather than causal agents.

When examining a data set like that of the hare–lynx cycle of Figure 13.4 it is necessary to keep in mind possible inaccuracies in the data. These are records of skins brought to sale in different parts of Canada—the data are based on historical research into the files of various trading organizations. Lynx and hare data, for instance, were collected from separate sets of records. One of the earlier published versions of the hare–lynx graph was found, for instance, to show hare numbers peaking slightly *later* than lynx numbers, a result that prompted a paper in a scientific journal with the title "Do Hares Eat Lynx?" (Gilpin, 1973). More historical research then showed that hare numbers did in fact slightly anticipate lynx numbers, putting hares back into the role of prey where they belong (Finerty, 1980). But this incident does show how nearly synchronous the highs of the two populations are.

Most ecologists now conclude that these small mammal cycles are not the result of coupled oscillations between predators and prey. The primary fluctuations are those of the prey species, and predator numbers merely track prey numbers. Obvious alternative hypotheses are that numbers of rodents and hares, the prey species, are coupled with something other than predator populations. The most likely candidates are diseases or the plant food of the herbivores.

A coupled oscillation between lemmings or hares and epidemic disease is, on the face of it, plausible. Lemmings and the rest are crowded at population highs, epidemics break out, the population is reduced to very low numbers, the disease dies out and population growth is resumed. But this hypothesis has been tested and found wanting. Signs of disease either cannot be found (Krebs, 1964) or are present in crowded lemming or hare populations at much less than epidemic proportions (Deevey, 1958; Elton, 1942). Moreover, data from hares suggest that different diseases are present in different periods of decline, a finding not consistent with the coupled oscillation hypothesis. Finally, we have experience of true epidemic disease in small mammals with the case history of myxamatosis of rabbits in Europe and Australia (Chapter 10) in which no cycles resulted. Natural selection worked to produce resistant rabbits, so that the rabbit population, once restored, was not again catastrophically reduced by the disease. It thus seems safe to conclude that small mammal cycles are not caused by a coupling with infectious disease.

The case for coupling with supplies of plant food, though less than convincing, is not so easily dismissed. The hypothesis has several forms, the simplest being that dense herbivore populations graze the vegetation so severely that plants are destroyed and the animals must starve. After the population dies off the vegetation begins to recover; the whole process working exactly like the classic coupled oscillation model of Lotka and Volterra. It can be shown that lemmings in particular heavily damage their range, grazing the ground bare in the summer and winter of a lemming high. But they do so only in patches. And the lemmings do not starve. When Krebs (1964) trapped and autopsied 4000 lemmings during a population die-off, the general appearance of the animals was healthy and most actually were fat.

But there are more ways of dying of malnutrition than outright calorie deficiency. Suppose the damaged range left lemmings short of some mineral or vitamin, or of protein? Might not such a diet have led to some subtle symptoms that were overlooked in Krebs' autopsies? Or might it not interfere with their later breeding success? For this hypothesis to stand two things must be shown; the impoverished quality of vegetation (as opposed to its total supply) must be demonstrated, and this failure in quality must then be linked to the deaths of the lemmings. The first of these has now been demonstrated by studies at Point Barrow (Pitelka, 1957; Mullen, 1969; Schultz, 1969). The overeaten vegetation of a high lemming year at Point Barrow has proportionately less protein, less calcium, and less phosphorus than in other years (Figure 13.8). But there is still no evidence linking this poor quality of the vegetation to either increased mor-

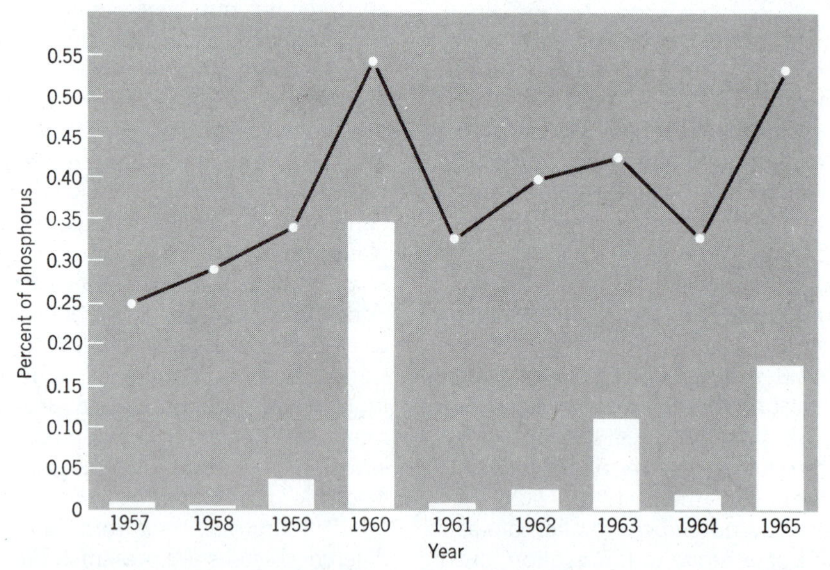

Figure 13.8 Correlation of phosphorus in forage with size of the lemming population at Point Barrow, Alaska.
The curve represents phosphorus and the bars represent the relative sizes of the lemming populations in the summers of the several years. The phosphate content of forage apparently drops following a lemming high, suggesting that the mineral is then being cycled back to the soil. (Redrawn from Schultz, 1969.)

tality of lemmings or to their failure to breed. There seems little doubt that the oscillations in lemming numbers at Barrow are accompanied by oscillations in the nutrient cycles, with a major portion of the nutrient reservoir being held in living things during the lemming highs and this same portion flowing back to the plants via the soil during lemming lows, but it cannot be shown that this shunting of the nutrients controls or even triggers the rise and fall of lemming numbers. For most ecologists the starvation hypothesis, even in its modified food-quality form, must go the same way as the predator and disease hypotheses. A mechanism more subtle than a simple coupled oscillation must be causing small mammal cycles.

Adaptations of Arctic Predators to Fluctuating Food Supplies

It is proper to look at arctic predators not so much as agents of population control, or as generators of cycles, but as animals cursed with a food supply that fluctuates widely. How do they cope with this and what do they do when the lemmings or the hares go away? Mammalian predators probably starve to death, particularly the smaller and less mobile predators like weasels. After a lynx high, gaunt animals, described as little more than living skeletons, have been seen on the tundra. Arctic foxes wander great distances, which is probably the best possible strategy when the lemmings fail: a long journey probably ends in death but staying put would make starvation certain. Probably all the mammals wander, which is really the only adaptation possible for them. Probably also most of them die of starvation at the end of a cycle. Some of the long-lived birds of the tundra are able to come to better arrangements.

Snowy owls (*Nyctea scandiaca*) respond to a tundra crawling with lemmings with a massive breeding effort (Pitelka *et al.*, 1955), laying clutches of as many as ten eggs and rearing most of the young that hatch. The result is that there are very many owls left to roam the tundra during the following winter, far too many for the dwindling food supply of vanishing lemmings. The owls wander away from a North that can no longer sustain them, and we see them in temperate countries as big white oddities. Perhaps most die there.

Pomarine jaegers (*Stercorarius pomarinus*) seem to be the best adapted predators of all. The adults feed themselves out at sea like other long-lived seabirds (Chapter 11) and are safe from hunger whatever the lemmings do. They use lemmings only to raise young and they breed only when there are enough lemmings to do so. They have the conservative attitude to size of clutch of many of their seabird kind but they respond to changes in lemming numbers with changes in size of breeding territory (Chapter 12). In a lemming high territories are small and the jaegers breed close together. They then take a heavy toll and can be seen killing lemmings with energy, apparently more even than they or their young can eat (Pitelka, 1955). Where they breed the jaegers must exert something of a stabilizing influence on lemming cycles, their effect being massively density dependent as they act with a type 3 functional response from a platform of stable numbers (Chapter 10).

Some of the larger mammalian carnivores like wolves may respond to lemming and hare highs by switching from their more usual food to the rodents. The same is true for hawks of various kinds. All these, like the jaegers, come in to hunt from a secure base of alternate food. No doubt they have a heavy impact on lemming and hare numbers, although this is not the impact of the coupled oscillation.

Multiple Components Hypotheses

Small mammal cycles can be viewed as community properties, allowing models that examine the effects of feedback between several components of the tundra ecosystem at the same time. In one model Finerty (1980) uses loop analysis to construct a scenario for the lemming cycle, largely dependent on the fact that the lemming environment can be seen to be patchy. Tundra, for instance, has good lemming range and poor lemming range. The good range will provide patches where lemming survival is highest at a crash. As recovery begins, numbers can build up in these patches of good habitat and migrants will leave for second-rate areas. If wandering predators are about, the second-rate regions with their new arrivals will provide good hunting grounds and they may be denuded of lemmings almost as fast as they arrive, thus making them near perfect population sinks.

Within the constraints of Finerty's model this system of generating population highs in good patches, even as lemmings are removed in second-rate patches, does result in oscillating numbers of lemmings. Whether the addition of more realism to the model, like selection for lemmings of different behavior or invoking other causes of mortality, would fail to dampen the oscillations is perhaps questionable. Furthermore is the matter of timing—because the fact that lemming highs are cyclic as well as periodic still must be accounted for in a multiple components model just as in more simplistic models. Finerty suggests that a timing mechanism may be found in the rate of regeneration of nutrients, since this is a gradual process that should proceed at a constant rate, though it is not easy to see how this hourglass process actually would work in practice. A further difficulty with all multiple components models is that many different species of small mammals cycle with roughly the same periods. The patches in tundra plant communities of good and bad lemming range, for instance, may be quite irrelevant to different herbivores. It must be true, however, that an accurate predictive model of any small mammal cycle will have to take into account all the multiple components of community structure and function.

Logistic with a Lag

The simple logistic equation can be modified to yield a population history that is cyclic about a mean at equilibrium. This makes it possible to imagine the primary oscillations of lemmings or hares to be intrinsic properties of their own populations. A uniform tundra without patches or predators would, on the assumptions of this model, still host fluctuating numbers of lemmings.

What is needed to make an isolated population oscillate is to introduce into the logistic equation (Chapter 6) a time lag, τ. We write

$$\frac{dN}{dt} = rN\left(1 - \frac{N_t - \tau}{K}\right) \tag{13.1}$$

May (1973, 1976) has shown that the stable equilibrium point predicted by the simple logistic equation gives way to stable cycles once the lag introduced exceeds by a sufficient margin the time it takes the population to respond to pressures at K.

This, like the multiple components hypothesis, is only a theoretical construct and it does not say that real animal populations should behave in this way. But it is easy to imagine causes of the necessary lags, perhaps physiological constraints of the animals, perhaps changes in the forage induced by feeding pressure, or perhaps life history phenomena. May (1976) found that setting the value of the lag constant τ to be 0.72 years generated a four-year population cycle that is hauntingly similar to the cycles of real lemmings (Figure 13.9).

The fit of May's model does not mean that the cycles in lemming numbers are explained by it. It does show, however, that stable limit cycles can be an inherent property of single-species populations. Indeed, oscillations of this kind seem so possible an outcome of real population growth that we might expect to find them to be widespread in nature.

The Stress Hypothesis

When it was first suggested that crowded animals might suffer from stress disease (Chapter 12) it was natural to seek to apply this mechanism to rodent cycles. Indeed, the first suggestion of shock disease in field populations was from studies on arctic hares of peak population years that went into "seizures" and died when caught (Green *et al.*, 1939).

Plenty of evidence suggests that lemmings behave oddly in the weeks before a population crash. In the folklore of Norway are the tales of

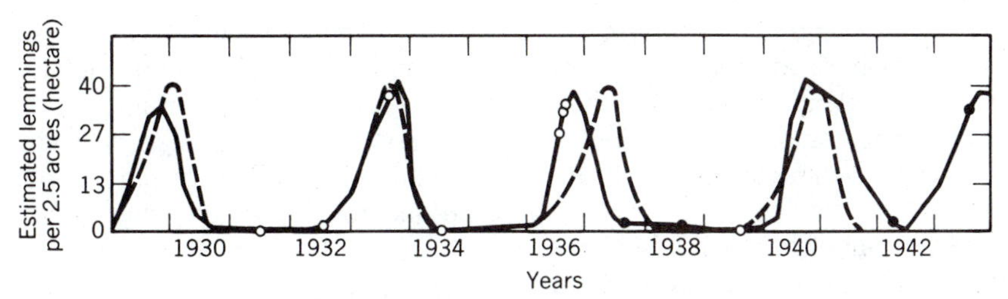

Figure 13.9 Logistic with lag plus lemmings.
The solid curve is the lemming census given in Figure 13.5. The dotted curve is the population history generated by May's (1976) logistic model for a single-species population incorporating a 0.72-year lag in response time. (From Finerty, 1980.)

those intermittent years when the lemmings were "on the march." They appeared in city streets, running about in the broad light of day. They were said to go mad and bite people. They were believed to make suicidal migrations, pouring down the Norwegian mountains like furry rivers, plunging to their deaths in the sea. If these tales were true, then the hypothesis that crowded lemmings suffer stress disease might seem plausible. Accordingly researchers have collected lemmings at times of population maxima and examined them for symptoms of stress disease. Krebs' (1964) autopsies of 4000 lemmings collected during a population die-off were particularly directed to looking for stress symptoms and involved weighing the adrenal glands to compare them with the glands of lemmings from other years. Swollen adrenals were used in some of the original laboratory work on stress as evidence for GAS disease (Chapter 12). But all the adrenals of Krebs' 4000 lemmings were normal and he saw no alternative signs of stress.

These autopsies suggested strongly that the lemmings of a high were perfectly normal and not in any sense "mad." What then of the tales of mass marches to the sea and the rest? Krebs undertook a little historical inquiry that showed how these tales came about, and in particular that mass marches to the sea have never happened. Lemmings of a "high" may be hundreds of times more abundant than at other times, and so people see them. Some lemmings always migrate when populations are dense, for this is part of their opportunist *r* strategy. In mountainous Norway, where lemmings live on mountain tops, all migration must be downhill, thus bringing numerous lemmings into the land of villages and farms where they are not seen in other years. This is all there is to the lemming "march." And the tales of mass suicide into the sea are fictional. Krebs could find no record of an eyewitness account, only long chains of hearsay evidence. The roots of this hearsay are clear enough. Coastal lemmings, at the time of migration across the snow and ice of their spring landscape, often are seen crossing pack ice and some must inevitably take wrong turns and drown. Furthermore, large numbers of lemmings often drown in the spring thaws that flood their burrows so that floating lemming corpses are a common sight. Thus all supposed abnormal behavior of lemmings is accounted for, either as the expected migratory behavior of an *r*-strategist in a variable environment or as fiction.

With lemmings at population highs being apparently healthy, and their behavior shown to be reasonable and adaptive, the stress hypothesis for lemming cycles loses its interest. The existence of the general adaptation syndrome in crowded lemmings, however, has since been demonstrated. Andrews (1968) measured the concentration of adrenocorticotrophic hormone (ACTH) produced by pituitary glands of lemmings in various stages of a lemming cycle. At a lemming high the pituitary was producing many times the amount of ACTH that it did in normal lemming years. There was no change in the visible appearance or weight of adrenals when lemmings crowded, but the glands did increase their output of ACTH all the same. The working of the GAS in crowded lemmings was demonstrated.

But that the GAS is invoked in crowded lemmings does not mean that lemmings are stressed to death; it merely means that lemmings respond to stresses of crowding with the appropriate mechanism, which we recognize as the GAS. Lemmings probably are not so much killed by the GAS as preserved by it. It is prudent to assume that shock disease is not an important contributor to lemming cycles (Krebs, 1964).

The Alternating Genotypes Hypothesis

Chitty (1958) suggested that cycles in small mammals might reflect alternating selection pressures, first favoring individuals with genotypes suited to crowded life and later individuals suited to low densities. In the form of the hypothesis developed by Krebs (1964) it is postulated that two strains of lemmings exist: good fighters who are poor breeders, and poor fighters who are good breeders. Essentially this is to say that different strategies for reproduction are suited to crowds or to low densities. In a crowd the only animals able to breed at all would be those so socially aggressive that they could hold dominance or territory by constant struggle. But although these animals might thus secure a chance to breed, their success might be minimal, having neither time nor energy to allocate to high fecundity or rearing of young. In a crowd, therefore, reproductive success would be low, but the few animals reared would all carry genes for aggression and poor fecundity.

At low densities territory or social breeding status would be easy to attain, letting the high fecundity animals, who were docile but good parents, breed. During times of low population, therefore, genes for docility and high fecundity would spread, even as population grew. When the crowd became dense once more, selection would favor aggressive individuals again. Work with microtine rodents has since shown that selection does work to favor individuals with different social traits according to population density as this hypothesis suggests (Krebs *et al.*, 1973). A tendency towards cycling populations could be one consequence of this.

The Role of Weather

A particular oddity of the arctic small mammal cycles is that they are synchronized over large areas. Ever since the cycles were first discovered this synchroneity has seemed to implicate weather in some way. Now that the autocorrelogram technique has demonstrated that the periodicities are indeed cycles, an external forcing function like weather would have particular appeal. The animals could be thought to be tracking, as well as may be, some cyclicity in the weather.

At one time the hare cycle was examined to see if it correlated with the well-known sunspot cycle, but the two do not in fact correlate, the one being of 10 years (hare) and the other of 11 years (sunspots). The two quickly get out of phase. And there are not other known cycles of weather that will do, particularly nothing like a four-year cycle. It is now quite certain that there is no cyclicity in the arctic weather acting as a forcing function for small mammal cycles.

Yet it is likely that weather does have a role in imposing regional synchroneity on the cycles. The lives of all parts of the tundra system—plants, rodents, and carnivores—must be affected by weather; weather covers large areas, and animals of large areas are in phase. There is no difficulty in this. But weather could have a more important part in the history of cycles as it determines and regulates life history phenomena.

Lemming Life History and Climate

The brown lemming (*Lemmus trimucronatus*) lives where summers are both cool and extremely short, generally having only six weeks that are frost-free. The winter is long, dark, and with prolonged intervals in which air temperatures never rise above freezing, and the ground is permanently frozen below depths of about 20 cm. In winter, of course, the ground is frozen right to the surface, and covered with snow. In spring the top 10 to 20 cm melts, but the ice underneath prevents drainage and the lemming environment changes from dry cold to cold flooding. Many lemmings actually drown in the spring in the puddles of meltwater that cover their range.

Lemmings must be adapted to surviving the long, dark winter and the flooding of spring, and they must make the best of the short summer weeks, after the spring floods are over and while the tundra plants are growing actively. They must also be able to cope with the high uncertainty of arctic weather, with a spring thaw that comes at different dates from year to year, with summers of different lengths, and with an unpredictable array of storms. There should certainly be much density-independent death in the lemming's world.

The essential strategy used by lemmings to cope with this hostile environment is one of high fecundity, short life, and habits of dispersal: the classical strategy of an opportunist or fugitive species. No lemming in the wild has been shown to live longer than a single year. But it is possible for one female lemming to have raised several litters in this short time, even granted that most of the time passes in the arctic night. A female can conceive when she is 27 days old and gestation time is 21 days, which means that from birth to birth can be as little as seven weeks. In a good summer a female can wean two litters, each having between four and nine young. But lemmings can also reproduce in winter, though more slowly. Their strategy is to pass the winter under the snow in tunnels that they make at the ground surface. Thick snow is very important to this strategy, since the temperature in subnivean tunnels is a function of the thickness of snow covering the tunnel. Temperatures may be only a little below freezing in the tunnels, even in midwinter. Below the snow, lemmings eat the stems and roots of tundra plants, using the brown vegetation as a natural hay.

An animal using a large-young strategy, with a generation time of seven weeks, and a clutch size of nine is capable of extremely rapid population growth. Even granted the difficulties of arctic living, this life style is bound to lead to dense local populations from time to time. The lemming's adaptation to this is migratory wandering, particularly in the spring when lemmings can be seen running in the open, a proceeding that grants fitness because there may be more chance of food to raise a litter elsewhere than in the crowded patch from which they came.

Refinements to the basic lemming life history should adapt behavior to the changing seasons. One refinement of importance to the cyclic history of whole populations turns out to be the use of environmental cues to stimulate breeding. If a lemming is to win most fitness she must make the best possible use of the short arctic summer for raising young. But half her reproductive cycle is taken up by gestation, the 21 days in which she must carry embryos before birth. A lemming would make best use of a summer if the time taken for socializing, mate choice, copulating, and carrying small embryos was taken out of late winter rather than early summer. Lemmings, therefore, should mate in late winter, ideally just three weeks before the onset of summer. Unfortunately summer comes at a different time each year, so that classic spring cues like length of day cannot be used.

Mullen (1969) showed that at Point Barrow the cue used to initiate breeding in late winter was an early spell of slightly warmer weather that anticipated the arrival of summer by about a month (Figure 13.10). There was no rhythm or cyclicity to this early warm spell, it was just the first random one of spring. The lemmings were turned on by it and mated, which is adaptive behavior that probably gains more fitness through producing young earlier in the summer than it loses through the occasional disaster when young are born too early.

If the cue to begin breeding is usually set by the weather, not only for lemmings but for other cyclic animals as well, a powerful synchronizing factor is revealed. All the animals within the sway of a patch of weather will have their life histories synchronized and the individual booms and busts of local populations will march in step. Huge areas of the arctic commonly lie under a single air mass

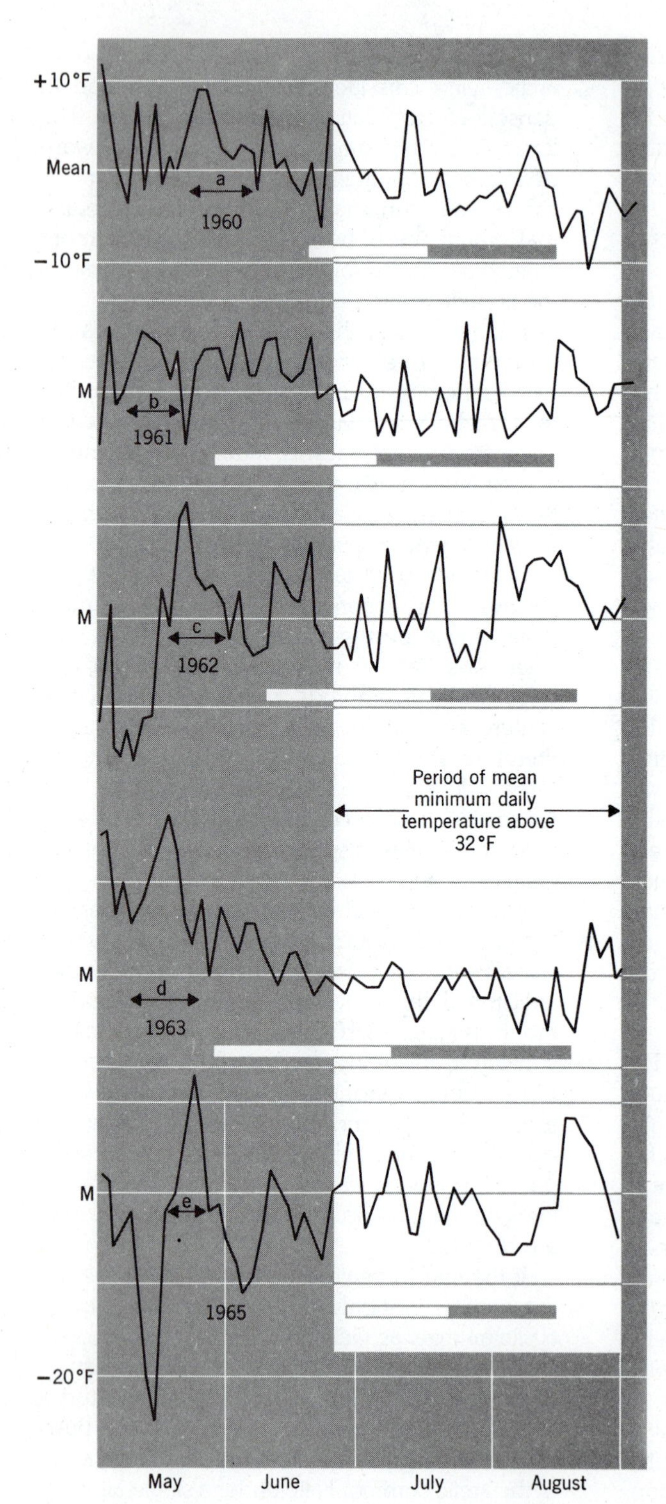

Figure 13.10 Deviations from the mean minimum daily temperatures in the spring at Point Barrow, Alaska.
Mullen suggests that the "warm" snaps marked a to e are cues that bring the lemmings into breeding condition. The grey horizontal bars were the observed breeding periods. These ended at about the same time each year, the different lengths being set by the onset of breeding. A broad correlation of the onset of breeding with the spring "warm" snap is suggestive. (Redrawn from Mullen, 1969.)

(Chapter 14). Use of temperature or some other parameter of weather as a cue for the onset of breeding, therefore, helps explain the synchroneity of population cycles over large areas.

Life History, Periodicity, and Timing

That lemmings are *r*-strategists inherently likely to experience rapid population growth in favorable circumstances seems clear enough. But it may be that all the herbivores at the base of food chains of cycling small mammal populations are likewise *r*-strategists adapted to strongly fluctuating environments. If the argument that local population events can be synchronized by weather has any merit, then it seems reasonable to expect simultaneous population growth to be frequent in all these species. To explain the cycles it is then necessary to explain the population die-offs and the timing.

Possibly the most important life history property contributing to population die-off is that these animals are naturally short-lived. Lemmings, for instance, apparently cannot live longer than about a year even in favorable circumstances. These short lives are, of course, predictable properties for highly fecund *r*-strategists (Chapter 11). But a consequence of short lives is that the failure of a breeding effort will have catastrophic consequences for a population as adults proceed to die without offspring to replace them. If whole populations have their breeding efforts synchronized by a cue from the weather, it must sometimes happen that the breeding efforts of all individuals living over a large expanse of real estate fail in synchrony. The result is the rapid or sudden decline of a whole population.

Once a population is in decline because a cohort of adults is not replaced by reproduction, then predators and other sources of imposed mortality will become particularly important (Chapter 10). The ratio of predators to prey becomes fatal to the few prey remaining and shortly to the predators also. Consideration of life history phenomena of short-lived *r*-strategists, therefore, leads to an understanding of why dense populations of large areas can vanish so quickly. The prime source of "mortality" probably is failure to breed; the adults that were unsuccessful at breeding die without replacements, and the dwindling survivors are then exposed to excessive predation.

Timing of the cycles is still the most puzzling thing about them. Synchronized periodicity seems understandable from the working of any or all of the mechanisms proposed that do not involve true coupled oscillations. But all these "uncoupled" mechanisms require timers if cycles are to result, and any timer we propose must work to produce either 4-year cycles or 10-year cycles with no other intervals apparently in fashion.

Recent attempts to find timers have dwelt on various aspects of forage recovery, either the supply of nutrients or actual plant growth (Finerty, 1980; Schultz, 1969). Replacement of vegetation and soil nutrients must be slow in the arctic or boreal ecosystems in which the cycles are most prominent. No mechanisms for generating 4-year or 10-year cycles from the slow recovery of vegetation seem to have been put forward, however.

It may be that the required timers are themselves included in the life history phenomena of these animals. If failure to breed coupled with short life is the ultimate cause of population die-off, as we suggest, then probable limits to the time taken for population collapse are established. The fecundity of the survivors (or immigrants) in the next breeding seasons, together with generation times, likewise set probable limits to the time required to build a new population high. Thus the total time required for population collapse followed by replacement is under constraint. The life histories perhaps provide the lags required in May's (1976) logistic with a lag model (page 485). The required "lag" is set by generation time and length of life responsible for the rates of population growth and collapse. If very short life and generation times produce a lag that generates 4-year cycles it is at least possible that the next longer life or generation times made possible by life in these highly seasonal environments yield the lag appropriate to a 10-year cycle.

A Lemming Farewell

A lemming high, like all temporary high populations of rodents, is an inevitable consequence of its opportunist life history, together with the fluctuating weather or resources on which it depends. As the lemming populations swell to their peak, the predators multiply; some disease may well appear; the lemmings begin to destroy their range; they are intolerant of low status animals, which must wander down into Norwegian valleys or onto the drift ice to drown at sea. All these things represent density-dependent death, and collectively they must kill many lemmings. Almost certainly these means of destruction will be sufficient to devastate the lemming populations unless something else happens to remove the lemmings first. The most likely such event is a simple failure of lemmings to replace themselves through an unsuccessful breeding season, in which case the predators and other agents of destruction will be in a race with old age to see who can carry off the lemmings first. But the lemming populations of a high cannot last, and they do not. Soon there are only a few survivors living in the best patches of habitat, overwintering and breeding in peace under the snow. The jaegers and the owls depart. The foxes and the weasels starve. The tundra grows up green again. And the journalists turn to embroidering other tales.

PART THREE

ECOSYSTEMS

PREVIEW TO CHAPTER 14

A vegetation map is also a map of climate. This was perhaps the first observation of the science that was to be called ecology. Climatologists tested the hypothesis that distinctive vegetation types called *formations* were set by climate by using formation maps as the basis of climatic maps. Later, ecologists came to understand the shapes of plants in the great formations as adaptations to prevailing temperature, humidity, or seasonality. The formations came to be called *biomes*, and the adaptations of animals in them were determined by both climate and the prevailing plants. Both climates and biomes grade into each other except where geography or the climatic system itself forms a physical divide. In eastern Canada the southern edge of the arctic air mass has a distinctive edge at different latitudes in summer and winter, yielding the disjunctions of the arctic tree line and the south edge of the boreal forest. The primary patterns of climate are caused by fluid flow on a spinning earth differentially warmed at the equator. Heat is distributed from equator to poles both by winds and in ocean currents. All fluid flow acquires angular momentum because of the spin of the earth, a result described as Coriolis force. As a result the earthly climate is divided into large patches or cells, wet where air ascends, dry where it descends. Ocean currents, driven by winds, are deflected by Coriolis force into great gyres around ocean basins, causing different climates where they approach continents and where they depart, thus superimposing local climatic patches on the global system. Finally, mountain ranges bar the way to moving air masses, dividing the patchwork of climate into still more parts. This final pattern is closely reflected in the pattern of biomes. But climate and biomes are transitory, the familiar patterns of the contemporary earth having developed in the 10,000 years since the last ice age. In glacial times sea level was one hundred meters lower than now, as water was withdrawn from the oceans to build ice sheets, and the climatic system was compressed towards the equator as arctic air masses waxed with the ice sheets. Plant communities and biomes were different in almost every part of the earth and modern biogeography results from building of new communities in the differently arranged earth that followed the ice age. The progression of biome types with latitude is roughly mirrored as a sequence of life zones up the sides of high mountains. Like the biomes, the life zones on mountains blend together and are not discrete communities. Individual species populations may be sharply disjunct with elevation on mountains as a result of character displacements but whole communities always merge except where such physical barriers as a cloud base are recognized by many species independently. Apparently discrete bands of vegetation visible on some mountains result partly from an optical illusion as the eye picks out noticeable layers and fails to see so clearly the regions of overlap. Maps of biomes made by ecologists essentially plot regions of common adaptation of plants and animals regardless of the relatedness of individual species. A quite different geography is used by evolutionary biogeographers, who map together into *realms* or *regions* plants and animals showing obvious signs of relatedness, regardless of adaptation.

Thus ecology, the most evolutionary of sciences, began with a conflict with evolutionists over the most useful way of making biogeographic maps. On a geological time scale the distribution of continents, sea, and climates has varied as continents have drifted. This continental drift has now been established by geological reasoning and the mechanism, plate tectonics, generally understood. Many species' distributions are explained as resulting from ancient movements of continents. But the distribution and abundance of life on the contemporary earth have largely resulted on a shorter time scale than plate tectonics, as species populations have moved in response to changing environments of glacial cycles. On time scales of a few thousand years the physical conditions for life are subject to almost constant change, requiring that the composition of communities change continually also.

CHAPTER 14

CLIMATE AND BIOGEOGRAPHY

A prime observation of biogeography is that vegetation of different parts of the earth looks different. Wet equatorial regions have *tropical rain forest*; temperate latitudes have *deciduous forest*; in more northerly places with long, cold winters are forests typified by coniferous trees, the *boreal*[1] *forest*; and further north still into the arctic are treeless expanses where plants are either low shrubs or herbs, the *tundra*. Simple descriptions of these vegetation types rely on plant *form*, whether tree or herb, whether the vegetation is layered or simple, whether trees are spreading or cone-shaped, seasonally deciduous or evergreen. The main vegetation types of continental regions, therefore, are called FORMATIONS.

The five formations from rain forest to tundra are arranged around the globe by latitude in rough bands from equator to arctic regions. Superimposed onto this pattern, however, are other *formations*. In hot, dry places are *deserts* or *savanna formations*, or *prairies* and *steppes*, and elsewhere are tough, evergreen bushlands like the chaparral of California called the *broad-leaved sclerophyll formation* or the *moist coniferous* forests of huge Douglas firs along the American northwest coast. Other formations can be recognized, but most of the vegetation of the earth can be assigned to one of the above GREAT PLANT FORMATIONS.

[1]In Greek and Roman myths *Boreas* was the god of the north wind.

The vegetation map of the earth included in most atlases is a map of formations (Figure 14.1). These maps do some violence to reality, for they show a world more neatly parcelled out between formations of plants than it really is. Map-makers draw boundaries at convenient places, taking advantage of an isthmus, mountain range, or tree line, where such exist, drawing lines where best they may across regions of transition. Nevertheless the maps do reveal that immense continental regions support vegetation of characteristic form. This is the first datum of community ecology, suggesting as it does that different rules of existence prevail in different parts of the earth, rules that first determine the forms of plants and consequently the terms of existence for animal life also.

THE CLIMATIC HYPOTHESIS FOR VEGETATION MAPS

It can be argued that the real beginning of modern ecology came in 1855 when Alphonse de Candolle[2] put forward the hypothesis that life

[2]Alphonse de Candolle, when in charge of the Paris museum collections, made the last attempt by one person to describe all plant species known to science, resulting in his *Prodromus* of 1824–1873. The formation data crossed his desk in the process.

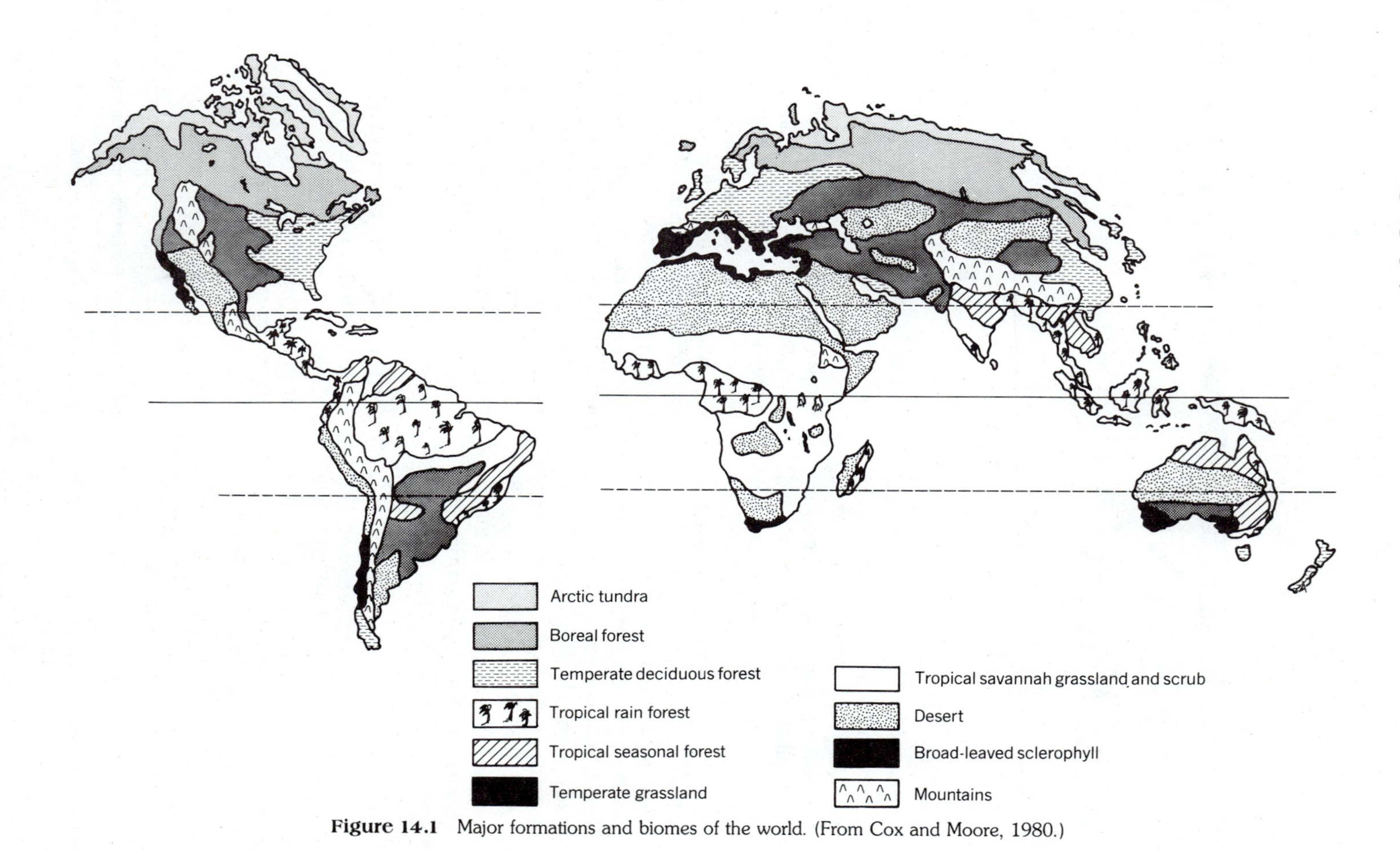

Figure 14.1 Major formations and biomes of the world. (From Cox and Moore, 1980.)

forms of plants, as described by a map of plant formations, were set by climate. He suggested that the most easily measured parameters of climate, moisture and temperature, should explain a formation map. Desert formations, and perhaps grasslands, were set by water shortage, but the latitudinal arrangement of other formations suggested temperature. Eventually de Candolle (1874) developed this hypothesis to the point of saying that there must be critical changes in the heat regimen at particular times of the year that accounted for the changes between one formation and another.

This refined form of de Candolle's hypothesis not only accounted (in a very general way) for different plant forms but also implied that transitions between one formation and another might be relatively abrupt. He predicted *critical changes in heat regimen at particular seasons*. This is an argument for environmental thresholds, allowing comparatively rapid switches from one region to another. The test of this hypothesis proposed by de Candolle was to see if the boundaries between formations shown on the vegetation map of the earth could be matched with ISOTHERMS (lines of equal heat), and he suggested what isotherms should be used. Data from world weather stations in 1874 were too few for this test to be tried, but de Candolle's isotherms make an interesting map when plotted using modern data (Figure 14.2).

A better test of the de Candolle hypothesis

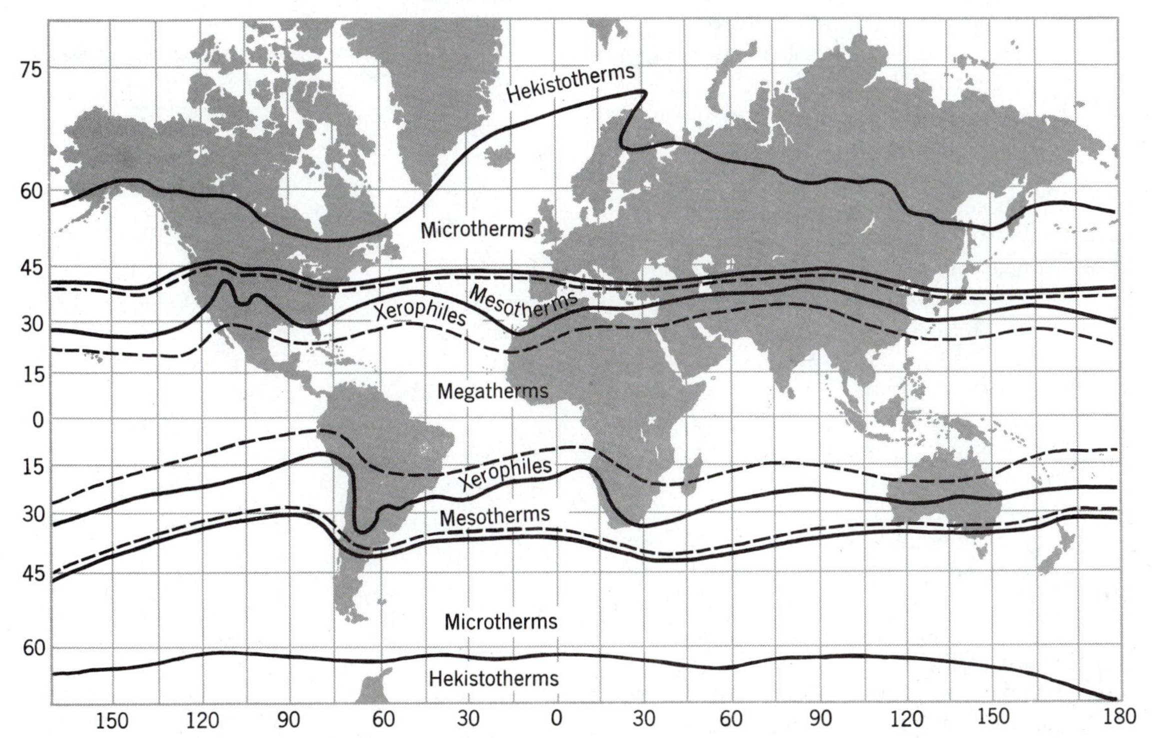

Figure 14.2 De Candolle's division of the earth according to the properties of plants. The broad extent of formations was thought to be set by temperature, except that shortage of water set the limits of plants of hot deserts (the xerophiles). De Candolle's ideas foreshadowed the attempts of ecological biogeographers like Allen to show that the forms of life on earth were distributed in latitudinal bands. The hypothesis that formation boundaries could be correlated with temperature and water led to the use of vegetation maps as the basis for climatic maps. The isotherms shown here are those suggested by de Candolle but are plotted from modern meteorological data.

came when Vladimir Köppen (1884, 1931) based climatic maps of the earth on vegetation maps. Köppen set out to classify climatic types in a world over much of which climate data were almost lacking. He used maps of the great plant formations as his base maps, then assigned to each plant formation the climate that seemed appropriate to it. De Candolle had ordered the formations according to their supposed heat-loving, or drought resistant, properties; rain forest plants were megatherms (most-heat), deciduous forests had mesotherms (middle-heat plants) *et cetera* (Table 14.1). Köppen elaborated on the temperature and precipitation required by each formation of plants, and called the climates A, B, C, *et cetera* (Table 14.2). He then took the boundaries shown on maps of plant formations and used those same boundaries to delimit climates, the result being the familiar map of global climates (Figure 14.3).

Thus Köppen, and the other founders of climatology, mapped the plants and called them climate. Maps of vegetation and climate in standard atlases, at least until satellite maps of weather were available, were the *same map*. But these climatic maps proved generally accurate, describing within tolerable limits the actual climate of different parts of the earth. The de Candolle hypothesis that the world was set out in a patchwork of different plant formations whose domains depended on temperature and precipitation, therefore, had been tested and found to be correct.

This exercise necessarily raises questions about the reality of the boundaries between formations or climatic regions shown on the maps. Climate should be fluid, so that there should be no discrete edges between its patches. Likewise formations should blend as the driving force of climate blends with the next climate over great distances. And yet it turns out that some patches of climate do have edges that oscillate around well-defined average positions, resulting in rapid transitions between plant formations. The match between fluctuating climate and migrating plants has been measured directly for the boreal forest of eastern Canada.

Table 14.1
Correspondence between Classifications of Life Forms and Climate

de Candolle	Raunkiaer Phytoclimate	Köppen Climatic Type
Megatherms	Phanerophytic	A
Xerophiles	Therophytic	B
Mesotherms	Chamaephytic	C
Microtherms	Hemicryptophytic	D
Hekistotherms	Cryptophytic	E

Table 14.2
The Köppen Climatic System

Zone	Symbol	Explanation
Tropical rain climate	1. Af	Tropical rain forest climate
	2. Aw	Savanna climate
Dry climate	3. BS	Steppe climate
	4. BW	Desert climate
Warm temperate rain climate	5. Cw	Warm, dry-winter climate
	6. Cs	Warm, dry-summer climate
	7. Cf	Moist temperate climate
Boreal or snow-forest climate	8. Dw	Cold, dry-winter climate
	9. Df	Cold, moist-winter climate
Snow climate	10. ET	Tundra climate
	11. EF	Perpetual frost climate

Symbols: B = inclined to be dry: S = steppe climate, W = desert climate. E = warm only in summer: T = tundra climate, F = frost climate. A, C, and D = location and timing of the dry season: s = main dry season in summer, w = main dry season in winter, f = perpetually moist (rain in every month).

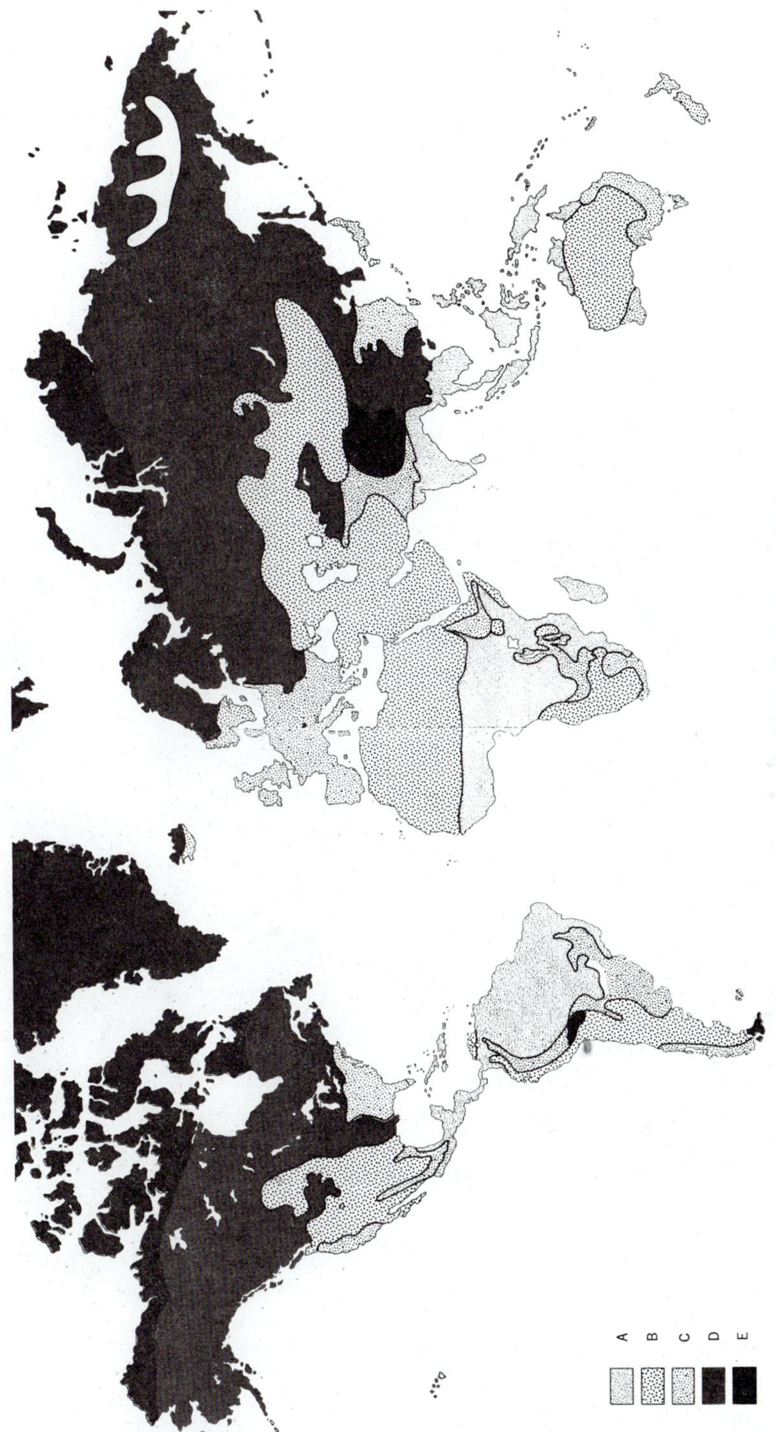

Figure 14.3 Climates of the earth mapped under the Köppen system.
Köppen worked on the hypothesis that the boundaries between different kinds of vegetation were set by climate. His climatic maps were essentially vegetation maps, and the similarity between this figure and a formation map (Figure 14.1) is obvious. But climatologists found the Köppen system useful, suggesting that the earth might really be divided into vague climatic regions each of which resulted in a special kind of vegetation.

BOREAL FOREST AND THE ARCTIC FRONTAL ZONE

Bryson (1966) mapped the average position of the *arctic frontal zone* over eastern Canada in both summer and winter. The zone marks the transition between the cold arctic air mass and warmer air coming in from the Pacific. The front moves, both with the random perturbations typical of all weather and more predictably with the seasons, going south in winter and north in summer. Bryson mapped the mean positions from a 10-year accumulation of measurements of air temperature and of wind speed and direction. He found that the frontal zone in summer coincided with the arctic tree line and that in winter it coincided with the southern edge of the boreal forest (Figures 14.4 and 14.5).

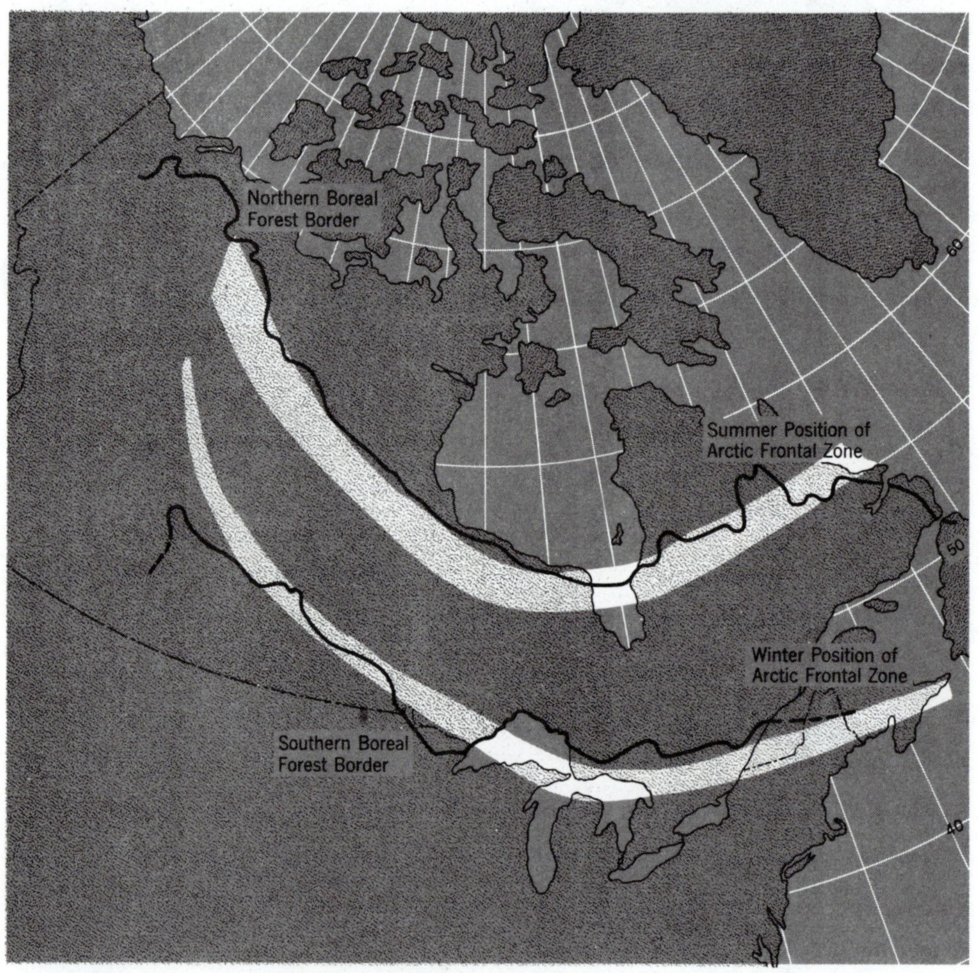

Figure 14.4 Forest boundaries and air mass fronts in eastern Canada.
The tree line and the southern edge of the boreal forest are perhaps the most distinct of continental vegetation boundaries. The plot shows that they closely follow the mean positions of the front of the arctic air mass at different seasons of the year. This study of Bryson provides direct evidence that extents of the Canadian boreal forest and tundra are directly influenced by the sway of an air mass. (Redrawn from Bryson, 1966.)

But how does vegetation detect the 10-year mean position of a front that is continually moving? The answer to this riddle probably lies in the slow generation time and rate of spread of the major units of the formations, the trees and other perennial plants. Ten years is a short part of the generation time of a forest. No doubt the formation boundaries are not really static but are waxing and waning at a rate that produces changes too slow to be noted in a human lifetime. Fluctuating boundaries are in dynamic equilibrium with more swiftly fluctuating fronts. What we see as a discrete boundary is just one temporary position in an always moving edge.

The really attractive thing about this investigation of arctic air in Canada is that it shows that climate can exist as a distinct patch roughly coincident with a patch of vegetation, just as botanists long suspected it must. The earth is covered with a patchwork quilt of air masses (Figure 14.6). This results in a patchwork quilt of plant shapes as the plants of each climate have shapes suited to balanicng heat budgets, conserving water, and fixing carbon dioxide in the different regimens of temperature and moisture (Chapter 5). This patchwork of climate results because the rotating earth is preferentially heated at the equator.

WINDS ON A SPINNING EARTH

The earth is a spinning sphere, the equator of which is just a few degrees away from being parallel with the plane of the earth's orbit around

Figure 14.5 The arctic tree line.
The forest on the right, the tundra on the left, and a good enough separation between them to please any map maker. In such places formations almost seem to face each other like rival armies at a disputed frontier. This part of the tree line is in Alaska. The tall trees are black spruce (*Picea mariana*), but there are several kinds of broad-leaved bushes in the forest also. The lines of small bushes in the tundra will be growing along drainage channels, and the open tundra, which looks grassy, will be humpy, a mixture of grass and sedge tussocks with prostrate woody plants like dwarf birch and blueberry.

Figure 14.6 The pattern of the earth's climate revealed by clouds.
The photograph was taken by an earth satellite crossing the equator over Brazil in November 1967. Sets of such photographs may show directly the patterns of climate that were earlier mapped by plotting the approximate areas of pieces of vegetation. Note particularly the mass of cloud that hangs over the tropical rain forest of the Amazon basin (NASA photograph).

the sun (Figure 14.7). This means that the intensity of irradiance per unit area is always less at high latitudes than it is at the equator. Low latitudes, therefore, get more heat, this being the primary inequality in energy supply that imposes a patchwork of climates on the earth.

The second inequality is in the distribution of day and night. All parts of the earth have exactly the same amount of night and day in the course of a year—six months of each. But at the equator light comes in alternations of a 12-hour day and a 12-hour night whereas at the poles in alternations of a six-month day followed by a six-month night: low latitudes have more constant heat. That low latitudes have more heat, and more constant heat, than high latitudes means that low latitudes always are warmer than the polar regions. This results in a mass transfer of heat from equator to poles in fluid flow, both by air movements and by water movements. A significant part of the heat transfer is actually done by movement in the oceans, since the specific

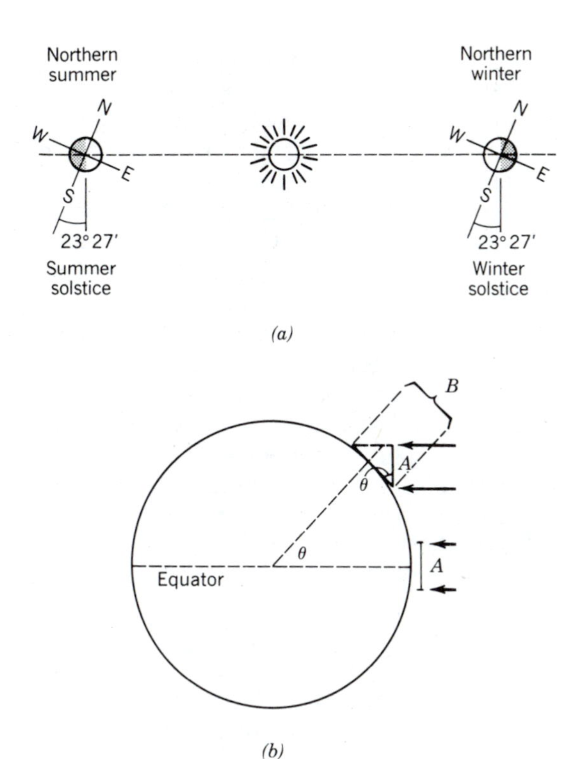

Figure 14.7 (*a*) Relative positions of earth and sun at summer and winter solstice. The sun is shown as effectively a point source. The plane of the earth's rotation is 23°27′ from being parallel with the plane of the orbit. (*b*) The effect of latitude on insolation received per unit area at the equinox (when the sun is over the equator). The sun's rays are assumed to be parallel. At high latitudes the same radiation received by area A is spread over the larger area B such that $B = A/\cos\theta$ where θ is the latitude in degrees.

heat of water is so high. But movements in the fluid air are more rapid, giving rise to the spectacular properties of climate.

Most of the energy of sunlight penetrates the atmosphere and strikes the surface of the solid or liquid earth, where the high-energy wavelengths of light are absorbed and re-emitted as radiant heat (Chapter 3). The primary heating of the atmosphere, therefore, is from below, from the ground and ocean surfaces. The warmed air rises; as it rises it expands; as it expands it cools.

Cooling of rising air as it expands is the reason why mountain tops are cold. The cooling results because packets of expanding air must do work to push aside other packets of air that are pushing back. Energy to do this work can come only from the air itself and it loses temperature accordingly. Since no external source of energy is involved the cooling is called ADIABATIC (from the Greek meaning "impassable"). The cooling is a simple function of altitude and the ADIABATIC LAPSE RATE in dry air is approximately 10°C per kilometer of elevation. In moist air the lapse rate is less because heat is gained as water condenses and the *adiabatic lapse rate in moist air* is about 6° per kilometer.

The major wind systems of the earth result from the fact that large masses of air around the earth's equator are forced to rise by bottom heating, causing air to rush in from high latitudes to fill the pending void. This of course lowers pressures in high latitudes, which are balanced by descending air. Winds represent the working of a heat-distributing engine. Heat is applied at the equator, the air moves up and then poleward, after which it descends and returns equatorward along the earth's surface. But this simple system is modified in two ways: by some properties of scale and by Coriolis force.

On the real earth the poleward-moving masses of upper air cannot penetrate to more than about 30° latitude north and south. This seems to be due to statistical properties of moving air masses, or what MacArthur (1972) calls "no good reason." The actual pattern that results is given in Figure 14.8. The atmosphere is thicker at the equator than at the poles and a second circulation cell is inserted between 30° and 60° north and south.

An object at the equator spins at about 24,000 miles a day. Farther north or south the object

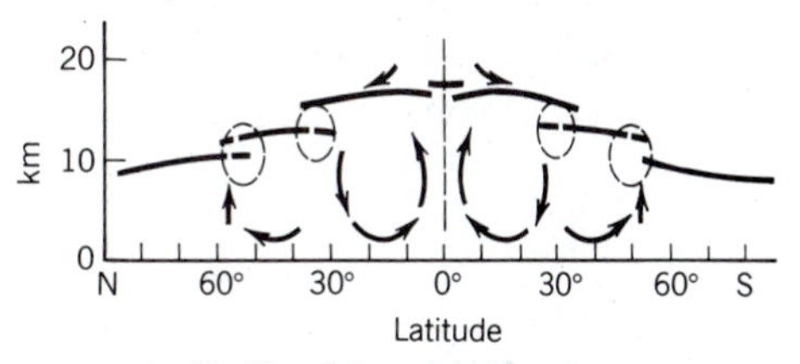

Figure 14.8 Profile of the atmosphere.
The dashed circles are tracks of the jet streams. (From Flohn, 1969.)

would be travelling more slowly as the distance to be travelled for one revolution of the earth is less. If you straddled a pole for 24 hours each foot would have moved perhaps one meter in that time. A more interesting latitude for climate dynamics might be 45°N, where the surface velocity is about 17,000 miles per day. A parcel of air moving due north from the equator starts with a powerful west to east momentum from its initial 24,000 mpd kick, and as it travels into regions where the surface moves more slowly it will veer in the direction of its superior momentum, to the right. This force that diverts the air through its own momentum is called CORIOLIS FORCE.[3]

A parcel of air travelling in the opposite direction, southward, in the northern hemisphere is being thrust into air masses that are moving west to east faster than itself with the result that it will drag its feet, conserving its slower angular momentum and being deflected west. So southward-travelling air in the northern hemisphere will be deflected to the right, just as is northward-travelling air. In the southern hemisphere the logic is the same but the directions are reversed. CORIOLIS FORCE DEFLECTS MOVING AIR OR WATER TO THE RIGHT IN THE NORTHERN HEMISPHERE AND TO THE LEFT IN THE SOUTHERN HEMISPHERE.

Air movements at tropical latitudes are now explained. The surface air that rushes to fill the equatorial void from the north is deflected to the right and becomes the NORTHEAST TRADE WIND. It meets similar air coming from the south that was deflected left, the SOUTHEAST TRADE WIND. But less land is present to slow air with assorted barriers in the southern hemisphere, where most of the time the winds pass over oceans. Accordingly the trades do not, on the average, meet at the geographical equator as you would expect, but somewhat to the north. The convergence, called the INTERTROPICAL CONVERGENCE ZONE (ITCZ), is displaced north of the equator, this being particularly visible over the Pacific Ocean. Along the ITCZ the heated air rises most rapidly, causing high cumulus clouds and the heavy rains typical of Caribbean islands and Hawaii.

Events at latitudes higher than 30° depend on the unexpected sinking of the upper air that closes the circle started at the equator (Figure 14.8). Impelled by the pressure from above, winds from 30° to 40° latitude set off poleward, but they are misdirected by strong Coriolis forces, to the right in the northern hemisphere and to the left in the southern hemisphere. The result is westerly winds in both hemispheres at about 40° latitude (Figure 14.9). In the southern hemisphere, latitude 40° has virtually no land to slow the winds, the result being the *roaring forties,* unceasing winds of high velocity that track the fortieth parallel, round and round the earth.

Upper atmosphere winds along both fortieth parallels are also westerlies, the JET STREAMS (Figure 14.8). The initial poleward movement in the upper air comes about because the at-

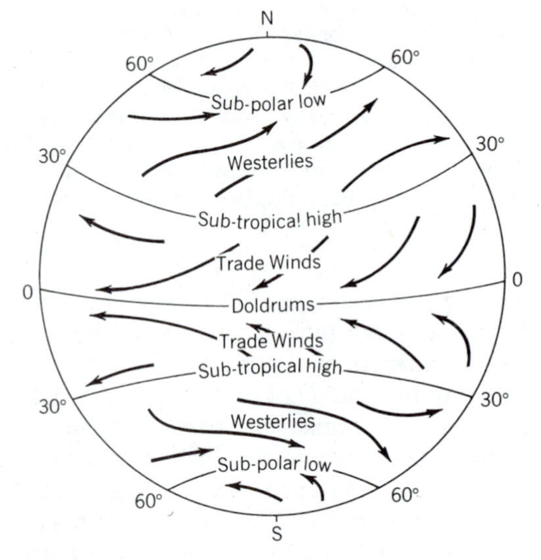

Figure 14.9 Winds at sea level.
Compare with Figure 14.8. The doldrums at the Intertropical Convergence Zone (ITCZ) are shown at the geographical equator as they would be on an ideal earth. The distribution of land and sea actually displaces the circulation system a few degrees to the north. (From Lamb, 1972.)

[3]Named after a 17th-century French mathematician, G.G. Coriolis. More accurately called *Coriolis effect* but the word "force" is in general usage.

mosphere is piled up high at the equator but is thinner farther north. Atmospheric pressure is higher at low latitudes, therefore, and upper air flows poleward to where pressure is lower. Again Coriolis force takes charge of this motion and drives it eastward in both hemispheres, leading to westerly jet streams. Placing an airliner in the jet stream when crossing the Atlantic from America to Europe cuts an hour or more off the flight time.

At the poles are the cold regions where the lack of sufficient irradiance has its own direct consequence for climate. One pole is covered with a floating ice pack and the other by glacial ice. Both sheets of ice act as reflecting mirrors, increasing the albedo and exacerbating the already low heating of the air. Dense cold air sits over the poles, therefore, flexing with the seasons and flowing out as east winds into subpolar regions, the cold east winds that northern Europe knows (Figure 14.9).

THE CIRCULATION OF THE OCEANS

The world oceans are stirred by the winds. A major process is the drive provided by the trade winds that thrust great masses of water before them, piling it up against continental dikes downwind. What this implies is neatly illustrated by the circulation of the North Atlantic, which begins as water is piled up in the shallow Caribbean Sea, where it is heated by the tropical sun. So effective is this piling up of water by the northeast trade winds that mean sea level is a meter or two higher on the Atlantic side of the isthmus of Panama than on the Pacific side.

The water piled up in the Caribbean escapes northward along the American coast as the GULF STREAM. But moving water is deflected by Coriolis force, just like moving air. The Gulf Stream veers right towards northern Europe and provides a warm-water bath for the island of Britain, which accordingly has a much warmer climate than places in Canada well to the south of it. The Gulf Stream goes on veering right to complete a gyre until the water is taken once more by the trades (Figure 14.10). A fact that shows up particularly clearly is the way ocean currents in the southern hemisphere follow the roaring forties in a perpetual circling of the globe.

One of the consequences for land climates of these circling currents is, of course, the warming of places like Britain. But even more interesting is what happens when currents turn back towards the equator along the lee of a continent. Consider the circulation of the North Pacific (Figure 14.10). The warm current flowing north and veering right is the Japan Current that bathes southern Alaska and British Columbia with water that is comparatively warm, which may even be warmer than the adjacent land during much of the year. These coasts accordingly have fog caused in just the same way as the "steam" coming from a warm bath. But then the ocean currents veer right again and down the coast of California, where Coriolis force turns them away from the coast and out to sea. What then happens is that water must be drawn up from below to take the place of that dragged away. This is cold water taken from the dark depths of the ocean. This surfacing of bottom water is called an UPWELLING. It promotes fisheries, because the water is nutrient-rich. But it makes coastal regions dry because the sea is usually colder than the land so that air cannot transfer moisture from the sea.

WHY AND WHERE IT RAINS

Rain falls when moist air is cooled. This will happen when high irradiance of the sea along the ITCZ raises moist air to great heights, resulting first in the condensation that causes cumulus clouds, then in rain. But air is also raised, and causes rain, when winds are deflected by mountain ranges. Figure 14.11 shows the progress of events. Air arriving from a sea journey rises up a mountain, cooling as it goes with the adiabatic lapse rate of 10°C km^{-1} until water condenses. A layer of cloud or rain forms high against the mountain side. The air continues to rise but at the moist air lapse rate of 6°C km^{-1}, crosses the mountains, and begins to descend. On the descent, air warms at 10°C km^{-1}. This warming,

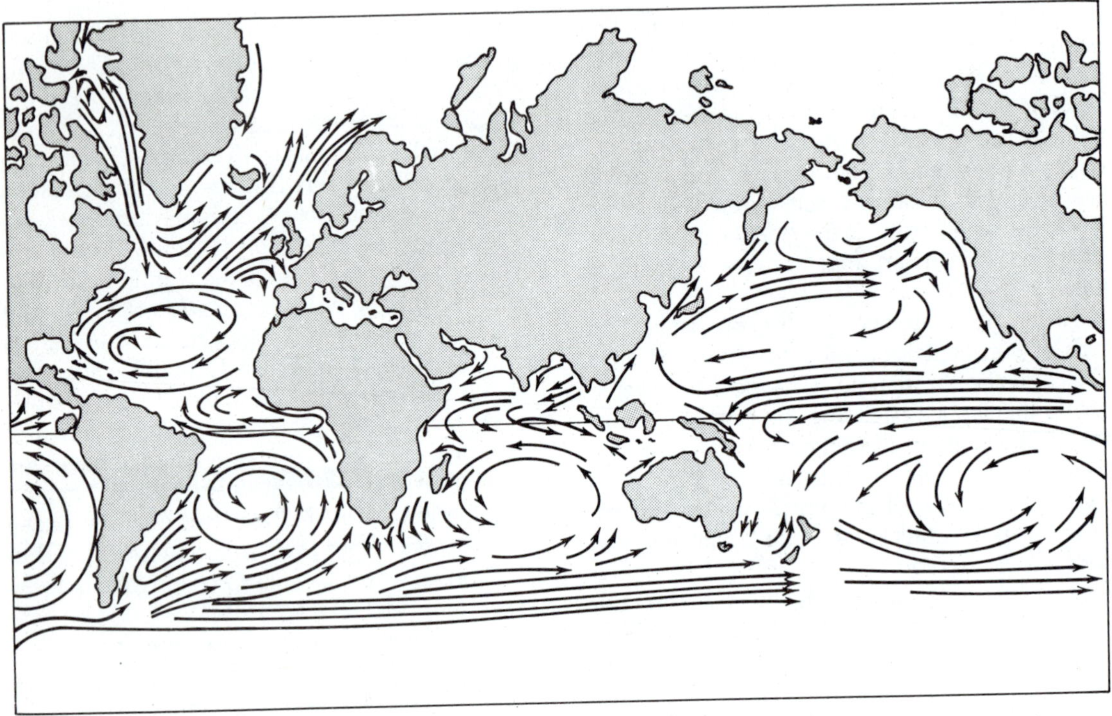

Figure 14.10 Surface ocean currents.
Winds and the Coriolis force move surface water in gyres around the ocean basins, clockwise in the northern hemisphere and anticlockwise in the southern hemisphere. Whether a current is approaching or departing a coast is decisive for the local climate. (From MacArthur and Connell, 1966.)

descending air will take up moisture rather than drop it, giving rise to a RAIN SHADOW on the lee side of a mountain.

Descending air should always absorb moisture rather than drop it. This means that a strip below the 30th parallels should be places without rain because the upper air descends there (Figure 14.8). Major deserts like the Sahara and in the American Southwest lie in this belt. Arizona is not only close to this latitude but is in the rain shadow of the Sierras.

The necessary ingredient for precipitation is rising air, but air can be prevented from rising over large geographic regions by a condition known as an INVERSION. An inversion is warm air floating over cold, a relationship that is physically stable (we say "inverted" because the usual pattern is cold air over warm, since the atmosphere is heated from below). The commonest form of inversion results when winds blow across cold ocean, because then the moving air is cooled from the bottom instead of being heated from the bottom. The cold winds blow on, bottom heavy, with little turbulence. They are actually comparatively dry, even though crossing ocean, because of the low temperature of the surface wind. Land under such an inversion can be desert.

The Galapagos Islands are desert because they are under an inversion, even though at the equator. The hot air rising from heated Galapagos rocks causes no more than a thin layer of *stratus* lying at the inversion (Figure 14.12). If the Galapagos Islands were only a few degrees farther north they would get rain from the ITCZ as does Hawaii, but they are merely in the path of stable,

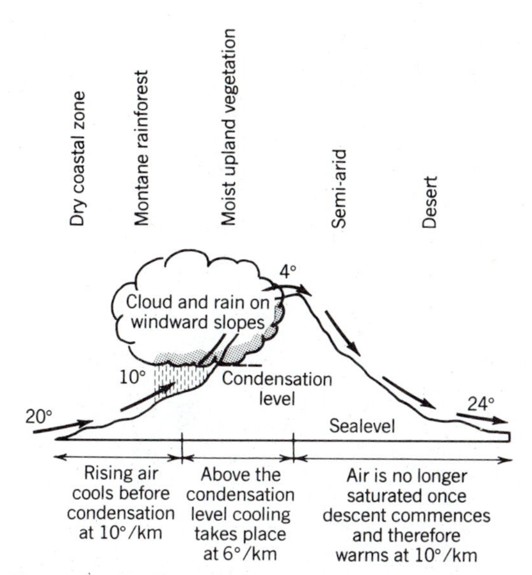

Figure 14.11 Precipitation along mountain ranges. The rain shadow downwind of a mountain results because the descending air, already dry, picks up moisture. (Based on a diagram of Flohn, 1969.)

Figure 14.12 Galapagos island under an inversion. The Island of Floreana (Charles) is low-lying, much of its surface being semiarid. The stratus cloud cap at the inversion may envelop the higher elevations, giving these a moist climate from fog-drip.

inverted air moving towards that convergence and so remain deserts. The most complete deserts known are on western coasts of continents as in northern Chile where cold upwellings along shores force inversions on the local winds, after which the cold, bottom-heavy air crosses hot land from which it actually tends to remove water.

Much of the vegetated landmass of the earth consists, in fact, of continents lying north of 30°N latitude, and in the path of the prevailing westerly winds. A mountain range, such as lines the west coast in America, puts the continental mass to the east in rain shadow. But the American interior gets rain all the same because it is under air masses that twist to the right under the direction of Coriolis force, bringing in moist air from warm seas to the southeast. Both the eastern United States and China are watered in this way, both lying north of the latitude where air descends for "no good reason" (Figure 14.8).

This is a simplified account of the circulation of the atmosphere and the causes of rain (see Lamb, 1972, for a full review) but even this elementary analysis shows that the earth does have blocks of climate of the kinds predicted by a vegetation map. Deserts lie in the southwest of continents and behind mountain ranges; lands along the ITCZ are wet, warm, and covered with rain forest; the northern edge of the boreal forest is the limit of cold, dense, polar air; and so on. Because the earth spins, and is heated unequally, its pattern of precipitation is as given in Figure 14.13. This is so fundamental a pattern to life on earth that geographers still find themselves producing maps of precipitation like this by using vegetation as their indices of precipitation. Cartographers still draw maps of plants and call the result "weather."

GEOGRAPHY AND CLIMATE OF THE ICE AGE EARTH: "IN THE BEGINNING"

For the ecological biogeographer "the beginning" was the last ice age. We can, of course, trace some patterns deep into the remoter past—

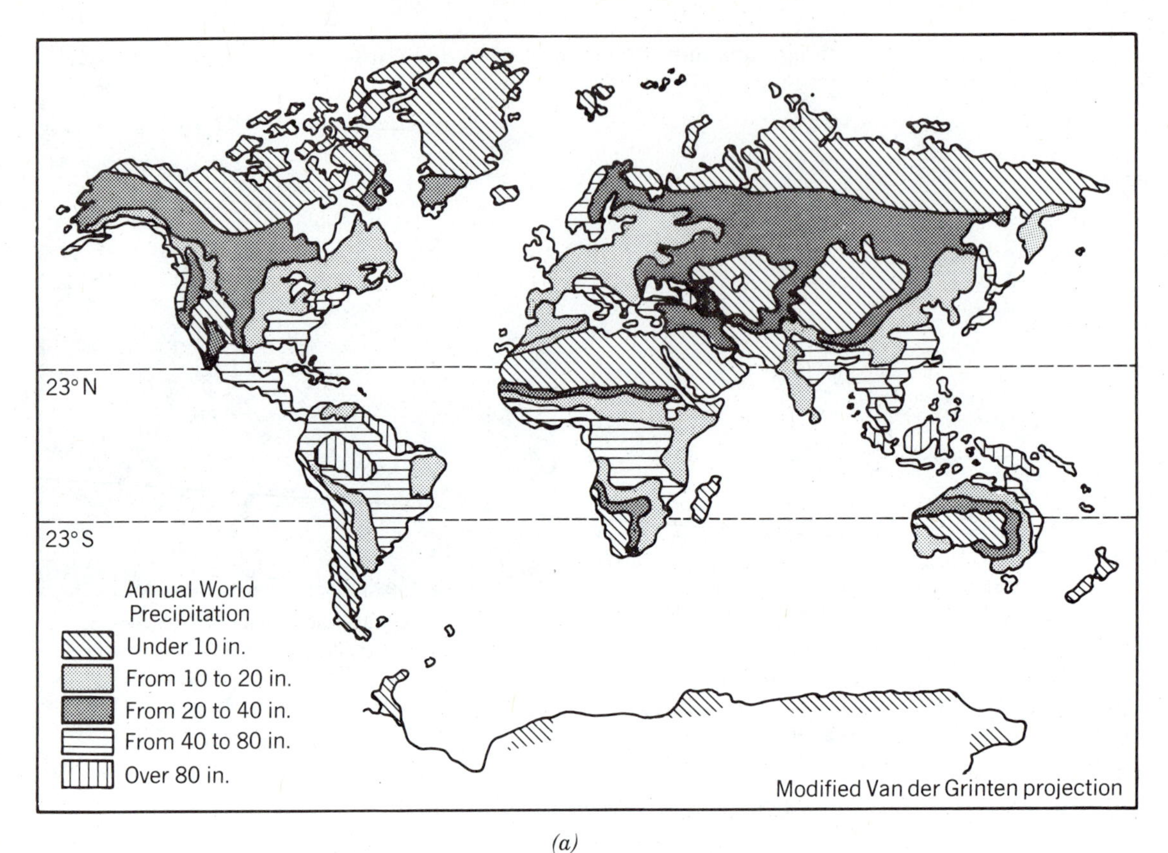

(a)

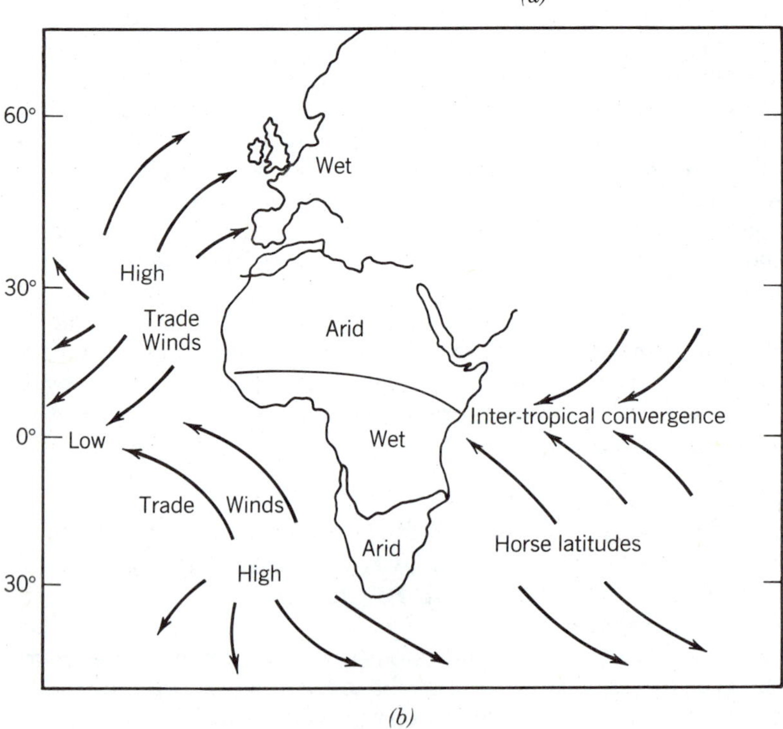

(b)

Figure 14.13 World precipitation.
(*a*) World map. The wettest and driest areas can easily be related to mountain ranges, the map of surface winds, and the map of ocean currents. (From MacArthur and Connell, 1966.) (*b*) Detail of causes of precipitation over Africa. (From Cox and Moore, 1980.)

but that is to talk of what happened in evolutionary or geological time. The distribution and abundance of life on the modern earth was either set in modern times or still holds the imprint of the last great geographic upheaval, the ice age.

The ice age ended only 10,000 years ago: a mere 400 human generations back, or perhaps 40 generations of trees from the climax forest. The ice age was recent, and its world was a very different place. Livable land was different, vegetation was different, and climate was different. Thick ice sheets covered the northern hemisphere down to roughly the fortieth parallel. These ice sheets represented a very large volume of water that was held in cold storage on the land instead of being allowed to run back to the sea. The first consequence of this is that world sea level was about 100 m lower during the ice age than it is now. This is called EUSTATIC lowering of sea level (as opposed to ISOSTATIC lowering caused by the land rising while the sea remains "static"). All shallow seas were drained, connecting continents and islands in novel ways. Alaska and Siberia were fused because the Bering and Chukchi seas flanking Bering Strait were drained. The island of Britain was part of continental Europe. Islands like the Galapagos were more than twice their present size. Many of the islands of the Indonesian archipelago were fused, some being joined to Australia and some to Asia. Changes like these transformed geography, rearranging the world with huge plainslands that have now vanished beneath the sea (Figure 14.14).

The ice had a direct effect on climate. The effect was not so much a cooling of the earth, for mean temperatures at low latitudes probably were only a few degrees different from those of the present day. What happened was that the circulation of the atmosphere was compressed southward as the polar zone of cold, high-pressure air extended south to the fortieth parallel along with the ice. This produced a domino effect on air masses the world over as local patterns of wind and rain changed, often in surprising ways. One likelihood, for instance, is that regular rains did not pour into the Amazon basin from the Atlantic as they do now so that there might have been less rain forest there, even perhaps patches of savanna. This is still a subject of debate and research (Colinvaux, 1979; Prance, 1982). But in Africa it is clear that the present wet climates of the rain forest regions were absent or displaced so that savannas or dry woodlands grew where now there are rain forests (Livingstone, 1974, 1982).

The first models of the climate of the ice age earth are now available and they generate climatic maps like that of Figure 14.15. The data on which the model for this map was built come from using fossils in cores of sediment from the deep sea to reconstruct temperatures at the sea surface, and hence both the circulation of the oceans and the atmosphere (Chapter 21). The map shows the base from which species migrated to fill communities of the contemporary earth over the last 10,000 years.

THE BIOMES AND THEIR CLASSIFICATION

In 1807 the naturalist and explorer Friedrich von Humboldt put together a classification of formations that assigned vegetation to groups typified by a common shape, such as spruce-tree shapes, palm-tree shapes, *et cetera*, and there have been other similar systems though none has come into general use. Later classifications attempted more formal descriptions. The principal characteristics of a piece of vegetation can be shown conveniently with a stylized sketch of it in profile and such drawings are still used widely in descriptive work. Figure 14.16 compares tropical rain forest, deciduous forest, and sagebrush desert in this way: the relative layering in these different formations is brought home clearly enough. This figure is designed to show the differences between disparate bits of vegetation but the same technique can be used to show other things. Figure 14.17 illustrates the conviction that plant formations grade one into another over long clines of territory: the figure includes most of the formations claimed by botanists for the earth and relates them one to another.

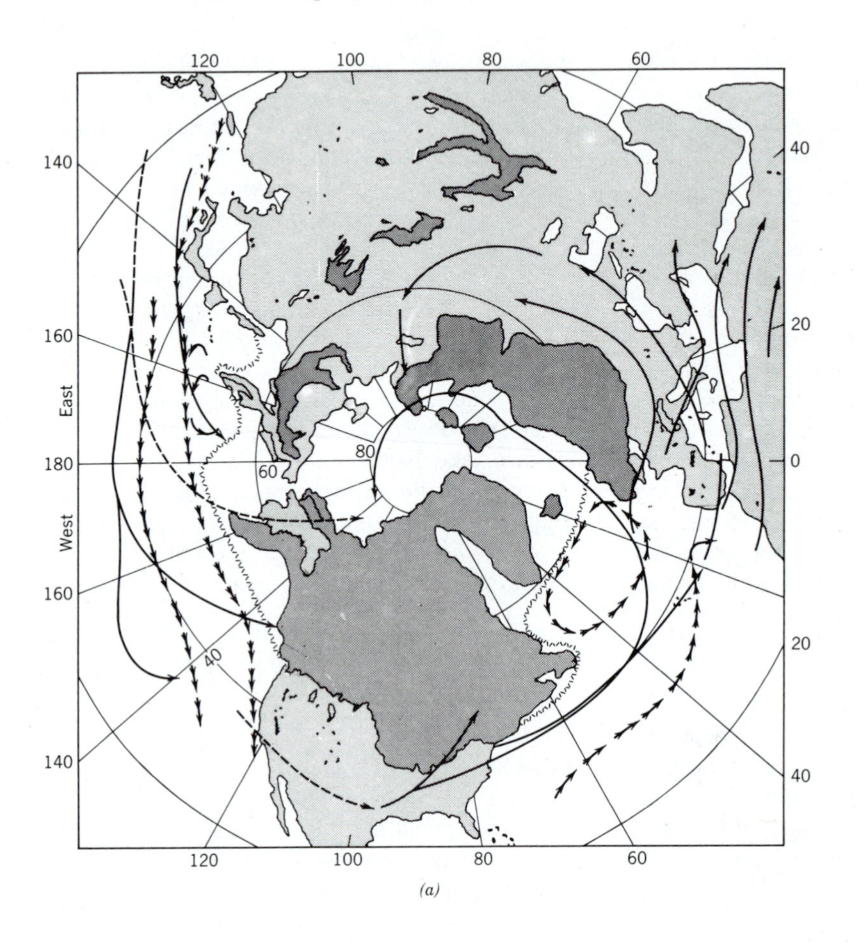

(a)

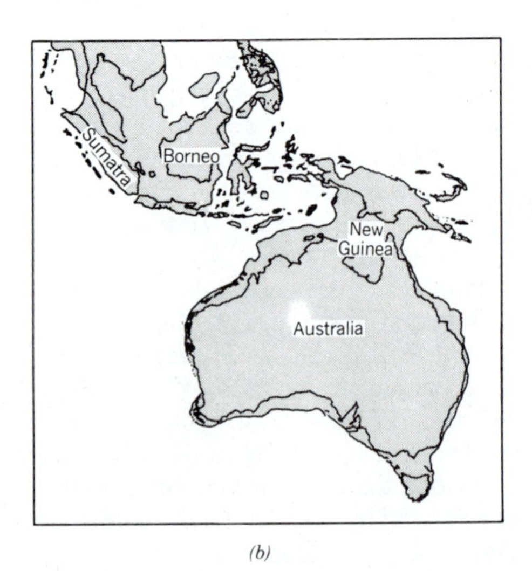

(b)

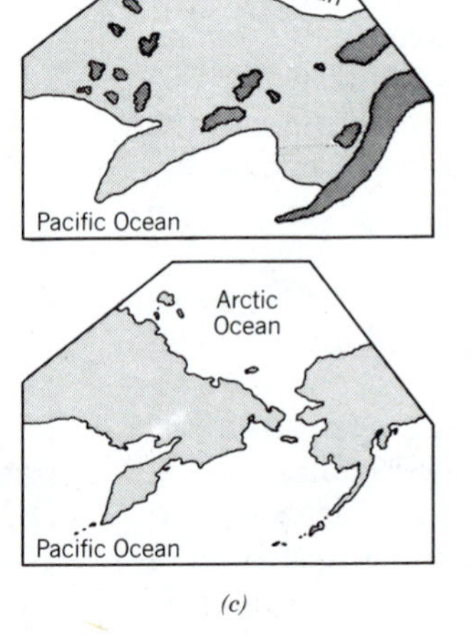

(c)

Figure 14.14 The ice age earth.
(*a*) Glaciation in the northern hemisphere at the last (Wisconsin-Würm) glacial maximum. (From Flint, 1971.) (*b*) Australasia in the last glacial maximum showing coastlines when eustatic lowering of sea level of about 100 m was in effect. (From Quinn, 1971.) (*c*) The Bering Strait region (Beringia) in glacial times and today. (From Pielou, 1979.)

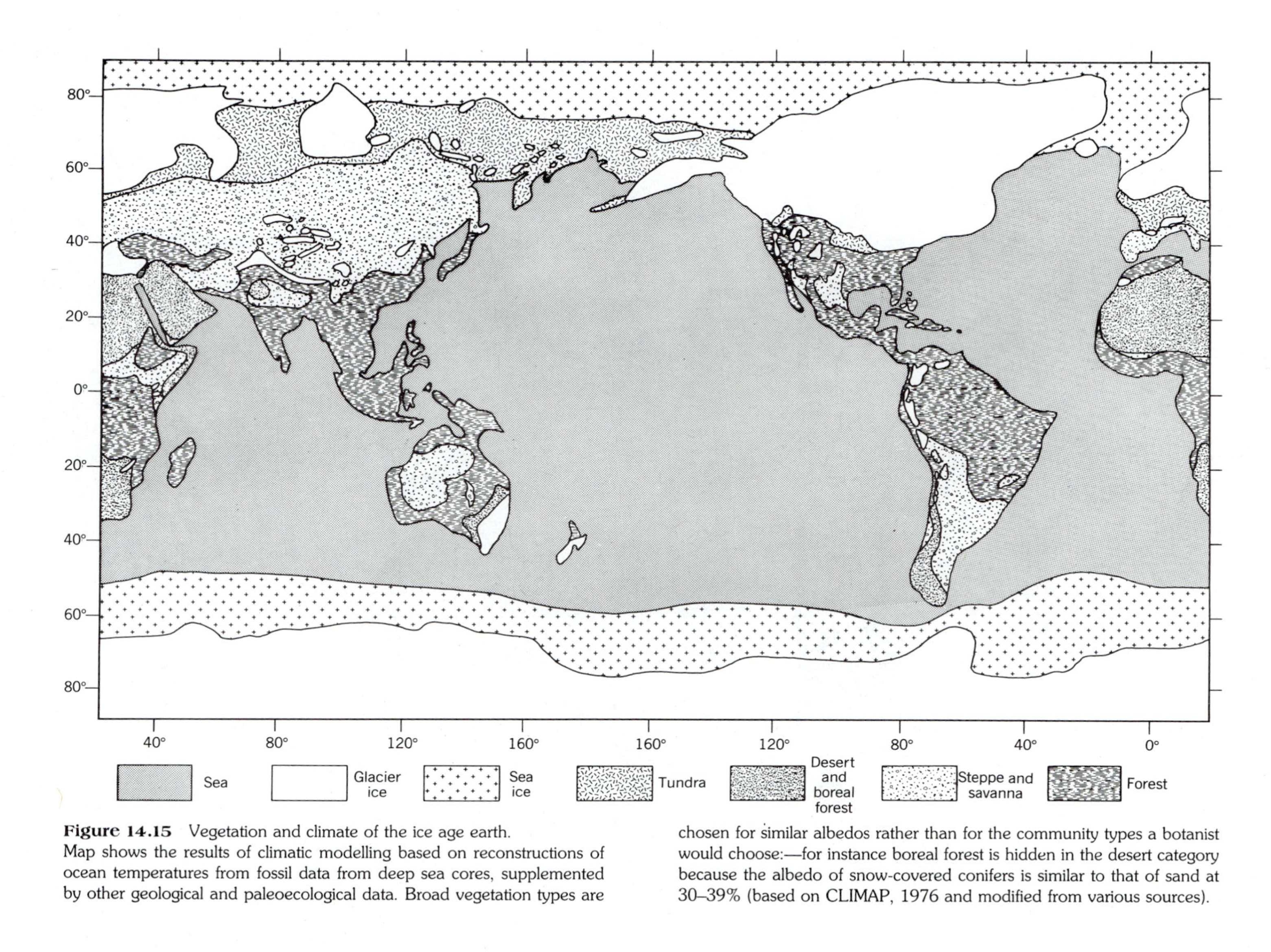

Figure 14.15 Vegetation and climate of the ice age earth. Map shows the results of climatic modelling based on reconstructions of ocean temperatures from fossil data from deep sea cores, supplemented by other geological and paleoecological data. Broad vegetation types are chosen for similar albedos rather than for the community types a botanist would choose:—for instance boreal forest is hidden in the desert category because the albedo of snow-covered conifers is similar to that of sand at 30–39% (based on CLIMAP, 1976 and modified from various sources).

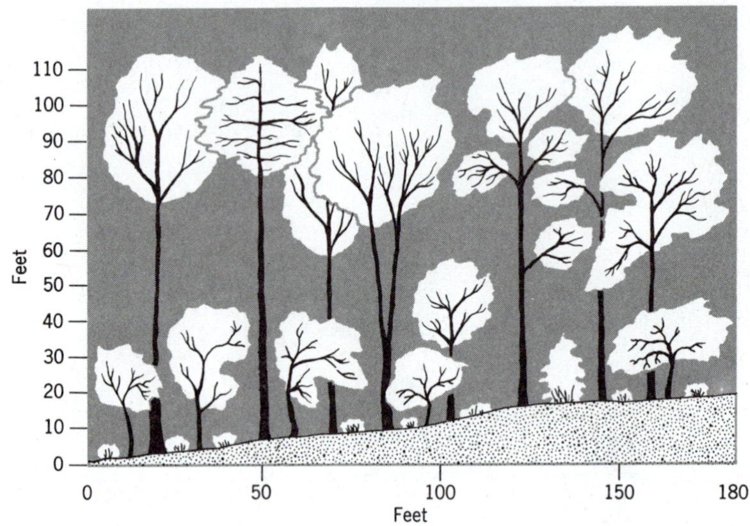

Figure 14.16 Elevation sketches of biome types. (Based on diagrams in Beard, 1955, and Billings, 1978.)

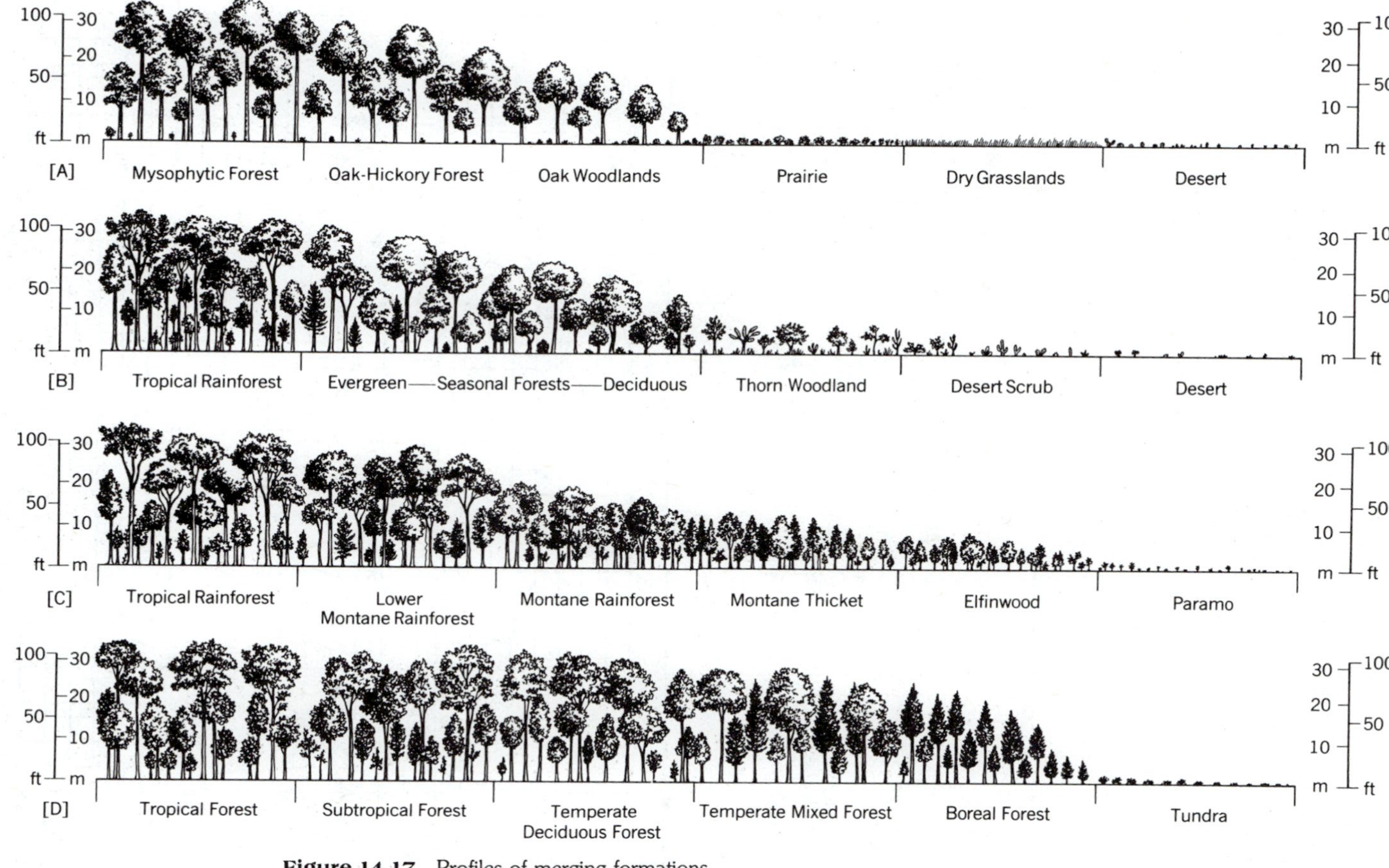

Figure 14.17 Profiles of merging formations.
(*a*) From moist Appalachian mountains westward to the desert. (*b*) Rain forest to desert in South America. (*c*) Up a tropical mountain in South America, from forest to treeless upland. (*d*) South to north from tropical forest to tundra. (From Whittaker, 1975.)

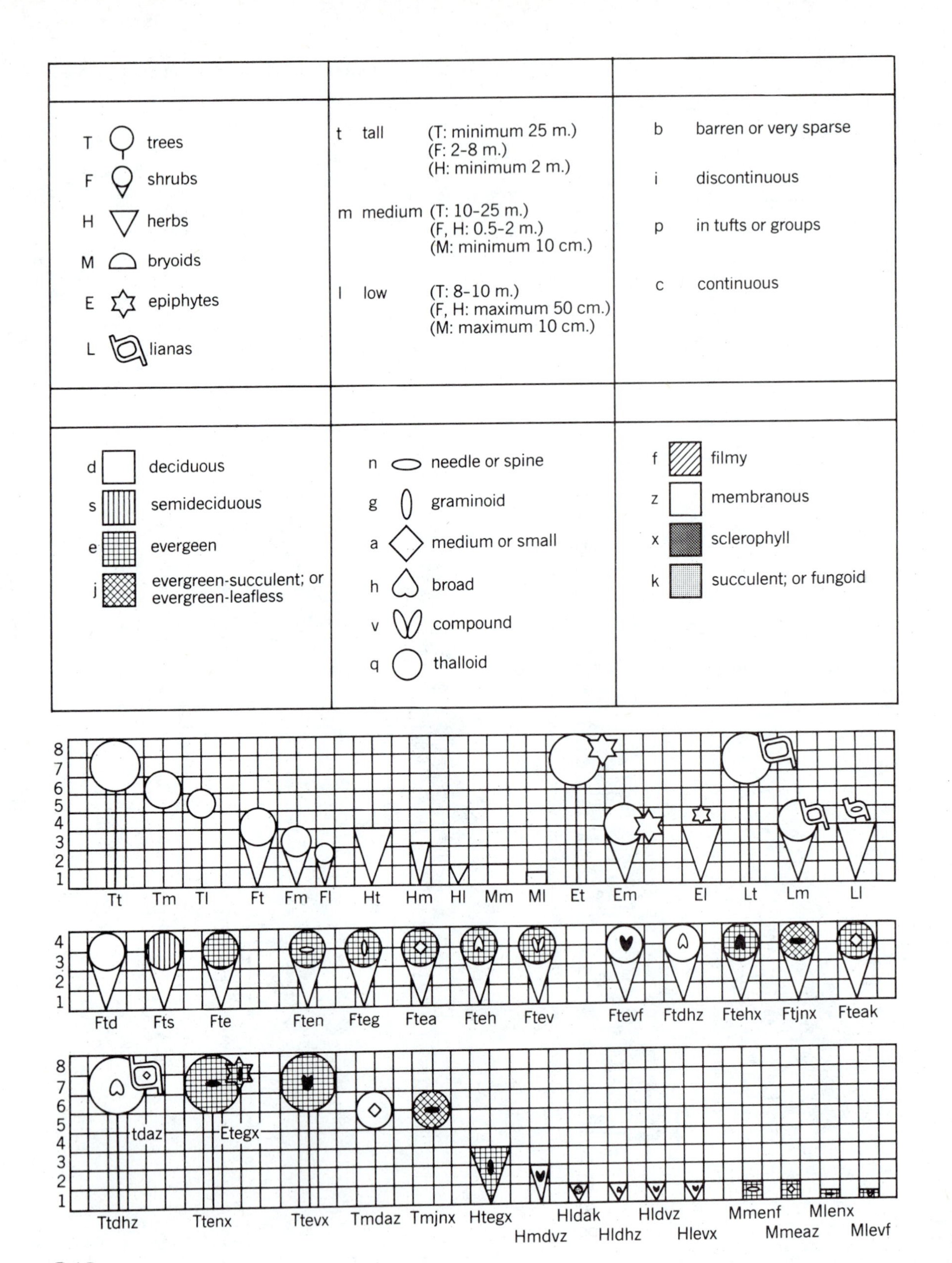

T trees
F shrubs
H herbs
M bryoids
E epiphytes
L lianas
t tall (T: minimum 25 m.) (F: 2-8 m.) (H: minimum 2 m.)
m medium (T: 10-25 m.) (F, H: 0.5-2 m.) (M: minimum 10 cm.)
l low (T: 8-10 m.) (F, H: maximum 50 cm.) (M: maximum 10 cm.)
b barren or very sparse
i discontinuous
p in tufts or groups
c continuous
d deciduous
s semideciduous
e evergeen
j evergreen-succulent; or evergreen-leafless
n needle or spine
g graminoid
a medium or small
h broad
v compound
q thalloid
f filmy
z membranous
x sclerophyll
k succulent; or fungoid
Tt Tm Tl Ft Fm Fl Ht Hm Hl Mm Ml Et Em El Lt Lm Ll
Ftd Fts Fte Ften Fteg Ftea Fteh Ftev Ftevf Ftdhz Ftehx Ftjnx Fteak
tdaz Etegx
Ttdhz Ttenx Ttevx Tmdaz Tmjnx Htegx Hldak Hmdvz Hldhz Hldvz Hlevx Mmenf Mmeaz Mlenx Mlevf

Profile sketches clearly illustrate the dramatic differences in form that are possible, but the analytical or intellectual power of such diagrams is limited. Attempts to make formalized profile diagrams are illustrated by that of Dansereau (1951) given in Figure 14.18. This system might well lend itself to computerized descriptions of vegetation for a more detailed analysis of how individual shapes are mixed.

The most quoted early attempt to classify life form in a functional way was introduced in 1905 by Raunkiaer (1934). The classification was based on the relative exposure of the perennating organ; the bulb, bud, or seed that got the plant through the most unfavorable season. Rain forest trees had their leaf buds boldly in the open. This was true of deciduous forest trees too, but many other plants in the deciduous forest hugged the ground or came up each spring from underground buds in rhizomes, bulbs, rootstocks, or corms. In the arctic, hugging the ground was the prevailing mode. So Raunkiaer reasoned that a plant's shape was a function of where its buds were placed and that the proportion of buds hidden in the ground should increase from the equator to the pole. This view seems somewhat simplistic now when heat budget, productivity, water stress, grazing protection, and other models are available to explain plant form (Chapter 5), but Raunkiaer's classification has left a heritage of terms still sometimes used in plant ecology. His system of classification can be set alongside the climatic classifications of de Candolle and Köppen readily enough because they recognize the same basic facts of geography (Table 14.1).

But experience showed that the most useful thing to do was simply to list and describe the principal vegetation types of the earth, including their climates and animals. The result was the BIOME.

The term *biome* was introduced to ecology by Clements and Shelford (1939) and has been used in various ways. Its modern usage is the result of slow change and a BIOME may now be defined as A BIOTIC COMMUNITY CHARACTERIZED BY DISTINCTIVENESS IN LIFE FORMS OF THE IMPORTANT CLIMAX SPECIES (Kendeigh, 1961). Reflection shows this definition to lack precision—we have to know what "important" means and what "climax" means. The word "biotic" creeps in to illustrate our conviction that we deal with both plants and animals, and yet the "important climax species" should certainly be a plant. But the definition is as good as we can do and our vagueness merely reflects the vaguer properties of the unit we seek to describe: a plant formation with the animals added in.

Ecologists talk much of "biomes" but they refer to abstractions. Infinite varieties of mixtures of animals, climate, and plants are possible (Figure 14.17). No section of this endless progression of change can be defined in a way that is clear-cut. Thus our definition is vague, but it is useful. We describe ideal types for our biomes and we use a list of these abstract ideal types as a measure against which we can compare real communities. Various ecologists produce various lists of biomes. Accepting that there should be eight principal BIOME TYPES on land as follows is conservative.

Figure 14.18 (opposite) Dansereau's symbols for describing vegetation.
A system to transform realistic profile diagrams of the kind shown in Figure 14.16 into a formal graphic language. Each descriptive term that a botanist normally uses is assigned a symbol and these are clustered on a shape drawn to scale on squared paper. Assuming that the scale used here is one division to 1 m, the plant at the left of the bottom row is a tree 8 m tall, deciduous (because the circle is not shaded) and with broad membranous leaves. Growing on it is a smallish liana, also deciduous. This graphic language could be used by field groups compiling data for a computer study. It has not come into wide use, however, partly for the reason that people are seldom ready to learn another's new language but more because it has been realized that refined description gives little help in answering grand ecological questions like why the plants are shaped as they are and why they live as they do. (Redrawn from Dansereau, 1957.)

Tundra

A biome type without trees or other tall perennial plants and where this absence of trees is associated with low temperature or short growing season (Figure 14.19). Lack of tall plants may be explained as a consequence of the requirements to balance a heat budget and to conserve moisture (Chapter 5). The principal region of tundra is the circumpolar lands north of the Arctic Circle but there are smaller areas in the southern hemisphere and a variant of tundra (so-called "arctic-alpine") occurs on high mountains at all latitudes. Circumpolar tundras have climates dominated by the polar zones of high pressure.

Figure 14.19 Alaskan tundra.
The upper photograph is a landscape on the Alaskan coastal plain within fifty miles of the Arctic Ocean; an herb tundra dominated by sedges with only low tussocks. The lower photograph is from Seward Peninsula within fifty miles of the tree line; high tussocks of *Eriophorum vaginatum* (a sedge known as cotton grass) are the prevailing life form with prostrate woody plants, including dwarf birches (*Betula nana*) between and below the tussocks.

Tundra soils are deeply frozen for all or much of the year, and are poorly drained and shallow. Climates and productivity are strongly seasonal. Animals hibernate, migrate in the colder season, or live as the lemmings do under the snow. Characteristic mammals are lemmings, hares, musk oxen, caribou, foxes, and wolves. Except for ptarmigan and a few predators and seabirds, tundra birds are migratory, using the short productive period for reproduction and surviving the rest of the year elsewhere; examples are jaegers, geese, shore birds, and insectivorous songbirds. Tundra communities are variable, probably reflecting different summer temperatures, length of growing season, and precipitation (Figure 14.19). Alpine tundras of high mountains do not look very similar to arctic tundras to the naturalist familiar with both. What they have in common is absence of trees and the sharing of many species, or at least genera, of plants and animals. Important causes of differences are the different lengths of growing seasons and the different irradiance, for a low-latitude mountain top may be almost without seasons and have many hours of sunlight each day and yet be cold enough, or physiologically dry enough, to be without trees. These sites are colonized by relatives of plants of cold arctic regions.

Coniferous Forest (Boreal Forest, Taiga)

Characteristic coniferous forest is the broad northern belt that fronts the tundra, usually called the *boreal forest*, after the Greek name for the north wind, or *taiga*, which is the Russian name (Figure 14.20). The climatic limits were discussed in Chapter 5. These forests exist essentially where winters are very cold, like those of the tundra, but where summers are longer, perhaps with a short period of warm continental weather. These climates lie along the line where cold, high-pressure air of the arctic meets air

Figure 14.20 Boreal forest in Canada.
Most abundant trees are white spruce (*Picea glauca*) and balsam fir (*Abies balsamea*).

travelling northward from about the fortieth parallel. Most of the forest trees are evergreen and needle-leaved, this being a design that apparently is suited to achieving useful working temperatures in temporarily productive periods at a minimum cost of maintenance (Chapter 5). Broad-leaved species within the forest are deciduous (maples), and some have special adaptations to restrict heat loss (for example, trembling and vertical leaves, as in *Populus tremuloides*; see Chapter 5). The evergreen gymnosperms that have prevailing adaptations of this biome type are flammable with the result that the boreal forest is subject to periodic fires. A burn–regeneration cycle in the forest is an important secondary characteristic. Apart from seed-eating rodents like mice and squirrels, the principal herbivores of older parts of the forest are insects, particularly caterpillars of Lepidoptera (butterflies and moths) and saw flies (herbivorous Hymenoptera). These herbivore populations are subject to irruptions when a section of forest has not been burned for a long time. The preponderance of insect herbivores in this seasonal environment causes abundant insects to be predictable in early spring, a property of the boreal forest that is used by many migratory songbirds as a resource for the rearing of young. Large mammalian herbivores are several species of deer and bears, all of which are adapted to make use of regenerating forest in burned areas for browse or fruits. Carnivores are essentially similar to, or the same as, the cold-adapted carnivores of the tundra: wolves, foxes, and medium-sized cats, like lynx and mountain lions. Patches of coniferous forest very like the boreal forest are found in the southern hemisphere but little southern land lies in the appropriate weather system so the extent of this southern forest is limited. More extensive are the *moist-coniferous forests* of the western coast of Canada and adjacent parts of the United States, a forest that is often described as a biome in its own right (Figure 14.21). These forests consist almost exclusively of very large, evergreen coniferous trees like Douglas fir (*Pseudotsuga*). These trees live in a climate that is cool and moist year-round from the influence of warm coastal seas against colder land. No completely satisfactory explanation seems available of why a coniferous evergreen strategy prevails under these conditions over a broad-leaved evergreen strategy, though the niceties of balancing heat budgets at a suitable operating temperature should be responsible (Chapter 5). Coniferous woodlands are known in other climates, like that of the south-eastern United States, but, except when on high mountains, these are usually transient, successional communities and not considered part of the boreal forest biome type.

Figure 14.21 Moist coniferous forest of northern California.
The trees are redwoods (*Sequoia sempervirens*).

Temperate Forest

This is the biome type to which the temperate deciduous forests belong, but there are variants even though the total range is not great (Figure 14.1). Temperate forests cover northern Europe, eastern China, the eastern and midwestern United States, a small area in South America, part of

New Zealand; in total not a large part of the globe. These are places of seasonal climates with cold winters but longer summers than in the land of the boreal forest. All patches are either adjacent to coasts, escape rain shadows, or are otherwise well watered in the growing season. In some places nearly all the trees are deciduous (Figure 14.22), but there may be mixtures of deciduous trees with evergreen conifers or evergreen broad-leaved trees like holly (*Ilex*). In the wetter or warmer parts of land covered with temperate forest, evergreen broad-leaved trees are a more important part of the forest. The predominant tree strategy, however, is to carry broad leaves and to drop them in winter, a strategy that can be understood in energetic and heat budget terms (Chapter 5). It should be clear that many individual solutions to manipulating heat budgets in seasonal environments exist as well as several strategies for overwintering. Evergreen and deciduous variants recombine differently over the range of the temperate forest. Possibly this biome type is less homogeneous than others, like boreal forest or tundra, because the climate is less homogeneous. Local weather depends on rains and air mass movements of midlatitudes where seasons are unpredictable. Temperate forest mammals are comparable to those of the boreal forest, but they are more abundant and include additional forms like badgers, wild pigs, and more species of squirrels and small predators. Bird species richness is high compared to that of the boreal forest, including both migrants that rear young on spring gluts of insects and many resident birds, some of these seed or fruit eaters (*frugivores*). The most important herbivores are insects. Amphibians (salamanders and frogs) are present whereas they are nearly, though not quite, absent from boreal forests. This distribution of amphibians may be the clearest evidence for a quantitative change in climate from boreal to temperate forest.

Tropical Rain Forest

Wet tropical places are productive year-round and are always green (Figure 14.23). The typical trees are broad-leaved, dicotyledonous plants of

Figure 14.22 Temperate deciduous forest in France. Most of the trees are beech (*Fagus silvatica*).

Figure 14.23 Tropical rain forest in the Amazon.
From a low-flying aircraft the uneven top to the forest and the great variety of tree species are noticeable.

wide taxonomic diversity, nearly all of which are pollinated by animals (insects, birds, or bats). The trees, therefore, have conspicuous flowers, unlike the wind-pollinated flowers of trees in deciduous and boreal forests. The pea family, Leguminoseae, is widely represented so that a significant percentage of the trees may be legumes. Most of these trees carry leaves year-round but they do replace leaves and some may shed leaves synchronously, thus being almost leafless for a short time. Buttress and stilt roots are common, adding to structural diversity in the forest. The forest also holds numerous tall, monocotyledonous plants, particularly palms, as well as tree ferns and other large cryptogams (Figure 14.24). These plants are usually of subordinate status in the forest, though in parts of the Amazonian rain forest the taller palm trees may reach to the canopy. Plants of the understory typically have very large leaves, suggesting a monolayer strategy to cope with the dim light there. Vines of many kinds (lianas) are present as well as many epiphytes, particularly orchids and bromeliads. Like the main forest trees, these have conspicuous flowers and use animals for pollination. Many rain forest plants also use animals to disperse seeds, so that fruits are large, succulent, or showy. This forest can be found in many subtle variants, depending particularly on the seasonality or evenness of the rains, but the essential properties for animal life are warmness and wetness. Most food production occurs many meters above the ground; indeed, one way of looking at a tropical rain forest is as a prairie standing on 30-m stilts. Heavy browsing animals of the ground are thus not numerous and are replaced by browsers that can climb, principally primates but also including sloths in South and Central America (Figure 14.25). A diverse array of fruit-eating and pollinating animals provides specialists on particular plant species. Climbing or flying amphibians, reptiles, and insects are in great variety. Important predators are large carnivorous birds that hunt the browsing animals of the canopy. A particular property of the biome type is the role played by animals acting as detritivores, partic-

Figure 14.24 Tropical rain forest in New Guinea.

Figure 14.25 *Alouatta seniculus*: herbivore of the canopy.
Most plant production in tropical rain forests is many meters above the ground and must be utilized by arboreal herbivores or omnivores like howler monkeys.

ularly ants and termites. As a consequence little litter collects on the forest floor and soils can be almost bereft of organic matter.

Tropical Savanna

Where it is warm year-round but with a very long dry season (several months) trees are stunted and grow spaced out, which allows grasses to grow between them (Figure 14.26). Broad belts of the resulting savannas flank the rain forests to the north and south where the equator bisects the continents of Africa and South America. This corresponds to the latitudes where air descends and begins its travels back to the ITCZ. Usually a gradual transition spans tree-studded savanna, dry woodland with grass, and tropical rain forest so that the boundaries of these biome types may be particularly difficult to map in detail. Sometimes, though, a sharp discontinuity forms because the savanna burns and the wetter forests do not. Fire, an important fact of the savanna environment, can set down a scorch line at the edge of the savanna so that, as in northern Nigeria, forest patches stand like cliffs out of the savanna. Tropical savanna grasses may be 1 to 3 m tall but they represent productivity within the reach of heavy grazing animals. Accordingly savannas are inhabited by herds of grazing mammals and their large mammalian carnivores, lions and other big cats, hunting dogs, jackals, and hyenas. The large mammals in turn provide living for large scavengers, so that savannas are characterized by vultures. The C4 photosynthetic pathway is widespread in this biome. Trees have small leaves, are multilayered (Chapter 3), and have flat tops; shapes that can be understood as representing a compromise between reducing the maintenance costs of unnecessary height with the requirement of being tall enough to reduce mammalian browsing from grounded animals. The result is the familiar appearance of tropical savannas (Figure 14.26), representing a total system of life starkly different from tropical rain forest and resulting from the simple fact that rain comes at wide intervals, though it may be ample when it does come.

Figure 14.26 Tropical savanna in East Africa.

Temperate Grassland

The prairies and steppes of the world lie in temperate, seasonal, and dry climates (Figure 14.27). A little wetter climate allows the deciduous forest to close over it; a little less rain in the dry season and bare ground between the grasses spreads until a desert results. This can be sensed by looking at the extent of prairies in North America in Figure 14.1 where boreal forest, deciduous forest, and desert are the boundaries. Boundaries from prairie to desert are extremely gradual but boundaries with forests can be distinct because of fire. Like the tropical savannas, the temperate grasslands burn and fire is an expected property of the environment. Unlike tropical savanna, however, the temperate grasslands are without trees. Perhaps we do not have a sufficient ex-

Figure 14.27 Temperate grassland.

planation for this. Lack of trees can be equated with drought, as was discussed when considering what sets a tree line on high mountains, the mechanism for which is now tolerably understood (Chapter 5). Scattered trees in tropical savanna presumably reflect the absence of winter, allowing an evergreen and broad-leaved strategy not applicable to prairies with winters. But the reality is that subtleties of climate produce a land free of trees and thus dominated by the inclined leaves of grasses (Chapter 3). Many of these grasses are C4 photosynthesizers. They are alternately green and brown with the seasons and have among them numerous species of low, flowering plants. A vital difference between these temperate grasslands and the treeless expanse of the tundra is that the ground is not frozen. Deep organic soils collect and a thick bed of organic matter is a prime parameter of this biome type. The fauna of a temperate grassland has much in common with that of tropical savanna: grazing ungulates, large carnivores, and scavengers. This can be seen most clearly in South Africa, where the temperate grassland, locally known as the "veldt," shares many animals with the tropical savannas of East Africa to the north.

Deserts

The ultimate desert with no water has no plants, merely bare rock or a sand sea of shifting dunes. But by the *desert biome* ecologists mean land too dry to support prairie or savanna, but with enough moisture to allow specially adapted plants to live (Figure 14.28). These deserts are not only dry but also hot. Typical perennial plants have

Figure 14.28 Vegetated desert in southeast United States.

swollen stems rather than leaves, with heavy cuticles, sunken stomates, C4, or even CAM, metabolism, and spiny defenses against large browsing animals, and live spaced out (Chapter 3). All these adaptations are for coping with heat stress, water stress, competition for water, or the necessity to avoid damage from herbivore attack when the energy costs of repairs are so hard to meet (Chapter 5). Where the mixture of heat and water stress is less severe, perennial bushes of the Chenopodiaceae or Compositeae form more or less regular arrays with bare ground in between. Numerous annuals grow briefly following rain and some perennial bushes may be opportunistically deciduous like the *Bursera graveolens* trees of the Galapagos (Figure 11.2). A particular property of desert life is an alternation of hot days with cold nights, resulting from the absence of clouds to impede irradiance by day or to stop the warm ground from radiating heat to the cold black body of space by night. Low productivity of plants, temperature extremes, and shortage of water narrowly restrict the life styles of animals. The low-energy systems of reptiles are well adapted to this regime (Chapter 4) and hot deserts accordingly have a wide variety of reptiles. Amphibians are almost absent because of lack of water. Mammals are restricted to taxa that produce concentrated urine or take refuge from the sun in burrows, or those with special tolerance to desiccation and with heavy fur insulation like that of camels (Chapter 5). Insect life histories tend to be strongly seasonal with diapausing eggs or adults able to survive the dry times. Plague grasshoppers or locusts are almost characteristic of deserts, like the Australian grasshoppers discussed in Chapter 8.

Chaparral (Broad-Leaved Sclerophyll)

The very distinctive life form of the vegetation of this biome type is shown in Figure 14.29. The strategy used by the plants that make up nearly all the vegetation is that of woody and dense bushes with permanent, thick, pubescent (hairy), leathery leaves. Neither trees nor grassland replace these bushes; they can, and frequently do, burn, but then they are replaced by their own kind. Many different families of flowering plants contribute species with this strategy and life form to *sclerophyll* vegetation in different parts of the world wherever the peculiar climatic pattern for which it is adapted occurs (Figure 14.1). The climate must be an alternation of hot, dry summers with cool, moist winters. These climates occur where cool ocean currents turn away from

Figure 14.29 Broad-leaved sclerophyll vegetation in southern France: the Maquis.

continents as they complete their gyres, causing cold upwellings along the coast (Figures 14.9 and 14.10). When this happens at middle latitudes with high irradiance, the land is hotter than the ocean in summer, preventing both clouds and rain; the result is known as a Mediterranean or Southern Californian summer. But lowered irradiance in winter makes the difference between land and sea temperatures less so that both clouds and rain are possible. The sclerophyllous bush strategy is apparently to grow during the winter rain and to sit tight in summer. Possibly trees are prevented from taking over by the hot summers, and grasses are outcompeted by the bushes that can overtop them in the cool winter. Certainly it seems that a hypothesis of both physical adaptation and competition is needed to account for the success of this peculiar life form. Seed-eating rodents and birds are common and a number of ungulates such as goats and deer can live in chaparral. These herbivore populations supported wolves or big cats in both the New and Old Worlds at one time.

These eight BIOMES or BIOME TYPES describe the most obvious kinds of vegetation on the earth, together with the constraints they impose on animal (consumer) populations and strategies. We emphasize again that each is an ideal type and that there are endless variations on these to be found in the long gradients of vegetation that cover the real world.

The present pattern of these biomes as shown in Figure 14.1 is, for the most part, less than 10,000 years old. Earlier than this the ice age earth had a different geography (Figure 14.14). Similar biomes should have existed, though in different places and perhaps with different mixtures of species in them.

It is possible to apply the biome concept to aquatic systems, though the concept loses much of its original meaning. The open waters of both lakes and oceans, for instance, support only planktonic plants and the food chains that come from them. Properties for life are set by the microscopic sizes of these plants and we can talk about the *planktonic biome.* Then there are different patches of coastline that can be called biomes; rocky shores with anchored seaweeds (*rocky-shore biome*), muddy estuaries without large seaweeds, coral (algal) reefs, and so on. In each of these places physical constraints of fluid flow and substrate determine the life forms of plants, and hence the strategies of consumers. The differences between them, however, are of a different order than those set by climate on the terrestrial earth. These biota are best discussed in the context of the physical systems of which they are a part.

LIFE ZONES ON HIGH MOUNTAINS

A climatologist's rule of thumb, called HOPKIN'S BIOCLIMATIC LAW, is that 1000 FEET OF ELEVATION IS EQUIVALENT TO 100 MILES OF LATITUDE. Unfortunately this does not translate into memorable numbers in the metric system. A result is that plants of high elevations are similar to those much farther north.

Figure 14.30 shows the vegetation spread up the side of San Francisco Mountain, near the Grand Canyon in Arizona. The names on the figure are names for the various zones given by C. Hart Merriam (1890) when he made the first biological survey of the region. A drive up an

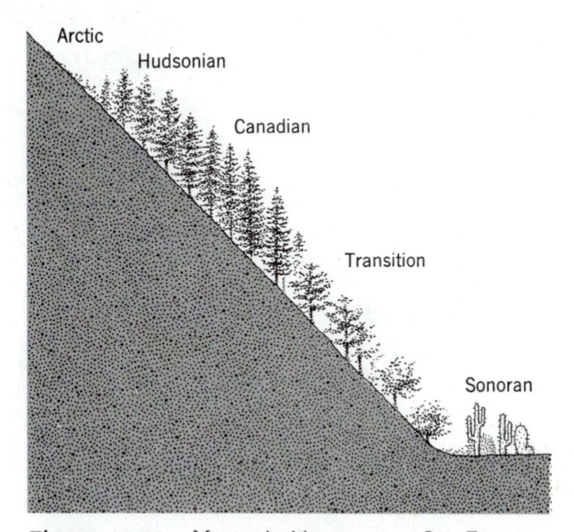

Figure 14.30 Merriam's life zones on San Francisco Mountain in Arizona.

Arizona mountain begins in the saguaro cactus desert (SONORAN LIFE ZONE) but, after only a few hundred meters of climbing, this is left behind. There follows a region of oak scrub, becoming gradually intermingled with pines, which Merriam called the TRANSITION LIFE ZONE. By 2000 m a real pine forest is reached, with streams and the rising fragrance of needles under foot, known as the CANADIAN LIFE ZONE. Higher still are Douglas firs and spruce trees, which Merriam called the HUDSONIAN LIFE ZONE after Hudson's Bay, though he might have done better to refer to British Columbia or the State of Washington. And a bit below 4000 m comes a tree line and higher still an alpine tundra, complete with a variety of species familiar enough to people used to travelling north of the Brooks Range or even in Greenland. This was the ARCTIC LIFE ZONE.

Merriam rose to be head of the U.S. Biological Survey and developed a system of mapping based on what he saw when climbing mountains. The result was the map of the United States given in Figure 14.31. It is clear enough that the

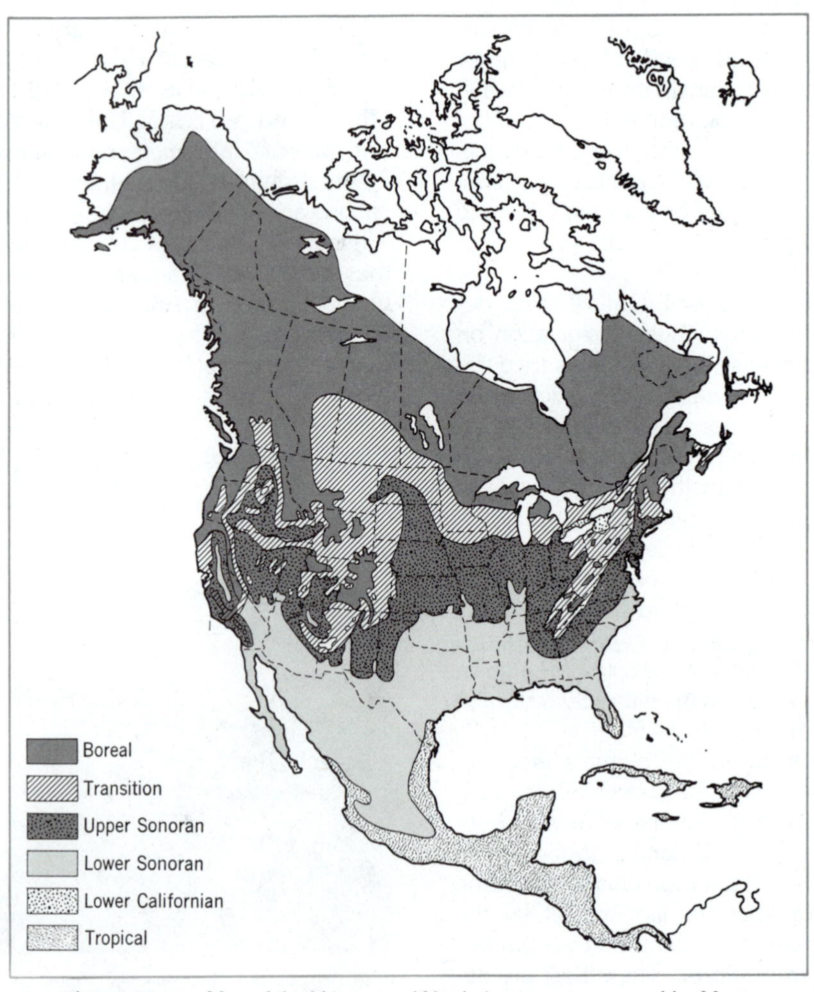

Figure 14.31 Map of the life zones of North America as proposed by Merriam.

vegetation of a continent has been constrained to fit that of mountains when the eastern deciduous forest appears as the *upper sonoran*, apparently because there are oak trees (of a sort) in both places. The spirit of the Arizona Territory had taken over the United States.

The life zone exercise does reflect a real distribution of life forms, however. Strategies and shapes suited to cold air masses, and which characterize *biomes*, are also suited to cold air on mountains and characterize *life zones*. The mountain vegetation, of course, comes in small or isolated patches, so that only a few of the species or forms of a biome are present. Some of the climates that set the life style cannot be expected on mountains, like the wet winters and hot summers of chaparral or the seasonal changes that let broad-leaved deciduous trees dominate much of the temperate forest. But some of the correlations between life form and climate were illustrated neatly enough by Merriam's exercise.

ZONE BOUNDARIES ON MOUNTAINS

Mountain sides have always had the special attraction that species distributions seem to be set within narrow limits. This idea seems clear enough to anyone who has seen a mountain from a distance and remarked at the bands of color their flanks display. Not only do the plants change with altitude but the zones of vegetation seem distinct enough to be sketched from across the valley. Up high is the tree line, and lower down are various color changes at the zonal boundaries, apparently all clear enough.

Figure 14.32 shows the vegetation bands on the side of the Galapagos island of Santa Cruz (Indefatigable) as seen from the sea. The photograph was taken on one of the rare days when the stratus cloud between the island top and the inversion (Figure 14.12) lifted to show the island clearly. The zones visible on the island, from sea level up, are the *coastal desert* zone, a *transition* zone of deciduous trees and bushes, a *forest* zone of evergreen trees supplied with moisture by being enveloped in clouds periodically, a *bush* zone of *Miconia robinsoniana* bushes mixed with tree ferns, and a *brown* zone of upland vegetation free of shrubs or trees. These zones, compressed into a climb of less than 1000 m, are due as much to moisture changes as to temperature, but they serve to show how definite zonal boundaries may seem to be.

Figure 14.32 Zonation on a Galapagos island.
Isla Santa Cruz (Indefatigable) as seen from the sea on one of the rare days when clouds are high enough for most of the island to be seen.

There are two sorts of boundaries that might appear on a mountain side:

1. disjunctions between the ranges of individual species and
2. disjunctions between whole communities.

These two theoretical possibilities would imply quite different mechanisms. Disjunct species populations offer no great difficulty for ecological theory, but disjunct communities would be another matter.

Disjunct Species Distributions

Bird populations have long been known to be very narrowly restricted on mountain sides so that the birds encountered during a climb keep changing. Since birds can fly where they like, this rigid segregation by altitude might seem unexpected but it can be explained comfortably enough as the result of behavior constrained during speciation by isolation, merging, and character displacement in the usual way (Chapter 7). A model of the stages in this process was put forward by Diamond (1973) for birds on New Guinea mountains and he was able to find actual bird distributions that represented most of the stages in the process. Figure 14.33 shows his model.

Stage 1. A single species occupies a broad range of elevations on all mountains.

Stage 2. The range is interrupted so that subpopulations occupy mountains at the eastern and western ends of the island with mountains in the middle left vacant. This is the geographical separation that we expect as the usual start of a speciation episode.

Stage 3. After long isolation the eastern and western populations have diverged so that it is likely that they would not freely interbreed if they were merged. Local selection pressures make the two populations occupy slightly different ranges of elevation.

Stage 4. The two populations spread so that their ranges abut but do not overlap: they are *parapatric* as opposed to being *allopatric* or *sympatric* (Chapter 7). Little interbreeding should occur, showing that evolution in isolation had separated the populations sufficiently.

Stages 5–8. Ranges overlap progressively with always a tendency for the populations to become sympatric. But in fact, where they occupy the same mountain, they divide it up by altitude as a form of character displacement. The end result is two species sympatric over the whole of New Guinea but allopatric on each individual mountain, with a sharp separation between altitudinal ranges.

Diamond found examples of each distribution proposed in the model from his field work in New Guinea, as shown in Figure 14.33. The end result is pairs of closely related birds on one mountain whose distributions are rigorously segregated into altitudinal bands. This is a perfect disjunct distribution along an altitudinal gradient. The disjunction is set not by any obvious physical limits but by competition. Such disjunct distributions of species along gradients are common.

On the Absence of Community Disjunctions along Gradients

Whole communities can be disjunct if physical habitats are disjunct; otherwise we expect communities to merge. To strike a discrete community boundary across a gradient would require all the many species of a community to have equal ranges, a result most unlikely to be brought about by individual selection. And yet ecologists have sometimes been presented with data that seem to require distinct communities. Data from mountain sides are again useful.

To the eyes of a spectator from across the valley, the sides of mountains truly seem to be banded with recognizable communities—life zones on an Arizona mountain or vegetation zones in the Galapagos (Figures 14.30 and 14.32). Some of these bands do have physical edges, like the cloud base, but others seem to strike across the gradient where no physical boundary exists. In the history of ecology these apparent zone

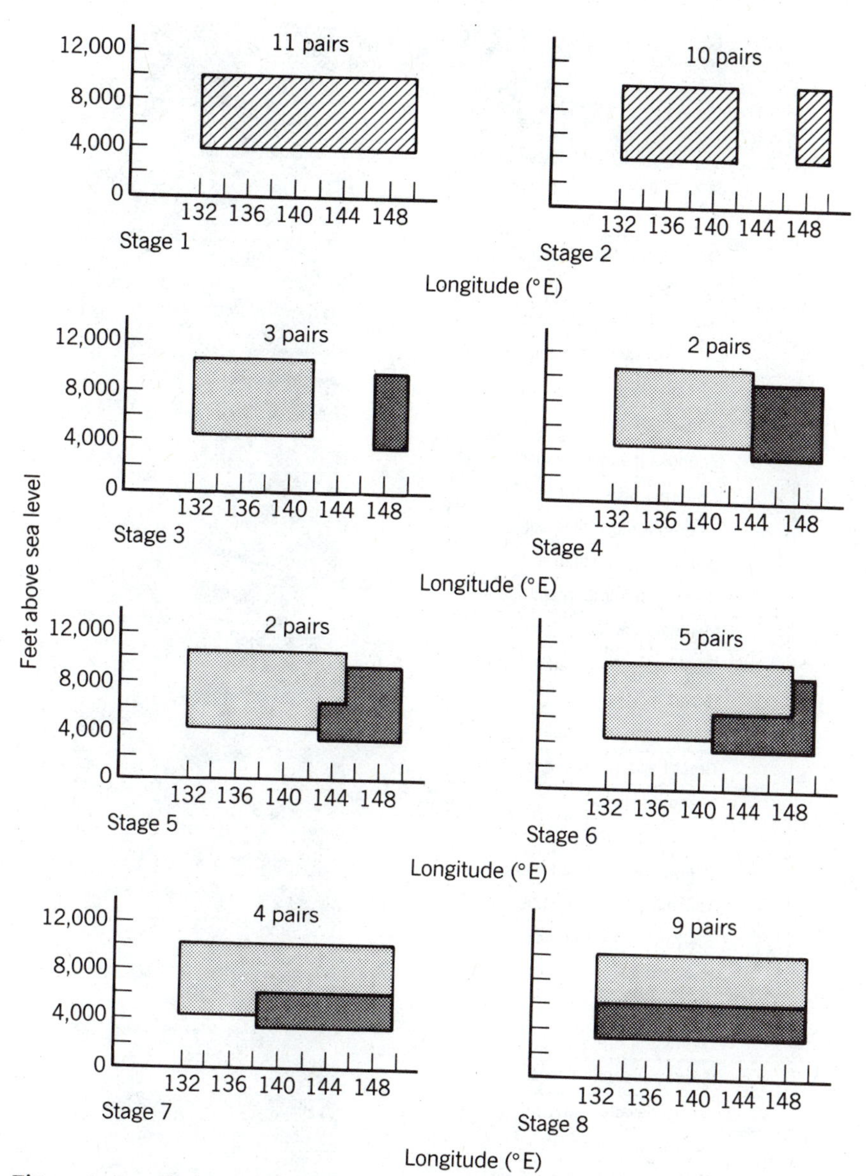

Figure 14.33 Bird zonation on mountains.
The diagrams illustrate the eight stages proposed by Diamond (1973) that result in species of birds with segregated distributions on mountain sides. For explanation see text. Numbers of pairs of birds representing each distribution found by Diamond in New Guinea are shown. (Modified from MacArthur, 1972.)

boundaries have given powerful arguments to those who postulated that discrete communities are able to maintain boundaries and disjunctions by biological processes. And yet the proposed disjunctions always turn out to be illusory on closer examination, unless a real physical boundary is found. Merriam's (1890) original data for mammals on California mountain sides show how these

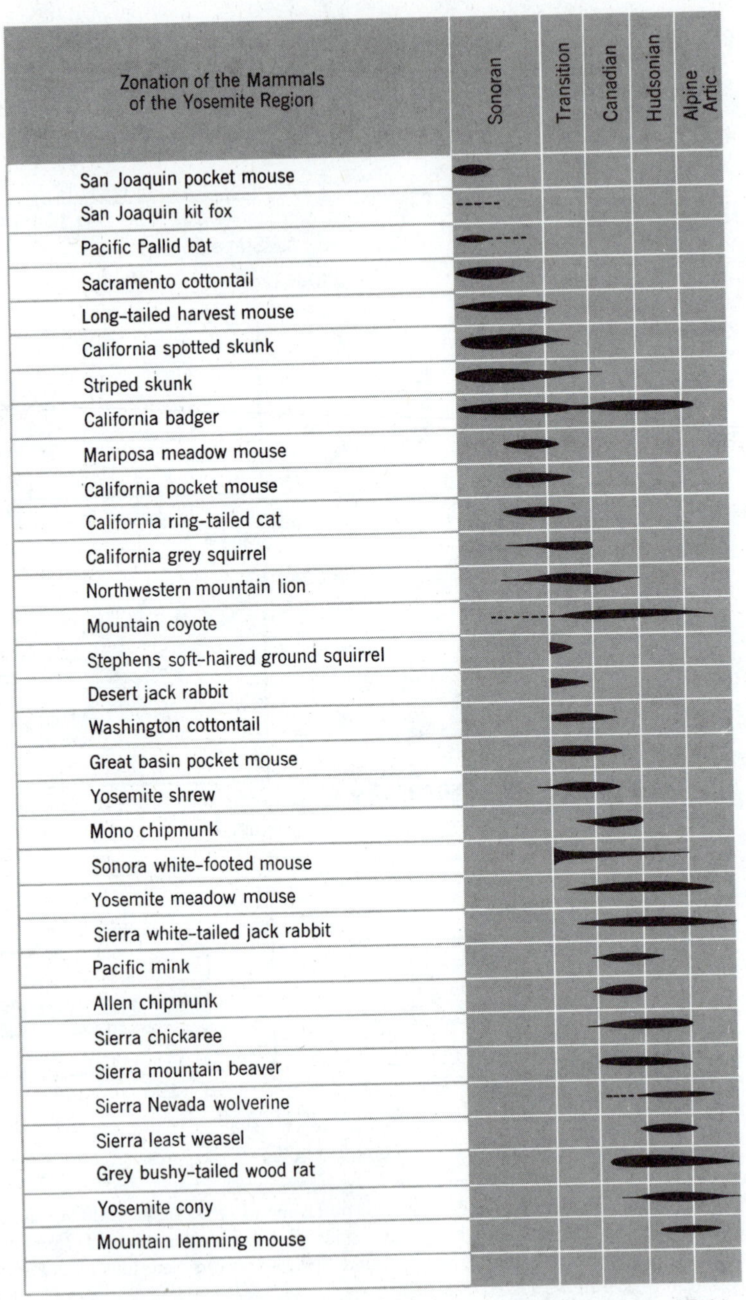

Figure 14.34 The ranges of mammals with altitude in the Yosemite region of California. Notice that the ranges of the animals overlap. There is a gradient, not a series of disjunct populations as Merriam postulated. (Redrawn from Allee *et al.*, 1949.)

distributions blend across the proposed boundaries (Figure 14.34).

Whittaker (1956) measured the distribution of individual tree species up and down mountain sides by setting out transects and measuring the percentage composition of trees in the forest at intervals. A revealing set of his data from the Great Smoky Mountains of eastern Tennessee is given in Figure 14.35. These are mountains that look to be nicely banded in the fall, having colors as different as the dark green of hemlock and the autumn tints of turning maple leaves, yet there are no disjunctions in the ranges of individual species let alone of whole communities. For each tree there is an optimum elevation at which it lives in peak numbers but the distributions overlap. There are actually no precise bands or edges in this system, though the human eye can pick out any of the peaks *a–d* in Figure 14.35 when viewing from across a valley and call it a band. The eye picks out prominent colors or textures and fits these together as bands in the way that we talk of bands of color in a light spectrum or rainbow. The typical "banded" vegetation of mountain sides is actually continuously merging in an ECOCLINE.

ECOLOGICAL AND EVOLUTIONARY BIOGEOGRAPHY: A HISTORICAL ASIDE

Some facts of biogeography were the essential data from which the process of evolution by natural selection was discovered. The animals or plants of any one place were more closely related in a taxonomic sense than they were to the animals or plants of more distant places. For instance, the rain forests of Africa, South America, and India all looked superficially alike and yet they had quite different species in them. Indeed, the species of the Amazon forest were more related to the species of Patagonia or the Argentine pampas than they were to the species of the African or Indian forests. Accordingly evolutionary biogeographers traditionally map the world into six REGIONS in each of which the animals and plants are closely related to each other (Figure 14.36).

This map of biogeographic *regions* may be compared with a map of *biomes* (Figure 14.1) or with an earlier and more general ecological map, that of Allen's REALMS (Figure 14.37).

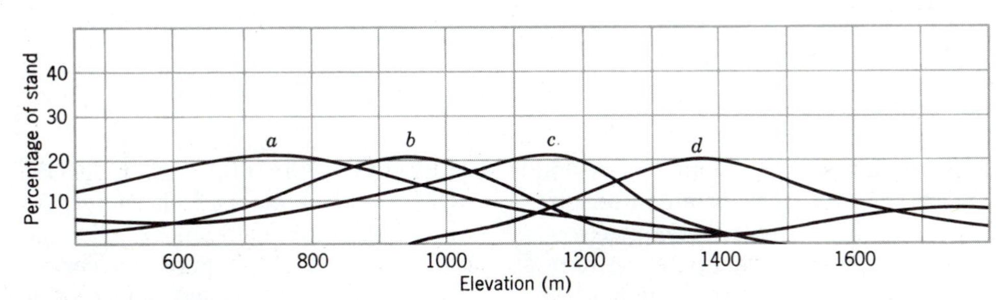

Figure 14.35 Distribution of tree species on a slope of the Great Smoky Mountains in Tennessee.
The percentage compositions of the forest were calculated from sample counts of trees more than one centimeter in diameter 4 feet from the ground. The forest was sampled at 100-meter elevation intervals. Four tree species out of the sample are shown here. There are no distinct vegetation belts, for the distribution of each species grades into that of its neighbors. A distant observer might well resolve points *a*, *b*, *c*, *d* as the middles of separate vegetation belts: *a*, hemlock; *b*, *Halesia monticola*; *c*, lime; *d*, mountain maple. (Modified from Whittaker, 1956.)

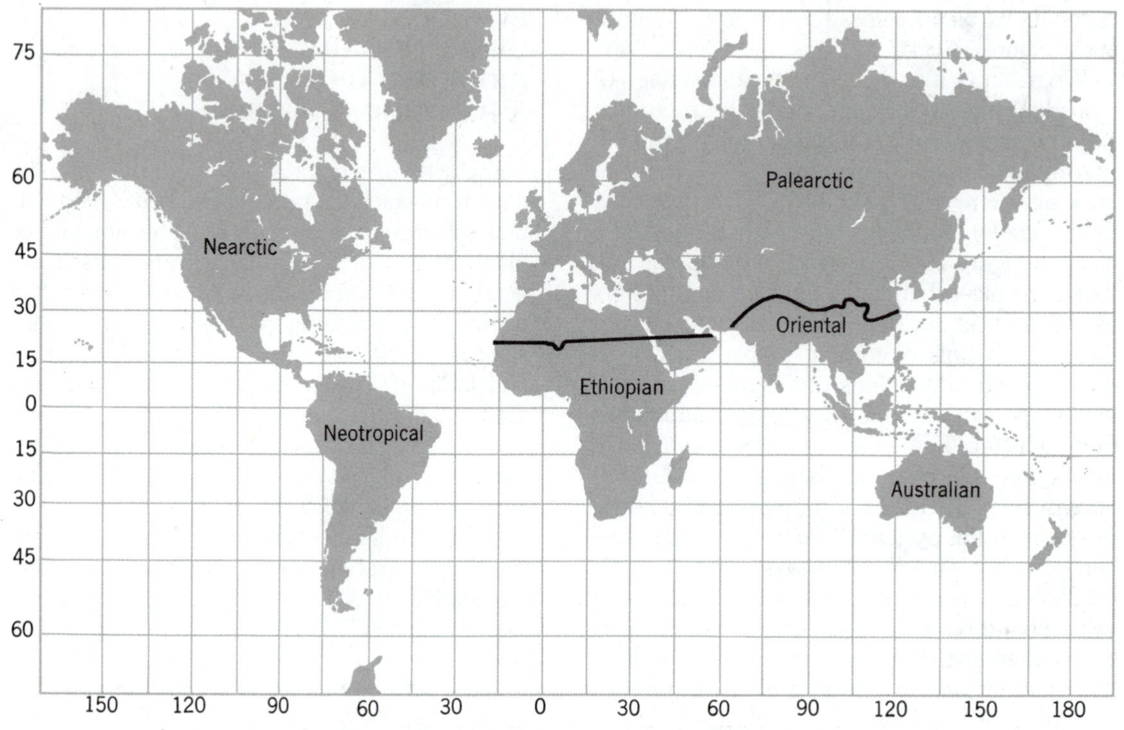

Figure 14.36 The zoogeographic regions of Sclater and Wallace. These regions were established because the birds and mammals of each continent or subcontinent were closely related to each other, even though they might have adapted to different ways of life. Use of this system by Wallace was one of the clues on which the theory of evolution was based.

The differences are striking. Biome maps and Allen's *realms* map recognize common adaptations to common environments: the *regions* map shows only relatedness or common ancestry regardless of form or environment. Both kinds of map are true in their own way. They classify different things.

In the early days of both ecology and evolutionary biology the different world views represented by the two systems of mapping came into conflict. On the one side were ecologists studying the effect of environment on form and life style. Allen (1871) stated this view of geography in the days when it was fashionable to write opinions as "laws," as THE LAW OF THE DISTRIBUTION OF LIFE IN CIRCUMPOLAR ZONES and he mapped his *circumpolar zones* as the *realms* of Figure 14.37. The influence of climate on the distributions he mapped is obvious enough. But on the other side were the evolutionists like Sclater (1858), who made the first map of evolutionary *regions,* and Wallace (1876), the co-discoverer of evolution.

When Wallace produced his great two-volume work on zoogeography he did so to reveal still more thoroughly the truth of evolution. He favored Sclater's classification of regions as doing this most clearly, and he applied it not only to birds but to all animals. As a first step, he found it necessary to demolish the "rightness" of Al-

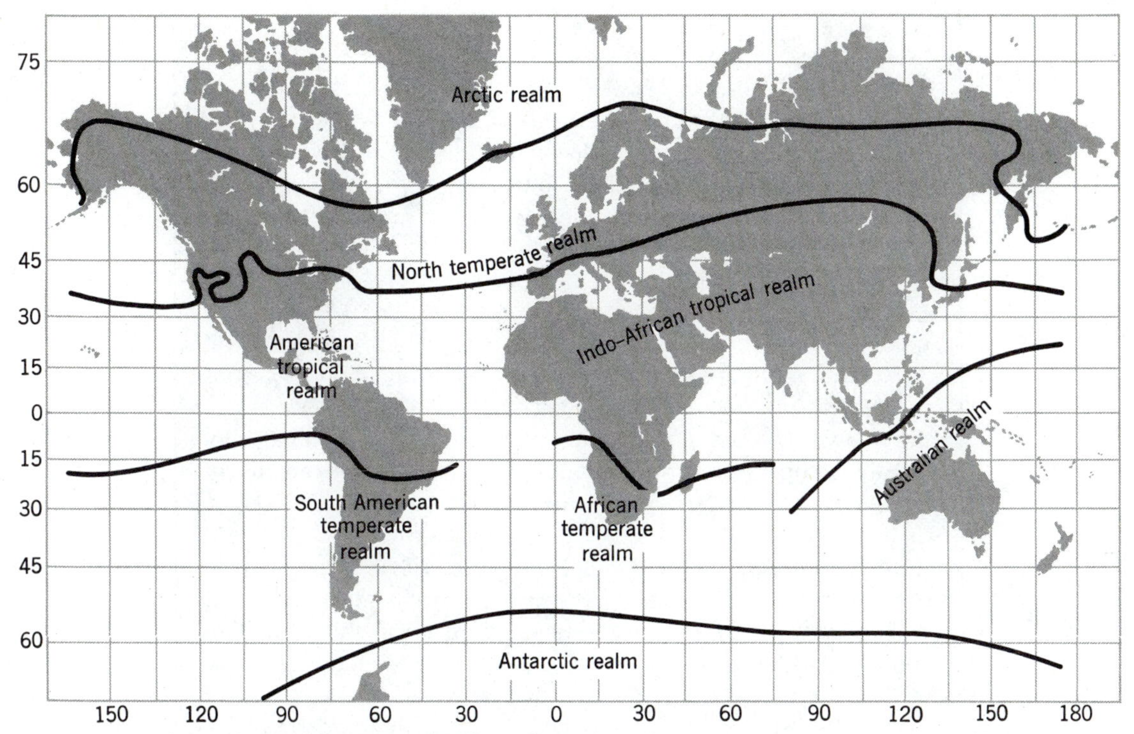

Figure 14.37 Allen's zoogeographic realms.
Each of these realms was erected to include birds and mammals that shared common adaptations. Like a map of formations, the main divisions fall naturally along lines of latitude. But Allen, working as a museum taxonomist as well as a geographer, was yet aware of the great taxonomic distinctness of the tropical continents, hence the separation of the tropical and south temperate realms into continental units.

len's ecological system and to do it in a way that would make the system seem ridiculous. Among his writings is a page of caustic comment along the lines of "Mr. Allen's supposed law is merely . . . etc." (Figure 14.38).

Wallace, of course, knew that the ecological classification was a logical possibility, that similarly adapted forms could be classified together, but this was in truth the reason he was so anxious to discredit the system. Classifying adaptations masked the evidence for evolution and was, indeed, the system long used by those who believed in special creation. For them, the toucans of South America were, in effect, the same as the hornbills of Africa, and hummingbirds and sunbirds were also more or less the same (Figure 14.39). Wallace, and those like him, spent their lives struggling to make certain that the great truth of evolution penetrated people's minds. There was a special creationist bishop hiding behind every page of an ecological classification, and the interest of these classifications was trivial compared with the importance of evolutionary truths which classification by family relationship revealed.

It is curious that ecology, the most evolutionary of subjects, should have had to begin with a conflict with evolutionists. The development of ecological thought was in fact held back for nearly half a century by it, until after the process of

Objections to the system of Circumpolar Zones.—Mr. Allen's system of "realsm" founded on climatic zones (given at p. 61), having recently appeared in an ornithological work of considerable detail and research, calls for a few remarks. The author continually refers to the *"law of the distribution of life in circumpolar zones,"* as if it were one generally accepted and that admits of no dispute. But this supposed "law" only applies to the smallest details of distribution—to the range and increasing or decreasing numbers of *species* as we pass from north to south, or the reverse; while it has little bearing on the great features of zoological geography—the limitation of groups of *genera* and *families* to certain areas. It is analogous to the *"law of adaptation"* in the organisation of animals, by which members of various groups are suited for an aerial, an aquatic, a desert, or an arboreal life; are herbivorous, carnivorous, or insectivorous; are fitted to live underground, or in fresh waters, or on polar ice. It was once thought that these adaptive peculiarities were suitable foundations for a classification,—that whales were fishes, and bats birds; and even to this day there are naturalists who cannot recognise the essential diversity of structure in such groups as swifts and swallows, sun-birds and humming-birds, under the superficial disguise caused by adaptation to a similar mode of life. The application of Mr. Allen's principle leads to equally erroneous results, as may be well seen by considering his separation of "the southern third of Australia" to unite it with New Zealand as one of his secondary zoological divisions. If there is one country in the world whose fauna is strictly homogeneous, that country is Australia; while New Guinea on the one hand, and New Zealand on the other, are as sharply differentiated from Australia as any adjacent parts of the same primary zoological division can possibly be. Yet the *"law of circumpolar distribution"* leads to the division of

F 2

Figure 14.38 Facsimile of a page from Wallace's *Geographical Distribution of Animals* (1876), in which he attacks the Allen ecological classification for obscuring evolutionary truths.

evolution by natural selection became the accepted working tool of biology. After that it was possible to study adaptation as an evolutionary process itself and concepts of niche or competition became ecological currency.

But perhaps the arrival of ecology as a respectable science was most clearly marked by the work of a Harvard chemist, L. J. Henderson (1913), who wrote a now classic book called *The Fitness of the Environment.* Henderson pointed out how well suited the terrestrial environment was to terrestrial life, in particular how the remarkable and unique physical properties of water at terrestrial temperatures were suited to the life of carbon chemistry (Figure 14.40). Before Henderson's book appeared, studies of the goodness

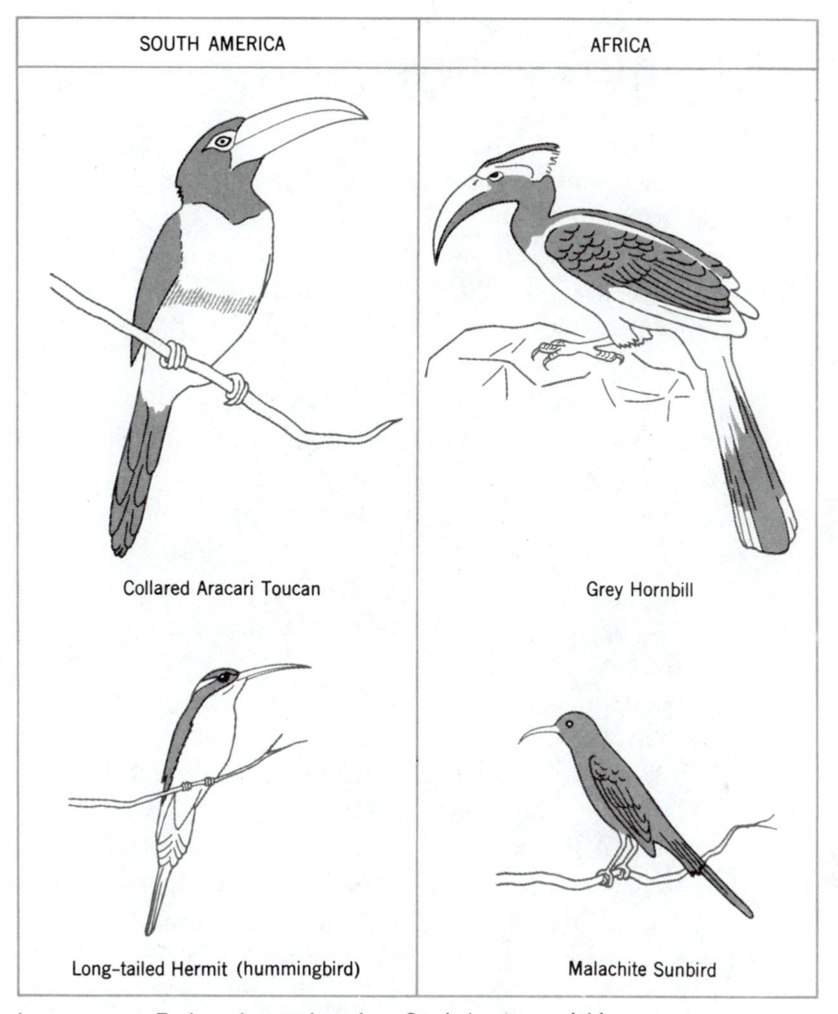

Figure 14.39 Ecological equivalents from South America and Africa.
Ecologists classify toucans and hornbills together, because they have similar Eltonian (class I) niches, but by evolutionary ancestry they are quite different.

of the environment, of the life-supporting properties of the earth, had a biblical and non-scientific ring to them. Nevertheless, animals and plants are adapted to their environments, so the converse must be true; that the environments are suited to them. Ecology studies the interaction between both; it must study the fitness of both.

CONTINENTAL DRIFT

Ecologists' understanding of geography has been changed by a new paradigm that has risen in geology, that of plate tectonics and sea-floor spreading. The continental masses of the lands, together with the continental shelves, are now to be thought of as plates of rock 700 kilometers

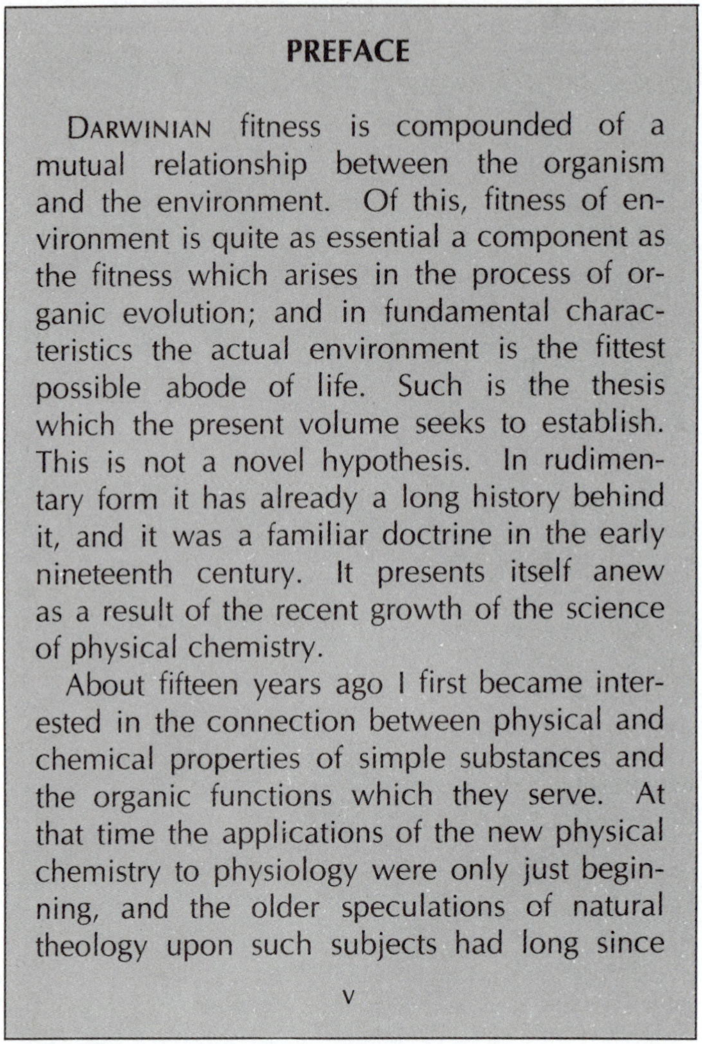

PREFACE

DARWINIAN fitness is compounded of a mutual relationship between the organism and the environment. Of this, fitness of environment is quite as essential a component as the fitness which arises in the process of organic evolution; and in fundamental characteristics the actual environment is the fittest possible abode of life. Such is the thesis which the present volume seeks to establish. This is not a novel hypothesis. In rudimentary form it has already a long history behind it, and it was a familiar doctrine in the early nineteenth century. It presents itself anew as a result of the recent growth of the science of physical chemistry.

About fifteen years ago I first became interested in the connection between physical and chemical properties of simple substances and the organic functions which they serve. At that time the applications of the new physical chemistry to physiology were only just beginning, and the older speculations of natural theology upon such subjects had long since

v

Figure 14.40 Facsimile of the first page of Henderson's book, *The Fitness of the Environment* (1913).
Henderson's statement that the adaptations of animals could not be properly studied without examining the qualities of the environments in which they lived was central to the establishment of much of modern ecological thought.

thick. These continental plates are nudged along by motions of the sea floor so that the continents drift on a scale of evolutionary time.

Where the continents drift away from each other there is a tear in the earth's crust, along the length of which hot rocks pour to fill the gap and form the lines of the mid-ocean ridges (Figure 14.41). Where the sliding plates collide there is the buckling of a mountain range as one plate slithers under the other. The line of a plunging plate (SUBDUCTION) yields a deep ocean trench on the side of it. As the descending plate is driven down, rocks melt and a line of volcanos marks its passing.

This general conception of plate tectonics is best described as a PARADIGM of science be-

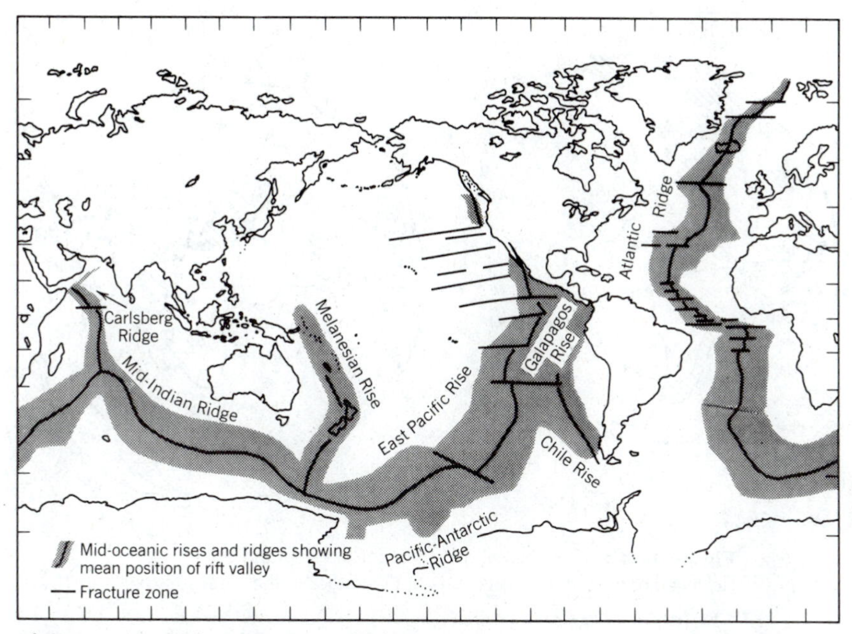

Figure 14.41 Mid-oceanic rises and ridges.
The rises and ridges are where new crustal rock is formed. Continents are moved away from the ridges.

cause it became important when the evidence for it was piled up to so visible a mass that virtually all geologists suddenly found it to be true (Cox, 1973; Kuhn, 1962). Plate tectonics explained so many things at once: the abyssal trenches; the island arcs of volcanos; the fitting shapes of distant continents; the lines of mountain-building or OROGENY; the high heat flux along the mid-ocean ridges; the intense magnetization of these ridges resulting from the rapid cooling of their lava in seawater; the long TRANSFORM FAULTS like the San Andreas.

The geological literature now provides us with a series of maps of the globe at different periods in the past (Figure 14.42). The maps are made on the assumption that each continental plate remains intact, unless there is evidence to the contrary, and the relative position of each plate is plotted from data on the direction of the earth's magnetic field recorded in ancient rocks. The age of each orientation is obtained by radiometric dating of the same rocks used for paleomagnetic determinations.

These maps are now raw material for biogeographers who can seek to explain modern patterns of distribution as, in part, resulting from the transport of ancestors on drifting continents. But formerly the evidence of biogeography itself was prime evidence on which a theory of continental drift could be built, and the possibilities of continental drift were as much a subject of biogeography as of geology.

Wegener[4] (1912, 1966) first put forward the theory of continental drift because he saw that the shapes of Africa and South America were so complementary that they could be fitted together to make a super-continent. He found some evidence that the two were once joined by matching rocks from either coast, but even more in the works of biogeographers who could not explain

[4]Alfred Wegener, a student of Vladimir Köppen in the years when the standard classification of climate was being made, explored remote places in quest for evidence of his theories of continental drift and meteorology. He was last seen on the Greenland ice cap, driving a dog team into a blizzard to fetch supplies for his winter camp (Georgi, 1934).

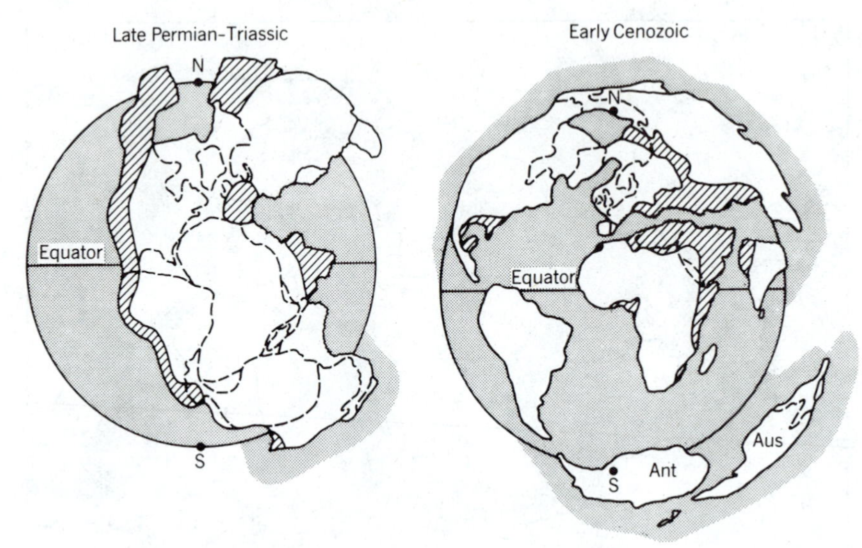

Figure 14.42 Continental drift: data for biogeography. Relative positions of the continents at two stages in the earth's history.

DISJUNCT DISTRIBUTIONS of species without postulating LAND BRIDGES across which ancestral populations once crossed. Most celebrated of the *land bridges* constructed on this logic was GONDWANALAND, the supposed bridge that connected Australia with South America via Antarctica. Gondwanaland seemed necessary to explain the otherwise puzzling occurrence of marsupials in Australia and South America but nowhere else in the world. Proponents of the land bridge said that this had sunk beneath the sea before placental mammals evolved in the Old World so that no placentals reached Australia. For Wegener, Gondwanaland was the old super-continent that had split in the middle before placental time, after which the parts drifted away.

Biogeographers argued over the validity of continental drift and land bridges for more than half a century. Was it *parsimonious*[5] to move large parts of the crust of the earth in order to get a fossil from one continent to another? Or was it more reasonable to postulate natural rafts for large animals and carriage in storms for small? Some land bridges were eminently believable, like the isthmus of Panama or a connection between England and France when sea level fell in an ice age, but the movements of continents across great oceans were not so easy to conceive. Darwin himself had rebuked persons who made land bridges "as easily as a cook makes pancakes."

Undoubtedly a biogeographer should try to find a way in which animals might have crossed the sea rather than expect land and animal to have crossed the ocean together. A sea crossing, for instance, certainly is required to get all the ancestors of the endemic fauna of the Galapagos Islands from South America, since modern geology of the sea floor shows conclusively that these islands never were joined to the mainland. The animals certainly rafted to the Galapagos, and rafts of flotsam can be seen today coming down the great Guayas river of Ecuador and heading out to sea in the general direction of the Galapagos. But some biogeographers once thought a Galapagos land bridge as likely as Gondwanaland.

A major work of biogeography was published just before the final triumph of continental drift

[5]The principle of parsimony states that, if several different explanations are all compatible with the evidence at hand, the simplest should be considered the most probable.

to show that all the known facts of zoogeography could be explained without any drifting at all (Darlington, 1957). But once the theory of continental drift rested on geological data, rather than requiring a whole continent to be boat to a fossil, Darlington (1965) announced he had become a drifter, "but not an extreme one." His position is the proper one for a biologist: if geological evidence shows that continents have moved, then continental drift is the likely explanation for disjunct distributions; but the argument should not be driven the other way to say that because the animals moved the continents must have moved also. Maps like those of Figure 14.42 now describe for ecologists the changing sets of the theater in which the evolutionary play of dispersion and speciation is performed (Hutchinson, 1965).

SUMMARY: THE SHIFTING COMMUNITY AND ECOSYSTEM STAGE

Recognizable communities and ecosystems exist because the surface of the earth is divided into different physical patches. The largest disjunctions of all are between land and sea, but even these are not permanent as the continents drift across evolutionary time. The next separate patches are set by climate, stamping a pattern that we recognize in our largest community units, the great plant formations, or biomes. Whether biomes are separate or blended depends on whether their controlling climates are separate or blended.

The permanence of biomes depends on the permanence of climate, and this is subject to frequent change. As recently as 10,000 years ago all the climates of the earth were displaced, or actually were different, as the globe emerged from the climatic upheaval of the last ice age. Equivalent changes of climate have been recurrent throughout at least the last two million years. It follows that biomes have been made and remade repeatedly, suggesting that communal relationships are in almost perpetual flux.

Communities and ecosystems are built as species enter or are removed from habitats defined by landscape and shifting climate. Communities are inherently subject to change. Ecologists often quote Hutchinson's (1965) aphorism about the ecological theater and the evolutionary play, saying that ecosystems set the stage for speciation. And yet for community studies, ecology contributes much of the play as well as the theater. Communities are shifting alliances of species, constantly being altered as physical habitats alter. If perpetual change forces the endless individual changes of evolution, change also drives continual alteration of communities made of shifting combinations of existing species.

PREVIEW TO CHAPTER 15

The interdependence of all living things with their physical environment, now implied by the term *ecosystem,* is an old concept in ecology but one that took a formal and central place only after intense study of plant communities. Ecologists, calling themselves *phytosociologists,* examined plant communities to see if they could be described as separate entities, as an earlier generation of biologists had done for special populations. *Associations* of plants in temperate landscapes could be identified by *ecological dominants* and *character species.* This could be done either by subjective identification of communities or by more quantitative means using *quadrat* samples and *species–area* curves. It was found, however, that these associations were set by habitats, at the edges of which the mix of plants changed rapidly, forming an *ecotone.* Classifying *associations* defined by species list (floristics) as sub-units of *formations,* defined by physiognomy, was possible but has not been valuable. Instead, the realization that habitats determined community composition allowed ecologists to emphasize ecosystem studies and the word *ecosystem* itself was a contribution of phytosociology. The importance of habitat was emphasized by ecologists who used ordination techniques to show that plants were distributed individually along environmental gradients, when neither community boundaries nor ecotones were present. *Continua* rather than *associations* were the norm except where physical boundaries were present. Within the continual change of species along gradients, however, biological interactions like competition or mutualism separated or grouped pairs or small clusters of species.

CHAPTER 15

VEGETATION, COMMUNITY, AND ECOSYSTEM

The first sustained studies of communities were attempts to identify and describe units of vegetation smaller than the plant *formations*. Within the eastern deciduous forest of North America, for instance, were oak–hickory woods, beech–maple woods, or woods where deciduous trees were mixed with the conifer hemlock. These woods were PLANT COMMUNITIES, distinctively different yet related as part of the TEMPERATE DECIDUOUS FOREST FORMATION. These observations raise immediate questions about how the individuals of communities are held together, essentially of how the communities are *organized*, and of how communities may be classified. Plants seemed to live in societies of sorts, inviting a sociology of plants. The first sustained efforts at community analysis came to be called PHYTOSOCIOLOGY.

Botanists first worked out formal methods of describing plant communities. These developed into methods of ORDINATION, in which the relationships of individuals within a community could be described. Community descriptions revealed important phenomena like *ecological dominance* and *ecological succession*. But most important was the development of the ECOSYSTEM CONCEPT. Phytosociologists found that the most important of the factors holding plant communities together was the physical habitat. The essential unit for study was the plant society and habitat combined. Botanists put a critical mass of scholars behind the idea of an integrated system of plants, animals, and the physical environment, and a botanist invented the term ECOSYSTEM to describe it.

Some plant communities are easily distinguished, like a STAND of forest trees—for instance, the *stand* of poplars down by a creek, or a *stand* of beeches and maples on a well-drained slope. But almost no statement describing a community can be separated from some mention of the environment as in these references to *stands*. This is reflected in Oosting's (1956) formal definition:[1] A COMMUNITY IS AN AGGREGATION OF LIVING PLANTS HAVING MUTUAL RELATIONS AMONG THEMSELVES AND TO THE ENVIRONMENT.

A first undertaking of phytosociology was to catalog visible communities, devising formal

[1] Oosting put the last phrase of this definition "and to the environment" in parentheses, thus acknowledging the views of those botanists who traditionally argued that communal associations of plants could be described without reference to the environment (see Oosting, 1956; Whittaker, 1962a).

methods of description on the way, then attempting to classify stands in a hierarchical system. Two prominent and opposed "schools" of phytosociology developed different methods, both of which have left a legacy in modern practice. The *Zurich–Montpelier* (Z–M) methods evolved for the complex vegetation of the south of France and those of the *Uppsala* school for the different task of separating communities from the less varied boreal forests of Sweden.

THE ASSOCIATION AND THE ZURICH–MONTPELIER (Z–M) SCHOOL

This approach is to decide in a purely subjective way what is a representative piece of vegetation or stand, and then to describe it according to a formal set of ground rules. The process is repeated in other stands and the separate descriptions are compared for similarities. Out of this exercise emerges a list of essential properties believed to be held in common by all stands of a community type, which was called the ASSOCIATION. Details of the procedure are as follows (Braun-Blanquet, 1932; Whittaker, 1962).

First choose a piece of vegetation believed to belong to a distinct type. The piece should be large enough to be representative of all facets of the community.

Second identify the ECOLOGICAL DOMINANTS subjectively. These are the most common or important plants; the oaks in an oak forest, the beeches and maples of a beech–maple woodlot, the *Festuca ovina* of a sheep's fescue pasture in Wales.

Third make a list of all the species of plant present in the chosen stand. This is not a census or sampling exercise. The botanist searches with diligence and a naturalist's skill to make sure that no single species is missed.

Fourth assign a set of importance values to each species on the list. These IMPORTANCE VALUES record whether a species is common or rare, small or large, dispersed or clumped. It is evident that these are all values that could, in principle, be measured in an objective and quantitative way. Some of the measures would not be easy, however, such as the number of grass plants in a meadow sample. Sometimes it is more convenient to measure the PERCENT COVER as the importance value in place of numerical abundance. But the Z–M method sidesteps this issue by assigning a COVER–ABUNDANCE INDEX, with "5" being the most important (covering more than three-fourths of the sampled area), "4" the next important (covering one-half to three-fourths of the sampled area), and so on down to "1." Dispersion is described with another 1 to 5 scale, the SOCIABILITY INDEX, as are CONDITION and VITALITY (Figure 15.1). It should be noted that all these indexes are assessed subjectively and depend on the good judgment of the botanist doing the work.

Fifth write the complete list of species found with the *cover–abundance* and *sociability* indexes after each name in that order, thus:

Zea mays 3.1

which describes the role of a corn plant in a typical field of corn (the corn community). The cover–abundance index of 3 means that corn plants cover one-half to one-third of the field (the rest being bare ground between the rows) and the sociability index of 1 means that the corn plants are evenly dispersed. The complete list written up in this way is called a RELEVÉ.

Sixth repeat the whole process at a number of other sites chosen for their apparent conformity to the type of community being described and tabulate the results (Figure 15.2).

Seventh compare all the *relevés* in quest of common denominators other than the ecological dominants. The dominants, of course, will be in every list because the botanist chose to sample only where the dominants grew. Now the botanist needs to find other, more humble plants that are faithful to the postulated community. Such faithful plants should be in every relevé. They are called the CHARACTER SPECIES.

Cover-abundance ratings according to following scale:

5 covering more than $\frac{3}{4}$ of the sampled area
4 covering $\frac{1}{2}$ to $\frac{3}{4}$ of the sampled area
3 covering $\frac{1}{4}$ to $\frac{1}{2}$ of the sampled area
2 with any number of individuals covering $\frac{1}{20}$ to $\frac{1}{4}$ of the sampled area, or very numerous individuals but covering less than $\frac{1}{20}$ of the area
1 numerous, but covering less than $\frac{1}{20}$ of the sampled area, or fairly sparse but with greater cover value
\+ sparse and covering only a little of the sampled area
r rare and covering only a very little of the sampled area (usually only 1 example)
(n.b. "+" is always spoken "cross")

Sociability is estimated for each species in terms of another scale:

5 in large solid stands; very dense populations
4 in small colonies or larger mats; rather dense populations
3 in small patches or polsters; distinct groups
2 in small groups or clusters or tufts
1 growing singly

Further symbols relating to the **condition** and **vitality** of the individual species, as those below, may be recorded beside the cover-abundance and sociability estimates.

oo - very poor and especially not fruiting (e.g., $+^{oo}$ or 2^{oo})
o - poor vitality (e.g. 1°)
g - germinating plant
Y - young plant
st - sterile
bu - budding
bl - blooming
fr - fruiting

no notation - normal growth
• - luxurious growth (e.g. 4•)
e - being driven out (by other plants)
d - dying
def - defoliated
dd - above-ground organs dead or dried out
s - present only as seed
- specimen collected

Figure 15.1 The descriptive indexes used by the Zurich–Montpelier school of phytosociology as interpreted by Benninghoff (1966).
This is an arbitrary set of indexes, but one that has had such wide use in the literature that it should be used in preference to ad hoc systems. For the purpose of pure description it is the most expedient method, but it is well to remember that the labor involved in using it may be very great and that such description might only be worthwhile when the object is to record for posterity the present condition of changing vegetation.

The exercise, summarized in Figure 15.3, is now complete. The investigator considers that an ASSOCIATION of plants has been demonstrated. It is characterized by the presence of both the *dominant* species and the *character* species. The association is then given a name based on the names of the dominants, like the *Beech–Maple Association* (there are various latinized systems of nomenclature also, some of them still in use, particularly in Europe) (Mueller-

Figure 15.2 Zurich-Montpelier system illustrated for Michigan cornfields.
The lists were each prepared from 4 × 4-meter plots, a size chosen intuitively, after which they were compared in quest of character species. The lists (relevés) are already long, despite the simplicity of a cornfield, suggesting how complex they might become for wild vegetation. (From Benninghoff, 1966.)

Total vegetation cover (5)	**75**	**90**	**95**	**75**	**70**	**40**	**10**	**60**	**40**	**25**
Total number of species	**9**	**12**	**5**	**8**	**9**	**5**	**7**	**7**	**9**	**11**
Relevé (list) number	**1**	**2**	**3**	**4**	**5**	**6**	**7**	**8**	**9**	**10**
Zea mays (planted)	3.1	3.1	3.1	3.1	3.1	3.1	3.1	3.1	3.1	3.1
Agropyron repens	r.1	+.3	+.2	r.3	r.1	•	+.1	2.2	+.1	+.1
Ambrosia artemisiifolia	+.1	1.2	•	r.2	+.2	r.1	•	+.1	1.1	+.1
Amaranthus retroflexus	r.1	•	2.2	r.1	1.2	r.2	r.1	•	1.1	r.1
Setaria glauca	r.2	+.2	•	•	•	1.2	r.1	2.3	+.1	+.1
Chenopodium leptophyllum	3.2	•	2.2	r.1	•	+.2	•	+.1	+.1	+.1
Daucus carota	•	2.3	•	•	r.1	•	r.1	•	+.1	+.2
Cirsium sp.	•	1.1	•	+.1	•	•	r.1	•	+.1	+.1
Chenopodium album	•	r.1	•	r.2	•	•	•	+.1	•	•
Portulaca oleracea	•	•	•	•	+.2	•	r.2	•	•	r.2
Digitaria sanguinalis	2.2	•	3.3	•	1.2	•	•	•	•	•
Panicum capillare	r.2	r.2	•	•	•	•	•	•	+.2	•
Plantago cf. major	•	r.1	•	•	•	•	•	•	•	r.1
Equisetum arvense	•	r.1	•	•	+.3	•	•	•	•	•
Xanthium chinense	•	•	•	•	•	•	•	2.2	•	•
Amaranthus sp.	•	r.1	•	•	•	•	•	•	•	•
Polygonum pennsylvanicum v. laevigatum	•	r.2	•	•	•	•	•	•	•	•
Taraxacum officinale	•	•	•	r.1	•	•	•	•	•	•
Panicum dichotomiflorum (typical)	•	•	•	•	+.3	•	•	•	•	•
Echinochloa crusgalli	•	•	•	•	•	•	•	•	•	r.1
Echinochloa pungens v. microstachya	r.2	•	•	•	•	•	•	•	•	•

Dombois and Ellenberg, 1974). Easy access to computers now makes possible the ordering of relevés with multivariate statistics but the sampling process itself remains subjective.

QUADRAT CENSUS: THE UPPSALA SOCIATION

The subjective element in the hunt for distinctive plant communities could be reduced by using a system of random subsamples when describing vegetation. These subsamples are called QUADRAT SAMPLES or simply QUADRATS. The essence of quadrat sampling is marking out small plots at random, then listing, measuring, or counting all the plants found in each plot. A formal method of identifying plant communities was developed in Sweden at the University of Uppsala even as more subjective methods were in vogue in the south of France (Du Rietz, 1929, 1930; Whittaker, 1962).

In the Uppsala method the species lists from series of quadrats are added and the results are plotted as a SPECIES–AREA CURVE (Figure 15.4).[2] As the area sampled increases with the number of quadrats added, so the expectation

[2] Species–area curves are used more widely now in the different context of diversity studies in which the goals are to understand how the number of co-occurring species in unit area comes about, see particularly Chapters 24 and 26.

of finding new species falls and the species–area curve tends to level off. This curve will never become truly horizontal, however, unless the whole earth is sampled at once, because increasing area always must bring in other habitats with different species. Yet reasonable rules can be written for deciding when a curve *inflects,* showing when all but chance plants of a locality are included. The area under the inflection point was called the MINIMUM AREA.

By this means a species list of co-occurring plants is derived and the smallest patch of ground on which this community can exist is defined. When the process has been repeated in a number of localities, the community lists can be compared to detect the regularity with which plants appear in the lists, expressed as PERCENT CONSTANCE. The community type, eventually to be called a SOCIATION, is then defined in terms of the *ecological dominants* and CONSTANT plants (those more than 80% constant) (Figure 15.5).

It will be evident that a piece of vegetation defined by *minimum area* and *constant species* in this way is likely to be much smaller than the piece of vegetation chosen for description by a follower of the Z–M school. Yet the goals of both methods are the same: to categorize fundamen-

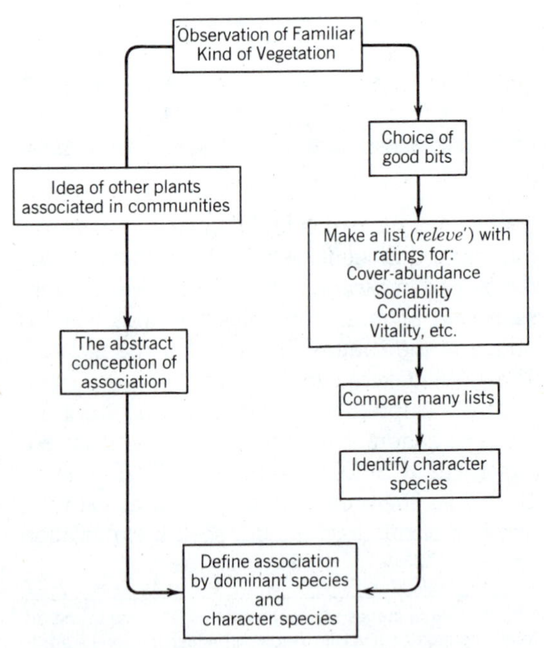

Figure 15.3 Procedure of the Zurich–Montpelier school of phytosociology.

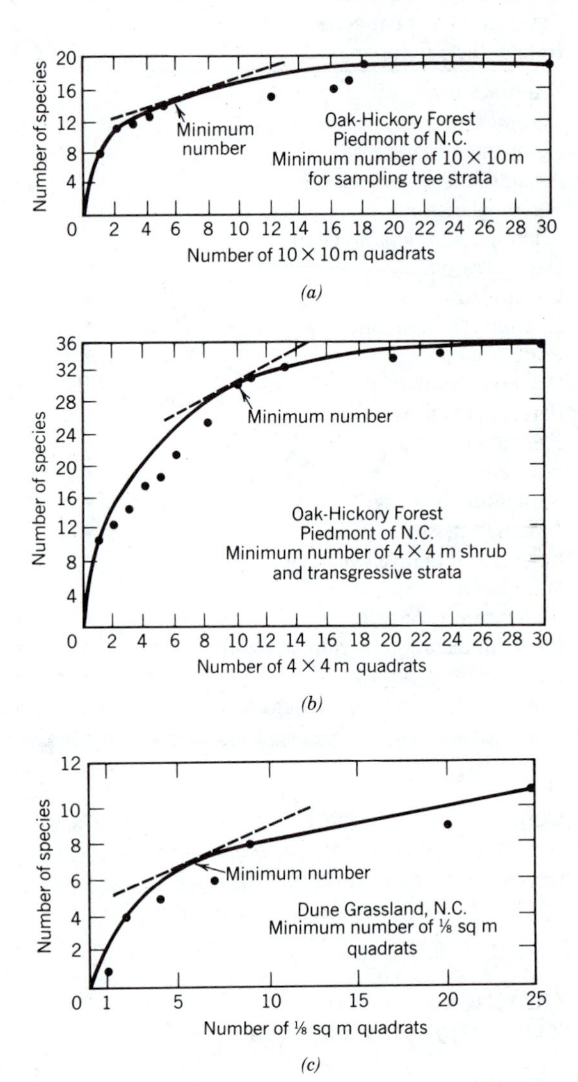

Figure 15.4 Species area curves for North Carolina vegetation.

Sizes of quadrats have been chosen in each example to be convenient. The "break" in the curve was chosen such that a further 10% increase in area would yield additional species equal only to 10% of those already present. This is a convention. Minimum areas are (*A*) (forest trees) 100 m^2, (*B*) (shrubs) 16 m^2, and (*C*) (grassland) 0.75 m^2.

tal community types. Both schools were in quest of fundamental *associations* of plants, and both talked of ordering and classifying ASSOCIATIONS. But since the actual pieces of vegetation were quite different, an International Botanical Congress ruled that communities defined by minimum area should be called SOCIATIONS to distinguish them from ASSOCIATIONS found by subjective means. It may be that little attention was paid to this ruling by botanists of the time because the term *association* continues to be used as a loose term to describe familiar bits of plantland, however this is described. Discussions of these matters can be found in Oosting (1956), Whittaker (1962), and Mueller-Dombois and Ellenberg (1974).

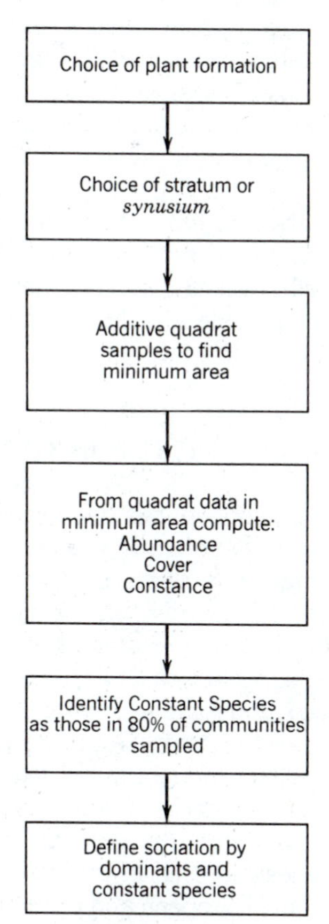

Figure 15.5 Procedure of the Uppsala school of phytosociology.

CLASSIFYING ASSOCIATIONS: THE INVENTION OF THE ECOSYSTEM

Many variants on the two opposed systems of Z–M and Uppsala were tried. Most were compromise methods that involved choosing stands in the Z–M manner, then using quadrats or other plot sampling techniques within the stands (Whittaker, 1962a). But all had in common a prime purpose that was taxonomic. The associations were to be classified into higher orders or subdivided into parts. Figure 15.6 shows some of the classifications proposed. At the left of the figure is a list of the subdivisions used in classical Linnaean taxonomy of species populations. The various schools of phytosociology produced their several names to be comparable. Uppsala saw their "sociation" as comparable to a species (but notice that they called this an "association" at first). Later both Uppsala and Zurich–Montpelier thought of "associations" as if they were biological families. Naturally enough, considering the variable properties of vegetation, the units sampled by botanical intuition or estimates of minimum area were of flexible size and so could be arranged to appear in different levels of the hierarchy.

A vital difference between the community classifications and Linnaean taxonomy appears when we consider the criteria used for separating units at different levels in the hierarchy. Linnaean-taxonomy groups related species together. The only act of differentiation is a species separation; the affinity between members of a phylum or class reflects a distant common ancestor, many acts of speciation ago. But the acts of separation of the higher orders proposed for communities are quite unrelated. The lowest unit is in fact a layer, called a *synusia* or *socion*. In plain language this means the bit of vegetation that happens to be made up of trees, or shrubs, or herbs. At the other extreme, the largest unit is the *formation*, the vegetable part of a biome set by climate. All the divisions in between depend on the botanical methods used to describe them, though they may also reflect large changes in habitat like a soil type. This is shown at the right of Figure 15.6.

Figure 15.6 Classifications of plant communities proposed by phytosociologists. For discussion see text. (Based on a compilation by Shimwell, 1971.)

<table>
<tr><th rowspan="2">Linnaen taxonomy</th><th colspan="6">Phytosociological classifications</th><th rowspan="2" colspan="2">Cause of separation</th></tr>
<tr><th>Uppsala 1921</th><th>Uppsala 1928–30</th><th>Uppsala 1930–35</th><th>Zurich–Montpelier</th><th>Rübel</th><th>Gams</th></tr>
<tr><td>Variety</td><td>—</td><td>Socion</td><td>Socion</td><td>—</td><td>Synusia</td><td>Synusia</td><td>Layering</td><td>Physiognomic</td></tr>
<tr><td>Species</td><td>Association</td><td>Sociation</td><td>Consocion</td><td>Facies</td><td>Sociation</td><td rowspan="6">Phytocoenosis</td><td rowspan="4">Botanical methods or physical habitat</td><td rowspan="4">Floristic</td></tr>
<tr><td>Genus</td><td>—</td><td>Consociation</td><td>Associon</td><td>(sub-association)</td><td>Consociation</td></tr>
<tr><td>Family</td><td>Association complex</td><td>Association</td><td>Federion</td><td>Association</td><td>Association</td></tr>
<tr><td>Order</td><td>—</td><td>Federation</td><td>Sub-formien</td><td>Alliance</td><td>Alliance</td></tr>
<tr><td rowspan="2">Class
Phylum</td><td rowspan="2">Formation</td><td>Sub-formation</td><td>Formien</td><td>Order</td><td rowspan="2">Formation</td><td rowspan="2">Climate</td><td rowspan="2">Physiognomic</td></tr>
<tr><td>Formation</td><td>Pariformien</td><td>Class</td></tr>
</table>

This analysis suggests that vegetation can be divided up by layers, or by life forms adapted to different climates, or by collections of species adapted to particular soils and patterns of drainage, *et cetera.* But these separations are all of different kinds and use different criteria.

Two widely used botanical terms speak to this dilemma of classification—*floristic* and *physiognomic.* FLORISTIC refers to the species composition of vegetation and PHYSIOGNOMIC to the shapes or structures of plants. Both layering and the life forms of formations are recognized by physiognomic characters but all the intervening units (associations and the like) are set by differences in floristics.

The phytosociological classifications are rather what we would get if we classified all flying animals together into the "flight formation" on physiognomic characteristics, and then separated them into mixes of insects, birds, and bats by specific differences; a sort of animal floristics. The result would be a most un-Linnaean and un-Darwinian classification. But something like this is what classifying plant communities as units of formations implied.

Braun-Blanquet (1932), a wise botanist of the Z–M school, revived the efforts of plant sociology to find and classify natural communities in a book that summarized the work of his school. He noted that when it was possible to make a clear decision about where one community ended and another began this was always because there was a change of habitat at that place, a change of soil, exposure, or moisture. The PLANT COMMUNITY WAS actually THE COLLECTION OF PLANTS THAT SHARED A HABITAT, and physical, not biological, factors were decisive in ordering communities. The observation that habitats were crucial determinants of plant communities was, of course, made independently by every botanist who went out to sample. All were equally aware that the fortunes of many a plant also depended on the animals that lived with it, to eat it, to pollinate it, or to carry its seeds. Even as attempts to classify vegetation met with indifferent success, these thoughts of habitats and animals led botanists to a different idea, that of the ecosystem.

THE CONCEPT OF THE ECOSYSTEM

The ecosystem concept has deep roots in ecology. Möbius (1877) often is given credit for the first formal statement of the idea, for which he used the term BIOCOENOSE (or *biocoenosis*). Möbius reflected on an oyster bank: the things the oysters ate, oyster predators, and the narrow set of physical requirements that must be met for an oyster bank to persist. Doubtless many naturalists before him thought in like ways about a community and its needs, but Möbius coined a formal term to define the unit. Even now the term *biocoenosis* is used in preference to *ecosystem* in central European and Russian writings.

In that same year Forbes (1877) proposed the

same idea with the name MICROCOSM.[3] Like Möbius, Forbes worked with an aquatic system, thinking of a lake as a self-contained unit, with its own plants, animals, and nutrient supplies all dependent on each other. A tradition of thinking of environment, nutrients, and food spread as naturalists everywhere arrived at positions close to those of Möbius or Forbes (Elton, 1927). But these ideas came to be stated as the central concept of ecology most clearly when botanists reached a general consensus on the nature of the *association*.

The one factor that determined the species list of every *association* was the physical habitat. The unit for study, therefore, must be the whole tangled mixture of plants, animals, and their physical surroundings. An idea was born of these studies and scholars groped for new words to describe it: the "LANDSCAPE UNIT" or RAUME, the BIOGEOCOENOSIS, the BIOCHORE, and the BIOTIC DISTRICT all had their devotees. Terms like "landscape" or "districts" show the influence of the terrestrial botanists on whose work this central theme of community ecology was based. And one of them, Tansley (1935), proposed the term ECOSYSTEM for the new idea (Figure 15.7).

Tansley's ECOSYSTEM would be considered to include all the animals, plants, and physical interactions of a defined space. Natural limits would be identified for the ecosystem under study, like the changes of soil or drainage defining plant associations, or the edges of streams and lakes. An ecosystem could be of any size depending on the communities to be studied; as small as the cup of a pitcher plant or as large as a biome or the whole biosphere.

Modern ecologists think easily of ecosystems in terms of energy flow, carbon flow, or nutrient cycles. But these concepts came later; the idea of energy flow, for instance, was not formally stated until seven years after Tansley clarified the phytosociological debate with his ecosystem (Lindeman, 1942).

[3]Forbes' essay was published in the Bulletin of the Peoria Scientific Association, and cannot have been widely read. It was rediscovered by later ecologists as the ecosystem concept was developing independently.

ORDINATION TECHNIQUES: CONTINUUM AND GRADIENT ANALYSES

The discovery that community boundaries are habitat boundaries leaves unanswered the possibility that some or all species are adapted to live together more than alone or with other species. Are, in fact, the familiar communities on a familiar habitat entirely chance aggregates in which the only common denominators are suitability to physical circumstance? Or do some species occur together because they are suited to each other in some way? These questions can be addressed by ORDINATION TECHNIQUES that measure the order in which individuals occur along any dimension of a community. The first *ordinations* were CONTINUUM ANALYSIS, which searched data ordered by biological similarity, and GRADIENT ANALYSIS, which searched for affinities along a series of related habitats. Different hypotheses of community organization are behind the two approaches.

1. The Discrete Community Hypothesis This states that plants exist in characteristic floristic assemblages (typically called associations), and that these assemblages have characteristic dominant and faithful species. A prediction of this hypothesis is that each assemblage will have recognizable limits or edges.

2. The Independent Plant Hypothesis This states that plants do not live compulsively in characteristic communities but that each species population has its distribution set independently. A prediction of the hypothesis is that there should be no recognizable edge to a community, but rather a continuous blending of stands.

The distributions expected by the two hypotheses are illustrated in Figure 15.8. The related schools of continuum and gradient analysis

THE ECOSYSTEM

I have already given my reasons for rejecting the terms "complex organism" and "biotic community." Clements' earlier term "biome" for the whole complex of organisms inhabiting a given region is unobjectionable, and for some purposes convenient. But the more fundamental conception is, as it seems to me, the whole *system* (in the sense of physics), including not only the organism-complex, but also the whole complex of physical factors forming what we call the environment of the biome—the habitat factors in the widest sense. Though the organisms may claim our primary interest, when we are trying to think fundamentally we cannot separate them from their special environment, with which they form one physical system.

It is the systems so formed which, from the point of view of the ecologist, are the basic units of nature on the face of the earth. Our natural human prejudices force us to consider the organisms (in the sense of the biologist) as the most important parts of these systems, but certainly the inorganic "factors" are also parts—there could be no systems without them, and there is constant interchange of the most various kinds within each system, not only between the organisms but between the organic and the inorganic. These *ecosystems*, as we may call them, are of the most various kinds and sizes. They form one category of the multitudinous physical systems of the universe, which range from the universe as a whole down to the atom. The whole method of science, as H. Levy ('32) has most convincingly pointed

[3] If this statement is applied to the individual organism, it of course involves the repudiation of belief in any form of vitalism. But I do not understand Professor Phillips to endow the "complex organism" with a "vital principle."

Figure 15.7 The first appearance of the word ecosystem. (From Tansley, 1935.)

both developed in the United States in the 1950s as a deliberate attempt to test these two hypotheses. It is probably fair to say that the investigators expected to confirm that the discrete community hypothesis was false.

Clearly a third hypothesis is possible: that some organisms have positive or negative affinities for each other, even if the community as a whole is largely a random collection of species sharing adaptations to common physical circumstance. At the level of vegetation analysis, this hypothesis also can be tested by ordination techniques, because affinities between species should appear as clumped distributions along continua.

Continuum Analysis

CONTINUUM ANALYSIS proceeds by ranking stands according to perceived relationships so that all stands within a forest appear in the list. The most dissimilar stands appear at opposite ends of the list, with a continual progression of variants in between. The list is said to define a CONTINUUM. Proponents of continuum analysis argued that the *continuum* was a more real-

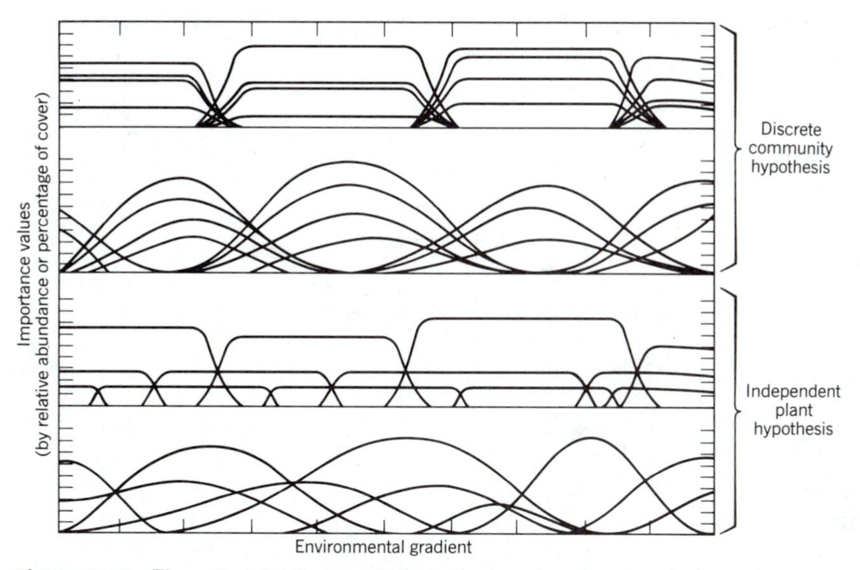

Figure 15.8 Theoretical distributions of plants along environmental gradients. Either (*A*) or (*B*) is required by the *discrete community hypothesis* of classical phytosociology. (*C*) and (*D*) are both consistent with the *independent plant hypothesis*. (Based on a figure of Whittaker, 1975.)

istic description of order in forests than was the *association*. They emphasized the blending rather than the discreteness of communities.

The original demonstration of continuum analysis used data from forest trees in Wisconsin (Brown and Curtis, 1952; Curtis and Mackintosh, 1950, 1951; Curtis, 1959; Cottam and Curtis, 1949, 1956). The importance values they used were relative abundance, frequency, and the measure of relative cover of trees provided by basal area, where

DENSITY is number of trees per unit area

FREQUENCY is percentage of plots within a stand in which individuals of a species are found and

BASAL AREA is a measure of area of trunks at breast height, then:

$$\text{Relative density} = \frac{\text{No. individuals of species } x}{\text{Total individuals of all species}} \times 100$$

$$\text{Relative frequency} = \frac{\text{Frequency of species } x}{\text{Sum of frequency values for all species}} \times 100$$

$$\text{Relative basal area} = \frac{\text{Basal area of species } x}{\text{Total basal area of all species}} \times 100$$

and

$$\text{Importance value of species } x = \text{Relative density} + \text{relative frequency} + \text{relative basal area of species } x$$

Within each stand sampled, the *importance values* for each species are averaged. In each stand the dominant tree, easily recognized in these northern forests, will certainly turn out to have the highest importance value. The stands are then ranked in order of importance values of the leading dominant as in Figure 15.9. That a real and valid separation is achieved by this procedure is shown by looking at the species involved in the arrangement in Figure 15.9: *Acer saccharum* (sugar maple) is a tree of the climax forest and

No. of Stands	Leading Dominant	Acer saccharum	Tsuga canadensis	Betula lules	Acer rubrum	Quercus rubra	Betula papyrifera	Pinus strobus	Pinus resinosa	Populus tremuloides	Quercus ellipsoidalis	Pinus banksiana
23	*Acer saccharum*	145	25	21	7	22	6	1	—	1	—	—
23	*Tsuga canadensis*	40	152	47	11	3	5	4	3	—	—	—
6	*Quercus rubra*	27	1	3	29	138	23	10	8	5	2	—
6	*Betula papyrifera*.......	48	8	7	27	16	108	19	1	29	1	—
19	*Pinus strobus*..........	12	6	2	24	12	12	150	39	9	5	—
9	*Pinus resinosa*.........	3	—	1	12	15	14	56	156	24	4	2
4	*Populus tremuloides*	11	—	—	10	29	34	14	19	140	—	—
4	*Quercus ellipsoidalis*......	—	—	—	5	7	1	11	9	9	103	34
10	*Pinus banksiana*.......	—	—	—	3	3	3	13	12	14	36	213

Figure 15.9 Sample of data used in continuum analysis.
The 104 stands were first identified by the dominant trees, yielding the nine kinds of stand. For each kind of stand the importance value for each tree species was then averaged. Finally, the stands were ranked according to the averaged important values. (From Brown and Curtis, 1952.)

Pinus banksiana is a pine of early succession stages, which are most unlikely to be found growing together.

Each species is assigned a CLIMAX ADAPTATION NUMBER on the basis of the position of the stand that it dominates in the rankings (Table 15.1). *Pinus banksiana* dominates stands at one side of Figure 15.9 and is given a rating of "1," *Acer saccharum* dominates stands at the other end and is given a rating of "10," and the other species are allotted numbers in between. It is essential to notice that no botanical assumptions are involved in allotting these numbers. Though the numbers are themselves arbitrary, they reflect a real ranking of the kinds of stands that are revealed by objective measurement.

A STAND INDEX NUMBER is now calculated for each stand. The *importance value* of each species present is multiplied by its *climax adaptation number* (CAN) and the totals are added (Table 15.2). The final act is to plot importance values for each species against the stand adaptation numbers for the stands in which it occurs, with results like that in Figure 15.10.

The results in Figure 15.10 are similar to those of the hypothetical distribution in Figure 15.8*D*.

Table 15.1
Climax Adaptation Numbers of Tree Species in Wisconsin
These numbers are derived directly from Figure 15.9 and depend only on the calculations of importance values. (From Brown and Curtis, 1952.)

Tree Species	Climax Adaptation Number
Pinus banksiana	1
Quercus ellipsoidalis	2
Quercus macrocarpa	2
Populus tremuloides	2
Populus grandidentata	2
Pinus resinosa	3
Pinus pennsylvanica	3
Quercus alba	4
Prunus serotina	4
Prunus virginiana	4
Pinus strobus	5
Betula papyrifera	5
Acer rubrum	6
Quercus rubra	6
Abies balsamea	7
Tsuga canadensis	8
Betula lutea	8
Fraxinus americana	8
Tilia americana	8
Ulmus americana	8
Ostrya virginiana	9
Fagus grandifolia	10
Acer saccharum	10

Each plant is distributed independently and communities cannot be recognized as discrete entities.

Gradient Analysis

In GRADIENT ANALYSIS plants are sampled on a smaller scale along steep and obvious environmental gradients like that of the side of a mountain. The use of this technique by Whittaker (1953, 1956, 1966) to show how vegetation blends on mountain sides was introduced briefly in Chapter 14.

Series of quadrats along a transect up a mountain are sampled to yield stems per hectare, percent cover, or grams of biomass per square meter. All of these can be used as importance values. It is then a simple matter to plot importance value against altitude with results like that in Figure 15.11.

Stretching plots out along so obvious a gradient as altitude is the simplest form of gradient analysis, but it has the disadvantage that many gradients are being sampled at once, for temperature, precipitation, soil type, wind speed, *et cetera,* all change with altitude. Whittaker's response to this was to examine lateral gradients within each 300-m section of transect (often a moisture gradient) at the same time as he sampled the altitude gradient. All parameters of altitude ought to be constant in the 300-m section so that Whittaker approaches the experimental ideal of varying one factor while keeping the rest constant.

The ranking and weighting procedure in *gradient analysis* is as follows:

First describe the stands, say, by calculating the relative density of all trees in each stand.

Second make a species list of all species found in all plots along the transect.

Third assign each species an arbitrary multiplier value on a scale of wetness and dryness (or other measure of gradient), for example, mesic (wettest site species) has multiplier of 0

submesic 1
subxeric 2
xeric (driest site species) 3

then, if 100 trees in a stand had

60 individuals of mesic species
30 individuals of submesic species
10 individuals of subxeric species
0 individuals of xeric species

Table 15.2
Procedure Used to Derive Stand Index Numbers in Continuum Analysis
(Data from Curtis, 1959, after Collier et al., *1973.)*

	Sampling Data from Stand								Importance Value (IV)		Climax Adaptation Number (CAN)		IV × CAN
Species	**Relative Density**		**Relative Dominance**		**Relative Frequency**								
Acer saccharum	30	+	35	+	25	=			90	×	10.0	=	900
Fagus grandifolia	25	+	35	+	30	=			90	×	9.5	=	855
Fraxinus americana	10	+	15	+	15	=			40	×	6.5	=	260
Ostrya virginiana	20	+	10	+	15	=			45	×	8.5	=	382
Carpinus caroliniana	15	+	5	+	15	=			35	×	8.0	=	280
	100		100		100								
									Stand index number			=	2,677

the MOISTURE INDEX for this stand is

$$\frac{(60 \times 0 + 30 \times 1 + 10 \times 2 + 0 \times 3)}{100} = 0.5$$

Figure 15.12 is an illustration of the results of this method from Shimwell's (1971) work in England. It deals with mosses in tiny transects instead of trees on mountains but it shows a typical result. The stands are ordered on the *x* axis by *moisture index* and arranged by gradient on the *y* axis. The curve follows a rough diagonal of the figure. There are no signs of the steps that would be expected if the communities were discretely separate.

It should be emphasized that for both continuum and gradient analysis the dice are not loaded in favor of blending. Individual stands *chosen for homogeneity* are ordered by an impartial statistical weighting, and the results are not a consequence of the method. The conclusion is that COMMUNITIES ARE AGGREGATIONS OF INDIVIDUAL SPECIES POPULATIONS EACH FINDING ITS LIMITS IN TIME AND SPACE. The species are not regimented together into standard communities. *D* in Figure 15.8 is the best approximation of reality.

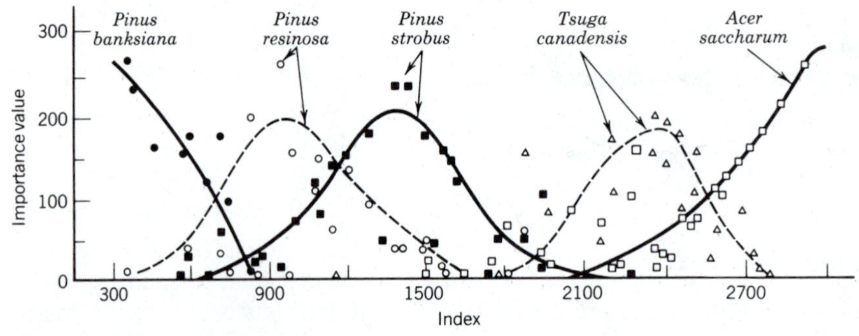

Figure 15.10 Gradient analysis of a continuum for the upland conifer–hardwood forests of northern Wisconsin.
Only the dominant tree species are shown here. (After Brown and Curtis, 1952.)

COMPETITION, MUTUALISM, AND DEPENDENCE IN COMMUNITY BOUNDARIES

Opposing views of phytosociologists revealed two principal kinds of community arrangement in plants. Community boundaries could be discrete (*A* or *B* in Figure 15.8), but only if physical boundaries were present. Species distributions then end in a narrow band of overlap called an ECOTONE. *Ecotones* always denote changes in the physical state of the habitat. Most commonly this is a change in soil, exposure, or moisture, but it can be set at the edge of burned ground or other clearing. If, to the contrary, physical conditions for life grade gradually species are assorted independently as in *D* in Figure 15.8.

There is, however, a third possibility. Small subsets of species can have ranges set or synchronized by biological interactions. Competition between species can set discrete boundaries even though the environment grades smoothly, as in *C* in Figure 15.8. An example is the segregation of bird species on mountain sides discussed in the last chapter. More importantly any interdependence between species will lead to synchronous ranges and thus to apparent community boundaries.

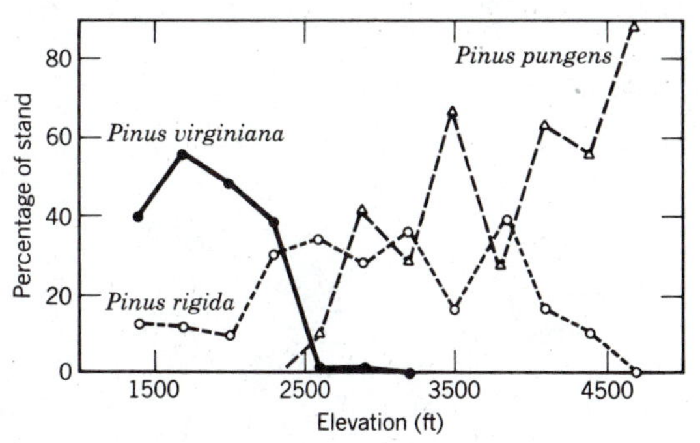

Figure 15.11 A Whittaker transect of an elevation gradient.
The transect is along dry, south-facing slopes in the Great Smoky Mountains of Tennessee. No boundaries separate the three community types that an ecologist is likely to distinguish along this gradient: *Pinus virginiana* forest at low elevations, *Pinus rigida* heath at middle elevations, and *Pinus pungens* heath at high elevations. (After Whittaker, 1956.)

The most complete dependence is MUTUALISM or SYMBIOSIS. The fungal and algal symbionts of lichens have identical ranges, unless one of the pair has a separate range as a free-living organism. Parasites and hosts also have synchronous ranges, unless the parasite has several hosts, or the host occupies ranges where the parasite cannot follow. Parasitic plants like Indian pipe (*Monotropa uniflora*) are distributed only where the host trees grow. Yet most mutualisms span trophic levels, as in the relationships between trees and ants who feed on special nectaries and preserve the trees from herbivores in return. Specialized pollinators are

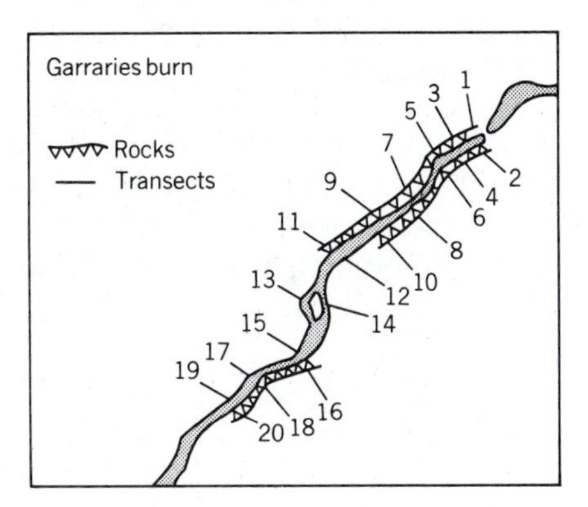

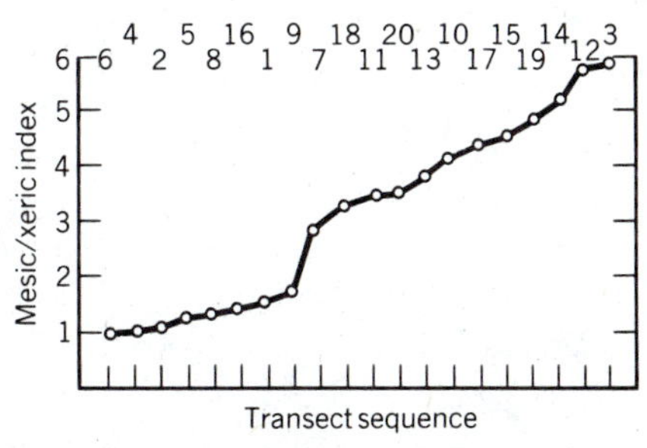

Figure 15.12 Composite transect of a moisture gradient in Scotland.
The communities sampled are of mosses growing along the Garraries Burn in Kirkudbrightshire. Twenty transects of moss communities beside the burn are equivalent to the 300-m slices used by Whittaker on American mountains. (After Shimwell, 1971.)

distributed with the plants that use them, specialized herbivores with the plants whose allelopathic substances they are able to detoxify, and mimics with their models (Chapter 25).

These patterns can be revealed by ordination techniques more sophisticated than those of classical phytosociologists, typically employing computers for cluster analysis or principal components analysis. Species are found to cluster together in portions of a general continuum, suggesting species interdependence.

CONCLUSION: ECOSYSTEMS AND COMMUNITY BUILDING

Attempts to describe and catalog plant communities occupied a large part of the total ecological effort in the formative decades of modern ecology. That ecologists in temperate latitudes found the description of *associations* relatively simple was due to the completeness of *ecological dominance* in these communities. A few physically prominent species were able to achieve such competitive advantage that the plants of relatively homogeneous habitats many hectares in extent could be grouped naturally with their superior competitor. Yet this relatively uniform species composition is always dependent on habitat uniformity. Discrete communities do not exist except where habitats are discrete.

Two parallel approaches in modern ecology result from these findings of phytosociology: ecosystem studies and studies of how species interact to produce the structures of communities. The ecosystem scientist seeks habitats with edges recognized by *ecotones*, then examines energy flow between trophic levels, productivity of whole vegetation, nutrient cycles, soil formation, and mass transport of substance within compartments of the system that has been defined. Meanwhile, the structure of the community is studied as resulting from the impact of one species on another, whether enemy, competitor, or friend, and community ecology is but an extension of population ecology. Community building is a game of species packing and the result of the game in any time and place is the temporary coexistence that we call a "community."

PREVIEW TO CHAPTER 16

The concept of the ecosystem invokes one of the better known overviews of ecology, that of communal relationships directed by a flow of energy. Conventional ecosystem diagrams show work being done by energy that cascades like falls of water, or that is directed down a network of pipes like water through plumbing. These hydraulic analogies can give misleading impressions of real ecosystems, however, because energy follows diverse pathways through real communities. Current thermodynamic theory does not let us produce useful models of ecosystem function from postulates about energy transfer, although the thermodynamic overview does give a satisfying understanding of the relative abundance of animals of different habits. Energetic studies of real ecosystems are empirical. Net production (not gross production) has been measured for many plant communities, and the data are sufficient to permit general investigations of the causes of particular levels of productivity in different ecosystem types. This work has progressed to the point where it is possible to make computer models that predict average productivity for broad regions of the earth from data of temperature and precipitation, showing that these two inanimate factors are the primary determinants of net production, at least on land. Net production may be measured by harvest techniques, though these are of limited use for forests because of the extensive work involved. It has been found to be possible to relate measures of tree size and age of some forests to productivity so that computer programs can be written that predict productivity of a forest from a few simple measures of tree height and stem diameter. Calibration of these models, however, is very laborious and the approach has been applied only to a few sites. Conversion of harvested dry mass data to calories turns out to be comparatively simple, since the compositions of most kinds of plant are similar. Measures of respiration of plants in the wild are possible, though difficult. Enclosing plants in chambers leads to uncertain results, but it is sometimes possible to make use of inversions, or numerous simultaneous measures of carbon dioxide concentration in the air by night and day, to derive the respiration and productivity of whole terrestrial ecosystems. Productivity of aquatic ecosystems is measured more easily because of the small size of the plants and the convenience of doing chemistry on ions in solution. The open oceans are, for the most part, unproductive deserts because they lack nutrients. There are very few data for animal production on land or sea and the principal, though very tentative, estimates are discussed. Possibly the most important of all conclusions from studies of ecosystem energetics is that by far the largest part of primary production is degraded by decomposers and detritivores rather than by herbivores.

CHAPTER 16

ECOSYSTEM ENERGETICS

An ECOSYSTEM can be defined as THE LIVING COMMUNITIES, TOGETHER WITH THEIR NON-LIVING ASSOCIATED FACTORS, PHYSICALLY CONSTRAINED OR DEFINED IN SPACE.[1]

In most studies the term *ecosystem* is reserved for a distinctive unit of landscape with geographical and geological continuity. Obvious examples are lakes, woodlots, estuaries, farm fields, or even cities. Smaller examples are hot springs, the lichen mat on the side of an alpine boulder, or the water in a pitcher plant. Very large ecosystems are the *biomes,* and we talk of the *tropical rain forest ecosystem, the tundra ecosystem,* and so on. In this usage an ecosystem merges into an ecosystem type. Strictly speaking ecologists study the tropical rain forest of South America, or Africa, or Malaysia, but we often talk of the *tropical rain forest ecosystem* when we describe properties that all rain forests hold in common. Likewise we talk about the "old field ecosystem" or the "lake ecosystem." This shifting from the particular to the general is a property of language itself.

The largest ecosystem of all we call the BIOSPHERE. This may be defined as ALL LIVING MATERIAL OF THE EARTH TOGETHER WITH NON-LIVING ASSOCIATE FACTORS, and this is the preferred usage in ecology. However, a subtly different meaning is often given to the word "biosphere," that of the earthly envelope containing life. This meaning speaks mostly of the inanimate world, of the atmosphere and ocean systems as the domain of life, rather than of the life itself. This second meaning appears most often in the resource literature rather than in primary ecological writings.

All ecosystems are so complex that the word "system" in its name can be an understatement. A SYSTEM IS ANY PHENOMENON, EITHER STRUCTURAL OR FUNCTIONAL, HAVING AT LEAST TWO SEPARABLE COMPONENTS AND SOME INTERACTION BETWEEN THESE COMPONENTS (Hall and Day, 1977). Possibly the Russian term BIOGEOCOENOSIS describes more accurately what is meant than our reference to systems. *Biogeocoenosis* literally means "the life-earth fusing together." All the major questions of ecology and evolutionary biology must invoke this concept of the interaction of life with its physical surroundings and with other life. As noted in the last chapter, far-seeing naturalists have long come to this realization for themselves, though the concept came to be stated in

[1]Definition derived from the discussion in Hall and Day (1977).

universally acceptable terms only when a critical mass of botanists found that terrestrial plant communities could be delimited unambiguously only by reference to the physical habitat. But one thing more was required for the ecosystem concept to take its central place in modern ecology. This was the CONCEPT OF ENERGY FLOW.

The energetic model of ecosystem function was introduced to ecology by Raymond Lindeman in his classic paper of 1942, "The Trophic Dynamic Aspects of Ecology." This was the paper in which the energy flow explanation of the pyramid of numbers was introduced (Chapter 2). Energy transfer along food chains was inherently inefficient, large predatory animals were rare, and much of the structure of ecosystems was a necessary consequence of energy degradation (Chapter 4).

This introduction of energetics to the ecosystem concept meant change in the ways that ecologists thought. Before then the emphasis was on physical impacts of habitats on communal life. Food chains and food webs were discussed by animal ecologists, but in the sense of Elton's (1927) *food cycle* (Chapter 2): it was physical transfer that was stressed. The concept of an energy flux being degraded as it passed through an ecosystem offered much more of a unifying principle for ecosystem studies.

This energetics view was taken up by the standard treatise on ecology of subsequent years (Allee *et al.*, 1949) and repeated in the following generations of college texts. Ecosystems were to be understood from the workings of the laws of thermodynamics: the first law that energy may be transformed but neither created nor destroyed, and the second law that no process involving energy transformation can occur spontaneously without dispersal of energy (Chapter 1). Cycling within an ecosystem, its physical structure, the modification of the habitat by life, and the abundance of animals or plants in each trophic level were to be understood by ecosystem energetics (Figures 16.1 and 16.2). This concept let a new dynamic into ecology as ecosystems were studied as functional units.

THE HYDRAULIC ANALOGY

A conceptual tool in modern ecology has been the hydraulic analogy (Figure 16.3). In this familiar ecosystem diagram energy cascades through the system: first a prism to represent the separation of light into that small portion used in photosynthesis and the larger part sent back to space, then the ever-narrower pipes connecting the trophic levels, the pipes to decomposers, and the total outgoing flux of heat energy that must balance the inputs of light. Understanding of ecosystem function, or comparing ecosystems, requires that the standing stock of potential energy represented for each compartment in the flow diagrams (Figures 16.2 and 16.3) be known, together with the energy inputs and outputs to each compartment. A very large research effort has gone into supplying representative numbers. The methods that must be used, and some of the results, are the subject of this chapter.

Figure 16.4 shows how hydraulic analogy diagrams can be used to compare six ecosystems. Standing stocks of energy are shown to be quite different between open water and a forest, which is hardly a novel observation, but it is something of a revelation to see that the respiration of a tiny crop of algae is comparable to that of the massive standing crop in a grass meadow. MacFadyen (1962) took this to mean that respiration is a function of surface to volume ratio, and so is relatively high for each tiny planktonic plant.

Hydraulic and other flow diagrams are thus tools for research and valuable practical guides. It is necessary, however, to put in a word of caution about their theoretical limitations. The analogy can be taken too literally so that real ecosystems are thought of as simple sets of second-law energy transformations between trophic levels such that a fairly straightforward application of thermodynamic theory would let us make predictive models of ecosystem function. The reality is different. In real ecosystems energy is not constrained to flow along set routes, but must be fed back through complex loops and pathways. Moreover, the energy transformations

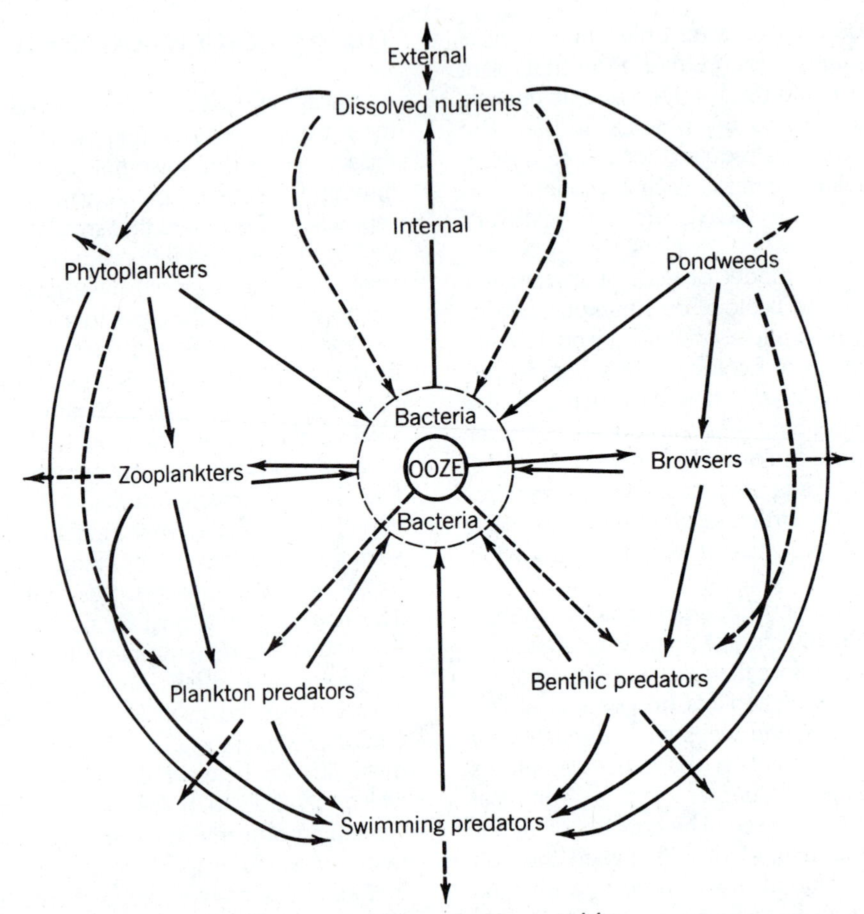

Figure 16.1 Lindeman's ecosystem diagram of a Minnesota lake.
Arrows represent flows of energy and nutrients. Broken arrows represent less certain pathways. The importance of bottom mud in a lake as a reservoir of nutrients and regulator of their supply is indicated by the central position of "ooze" in the diagram. (From Lindeman, 1942.)

measured as respiration record work done in more ways than just the transfer of energy from one potential state to another. The second law of thermodynamics itself offers little more than analogy when describing this energy cascade through an ecosystem (Slobodkin, 1962).

Work with the energetics of real ecosystems must be empirical rather than theoretical. A theory of thermodynamics adequate to predict energy pathways and energy stocks in ecosystems does not exist, and may forever be impractical. Instead we measure where energy actually goes and use this information as a source of enlightenment about ecosystems. For instance, one important realization coming from modern work is that most of the energy fixed by plants in most ecosystems goes to decomposers. The plants themselves respire some of their energy and herbivores tap some more. But most goes to rotting organic matter. On the face of it a system of plants and animals depending on energy that lets the larger part of the energy pass them by has curious theoretical properties. Yet this energy of dead biomass often is vital to the maintenance

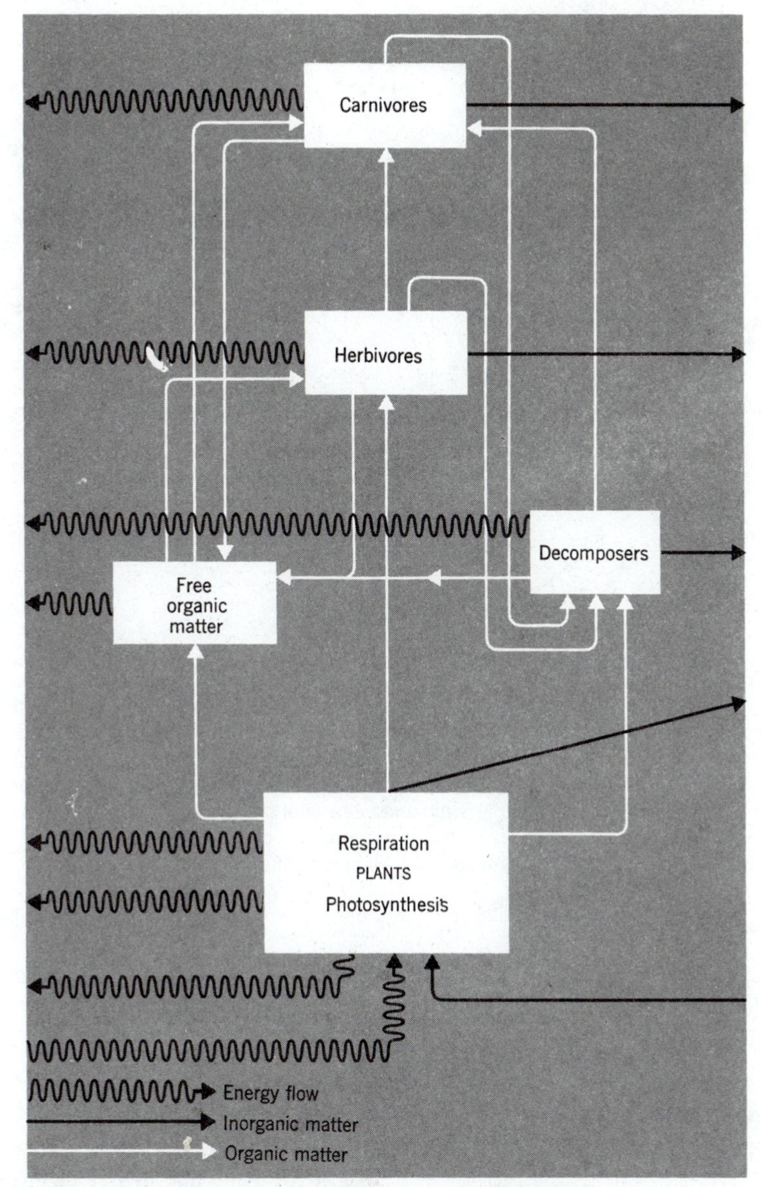

Figure 16.2 Theoretical energy flow scheme in an isolated ecosystem.
In this view of an ecosystem the various trophic levels, together with corpses and decomposers, are seen as discrete entities that one can draw as boxes. Energy enters by way of the plants, then cascades from box to box, being radiated from the system the while. Matter circulates as energy cascades. Such a diagram forms a useful conceptual device, and one that suggests what a practical ecologist should measure, but it is important to remember that nature does not come packaged in such discrete boxes. The animals, plants, and decomposers of real systems often have many and changing roles to play. (Redrawn from MacFadyen, 1962.)

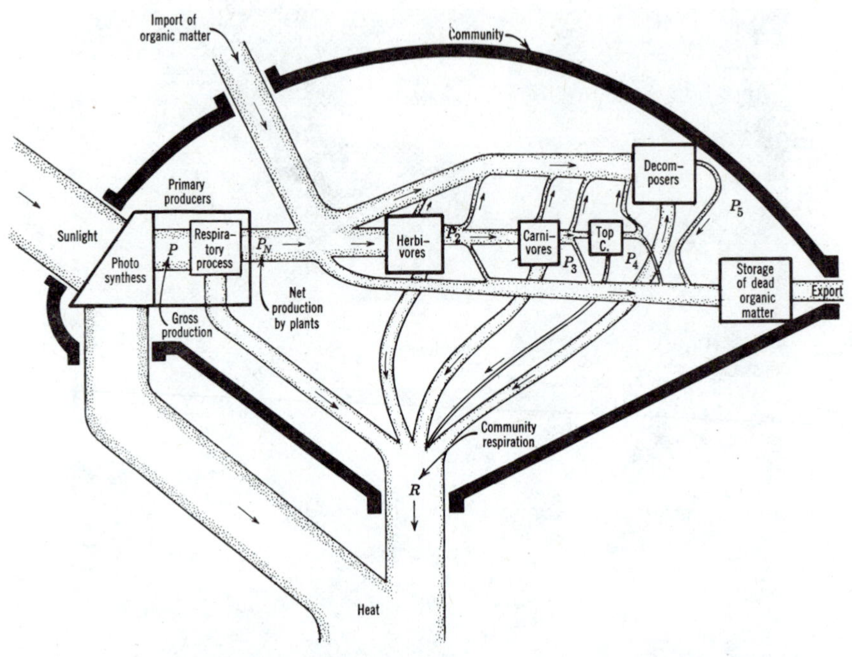

Figure 16.3 The hydraulic analogy of H. T. Odum.
In this analogy the energy cascade is imagined as being channeled through pipes whose thickness is proportional to the rates of energy flow. A prism placed at the entrance (or some hydraulic equivalent of a prism) deflects most of the sunlight from the community to represent that proportion of incident light not used in photosynthesis. From then on the degradation of energy at each trophic level is shown by pipes running to the heat outlet. The diagram is an excellent illustration of the ecosystem concept, but is poor as a practical tool because energy in real systems is not channeled and regimented as if by pipes. (After Odum, 1956.)

of ecosystem stability by preventing erosion and serving as a nutrient reserve (Chapter 17).

MEASUREMENT OF ECOSYSTEM PRODUCTIVITY

We cannot label energy to trace its passage through an ecosystem; nor have we voltmeters to record its potential. All we can do is to measure biomass and respiration. Biomass and respired carbon can be expressed fairly readily as the calories they represent, but rates of energy transfer must be computed from time-series measurements of changing biomass (Chapters 2 and 4). Designs for productivity measures must address the following four questions:

1. What are the best measures of biomass (or respiration) to make?
2. How can measures of biomass be converted into *rates* of biomass production?
3. How shall units of biomass be converted into units of energy?
4. Can estimates of net production usefully be converted into units of gross production and, if so, how?[2]

Studies in ecosystem energetics critically depend on finding the right answers for the first two questions and much effort has been devoted to devising suitable methods for ecosystems of different types. Converting to units of energy

[2]Gross and net production are defined in Chapter 4.

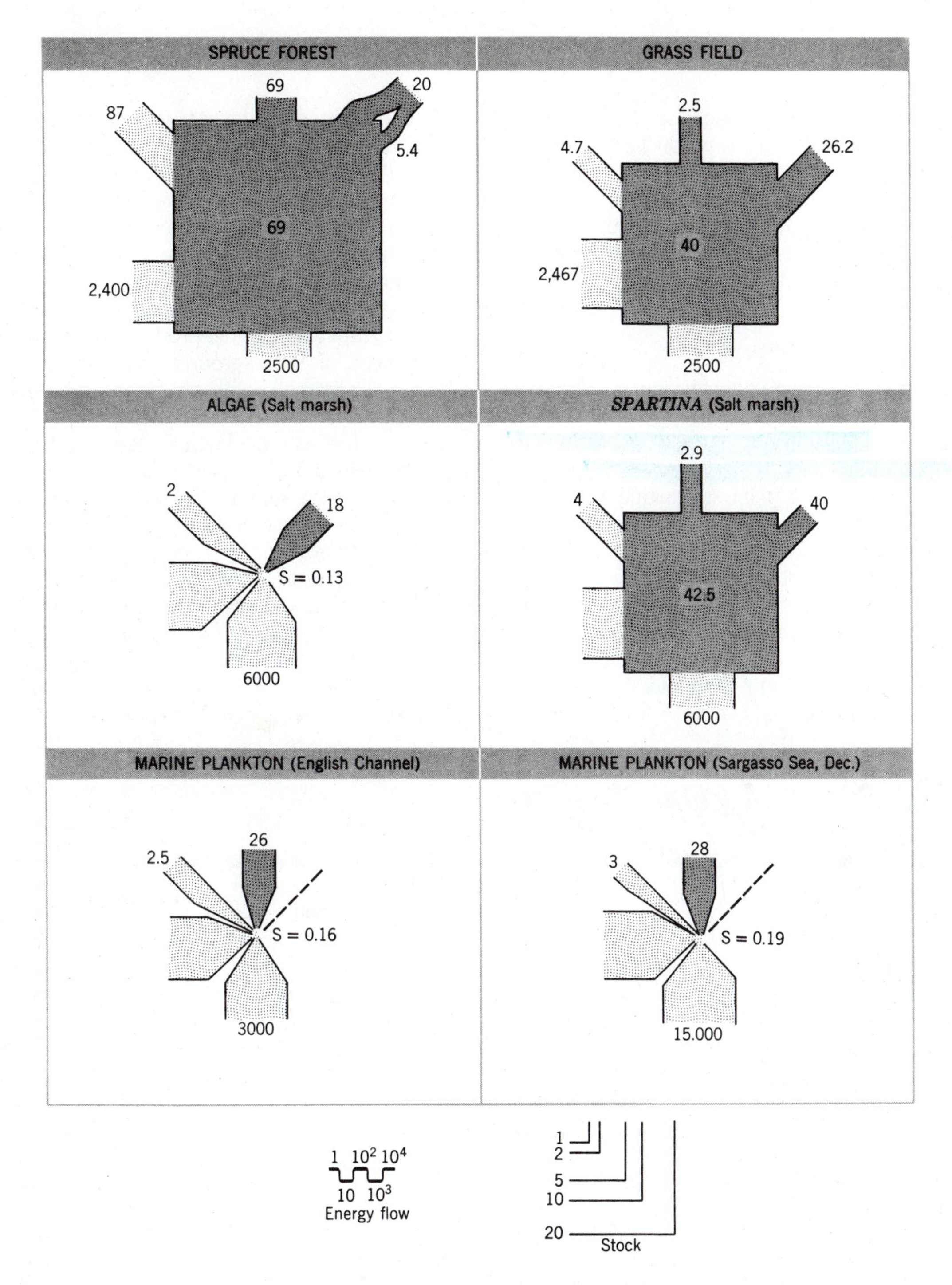

Figure 16.4 Hydraulic analogy diagrams to compare plant systems.
Units are Calories × 10^6 per hectare. Mean stock is proportional to the area of the central square. Such diagrams afford a striking way of presenting data, of showing such things as how similar productivities may be achieved by quite different standing crops under different conditions. They also reveal some of the uncertainties in productivity data, since it seems doubtful to some ecologists that a spruce forest respires at 20 times the rate of a salt marsh, or that consumption exceeds decomposition in a spruce forest as suggested by the diagrams. (Redrawn from MacFadyen, 1962.)

(question 3) is more straightforward. Question 4, on conversion to gross production, is usually nearly insoluble. Accordingly most ecosystem studies concentrate on measuring, comparing, and modelling net production from measures of biomass. We now describe some of the principal methods used.

Harvest Methods

The simplest measure of net production is to collect what grows, dry it, and weigh it—a simple harvest. Harvest in unit time thus gives the rate of net production, or the *net productivity* (Chapter 3). Yet applying this technique to natural vegetation can become complex and involves multiple harvests. This is because wild vegetation seldom grows synchronously. The standing crop in any field or forest nearly always includes biomass from previous years or seasons so that at least two measures are needed: standing crop at the start of a growing period and standing crop at the end. These data can come from measures of quadrat samples taken at intervals through the growing season.

Herbaceous vegetation can be sampled by clipping all above-ground vegetation from the sample quadrats, being careful at the same time to collect plant parts that have died and become litter in the interim. It is then necessary to take core samples of soil to be extracted for roots in order to arrive at total net production. If only the production of above-ground plant matter is required, then it may be possible to avoid sampling of roots. This might be possible, for instance, if the objective was confined to learning the net productivity available to the grazing or browsing herbivores of a system. It would then be necessary only to control for production taken by herbivores, probably by fencing herbivores away from the sample plots. These methods are straightforward and effective when used on herbaceous vegetation, but they can be very time consuming. Table 16.1 gives some typical results.

Meticulous sequential harvest is the most ac-

Table 16.1
Net Primary Productivity by Harvest Techniques
(Data compiled by Whittaker and Marks, 1975, from various authors.)

Communities and Species	Total ($g/m^2/yr$)	Stem and Branch Wood (%)	Leaves and Twigs (%)	Fruit and Flower (%)	Root System	
					Rhizomes (%)	Roots (%)
Wheat	294	—	53.0	29.4	17.6	
Barley	242	—	46.6	35.5	17.9	
Zea mays (maize, high-yield)	1935	16.8	17.1	61.4	4.6	
Helianthus annuus (sunflower)	3213	37.5	17.6	36.0	8.9	
Arctic tundra, all species						
Production	100	2	28	—	70.	
Populus (7-year-old poplars)						
Production	226	48.6	36.7	—	14.7	
Blanket bog						
Calluna vulgaris	351	10.8	37.0	—	52.2	
Eriophorum vaginatum	221	—	78.2	1.4	8.6	11.8
Empetrum and others	26	12	38	—	—	(50)
Sphagnum, other bryophytes	47	—	100	—	—	—
Lichens	3	—	100	—	—	—
Total bog	648	6.3	56.0	0.5	37.2	

curate method for measuring net productivity of forests also, but the task is likely to be ruinously laborious. Stem thickness, shoot elongation, fruit biomass, flower biomass, and leaf mass must all be measured at repeated intervals, to say nothing of root thickness, root elongation, and root-hair biomass. Furthermore, the measure of dead parts (litter) produced and of herbivore damage can be difficult. *Litter traps* must be used to collect representative samples of litter. Protection from herbivores, which may be mostly beetles or caterpillars, can be very tricky indeed. Nets over branches, perhaps coupled with chemical fumigation, offer one solution. The details of the methods have to be worked out for any particular patch of forest to be examined. The work required by such a sampling program can be considerable, but it has been possible to extrapolate from good harvest data collected on a few patches in ways that allow less time-consuming measures to be used on other forests.

The Mean-Tree Approach

Forests are made up of distinct, and easily distinguished plants, unlike the vegetation of herbaceous or shrubby layers. It ought, therefore, to be possible to identify typical trees, measure the productivity of these individuals, and then extrapolate to the whole forest. This turns out to be practicable for many artificial plantations but not for wild forests.

Productivity can be measured if a series of plantations of different ages is available; say, plantations started thirty, forty, and fifty years ago. In each stand the trees will be similar so that frequency distributions of height, or basal area, or stem thickness, or cover, *et cetera,* will be steep bell-shaped curves. It is fairly easy to extrapolate from measures of a few trees to the whole stand, thus yielding a measure of standing crop. And once standing crop data are available for the stands of different ages it is a simple matter to calculate growth over time, and hence productivity. The biggest correction that need be applied to these measurements is for dead branches and leaves shed between sampling dates, but this can be estimated with litter traps. Estimates of loss to herbivores or from shed roots are more difficult, yet clever sampling can provide tolerable estimates of these contributions to total productivity.

The PLANTATION SERIES has provided us with some of our most reliable estimates of forest production (Ovington, 1956, 1957, 1965; Kira and Shidei, 1967), yet even in plantations the MEAN-TREE approach carries a large error, estimated by Baskerville (1965) to be 25% and 45% for plantations of balsam fir (*Abies balsamea*). The reason for this is that tree geometry is so complex, and individual variation so great, that simple properties of the trees that we choose to measure have no constant relationship to one another. All we can measure easily are parameters like trunk thickness or tree height and these do not have a constant relationship to total tree mass, even within the individuals of a plantation.

It is tempting to seek to apply the *mean-tree* approach to wild forests. Unfortunately wild trees vary too much in shape, size, and individual dimensions. If errors can be up to 45% in plantations of balsam fir, the errors in natural forests will clearly be so large as to make the exercise meaningless.

Dimension Analysis

Although trees vary so much in the relative lengths and widths of their parts, it yet seems certain that there must be some relationship between tree size and productivity. Measures of tree sizes throughout the forest then should also be measures of productions and the problem becomes one of a suitable measure of size. This problem is solved by the technique known as DIMENSION ANALYSIS. It depends on the principle of *allometric growth.*

As structures are made larger, the relative proportions of their parts must change. This is a well-known principle of engineering, one that requires the redesign of all the parts of, say, an airliner as its capacity is increased from 10 passengers to 300. A mature forest tree accordingly is built differently from a young sapling, just as an old mare is built differently from a foal (Figure 16.5). The process of changing shape with size

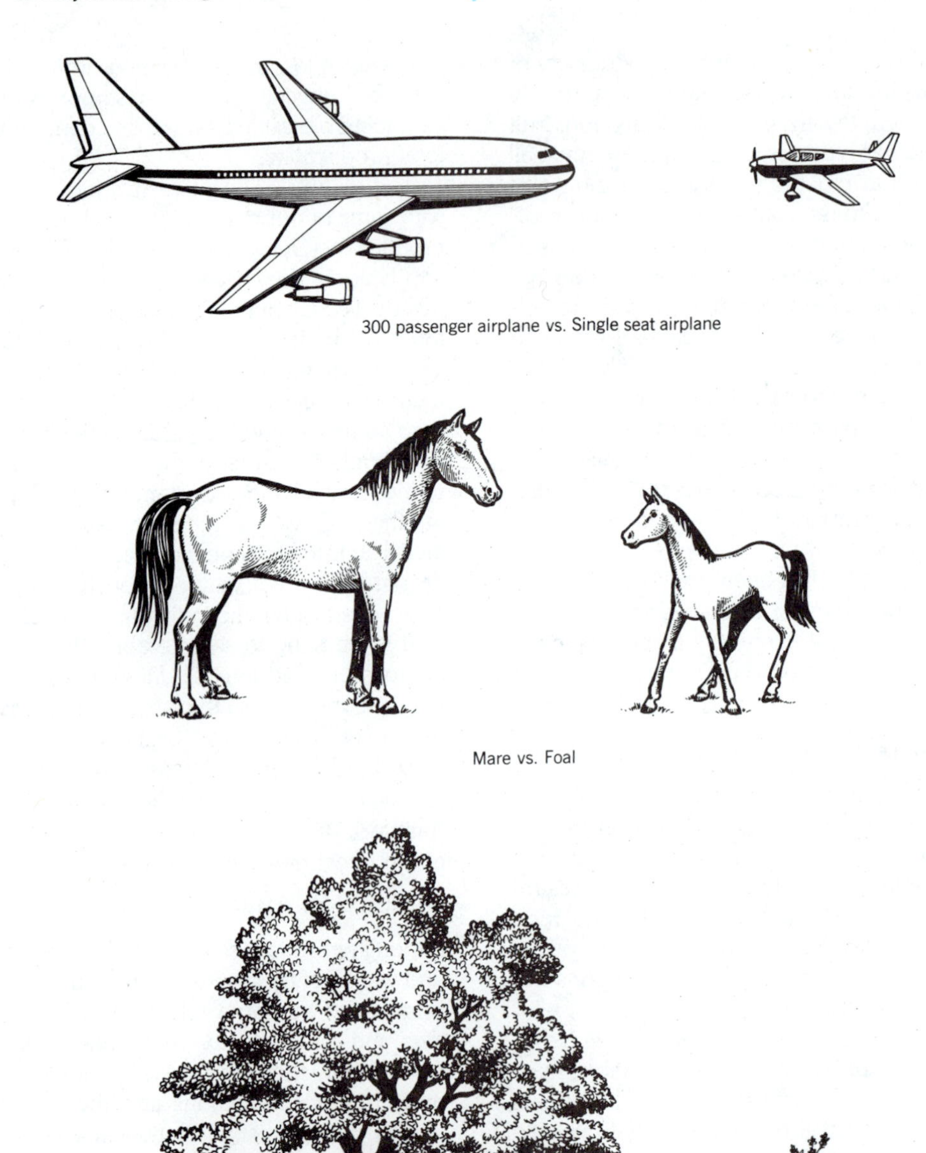

Figure 16.5 Examples of allometric growth.
Small objects have different proportions to large objects of the same kind. Regression equations can relate linear dimensions to mass and productivity of growing trees to allow computation of standing crop and net primary productivity of forests.

is called in biology ALLOMETRIC GROWTH (Huxley, 1932; Thompson, 1942).

Allometric growth implies that the relationship between two changing dimensions is curvilinear. Thus if

y is the height of a tree and

x is the thickness of the stem

the relationship between height and stem thickness might well be approximated by

$$y = ax^B \quad (16.1)$$

and

$$\log y = A + B \log \quad (16.2)$$

where B is a slope constant describing the relationship between these two fundamental measures of tree size.

Dimension analysis proceeds by constructing regression equations that will relate easily measured parameters of individual trees to tree size (leading to estimates of standing crop) and to individual production (leading to estimates of ecosystem productivity). The first step is to collect numerical data with which to construct the regression equations and to identify the measurements that will prove most useful. This is done by cutting down sample trees, measuring them, and estimating dry matter. Dimension analysis, therefore, is based on an initial harvest done in a way that will allow realistic extrapolation. The second step is to make the simple measurements indicated on an adequate sample of trees in the forest to which the regression equations can be applied (Whittaker and Marks, 1975).

The first complete measurements of net productivity of a forest using dimension analysis were made on sample patches of the eastern deciduous forest of North America at Brookhaven, New York, and at Hubbard Brook, New Hampshire (Whittaker and Woodwell, 1968; Harris *et al.*, 1973). The forest was sampled by quadrats. Every tree more than 1 cm in diameter at breast height in each quadrat was measured for dimension analysis and all plants of smaller size were sampled by direct harvest.

For each tree the following were measured:

Diameter at breast height (*DBH*)

Height

Bark thickness (by boring)

Mean current wood growth (by boring to measure and count tree rings)

Age (by counting tree rings in borings)

These data were sufficient to allow calculation of standing crop and productivity from the regression equations based on a much larger set of measurements on the specimen trees that were harvested. Table 16.2 lists a number of calculated stand dimensions that are used routinely in the work. Terms like "basal area" and "EVI" tend to become jargon of the productivity measurer's trade, but they are all calculated from the set of simple measures given above.

The fit of the regressions between stem thickness and productivity is illustrated by data in these studies given in Figure 16.6. These data show nicely how dimension analysis does allow biomass and production to be measured in forests. The work, however, is extremely time consuming. In effect one measures a patch of forest by harvesting samples of everything small enough

Table 16.2
Calculated Dimensions Used in Dimension Analysis

Basal area ($BA = \pi DBH^2/4$) of the stem, and of wood only (*BAW*) at breast height
Parabolic volume of the stem (*VP* = one-half basal area times tree height *H*) for the stem wood plus bark, and for stem wood only (*VPW*)
Conic surface (*SC* = one-half breast height circumference × height) for the stem, and for stem wood (*SCW*)
Basal area increment (*BAI* = mean annual increase in wood area at breast height during the past 5 or 10 years)
Estimated volume increment ($EVI = BAI \cdot H/2$)

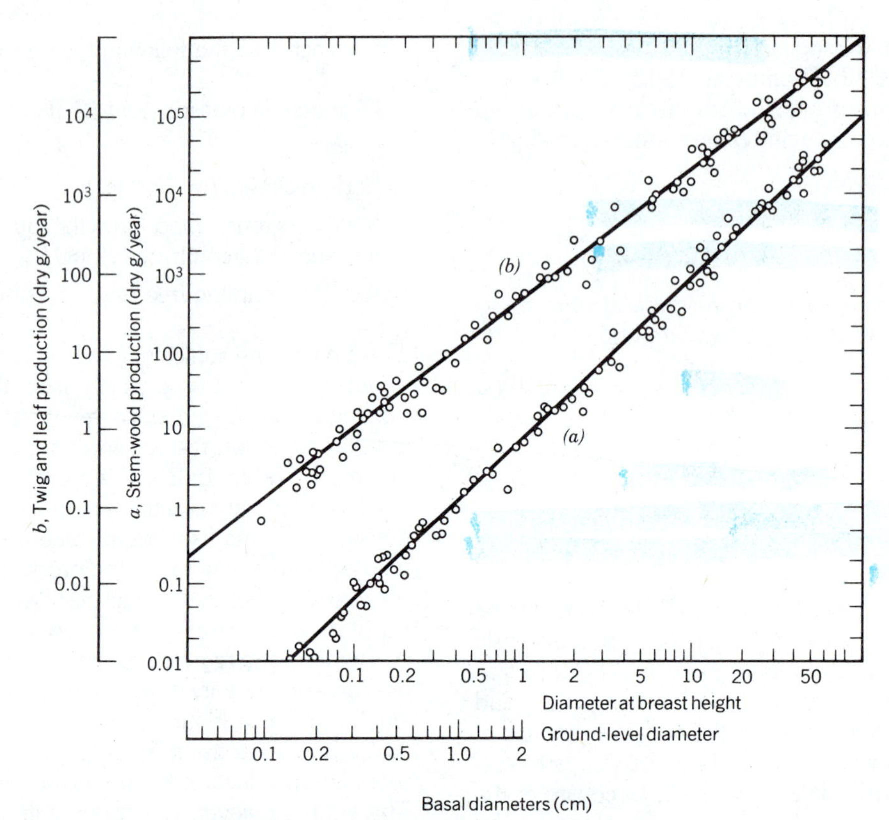

Figure 16.6 Regressions comparing stem thickness with productivity.
Data are from many species, ranging from small shrubs to large trees. As might be expected, individual productivity is a function of individual size, regardless of taxonomic affinity. The regressions show that stem thickness is a good predictor of both size and productivity. (From Whittaker and Woodwell, 1968.)

to harvest, but harvesting only a smaller proportion of the larger trees, and extrapolating to the remaining trees by regression equations. As with other harvest work, litter and destruction by herbivores also must be measured, so that the total labor can be near overwhelming. It will be noted that counting tree rings is an integral part of the procedure so the method cannot easily be applied to forests in non-seasonal climates (tropical rain forest) where trees do not make rings.

Converting Dry Weights to Calories

Table 16.3 describes the caloric content of the principal compounds making up organic matter. One way of converting dry weight estimates to measures of energy is to perform chemical analyses of sample parts of the plants and to calculate caloric contents using the data in this table. Scrutiny of sample results in Table 16.4 reveals that collective plant parts of such widely different taxa as beech trees, spruce trees, grass plants, and legumes have very similar total caloric contents. It is this fact that gives ecologists confidence in using the more simply derived estimates of dry matter when comparing the vegetation of one place with another. It is, in fact, usual in field or ecosystem studies to express productivity as grams of dry matter per unit time (or sometimes as grams of carbon per unit time

Table 16.3
Caloric Content of Chemical Compounds
Estimates of productivity in grams of dry weight can be converted to calories by chemical analysis when the caloric equivalents of the components are known. (Data compiled by Lieth, 1975b, from various authors.)

Compound or Matter Class	kcal/g
Starch	4.18
Cellulose	4.2
Saccharose	3.95
Glucose	3.7
Raw fiber	4.2
N-free extract	4.1
Glycine	3.1
Leucine	6.5
Raw protein	5.5
Oxalic acid	0.67
Ethanol	7.1
Tripalmitin	9.3
Palmitinic acid	9.4
Isoprene	11.2
Lignin	6.3
Fat	9.3

where carbon is assumed to be 50% dry weight) instead of the more precise estimate of calories per unit time.

Carbon Dioxide Exchange in Terrestrial Ecosystems

The use and difficulty of gas exchange for measuring productivity was introduced in Chapter 3. If the CO_2 absorbed by the plants of an ecosystem in unit time by day could be measured this would be a direct measure of NET PRIMARY PRODUCTIVITY. A similar measure for the whole community (animals included) would be smaller by the flux of animal respiration, and would yield NET ECOSYSTEM PRODUCTIVITY. Adding to the measure of net primary productivity an estimate of respiration of CO_2 by the plants in the dark would give an approximate estimate of GROSS PRIMARY PRODUCTIVITY. Measures of productivity by measuring CO_2 flux continue to be made, either of small subsamples or of whole communities.

Figure 16.7 shows the design of a small chamber system to measure CO_2 exchange of a forest. A glance at this design should leave no doubt of the complexity of the operation and some concerns will immediately be apparent. The photosynthesizing and respiring tissues, for instance, must be totally enclosed, which is likely to affect their function. Then it is found that fluctuations of CO_2 concentration within all but the smallest enclosures are large, making the estimates of average CO_2 concentration difficult. Very small chambers, each enclosing a single twig, have been found to be necessary (Woodwell and Botkin, 1970). When the measurements are made, it is still necessary to extrapolate from the small samples in the chambers to the whole community. This method has accordingly not been widely used.

Very large chambers that enclose complete sections of vegetation are, at first sight, an attractive possibility. A chamber enclosing, say, a complete section of forest as in a gigantic greenhouse should yield estimates of the total gas exchange of the ecosystem. Measurement would be as simple in the dark as by day so that gross ecosystem production should be delivered. Unfortunately the practical difficulties are near overwhelming. H.T. Odum (1970) enclosed a section of rain forest in a huge plastic chamber, then pumped air through and measured CO_2 in an array similar to that of the small chamber work (Figure 16.8). The study does not seem to have been repeated elsewhere.

On rare occasions natural phenomena may enclose an ecosystem. When an atmospheric inversion descends low over a forest, cold air is injected under warm air and trapped, almost as if under the glass of a greenhouse roof. This condition occurs frequently in the forest near the Brookhaven laboratory and has been used for measures of forest respiration. A series of gas sampling stations were built on towers rising through the trees so that CO_2 concentrations could be measured throughout an inversion event lasting, perhaps, for days. Figure 16.9 shows the direct results from one such episode. The change

Table 16.4
Chemistry and Caloric Content of Representative Plants
These data show the similarity of plant biomass across a large spread of taxa. It is this similarity of composition of organic matter that makes comparisons of productivities expressed as rates of dry mass production useful. (Data of Runge, 1973.)

		Beech Wood		Spruce Wood		Grass		Legumes	
Biologic Material	Average Caloric Value	Wt (%)	kcal Contributed to Total	Wt (%)	kcal Contributed to Total	Wt (%)	kcal Contributed to Total	Wt (%)	kcal Contributed to Total
Crude fiber	4.2	—	—	—	—	30.6	1.28	20.1	0.84
Lignin	6.3	22.7	1.43	28.0	1.76	5.0	0.32	5.0	0.32
Cellulose	4.2	45.4	1.91	41.5	1.74	—	—	—	—
N-free extract	4.1	—	—	—	—	45.1	1.85	40.9	1.68
Woodpolyoses	4.1	22.2	0.91	24.3	1.00	—	—	—	—
Resins and fats	8.8	0.7	0.06	1.8	0.16	3.0	0.28	6.2	0.58
Crude protein	5.5	—	—	—	—	9.4	0.52	19.9	1.09
Rest fraction	5.5	7.4	0.41	3.2	0.18	—	—	—	—
Ash	—	1.6	—	1.2	—	6.9	—	7.9	—
Total (kcal/g)			4.72		4.84		4.25		4.51

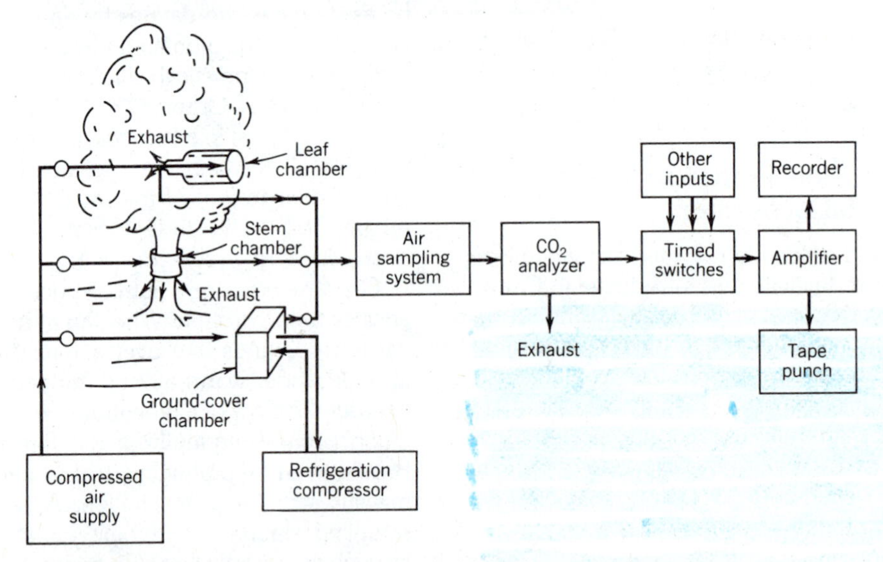

Figure 16.7 Small chamber measurement of CO_2 exchange in a forest.
In this analysis air must be driven through the chambers at a known rate, the temperature must be measured and controlled, and the gasses mixed completely and subsampled accurately. (From Woodwell and Botkin, 1970.)

Figure 16.8 Odum's "sleeve" in a Puerto Rican rain forest.
A portion of the forest is enclosed in this open sleeve. Air flows through it from top to bottom, and the content of carbon dioxide in the inlet and outlet streams is continuously monitored. Carbon dioxide lost by day added to carbon dioxide gained by night gives a measure of the gross primary production of this portion of the forest. The disadvantage of the method is that the sleeve may produce conditions so unnatural that the measure may not be applicable to undisturbed forest.

in CO_2 concentration in the forest between day and night shows up particularly clearly. In Figure 16.10 these data are reduced to estimates of total respiration regressed as a function of temperature. Ecosystem respiration was computed from the CO_2 flux at 2104 g dry matter per square meter year, a figure in keeping with measures by harvesting and dimension analysis.

It is theoretically possible to monitor the flux of gasses between and over the trees of a forest without any confinement. What is needed is a set of sensors arranged on a vertical mast projecting through the vegetation, when the simultaneous readings may be integrated to yield a total carbon dioxide flux. Practical difficulties mainly concern calculating the correct transfer efficiencies. (Lemon, 1969; Lemon *et al.*, 1970; Woodwell and Dykeman, 1966; Baumgartner, 1969; Allen *et al.*, 1972).

Gas Measurement in Aquatic Systems

In open water two circumstances make the measurement of productivity easier than on land:

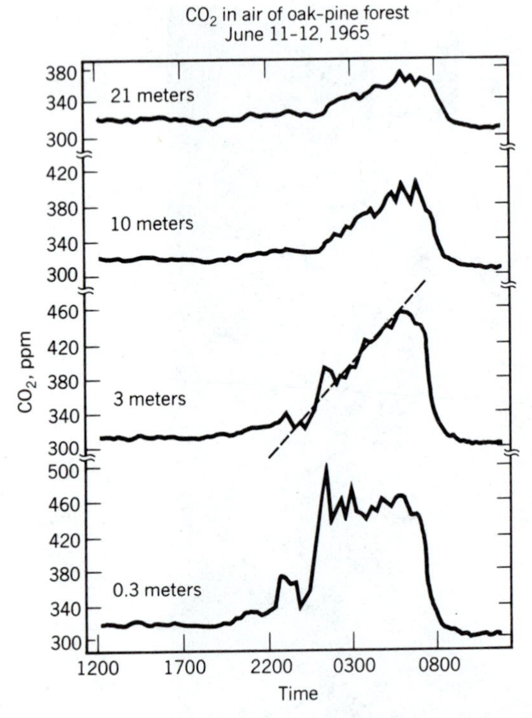

Figure 16.9 Carbon dioxide concentrations in a forest under an inversion.
The data are for an eastern American deciduous forest at Brookhaven. Air is trapped over the forest by inversions lasting several days. (From Woodwell and Botkin, 1970.)

plants of the open water are small and water is a simpler medium for chemical determinations than air.

The LIGHT AND DARK BOTTLE METHOD was first invented for use in the sea by Gaarder and Gran (1927). They put seawater containing planktonic algae into bottles, one of clear glass and the other covered with black paint, then they lowered the bottles over the side of their ship and left them in the sea for a measured time. When the bottles were recovered, the oxygen in each was measured by titration with potassium permanganate (the Winkler method). The light bottle had *gained* oxygen, due to photosynthesis, and the dark bottle had *lost* oxygen, due to respiration. The oxygen gained in the light bottle added to the oxygen lost in the dark bottle provided an estimate of the gross primary production of the time interval.

Although the light and dark bottle method is most accurate it has been replaced for routine work by a method based on the uptake of labelled carbon in photosynthesis. A sample of lake or ocean water is placed in a clear bottle with bicarbonate labelled with ^{14}C and the bottle is dangled where the sample was taken. Algae incorporate ^{14}C as a function of the rate of photosynthesis. The algae can be filtered out, fixed, and the labelled ^{14}C measured with a radiation counter (Steeman-Nielsen, 1952; Strickland and Parsons, 1972). Estimates of productivity by this method are always lower than those by the light and dark bottle method, probably because the algae respire some of the labelled CO_2 back to the water. The method apparently gives an estimate that is closer to net production than to gross production. It has, however, the benefits of convenience and standardization, letting productivities of different water bodies be compared. For remote field work the method is particularly valuable because the dried filters can be preserved for counting later.

A quite different approach is to work from estimates of algal standing crop provided by measures of chlorophyll. Algae are filtered from a known volume of water, their chlorophyll is extracted in acetone, and its concentration is measured in a spectrophotometer. Standing crops can be measured rapidly on large numbers of samples in this way. Moreover, samples may be stored for later measurement in the laboratory so that many samples can be taken rapidly in the field and treated at leisure, which is a great convenience. It is then necessary to estimate productivity from knowledge of chlorophyll concentration and light intensity. This requires empirical calibration of the standing crop data with similar concentrations of live algae whose productivity in the field is measured with the light and dark bottle technique. This has been found to be straightforward so that chlorophyll assay is now one of the preferred methods of measuring aquatic primary productivity.

A peculiar virtue of all these methods (except uptake of ^{14}C) is that they can be applied to

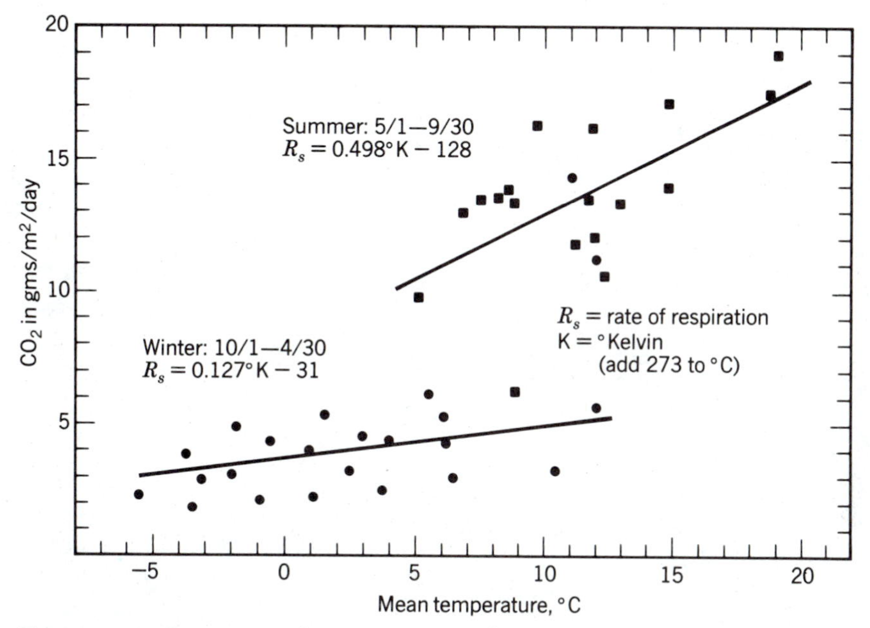

Figure 16.10 Respiration under an inversion as a function of temperature.
The base data are those given in Figure 16.9, temperature being the average of measures at 10 m and 3 m. R_s = rate of respiration in grams of CO_2 per square meter per day. (From Woodwell and Dykeman, 1966.)

whole bodies of water, not just to subsamples. The productivity of a whole pond or small lake ecosystem, for instance, can be measured by estimating dissolved oxygen or pH by day and by night. Typically the measurements would be made every two hours or so around the clock, leading to results like those in Figure 16.11. Another variant applied to moving water is to make the measurements upstream and downstream of the community to be studied. This was done, for instance, in the classic measure of the productivity of a coral reef at Enewetak by Odum and Odum (1955), where water flows constantly in one direction over the submerged reef between two islets of the atoll (Figure 16.12).

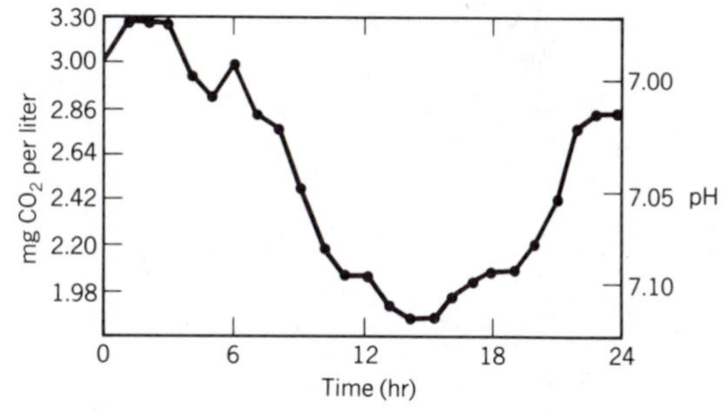

Figure 16.11 Diurnal changes in gas concentrations in an aquatic system.
The data are for New Hope Creek in North Carolina, where both CO_2 concentrations and pH were measured. (From Hall and Moll, 1970.)

ECOSYSTEM PRODUCTIVITY

Net primary productivity has now been measured in fairly reliable ways for samples of all types of ecosystem. Table 16.5 gives estimates for the principal ecosystem types. In the table these estimates are also applied to the areas of each ecosystem type to compute the tonnage of net production, and the net energy, that each contributes to the energetics of the whole earth. This is not

Table 16.5
Annual Primary Production of Principal Regions of the Earth
Dry biomass has been converted to carbon on the assumption that biomass is 45% C. (From Whittaker and Likens, 1973.)

	Means		Biosphere Totals			
Ecosystem Type	Mean Net Primary Productivity (g C/m^2/yr)	Mean Plant Biomass (kg C/m^2)	Area (10^6 km^2 = 10^{12} m^2)	Net Energy Fixed (10^{15} kcal/yr)	Total Plant Mass (10^9 metric tons C)	Total Net Primary Production (10^9 metric tons C/yr)
Tropical rain forest	900	20	17.0	139	340	15.3
Tropical seasonal forest	675	16	7.5	47	120	5.1
Temperate evergreen forest	585	16	5.0	31	80	2.9
Temperate deciduous forest	540	13.5	7.0	39	95	3.8
Boreal forest	360	9.0	12.0	46	108	4.3
Woodland and shrubland	270	2.7	8.0	23	22	2.2
Savanna	315	1.8	15.0	42	27	4.7
Temperate grassland	225	0.7	9.0	18	6.3	2.0
Tundra and alpine	65	0.3	8.0	5	2.4	0.5
Desert scrub	32	0.3	18.0	6	5.4	0.6
Rock, ice, and sand	1.5	0.01	24.0	0.3	0.2	0.04
Agricultural land	290	0.5	14.0	37	7.0	4.1
Swamp and marsh	1125	6.8	2.0	20	13.6	2.2
Lake and stream	225	0.01	2.5	6	0.02	0.6
Total land	324	5.55	149	459	827	48.3
Open ocean	57	0.0014	332.0	204	0.46	18.9
Upwelling zones	225	0.01	0.4	1	0.004	0.1
Continental shelf	162	0.005	26.6	43	0.13	4.3
Algal bed and reef	900	0.9	0.6	5	0.54	0.5
Estuaries	810	0.45	1.4	11	0.63	1.1
Total oceans	69	0.0049	361	264	1.76	24.9
Total for biosphere	144	1.63	510	723	829	73.2

Figure 16.12 Reef productivity laboratory.
The windward reef at Enewetak Atoll has ocean water flowing rapidly over the barely submerged reef between islets of the atoll and into the lagoon. Measuring oxygen concentrations on the ocean and lagoon sides of the reef allowed calculation of productivity of the whole reef ecosystem.

Table 16.6
Relative Productivity of Ecosystems
(For details see Table 16.5.)

Productivity	Ecosystem Type	Postulated Causes
HIGH >350 g $C/m^2/yr$	Forests Marshes and estuaries Reefs	Wet, warm, and anchored ↑
MIDDLE 70–350 g $C/m^2/yr$	Grasslands Upwellings Lakes Agriculture	← Combinations →
LOW <70 g $C/m^2/yr$	Tundra Deserts Oceans	↓ Cold Dry Lack nutrients

quite the same as an energy budget for the biosphere since this would require gross primary productivity, for which we have many fewer data (Chapter 3). But the results are certainly closely proportional to the contribution of the biota to a true energy budget.

The truly productive places are forests, marshes, estuaries, and reefs. The unproductive places are tundras, deserts, and oceans. Other systems, including agriculture, fall in between (Table 16.6). Some of these findings must always have been expected: high productivity in wet, warm forests, for instance, and low productivity in deserts. Outline postulates to explain productivity are thus readily available, and are offered in Table 16.6. Not all the conclusions are self-evident, however. People are still surprised to learn that the oceans are biological deserts and that the average productivity of agriculture is less than the typical forest.

These patterns of ECOSYSTEM PRODUCTIVITY are the summed results of many individual plants all doing as best they can in local physical circumstances. And yet the basic machinery of energy transformation in all plants is closely comparable: the different kinds vary mostly in the engineering that adapts their photosynthetic apparatus to local environmental constraints. These properties of individual plants as energy transformers were reviewed in Chapter 3.

Different ecosystem energetics are to be understood from habitat or system constraints, not from considerations of plant design. Forest ecosystems have high productivity because they are wet and warm, not because they have trees; prairies have low productivity because they are dry, not because they have grasses; lands beyond the arctic circle have still lower productivity because they are cold, not because they are covered with tundra.

Productivity as a systems property is nowhere better illustrated than in agriculture. We ought to expect that many forms of agriculture have lower productivity (as an ecosystems scientist calculates productivity) than the wild ecosystems replaced by agriculture (Table 16.5). This is because the agricultural crop has a lesser occupancy of the land than had the original vegetation. These constraints were discussed in Chapter 3 when examining the original calculations for the low assimilation efficiency of corn. Before sowing, and after harvest, a corn field is merely bare habitat, producing nothing at all, so that the average productivity over the growing season is low. Something like this pattern occurs in wild vegetation in strongly seasonal environments, though all wild communities manage more complete cover for longer times. In a temperate forest, for instance, spring flowers cover the ground before trees carry leaves, and well before nearby fields are covered with crops. This represents production that is denied an agricultural ecosystem.

The best agriculture does much better than the averages of Table 16.5, being comparable to the highest productivity given in the table for any ecosystem. But we have better data for agriculture than for forests. A prudent ecologist would expect the best production from the best forest to exceed the best of agriculture simply because the ground is covered more completely for longer. Possibly tropical systems of farming that keep the ground covered with many kinds of crops at once would produce as well as wild vegetation on the same site, otherwise agriculture starts at a disadvantage.

There are, of course, many habitats so changed by people that the local productivity of agriculture is actually higher than that of the displaced plant communities. The most obvious examples are irrigated deserts. The increased productivity, however, is a result of the change in the nonliving parts of the ecosystem. Productivity might be improved still further by letting wild vegetation cover the site while maintaining the irrigation. Farmers do not do this because they are not interested primarily in "productivity" but in agricultural yield of edible portions of crops.

TESTING THE MOISTURE–TEMPERATURE HYPOTHESIS

The general hypothesis that global patterns of terrestrial production are set by climate has been tested in two ways, by seeking correlations between productivity, temperature, and precipitation and by ecosystem modelling.

Data from dry grasslands show that productivity is closely correlated with mean annual precipitation, the relationship being net production of 1 g m^{-2} y^{-1} for each 1 mm of precipitation (Walter, 1962). There are other curbs on productivity in grasslands, but the data cannot refute the hypothesis that water has a paramount importance over large areas.

A more general test of the precipitation and

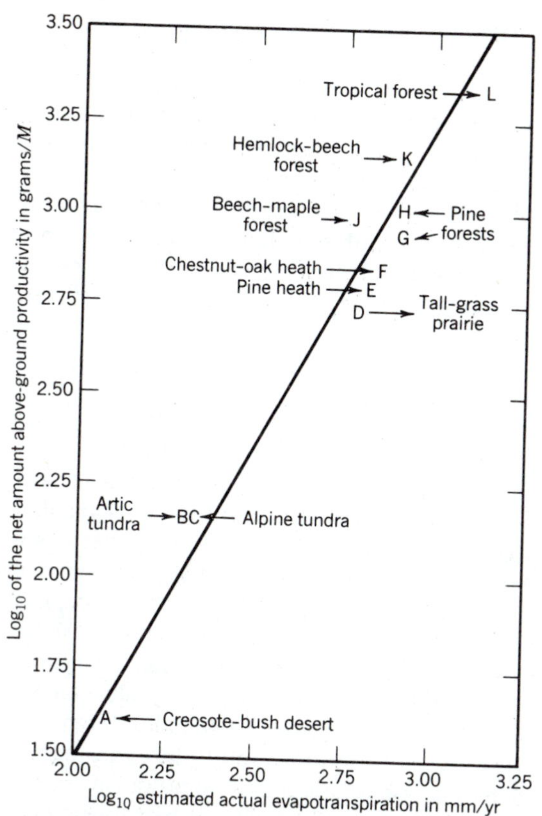

Figure 16.13 Correlation test of the hypothesis that precipitation and temperature determine productivity. Evapotranspiration (precipitation less run-off) depends on precipitation and temperature. The range of productivities is shown in Table 16.5. (From MacArthur and Connell, 1966, data of Rosenzweig, 1962.)

temperature hypothesis is the correlation that is found between productivity and evapotranspiration (Rosenzweig, 1962; Figure 16.13). EVAPOTRANSPIRATION is the excess of precipitation over run-off, and is, therefore, the mass of water that is returned directly to the air as vapor. In well-vegetated places we know that a large part of *evapotranspiration* is in fact channeled through the transpiration streams of plants, only the smaller portion being returned to the air as direct evaporation from passive surfaces. Figure 16.13 shows how close is the correlation between evapotranspiration and productivity. Various ecosystems included in the figure show their rankings in the productivity stakes as in Table 16.5. Since rate of transpiration is a function of temperature as well as of precipitation, the correlation is a test of the combined temperature–precipitation hypothesis.

It would be possible to produce a large number of individual examples of successful correlations like these between productivity, temperature, and water, but the most persuasive is probably the attempt to model regional productivity of the whole earth from climatic data. The pioneering attempt was made by Lieth (1975a), who used functions correlating productivity with temperature or precipitation and then applied these to global climatic data to generate a global production map with a computer.

The first step in the Lieth model was to plot production data from a range of ecosystems against mean annual precipitation (Figure 16.14) and mean annual temperature (Figure 16.15; Whittaker and Likens, 1973). These data themselves show the strong underlying correlations of productivity with wet and warmth. Curves are fitted mathematically to the data in Figures 16.14 and 16.15 and the expressions for these curves were used for the computer model. For each of a series of reference points on the globe, mean annual precipitation and temperature were determined independently. From the curve equations, the model then generated separately the productivity expected if temperature or precipitation at each place were the deciding factor, providing two independent estimates of productivity for each point. The model then chose whichever estimate was the lowest and accepted this figure as the predicted productivity of that place. This procedure of choosing the most conservative estimate invokes the basic ecological concept of limiting factors (Chapter 2). Figure 16.16 is the final productivity map of the globe that results. The productivity data of this map are closely similar to the direct measurements shown in Table 16.5.

This mapping shows the use of the ecosystem concept in its hypothesis testing role. Modelling is the prime tool. Initial postulates are applied to discrete data in order to derive numerical values

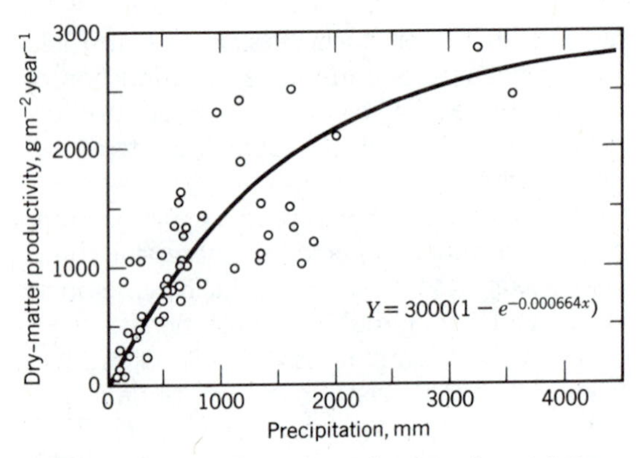

Figure 16.14 Productivity as a function of precipitation. The function Y for the fitted curve was used to generate production from mean annual precipitation in the Lieth model. (From Whittaker and Likens, 1973.)

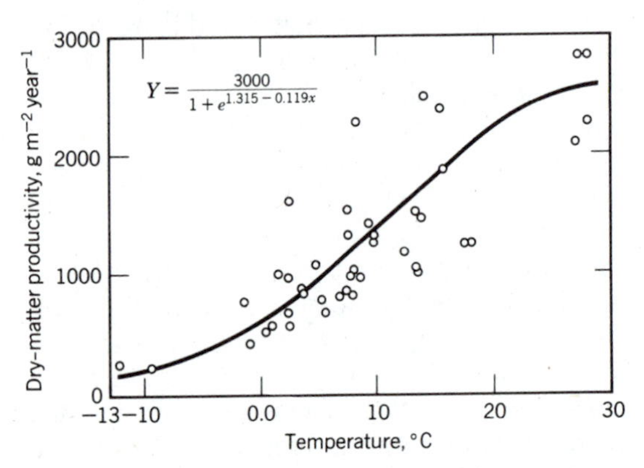

Figure 16.15 Productivity as a function of temperature. The function Y for the fitted curve was used to generate production from mean annual temperature in the Lieth model. (From Whittaker and Likens, 1973, using data of Lieth.)

on which the model can be built. If the original postulates are correct, then a model built on these numerical values ought to mimic nature. When the final fit is good, as in the Lieth productivity model of Figure 16.16, then it is permissible to assume that the original postulates had value: the postulates were not "falsified" by the exercise and we may accept them for future refinement. The ecosystem concept, therefore, lets us apply, through the medium of modelling, the classic scientific method to very large scale natural phenomena.

One weakness that has to be considered is the contribution of circularity to the results. The original numerical basis of the model depends on correlating mean annual precipitation and temperature with ecosystem productivity. Thus the model, in part, merely recovers productivity data on which it fed. Yet this may not be so serious in practice with so large a model, because many more data were generated than were used to build the model in the first place. To this extent the model, and others like it, are truly predictive.

It should be noted that detail is lost in going for the large scale. The global model derives productivity from temperature and precipitation alone, despite the fact that we know that many other factors regulate productivity at any one place. Nutrient supply, soil type, seasonality, plant age and plant health, grazing pressure, all these and more can influence productivity. Yet on the scale of regional ecosystems, essentially a biome scale, the influence of precipitation and temperature seems to be paramount. In every biome there must be variation in local productivity, but the size of a biome is so great that these local variations cancel out.

PRIMARY PRODUCTION IN THE SEA

Perhaps the most striking aspect of the world productivity figures is the poor performance of the open sea. The average for the oceans of 57 g C m^{-2} y^{-1} given in Table 16.5 is about one-sixth of the average for agriculture and not much better than one-twentieth of the average for a rain forest. Table 16.7 gives a number of regional measures of ocean productivity, showing that there is variation in the open sea but that very low productivities are typical. The blue waters of the warm tropics are particularly unproductive, being truly comparable to semideserts like the Sonoran of Arizona. This is contrary to the popular idea that the oceans are grand resources waiting to be tapped. This popular misconception comes from a variety of social processes,

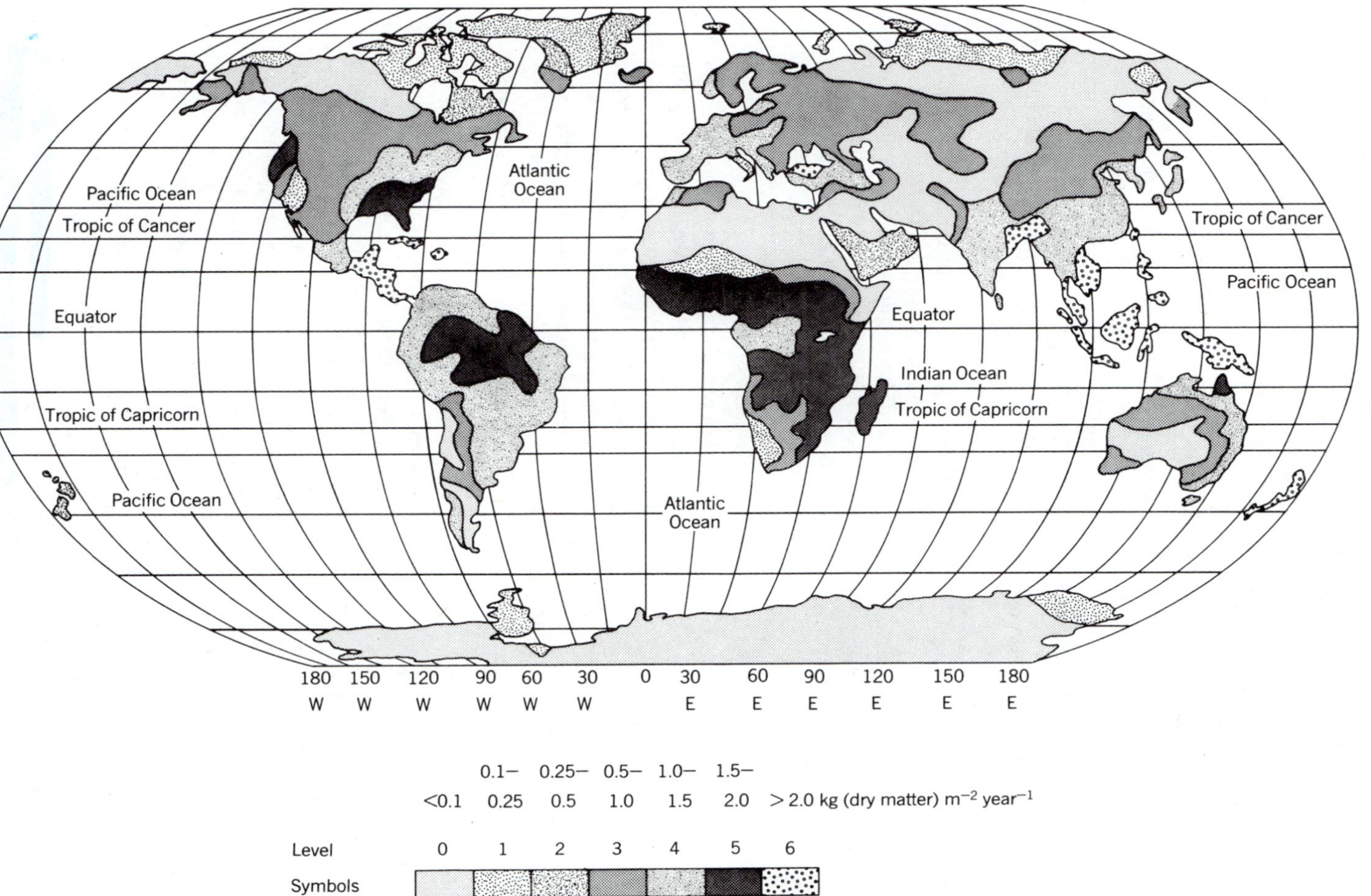

	<0.1	0.1–0.25	0.25–0.5	0.5–1.0	1.0–1.5	1.5–2.0	>2.0 kg (dry matter) m^{-2} $year^{-1}$
Level	0	1	2	3	4	5	6
Symbols							
Frequency (Frequency distribution of data-point values in each level)	59	97	152	262	180	146	105

Figure 16.16 Lieth "Miami" model of primary productivity (1971).
The model uses data of temperature and precipitation to compute net production from the relationships in Figures 16.14 and 16.15, then accepts the lower of each pair of estimates for mapping. (Model by H. Lieth, E. Box, and T. Wolaver, from Whittaker and Likens, 1973.)

Table 16.7
Rates of Primary Production from Various Parts of the World Oceans
Data for small planktonic plants are all derived from the ^{14}C uptake method (compiled by Strickland, 1960) from various sources. Data for large benthic plants include various harvest and gas exchange measurements. (Halimeda *data from Hillis-Colinvaux, 1980; remainder compiled by Bunt, 1975, from various authors.)*

	Location	Grams Carbon m^2/day
Small Planktonic	English Channel	0.50
	North Sea	0.1–1.5
	Danish coastal (August)	0.70
	Danish coastal (March)	0.30
	Danish coastal (December)	0.01
	Western Barents Sea (arctic water)	1.30
	Western Barents Sea (Atlantic water)	0.275
	Mediterranean	0.03–0.04
	Eastern Atlantic (15 miles offshore)	1.0
	Eastern Atlantic (200 miles offshore)	0.15
	Sargasso Sea	0.04–0.05
	Pacific off Ecuador (fishery)	0.5–1.0
	Equatorial Pacific (Fall)	0.01
	Equatorial Pacific (Spring)	0.10–0.25
	Sea of Japan	2.0
	Off Southwest Africa (inshore)	0.5–4.0
	Arctic Ocean (ice island)	0.024
Large Benthic	Kelp beds	1.65–7.90
	Intertidal (brown algae)	20.00
	Tropical seagrasses	5.8
	Codium (green, noncalcareous)	12.90
	Halimeda (green, calcareous)	2.5
	Lithothamnion (red, calcareous)	0.66
	Blue-green filamentous	0.65–2.15

including wishful thinking, the thought of the oceans being so big, political misrepresentation, and distant echoes of false claims about the efficiency of algae as producers of energy.

Limits to Ocean Production

Neither the supply of water, nor temperature, can explain low ocean productivity as they do for so much of the land. Some of the most unproductive of all are tropical oceans—blue, warm, and wet, just offshore from a tropical rain forest and yet a desert (see data for the equatorial Pacific, Table 16.7).

At first sight a hypothesis of ocean production limited by light has some attractions, since light is reflected from the sea surface and some is absorbed by clear water. But the data show that light is far from limiting. Only about 5% is backscattered or reflected (Ryther, 1959) and the short wavelengths used in photosynthesis are absorbed slowly (Chapter 3). The blue color of tropical oceans is testimony to this fact since the blue light penetrates deep into the plant-free water, from where some is scattered back to make the sea glow blue. The color of the open oceans, therefore, is the color of surplus light that is not used in photosynthesis because insufficient plants are present to use it.

Remaining is a general hypothesis of resource limitation. Carbon is most unlikely to be a limiting resource because ocean plants are able to use carbon from the bicarbonate–carbonate solution system, as well as the limited supply from the low tension of dissolved CO_2 itself. All other dissolved nutrients, however, can be scarce.

All the nutrients needed for plant growth are present in seawater in low concentrations. But plants must live where there is light, effectively in the top 100 m, and it is only the nutrients of these surface waters that are available to plants. This supply can be depleted quickly. Furthermore the open ocean does not mix very freely, being layered by temperature and salinity, so that mixing between surface and depths is impeded (Chapter 21). This means that nutrients removed from the surface zone of light and life (EUPHOTIC ZONE) cannot easily be replenished from the vast hoard of the depths. Finally, the dead bodies of plants or the animals that have eaten them fall through the sea, representing a continual transport of nutrients out of the euphotic zone.

The low concentration of nutrients in surface waters is compounded by the fact that the oceanic plants are microscopic. This means that it is not possible for plants to build reservoirs of nutrients in their bodies. Big plants should be able to store nutrients, thus building reserves in the surface waters. This is effectively what is done on land by rain forest trees that live in nutrient-poor soils: they collect nutrients with great assiduity, then store them in their large reserves of biomass (Chapter 17). But this is not possible in the open sea because the plants are so small. Part of the low productivity of the sea, therefore, does turn out to be a property of the plant engineering of the place. But that engineering (small size) is itself a function of the fluidity of the medium and hence a property of the system (Chapter 21).

The working *hypothesis for the low productivity of the ocean*, then, *is that low nutrient concentrations in the euphotic zone are the primary cause.* Concentrations remain low because transport of fresh nutrients from deep water is usually impeded and because the small size of oceanic plants prevents nutrients from being stored. This hypothesis is potentially falsifiable since it yields definite predictions:

1. Productivity in the open sea will be highest where there is strong circulation of deep water coming to the surface.

2. Nutrient concentrations should increase with depth, particularly as depths greater than the euphotic zone are reached.

3. Productivity will be high whenever local circumstances permit the growth of larger plants in the sea.

Ample data are available to test all three of these predictions.

The first prediction implies that a circulation map of the world oceans is also a crude map of productivity and the data confirm this. Figure 16.17 and Table 16.8 show data for the world oceans compiled by Russian oceanographic ships over many years (Koblentz-Mishke, 1970). Similar data are available from other countries, of course, but the Russian data are both voluminous and collected with standardized methods by all ships of their oceanographic fleet. The productivity map may be compared with the map of ocean currents given in Figure 14.10 (Chapter 14). Where currents diverge, so that water rises from below, are regions of high productivity. The water in the middle of the great subtropical gyres, though warm, has the lowest productivity of all, much lower than cold subpolar water where currents part because of winds and Coriolis force. Shallow shelf and coastal regions, where currents and storms can stir water to the very bottom, are the most productive waters of all, regardless of latitude or temperature. Seasonal mixing data also confirm the importance of nutrient transport to surface productivity. This is discussed for lakes in Chapter 21.

Figure 16.18 shows the distribution of phosphate, nitrate, and silicate ions with depth in the three major oceans of the world (Richards, 1968). The clines of nitrate and phosphate in the surface waters (prediction 2) are particularly striking. Silicate is important to diatoms, whose skeletons are made of silicates. The importance of diatoms to the total phytoplankton flora can vary with many factors, which probably explains why there is more variation in the silicate curve from ocean to ocean than there is in the other two. But the distribution of all three essential ions with depth throughout the oceans upholds the second prediction clearly enough.

The third prediction, concerning the effect of plant size, also is abundantly upheld by the data (Table 16.7). Large plants of the sea are confined to shallow water where they can anchor, but they are always much more productive than the small plants of open water. In colder waters large plants take the form of seaweeds like the kelps, the intertidal browns and reds, or greens like sea lettuce. In warm waters the large plant biomass is associated with calcareous reefs. Plants may have large calcareous skeletons of their own, like *Halimeda* and *Penicillus* (Figures 11.13, 3.24,

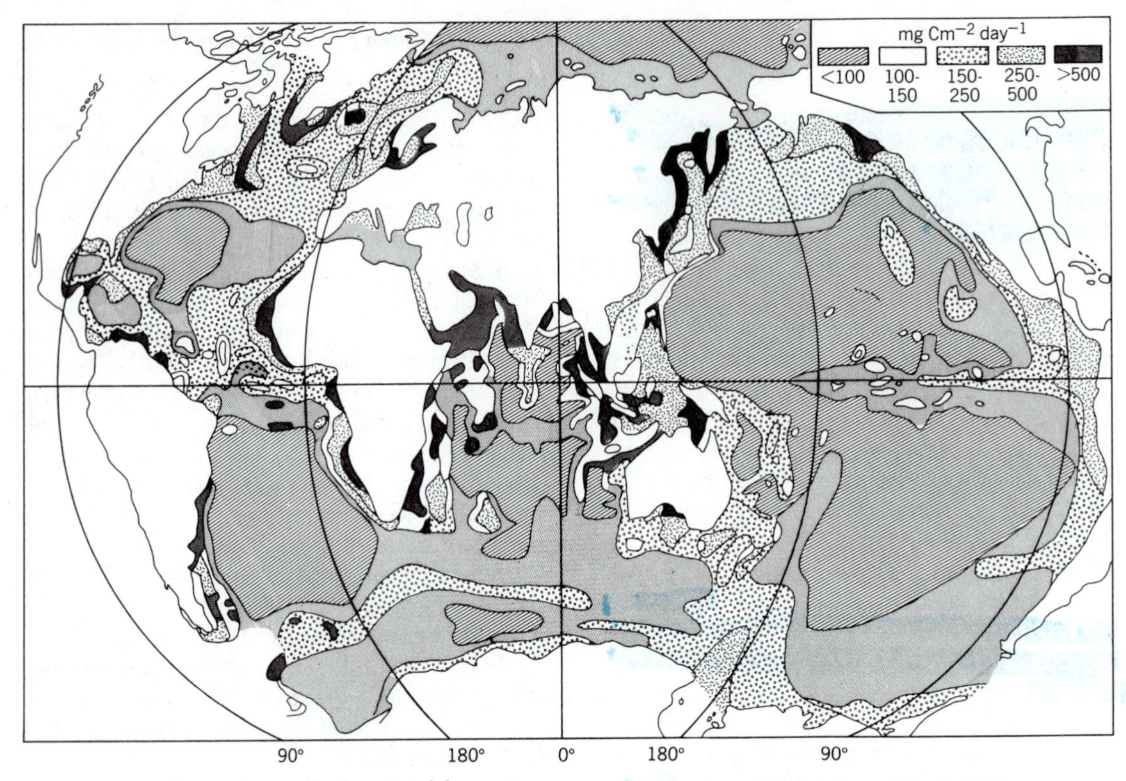

Figure 16.17 Productivity of the oceans.
The map is based on data gathered by Russian research ships using standardized methods. The five productivity regions mapped are described in Table 16.8. (Based on Koblentz-Mishke *et al.*, 1970.)

Table 16.8
Productivity of Regions of the World Oceans
These data are based on many ^{14}C productivity measurements by Russian research ships. They have the particular merit that methods were standardized, allowing regional comparisons with confidence. The data are mapped in Figure 16.18. (From Koblentz-Mishke et al., *1970.)*

	mg C/m²/day			
Region of Ocean	**Mean**	**Range**	**% Ocean**	**Yearly Tonnage (10 tons C/yr)**
1. Blue waters of subtropical gyres	70	<100	40.4	3.79
2. Transitions	140	100–150	22.6	4.22
3. Equatorial divergence and subpolar	200	150–250	23.6	6.31
4. Inshore waters	340	250–500	10.6	4.80
5. Shallow shelves	1000	>500	2.9	3.90
Total world oceans				23.0

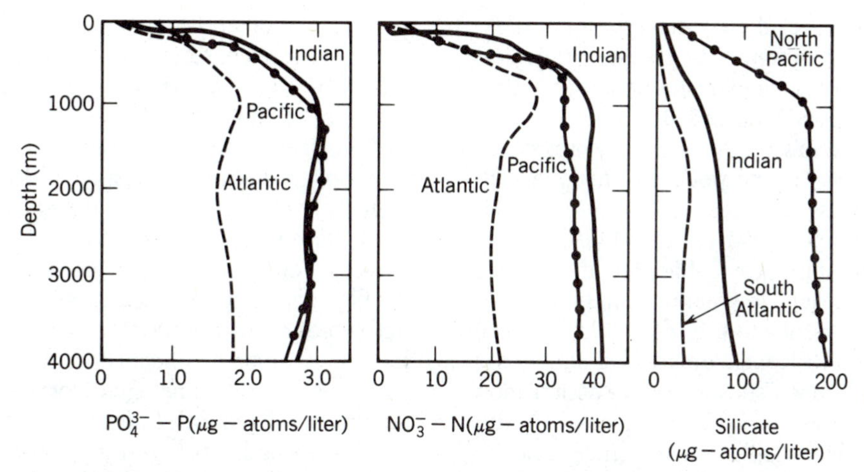

Figure 16.18 Nutrient concentration with depth in three oceans.
The data show that concentrations of essential nutrients are low in surface waters where light permits photosynthesis. This is predictable if ocean plants are nutrient-limited. (From Richards, 1968.)

and 3.26), or they may exist as filamentous forms embedded in the reef matrix, or as zooxanthellae living symbiotically with coral animals. But these arrangements, like the large thalli of cold-water seaweeds, have the effect of making the plants part of a large, fixed, physical mass. In all these circumstances marine plants allow high ecosystem productivity, comparable to that of the best sites on land and an order of magnitude better than that achieved in the open sea, and they do so despite the fact that they are bathed with the same water as plants in the nearby sea.

It is, therefore, a well-established hypothesis that productivity of marine ecosystems is set by the solution or delivery of dissolved nutrients, particularly phosphate, nitrate, and silicate. Coastal waters and shallow banks have relatively high productivity because they are without stagnant depths to which nutrients can be lost. These regions, accordingly, are turbid with algal life that gives them a characteristic green color. Upwellings in the open ocean are also relatively productive, because they are provided with a continual conveyer of nutrients from below. Upwellings, too, are green. But the great mass of the oceans is blue because surface nutrients are lost to the depths, preventing a growth of plants to color the water. And the pattern of productivity in the sea is set by the circulation of the oceans, not by temperature or climate as is the productivity of the land.

Reasons for Mistaken Beliefs in Productive Oceans

Ecological analysis shows quite definitely that the world oceans as a whole are unproductive deserts. Regions of upwellings supporting productive fisheries are exceptions to a rule of unfarmable desert. And yet people have been led to believe that the oceans are a vast, untapped resource waiting for us like a last frontier. Probably most of this public misconception came from realization that the ocean was large and that we had not yet farmed it. The analogy of little fishing boats as the equivalent of stone-age hunters is powerful, suggesting that modern technology could extract far more food. In fact the hunting technology of the modern fishing fleets appears to be more than adequate to deplete fish stocks. Since the real problem is one of poor primary production because of low nutrient status, the farming option is likely to remain wishful thinking. The ocean system has powerful chemical regulating devices keeping nutrient concentrations forever low (Chapter 18).

Probably the misconception was reinforced by a mistaken belief in the powers of algae as primary producers. A decade or so ago popular articles sometimes referred to ALGAL CULTURE as one of the ways of feeding the world's hungry. This belief came about from extrapolating laboratory results to the real world in a way that was invalid. Algal cultures in dim light in the laboratory had been shown to have assimilation efficiencies as high as 20% (Chapter 3). A culture working like that under the sun would be an attractive proposition, suggesting that an ocean full of algae should be highly productive. But of course the algae would not be 20% efficient under the sun. Those high efficiencies of the laboratory were all in dim light. Corn can be 20% efficient in dim light too. The truth is that algae in bright light are no more and no less efficient than other plants (Ryther, 1959; Chapter 3).

ANIMAL PRODUCTION OF LAND AND SEA

SECONDARY ECOSYSTEM PRODUCTIVITY is much less well known in all ecosystems than is *net primary ecosystem productivity.* Table 16.9 gives the estimates of NET SECONDARY PRODUCTION for the biosphere compiled by Whittaker and Likens (1973). These estimates are not based on measurement but are calculated from data for net primary productivity according to assumptions about the eating and digestive abilities of animals of the various places. The procedure was to begin with plant production and to take a percentage of this as eaten by animals in each ecosystem, based on what seemed reasonable to those familiar with the appropriate plants and animals. The potential for error can be illustrated by reflecting on how little is known about the herbivore take in tropical rain forests, and yet the compilers of the table had to be bold and chose a figure (7%) as the portion of rain forest that gets eaten. The percentage of plant productivity taken by animals gives the data in the "herbivore consumption" column, which then has to be corrected to allow for animal defecation. The compilers took the best data for animals typical of each ecosystem to arrive at an *assimilation efficiency* (Chapter 4) for each ecosystem. From this is calculated the *net secondary production* in the last column.

Very noteworthy about these estimates are the high consumption efficiencies assigned to the herbivores of marine systems, which are thought to be five times as effective at cropping ocean plants as are terrestrial herbivores at cropping land plants. Because of this, the estimate shows the oceans as having nearly as much total secondary production as the land, even though the total primary productivity of the oceans is only a third of that of the land. Something like this may be the truth of the matter, but there can be room for doubt. The whole issue of the effectiveness of grazing zooplankton at eating down planktonic plants is one of current debate. If it turns out that animals cannot in fact consume 40% of the primary productivity of the oceans, or 15% of reefs and anchored algae, as Table 16.9 assumes, then we shall find that secondary production does not compare nearly so favorably with that on land as some present estimates suggest. It is important to realize that this issue is not yet resolved and that conclusions still lean as much on "estimate" as on measurement.

The issue of ocean secondary productivity is important for practical economic reasons because it determines our estimates for potential yields of fish. Few commercial fish are herbivores, most feeding higher on food chains. The most-quoted estimates for potential yields of world fisheries are those of Ryther (1969) (Table 16.10). Ryther began with primary productivity estimates that were partly based on preliminary results of the Russian measurements and that were close to them (Table 16.8). He then estimated how many trophic levels there must be in each ocean province between plant and harvestable fish. Food chains should be long (five links) in the open unproductive oceans but only between one and two links in the more productive areas (Table 16.10). Ryther then assigned conversion

Table 16.9
Annual Secondary Production of Principal Regions of the Earth
These estimates are very approximate, being based on estimates of animal consumption of plants to compute herbivore consumption, and of assimilation efficiencies to compute gross secondary production. Some of the estimates are little more than intelligent guesses, probably being too high. (From Whittaker and Likens, 1973.)

Ecosystem Type	Total Net Primary Production (10^9 metric tons C/yr)	Animal Consumption (%)	Herbivore Consumption (10^6 metric tons C/yr)	Net Secondary Production (10^6 metric tons C/yr)
Tropical rain forest	15.3	7	1100	110
Tropical seasonal forest	5.1	6	300	30
Temperate evergreen forest	2.9	4	120	12
Temperate deciduous forest	3.8	5	190	19
Boreal forest	4.3	4	170	17
Woodland and shrubland	2.2	5	110	11
Savanna	4.7	15	700	105
Temperate grassland	2.0	10	200	30
Tundra and alpine	0.5	3	15	1.5
Desert scrub	0.6	3	18	2.7
Rock, ice, and sand	0.04	2	0.1	0.01
Agricultural land	4.1	1	40	4
Swamp and marsh	2.2	8	175	18
Lake and stream	0.6	20	120	12
Total land	48.3	7	3258	372
Open ocean	18.9	40	7600	1140
Upwelling zones	0.1	35	35	5
Continental shelf	4.3	30	1300	195
Algal bed and reef	0.5	15	75	11
Estuaries	1.1	15	165	25
Total oceans	24.9	37	9175	1376
Total for biosphere	73.2	17	12433	1748

efficiencies for the transfer of energy between trophic levels. These efficiencies are not precisely defined. The efficiency used at the bottom of the food chain is between net production of plants and gross production of animals, and so is not an ecological (Lindeman) efficiency (which is the ratio of gross to gross, see Chapter 4). The efficiencies used for higher trophic levels apparently are ecological (Lindeman) efficiencies. These efficiencies used by Ryther are of the order originally calculated by Lindeman himself and are now known to be too large in many cases (Chapter 4). These considerations suggest that Ryther's estimates may be optimistic. Until convincing measures of consumption at each link in the food chains are made, however, we are not in a position to improve upon Ryther's calculations.

Table 16.10
Ryther's Estimates for Fish Production of the World Oceans
These estimates are calculated from measures of net primary productivity. There is large uncertainty about the numbers of links in the food chains and even more about the conversion efficiencies used. The fish production estimates are probably too large. (From Ryther, 1969.)

Province	Percentage of Ocean	Area (km^2)	Mean Productivity (g $C/m^2/yr$)	Total Productivity (10^9 tons C/yr)	Primary Production [tons (organic carbon)]	Trophic Levels	Efficiency (%)	Fish Production [tons (fresh wt.)]
Open ocean	90	326×10^6	50	16.3	16.3×10^9	5	10	16×10^5
Coastal zone[a]	9.9	36×10^6	100	3.6	3.6×10^9	3	15	12×10^7
Upwelling areas	0.1	3.6×10^5	300	0.1	0.1×10^9	$1\frac{1}{2}$	20	12×10^7
Total				20.0				24×10^7

[a]Includes offshore areas of high productivity.

Table 16.11
World Production of Detritus
Detritus production is calculated from estimates of net primary production by assuming how much is eaten by herbivores, et cetera. (From Reiners, 1975.)

Ecosystem Type	World Net Primary Production (dry weight) (10^9 tons/yr)	Percentage of Production to Detritus	World Detritus (dry weight) (10^9 tons/yr)	Carbon (10^9 tons/yr)
Swamp and marsh	4.0	90	3.6	1.8
Tropical forest	40.0	95	38.0	19.0
Temperate forest	23.4	95	22.2	11.1
Boreal forest	9.6	97	9.3	4.6
Woodland and shrubland	4.2	80	3.4	1.7
Savanna	10.5	60	6.3	3.2
Temperate grassland	4.5	50	2.2	1.1
Tundra and alpine	1.1	95	1.0	0.5
Desert scrub	1.3	95	1.2	0.6
Extreme desert	0.07	97	0.1	0.03
Agricultural land	9.1	50	4.6	2.3
Totals	107.8		91.9	45.9

PRODUCTIVITY OF DECOMPOSERS AND DETRITIVORES

Most of the energy of net primary production goes to rot. Herbivores do not eat most plants; bacteria and fungi do. This is common knowledge to naturalists and it is given a certain formality in the estimates of animal consumption used for the secondary productivity estimates (Table 16.9). When, for instance, an ecologist assumes that 7% of the net production of a rain forest goes to the consumers there must be the balancing assumption that 93% goes to rot. Even the optimistic assumptions of the take of plants by herbivores in the sea still leaves more than half for the decomposers. On land, there is no doubt that energy degradation by decomposers is far more important than energy degradation by all the animals combined.

But perhaps a slight eating of words is necessary here, because many of the organisms that dispose of the *detritus* are in fact animals. Termites in the tropics and earthworms in temperate regions are obvious examples, but there are many others. Fiddler crabs (Figure 16.19) are one of the more engaging examples, and there are many benthic animals of lakes and the sea who find much of their energy as detritus. It is because the energy flux by-passing the herbivores is so large that there are so many scavenging animals. These are properly called DETRITIVORES and they are immensely important in the economy of most ecosystems. Together with the true DECOMPOSERS (the bacteria and fungi) they undertake most of the non-plant biological activity of all ecosystems.

Figure 16.19 *Uca* spp: display by a detritus feeder. Fiddler crabs feed on the detritus collecting on mud flats.

Since plants are producing detritus continuously, a flux of this production passes through all ecosystems. Reiners (1973) has estimated the total flux for terrestrial ecosystems using a number of alternative assumptions, his extreme results differing by factors of less than two. Table 16.11 gives a representative sample of his results. He begins with standard estimates of net primary productivity, as given in Table 16.5, provides his own estimates of what animals might eat, calls the remainder detritus, and calculates accordingly. Forests are the great producers of detritus, as one might expect. The existence of numerous termites in tropical rain forests is understandable enough.

PREVIEW TO CHAPTER 17

Cycling of nutrients in solution within ecosystems is second in importance only to the transformation of energy on which the running of ecosystems depends. Even without life nutrients would be maintained at low levels in habitats. There would be inputs from rain, weathering, and the air; storage in water and minerals of surface rocks; outputs in drainage water; and a tendency towards a steady-state nutrient supply. Vegetation regulates the nutrient supply by modifying each part of this system. Plants increase storage by adding to the habitat both living and dead organic matter. They increase the nutrient input by deep roots and by extraction from the air. Complete plant cover restricts nutrient output by regulating the concentration of nutrients in soil water. Moreover vegetation alters the flow of water through the ecosystem by diverting a very large part of the flux from a liquid flow through the soil into a transpiration stream that releases water vapor to the atmosphere. Since vegetation regulates both the flow of water and the concentration of dissolved nutrients, it exerts a tight control on the flux of nutrients. The energy with which the work of diversion by transpiration is done comes not from energy fixed by photosynthesis but from energy absorbed by the plants as heat, thus increasing their effective efficiency as energy users by a large amount. Terrestrial ecosystems can be monitored by the watershed approach where all outputs in solution converge on a single stream. Watersheds can be manipulated experimentally. Some tropical forests have nutrient cycles unlike those of temperate regions. Except on flood plains or young volcanic soils, tropical habitats have few nutrients stored in soil minerals or as dead organic matter. Instead the main reservoir of nutrients is in the bodies of the living plants themselves. Input of fresh nutrients to this reservoir via dust, rain, or weathering is slight and the reservoir is maintained by an efficient root system, associated with symbiotic fungi, that recaptures nearly all nutrients lost to the living plants by death or decay. Typical tropical rain forest ecosystems are thus unstable, in that destruction of the forest means a disruption of the nutrient cycles and a loss of nutrients that cannot immediately be replaced. The different nutrient budgets of temperate and tropical forests are determined by physical factors of the habitat, notably temperature. Nutrient budgets in other biomes are also dependent on the physical environment. Grasslands build nutrient reservoirs in the soil in the manner of temperate forests. Tundra soils contain nutrients but vegetation is dependent on the rate at which these can be released under the prevailing low temperatures. Human civilizations arise only where natural nutrient budgets leave large reserves in the soil, either in temperate habitats or in such locally fertile tropical habitats as flood plains.

CHAPTER 17

ECOSYSTEM CHEMISTRY AND THE CYCLING OF DISSOLVED NUTRIENTS

The cosmic abundance of the elements is quite different from the abundance of elements in living things. Burning any sample of life—a tree, a mouse, or a box full of bacteria—yields ashes that are remarkably similar to each other chemically, yet all with elemental compositions quite unlike the mineral crust of the earth or a solar flare (Table 17.1). Living things, therefore, select from the elements about them. They have little use for really abundant elements like silicon (except diatoms), aluminum, magnesium, and iron, but concentrate the relatively rare elements phosphorus, potassium, and calcium.

Life does its collecting of most elements in solution. Carbon, of course, enters ecosystems as gas and thence is transformed along food chains as solid compounds of reduced carbon. Nitrogen can be imported as a gas by nitrogen-fixing bacteria and later vented back to the atmosphere. Likewise, oxygen is imported as a gas used for respiration, then re-exported to the atmosphere by the photolytic destruction of water in photosynthesis. These processes can be seen to regulate the atmosphere bathing terrestrial ecosystems (Chapter 19), but most of the elements used by ecosystems are acquired in solution. Each organism in an ecosystem, therefore, processes water and manipulates solution equilibria. The effects of these individual efforts are combined because all organisms have the same common end of concentrating generally the same chosen few elements and rejecting the rest.

Seven ions are processed in large masses by ecosystems, and it is the cycling of these seven that has been most studied by ecologists. They are the CATIONS calcium, sodium, and potassium (Ca^{++}, Na^{+}, K^{+}) and the ANIONS nitrate, phosphate, sulfate, and carbonate (NO_3^{--}, PO_4^{--}, SO_4^{--}, CO_3^{--}). Much energy has to be degraded to concentrate these ions, and to keep them concentrated. Furthermore, the fact that the ions are all soluble in water means that all must be kept against parts of the hydrologic cycle that tend to carry them away.

The biota has a relatively tiny flux of energy at its command to do this work—some small portion of the 1 to 2% of solar energy trapped by the photosynthesis of an ecosystem—but against this is set the huge energy flux that goes to drive the hydrologic cycle. A very large portion of the total solar energy incident on any ecosystem (except a desert) goes to heat or transport water. The biota can answer only with the puny flux of photosynthetic energy. The energy odds are thus heavily weighted against the life forms of an ecosystem. Nevertheless, the collective activities of individual organisms in some ecosystems do modify hydrologic cycles, imposing order and stability to a degree that seems quite unlikely in view of the apparent weighting of the energy scales towards entropy and chaos.

CYCLES AND STATES WITHOUT VEGETATION

The hydrologic cycle for a world without vegetation is shown in Figure 17.1. The essence of this system is that water moves between large reservoirs formed by the oceans, surface water (including glaciers), and underground water. The actual proportion of water in transit at any one time is very small (Table 17.2) since all rivers and the atmosphere together hold only 0.0011% of the water of the biosphere. The rest is held in the various reservoirs, particularly oceans, glaciers, and deep in the ground. Only soil water is transported much by life, and this minor reservoir holds only 0.005% of the total reserve.

Even in a plantless world some water would be retained in the surface soil; as the ground was wetted, as pore spaces in the soil were filled with water held by the surface tension effect in capillaries, and as some water was absorbed on minerals, notably clays (Chapter 20). The amount of water actually held in the soil of a plantless

Table 17.1
Relative Abundance of Principal Elements
Data as percentage dry weight. Human and alfalfa plant (Medicago sativa) *data from Rankama and Sahama (1950), remainder from Zajic (1969).*

Adult Human	Alfalfa	Element	Solar Atmosphere	Total Earth	Crust of Earth	Seawater
6.60	5.54	Hydrogen	53.0	Trace	0.14	10.8
		Helium	42.0	Trace	0.00000003	0.0000000005
		Boron	Trace	Trace	0.0003	0.0005
48.43	45.37	Carbon	0.012	Trace	0.03	0.003
12.85	3.30	Nitrogen	0.031	Trace	0.005	0.00005
23.70	41.04	Oxygen	4.7	28.0	46.6	87.5
0.65	0.16	Sodium	0.0024	0.14	2.8	1.05
		Magnesium	0.043	17.0	2.1	0.13
		Aluminum	0.0031	0.4	8.1	0.000001
		Silicon	0.029	13.0	27.7	0.0003
1.58	0.28	Phosphorus	Trace	0.03	0.12	0.000007
1.60	0.44	Sulfur	0.014	2.7	0.05	0.09
0.45	0.28	Chlorine	Trace	Trace	0.02	1.90
0.55	0.91	Potassium	0.00033	0.07	2.6	0.04
3.45	2.31	Calcium	0.0036	0.61	3.6	0.04
		Vanadium	0.000031	Trace	0.02	0.0000002
0.10	0.33	Manganese	0.00086	0.09	0.10	0.0000002
		Iron	0.167	35.0	5.0	0.000001
		Cobalt	0.00034	0.20	0.002	0.00000005
		Nickel	0.0029	2.7	0.01	0.0000002
		Copper	0.000058	Trace	0.01	0.0000003
		Zinc	0.00021	Trace	0.01	0.000001
		Molybdenum	Trace	Trace	0.0015	0.000001
		Iodine	Trace	Trace	0.00003	0.000006

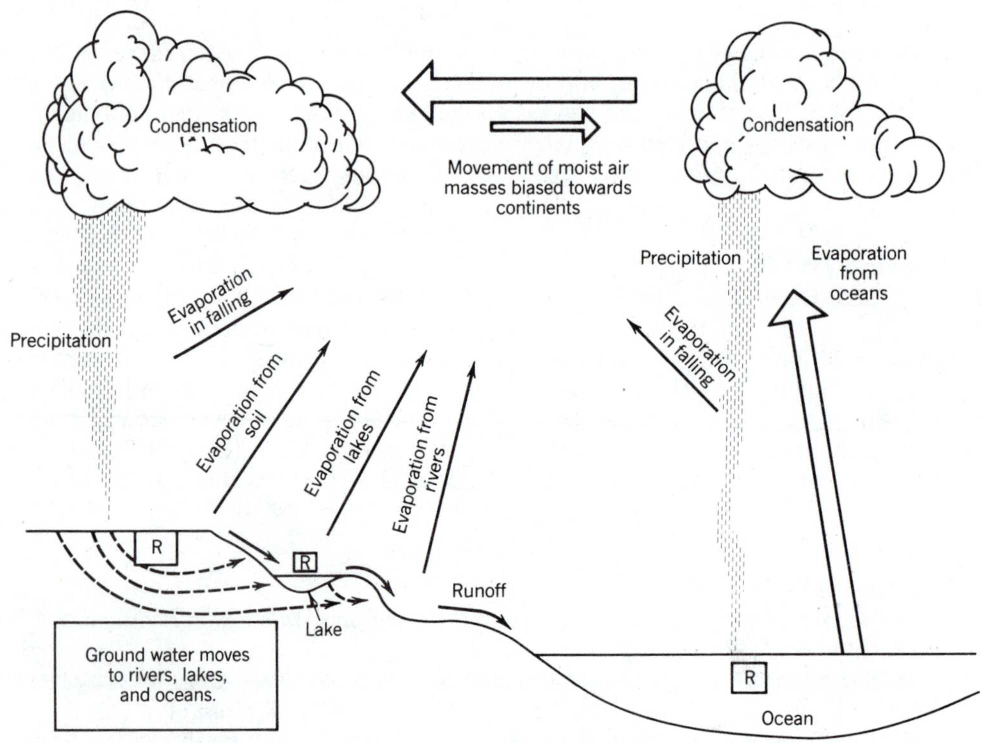

Figure 17.1 The hydrologic cycle in the absence of vegetation.
Three principal reservoirs of liquid water are marked (R). These reservoirs are the oceans, lakes, and the pore spaces of soils and surface rocks. The diagram omits the second largest of the earth's water reservoirs, glacier ice (see Table 17.2).

Table 17.2
Water Budget for the Biosphere
From Strahler (1969), compiled from data of R. L. Nace.

Reservoir	Volume Water ($\times 10^6$ km^3)	% Total
World oceans	1322.0	97.21
Glacier ice	29.2	2.15
Ground water below soil	8.4	0.62
Soil water	0.067	0.005
Fresh-water lakes	0.125	0.009
Inland seas and salt lakes	0.104	0.008
Rivers and streams	0.001	0.0001
Atmosphere (clouds and vapor)	0.013	0.001

land would depend on the frequency and amount of rain and on the drainage. Another way of saying this is that the mass of water usually present would be a function of the rate of water input and the rate of water loss. A STEADY STATE would tend to be established with water constantly entering and leaving the soil reservoir. In seasons of high rainfall the steady-state volume of water in the soil would be higher than in seasons of low rainfall. With intermittent rain the reservoir would fluctuate.

Water entering and leaving a patch of mineral soil will contain dissolved nutrients,[1] whether these were acquired from the air or by solution of min-

[1]Ecosystem nutrients may be defined as ions needed by plants in solution. This definition excludes carbon, oxygen, and water itself.

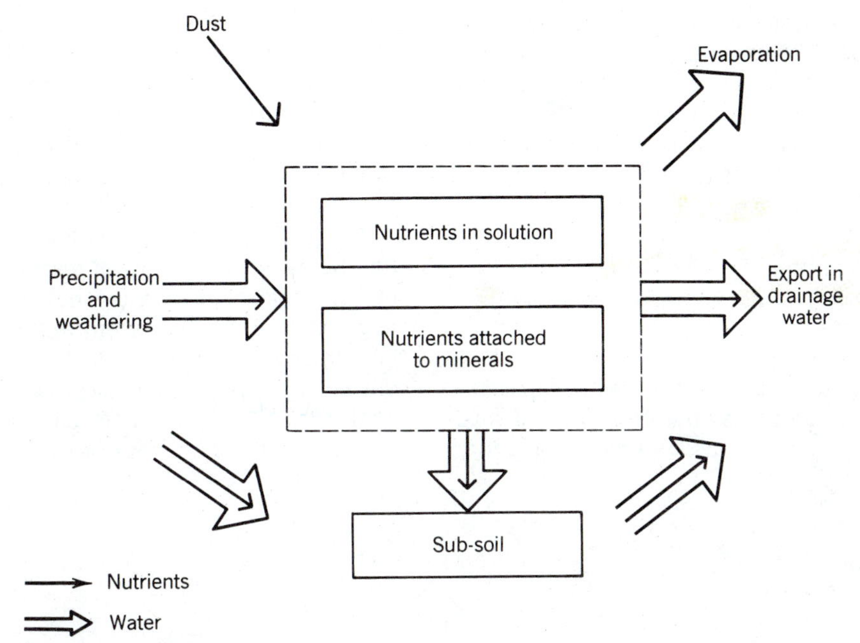

Figure 17.2 Water and nutrients to a lifeless habitat.
If precipitation is roughly constant, the nutrient reservoir will tend to a steady state, even without plants. The potential habitat will hold nutrients in simple solution in its nutrient reservoir and absorbed or adsorbed to minerals.

erals in passage through the ground. The reservoir of water in the soil will, therefore, also be a nutrient reservoir (Figure 17.2). Where rainfall is constant enough for a steady state to be established there will be a *steady state reservoir of nutrients.* The actual nutrient reservoir will depend on the nutrient-retaining power of the soil minerals, and this will not be the same as the water-retaining powers. Many nutrients, notably cations, are selectively adsorbed by clays or other minerals in the soil.

When plants are added to the habitat, three essential changes are made. First, water is pumped through transpiration streams instead of being allowed passively to evaporate or flow away. Second, the nutrient concentration of soil water is actively changed by the plants acting as ionic pumps. Third, the nutrient reservoir is changed by the addition of both living and dead biomass. These three processes occur in different arrangements in habitats of different biomes. Together they account for the apparently ordered storage and cycling of nutrients in many ecosystems.

NUTRIENT CYCLING IN A TEMPERATE ECOSYSTEM

In a typical vegetated habitat in the temperate parts of the earth the actual nutrient reservoir is augmented in various ways. Instead of just the dissolved nutrients and those attached to the minerals of surface rocks of the unvegetated habitat, temperate ecosystems hold the following:

1. nutrients in living plants
2. nutrients in living animals
3. nutrients in detritus
4. nutrients in soil humus.

In addition to these supplies, extra nutrients are held in the soil that results from covering the land with plants. Finely divided mineral matter is mixed into aggregates with soil humus, and the finely divided mineral matter may itself be increased in the soil-forming process (Chapter 20). The vegetated habitat thus also holds

5. nutrients adsorbed to soil humus

6. nutrients adsorbed to additional clay minerals.

Each of these reserves represents extra capacity. Even if the input and output of water and nutrients to and from the habitat were roughly the same as in a plantless plot, the steady-state supply in a vegetated habitat would be larger. But in fact the plants increase the inputs also. They extract nutrients from water deep in the subsoil and provide a substrate for the bacterial fixing of nitrogen. The flux of nutrients through such a temperate system is illustrated in Figure 17.3.

In Figure 17.3 the "plantless soil component" represents the entire habitat reserve of nutrients of the original unvegetated habitat. By far the larger part of the ecosystem nutrient reserve is held in those compartments of the system that did not exist in the lifeless place. And yet the work done by organisms to amass this ecosystem reserve need not have been much, since the

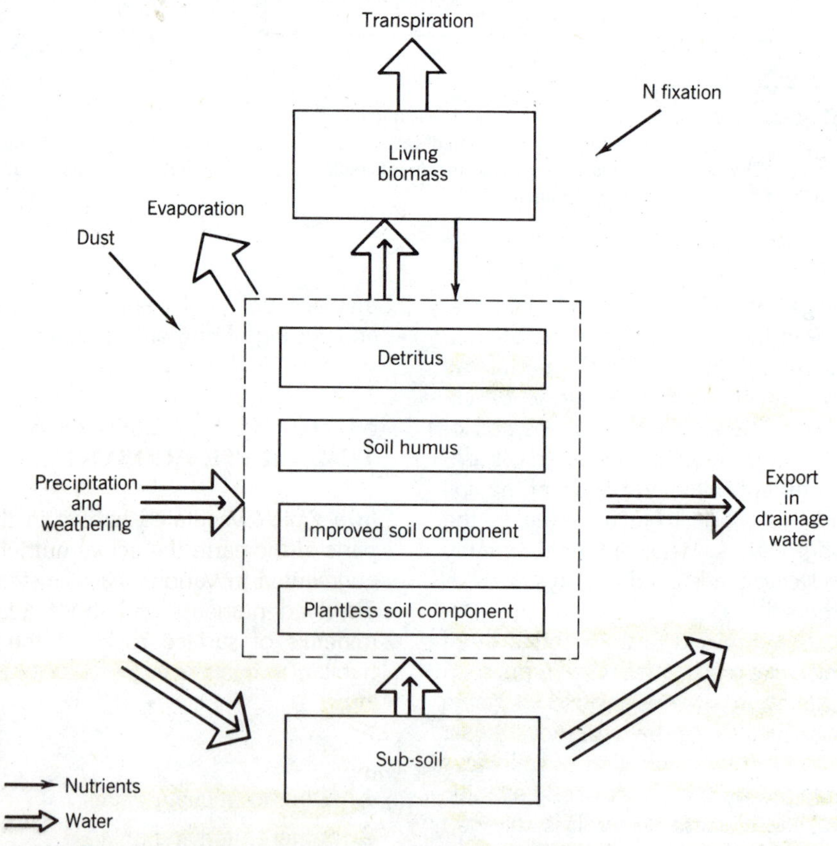

Figure 17.3 Nutrient flux and storage in a temperate ecosystem.
The system differs from the unvegetated habitat of Figure 17.2 by having vastly increased storage capacity for nutrients and by diverting the flow of water through the transpiration stream.

larger part of the reserve is detritus or the product of detritus. This is the mass of net primary production that is not eaten by herbivores but left to rot (Chapter 16). Much of this material has been made as indestructible as possible by the plants, very largely as a defense against herbivores. Cellulose, lignin, and even the chitin of animals is jettisoned and buried, and is very slow to rot. It is this produce of the living part of the system that has so dramatic an effect on the nutrient reservoir.

Thus we can conclude that one of the most important properties of a terrestrial ecosystem, the enrichment of a habitat, is largely an accident. It is a side-effect of selection for plants and animals that they be as indestructible as possible in their own self-interest. Natural selection gives individuals cellulose and the rest to promote fitness, but the contribution of these persistent parts of corpses to the habitat profoundly improves the place for others who come after. This looks like altruism, but is not. The ecosystem appears to grow more fertile through some inner design, but this also is not true. The happening is accident, though it is nonetheless important.

Yet the indestructible remnants of the biota in the litter do not themselves represent all the extra nutrient reserve: cellulose and other polysaccharides, for instance, are mere arrangements of C:H:O and are not usually nutrient-rich. Their importance lies partly in providing soil structures to which nutrients can be attached. In this sense, the organic parts of soils represent a storehouse that must be filled. This storehouse receives nitrates as nitrogen-fixing bacteria work in the roots of many plants, but a major flux of dissolved nutrients also is imported to the reservoirs from deeper in the soil or from percolating water via plant roots. This nutrient-bearing water is physically pumped to the surface by plants in their transpiration streams, and the transpiration pump is not powered by the energy of reduced carbon compounds fixed in photosynthesis but by heat energy absorbed directly from the sun.

In a closed piece of vegetation like a temperate forest, a very large portion of the total incident energy that is neither reflected nor used in photosynthesis goes to evaporate water, thus working the transpiration pump. We have a striking calculation of the actual percentage of solar energy that may be used for life-controlled pumping in this way from the Hubbard Brook watershed study (see below), where calculations of the total water budget of an ecosystem were made (Gosz *et al.*, 1975, 1976; Bormann, 1976). Both the water received by precipitation and that lost in drainage channels were known, so that the difference represented the loss by evaporation. Nearly all this evaporative water loss in a closed forest actually passes through plants as transpiration. The energy required to evaporate this mass of water is easily calculated. At Hubbard Brook it represented 45% of the solar radiation received by the forest during the growing season (Table 17.3). Covering bare ground with a forest, therefore, can result in a very large appropriation of energy by life outside the mechanism of photosynthesis and in ways that permit the pumping of nutrients into living and humic reservoirs.

Two processes thus underlie the changes from the plantless state brought about in nutrient reserves by vegetation, and each of these processes uses a different portion of the total solar energy budget. First is the creation of fresh reservoirs by the plants (Figure 17.3). Since these reservoirs are mostly different fractions of organic matter, living or dead, the energy flux used to make them comes from photosynthesis. Second is the pumping of water through these reservoirs and its discharge into the atmosphere above the plant canopy. This process serves to import dissolved nutrients, with the final result that the reservoirs are charged. The energy flux used is absorbed heat energy, and may constitute nearly half of the total solar flux (Table 17.4).

NUTRIENT BUDGETS OF FORESTS: DIRECT MEASUREMENTS OF GAINS AND LOSSES

It is possible to calculate the nutrient budget of a piece of vegetation by measuring net production, the nutrient content of biomass, and the various nutrient losses from the plants. The losses will reach the soil in dead plant parts, in animal droppings, and in rainwater that washes through the canopy and down the stems. We can say that:

$$N_u = N_r + (N_l + N_w + N_{sf}) \quad (17.1)$$

where

N_u = nutrient uptake
N_r = nutrient retained in biomass (nutrient *increment*)

Table 17.3
Energy of Gross Production and Evapotranspiration at Hubbard Brook
Nearly twenty times as much energy is used to pump water by the transpiring forest as is transduced in photosynthesis. Data are for a second growth forest of sugar maple, beech, and yellow birch on a watershed in New Hampshire (Hubbard Brook, watershed No. 6). (Data from Gosz et al., *1976.)*

	Solar Flux Received (kcal/m²)	Gross Primary Production (kcal/m²)	% Solar Flux into Production [Ecological (Lindeman) Efficiency]	Flux of Energy Used to Evaporate Water (kcal/m²)	% Solar Flux Used to Evaporate Water
				Total: 288,990	Total: 20%
Budget for year	1,485,700	10,400	0.7%		
				Plants: 200,070	Plants: 14%
Budget for growing season	471,600	10,400	2.2%	200,070	42%

Table 17.4
Partitioning of Solar Energy in the Nutrient System of a Forest
This is a likely budget for a temperate forest ecosystem during the growing season, based on the data from Hubbard Brook (Gosz et al., *1976). The budget is only for energy actually striking the forest and does not take into account the fact that the vital transport of water into the system is powered by energy absorbed outside the system (notably by the oceans) and thus greatly exaggerates the importance of the vegetation in energetic terms.*

Energy Capture in Forest Ecosystem	Effect on Nutrient Cycles	% of Incoming Radiation
Photosynthesis	Construction of reservoirs, humus, biomass, litter, and soil structure	2%
Absorbed in living part of ecosystem as heat	Drives transpiration system and pumps nutrients into reservoirs	42%
Absorbed in non-living part of ecosystem as heat	Drives hydrologic cycle and weathering processes	46%

N_l = nutrient returned in litter (includes leaves and droppings)
N_w = nutrient washed off canopy in rain (throughfall)
N_{sf} = nutrient washed back to soil in *stem flow*.

NUTRIENT RETAINED can be measured by monitoring *net production* by harvest (Chapter 16) and making chemical analyses of representative stems, twigs, *et cetera*. NUTRIENT RETURNED is measured in a similar way by monitoring the rate of fall of leaves, branches, caterpillar droppings, *et cetera*, with litter traps. THROUGHFALL and STEM FLOW NUTRIENTS are calculated by measuring water falling from the canopy and down stems, then subtracting the chemical data for the local rainwater. The term $(N_l+N_w+N_{sf})$ is then known and is referred to as NUTRIENT RESTITUTION or NUTRIENT LOSSES by different authors. Then all has been measured to solve:

$$\text{Nutrient uptake} = \text{Nutrient retained} + \text{Nutrient losses} \quad (17.2)$$

A number of nutrient budgets of this kind are now available for forests of northern Europe and European Russia.

Data for Belgian oak forests are given in Figure 17.4 and Table 17.5 (Duvigneaud and Denaeyer-De Smet, 1970). Notice that two sets of analyses were made on the soils: a set of ash analyses to yield estimates of total contents of nutrients in the top 40 cm of soil per hectare and some estimates of EXCHANGEABLE NUTRIENTS that can be leached from the soil in the laboratory. The *exchangeable nutrients* are those that are easily removed in aqueous solution and thus presumably readily available to plants, though this is not to say that plants may not be able to get at some of the nutrients that are more firmly held in the mineral or humic matrix. The results (Table 17.5) show how small a portion of the available nutrients is used by the forest in a year.

A reasonable interpretation of these data is that the trees of a Belgian oak wood do not have their productivity limited by the supply of the various nutrients that were measured. This is in keeping with the general conclusion that photosynthesis is usually restricted by water or temperature acting on a system designed for the constraints of a chronic shortage of carbon dioxide (Chapters 3 and 16). In this connection, it is noteworthy that the ecological (Lindeman) efficiency over the growing season was 2.2% (Table

Table 17.5
Nutrient Budget of a Belgian Oak Wood Compared with Soil Reserves
These are data for the same oak wood illustrated in Figure 17.4. The data suggest a frugal use of the potential reservoir of nutrients held in the soil. This is in keeping with the observation that soils of temperate latitudes tend to have large reservoirs of nutrients in organic matter and clays. The high calcium and magnesium reserves reflect the fact that the soil is a **rendzina** *developed on limestone. (Data from Duvigneaud and Denaeyer-De Smet, 1970.)*

	K	Ca	Mg	N	P	S
In plants (kg/ha)						
Retained (increment)	16	74	5.6	30	2.2	4.4
Returned (losses)	53	127	13	62	4.7	8.6
Uptake	69	201	18.6	92	6.9	13
Soil (top 40 cm)						
Exchangeable (kg/ha)	157	13,600	151	—	—	—
Total (metric tons/ha)	2.68	133	6.46	4.48	0.92	—

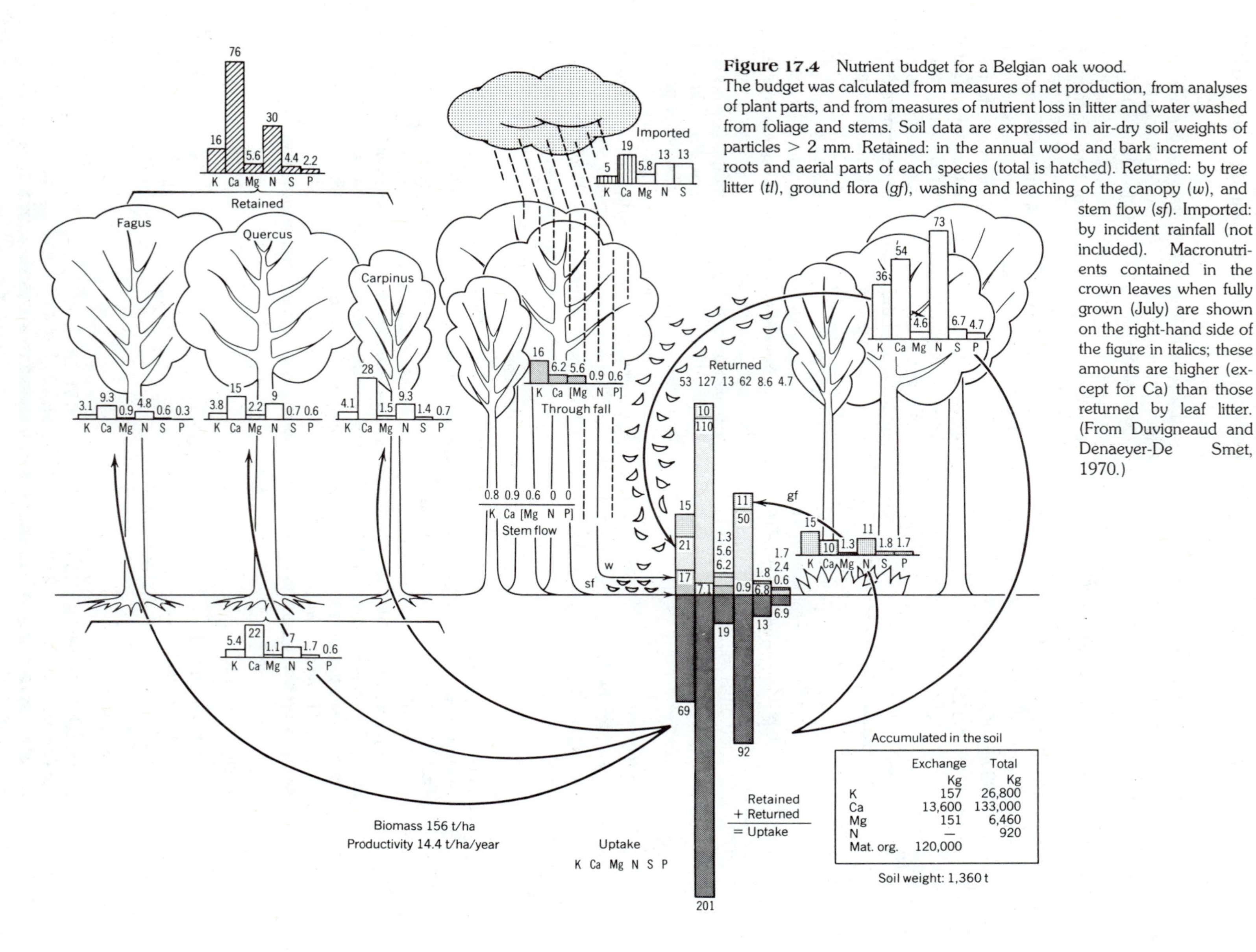

Figure 17.4 Nutrient budget for a Belgian oak wood. The budget was calculated from measures of net production, from analyses of plant parts, and from measures of nutrient loss in litter and water washed from foliage and stems. Soil data are expressed in air-dry soil weights of particles > 2 mm. Retained: in the annual wood and bark increment of roots and aerial parts of each species (total is hatched). Returned: by tree litter (*tl*), ground flora (*gf*), washing and leaching of the canopy (*w*), and stem flow (*sf*). Imported: by incident rainfall (not included). Macronutrients contained in the crown leaves when fully grown (July) are shown on the right-hand side of the figure in italics; these amounts are higher (except for Ca) than those returned by leaf litter. (From Duvigneaud and Denaeyer-De Smet, 1970.)

17.4), an efficiency comparable to that of temperate agricultural crops on good, well-fertilized land.

This demonstration of a very large and redundant reserve of nutrients under temperate forest partly shows why western-style agriculture works. Soils in temperate latitudes under wild vegetation accumulate nutrients so that soil reservoirs are maintained with stores far beyond the immediate needs of the plants. Farmers are able to draw on these reservoirs that remain even when the primeval forest has been removed. A number of studies in northern Europe and the Soviet Union allow nutrient budgets of different forests to be compared (Table 17.6). The data suggest rather strongly that the species of plants growing on the habitat are more important to the relative cycling of different nutrients than the geographical position of the site. Different species of tree have different nutrient requirements, a common-sense conclusion but it is nice to have it backed by real data.

THE WATERSHED IN THE STUDY OF NUTRIENT CYCLES

The nutrient budgets described so far are limited, since they give only rates of turnover for nutrients in living plants, together with an estimate of the masses stored in the ground. They are not budgets for the whole ecosystem, which would require knowing inputs and outputs to all parts of the system, including rates of turnover in the soil reservoir. But in some special circumstances it has been possible to make a complete nutrient budget for an intact ecosystem. We can measure nutrient inputs that reach the ecosystem via incoming water and dust, outputs that leave the system in drainage waters, turnover rates in the living plants by direct measure, and turnover rates in soils by subtraction. The technique is to study the ecosystem occupying a whole drainage catchment, or WATERSHED,[2] where the limits to the ecosystem can be defined as the edges of the drainage divides. The inputs and outputs to the whole watershed can then be measured directly with sufficient ingenuity and effort. The technique first made news in ecology with its use at Hubbard Brook in New Hampshire (Likens *et al.*, 1967, 1970, 1977, 1978; Likens and Bormann, 1972; Bormann, 1976; Bormann and Likens, 1967; Bormann *et al.*, 1974, 1977; Gosz *et al.*, 1975, 1976; Burton and Likens, 1973; Siccama *et al.*, 1970; Whittaker *et al.*, 1974).

Figure 17.5 is a map of the Hubbard Brook experimental forest showing watershed ecosystems used for measurements. Each ecosystem is defined by heights of land and is drained by a tributary stream of the Hubbard Brook. All watersheds are underlain by impermeable bedrock, an important feature because it means that virtually no water or nutrients depart from the ecosystems by deep penetration and groundwater flow. Monitoring the flow of water in the drainage stream then accounts for all the liquid water leaving the ecosystem. This is done by damming each stream and making the water pass over a notched weir (Figure 17.6). Regular sampling of this water yields a measure of all the nutrients leaving each ecosystem in solution.

Since the watersheds do not have streams flowing into them, the only sources of nutrients to balance those lost in the drainage streams are rain, dust, and weathering of the surface rocks (the *regolith*, Chapter 20). Rain and dust are measured by rain gauges and traps set out at a rate of one for each 12.9 hectares. A complete soil survey was made of each watershed. Net primary production of the forest is monitored by the technique of *increment analysis* (Chapter 16) and the vegetation is described by standard phytosociological methods (Chapter 15). The results of this work on each watershed are direct measures of *water input and output, nutrient input and output*, and *nutrient retention in biomass and soil*. These data are sufficient to allow calculation of other rates of nutrient and water flux to provide a complete description of the cycling of dissolved nutrients in a temperate hardwood forest (Likens *et al.*, 1977).

The valley bottoms of the watersheds have

[2]The term "watershed" has different meanings in American and English usage. The American "watershed" is the English "catchment area." The English "watershed" is the American "height of land" or "watershed divide."

Table 17.6
Nutrient Budgets for North European Forests
Compiled by Duvigneaud and Denaeyer-De Smet (1970) from data of various authors.

Country	Forest Type	K (kg/ha/yr) Lost	+ Retained	= Uptake	Ca (kg/ha/yr) L	+ R	= U	N (kg/ha/yr) L	+ R	= U	P (kg/ha/yr) L	+ R	= U
Belgium	Oak–ash	54	21	75	76	42	118	79	44	123	5.4	4.0	9.4
Belgium	Mixed oak	35	16	51	120	74	194	61	30	91	4.1	2.2	6.3
USSR	Oak	46	27	73	81	20	101	48	56	104	12	3	15
Germany	Beech	8.4	8.8	17.2	69	26	95	29	16	45	3.5	1.2	4.7
U.K.	Birch on marsh	25	3	28	34	10	44	48	8	56	3.6	0.5	4.1
USSR	Spruce	5	6.7	11.7	64	22.5	86.5	40	20.6	60.6	3.3	1.8	5.1
Germany Germany	Spruce Pine	4.4	2.4	6.8	35.6	9.0	44.6	22.2	12.1	34.3	2.1	0.9	3.0

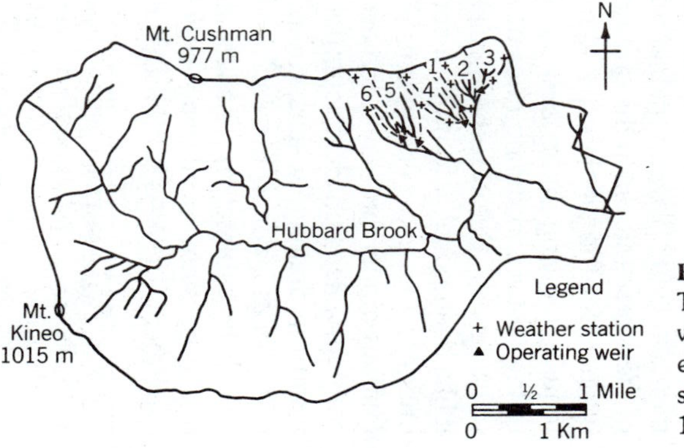

Figure 17.5 Watershed ecosystems at Hubbard Brook.
The six watersheds used for nutrient studies are shown by dashed lines. Each little watershed has its own stream and is divided from the next by steep terrain. A weir on each stream (indicated by a triangle) provides a site to sample outgoing water. Weather stations (crosses) are sites for sampling incoming water. (From Bormann and Likens, 1971.)

Figure 17.6 Notched weirs at Hubbard Brook.
The outlet of each sampled watershed is made to flow over a dam or weir where the volume of water coming down, its load of dissolved nutrients, and its sediment load can be monitored. The upper photograph shows the devegetation experiment.

forests that can be assigned to *beech–maple associations* (Chapter 15), being dominated by sugar maple *(Acer saccharum)* and American beech *(Fagus grandifolia)*. There is also yellow birch *(Betula alleghaniensis)* and, especially towards the tops of the ridges, red spruce *(Picea rubens)* and balsam fir *(Abies balsamea)*. The soils are alfisols (gray-brown podzolic), with well-leached "A_2" horizons (Chapter 20). The forest is second growth but the land has not been farmed so that the soils are virgin with well-developed structures. A cover of snow and litter prevents the soils from freezing despite the cold New England winters. Data from this forest, therefore, should be roughly comparable to that of European forests and the temperate deciduous biome generally.

The sampling procedure at Hubbard Brook is summarized in Figure 17.7. The important revelations of these measurements are as follows:

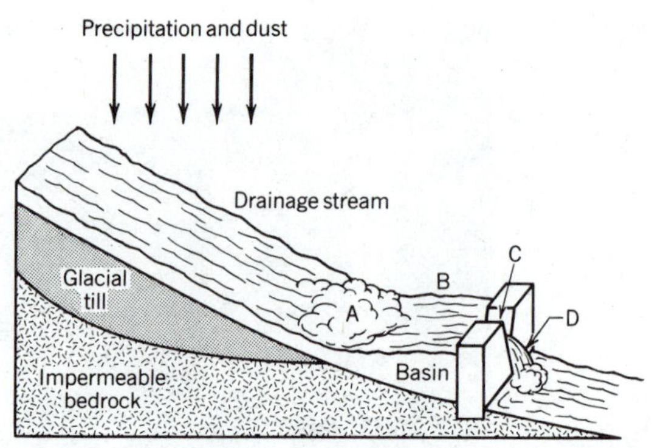

Figure 17.7 Nutrient measurements at Hubbard Brook. Nutrients in precipitation and dust are measured in grids of collectors. Samples for nutrients in solution leaving the ecosystem are taken at A, before the water enters the settling basin in water backed up behind the notched weir. Nutrients leaving the ecosystem as particulate matter are collected as they settle in the basin at B, and by millipore filtration and screening of samples going over the weir at D. Rate of water loss is measured at C. (Modified from Bormann *et al.*, 1974.)

Evapotranspiration Is Constant

Evapotranspiration is measured easily as the difference between precipitation and water leaving each watershed over its dam. Over five years, precipitation varied between 95 and 142 cm but evapotranspiration only between 46 and 52 cm (Table 17.7). One conclusion to draw from this result is that the ecosystem regulates the amount of water pumped through the ecosystem reservoirs, a narrowly defined volume being pumped each year. The intact forest, therefore, resists being perturbed by weather within rather wide limits.

Concentration of Dissolved Nutrients at Outlet Is Constant

The ionic content of the water coming over the weir was always nearly the same, despite the fact that the mass of water coming down varied widely.

Table 17.7
Precipitation and Evapotranspiration at Hubbard Brook
Although annual precipitation varies widely the evapotranspiration is nearly constant. (From Likens et al., *1970.)*

Water Year (June 1–May 31)	Precipitation (cm)	Stream Outflow (cm)	Evapotranspiration (cm)
1963–64	117.1	67.7	49.4
1964–65	94.9	48.8	46.1
1965–66	124.5	72.7	51.8
1966–67	132.5	80.6	51.9
1967–68	141.8	89.4	52.4
Average	122.2	71.8	50.4

A slight negative correlation between sodium and water volume and a slight positive correlation between potassium and water volume are trivial compared with the changes in the mass of water coming down. It follows that the ecosystem regulates the concentration of ions in the outgoing water. The outgoing *mass* of solutes is not controlled, because the mass lost varies directly with the water flowing through, and hence the precipitation. Only the *concentration* is under control. What is revealed is a chemical buffering of the ecosystem water. This could be through passive soil chemistry or it could be a function of life processes. Experimental removal of a forest (see below) in fact shows that life processes are decisive for this buffering.

Particulate Losses from the Intact Ecosystem are Small

Data comparing nutrient loss in solution with losses in particulate matter are given in Table 17.8. The mass of nutrients lost as particles is a small fraction of the loss in solution. Erosion losses in the intact forest ecosystem, therefore, are close to negligible.

Table 17.8
Dissolved and Particulate Nutrient Loss at Hubbard Brook
When the forest vegetation is intact, the loss of nutrients in solids carried by drainage streams is small compared with the loss in solution. Data are the results of analyses at the weir of watershed number 6 between 1965 and 1967. (From Bormann et al., 1969.)

	Percentage of Total Losses		
	Particulate Matter		
Element	Organic	Inorganic	Dissolved
Calcium	0.7	1.8	97.5
Magnesium	2.0	3.7	94.3
Nitrogen	5.9	[a]	94.1
Potassium	0.5	17.5	82.0
Sodium	0.0	2.8	97.2
Silicon	[b]	18.8	81.2
Sulfur	0.1	0.1	99.8

[a]Not measured, but very small.

[b]Not measured.

Weathering Equals or Exceeds the Nutrient Leak

The loss of nutrients in solution generally exceeds the input by precipitation and dust by a wide margin (Table 17.9), yet the ecosystem cannot be suffering a net loss of nutrients or it would be impossible for the concentration in outgoing water to remain constant year after year as it does. It follows, therefore, that the nutrients lost are balanced by a nutrient input from other than dust or rain. These balancing nutrients can come only from weathering of rocks. This is the

Table 17.9
Precipitated Nutrients and Stream Outflow at Hubbard Brook
The data are for watershed number 6 in the year 1967–1968. With the exception of nitrogen, the nutrient input does not equal the nutrient output, thus demonstrating that weathering is the most important source of nutrients to the ecosystem. The data do not of themselves show that nitrogen is gained by the reservoirs of the system since the balance of combined nitrogen may be broken down within the system and released to the atmosphere. Other data, however, do show nitrogen accumulation (Table 17.10). (From Likens et al., *1970.)*

Nutrient	Precipitation Input	Stream Outflow	Net Loss or Gain
Water	171.8	89.4	—
Ca	3.0	12.2	−9.2
Mg	0.8	3.4	−2.6
K	0.8	2.4	−1.6
Na	1.8	8.8	−7.0
N (NH_4)	2.6	0.2	+2.4
N(NO_3)	5.2	2.8	+2.4
S(SO_4)	16.0	19.3	−3.3
Cl	5.2	5.3	−0.1
C(HCO_3)	[a]	0.5	−0.5
Si(S_1O_2)	[a]	17.0	−7.0

[a]Not measured, but very low.

best demonstration we have of the importance of deep weathering to the maintenance of nutrient reservoirs in temperate ecosystems. Presumably the nutrients that go into the ample soil reserves in Belgian oak forests and elsewhere, as well as at Hubbard Brook, come ultimately from deep weathering within the ecosystem itself.

Combined Nitrogen is Stored Preferentially

Less combined nitrogen goes over the dam than comes in as dust or rain (Table 17.9). This is a remarkable statistic because we know very well that the input of combined nitrogen from the atmosphere is always relatively low. Farmers must use nitrogen fertilizer and many wild plants rely on nitrogen-fixing bacteria; evidence enough that combined nitrogen falling from the sky is of little importance to an ecosystem. And yet even this small input from dust and rain is more than the losses in solution at Hubbard Brook. Combined nitrogen, of course, may be lost in more important ways than in solution, since nitrogen gas may be vented directly to the air. However, the breakdown of nitrate to nitrogen gas requires reducing conditions (Chapter 19) and is not likely to be an important process at Hubbard Brook. It seems likely that most of the combined nitrogen is being stored in biomass, something made particularly likely by the fact that biomass is known to be increasing at Hubbard Brook (Whittaker *et al.*, 1974).

EXPERIMENTAL REMOVAL OF A HUBBARD BROOK FOREST

The basic Hubbard Brook data suggest a number of functions for the living parts of the ecosystem. Notable among these are the regulation of nutrients in outgoing water, storage of combined nitrogen, and control of run-off up to the volume of transpired water. These conclusions can be stated in the form of hypotheses that could be tested by the relatively simple experiment of killing all life on one of the watersheds. With six watersheds to choose from it was possible to keep some as control while manipulating others.

On watershed 2, the forest was clear-cut and the land was sprayed with herbicide for three years to prevent all regeneration (Likens *et al.*, 1970; Bormann *et al.*, 1974). On watershed 4, some strips of forest were cut and other strips were left intact (Burton and Likens, 1973). Studies of regeneration in other New Hampshire clear-cut forests were made for comparison (Marks and Bormann, 1972). The standard measurements of the study continued to be made at each weir of each watershed. Needless to say, there were dramatic differences between the measurements for forested and treeless watersheds. When the trees went, more water came down as run-off, the water was warmed as it rolled unshaded under the sun, and there was erosion and an increased leak of nutrients from the ecosystem. But the patterns of these events held a number of revealing surprises. The principal revelations of these experiments in ecosystem destruction are as follows:

Run-off increased by about 30%. This was expected because the transpiration pump that loosed water into the air had been removed. Much of the old water that was originally transpired now ran into the streams.

Water temperature increased by several degrees Celsius. This was particularly apparent in the strip-cutting experiment where water warmed and cooled as its parent stream crossed strips. A secondary heating effect is that snow banks melt earlier in cleared sections.

In the first two years following clear-cutting the big nutrient loss was in solution. This was an exciting discovery. We expect *erosion* losses, but what we find first is an escalation of *solution* losses. Over the first two years the ratio of net loss in solution to net loss as solids increased nearly four times. Killing the plants has the effect of increasing the concentrations of nutrients in the flushing water, and the effect is almost immediate. This is a clear demonstration that the concentration of nutrients in water leaving an intact forest ecosystem is in fact controlled by the plants themselves. The living ecosystem regulates the leak.

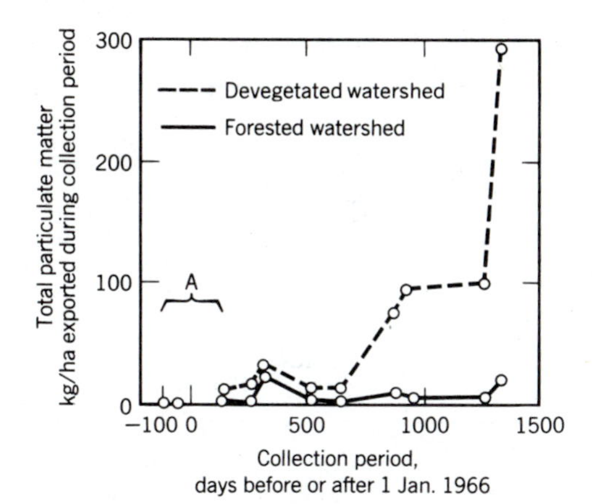

Figure 17.8 History of erosion in a devegetated watershed.

Particulate losses from the cleared watershed at Hubbard Brook were comparable to those from an intact watershed for the first two years, after which they rose exponentially. One watershed was devegetated in time interval A. (From Bormann *et al.*, 1975.)

Table 17.10

Nutrient Losses in the Clear-Cut Hubbard Brook Watershed

The largest increase in losses was in the anion nitrate. Lesser increases in each cation are probably a consequence of the nitrate flux. (From Likens et al., *1970.)*

	Metric Tons/km²/yr			
	1966–67		1967–68	
	W2	*W6*	*W2*	*W6*
Ca^{++}	−7.5	−0.8	−9.0	−0.9
K^+	−2.3	−0.1	−3.6	−0.2
Al^{++}	−1.7	−0.1	−2.4	−0.3
Mg^{++}	−1.6	−0.3	−1.8	−0.3
Na^+	−1.7	−0.6	−1.7	−0.7
NH_4^+	+0.1	+0.2	+0.2	+0.3
NO_3^-	−43.0	+1.5	;62.8	+1.1
SO_4^{--}	−0.5	−0.8	0	−1.0
HCO_3^-	−0.1	−0.2	0	−0.3
Cl	−0.1	+0.2	−0.4	0
SiO_2(aq.)	−6.6	−3.6	−6.9	−3.6
Total	−65.0	−4.6	−88.4	−5.9

Erosion of solids began to increase significantly only after a 2-year lag. This result is illustrated in Figure 17.8. The conclusion is that the once-forested landscape retains its relative immunity to erosion long after the trees are killed, presumably until bacterial decomposition destroys the litter and organic fabric of soil and of the stream bed.

Nitrate was the nutrient to show the largest losses. The concentration of nitrates coming over the weir of the clear-cut watershed went up over 40 times whereas losses of other nutrients were mostly increased by only 2 or 3 times (Table 17.10). This is a dramatic test of the hypothesis that combined nitrogen is being retained and hoarded by the intact forest. Denitrifying bacteria must be liberating combined nitrogen to the soil water as nitrates rapidly, but on the death of the forest these nitrates are no longer scavenged by plants. Probably the reservoir of combined nitrogen that has been built up by nitrogen-fixing bacteria over the lifetime of the forest is leaked from the site in the first two or three years after destruction of the forest.

The loss of other nutrients was a consequence of nitrate losses. In solution, the anion NO_3^{--} must be balanced by a divalent cation or two monovalent cations. Unless a pH change occurs that is sufficient to balance NO_3^{--} with H^+ the loss of nitrates from the system inevitably means a comparable loss of cations. This is readily appreciated from Table 17.10.

Revegetation is rapid where it is not prevented by herbicides. The early succession shrub or tree, wild cherry *(Prunus pennsylvanica)*, grows rapidly on a clear-cut site. In three years it forms a closed canopy and is quickly accumulating nutrients. This rapid growth presumably benefits not only from the absence of shade but also from the rich supply of nutrients in the soil water.

Algae bloom in the drainage streams. This can be attributed to the increased flux of dissolved nutrients.

The mutual dependence of these various effects is illustrated in Figure 17.9.

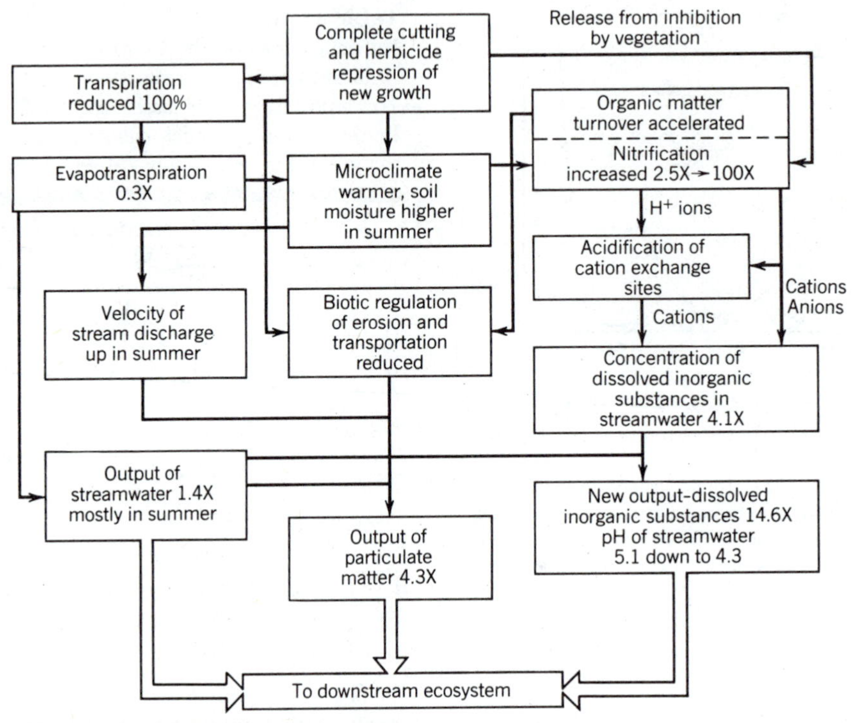

Figure 17.9 Interactions triggered by deforestation of Hubbard Brook watershed. Ecologic, hydrologic, physical, chemical, and erosion consequences of removing the vegetation based on data of the first two years. (From Bormann *et al.*, 1974.)

INTERIM CONCLUSION: NUTRIENTS, ENERGY, AND STABILITY IN TEMPERATE FORESTS

That nutrients should be cycled is one of the more basic ideas that went into the ecosystem concept. The community shares a habitat and hence the nutrients that are there. But more than this, there has always been the idea that the living system managed things so that the habitat was materially different from the bare ground that once was. Studies of nutrient budgets give strong support to this general idea of an improved habitat, at least where trees grow in temperate latitudes.

Nutrients do collect under temperate forests, and it is the living trees themselves that are responsible for the collecting and retention of these nutrients. Furthermore, the living vegetation regulates the passage of water and controls the chemistry of that water. Rapid changes in nutrient flux, water flux, and even temperature are prevented by the plant community. It is therefore proper to say that temperate forest ecosystems promote HOMEOSTASIS in many respects and resist perturbation. Life on a site promotes physical stability of the system.

One of the most interesting properties of the forest is the way in which command of a relatively small flux of energy by the plants results in so much control. Photosynthesis yields the biota only 1 to 2% of incident energy. The trick that makes this small energy flux so potent is to use it for physical construction. A temperate forest controls wind, water, and soil movement by what amounts to mass engineering, and the forces of chaos in a habitat are checked as the floods of a great river are checked—by a few subtle engineering works. Like the buildings and infrastructure of a human city, the constructions of a temperate forest confer physical stability on the system they support.

A prominent viewpoint in ecology has been to look for high efficiency of energy transfer between organisms as the refinement of life process. It is this view that lies behind much of the work seeking to calculate ecological (Lindeman) efficiencies (Chapter 4). But the remarkable physical stability achieved by a temperate forest largely results because energy transfer is NOT efficient. Herbivores are not able to win a large fraction of the net productivity of plants of the temperate forest. Production goes instead to bolster the physical structure of the ecosystem. From the ecosystem viewpoint, therefore, it might be appropriate to call a plant "efficient" if it budgets a large part of its total energy to making dead biomass.

Perhaps the most exciting revelation for ecosystem energetics that comes from these studies is that vegetation actually wields far more of the total incident energy than it can win by photosynthesis. Trees of a temperate forest can use 40% or so of incident solar energy to pump water through their own bodies. Forest trees are actually heat engines, running the cruder parts of their machinery with direct solar power. It is because a forest is a ranked array of these heat engines that much of the stability of temperate landscapes results.

These ecosystem properties of temperate forests can be understood as resulting from adaptations of individuals. The shape and structure of each temperate tree can be explained fully as suited to individual fitness: selection favors massive trunks made of resistant polysaccharides because the owners compete well, not because the dead trunk holds the soil together. It just happens that when the life strategies of forest trees are followed in temperate latitudes, then structured soil and other ecosystem properties so elegantly revealed at Hubbard Brook follow.

FOREST IN DIFFERENT CIRCUMSTANCES: THE WET TROPICS

A tropical forest ecosystem has a pattern of nutrient cycling and physical stability that is very different from that of a temperate forest. The critical differences are revealed in the soils, which are deficient in nutrients. Many tropical soils are red instead of brown, and very deep (*oxysols* or red lateritic soils, Chapter 20). If these soils are examined to see what factors contribute to their lack of nutrients, the following facts emerge readily:

1. The fine particles of the soils include few of the large-lattice clay minerals that absorb cations to their surfaces. Clays like montmorillonite that perform this service in temperate soils are largely lacking, being replaced by amorphous fine particles or smaller clays that are less chemically active.
2. There is extremely little organic matter in the soils on which anions can be affixed. Below the top few centimeters, in fact, the soil holds virtually zero organic matter.
3. Surface litter or detritus is scanty and thin.
4. The effects of solution from percolating water can be detected many meters deep, far deeper than in temperate soils.

These observations lead to a remarkable conclusion: tropical rain forest can live on soils that, by any scale of agriculture, are infertile. The most productive ecosystem known, the forest that supports the largest biomass, exists where the soil reservoir of nutrients is trivial by the standards of nearly all temperate ecosystems.

Not only are the nutrient reserves of this ecosystem low, but the inputs are low too. The fact that the soil has been leached to great depths by percolating water means that the supply of nutrients in ground water will be low. On flat land like the Amazon basin where most rain forest exists few nutrients are weathered from rocks upslope. Nor is it likely that tropical rain and dust bring in more nutrients than do temperate rain and dust. Dispassionate analysis, therefore, shows that the only large reserve of nutrients in a tropical forest lies in the trees themselves. Only with a thorough and rapid cycling of nutrients, coupled with a very long residence time in the trees, can the forest survive. The essentials of this system are shown in Figure 17.10.

Two major questions arise from this analysis: The first is, "How does the tree strategy yield a system with so few reserves in the tropics whereas the same strategy yields ample reserves in temperate regions?" This is a systems question. Individual trees seem to behave in rather similar ways in the two places and yet different ecosystem properties result. The second question is, "How do the individual trees of the tropical rain forest adapt to nutrient shortage so that the forest persists?"

On the Nutrient Structure of Ecosystems in the Wet Tropics

The explanation for the low nutrient reserves in rain forest soils depends on temperature. High temperatures, and lack of a cold season, mean that decomposition is far more rapid. Furthermore, a characteristic of tropical forests is the rich fauna of detritivores whose existence is probably itself temperature dependent. Termites, together with detritivorous ants and other insects, dispose of litter, even tree trunks, with extreme rapidity. In addition the true decomposers, fungi and bac-

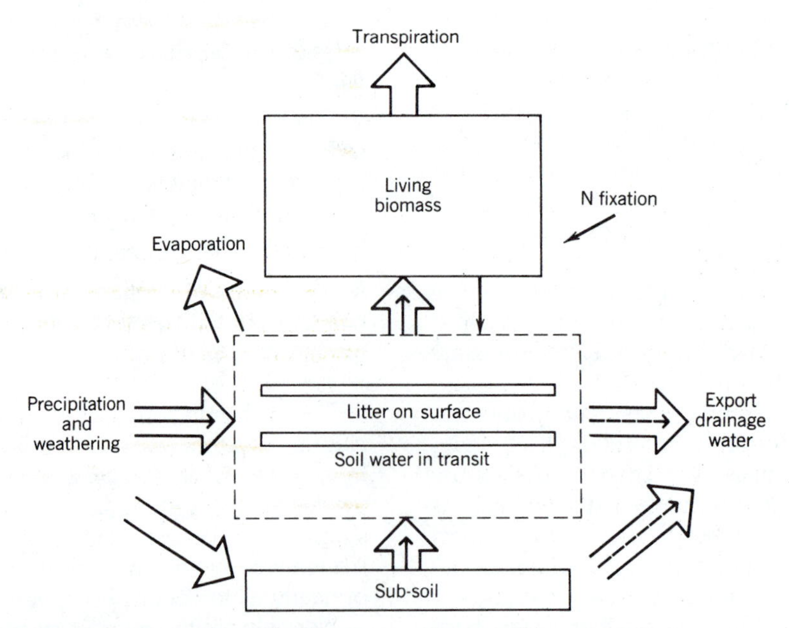

Figure 17.10 Nutrient flux and storage in an ecosystem of the wet tropics. The system has even less soil storage capacity than the unvegetated temperate ecosystem of Figure 17.3. It does, however, have water pumping and deep nutrient retrieval systems of the temperate forest of Figure 17.4.

teria, remove humic substances completely from the soil surface. Soils have minimal anion-holding powers as a result.

Lack of organic matter may be particularly serious in not allowing the retention of nitrates in litter in the way that the Hubbard Brook study showed was so spectacular a feature of temperate soils. But high temperatures also have direct effects on the mineral soil in precluding the formation of large-lattice clay minerals (Chapter 20). The inability of the soils to hold cations, therefore, is a direct consequence of temperature and is a property of the physical system.

On Adaptations of Tropical Trees to Nutrient Shortage

For rapid and effective cycling of nutrients the essential adaptation of an individual rain forest tree is a high capacity to take up nutrients. Rain forest trees must be able to sweep the water passing their roots clean of dissolved ions. They do not, as it were, bring fresh tree technology to this task but they certainly possess the full array of uptake systems owned by trees everywhere. In particular, it has been found that they have prominent symbiotic relationships with fungi called VESICULAR ARBUSCULAR MYCORRHIZAE (VAM). Various soil fungi are important to nutrient cycles by releasing nutrients from organic complexes and by extracting dissolved ions from very dilute solutions (Wicklow and Carroll, 1981). The VAM symbiotic fungal hyphae penetrate root hairs and derive energy by ingesting cell contents. The roots, however, are able to absorb nutrient solutions from the fungus so they pay with lost cell sap for the services of an excellent ion pump that lets root hairs sweep adjacent water virtually clean of nutrients. VAM are not a peculiarity of rain forest trees since they are found on some roots in virtually all terrestrial ecosystems, but they seem to be ubiquitous in the wet tropics. There is evidence that the adaptation is refined in the rain forest to the point that the fungal symbiont may be in direct contact with the decomposing litter, so that nutrients can be transferred from detritus to tree roots directly. The evidence for this comes from a study near Manaos in the Brazilian Amazon where rain forest trees grow in extremely poor sandy soils. Went and Stark (1968) found that rootlets of trees in the top 15 cm of soil were directly connected to the thin litter of the surface through fungal mycorrhizae. It looks to be a beautiful adaptation that prevents any danger of loss of nutrients by not permitting them to reach the soil water at all.

All this is no more than a thorough application of nutrient-absorbing systems theoretically available to all trees. But on an old tropical oxysol every individual tree is on short rations for nutrients, whereas the trees of a temperate forest sometimes have nutrients in abundance. It is predicted that the actual network of fine roots put out by a rain forest tree is finer than that of a temperate tree, though it is doubtful if data exist to test this prediction.

On top of these refined adaptations there is one physical circumstance that aids the tropical trees—they pump more water. The adaptations of tropical trees to nutrient shortage, therefore, are to pump much water and to sweep the water they do not pump clean of useful ions, making the best possible use of symbiotic fungi in the process. The best hard data we have for the effectiveness of these ion pumps is that water leaving a tropical rain forest in streams may have electrical conductivity close to that of distilled water (Walter, 1973). Figure 17.11 compares the nutrient cycles of tropical and temperate forests.

The overriding characteristic of the nutrient cycle in the wet tropics is the absence of any significant reserve outside the biota. There may, of course, be exceptions where local accidents of landform or parent rock supply additional nutrients. This is so on young volcanic soils, where rapid weathering of nutrient-rich rock supplies the ecosystem with a large input of nutrients. The same benefit is given ecosystems on tropical flood plains, where the annual floods bring nutrient-rich sediments from distant ecosystems. But on many wet lowlands of the tropics with oxysols (lateritic soils) the prominent ecosystem characteristic is few nutrients outside the living biomass.

This has important implications for ecosystem

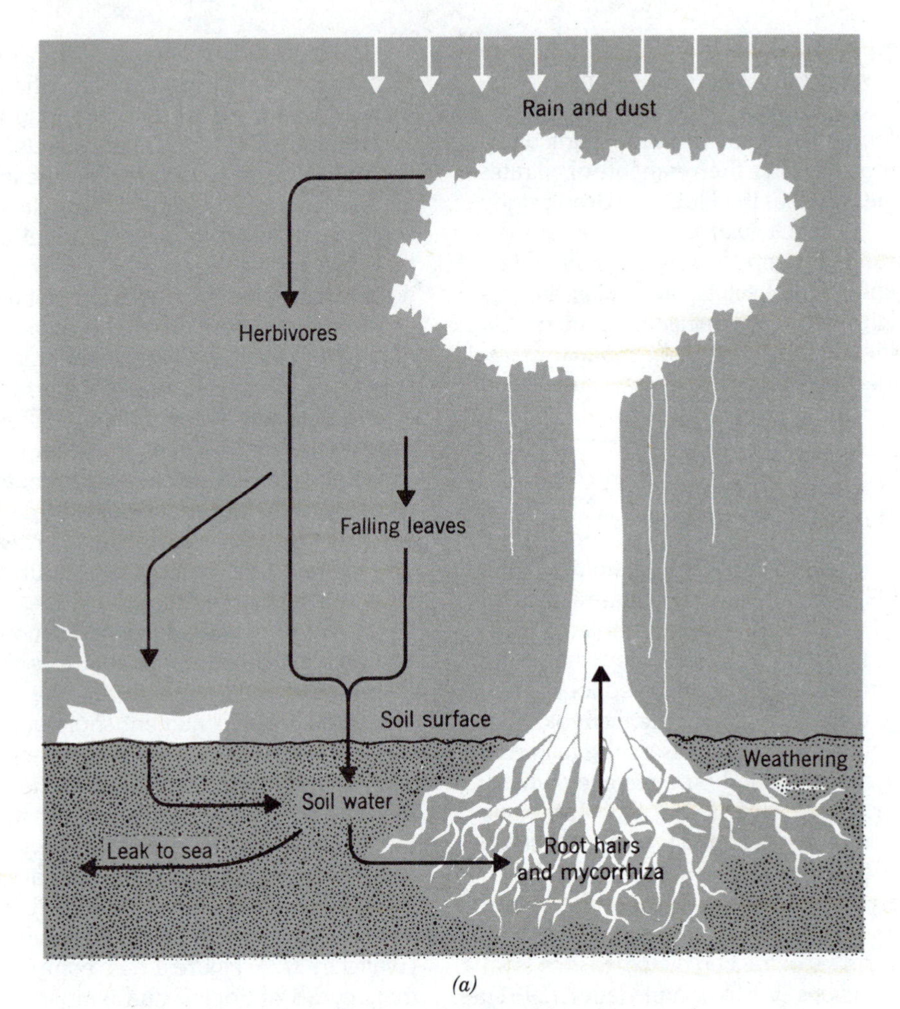

(a)

Figure 17.11 Nutrient cycles in temperate and tropical forests compared.
In some lowland tropical rain forests (*A*) the principal nutrient reservoir is in the trees themselves. Nutrients entering the soil water by decomposition of plant or animal remains are quickly retrieved by a fine network of root hairs and mycorrhizae. This retrieval system may be so efficient that the loss of nutrients in the drainage waters is small enough to be offset by what fresh supplies are provided in rain water or from ancient weathered rocks. Farming in such places destroys the living nutrient retrieval system with a consequent loss of the nutrient reserves in the drainage water.

stability. In one sense a tropical rain forest seems to be a stable system, for it can persist without visible change for long periods of time. The biota are self-perpetuating; hoarding nutrients, producing year-round, controlling run-off, and regulating the passage of water through the transpiration mechanism. As long as the climate does not change this ecosystem persists and is, in a sense, stable.[3]

NUTRIENT CYCLES IN OTHER BIOMES

In grasslands the pattern of nutrient cycling has much in common with temperate forests in that there tends to be a large soil reservoir and a

[3]Ecosystem stability in this sense could be defined as the ability to persist indefinitely in the face of minor perturbations (see Chapter 26).

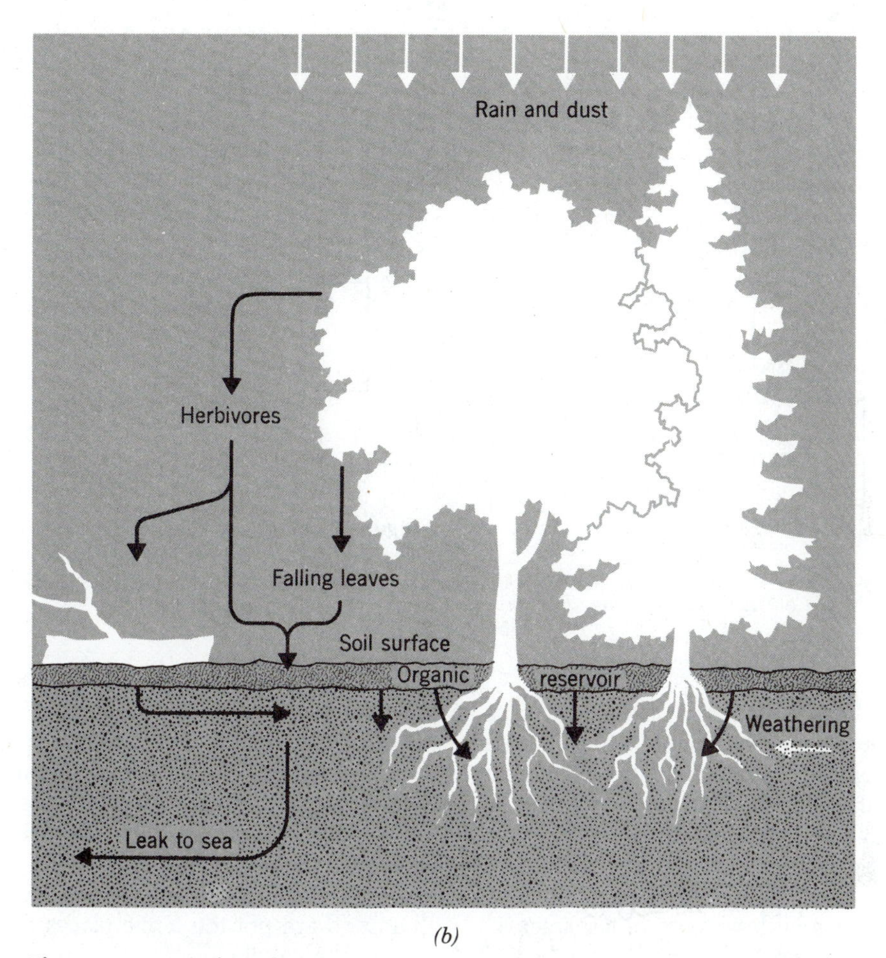

(b)

Figure 17.11 (continued)
In temperate forests (*B*) the main nutrient reservoir is in the soil itself. This soil is not subjected to the intense leaching of soils of the wet tropics and it holds much organic matter that, in turn, holds nutrients. Nutrient cycles are slower, but there is still a steady-state supply. Wise agriculture may maintain this steady state in the agricultural ecosystem.

lingering supply in litter and moribund parts. Figure 17.12 shows a sulfur cycle worked out for Australian grasslands by using ^{35}S as an isotopic tracer. The soil reservoir is huge by the standards of plant needs. Farmers of this ecosystem supply sulfur as fertilizer though, because sheep impose an output of sulfur on the system when this is carried off in wool. Figure 17.13 shows the phosphorous cycle on Indian grasslands, again revealing a large soil reserve. This pattern is to be expected in all natural grasslands that build organic soils (Chapter 20).

In tundra biomes, temperature tends to impose a different constraint on nutrient cycles. Low temperature and frozen ground make the nutrients of soil and litter slow to be mobilized. In a tundra, therefore, the rate at which nutrients can be mobilized from the frozen ground in the spring is crucial to the plant supplies. In extreme arctic sites the ability to mobilize nutrients from within the living tissue of the plant may be important and plants are adapted to retain nutrients in stem or root tissue that persists for years (Dowding *et al.,* 1981). The total nutrient reserve in a tundra soil may appear to be redundantly large, almost on the scale of a temperate

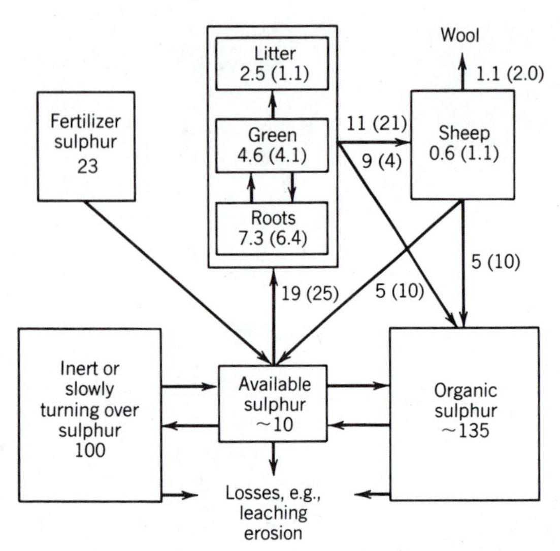

Figure 17.12 Sulfur cycle in Australian grasslands. Data estimated by use of ^{35}S tracer technique. (From Till, 1979.)

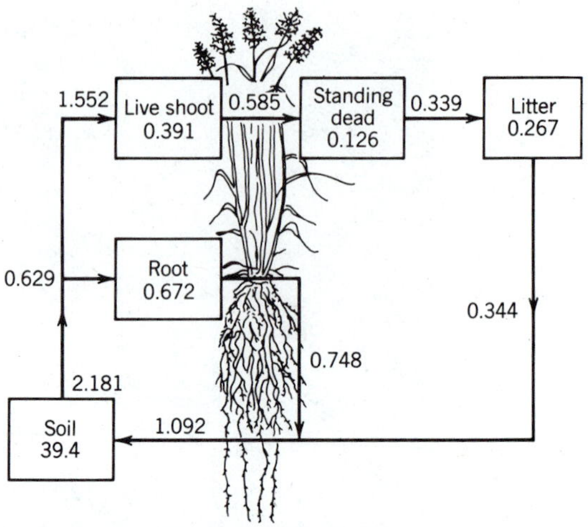

Figure 17.13 Phosphorus cycle in Indian grasslands. Numbers in boxes are mean standing crops of P in g m^{-2}. Numbers on arrows are net flux rates of P in g m^{-2} d^{-1}. (From Yadava and Singh, 1977.)

forest (Table 17.11). Yet the mobilization of these reserves may be so difficult that tundra plants are nutrient-limited. Evidence for this is the rapid tundra growth following destruction of habitat by lemmings (Chapter 13), when nutrients long held in moribund plant tissue are returned to the soil via lemming excrement.

Nutrient cycles in deserts are of less interest, unless water is considered a nutrient. In seasonal dry places like chaparral, soil reservoirs are likely to accumulate during moist times and persist, provided the plant cover remains intact. These ecosystems can show some of the instability of wet tropics, however, if completely cleared. Some lands of this biome around the Mediterranean Sea that have been farmed for thousands of years certainly have lost the nutrient reserves they once had and are not the fertile places they were in classical antiquity.

One practical thought emerges from studies

Table 17.11
Arctic Nutrient Reservoirs

The data are for total nutrients in live plants, dead plants, and total soil and plants combined to a depth of 20 cm. It is evident that in tundra and subarctic environments the soil reserve is larger than plant demand. (IBP data compiled by Dowding et al.*, 1981, from various sources.)*

	N (g/m^{-2})			P (g/m^{-2})			K (g/m^{-2})			Ca (g/m^{-2})		
	Live	Dead	Total	Live	Dead	Total	Live	Dead	Total	Live	Dead	Total
Alaskan arctic coast	9.1	12.8	960	0.8	0.8	63.2	1.3	0.5	15.8	3.0	5.4	—
Siberian hummock tundra	41.1	7.9	—	5.1	0.9	9.2	5.5	1.1	14.2	34.1	1.3	—
Antarctic tundra	14.0	10.3	265	1.9	1.0	8.2	10.2	3.0	17.8	1.8	3.0	14.8
Norwegian willow thicket	19.4	10.1	591	3.4	1.4	373	7.0	1.7	232	16.8	9.3	986

of nutrient cycles in different biomes: the success of agriculture is a biome property. The agriculture of Western Civilization depends on a large nutrient reservoir in the soil. This is present under temperate deciduous forest and temperate grasslands. It is for this reason that civilizations arise in these temperate places, not because people like the weather there. Comparable civilizations are not possible over most of the wetter parts of the lowland tropics, like the Amazon or Congo basins. In the tropics, civilization is associated with the deltas and flood plains of the great rivers: the Mekong, the Ganges, and the Nile, where nutrients are replaced annually. The ancient civilization of Egypt flourished before the North Europeans learned to cut down their forests, not because Egypt was a nicer place in which to live, but because the Nile was turbid with silt every year.

PREVIEW TO CHAPTER 18

The oceans are remarkable for containing a fairly strong solution of sodium chloride despite the fact that contributing rivers contain various mixtures of potassium, calcium, and sodium carbonates and sulfates. It seems that ocean chemistry becomes dominated with the cation sodium because other major cations contributed by rivers are removed, either in estuaries or in shallow seas. Potassium is removed in the synthesis of the clay illite and deposited in mud. Calcium is removed by living things and deposited as carbonate rock, eventually ending up as limestone. The sodium becomes balanced by the anion chloride, which seems to be contributed to the ocean from volcanic vents, particularly along the mid-ocean ridges. But a full explanation of ocean chemistry requires a mechanism to remove most of the other elements of the periodic table that should have been contributed over geologic time by rivers. The best hypothesis to explain this removal is that clay minerals, flocs of humic acids, and the corpses of small marine animals and plants selectively remove different elements and carry them down to the ocean mud. Benthic animals living in burrows in estuaries probably help refine ocean chemistry as they pump water from the oxic sea into chemically reduced mud and back into the sea. Water flowing through the burrows gains and loses solutes as its oxidized state changes, the whole system working as an ion pump. The best evidence that ocean chemistry is constant over geological time comes from comparing the sodium budget of the ocean with the radiometric age of the oceans. Rivers could have contributed all the sodium now in the ocean and its sediments in 100 million years, but the oceans are at least 500 million years old. We conclude that sodium is cycled between land and sea with a steady-state concentration resulting in the ocean. Chloride, on the other hand, is conserved in the oceans, the small input from volcanos being balanced by small losses to the earth's crust and in sediments.

CHAPTER 18

BIOGEOCHEMISTRY I: SOLUTES IN THE OCEANS

The sea is salty water. The air is a mixture of oxygen and nitrogen, with just a little bit of carbon dioxide. These are overriding properties of the global surface. Neither the condition of the sea nor of the air seem intuitively likely and a planetary scientist told only that the earth was a sphere of typical cosmic chemistry orbiting a minor star would be most unlikely to guess at the true composition of sea or air. Certainly the other planets of the solar system give us no hint of the life-giving surface of the earth, since they are either airless expanses of rock, hammered by ionizing radiations from space, or are covered with poisonous, reducing brews of methane or sulfuric acid and the like. Yet the maintenance of the unlikely chemistry of earth's atmosphere and oceans is essential to life.

Animals and plants are intimately involved in the maintenance of both sea and air, so much so that the predominant hypothesis to explain the chemistry of the air relies on life processes for the release and maintenance of both principal gasses, oxygen and nitrogen.

QUESTIONS OF OCEAN CHEMISTRY

The sea is fed by rivers, and all the rivers draining into the sea are solutions of crustal rocks in water. They contain nutrients leaked from the imperfect cycles of every terrestrial ecosystem, and they also hold solutes of crustal rocks that are largely ignored by the living parts of ecosystems. Rivers can be viewed as devices for the mass transport of surface land to the bottom of the sea, either by solution or as sediment.

A most elementary reflection from this fact is that the sea ought to be getting saltier. We do know, of course, that the sea is salty, and it is a reasonable hypothesis to suggest that this saltiness is connected with the continual import of salt to the sea by rivers. A first hypothesis for the salt oceans, therefore, is that they have been getting steadily saltier throughout geological time, and will continue to get saltier still through whatever span of geologic time is left.

But this first hypothesis faces difficulties, an obvious one being that the sea holds a very special salt solution. The oceans are a remarkably pure solution of sodium chloride, considering the assorted chemicals added yearly by rivers. Seawater is 35 parts per thousand (ppt) NaCl, with most of the rest of the periodic table thrown in, mostly in small or trace amounts (Table 18.1). But none of the world's rivers are dilute versions of the sea (Table 18.2). There are thus two basic questions posed by the salt oceans: "Are they getting saltier?" and "How did the oceans come to be a solution of sodium chloride rather than some more typical mixture of terrestrial solutes?"

Table 18.1
Chemistry of Seawater
Data are adjusted for a salinity of 35 ppt NaCl. Elements are arranged in order of atomic weight (From Turekian, 1968.)

Element	Symbol	Micrograms per Liter	Element	Symbol	Micrograms per Liter
Hydrogen	H	1.10×10^8	Molybdenum	Mo	10
Helium	He	0.0072	Ruthenium	Ru	0.0007
Lithium	Li	170	Rhodium	Rh	
Beryllium	Be	0.0006	Palladium	Pd	
Boron	B	4,450	Silver	Ag	0.28
Carbon (inorganic)	C	28,000	Cadmium	Cd	0.11
(dissolved organic)		500	Indium	In	
Nitrogen (dissolved N_2)		15,500	Tin	Sn	0.81
(as NO_3^-, NO_2^-, NH_4^+)		670	Antimony	Sb	0.33
Oxygen (dissolved O_2)		6,000	Tellurium	Te	
(as H_2O)		8.83×10^8	Iodine	I	64
Fluorine	F	1300	Xenon	Xe	0.047
Neon	Ne	0.120	Cesium	Cs	0.30
Sodium	Na	1.08×10^7	Barium	Ba	21
Magnesium	Mg	1.29×10^6	Lanthanum	La	0.0029
Aluminum	Al	1	Cerium	Ce	0.0012
Silicon	Si	2900	Praseodymium	Pr	0.00064
Phosphorus	P	88	Neodymium	Nd	0.0028
Sulfur	S	9.04×10^5	Samarium	Sm	0.00045
Chlorine	Cl	1.94×10^7	Europium	Eu	0.0013
Argon	Ar	450	Gadolinium	Gd	0.00070
Potassium	K	3.92×10^5	Terbium	Tb	0.00014
Calcium	Ca	4.11×10^5	Dysprosium	Dy	0.00091
Scandium	Sc	<0.004	Holmium	Ho	0.00022
Titanium	Ti	1	Erbium	Er	0.00087
Vanadium	V	1.9	Thulium	Tm	0.00017
Chromium	Cr	0.2	Ytterbium	Yb	0.00082
Manganese	Mn	0.4	Lutetium	Lu	0.00015
Iron	Fe	3.4	Hafnium	Hf	<0.008
Cobalt	Co	0.39	Tantalum	Ta	<0.0025
Nickel	Ni	6.6	Tungsten	W	<0.001
Copper	Cu	0.9	Rhenium	Re	0.0084
Zinc	Zn	5	Osmium	Os	
Gallium	Ga	0.03	Iridium	Ir	
Germanium	Ge	0.06	Platinum	Pt	
Arsenic	As	2.6	Gold	Au	0.011
Selenium	Se	0.090	Mercury	Hg	0.15
Bromine	Br	6.73×10^4	Thallium	Tl	
Krypton	Kr	0.21	Lead	Pb	0.03
Rubidium	Rb	120	Bismuth	Bi	0.02
Strontium	Sr	8,100	Radium	Ra	1×10^{-7}
Yttrium	Y	0.013	Thorium	Th	0.0004
Zirconium	Zr	0.026	Protactinium	Pa	2×10^{-10}
Niobium	Nb	0.015	Uranium	U	3.3

Table 18.2
Sea and River Water Compared
The table shows the more abundant elements in oceans and typical river water. It is evident that simple concentration of river water does not produce seawater. (From data in Garrels et al., 1975.)

	ppm (mg/l)	
	Sea	Rivers
Chlorine	19,000	7.8
Sodium	10,500	6.3
Magnesium	1,300	4.1
Sulfur	904	5.6
Potassium	380	2.3
Calcium	400	15.0
Silicon	2.9	6.1
Bromine	65	0.02
Strontium	8	0.07
Aluminum	0.001	0.4
Boron	4.5	0.01
Fluorine	1.3	0.1
Phosphorus	0.09	0.02
Iron	0.003	0.67
Manganese	0.002	0.007
Zinc	0.002	0.02

Research on these questions suggests that ocean chemistry is constant, and is maintained by a GEOCHEMICAL CYCLE and by biologic and sedimentary processes in the sea itself.

The Sodium Cycle and the Age of the Oceans

The simplest possible model for ocean chemistry is to start with fresh water in the ocean basins, and for the water to have become steadily saltier throughout geological time as the endless washing of rain and rivers brought down salt from the eroding rocks of the land. If the process has gone on at a roughly constant rate since the seas were first formed, as seems reasonable, then an age for the oceans can be calculated by working out the annual tonnage of salt dumped into the sea by all the rivers of the world and by dividing this into the total tonnage known to be in the sea. The age long ago arrived at by this method was 100 million years.

This age of 100 million for the oceans was generally accepted by geologists until the 1940s. In those days, therefore, geochemists generally believed that the oceans were getting saltier, and it was only with the advent of radiometric dating of rocks that this notion seems seriously to have been questioned. But radiometric dates then revealed that rocks holding recognizable marine fossils could be found that were at least 500 million years old. Either the assumption that the sea was getting saltier was wrong or the old calculations were off by a factor of five.

Livingstone (1963a, b) repeated the classic calculations for the expected age of the oceans on the assumption that the sea was accumulating salt and confirmed that this procedure produced the traditional conclusion of oceans 100 million years old. Obviously there must be error in the method. If all the sodium in the world oceans could be accounted for in just 100 million years of river transport, where is the sodium that was put in the oceans during the first 400 million years of oceanic existence? Livingstone estimated all the storage places outside the actual water of the seas—sediments, sedimentary rocks, and the like—in a quest for the missing sodium, but even the most generous estimates of these depositories could not account for the loss. The calculations are as follows:

Sodium carried by rivers to the sea during 500 million years of post-Cambrian time	119.4×10^{15} tons
Sodium dissolved in world oceans	14.1×10^{15} tons
Sodium in deep-sea sediments	5.1×10^{15} tons
Sodium in other suboceanic sediments	5.4×10^{15} tons
Sodium in old sedimentary rocks now raised out of the oceans	2.6×10^{15} tons
Known reserves of rock salt	0.4×10^{15} tons
Total sodium accounted for	27.6×10^{15} tons

This still leaves three-quarters of the missing sodium unaccounted for. Some may be present as deep-sea salt domes not included in Livingstone's calculations, but probably not enough to alter the conclusion significantly.

Livingstone's explanation of these facts, now generally accepted, is that no missing deposit of sodium remains to be found, rather a smaller mass of sodium is recycled between land and sea. As much sodium is returned to the land each year as travels to the sea in rivers, and the oceanic concentration of sodium is maintained as a steady state. The return mechanism is in two parts: return through the air of sodium picked up in spray to be carried by wind, and sodium in sediments returned to the land as sedimentary rock. Computation of sodium in rain water the world over shows that this accounts for half the sodium found in rivers. Emergence of crustal rocks from the sea must account for the rest.

Little doubt remains that the cycle of sodium between land and sea is largely closed, so that the sodium in present-day rivers has been in the sea before (Figure 18.1). One obvious complication is that some sodium in sediments will be carried down into the deep ocean trenches by sea-floor spreading (Chapter 14), probably to be balanced by sodium injected to the land in volcanic rocks.

This analysis of sodium budgets suggests that the chemistry of the oceans may be constant having reached something very like a steady state. This suggestion is of high interest to ecologists because it means that we need not think of biota of past epochs as living in seas unfamiliar to us.

The obvious way to test the hypothesis that ocean chemistry and salinity have been constant since, say, Cambrian time at the start of the fossil record is to look for chemical data in sedimentary rocks. These are often hard to interpret because of the changes that take place between sea, sediments, and rock, and because of what can happen to a rock in its long history on the land. The best that can be done is to compare concentrations of some elements in young and old rocks of similar type. These data are still somewhat equivocal, though the preponderance of the evidence, as the lawyers say, is consistent

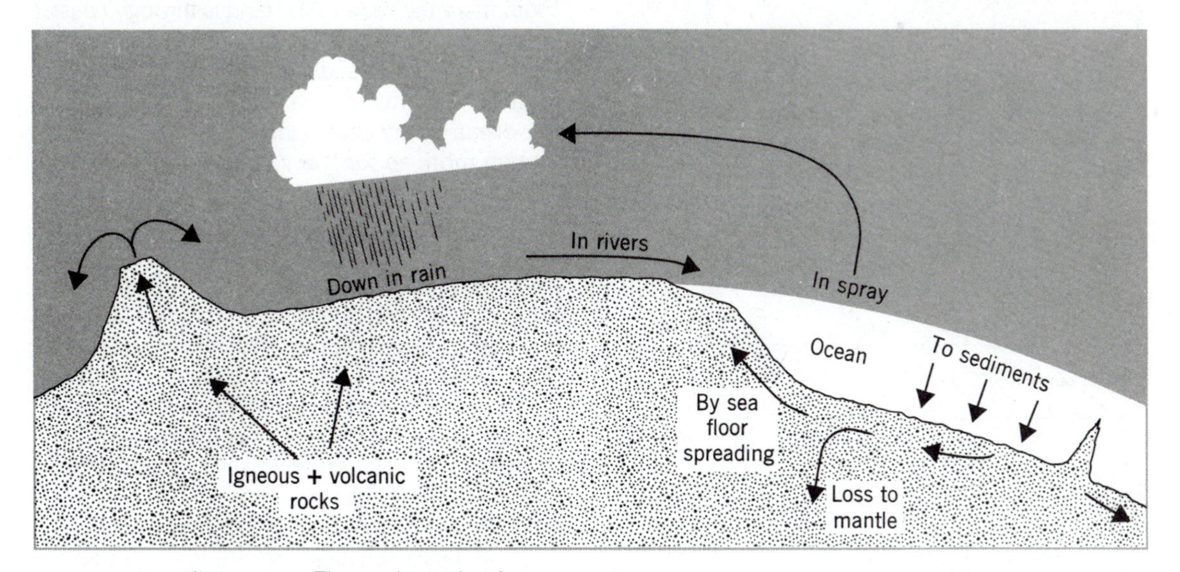

Figure 18.1 The geochemical cycle.
Minerals are cycled between the oceans, the land, and the crust of the earth. Rivers carry soluble minerals to the sea but wind and rain carry some back and most of the rest are probably returned by crustal movements. Minerals in sediments are moved across the ocean bottoms by sea-floor spreading from the mid-ocean ridges (right of figure) to the continents. Loss to the mantle in the deep ocean trenches may be made good by the extrusion of fresh rocks.

with an ocean of near constant chemistry over at least the last 500 million years (Garrels and Mackenzie, 1971). One of the nicest sets of data comes from analyses of the boron content of the clay mineral *illite* in similar rocks spanning not just the rocks of the fossil record but the better part of 3000 million years (Figure 18.2). The boron content of the illite is so constant that there must be a strong presumption that each illite sample collected its boron from solutions of comparable chemistry.

It is now parsimonious to postulate that the oceans have had near constant chemistry since at least the start of Cambrian time, and throughout the span of most of the fossil record. Many trace elements in the sea must certainly fluctuate, not only with time but across oceans at the same time, and the elements concentrated within organisms, like phosphorus or silicon, will always vary from place to place. These local variations, however, proceed within limits that have always been the same. The steady state required in this hypothesis is maintained partly by the geochemical cycle revealed by the study of sodium and partly by chemical or biological processes within the ocean basins themselves.

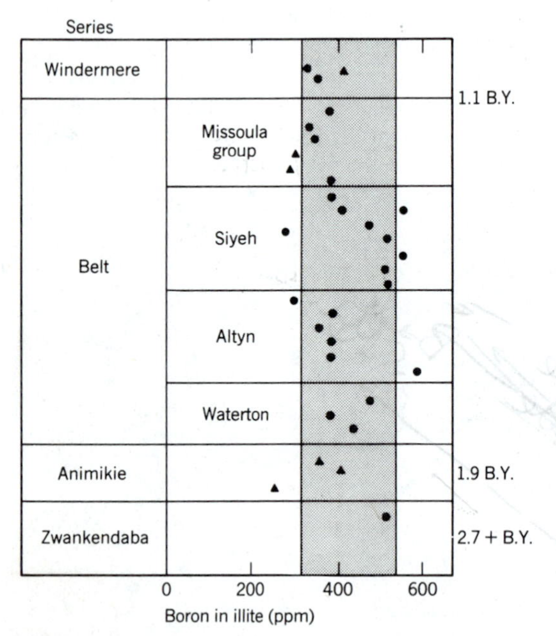

Figure 18.2 Evidence of boron for constant ocean chemistry.
Data are parts per million (ppm) of boron in the clay illite within carbonate rocks and shales. The diagram covers only pre-Cambrian time but analyses of similar rocks of the last 500,000,000 years are all in the range shown by the shading. The data suggest that illites of all ages have collected boron from similar solutions. (From Reynolds, 1965, and Garrels and Mackenzie, 1971.)

The Geochemical Cycle and the Special Case of Phosphorus

Any element or mineral that can be dissolved, precipitated, or moved in suspension will be subject to geochemical cycles on the scale of sodium. They will be weathered from rocks and cycled or held in ecosystems for characteristic RESIDENCE TIMES. They will be carried to the sea in rivers, though possibly short-circuited in the deposits of an estuary. But they find their way to deposits eventually; as solid, in interstitial water, as precipitate, or adsorbed to a particle. From there the road to the land is through coastal emergence, or the slow conveyor of plate tectonics, until a new sedimentary rock is emplaced for fresh erosion to begin. A short-cut to the land via ocean spray and wind is always available, though more so for some elements than others. The process is complicated by the weathering of igneous rocks and by subduction of deposits deep into the earth through ocean trenches (Figure 18.1). Processes are slow and *residence times* long, up to the 100 million years taken by seafloor spreading to move deposits from a mid-ocean ridge to a subduction trench.

Phosphorus is of particular interest to ecology because of its rarity in the crust of the earth and its extreme concentration in living things (Chapter 17, Table 17.1).

Some of the best data on the phosphorus cycle are from lakes, which serve well enough as miniature oceans. Excess phosphorus from agricultural fertilizer or detergent causes dense, green blooms of plant growth, the familiar condition of polluted lakes. The algae have taken up the excess phosphorus rapidly and produced a large standing crop of biomass. It is now known that these blooms of algae will disappear if the input of fresh phosphorus ceases. The whole process has been followed experimentally.

An early experiment (Smith, 1948) used one and a quarter tons of phosphatic fertilizer on an unproductive small lake in Canada, sufficient phosphorus to increase the concentration of the lake by 2500%. But the actual increase was only 67%, showing that most of the fertilizer was taken up by plants or vanished into the lake mud. The detailed history of phosphorus in *enrichment experiments* like this is now known from using radioactively labelled phosphorus, the pioneering version of which was by Hutchinson and Bowen (1950). Within hours the labelled phosphorus appears in tissues of algae and rooted aquatic plants, and within days it begins to appear in the mud. Some of the phosphorus attached to bottom mud goes back into solution at times of low oxygen tension and is brought back to the plants by circulating water (Chapter 21; Figure 18.3). Only a thin layer of sediment at the mud/water interface can release phosphorus back to the water, however, even during episodes of chemical reduction when oxygen tensions are low. Most of the phosphorus is buried permanently (Chapter 21).

It turns out, therefore, that the greed for phosphorus by planktonic plants accelerates the race of phosphorus to the mud. Phosphorus is not allowed to remain in solution but is instantly removed into a plant body. But the plants are tiny and soon die, perhaps to be eaten by an animal but perhaps to go straight to the mud. The residence time of phosphorus in the ecosystem of an open lake may, therefore, be measured in weeks or days. Some can return to the open water by resolution, but only a little. The rest is sediment.

A parallel process proceeds in the oceans, but there the export of phosphorus to the mud may be even more final because the export may be over a vertical distance measured in kilometers. Furthermore the chance of chemically reducing conditions on the ocean floor is much less than on the floor of a lake, so that phosphorus consigned to the mud is likely to remain in the mud.

But there is a curious and specialized short-circuit of the global phosphorus cycle because

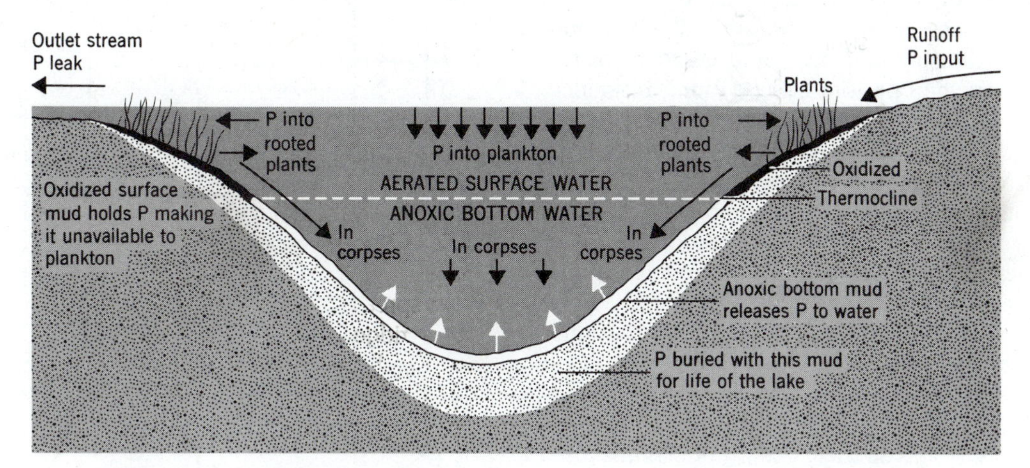

Figure 18.3 The phosphorus cycle in a lake.
Phosphorus enters all lakes continuously in run-off water, in inlet streams, and from the air. Phosphorus is also continuously lost to the lake in the outlet streams and by incorporation in lake mud. When a lake has anoxic bottom water, as when a lake stratifies in summer (see Chapter 21), the top few millimeters of mud are chemically reduced, a condition that allows the mud to release phosphorus back to the water. The bottom water thus becomes phosphorus-rich. Stirring of the lake by winter storms brings the phosphorus-rich water to the surface, completing an annual cycle, and fertilizing the lake for a spring plant bloom.

of the activities of fish-eating birds, now supplemented by human fishers. Vertebrate bone is phosphate-rich so that fisher birds and fisherpeople return much phosphorus to the land each year. Hutchinson (1950) calculated the scale of this return of guano and fishmeal to the land. He found that birds and people combined return about 70,000 tons of phosphorus to the land each year, but that the world's rivers send 14,000,000 tons to the sea. Fishing, therefore, is not important to the global phosphorus cycle.

Most of the annual 14,000,000 ton flux to the sea probably goes through food chains and into the mud in the first year. Much will travel only a little past an estuary or continental shelf, where phosphorus-rich muds will collect under relatively productive waters. But a little will journey in ocean currents until carried again to the surface in an upwelling, beneath which it will find its resting place for the next geologic epoch or so. The principal phosphate deposits for mining, like those now fought over in northwest Africa, are probably the graveyards of ancient ecosystems poised over upwellings.

There is thus the curious paradox that greed of plants for phosphorus leads to very short residence times in the open water of lakes and oceans, but on land this same greed of plants can have the opposite effect of long residence times. In a tropical rain forest, for instance, a phosphate ion once taken into an ecosystem may well remain as long as the ecosystem persists, remaining in a tree for centuries, passed to another tree in days or hours, and so on. Nutrient residence times can be very long when plants are both large and long-lived.

The organic soils built by plants of the land also promote long residence times for phosphates as these are held in humic complexes (Chapter 17). In addition, iron and aluminum minerals of soils strongly adsorb phosphate ions so that an equilibrium is established between the concentration of phosphate ion in the soil water and that in the soil minerals. Such a system can buffer the phosphate supply to the plants, helping to promote a steady-state flow of the mineral into and out of the plant community.

The global phosphorus cycle illustrates nicely the components of all such cycles, as they span more than 100 million years (Figure 18.4):

1. long wait in sedimentary rocks
2. weathering and cycling in terrestrial ecosystem
3. almost instantaneous passage through aquatic ecosystem
4. long wait in ocean sediments
5. long wait during uplift and metamorphosis of rocks.

Probably every element has its own unique fine-

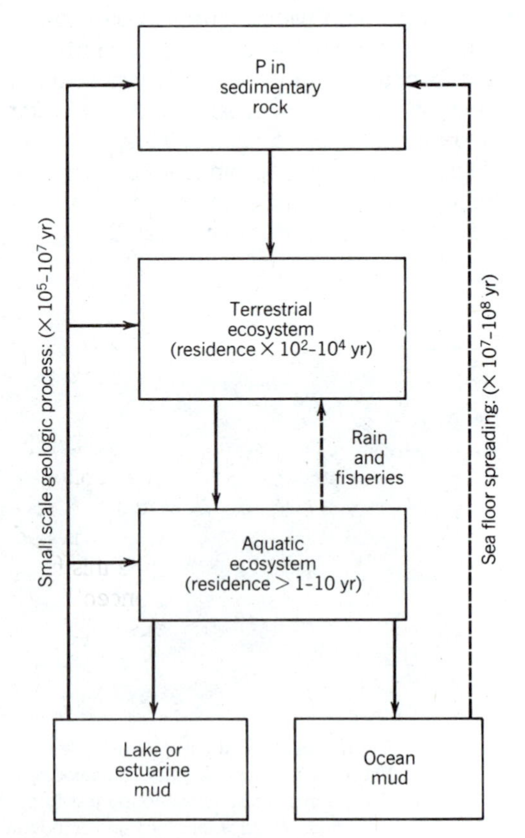

Figure 18.4 The global phosphorus cycle. Residence times and transfer times are order of magnitude guesses. Transfer back to land from the deep oceans is problematical.

tuning of these stages, particularly when in the power of an ecosystem.

Why the Sea Is Salt

The details of the mechanism that turns river water into a solution of sodium chloride plus lesser elements are far from known, although the kinds of mechanisms involved can be identified. River water is the product of dissolving crustal rocks in the dilute carbonic acid of rain. Important minerals of the rocks are FELDSPARS like

potassium feldspar	$KAlSi_3O_8$
anorthite	$CaAl_2Si_2O_8$
sodium feldspar	$NaAlSi_3O_8$

The resulting solution is rich in potassium, sodium, and calcium but poor in chlorine—the typical composition of river water (Table 18.2). For this solution to be changed to that of seawater, potassium and calcium must be removed and chlorine added. In general it may be said that potassium is removed on mud, calcium is precipitated as carbonate, and chlorine is supplied from the mid-ocean ridges.

Potassium Removal

Clay minerals of mud are supplied to rivers and the oceans as a by-product of the solution of feldspars. A clay mineral like KAOLINITE, $Al_2Si_2O_5(OH)_4$, is made from the insoluble residue of feldspar and is carried to the sea in suspension. It is a flat, plate-like mineral, charged on all surfaces. To thrust minerals like this from the dilute waters of rivers into the concentration of ions found in seawater is to cause complex rearrangements of their structures. It is of particular importance that kaolinite reforms into the potassium-based clay ILLITE. This chemistry is subtle, slow, and poorly understood. As MacIntyre (1970) has it, "Graduate students who study these reactions invariably leave before the reactions are complete." But there is little doubt that the reactions take care of the potassium donated to the sea. The resulting illite takes potassium to the bottom and holds it in the mud.

Calcium Removal

The fate of calcium is possibly more satisfying to an ecologist because the biota are involved. Calcium is deposited in shallow oceans as calcium carbonate, particularly in the shells or skeletons of plants and animals, most spectacularly in coral reefs. The emphasis is on the word *shallow* because calcium carbonate redissolves under high pressure, effectively at depths below 4000 m. There is no donation of calcium as carbonate to the muds of the deep sea, therefore, and any settling skeletons that penetrate there return to solution.

Like everything about ocean chemistry, the chemistry of precipitating calcium carbonate is not simple (Milliman, 1974). Inorganic precipitation is possible, as when cold water saturated with calcium carbonate rises from the depths and is heated at the surface. This is a well-known phenomenon in parts of the Caribbean where the surface of the sea turns milky white from the small crystals of the carbonate mineral ARAGONITE. Some of the mass of carbonate in coral atolls may be inorganic aragonite also, probably deposited alongside the plant skeletons in the lagoons, behind the framework set by coral animals (Hillis-Colinvaux, 1980). But the great mass of calcium carbonate deposition seems to be by animals or plants, and has been so since the start of the fossil record. Life disposes of calcium, therefore, leaving sodium in lonely state as *the* alkali metal of the oceans.

Chlorine Enrichment

Two sources of chlorine are known—volcanos and mid-ocean ridges. Chlorine appears in the form of hydrochloric acid, HCl: as gas from volcanos but as the dissolved acid in juvenile waters from the submarine volcanos of the mid-ocean ridges. Either way, chlorine enters the ocean system as aqueous hydrochloric acid. It then reacts with bicarbonate ions, producing water, carbon dioxide, and chloride ions. The carbon dioxide returns to the atmosphere leaving the chlorine to take the place of the bicarbonate as a negative valence, balancing sodium.

Requiring chlorine to enter the sea from volcanic vents almost certainly implies a lesser annual flux of chlorine than there is of sodium, potassium, and calcium. For the sea to remain sodium chloride, therefore, it seems certain that chlorine must remain in the sea even as sodium and the others are removed in the geochemical cycle. This is, indeed, what seems to happen. The reactions that deposit clays or carbonates do not deposit chlorides, and the minerals of sediments are chloride-poor. It may be proper to say that the ocean conserves chlorine even as sodium is passed backwards and forwards between land and sea. But it is also advisable to say that these are still early days in our understanding of the chemical stability of the oceans.[1]

Scavenging Other Elements from the Sea: Mud and the Biota

Not only is the sea salty, but the ocean solution is almost nothing but salt, since the rest of the periodic table is present in only small amounts (Tables 18.1 and 18.2). Since rivers carry most of the periodic table continuously to the oceans, mechanisms must exist to sweep the ocean clean again.

Geochemists have looked for the places where the sweeping occurs by using uranium as a tracer. URANIUM and its decay products (the URANIUM DAUGHTER SERIES) are particularly useful for this, since the decay of the parent into the daughters at known rates provides clocks that time many of the processes (Figure 18.5). These studies show that the process of sweeping goes on at every stage in the journey of minerals from land to sea: in soils, streams, rivers, estuaries, and in the deep sea itself (Turekian, 1977).

First among the sweeping mechanisms is attachment to clay minerals in the same way that potassium is attached to illite. Other elements are attached to the long-chain organic molecules in solution called HUMIC ACIDS (Chapter 20). Both resulting complexes are precipitated in the high-electrolyte medium of the oceans. Additional

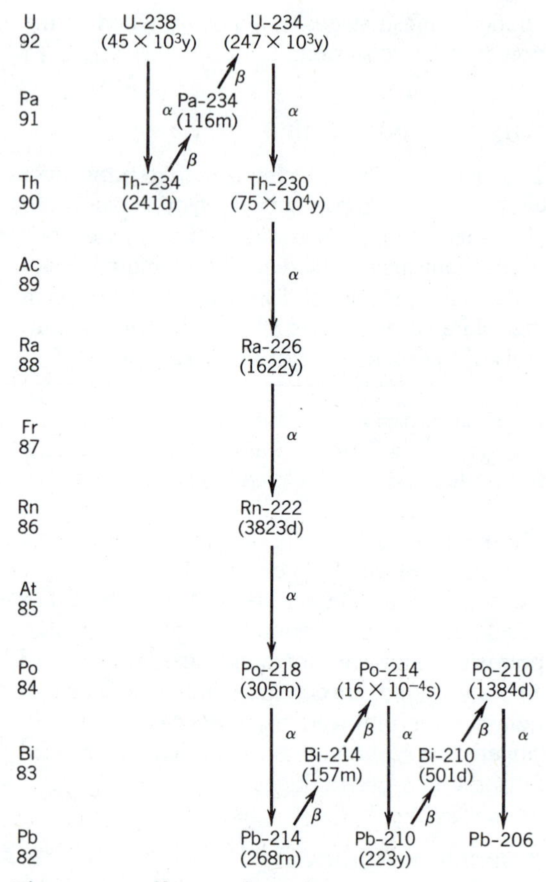

Figure 18.5 Uranium decay chain.
Elements of the uranium daughter series serve oceanographers both as tracers representative of heavy metals and as clocks. The half-life of each element is given in days, months, or years. (From Turekian, 1977.)

elements may be removed from solution as bacterial flocs, which are known to remove even such esoteric molecules as DDT from water and which are probably able to scavenge more mundane elements as well (Leshniokowsky *et al.*, 1970). Algae extract phosphorus, copper, and other minor elements. As these bacterial flocs, algal corpses, and minerals drop to the sea floor, they leave behind an ocean containing little more than sodium chloride.

The living parts of ecosystems, particularly in estuaries, probably serve to refine this process of sweeping elements from the sea. This is be-

[1]This account draws on Sillen (1961); Ronov (1968), Horne (1979); MacIntyre (1970); and Garrels and Mackenzie (1971).

cause respiration controls the REDOX POTENTIAL of bottom mud, and because burrowing animals pump solutes in and out of the bottom mud. Many ions may be dissolved from chemically reduced mud but are reprecipitated on oxidized mud. In estuaries, and on the sea floor, mud under a thin surface layer is reduced from the oxygen demand for respiration, as attested to by the typical gray and smelly mud in shallow estuaries. But the overlying water usually holds free oxygen and the thin top layer of mud consequently is oxidized.

A dense BENTHOS of burrowing animals lives in the surface mud; tunnelling, or occupying tunnels open to the water. Animals pump oxygenated water deep into the reduced mud through these tunnels, and they pump anoxic water out. This means that each burrowing animal is also a chemical pump. It ejects water in which metals and other ions have been redissolved. This ejected water is loaded particularly with iron and manganese. Close to the sediment surface, and in the open water, the iron and manganese are certain to be reoxidized, reprecipitated, or again fixed on settling particles. The animals of the benthos thus run a chemical cycle that sends active ions in and out of the open water. It is likely that included in the consequences of this pumping are the building of manganese and silica nodules in the deep sea.

Summary: The Maintenance of Ocean Chemistry

The oceans as ecosystems have a chemical composition that is essentially invariant. The concentration of the principal cation, sodium, is maintained by cycling between sea and land. The other principal cations of fresh water, potassium and calcium, are removed by precipitation: the potassium on clay minerals and the calcium through biogenic processes. Concentration of the principal anion, chlorine, is not so dependent on cycling; chlorine is contributed to the oceans through volcanism and conserved in balance with sodium. All other ions are maintained at low concentrations from the sweeping action of settling clays and organic debris. Benthic ecosystems probably are important in regulating concentrations of many elements through manipulating redox potentials and hence solution equilibria.

It should be noted that the concentrations of nutrients in surface ocean water above the thermocline (Chapter 21) are dependent on biological demand and will fluctuate with changes in productivity.

PREVIEW TO CHAPTER 19

The three principal gasses of the air, nitrogen, oxygen, and carbon dioxide, are maintained by biological processes, at least in part. The gas least under biological control is carbon dioxide, strange though this may seem. The low concentration of CO_2 in the air is maintained by solution in the oceans, which act as a giant buffer to the concentration of CO_2 in the air, and the oceans are themselves buffered by deposition of carbonates within shallow seas. This elaborate buffering keeps the concentration of carbon dioxide in the air very low, but the buffer is rate-limited by the time taken for the oceans to mix. Because of this, release of carbon dioxide into the air by burning fossil fuels and clearing forests has resulted in a measurable increase in atmospheric carbon dioxide over the last few decades. Studies of the isotopes of carbon in both the air and in tree rings have demonstrated that this long-term rise in atmospheric carbon dioxide includes contributions from both fossil fuels and the biota as expected. The oxygen of the atmosphere—ocean system must originally have been contributed by processes that could override massive sinks present on a reducing earth. Evidence of this process is preserved in the banded ion formations deposited more than 2000 million years ago. An important process for producing oxygen in those early days was the dissociation of water vapor in the upper atmosphere caused by ionizing radiations from space. It is likely that photosynthesis was producing oxygen at the same time, though this is still somewhat a matter of dispute. Any production of oxygen by photosynthesis must be accompanied by the removal of reduced carbon from the system and it is not certain where this carbon might be found. After free oxygen became present in the air, however, it seems likely that photosynthesis was the primary oxygen producer because the presence of an ozone shield would lower the rate at which water would be dissociated in the upper atmosphere. The carbon necessary to account for the mass of free oxygen now present in the atmosphere has not been accounted for, giving rise to what is known as "the missing carbon problem." Very likely the missing carbon is present as finely divided graphite in crustal rocks. The most convincing hypothesis for the actual regulation of the oxygen of the atmosphere calls upon the activities of sulfate-reducing bacteria. These release both oxygen and hydrogen sulfide from sulfate in deposits of mud, the hydrogen sulfide released to the atmosphere is then reoxidized, the whole mechanism providing a feedback regulatory device. Nitrogen is unique because it has low solubility in the oceans and is chemically inert. The large nitrogen reservoir, therefore, is the air and the only physical output of nitrogen from the air to balance the input from volcanos is the synthesis of nitrates in electrical storms. This loss of nitrogen from the air is made good entirely by biological processes, as nitrate-reducing bacteria use nitrate as a hydrogen acceptor in oceanic or swamp mud devoid of free oxygen. This process also offsets the loss of nitrogen due to biological nitrogen fixation and serves to regulate the nitrogen supply of the atmosphere. The oxygen and nitrogen concentrations in the contemporary atmosphere are so large and so well regulated that they cannot seriously be influenced by human activities, contrary to some of the postulates of environmental damage that have received wide publicity. Rising concentration of carbon dioxide, however, may have some climatic effects but we have too few data to decide what these might be or to assess their magnitude.

CHAPTER 19

BIOGEOCHEMISTRY II: THE MAINTENANCE OF THE AIR

The most startling property of the air is that it is oxidizing; it holds free oxygen. This despite the fact that the surface rocks of the crust of the earth are all chemically reduced, except where they come into actual contact with the air. Moreover, it is known that volcanos continue to put fresh, partly reduced gasses into the atmosphere (Figure 19.1). Every other planet examined so far has an atmosphere of reducing gasses; only earth has an oxidizing air. A cynic observing us from orbit around a distant star might well suggest that we ought spontaneously to combust. At times, of course, some of our ecosystems do.

Apart from water vapor, the air is made up of nitrogen (79%), oxygen (21%), and carbon dioxide (0.03%), with other gasses present in trace amounts only. Nitrogen, oxygen, and carbon dioxide all are manipulated by individual organisms in ecosystems. Over geologic time, it seems likely that the concentration of all three in the air has either been determined by life or at least drastically altered by life.

It seems very likely that the mixture of gasses in the air has been roughly constant throughout what might be called ecological time. With more oxygen, spontaneous combustion would increasingly be likely. Much more carbon dioxide would be toxic. Life on earth seems organized to cope with the present mixture, suggesting that something close to our present air has persisted since at least the Cambrian epoch, 500 million years ago.

REGULATION OF ATMOSPHERIC CO_2

In the 1980s nobody can think of atmospheric CO_2 without knowing the haunting sound of en-

Figure 19.1 Raw material for the atmosphere. Mayon volcano in the Philippines spews gasses into the atmosphere. Volcanic gasses are the major source of carbon, nitrogen, and sulfur for the atmosphere over geologic time.

vironmental debate. The dangers of burning coal, the greenhouse effect, air pollution; these are the phrases that come to mind. And yet atmospheric carbon dioxide *ought* to be as well buffered as any part of a terrestrial system we can easily imagine.

The air is in direct contact with the sea, and there is free exchange of CO_2 at the air–sea interface. The oceans hold CO_2 in solution in the various oxide species equivalent to about fifty atmospheres' worth. The oceans, therefore, should act as an enormous shock absorber for the air against which humanity's puny efforts at burning coal and oil should have little effect.

Moreover, the oceans are themselves shock-absorbed; by the mechanism, largely under biological control, that deposits excess dissolved CO_2 as carbonate rock. A rough estimate of the carbonate rocks of the earth suggests that they represent about 40,000 atmospheres' worth of CO_2. So we have a potentially enormously powerful buffering system of oceans with carbonate sediments behind the scanty mass of CO_2 held in the air (Figure 19.2).

These facts were pointed out by G. E. Hutchinson in 1949 in a seminal paper analyzing, among other things, the possibility of serious perturbation of atmospheric CO_2 by industrial activity. Hutchinson's conclusion was that it was unlikely that significant changes in atmospheric CO_2 could result from industrial activity because the rate of release of CO_2 from industrial chimneys and vehicle exhausts should be too low to swamp so massive a shock absorber. Interestingly for future debates, he suggested that, if changes in the CO_2 of the air were to be detected, they would probably result from cutting down of forests and the draining of bogs rather than from industrial activity. This was because a more massive and sudden release of CO_2 might be effected by these agricultural means than by mining and burning coal.

These conclusions of Hutchinson (1949) are surely what must come from a first look at the

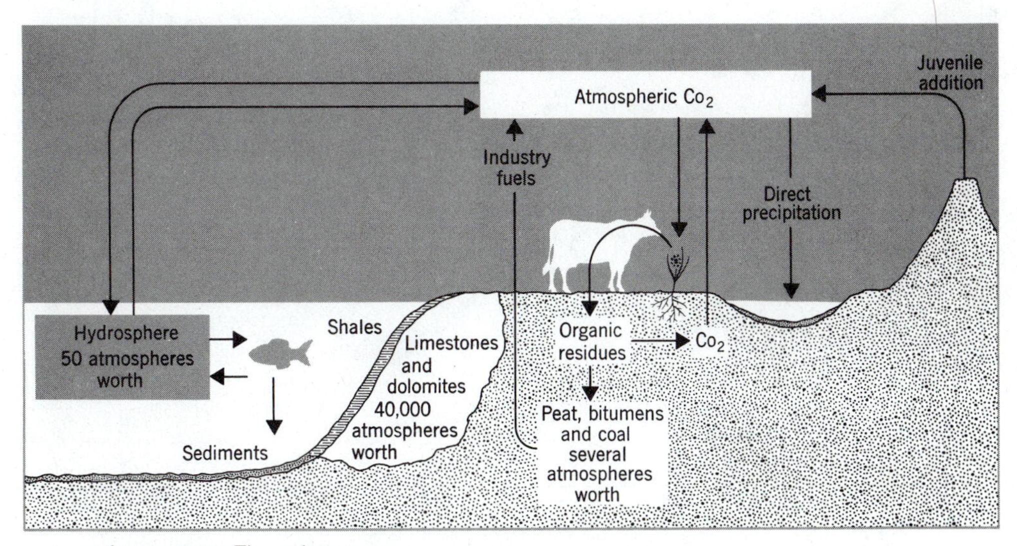

Figure 19.2 The carbon system.
There is apparently a steady-state supply of carbon to the biosphere, the input being from volcanos and the sink being carbonate rocks. Within this steady state there are three orders of cycling: a cycle between land and sea, through crustal folding and solution of limestones and dolomites, which has a mean time span of geologic epochs; a cycle involving reduced organic deposits, which is only now being closed by human activity; and the short-term cycles between animals and plants.

geochemistry of carbon, for it is hard to imagine any system so thoroughly buffered. And yet we now know that the conclusion was partly wrong and that there is in fact a sustained and significant secular rise in the carbon dioxide of the air.

The most spectacular data come from air chemistry measurements at the observatory at Mauna Loa in Hawaii. Hawaii is well-sited for taking these measurements because the islands are in the middle of the Pacific amid well-mixed oceanic air. There are no continental industries nearby whose chimneys could produce purely local effects. The data show that CO_2 concentration in the air over Hawaii has been rising, and at an increasing rate, ever since the measurements began (Figure 19.3). Less complete data from Antarctica show a comparable rise, and other studies confirm that an increase in atmospheric concentration of CO_2 has been proceeding for about a century (Bacastow and Keeling, 1973; Stuiver, 1978).

Despite these rather alarming data that the content of CO_2 in our air is rising, the original arguments about the potential buffer of oceans and rocks remain sound. In the long term, the CO_2 of the air is kept a rare gas by the buffering system. The long-term fate of carbon is a one-way journey from volcano to limestone, with the chance of some future wanderings in solution as part of the geochemical cycle (Figure 19.4).

But in the short term, as we now know, CO_2 can vary. Perhaps the aspect of the Mauna Loa data most revealing of this is the annual fluctuation (Figure 19.3). CO_2 is pumped in and out of the atmosphere with a yearly rhythm, one inhalation and one exhalation every year. This is because most of the land surface is in the northern hemisphere and most plant production is on

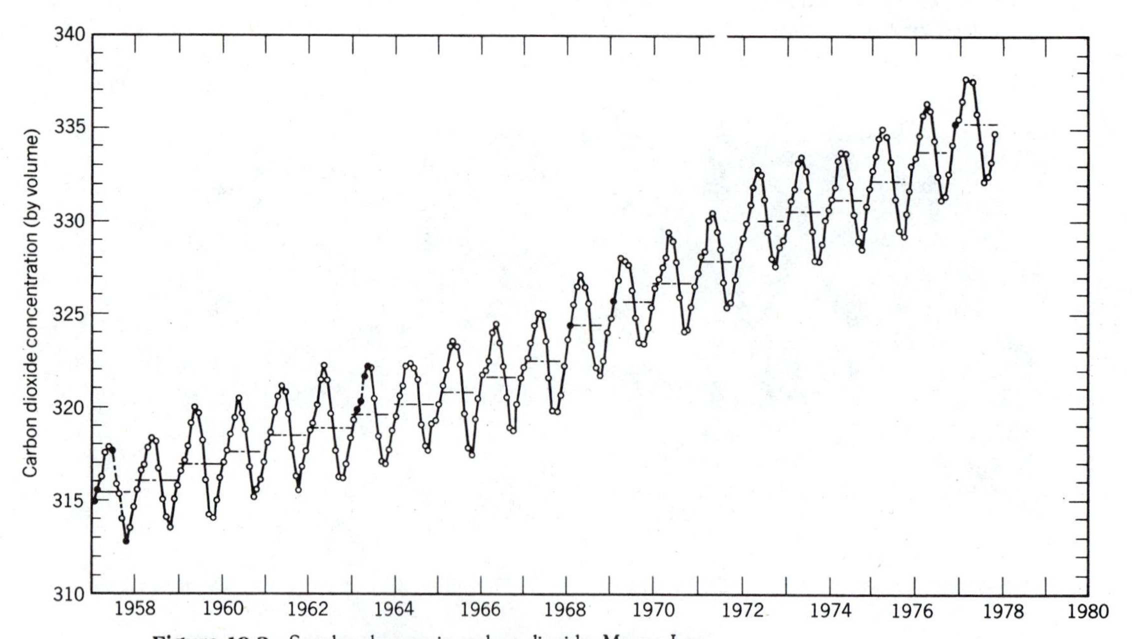

Figure 19.3 Secular changes in carbon dioxide: Mauna Loa.
The data are concentrations of CO_2 in the air at the Mauna Loa Observatory, Hawaii. Circles are observed monthly average concentrations. The curve is a least-squares fit to these averages based on an equation derived from the empirical data. Peaks in the curve reflect the respiration of northern ecosystems during winter. Troughs reflect the excess of photosynthesis over respiration of northern ecosystems in summer. The upward trend of the curve is a direct demonstration that the carbon dioxide concentration of the air is rising. (Data from L. Machta, NOAA Air Resources Laboratories, 1979.)

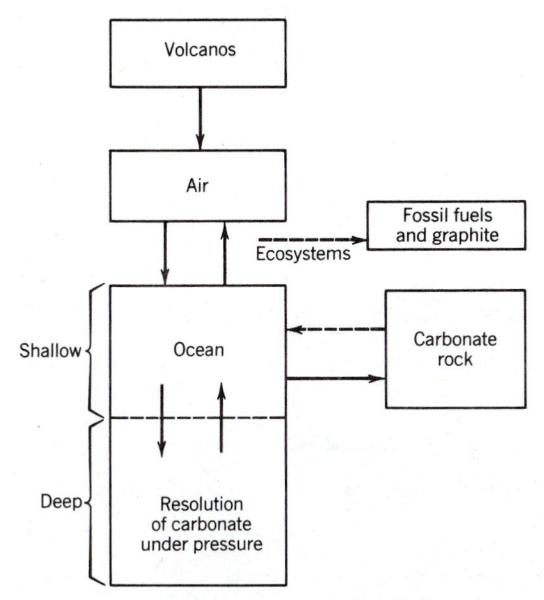

Figure 19.4 Carbon system in geologic time.
The main feature is continuous production by volcanos that is disposed of by incremental storage as carbonate rock. Because carbonate minerals dissolve in the deep sea under pressure, there probably is little return to the mantle in subduction trenches. The surface of the earth now stores the CO_2 production of all geologic time. The CO_2 stored as organic compounds is trivial compared with that stored as carbonate.

the land. In the northern summer, vegetation takes in more CO_2 than it respires; in the northern winter, vegetation respires more than it takes in. The concentration of the gas in the entire atmosphere changes in synchrony with the production and respiration of the ecosystems of northern lands (Figure 19.5).

It is important to notice that seasonal and longer-term changes in CO_2 are noticeable only because the gas is at such low concentration in the atmosphere. The acutal masses of gas involved are small compared with the mass of the atmosphere as a whole. Similar mass changes in the oxygen or nitrogen of the air would not be detectable by our instruments. Add a gram to a gram and the result is noticeable, but add a gram to a ton and nobody knows the difference.

We have a short and convincing answer as to why the carbon buffer system is less than instantaneous in its action. This is that the world oceans stir only slowly. It is one thing to dissolve excess CO_2 in waves, but quite another to carry the solute to the depths for equilibrium of the entire ocean with the air to be achieved. Direct evidence that slow mixing of the oceans is the cause comes from radiocarbon dating of water from the deep sea, for often this water turns out to have last been exposed to fresh carbon more than a hundred years ago and some samples of bottom water are a thousand years old.

CARBON BUDGETS AND THE ATMOSPHERIC RISE OF CO_2

Since we now know that the concentration of CO_2 in the air is rising, it is natural to suspect that industrial use of fossil fuels is responsible. Yet it is equally plausible that clearing forests or bogs is to blame, as Hutchinson (1949) suggested before we even knew that there was a long-term rise. But for that matter, why not an increase in the emission of volcanos, or some disturbance in the oceans? When the Mauna Loa observations at last gave us incontrovertible evidence that CO_2 really was rising, all these possibilities and more had to be explored.

Considering all possible causes of the CO_2 rise requires drawing up a budget of all possible sources and sinks for CO_2. From the point of view of the atmosphere a SOURCE IS ANY RESERVOIR THAT MAY CONTRIBUTE CO_2 TO THE AIR and a SINK IS ANY RESERVOIR INTO WHICH ATMOSPHERIC CO_2 MAY DRAIN. Sources and sinks of atmospheric CO_2 are listed in Table 19.1 and Figure 19.6.

Over geologic time the important source is volcanos as we have seen, and the output of volcanos is taken up by carbonates. Any change in volcanism should have short-term effects on CO_2 in the atmosphere but direct observation of volcanos makes it certain that volcanic events have not been important in the present CO_2 rise. Our data are good enough to declare that volcanos have not emitted significant extra CO_2 in the recent past but there could have been big changes in CO_2 in the air due to volcanism in the more distant past.

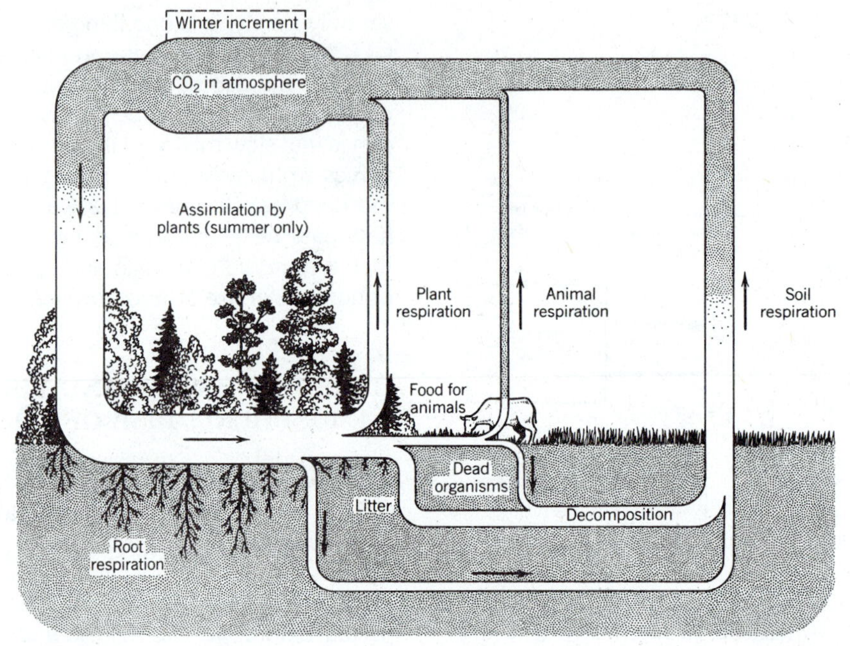

Figure 19.5 Ecosystem cycling of CO_2.
Notice that plants assimilate only in summer but that respiration proceeds year-round. The "pulse" in the Mauna Loa data of Figure 19.3 reflects the seasons in the northern hemisphere.

Table 19.1
Sources and Sinks of Atmospheric CO_2
Reservoirs in parentheses are probably of little quantitative importance. All reservoirs in the bottom half of the table can act as both sources and sinks.

Sources	Sinks
Volcanos	Carbonate rock
(Synthesis of ^{14}C from ^{14}N)	(Decay of ^{14}C to ^{14}N)
	(Descent to mantle)
	Graphite
Oceans	Oceans
Fossil fuels	Fossil fuels
The biota (living plants and animals)	The biota (living plants and animals)
Detritus	Detritus

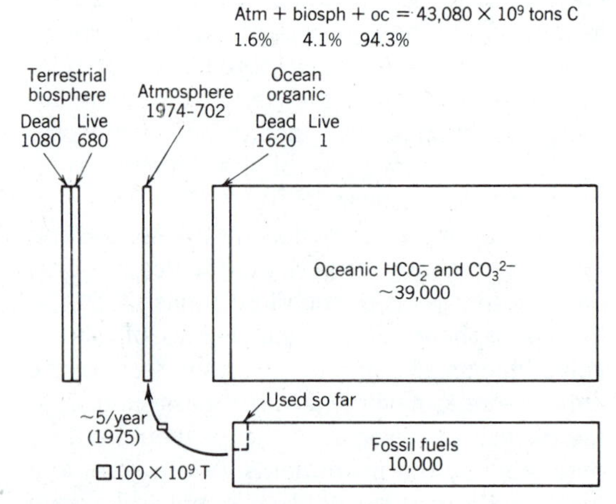

Figure 19.6 Major reservoirs of the contemporary carbon cycle.
Figures are in units of 10^9 tons of carbon. Reservoirs are drawn to scale. (From Stuiver, 1978, based on data of C. D. Keeling and R. M. Rotty.)

We are left with fossil fuels, the organic parts of ecosystems, both living and dead, and the oceans as possible culprits (Table 19.1, Figure 19.6). These three possibilities will be reviewed briefly and then the isotope evidence that we can use to choose between them will be described.

Fossil Fuels as a Source

The total reservoir of fossil fuel carbon at the start of the industrial revolution was about one-quarter of the reservoir in the oceans (Figure 19.6). Of a total reservoir of about $10{,}000 \times 10^9$ tons C in petroleum, peat, coal, and shale deposits we have released about 180×10^9 tons in the last century (Stuiver, 1978). This compares with a present atmospheric reservoir of about 650×10^9 tons. The amount released from fossil fuels so far, therefore, is equal to 28% of the present atmospheric reserve, certainly sufficient to account for the increase shown by the Mauna Loa data even if the other sources had not changed.

It is worth noting that most of the $10{,}000 \times 10^9$ tons of fossil fuel in the global reservoir is out of our reach as fuel, being in the more diffuse or inaccessible shale and coal deposits. The most that we could release is perhaps several times what we have burned so far, possibly as much as three or four atmospheres' worth. So far, however, we have burned only about a quarter of one atmosphere's worth.

The Biota: Carbon Sink or Source?

Clearing forests, burning peat, draining marshes, ploughing prairies: all these release CO_2 that has been stored in ecosystems. Usually CO_2 released in these ways is balanced by CO_2 captured by regrowth so that the carbon cycle is closed (Figure 19.5). But it is quite reasonable that modern humanity has been clearing the growth faster than it can be replaced.

Yet it is also true that photosynthesis tends to be carbon-limited (Chapter 3). Increasing the CO_2 concentration of the air should increase the rate of photosynthesis in all of the more productive ecosystems. As the CO_2 of the air rises, therefore, we should expect the biota to fix carbon more rapidly, acting like the oceans as a CO_2 shock absorber. This is not just a theoretical construct, or one known only on a laboratory scale, because field crops have been shown to increase their productivity when grown under plastic tents supplied with extra CO_2. Recent work has demonstrated increased ring growth in trees, apparently due to CO_2 enrichment (LaMarche *et al.*, 1984).

Furthermore, the fields and forests of a humanized world are all young fields and forests. We cut down the climax and allow secondary succession (Chapter 23). These young plant associations are those of most rapid growth, making it certain that the flux of CO_2 into the biota has been maximized. Thus the prime ecological view of CO_2-enriched air is that it should sponsor plant growth tending to return the air to normal.

An ingenious use of the Mauna Loa data suggests that extra uptake of CO_2 by photosynthesis is negligible, however (Hall *et al.*, 1975). If there had actually been a net return of carbon to the biota in the last few decades then this should be reflected in the Mauna Loa data (Figure 19.3) as the northern ecosystems took in more carbon with each successive summer. Rigorous analysis of the data reveals no systematic change in the amplitude of the fluctuations, showing that any tendency of the biota to act as a sink for fossil fuel carbon through increased productivity was slight.

The other obvious way to decide whether the biota are now sink or source is to make a budget of organic gains and losses from the world ecosystems. This is a difficult undertaking, one fraught with assumptions that possibly are perilous. A cost accounting by Woodwell *et al.* (1978) used the net productivity estimates of world ecosystems of Whittaker and Likens (1973), which are described in Chapter 16. Starting with these regional estimates for net productivity they then made the following assumptions:

(i) World forests are being cleared at a rate of about 1% per annum.

(ii) The response of photosynthesis to increased concentrations of CO_2 is linear.

Both these assumptions are reasonable. The scanty data we have from the Amazon region, where the largest remaining forests are, suggest that a 1% reduction is conservative, and the exploitation of other tropical countries with growing populations is at least as great. Only in the northern temperate regions are forests coming back (in New England, for instance). A linear response by photosynthesis to CO_2 concentrations is probably generous.

In these calculations the carbon released by cutting and clearing exceeds the carbon fixed by contemporary photosynthesis (Table 19.2). The biota turn out to be a source. Woodwell *et al.* (1978) claimed to be conservative when they concluded that the biota now contribute between 2×10^{15} and 8×10^{15} grams of carbon to the atmosphere each year. This range spans the present contribution of fossil fuel carbon from industry (5×10^{15} g C yr^{-1}). This conclusion, however, is vulnerable to attacks on the assumptions. If the world forests are being cleared at a rate other than 1% a year the conclusions fail; if the regrowth of forests has been underestimated the conclusions again fail (Bolin, 1970).

The Oceans: Carbon Sink or Source?

We naturally tend to think of the oceans as a carbon sink. The oceans are the great shock absorber, and over the course of geologic time they do usually serve as a sink. but the fact remains that fifty atmospheres' worth of CO_2 are held in the oceans in intimate proximity to the air. Anything that changed the solution chemistry at the ocean surface might easily convert the oceans temporarily from sink to source.

The CO_2 regimen in the oceans as a whole is extremely complicated such that no detailed simulation of the system is possible yet (Gieskes, 1974; Broeker, 1971, 1973; Garrels and Mackenzie, 1971). But it is clear that the concentration of carbon oxide species in the surface waters is influenced by the concentration of CO_2 in the air, by temperature, and by ocean chemistry as this is directed by biological processes.

Of particular importance to solution equilibria is the concentration of the bicarbonate ion, HCO_3^-. This ion may be formed directly by solution of CO_2 thus

$$H_2O + CO_2 + CO_3^{--} \rightleftharpoons 2HCO_3^-$$

But a second important source of bicarbonate is rivers that bring the ion down to the sea in solution. Bicarbonate is formed as a result of the aqueous weathering of crustal rocks in the presence of gaseous CO_2 in various reactions of which the following, for magnesium silicate minerals, is an example:

$$2CO_2 + 3H_2O + MgSiO_3 \rightarrow Mg^{++} + H_4SiO_4 + 2HCO_3^-$$

Table 19.2
Carbon Budget for the Biosphere.
These are the results calculated by Woodwell et al. *(1978). See text for assumptions of the calculations.*

		× 10^{15} g C/yr			
	Area (10^6 Km)	Total Net Primary Production	Forest Clearing and Harvest	Increase in Net Ecosystem Production	Net Release (total effect of human activity)
Tropical forests	24.5	22.2	4.4	0.9	3.4
Temperate forests	12.0	6.7	1.7	0.3	1.4
Boreal forests	12.0	4.3	1.0	0.2	0.8
Other vegetation (including agriculture)	98.5	19.2	0.3	0.1	0.2
Detritus and humus	147	—	—	—	2.0

This flux of riverain bicarbonate into the ocean appears to be the main reason that seawater is alkaline. The pH of seawater is weakly buffered around pH 8 by the carbon oxide solution system

$$CO_2 \rightleftharpoons HCO_3^- \rightleftharpoons CO_3^{--}$$

This system must be sensitive to demands on particular carbon oxide species within the ocean. Photosynthesis, for instance, removes bicarbonate so that increased productivity should draw CO_2 into the sea and reduced productivity should release CO_2. Fixing of the electrically neutral form of silica known as OPAL by diatoms and other organisms also influences the reaction as more bicarbonate is needed to replace the valence of the lost silicate ion. Through reactions like these, therefore, biological activity in the sea can affect the solution equilibria of carbon oxides. Since the full complexity of this system is not known, it remains possible that the relative state of the oceans as sink or source can change as productivity changes (Broeker, 1971, 1973).

But ocean temperature is probably more important to the retention of CO_2 because the solubility of CO_2 is strongly temperature dependent. Solubility falls as water warms. A rise of mean water temperature of 0.1°C will increase atmospheric CO_2 concentration by 10 ppm (Machta, 1973). This means that the oceans become a net source of CO_2 if they warm while other parameters remain the same. If the earth is warming slightly, from the results of human activities or from natural climatic change, then the oceans may be contributing CO_2 to the air. It is possible that warmer geological periods always have higher concentrations of CO_2 in the air in this way.

The Carbon Isotope and Tree-Ring Evidence

The most persuasive evidence for the cause of the secular rise in CO_2 comes from the studies of carbon isotopes in air, water, fuels, and the biota. Ratios of the different isotopes tell us the origin of any carbon sample, whether fossil fuel, the contemporary air, or organic matter. Moreover, we can take samples of "air" from different years over the past century or two by cutting out rings of trees and using the carbon isotopes in the resulting dated slivers of wood as vouchers for the past atmosphere. In this way it has been possible to trace in outline the course of the present CO_2 enrichment. A general conclusion is that both fossil fuels and cleared forests have contributed to the CO_2 rise, but that the oceans probably are still acting as a sink.

Carbon occurs naturally as three isotopes—^{12}C, ^{13}C, and ^{14}C. Both ^{12}C and ^{13}C are stable permanent parts of the earth's elemental species, the one abundant, the other rare. But ^{14}C is produced in the upper atmosphere by cosmic bombardment of nitrogen and it decays back to nitrogen with a half-life of about 5700 years. Fossil fuels, being organic matter that has been sequestered for millions of years, contain no ^{14}C; the atmosphere does contain ^{14}C.

One of the consequences of our burning fossil fuel these last one hundred years has been a dilution of atmospheric ^{14}C with extra ^{12}C, because none of the CO_2 coming from fossil fuels contains ^{14}C. This dilution is known as the SUESS EFFECT after its discoverer, Hans Suess (Revelle, 1965; Suess, 1955). We can measure the actual reduction in ^{14}C of the contemporary air from the Suess effect by taking samples of wood from before the industrial revolution and measuring the $^{12}C/^{14}C$ ratio. The actual ^{14}C reduction is between 1.7 and 2.3% (Broeker, 1971). This reduction lets us calculate that between one-third and one-half of fossil fuel carbon so far burned is still in the air. The rest, if not in the biota, must be in the world oceans.

During photosynthesis, plant enzymes discriminate against the unusual carbon isotopes, both ^{14}C and ^{13}C. These are taken up and synthesized into organic compounds but proportionately less easily than is ^{12}C. The ratio of ^{13}C to ^{12}C in a plant, therefore, is different than the ratio found in the air on which that plant draws. Chemical processing of organic matter into fossil fuels over geological time also skews the isotope ratios. The result of this is that fossil fuels, and different contemporary organic materials, have different and characteristic ratios of the stable

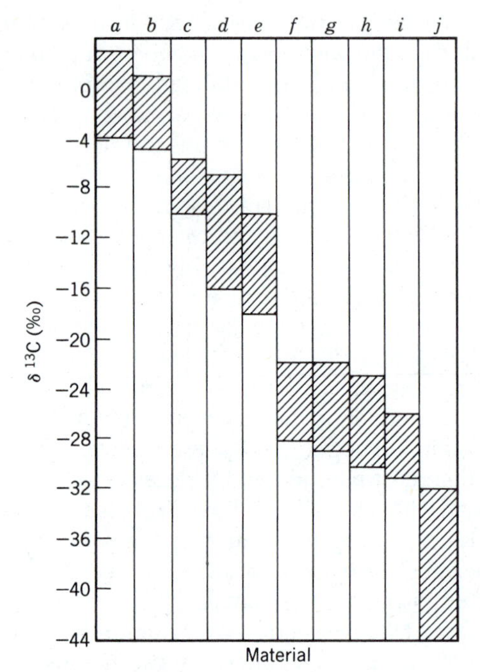

Figure 19.7 Excess ^{13}C (δ ^{13}C) over ^{12}C of various carbonaceous materials.
Because plants discriminate between the isotopes in photosynthesis, and because fossil fuels have characteristic isotope ratios, it is possible to identify the source of CO_2 used in photosynthesis in the past from pieces of wood. *a,* Carbonates; *b,* volcanic CO_2; *c,* marine animals; *f,* coal; *g,* land plants; *h,* liquid hydrocarbons; *i,* organic sediments; *j,* natural gas. (From Farmer and Baxter, 1974.)

isotopes ^{12}C and ^{13}C as well as of ^{14}C (Figure 19.7).

Yet the actual ratio of isotopes fixed by a plant in photosynthesis must be a function of the ratio in the air from which the plants were drawing their supplies. Several investigators have now taken trees of known age, extracted samples of tree rings, and measured the carbon isotope ratios of each (Farmer and Baxter, 1974; Stuiver, 1978). With all three isotopes known for a time series of samples we can proceed as follows, following the analysis of Stuiver (1978):

(i) Measure actual lowering of ^{14}C content for each tree-ring increment in the time series and calculate the actual contribution of fossil fuel carbon to each increment.

(ii) Calculate from the results of (i) the lowering of ^{13}C for each increment due to fossil fuels.

(iii) Adjust each ^{13}C measure for the contribution of fossil fuels, then any trend remaining is due to the flux of carbon from the biota to the atmosphere.

Figure 19.8 gives the results of Stuiver's (1978) calculations. There has been a more or less progressive fall in the ^{13}C concentration over the last century, with a particularly sharp decline between 1880 and 1930. This decline is due to the

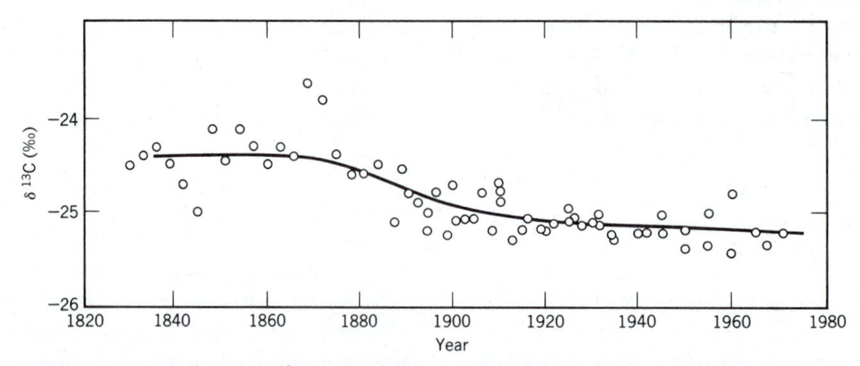

Figure 19.8 Reduction in ^{13}C in atmosphere resulting from reduction in biota.
Data are residual reduction in ^{13}C after allowing for the effect of fossil fuels. The modern reduction in atmospheric ^{13}C shows that the biota are acting as a source of CO_2 rather than a sink. The curve is a visual aid drawn by hand. (From Stuiver, 1978.)

dilution of the atmospheric carbon pool with ^{13}C-deficient carbon released from the biota.

Stuiver (1978) concludes that about half the increment in CO_2 floating about in our air comes from fossil fuels and the other half comes from the biota. On this analysis, therefore, the biota are at present a *source* of CO_2 and the old belief that plants would compensate for our burning of fossil fuels with extra photosynthesis is found to be wrong.

The following tentative conclusions seem warranted:

1. The mass of CO_2 in the atmosphere is rising.

2. Human activities are the cause of this rise.

3. Both burning fossil fuels and the clearing of forests contribute to this rise.

4. The rise is possible because the rate at which the oceans can take up CO_2 is kept low by the slow mixing of the oceans.

5. Unless ocean temperatures change, the extra CO_2 eventually will be absorbed by the oceans, allowing the atmospheric concentration to return to what it was before the industrial revolution.

6. Changes in mass of CO_2 in the atmosphere comparable to that now being caused by human activity probably occurred in the past with global changes of temperature or displacement of continents.

HOW THE AIR GOT OXYGEN

For free oxygen to be put into the earth's atmosphere all the primeval oxygen sinks had to be swamped. Surface rocks were all chemically reduced, as indeed they still are beneath the thin film of oxidized material that has since been draped over the earth. The primeval atmosphere was made of reduced gasses from the original "outgassing" of the earth's beginning. Chemical reduction was the universal norm and the chances of a molecule of oxygen remaining free over this original earth were those of the proverbial snowflake in a very hot place.

Two kinds of mechanism are known that pumped oxygen into this reducing world. One is purely physical, the dissociation of water vapor in the upper atmosphere, with the loss of hydrogen to outer space. The other is photosynthesis. Most geochemists think that photosynthesis has been the more important.

DISSOCIATION OF WATER is the more straightforward mechanism. The upper atmosphere is exposed to ionizing radiations from the sun and these radiations provide the energy to split water into molecular oxygen and molecular hydrogen. Hydrogen is then lost to outer space as the lighter gas. This mechanism is in place still, though working effectively only above the ozone shield. The beauty of the mechanism from the geochemist's point of view is that the hydrogen is removed from the earth system entirely; the mechanism is a real whole-earth oxidizer.

PHOTOSYNTHESIS in its various forms also uses solar energy to release oxygen, typically from water. But the oxygen release of photosynthesis is less final an act because the other product of the reaction is still around. Photosynthesis balances the oxygen released with reduced carbon compounds that are the object of the synthesis. Photosynthesis, therefore, can yield free oxygen only by burying carbon.

If all the oxygen used to oxidize the primeval gasses and the exposed parts of the earth's crust, together with all the oxygen in the air, came from photosynthesis, then there must be an enormous mass of reduced carbon buried in the crust of the earth. The carbon required is far larger than all the known reserves of fossil fuels, including oil, coal, natural gas, and oil shale. These combined cannot even balance the mass of oxygen in the air, let alone the far larger mass that must have been used to oxidize minerals before any oxygen got into the air at all. The PHOTOSYNTHESIS HYPOTHESIS, therefore, has a MISSING CARBON PROBLEM.

Free oxygen could have resulted from photosynthesis only if the crust of the earth still holds the reduced carbon that was produced at the same time. Those geochemists (the majority) who accept photosynthesis as the major of the two sources of free oxygen solve the *missing carbon problem* by appealing to finely divided graphite in sedimentary rocks or to massive reserves of natural gas as yet undiscovered. These solutions are perfectly plausible. But, when accepting the *photosynthesis hypothesis*, it is proper to be clear that this does require that there be reserves of reduced carbon in the crust of the earth that have not yet been measured.

Proponents of the *dissociation* and *photosynthesis* hypotheses are still in conflict about the initial rise in oxygen before the start of metazoan evolution. This was the period between 3 G yr and 1 G yr ago (three billion and one billion; G = *giga* = 10^9). The record of sediments over this long span of time includes direct evidence of oxygen entering the atmosphere and the first evidence of unicellular life. There can, however, have been very little free oxygen in the air until the end of the epoch. This was the period, the enormous period, needed to swamp the oxygen sinks of the crustal rocks and primeval atmosphere before free oxygen could appear in the air.

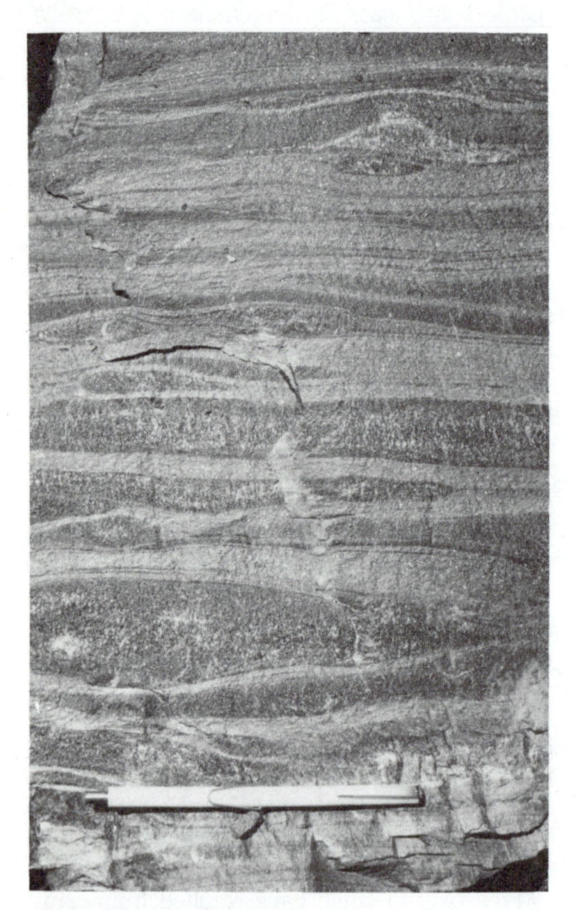

Figure 19.9 Banded iron formation.
The finely layered deposit could not have been emplaced by any process continuing on the present earth but required oceans saturated with ferrous iron solutes and subject to repeated infusions of oxygen at short intervals. Scale object about 18 cm long.

The evidence that oxygen was being released in vast quantities in these early days is at its most spectacular with the banded iron formations (Figure 19.9). These are deposits of ferric oxides of a kind most certainly not deposited anywhere on the contemporary earth. They imply oceans, or at least water, with high concentrations of dissolved ferrous compounds that are supplied with a source of oxygen. Ferric compounds would then be precipitated. The iron itself was thus a very large sink for oxygen, but in the reducing world in which ferrous oceans were possible there would have been many other mineral sinks for oxygen also. There was no oxygen in the air, yet oxygen was being produced rapidly.

Advocates of the *photosynthesis hypothesis* for these early days draw on the evidence of fossils and on properties of the banded iron formations themselves (Cloud, 1972, 1973, 1976; Marguilis *et al.*, 1976; Schopf, 1975). The fossils show that life, probably photosynthetic life, was already well established. Most notable among the fossil evidence are the STROMATOLITES, the carbonate relics of ancient algal reefs. The stromatolites leave little doubt that algae rather like the present blue-greens were building reefs of carbonates in shallow seas of the time, rather as other algae and coral animals are building reefs in present-day oceans. Photosynthesis, therefore, is demonstrated and, being present, becomes a parsimonious hypothesis for the release of oxygen demonstrated by banded iron.

The bands in the banded iron formation are the second prop of the photosynthesis hypothesis, because these look like seasonal bands.

Certainly whatever caused the banding had its influence over large pieces of real estate at the same time, making seasonal change a likely candidate. Presumably the algal reefs produced less oxygen in winter than in summer so that the bands are annual. The *photosynthesis hypothesis*, therefore, explains the complete set of phenomena of those early days, at least in a qualitative way. Its one weakness is its requirement of the missing carbon. Where is the "banded carbon" formation to balance the banded iron?

Advocates of the *dissociation* hypothesis suggest that conditions for upper atmosphere dissociation were so good in those early days that the process was decisive despite the evidence for photosynthetic life being present also (Towe, 1978). In the world before free oxygen there would be no ozone in the upper atmosphere to act as a radiation shield. Water vapor would be exposed to radiation even more than now so that the rate of dissociation might be high. Estimates of the actual rate of oxygen production by this means are bound to be sensitive to assumptions like the rate of atmospheric mixing, the density of water vapor, the rate of hydrogen escape, *et cetera*, and can never be decisive. But models allowing very significant oxygen production by photolytic dissociation in the anoxic air are reasonable, and oxygen production by this method has no missing carbon problem.

But the dissociation hypothesis does have to account for the bands in the iron. This it can do by invoking seasonal mixing, or another physical process that is triggered by earthly seasons. The hypothesis sees oxygen produced in the upper atmosphere at a steady rate as hydrogen is lost, but only being pumped into the world ocean with seasonal events of the global climate.

As developed by Towe (1978) the dissociation hypothesis can do away with photosynthesis entirely as a producer of oxygen at this early period. The stromatolites and other fossils record the presence of photosynthetic life clearly enough, but the ancient photosynthesis need not have released oxygen. It could be that only photosystem I was used by the early plants, and this does not release oxygen. Oxygen is released only in photosystem II (Chapter 3). Towe suggests that this need not have evolved until the end of the time of the banded iron formation, about 2 G yr ago. Much of the missing carbon problem is thus neatly shelved.

Balanced judgment probably has it that both the dissociation and photosynthesis mechanisms were working forcefully in the time of the banded iron. But 2 G yr ago banded iron deposition ceased and the geological record begins to supply widespread evidence of an atmosphere that holds some free oxygen, notably the evidence of "red beds," the color of iron and aluminum oxides. Between 2 G and 1 G yr ago this atmosphere with free oxygen acquired more and more of the gas until the final concentration of 21% by volume was reached. Oxygen production must still have been very high because eroding crustal rocks still provided a very large oxygen sink that must be swamped before extra oxygen could remain in the air. Over this period there seems little doubt that photosynthesis was the main source, because the earth now had an ozone shield. The most probable reconstruction is that the free oxygen of the present air was made by plants of ecosystems, but that the ancient crustal sinks of the reduced earth might have been filled partly by the inanimate dissociation mechanism first.

How Oxygen Is Maintained at 21% by Volume: The Sulfur Cycle

The free oxygen of the contemporary atmosphere must be maintained at a near steady state by a system of sources and sinks comparable to those that maintained the CO_2 concentration. There is an important difference between these two systems, however, in that the reservoir of oxygen (21%) in the atmosphere is very much larger than that of CO_2 (0.03%). All the oxidation and reduction reactions of the carbon cycle are sinks and sources for atmospheric oxygen and, to this extent, the fate of oxygen is linked

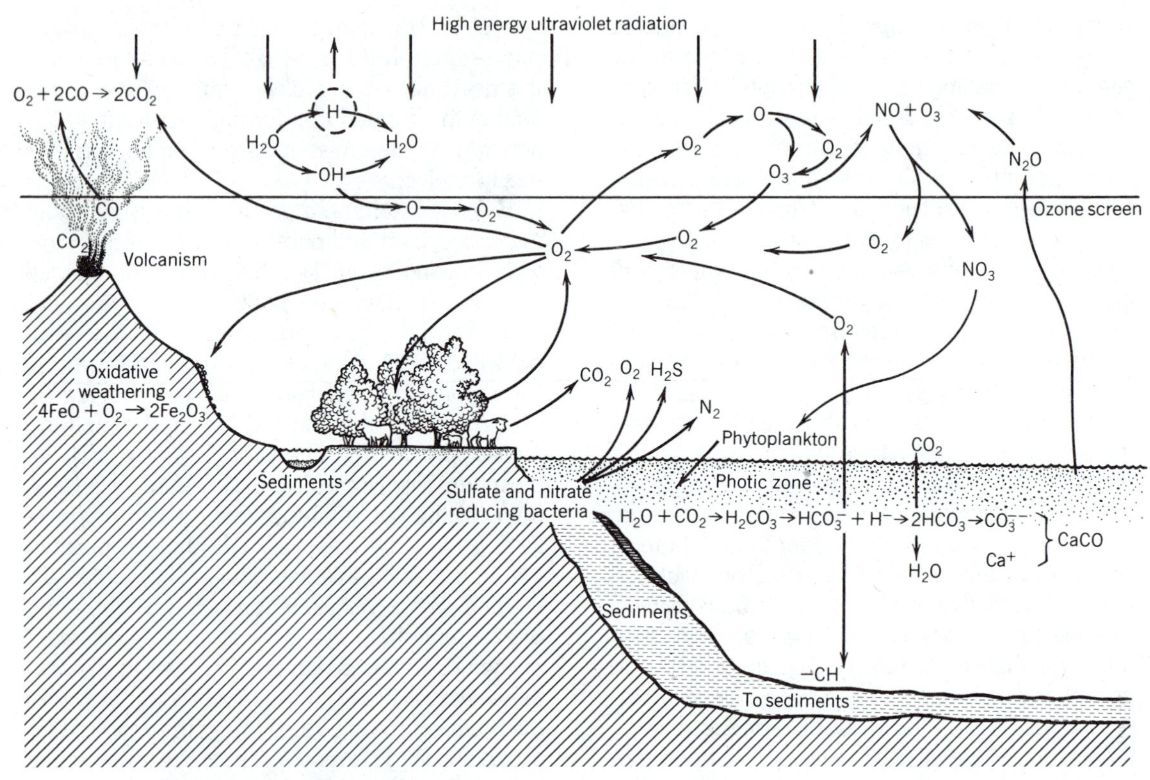

Figure 19.10 The oxygen cycle.
The diagram identifies important pathways in the oxygen cycle. It is important to remember, however, that the flux of all the processes illustrated is small compared with the volume of O_2 present in the air. (Based on a figure of Cloud and Gibor, 1970.)

to that of carbon. But the carbon–oxygen cycle is only a small part of the oxygen system of the biosphere (Figure 19.10).

Important sources and sinks of oxygen are listed in Table 19.3. Photosynthesis as a source is very nearly balanced by respiration and the dissociation mechanism is thought now to be trivial. The ocean surface generally is saturated with oxygen. The slight excess production of oxygen from dissociation and from burial of organic matter in ecosystems may just be sufficient to balance the sink of oxygen used in what is left of the process of oxidizing iron minerals of crustal rocks and volcanic gasses and in the production of marine carbonates (Garrels *et al.*, 1975). If these were the only sinks and sources then something close to a steady-state supply of oxygen for the atmosphere is easily conceived, but

Table 19.3
Sources and Sinks of Atmospheric Oxygen
Fossil fuels are ignored as a sink (when burned) because they are probably of negligible importance. Fossil fuels and other organic deposits as a sink are allowed for when listing photosynthesis as a source.

Sources	Sinks
Escape of hydrogen from water to space	Oxidation of crustal minerals
Sulfate-reducing bacteria	Sulfur bacteria
Nitrate-reducing bacteria	Nitrogen-fixing organisms
	Volcanic gasses
Oceans	Oceans
Photosynthesis	Respiration

larger sinks and sources result from the oxidation and reduction of sulfur and nitrogen (Table 19.3). These are the processes most necessary to the maintenance of our contemporary air.

The nitrogen cycle may well be internally balanced. It is discussed in the next section. The complicated sulfur cycle may not only be balanced in itself, but may also be the final regulator that sets the atmospheric concentration of oxygen at 21%, no more and no less.

Sulfate-Reducing Bacteria

Spontaneous oxidation of sulfide minerals and SO_2 from volcanos is potentially a large sink of both oxygen and sulfur. The sulfates produced are stable and soluble ions. They are cycled in ecosystems, where they are treated by living things as resources to be conserved. Inevitably they leak from the ecosystem, into the drainage channels and thence to the sea to be dropped into its mud. This process might be expected to result in accumulations of sulfate rock just as carbonate rock accumulates; sulfur stone instead of limestone. The reason that this does not happen is that specialized bacteria take oxygen from sulfate in anoxic mud.

Sulfate-reducing bacteria like *Desulfovibrio* and *Desulfotomaculo* live in places without free oxygen, typically reduced mud in bogs, lakes, or estuaries. These are places where energy as dead organic matter abounds but oxygen for respiration is absent. Sulfate-reducing bacteria respire by using the oxygen of sulfate ions as the required electron and hydrogen acceptor:

$$H_2SO_4^{--} + 2C \rightarrow 2CO_2 + H_2S$$

This self-interested undertaking of the bacteria releases hydrogen sulfide gas and CO_2 into the air or to the overlying water column. Marshes smell of sulfur; and oxygen starts its journey back to the air.

The CO_2 released by the sulfate-reducing bacteria into the deep sea has a special significance in that its release is accompanied by a flux of nutrients leaving the reduced mud in solution in the usual way. When these nutrients and CO_2 are transported together as a package into lighted water, the CO_2 can be expected to be reduced in photosynthesis and free O_2 released. The return to the air of oxygen once used to make sulfate out of sulfide is now complete (Deevey, 1970; Kellog *et al.*, 1972).

No other ways are known in which oxygen can be released from sulfates. It is hard to escape the conclusion that the oxygen concentration in the atmosphere would be less than its steady-state figure of 21% were it not for these bacteria. The sulfate-reducing process is possible only in anoxic mud. Ecosystems of swamps and fertile waters, therefore, are essential to the maintenance of the biosphere.

Other bacteria use sulfur in their daily lives, though they do not have consequences for the oxygen budget of the earth like those of the sulfate-reducing bacteria. Some even reverse the work of the sulfate reducers by oxidizing sulfide back to sulfate. An outline of the global sulfur cycle is given in Figure 19.11.

Sulfur Bacteria as Oxygen Regulators

Oxygen production in the sulfur-sulfate cycle in the oceans is thought to be the regulator of atmospheric oxygen at 21% by volume. This belief comes about because the oxidation and reduction of sulfur has an essential feedback between land, sea, and sediments that must be necessary for a regulator.

It is known that a large flux of sulfur comes out of seawater back to land. The data supporting this conclusion are measures of sulfate entering the sea in rivers showing this flux to be many times larger than can be accounted for by weathering. If these river sulfates have not been weathered out of rock, they must come from the air. The air in turn must receive its sulfur from volcanos or the sea. Since the contribution of volcanos is trivial, it follows that a large annual flux of sulfur goes from the sea to the air, thence into rivers as sulfate and once more back to the sea.

It can be shown that most of the sulfur reaching the air from the sea does so via biogenic

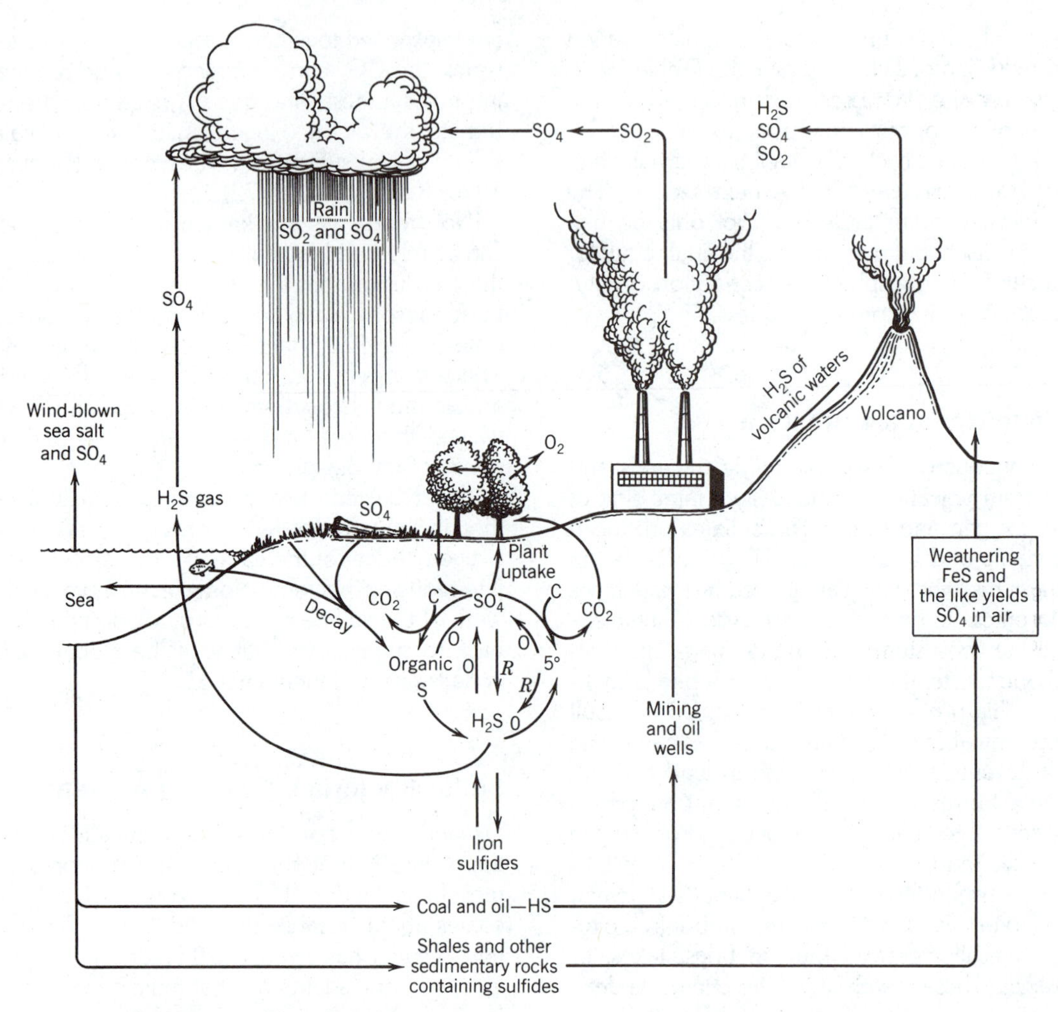

Figure 19.11 The sulfur cycle.
Ecosystems are provided with sulfates, which they cycle, from the oxidation of the products of weathering. In marine and fresh-water muds, sulfur is cycled between sulfate-reducing bacteria, *Desulfovibrio* and *Desulfotomaculo,* which reduce sulfates to S and H_2S, and sulfide-oxidizing bacteria, such as the photosynthetic green or purple sulfur bacteria and the chemosynthetic colorless bacteria (Leucobacteriaceae), which oxidize sulfides. H_2S returned to the atmosphere is spontaneously oxidized and delivered by rain. Sulfides incorporated in fossil fuels and sedimentary rocks are eventually oxidized following human combustion or crustal movements and weathering. Primary production accounts for the incorporation of SO_4 into organic matter, and anaerobic and aerobic heterotrophic microorganisms transform organic S to H_2S and SO_4, respectively.

processes, a conclusion based on the isotopic composition of atmospheric sulfur (Erikson, 1959, 1960). This makes it virtually certain that the prime source of this sulfur is sulfate-reducing bacteria in oceanic mud.

The complete ocean–atmosphere–land sulfur cycle, therefore, can be deduced (Figure 19.12). Sulfate concentration in the mud regulates the rate at which sulfate-reducing bacteria vent H_2S to the water column and the atmosphere. This

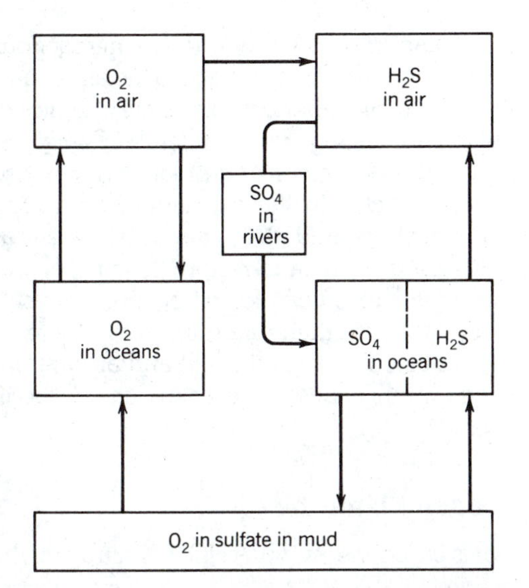

Figure 19.12 Oxygen regulation: the sulfur hypothesis. Oxygen leaves the air to oxidize sulfides emitted from the oceans. Oxygen is released from these sulfates by sulfate-reducing bacteria in mud. This process may be a prime regulator of oxygen concentration of the oceans, and hence of the air. The sulfur system is itself regulated by release of hydrogen sulfide back to the air.

Table 19.4
Sources and Sinks of Atmospheric Nitrogen
The biological sinks and sources are by far the most significant.

Sources	Sinks
Volcanos	Synthesis of nitrates in electric storms
Reduction of nitrates by bacteria	Biological nitrogen fixation
Oceans	Oceans
Decay of ^{14}C	Formation of ^{14}C

Table 19.5
Nitrogen in Air, Oceans, and Contributed by Volcanos
(Data of Delwiche, 1970.)

N_2 in atmosphere	3.8×10^{15} tons
N_2 in oceans	2.0×10^{13} tons
Annual input of N_2 from volcanos	2.0×10^{5} tons
Annual output of N_2 from oxidation in electric storms	4.9×10^{9} tons

flux of H_2S regulates the concentration of sulfate in rain water, which determines the concentration in rivers. Finally, the concentration in rivers regulates the concentration in ocean mud, completing the feedback system. Oxygen emission by sulfate-reducing bacteria, and hence atmospheric oxygen, is regulated as a by-product of the sulfur cycle.

THE REGULATION OF NITROGEN IN THE AIR

The almost inert gas nitrogen makes up 79% of the atmosphere by volume. The prime source seems to be volcanos that have been emitting small quantities of nitrogen throughout geologic time. Since the gas has a low solubility in water most has collected in the air itself. The main sinks and sources are given in Table 19.4.

The mass of nitrogen held in the oceans is trivial compared with the mass in the air (Table 19.5). This is quite different from the circumstance of carbon dioxide and oxygen, where the oceans are principal reservoirs. For nitrogen the atmosphere is the principal reservoir (though the total mass in the biota may be large also), so the oceans are unimportant as a buffer. The annual input of volcanos is small, though large enough to suggest that the atmosphere might continually be enriched with nitrogen throughout geologic time without corresponding sinks (Table 19.5). But there are two very significant sinks, both directing nitrogen into the stable oxide nitrate. One is nitrate synthesis in electrical storms and the other is nitrogen fixation in ecosystems.

The manufacture of nitrates by storms and organisms represents a sink for atmospheric nitrogen that is potentially bottomless. Nitrogen could be removed from the air continuously until both nitrogen and oxygen ended up as nitrate rock, destroying the atmosphere as we know it and ending life on earth. In the face of these cumulative drains it appears that the nitrogen concentration of the air can be maintained only by the nitrogen-return functions of ecosystems. These nitrogen-return functions arise because some bacteria use nitrate ions as hydrogen acceptors in their oxidations of organic matter, just as other bacteria use sulfates for the same purpose. Nitrogen in the air, therefore, like carbon dioxide and oxygen, is maintained and moderated by living systems.

Nitrogen Fixation: The Traditionalist or Farmer's View

Farming increases the leak of combined nitrogen from a habitat. Clearing land and letting soil humus rot always increases the leak of combined nitrogen, as it does for the other soluble nutrients. In the wet tropics this loss of nitrogen is part of the catastrophic cost of cutting down a rain forest. Even in a temperate habitat the loss can be dramatic as the Hubbard Brook clear-cutting experiment (page 435) showed, because the loss of nitrate two years after clear-cutting was an order of magnitude greater than the loss of any other ion. Regular farming increases the leak still further by carrying protein-rich crops from the land, passing them through the digestive tract of various bipedal and quadripedal animals, and burying the nitrogenous remains in privy, lake, or sea.

Old-style farming was spared paying the crippling cost for this prodigality by the activity of NITROGEN-FIXING BACTERIA during years when land was left fallow. New-style agriculture is impatient of the delays inherent in waiting on the bacteria and employs an industrial synthesis to the same end to such an extent that the flux of nitrogen from the air to soluble ion in land and sea from industrial process is at least as large as that worked by the nitrogen-fixing bacteria of the whole world put together (Figure 19.13).

The traditionalist view has it that nitrogen-fixing organisms are friends of mankind. It is known that their work is continually being undone by other organisms that break down nitrogenous fertilizer and void nitrogen back to the air. It must follow that these other organisms are enemies of mankind for eating into our fertilizer supplies. And yet the nitrogen-fixing organisms can also be seen as being in the business of destroying the atmosphere and the nitrogen releasers as grimly opposing them to restore the air. The potential major reservoirs for nitrogen on the surface of the earth, therefore, turn out to be fertilizer (combined nitrogen in soil) and air. Various bacteria arrange how much nitrogen will be in each.

Nitrogen Fixation

What is called the ACTIVATION of nitrogen, the reaction

$$N_2 \rightarrow 2N$$

requires much free energy. In the atmosphere it appears to result only from lightning discharges, when a series of alternate reactions take place that end up with nitrogen oxides and, in the presence of water, with nitric acid (Ferguson and Libby, 1971). The nitric acid comes down with rain, reacts with basic minerals, and becomes a nitrate input to the ecosystems below. As a practical source of nitrates to ecosystems it is generally of small importance (Figure 19.13) but it may provide a significant part of ecosystem nitrate inputs in exceptional, barren, stormy places like Antarctica (Schofield and Ahmadjian, 1973).

In biological NITROGEN FIXATION the energy for the activation of nitrogen is supplied by the oxidation of organic matter and the process is facilitated by an enzyme, *nitrogenase*. Once N_2 has been activated to 2N at a heavy cost in energy it is combined with hydrogen to make ammonia, an exothermic reaction that pays back a small portion of the original energy cost. Ammonia, dissolved in water as the ammonium ion NH_4^{--}, is then available for bacterial and plant metabolism and has entered the biosphere.

We do not know how nitrogenase works, or even whether the whole apparent cost of 147 kilocalories per mole (160 kc less 13 kc; Figure 19.13) is actually paid by a nitrogen-fixing or-

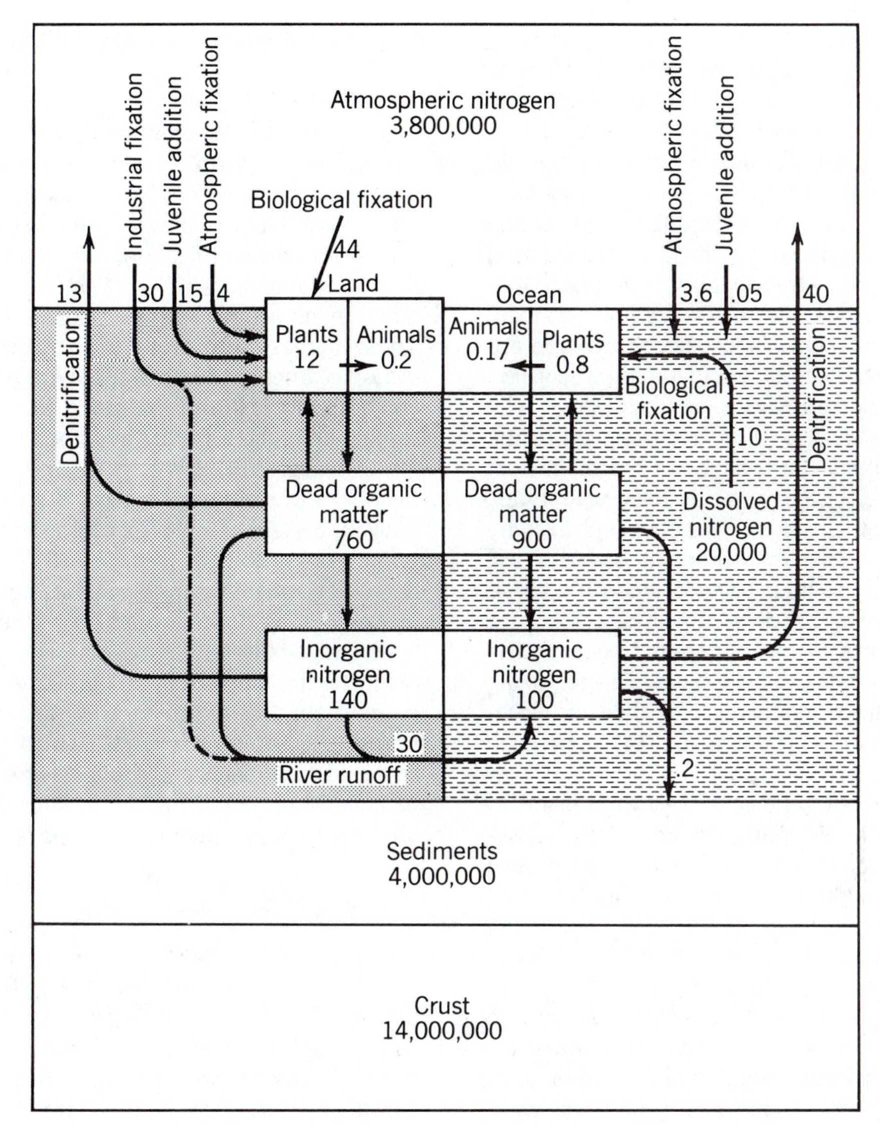

Figure 19.13 Nitrogen in the biosphere: sinks, sources, and transfers. Figures in reservoirs are $\times 10^9$ metric tons. Figures for transfer rates (arrows) are $\times 10^6$ metric tons per year. There are large uncertainties in the estimates of transfer rates, some of which may be off by as much as a factor of 10. The very large size of the atmospheric reservoir compared to the transfer rates is, however, apparent. (From Delwiche, 1970.)

ganism, since enzyme function can involve penetrating activation barriers rather than breaking them by main force. Industry uses the force method in the Haber process of ammonia synthesis, the basis of both the fertilizer and explosive industries. This process involves high temperatures and pressures as well as a metal catalyst, so a fortune must await the chemist who finds out how to do it the enzyme way.

Nitrogen-fixing organisms are all single-celled prokaryotes—various bacteria and blue-green algae. The nitrogen-fixing ability is not possessed by any multicellular plants or animals. It is, therefore, possible to find the very odd circumstance of large organisms dying for want of nitrogen even though they are bathed in a fluid that is 79% nitrogen gas. Perhaps prokaryotes possess the nitrogenase mechanism as an inheritance from ancient days before the air acquired oxygen and when different chemical pathways were required for life but, if so, why has it not been possible to evolve the mechanism since? We have no idea.

Root nodules in which bacteria live are apparently the best stratagem that natural selection has been able to provide for large plants. The plants pay a ransom in energy supplied to the bacteria and thus get combined nitrogen back at a greater cost than they would have paid if they had made it themselves. This cost explains why legume crops take their nitrogen from the soil rather than from their bacteria when the soil is fertilized (Child, 1976). A similar stratagem for obtaining nitrogen has apparently been undertaken for termites, which have nitrogen-fixing bacteria in their guts along with their cellulose-digesting protozoa (Benemann, 1973; Bresnak *et al.*, 1973).

It is likely that the failure of all large plants to fix their own nitrogen has important consequences for the atmosphere. Were all plants able to do so cheaply it is easy to imagine the ecosystems of the world adopting the sort of attitude to utilizing resources that people have: take it and throw it away. Nitrogen from the air might be cheaper than nitrogen from the soil and the consequence of this would certainly be that there would be a different steady-state mixture of gases in the air as both nitrogen and oxygen were withdrawn to the nitrate reservoir in "nitrostone."

A complex series of modifications of combined nitrogen as it cycles around ecosystems depends on the fact that each stage in the reaction series—amino acid, ammonia, nitrite, nitrate—yields free energy. Groups of bacteria specialize in sponsoring each reaction in this series, using the free energy won as a prime energy source. The bacteria are amino acid, ammonia, or nitrite *feeders*, nicely illustrating the ecological principle that IF FREE ENERGY IS AVAILABLE THROUGH A CHEMICAL REACTION SOME ORGANISM HAS PROBABLY EVOLVED TO USE IT.

Figure 19.14 illustrates the principal reactions in this series of bacterial oxidations and reductions of various forms of combined nitrogen and compares the free energy of the various undertakings with that released in respiration or required for nitrogen fixation. A point to notice is the regrettable names given to many of the reactions: "nitrification," "denitrification," and the like. These names were given to processes discovered piecemeal in the biological long-ago. They are both illogical and confusing but they keep turning up in the literature and need to be handy for reference. What is important is not the names, neither of process nor bacteria, but the fact that series of bacteria use nitrogen compounds as their prime energy sources.

In wet anaerobic mud, nitrates can be used by bacteria as hydrogen acceptors, just as sulfates can be used (page 471). These bacteria complete the nitrogen cycle that moves nitrogen to and from the air. We know of no other mechanism to complete the cycle, which is thus essential to maintenance of the present steady-state balance of gasses in the air. Figures 19.13 and 19.15 summarize the whole nitrogen system.

The Catastrophe Hypotheses

The most lurid of the scares put forward by "environmentalists" of the late 1960s were those that suggested we could destroy the atmosphere as a life support system. In its most spectacular form, the catastrophe hypothesis suggested that we might so interfere with oxygen sources that the air would be critically depleted of this essential gas. The argument, at first sight, had a certain plausibility because of the growing awareness of geochemists that oxygen may have been put into the atmosphere by life processes in the first place and that it is maintained by life processes now. With this plausible start the argument went as follows:

(i) Oxygen in the air is maintained principally by photosynthesis.

(ii) Since the oceans cover more than two-thirds

Figure 19.14 Energetics of the nitrogen cycle.
Steps 1 through 6 yield free energy and are undertaken by the various bacteria for that reason. The production of N_2O in the denitrification of step 1 may release the gas to the atmosphere with consequences for the ozone cycle of the upper atmosphere. The costs of all the bacterial operations of steps 1 through 7 have to be met eventually by green plants, either by metabolizing nitrates or by "feeding" nitrogen-fixing bacteria. Nitrogen fixation (steps 7 to 8) requires free energy. The energy yield of respiration (9) is given for comparison. (Modified from Delwiche, 1970.)

Reaction	Energy Yield (Kilocalories)
DENITRIFICATION	
1 $C_6H_{12}O_6 + 6KNO_3 \rightarrow 6CO_2 + 3H_2O + 6KOH + 3N_2O$ Glucose; Potassium Nitrate; Potassium Hydroxide; Nitrous Oxide	545
2 $5C_6H_{12}O_6 + 24KNO_3 \rightarrow 30CO_2 + 18H_2O + 24KOH + 12N_2$ Nitrogen	570 (per mole of glucose)
3 $5S + 6KNO_3 + 2CaCO_3 \rightarrow 3K_2SO_4 + 2CaSO_4 + 2CO_2 + 3N_2$ Sulfur; Potassium Sulfate; Calcium Sulfate	132 (per mole of sulfur)
AMMONIFICATION	
4 $CH_2NH_2COOH + 1\frac{1}{2}O_2 \rightarrow 2CO_2 + H_2O + NH_3$ Glycine; Oxygen; Ammonia	176
NITRIFICATION	
5 $NH_3 + 1\frac{1}{2}O_2 \rightarrow HNO_2 + H_2O$ Nitrous Acid	66
6 $KNO_2 + \frac{1}{2}O_2 \rightarrow KNO_3$ Potassium Nitrite	17.5
NITROGEN FIXATION	
7 $N_2 \rightarrow 2N$ "Activation" of nitrogen	−160
8 $2N + 3H_2 \rightarrow 2NH_3$	12.8
RESPIRATION	
9 $C_6H_{12}O_6 + 6O_2 \rightarrow 6CO_2 + 6H_2O$ Carbon Dioxide; Water	686

of the globe it is reasonable to conclude that it is ocean photosynthesis that maintains our oxygen.

(iii) The world oceans are being polluted already. A major spill of a tanker conveying herbicides like 2,4-D and 2,4,5-T or picloram was inherently likely (perhaps on its way to Viet Nam). A few such spills would kill most of the plants in the sea.

(iv) With the plants of the sea dead the air's oxygen begins to run down without being replaced.

This hypothesis was always absurd on the grounds of one unjustified inference and two massive errors of fact.

The unjustified inference was that it would be possible to kill all the plants of the sea with a

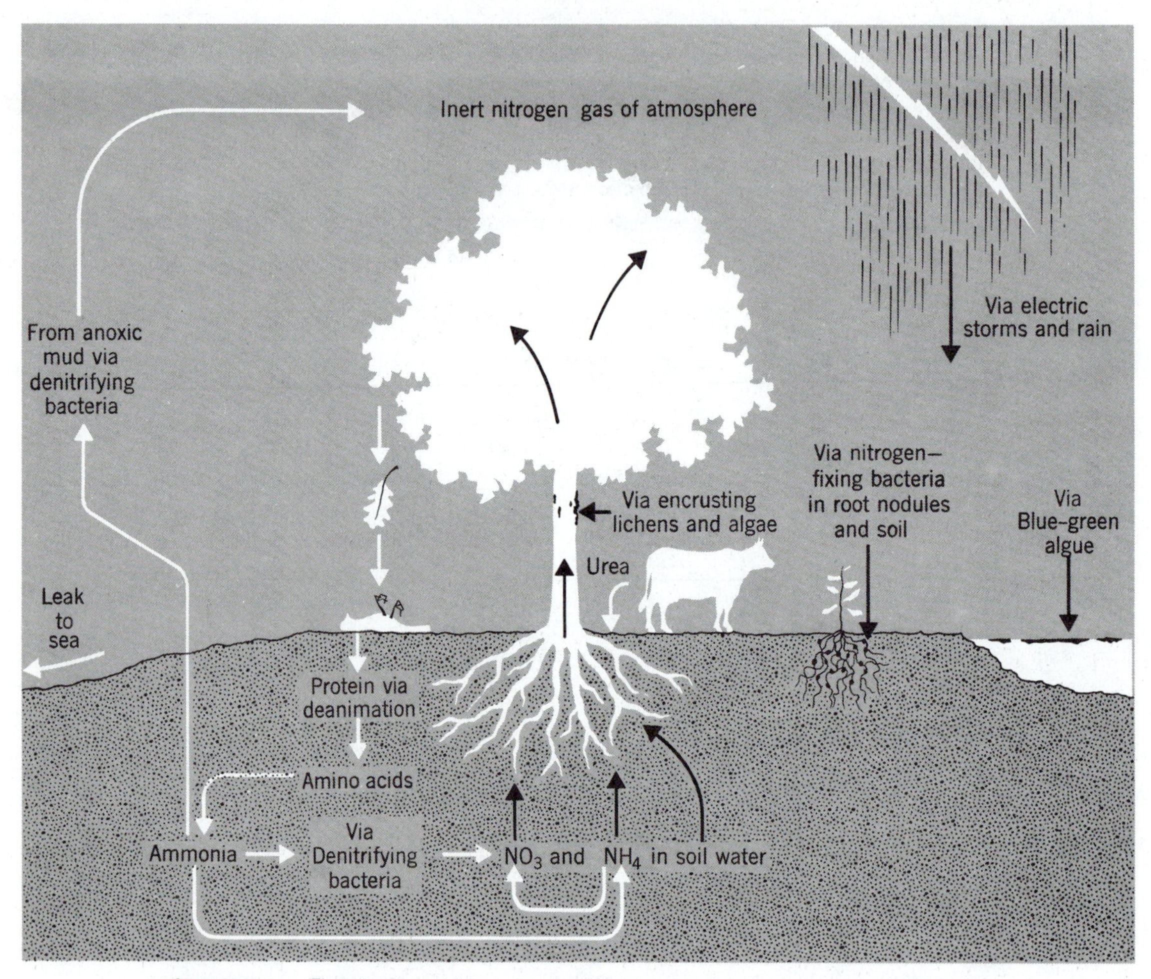

Figure 19.15 The nitrogen cycle.
Nitrogen is cycled through the ecosystem in combined form either as oxides or reduced as in ammonia, amino acids, or proteins, and the like. There is a continuous leak from the system in flowing waters that must be made good by fresh supplies from the atmospheric reservoir. Many kinds of microorganisms fix nitrogen, principally by reducing it to ammonium compounds, and there may be additional supplies (as oxides of nitrogen) in rain water that have been synthesized in electric storms. Nitrogen is returned to the atmospheric reservoir by denitrifying bacteria living in reduced muds of fertile waters and bogs. Black arrows are paths of nitrogen synthesis and white arrows are paths of nitrogen release from more complex molecules.

herbicide spill or two. Ocean plants are tiny phytoplankton, the populations of which are so large as to defy easy comprehension. Rapidly reproducing populations of enormous size include very great genetic diversity such that strains resistant to spilt chemicals are certain. In other words, natural selection would cope with herbicides spread on phytoplankton even if the virtually impossible goal of spreading and maintaining a concentration of the stuff in the world's oceans was achieved.

But the two massive errors of fact are more

Table 19.6
The Comparative Production of Land and Sea
The fact that the oceans are unproductive deserts means that the photosynthetic yield of oxygen for all the world oceans is less than that of the land, despite the larger size of the oceans. (Compiled from data of Whittaker and Likens, 1973.)

	Net Production (10^9 tons/yr)	Estimated Respiration (% gross production)	Gross Production (10^9 tons/yr)	Percentage of Total Global Gross Production
Land	109	60	272	75
Sea	55	40	92	25
Total	164		364	100

immediately fatal to the oxygen catastrophe hypothesis. The first of these is to think that the oceans are the principal source of photosynthetic oxygen. They are not. The oceans are unproductive deserts (Chapter 16) that, although covering much of the globe, contribute only a quarter of the total plant production (Table 19.6). Killing all ocean plants would interfere with no more than a quarter of the photosynthetic oxygen return.

The even more massive error of fact was to conclude that the annual oxygen return from photosynthesis was large compared with the amount of oxygen in the air. In truth the volume of free oxygen is so large that the contribution of contemporary photosynthesis is close to trivial. The geochemist Wallace Broeker (1970) deserves the honor of most clearly scotching this nonsense by examining the oxygen flux and reserves over a typical square meter of the surface of the globe. 60,000 moles of oxygen rest on every square meter of the earth's surface. But photosynthesis on each square meter produces only 8 moles in a year; 8 out of a total reserve of 60,000! If we killed all the plants on earth we should starve, but there would be plenty to breathe while we starved.

To ecologists generally it has long been known that the atmospheric gas with which we can and do tinker in significant ways is not oxygen but CO_2. The Mauna Loa data (Figure 19.3) are the most spectacular evidence and the data are reviewed earlier in this chapter. The reason human activities are significant to the level of CO_2 in the air is, of course, that the gas is present in such low concentration. It is likely, however, that the scale of the changes we are now causing is within the range of similar events in the past. These changes will definitely have consequences for life on earth, ours included with the rest, but how serious these changes will be we do not yet know.

The effects of rising CO_2 will depend on how concentrated the gas becomes finally. To predict what this final concentration will be we need data on fossil fuel consumption in the future, atmosphere mixing rates, changes in the biomass of the world, and, above all, trends in ocean temperature. Models are being made on the basis of various estimates for these parameters, some of them suggesting perhaps a doubling of atmospheric CO_2 sometime in the next century (from 0.03% to 0.06%). To choose between these models we need many more data on the atmosphere and oceans, past and present. The most important consequence of the enrichment, whatever it is, probably will be local changes in climate. Atmospheric CO_2 absorbs infrared radiations from the ground, so that a higher concentration of the gas increases air temperatures (THE GREENHOUSE EFFECT). The details of

Figure 19.16 The chemical boundaries.
Exchange of gas and spray across the ocean surface is one regulator of the chemistry of both. Even more profound is the exchange of gas and solutes across the boundary between the mud surface and the oceans. In the mud life has its most controlling influence, even undoing the synthesis of the primary gasses oxygen and nitrogen, so constantly undertaken by lightning.

local climatic change from CO_2-induced warming are a subject of debate and uncertainty, but it is expected to be considerable.

CONCLUDING REMARKS: LIFE IN THE MAINTENANCE OF THE AIR

In the creation and maintenance of the air the role of life is certain. It seems safe to say that the atmosphere would contain much less free oxygen were it not for photosynthesis. Moreover, the best explanations we now have for the regulation of the mix of nitrogen and oxygen in the air depend on the activities of bacteria that use sulfate and nitrate as hydrogen acceptors in reduced mud. The flux of oxygen, nitrogen, and sulfur into the oceans from the activities of these organisms maintains the air that the green plants build (Figure 19.16).

PREVIEW TO CHAPTER 20

Soils are formed in the loose material at the surface of the earth by processes working from above, chiefly through the agency of percolating water. This results in characteristic series of layers in the soil called "horizons." The surface "A" horizon collects organic matter from the vegetation and is, at the same time, denuded of both organic matter and soluble minerals as these are washed to lower horizons and to drainage waters. Deeper in the soil profile, mineral and organic materials from above may be redeposited in a "B" horizon where new minerals, like clays, may also be synthesized. Under these B horizons of accumulation and synthesis lies the surface rock, probably fractured but otherwise unaltered and called the soil "parent material" or "C" horizon. In any one climate or vegetation type the parent material will vary in conformity with the underlying pattern of rock outcrops, but the essential properties of the upper two groups of layers remain constant. Soil profiles are more dependent on climate and vegetation than they are on rock type. Different parts of the world and different vegetation types are found to have characteristic profiles of horizons. Early maps correlated these zonal soils with vegetation and climate, a correlation that was one of the early prompters of the ecosystem concept. *Great soil groups* are the rough equivalents of *plant formations.* A few names for them, like *podzol* for the layered soil under boreal forest, have long had international recognition but definitions are loose. These names have been partly replaced in the United States with an official classification of the U.S. Soil Survey. Within any one biome or larger ecosystem type, the parent rock or local topography specifies the particular local properties of a soil from among the regional properties set by vegetation and climate. The most widely used local unit is the *soil series,* named for a type locality and defined by physical properties. The smallest sample of a soil series for descriptive purposes is the *pedon.* Where soil series change continuously with slope or drainage, so that variants are repeated, the collection of related soil series is called a *catena.* Classifying soil series into related groupings is difficult and comparable to classifying plant communities, or whole ecosystems, into related groupings, but detailed sets of rules are in use to guide soil surveyors. An intriguing general observation is that tropical soils tend to be red, whereas soils of high latitudes tend to be brown or gray. It is likely that organic complexes remove red sesquioxides of iron and aluminum from temperate soils, allowing gray silica to accumulate. Soil color, therefore, may depend on life processes within ecosystems.

CHAPTER 20

THE SOIL

A part of any terrestrial ecosystem is the soil in which the plants grow. Soil to an ecologist is not something to do with geology but is a thin layer of the earth's crust that has been remade by life and weather. Characteristic ecosystems have characteristic soils, with their own mixtures of organic matter and their own layering. The study of soils is a discipline in its own right, PEDOLOGY, from the Latin word meaning a "foot."

Soil, to an ecosystem, is a nutrient delivery system, a recycling system, and a waste-disposal system. For plants, soils are sites of germination, support, and decay. For animals, soils are a refuge, a sewer, or a whole habitat. For decomposers, soils are a resource.

Soils of different ecosystems or climates have special properties. They may be of strikingly different colors, regardless of the parent rock, like the red of the ground in the lowland tropics as opposed to the prevailing browns of temperate latitudes. Or the topsoil may be ashy white. Prisoners held in Eastern Germany during the Second World War found that an extra hazard in tunnelling out of camps was disposing of colored subsoil on top of the whitish surface soil of the compound, where the ashy surface of a trampled *podzol* would show up the tell-tale signs of a tunneller's spoil. These different colors and properties of soil must be understood as weather, vegetation, and rock act together to produce this complex system we call "soil."

PEDOLOGY thus turns out to be a subject with some remarkable natural phenomena to explain. Why is the ground red in the tropics, brown under most of temperate agriculture, and perhaps ashy white when the surface litter is scuffed away under the northern pines? Why are some soils fertile but others not? And why should the top meter of the earth hold different minerals from the rock underneath?

Inevitably part of the approach of the soil scientist is to seek to classify soils, a task that seems simple enough at first sight but that is difficult in practice. Nearly as many soil classifications exist as there are national soil surveys. A multitude of classification may speak partly to rapid growth of pedology in the last decades, for the science is very active, but a more fundamental cause is the complexity of soil itself, where processes as disparate as climate, volcanism, and migrating plants can converge to produce a result impossible without all three. These converging processes are revealed most clearly when the soil is seen in section, showing what is called a SOIL PROFILE.

THE SOIL PROFILE

The side of a trench in any well-vegetated place reveals a succession of layers in the soil. At the surface is a litter of dead or rotting plant parts, with underneath one or more distinctly different layers that separate the surface litter from the subsoil a few feet down. Sometimes these layers are sharply distinct, other times they merge gradually into one another, but always they have been formed within the subsoil by weathering processes working down from the surface. The layers are not strata in the geological sense because they are not separate deposits. It is convenient to have a special word to describe them and they have come to be called HORIZONS by soil scientists.

SOIL HORIZONS form as rotting plant parts mix with the upper layers of the mineral soil and as drainage water percolates down through the litter to work a slow washing and chemistry on the lower horizons. The thickness of earth affected by these processes constitutes the soil. The subsoil underneath is the earth from which the soil was made and is, therefore, called the "PARENT MATERIAL." This layered appearance is called the SOIL PROFILE, which is THE SET OF SOIL HORIZONS BETWEEN THE UNDIFFERENTIATED PARENT MATERIAL AND THE SURFACE LITTER (Figure 20.1).

A technical term almost synonymous with *parent material* is REGOLITH. This is the top layer of unconsolidated material that covers crustal rocks. The *regolith* is the loose debris and it may be of widely varying thickness; negligibly thin at places where rocks outcrop or many meters thick in other places. Perhaps the characteristic way of forming a regolith is from the underlying rocks when these were shattered by expansions and contractions of changing temperature over long spans of geologic time. Or the regolith may be a deposit of wind-blown material (called LOESS) that buries crustal rocks, or of water-borne ALLUVIUM, or of the debris of surrounding hillsides that has been slowly worked down the slope (COLLUVIUM). In any of these ways the regolith provides parent material out of which soil is formed.

We recognize three kinds of soil horizon starting from the top: A, B, and C horizons. It is usually easy to separate a soil into these three horizons just by looking at it, although there may be many subdivisions and finer layers present. The principle of this simple classification is that:

A horizons have *lost* material from leaching, though gaining organic matter as deposit.

B horizons have *gained* material from leaching and by synthesis *in situ*.

C horizons are the *unaltered* parent material.

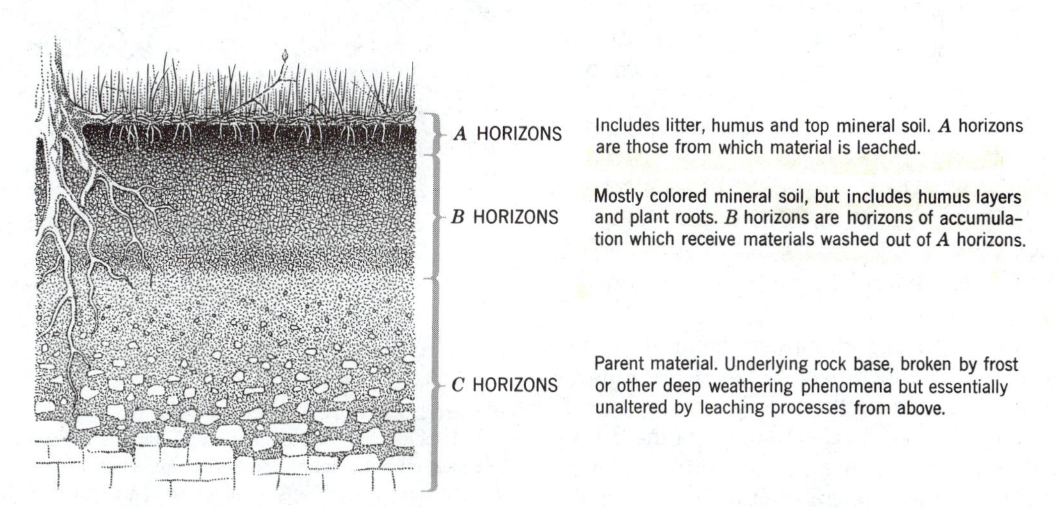

Figure 20.1 Idealized soil profile.
There may be several subhorizons in each of the main horizons and roots may penetrate them all, even deep into the parent material. There may be colored bands in each horizon, as some of the later figures in this chapter show. Finally, the boundaries may be elusive. But the functional separations are clear enough: material is removed from "A" horizons, material is added to "B" horizons, and "C" horizons are essentially unaltered by processes acting from above.

At the top of the soil profile (Figure 20.1), underneath the litter of leaves, the mineral soil is colored and structured by the organic particles mixed into it by soil animals such as earthworms, or by roots, and by various organic materials produced by decomposition and collectively called HUMIC ACIDS. The percolating water, which is a solution of various substances washed out of the litter, dissolves anything soluble in the surface of the mineral soil and carries it down to deeper horizons. Percolating water also removes finer particles from the top horizons and carries them physically downward. The horizons at the top of a soil profile that are being continually denuded in this way are called the A HORIZONS of the soil. We may, therefore, define the A HORIZONS as THE SET OF LAYERS AT THE TOP OF THE SOIL PROFILE THAT TENDS TO LOSE MATTER TO THE SOIL WATER OR TO THE LAYERS BELOW. "A" horizons are sometimes called ELUVIAL horizons.

Underneath the A horizons is a second set of horizons that has caught matter washed down from the top, such as fine particles trapped in the spaces of the deeper soil as in a filter, or solutes redeposited over the immense surfaces of that filter bed. The detailed chemistry of this process of redeposition is not always known, but the fact that it occurs is clear enough. The result is horizons in which there has been redeposition. These B HORIZONS may be defined as THE MIDDLE GROUP OF HORIZONS IN WHICH THERE IS REDEPOSITION OR ENTRAPMENT OF MATTER BROUGHT DOWN FROM ABOVE. "B" horizons may also be called ILLUVIAL horizons.

A point to notice about the chemical events leading to the enrichment of the B horizons is that these involve synthesis as well as transport and redeposition. The clay particles of the B horizons, for instance, are not just broken down fragments of rock minerals but are plate-like aluminosilicates formed *in situ*. The structures of common clay minerals are given in Figure 20.2.

The lowest C horizons are the PARENT MATERIAL or CHEMICALLY UNALTERED REGOLITH IN WHICH PHYSICAL CHANGES MADE BY SOIL-FORMING PROCESSES ARE MINIMAL. When the regolith is thin, it is, of course, possible for this to be entirely transformed into the A and B horizons of a soil, the basement stratum of which is not the C horizon but the underlying rock or even a different geologic formation. Underlying rock unrelated to the soil, but in close contact with the soil profile, is sometimes called the D HORIZON.

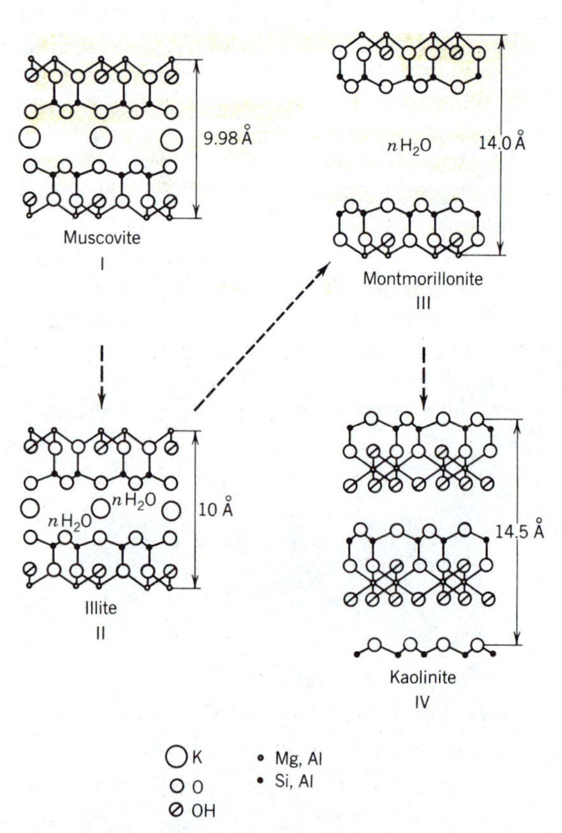

Figure 20.2 Structures of clay minerals.
The minerals are silicate lattices with various hydrations of magnesium, aluminum, and potassium. They may form in large plates of which only small sections are shown in edge view. Molecular water is held between the plates of the lattice of illite and montmorillonite, which makes those clays of prime importance in soils. The four lattices are shown in the sequence in which they can be weathered from the primary mineral muscovite (a mica). Lattice widths are given in angstroms. (From Thomas, 1974; Huckel, 1951.)

It will be evident that the thickness and complexity of a soil profile is a function of time and weather, as well as of the plants that grow there, themselves functions of time and weather. In a desert, for instance, the soil profile may be just a thin surface layer slightly added to by plants and slightly washed by infrequent rain, with underneath this the parent material. This soil might be called an A–C soil, to record the fact that there was no discernible B horizon of redeposition or clay synthesis.

In well-watered, well-vegetated places we should expect a thick B horizon where mineral synthesis is active. An array of clay minerals in this B horizon would serve to collect nutrients and regulate their supplies (Chapter 17; Figure 20.3). If the soil is in a latitude where earthworms can live, then the array of minerals synthesized in the B horizon will have been moved physically up through the soil, left on the surface in worm-casts, and churned back into the A horizons.

The soil profile, therefore, is an instant indicator of important ecosystem processes. These processes proceed with remarkable independence of the parent rock. It is true that quite different mineral substrates, say, alluvial clay as opposed to a granite regolith, do affect the soil profile

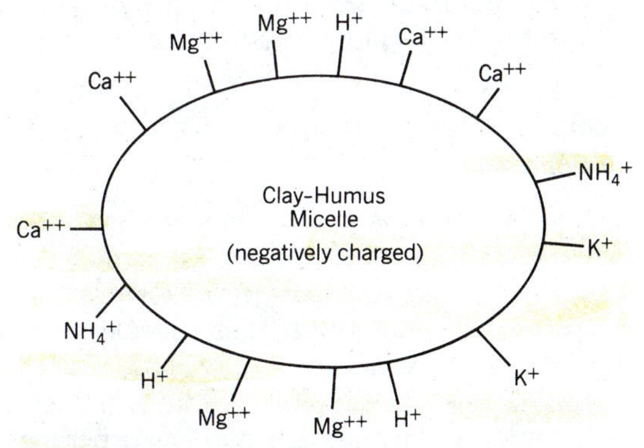

Figure 20.3 Clay–humus micelle.
Complexes of fragments of clay lattice and humic substances with net negative charges form in soil. At low pH the charges will be neutralized by protons (H^+) but these can be displaced by cations. Divalent ions like Ca^{++} and Mg^{++} bind strongly to the micelles, giving soils the power to retain nutrients.

to some extent, but the influence of mineral rock is not nearly so important as someone approaching the study of soils for the first time would expect. The soil profile peculiar to a particular climate or biome often may be expressed regardless of the type of crustal rocks.

In every soil we find traces of two quite different origins. Always there will be the imprint of geology, the traces of fabric and mineral array that are derived from a particular rock type, but these traces of the mineral origins will always be overlaid by a soil profile left in the ground by forces working from above. This profile is the true reflection of the soil, the product of organic and mineral synthesis that has very little to do with crustal rocks.

PRIMARY CLASSIFICATION: THE GREAT SOIL GROUPS

Just as the first botanists who had the means to travel the globe found that the forms of plants changed from place to place, so the first soils people who travelled with spade in hand found that the forms of soils varied from place to place. The pedologist's "formation" is called the GREAT SOIL GROUP. It happened that the first inspired digging of holes was in Russia, where the Tsarist government thought it would be worthwhile to classify soils to provide an objective basis on which the land could be taxed. The de Candolle of the soil world was V. V. Dokuchaiev (Muir, 1961; Coffey, 1912). As a result, soil names in wide use around the world are Russian names. These names are used on FAO/UNESCO soil maps and in most national soil surveys, and were also the standard in the United States until the 1960s. The U.S. Soil Conservation staff developed an alternative system in 1960, in which the *great soil group* was replaced with a somewhat different unit, the *soil order,* and the Russian names were replaced with new names invented from Greek and Roman roots (Table 20.1).

On a world scale, the great soil groups are the most easily mapped of any class in a soil classification. They are to pedology what the biome is to ecology and we can talk of the typical profile

Table 20.1
Soil Orders Recognized by U.S. Soil Survey Manual
These orders do not correspond directly with great soil groups, though there is a rough correspondence. Other orders not in the list are recognized.

Name	Derivation	Classic Great Soil Group
Entisol	Nonsense symbol	Azonal (simple profile or purely local as water-logged (gley) soils)
Inceptisol	L. *inceptum,* beginning	Brown forest soils, some poorly drained (gley) soils
Aridisol	L. *aridus,* dry	Desert soils
Mollisol	L. *mollis,* soft	Chernozems and prairie soils
Spodosol	Gk. *spodos,* wood ash	Podzols
Alfisol	Nonsense symbol	Gray-brown podzolic
Ultisol	L. *ultimus,* last	Red-yellow podzolic soils
Oxisol	F. *oxide,* oxide	Latosols and laterites
Histisol	Gk. *histos,* tissue	Bog soils

of a soil in one of the great groups as we can talk of the characteristic profile of vegetation in a biome. It is now often argued, particularly by soil surveyors in the U.S., that dividing soils into classes to correspond to the great soil groups is not a useful basis for a detailed taxonomy. This may be so, though many soil surveyors of other countries continue to argue otherwise. But the first step in understanding the natural history of soils must surely be a recognition of taxa as broad and recognizable as the biomes themselves. Biomes are of little use as units for a formal taxonomy of plants, but biomes have shown their value for the ordering of ecological knowledge. Great soil groups should be known and used in the same way that plant formations or biomes are known and used.

Color, and color-banding of the soil profile, distinguish the great soil groups in the way that plant shapes distinguish the biomes. Podzols have their bleached ashy layer under the litter surface, and the dark red and black bands of iron and humus lower down. Under the temperate deciduous forest, soils are tinted in rich browns. Deep soils under tropical rain forest are red, some of them vividly so. Soils under prairie and steppe may be the color of chocolate, though sometimes flecked or banded with white carbonates. Only color plates, or the real thing, can show properly this rich variety of color.[1] This color is spread on a biome scale under the veneer of vegetation and debris that hides the thick paste of pigments that make up the ground underfoot. The following sections describe the more important of the great soil groups. Traditional names, some of them with Russian roots, are given first for each and the U.S. Soil Order names of their nearest equivalents (Table 20.1) follow. Whatever the final global consensus on nomenclature may turn out to be, familiar terms like *podzol* or *laterite* are unlikely to go out of use.

Podzols (Russian—ash earth): Spodosol

In a well-chosen site under boreal forest the soil profile is richly banded. At the top is an edgeways view of the fragrant carpet of brown needles typical of the boreal forest. These needles retain their shape. They are the A_{01} horizon (organic horizon of litter that retains its form). Immediately under the needles is a thin layer, perhaps 1 cm thick, of black, slimy, formless humus called

[1]A vivid impression of the diversity of soil profiles is given in Kubiena's *Soils of Europe* (1953), where more than fifty sensitive watercolor paintings of soil profiles are reproduced.

MOR, the A_{02} horizon. Then comes a horizon of mineral soil stained dark with humus (A_1 horizon) and under this a striking pale gray or even white horizon that has evidently been bleached, the A_2 horizon. It is this bleached horizon that gives a podzol its name because ploughed podzols look as if they have had ashes scattered over them.

The B horizons of podzols are nearly as spectacular as the A horizons, sometimes including an "iron pan" layer stained red with ferric oxide and another layer stained bluish-black with organic matter. Under this chromatic array will be the C horizon, perhaps with a quite different color given by minerals of the regolith and differentiated from the A and B horizons by the processes of bleaching and chemical redeposition. People who have never seen a podzol are urged to dig a hole the first time they get into the boreal forest; it is aesthetically satisfying.

No other soils are as prettily banded as podzols, which fact has resulted in hypotheses to explain the banding being generally included in the podzol description. The standard hypotheses rely on observations of the soil fauna. There are no earthworms to be found in podzols; earthworms do much digging; if the podzol had been dug-over the pretty horizons would all have been mixed up to form the hue of mud; therefore the presence of the pretty bands reflects the absence of earthworms. A bit circular but probably true. The ecological question of why there are no earthworms in podzols is, perhaps, not so easily answered.

Podzols represent the result of soil-forming processes in temperate or cool regions in their most extreme expression (Buckman and Brady, 1969; Russell, 1953; Kubiena, 1953; Rankama and Sahama, 1950). The principal observations are:

(a) There is an accumulation of organic matter at the surface.

(b) The percolating soil water has to pass through a layer of decomposing organic matter where it acquires many humic solutes and acquires a low pH (typically pH 4).

(c) An upper horizon of mineral soil is partly bleached. The pale color is found to be due to the fact that brightly colored iron and aluminum oxides have been removed, leaving the horizon enriched with silicates.

(d) Iron and aluminum oxides and hydrates (the so-called sesquioxides) together with organic colloids are collected in the B horizons. This enrichment of iron and aluminum has apparently resulted from a process of transport and redeposition.

Tundra Soils: Included with Entisol, No Direct Equivalent

The profile of a tundra soil is given in Figure 20.4. The regolith is permanently frozen (PERMAFROST) and the complete soil profile is very wet for most of the year and is completely frozen for many months each winter. There are no earthworms. Organic matter accumulates often under the tussock life forms of tundra sedges and grasses. A thick organic cover insulates the frozen mineral soil from the summer sun so that a long-established soil may not thaw down to more than a few centimeters within the mineral parent

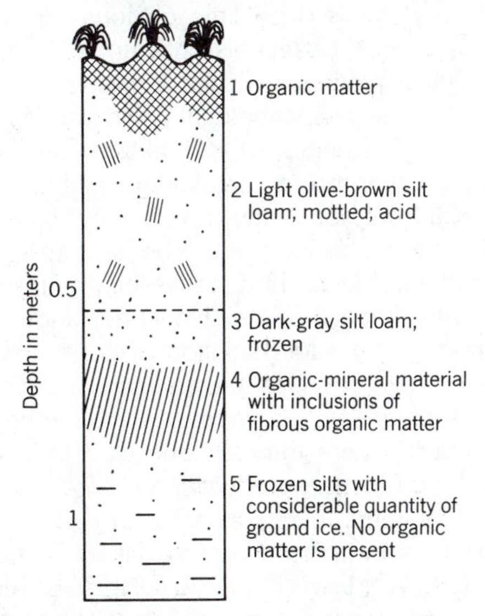

Figure 20.4 Profile of tundra soil.
Diagram of typical tundra soils of northern Alaska. Surface material is buried by movements of the freeze–thaw cycle more than by leaching in solution. (From Tedrow, 1965.)

material. In these circumstances it is not surprising to find that soil horizons are poorly developed. Possibly one way of looking at a tundra soil is to think of it as an incipient podzol prevented from developing by ice.

On arctic sites so well-drained that permafrost is absent (gravel beach ridges, *et cetera*) podzolic soils like the brown arctic soil (Tedrow and Hill, 1955) develop, but these are local peculiarities.

Gray-Brown Podzolic Soil: Alfisol

These are the soils that underlie the brown fields of classical agriculture in Western Europe and New England, land that was originally forested with hardwoods. At the top is mulchy leaf litter, rotting underneath into humus that is intimately mixed with the surface mineral soil and called MULL. Typically *mull* is less acidic than *mor,* though, as in all things to do with soils, an infinite series of in-between types exists. The mixture of organic matter with mineral soil spans perhaps 10 or 20 centimeters of dark, fertile-looking earth, quite like the surface of a ploughed field or a garden. Earthworms, which live on the nearly neutral leaf litter, have done the digging and mixing. There is often little obvious sign of a bleached layer under this dark upper horizon, although a reddening of the B horizon farther down tells of materials brought down from above. The colored minerals of the B horizon have been washed down in the percolating water and redeposited but not nearly to the same extent as under the needle litter of the boreal forest, and this more gentle leaching leaves the profile looking unlike that of a podzol. Under deciduous forest there is a gentle gradation of horizons with subtle changes of hues: brown litter, a dark mixture of mineral soil and humus, the deep brown B horizon, where minerals and clay have collected, and thence, gradually, the parent material. In its finest development, as in parts of Europe where it was first described, this soil is known as a BROWN EARTH or BRAUNERDE. ALFISOLS conform to the four points of observation listed for the true podzols (SPODOSOLS) above. It is tempting to suggest that they represent what would happen when diggers (earthworms) work over a podzol as it is forming so that the layers are blended with one another.

Tropical Lateritic Soil: Oxisol

Laterites are the thick, red, and heavily weathered and altered strata of the tropical ground. Theirs is the red color of much of the wet tropics. They are defined by mineralogy and gross structure, particularly in that they have been enriched with oxides and hydrates of iron and aluminum left behind and synthesized while other elements of the original rock are leached away.

The actual chemistry and mineralogy of laterites are extremely complex and not completely known (Russel, 1953; Mohr and van Baren, 1954; Thomas, 1974). They may have had a geological history of intermittent formation and erosion on ancient peneplains,[2] though some may be much younger, and it is not uncommon to find them to be 10 to 25 m thick. One kind can be dug wet, shaped, and allowed to dry into excellent durable bricks in the sun; hence the name *laterite* (*later* means "brick" in Latin).

Laterites, because of their age, thickness, and the variety of processes that go into their formation, can be said to be part of the geology of an area. But plants grow in them and they weather from the top; so they are also soils, hence LATERITIC SOILS. The U.S. term OXISOL clearly separates the soil interest from the geological. The profile, for as deep as an ecologist is likely to dig, consists of a few centimeters of leaf litter (A_0), an A horizon identified mostly by a slight browning of humus, and the red array of weathered residue consisting of the synthesized iron and alumina complex. It is difficult to decide where the A and B horizons divide.

Important aspects of an oxisol for an ecologist to notice are that it is extremely poorly supplied with plant nutrients, it has very little organic matter either on or in it, and it is red. The red color itself reflects the fact that iron and aluminum have been preserved at the expense of white and gray

[2]A *peneplain* (Latin for *almost* plain) is a dissected but flat countryside, the result of prolonged erosion by rivers that has removed high relief.

silicates, apparently the opposite of the process that leaves podzolic soils both gray and silicate-rich.

Chernozems (Russian for Black Earth), Prairie Soils (Both Mollisols), and Desert Soils (Aridisols)

Under grasslands, such as the Russian steppe or the American prairie, is a more striking soil. Here rainfall is limited, and there is rarely much water to percolate through the soil. Despite the low rainfall, there may yet be heavy showers, providing enough water at one time to penetrate to considerable depths before evaporation at the surface draws it back. A profile forms within the depth reached by the percolating water, but with some peculiar characteristics. Dead grass parts decompose slowly so that a thick, largely organic layer builds up at the top of the soil, forming the famous black earth of the wheat lands. This black peaty layer grades gently into mineral soil below and is mixed with it by burrowing animals, although this mixing is not done so thoroughly as in the wetter soils under deciduous forest. The result is a deep black soil, turning gray towards the bottom, although still dark, and then an abrupt termination when the line of lowest water penetration is reached. Here at the "water-line" there may be a mineral band, often white with carbonates, where the drying water has left its load of dissolved matter. Below this, and sharply distinct, is the parent material.

A PRAIRIE SOIL is the more organic result of this soil process. When calcium carbonate deposits are clearly visible so that a separate horizon (C_{Ca}) is visible, the profile conforms to the original Russian description for a CHERNOZEM. Versions of this soil in the American Southwest are called CALICHE, after the local name for the thick carbonate deposits in them.

Where rainfall is very slight (hot deserts) this process is so attenuated that a B horizon of redeposition can scarcely be recognized and a simple DESERT SOIL having only A and C horizons results.

All three of these soils of dry places have in common that they are not washed completely through by percolating water. There must obviously be places at the edge of the range of these great soil groups where the rainfall is such that sometimes the soil is flushed through whereas at other times it isn't so that a solid distinction between the two types of soil is not possible. But, for convenience, pedologists talk of PEDALFERS, meaning soils that are flushed through, and PEDOCALS, meaning soils of dry (usually hot) places where the soil water is pulled back from above by evaporation. This separation is used as the basis of some soil classifications and is a consideration in most.

Tropical Black Soil or Margalitic Soil (Included in Mollisols)

In parts of the tropics the ground surface appears black instead of red. These are fertile soils, and unlike those on laterites they are base-rich and contain clay minerals of a type (montmorillonite) associated with the weathering of basic materials. They are often referred to simply as "black soils." The name "Margalitic soil" is constructed from the Latin *marga,* meaning "marl," which is a mixture of lime and clay used since ancient times as fertilizer. There seem to be many varieties of process leading to the black end-point, but all seem to occur on young, base-rich rocks (often fresh volcanic ash or lava) and in the less humid parts of the tropics (Mohr and van Baren, 1954; Eyre, 1968; Walter, 1973).

Table 20.2 describes the profile of one black soil from India. There is a striking uniformity of the appearance of all horizons in this soil, a uniformity that was found to extend to the chemical constituents in analyses of samples from six depths. This pattern suggested that it was unsafe to interpret the horizons into the A–B–C scheme and the investigator instead numbered them. It is likely that the black color of these soils is due to fine coatings of almost insoluble, unusual compounds of carbon and nitrogen, although the organic content of the soils is still very low by temperate standards (Mohr and van Baren, 1954).

It is particularly noticeable that the lowest horizon is calcareous and, in some profiles, there may actually be carbonate concretions. This,

Table 20.2
Tropical Black (Margalitic) Soil Profile
Soil is from near Bombay, India. A pronounced dry season results in a black soil well supplied with nutrients. (From Mohr and van Baren, 1954.)

Depth of Horizon in cm	0–15	15–30	30–45	45–60	60–75
Clay in %	55.75	57.50	56.75	54.50	56.00
Silt in %	15.75	16.00	14.50	11.75	20.50
$CaCO_3$ in %	9.42	11.68	8.86	7.90	6.15
pH–H_2O	8.94	8.81	9.01	9.01	9.01
Humus	1.32	1.26	1.22	1.21	1.19
Adsorption Capacity in m.e./100 g soil	65.25	65.91	63.18	53.07	66.29
Ca in % of A.C.	80	75	77	74	67
Mg in % of A.C.	14	18	18	20	22
K in % of A.C.	4	4	2	3	2
Na in % of A.C.	2	3	3	4	9

coupled with the fact that they often occur in places with both wet and dry seasons, lets them be classified as *pedocals*. Many of the productive soils of India are of this kind.

ON THE CORRESPONDENCE OF MAPS OF SOIL PROFILE, VEGETATION AND CLIMATE

The kind of map of "soils of the world" that appears in every school or family atlas is a rough plot of the great soil groups (Figure 20.5). These maps have a very strong similarity to maps of climate or vegetation in the same atlas (Figures 14.1 and 14.3). This need cause no surprise—if what is mapped is the prevailing soil profile and if the soil profile results from weather and climate. Vegetation, climate, and soil; the three are logically linked.

And yet there is room for a word of caution because the soil map in the atlas has been drawn from a vegetation map. It was noted in Chapter 14 that early climate maps were made by mapping plants and calling the result a climate map. Likewise cartographers map plants and call them soil. In truth, there is scarcely any other way of making a large-scale soil map than this because a soil profile can only be seen after digging. Once the soil profile of a region is determined from sample diggings, the plants must be used as indexes of the territory covered by that kind of profile. Even on a small-scale this is necessary, and soil surveyors draw their maps from air photographs, which typically show vegetation, not soil.

Early maps of world soil were not so much sets of data as statements of a hypothesis: the hypothesis that vegetation and climate determine the soil of any particular place. This hypothesis is illustrated by a much-copied chart (Figure 20.6) of the geographers Blumenstock and Thornthwaite (1941). It shows the earth nicely parcelled out into regimens or systems. If the plant formations are imagined as set out between mappable frontiers in the nation-state model, the world is set into neat system-cells. It was an idea much in the minds of ecologists when the concept of the ecosystem was invented. It is, however, a poet's view of the earth, holding a core of fundamental truth but overlooking many a contradictory detail.

Mapping world vegetation as a guide to mapping soils probably is less satisfactory than as a guide to climate. A true map of even the most distinctive of the great soil groups would not match boundaries with plants nearly so well as does climate. Perhaps the most distinctive of all great

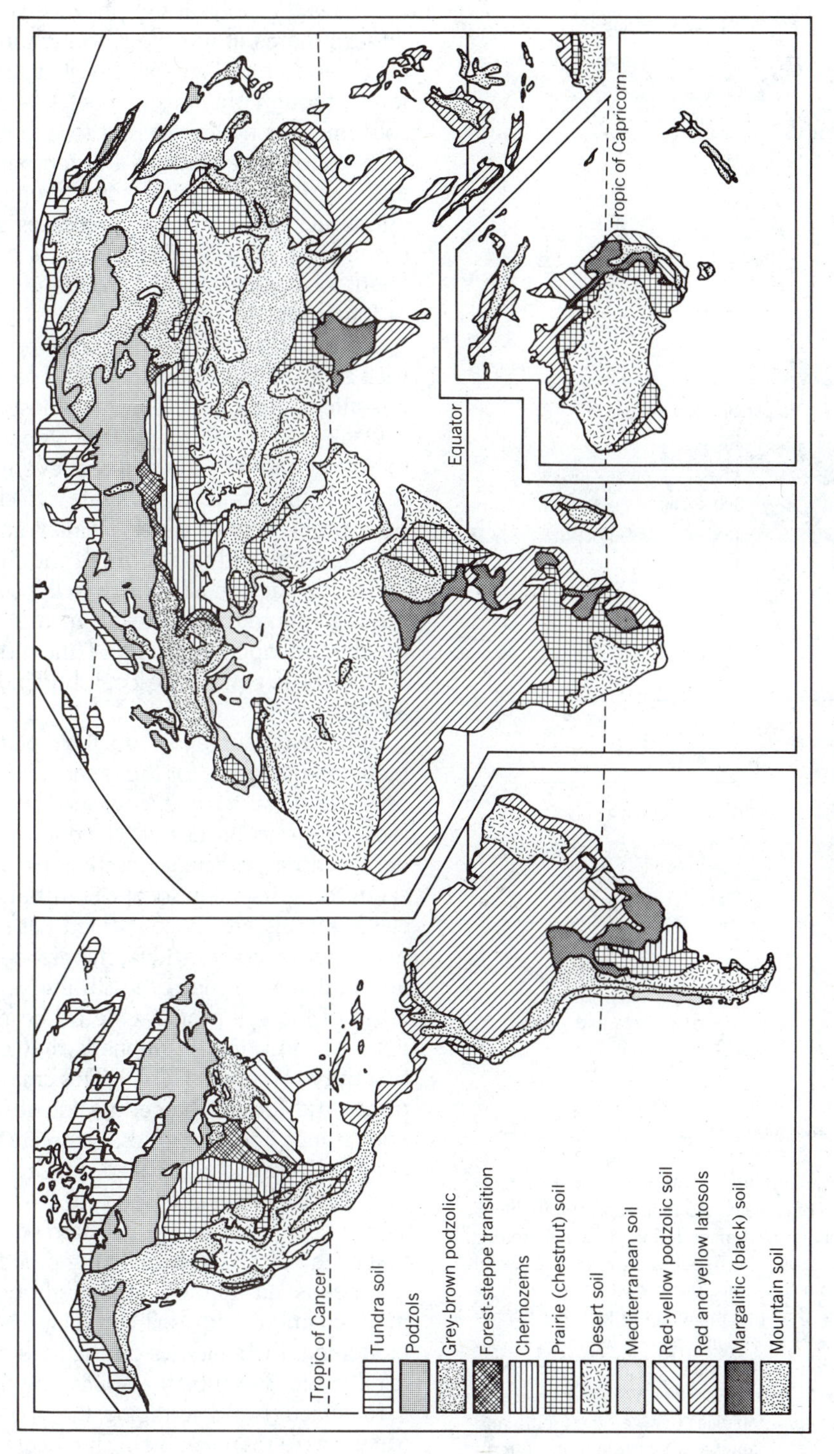

Figure 20.5 World soil map. (Based on Bridges, 1978.)

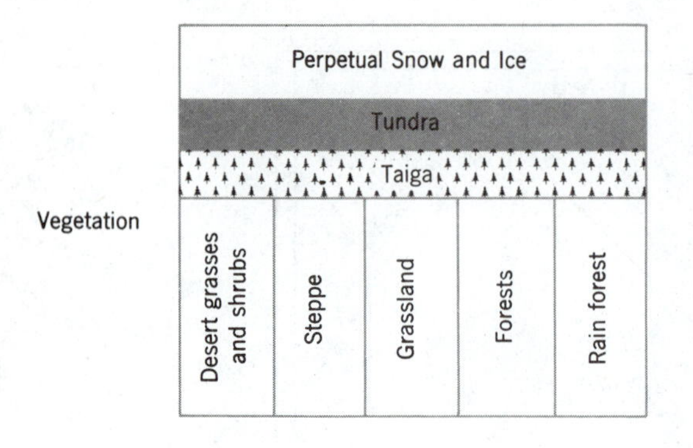

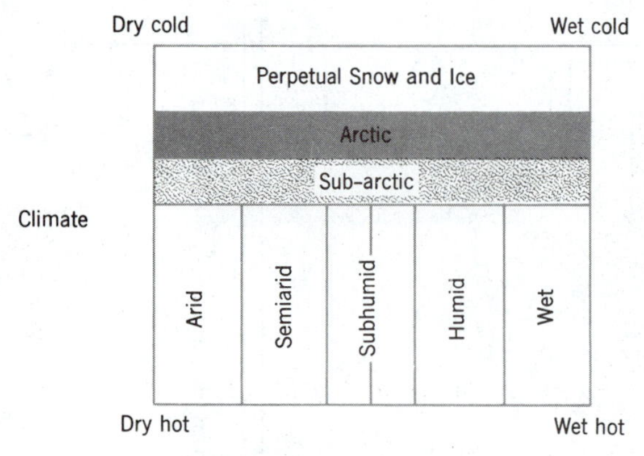

Perpetual Snow and Ice
Tundra Soils
Podzols
Soils
Desert soils
Chestnut and brown soils
Chernozems and prairie soils
Podzols
Gray-brown podzolic soils
Red and yellow podzolic soils
Latertic soils

Figure 20.6 Conventional climate, vegetation, and soil diagrams.
These diagrams summarize the hypothesis that mapping formations yields maps of climate, and that maps of formations and climates combined yield maps of the great soil groups. As a grand generalization the hypothesis provides a useful way of thinking about the earth. But it is well that we draw the earth as square, thus accentuating the fact that we are being idealistic. Few of the boundaries shown in these diagrams can actually be found on the ground. (*Taiga,* which appears in the top diagram, describes the belt of more or less open forest that commonly lies between the tundra and the closed boreal forest; many would include it in the term "boreal forest" when speaking in very general terms.)

groups is the podzol: yet podzols can be found under tundras, if the site is well-drained (Figure 20.7), and under heaths as well as under boreal forest, though only in small patches. And "podzols" merge on all sides into soils that are "podzolic" and these span all the temperate forests and even some tropical soils. Tropical soil scientists talk of "podzolic" processes in rain forests. A true map of all the podzols of the world might look quite different from any known map of formations or biomes.

Local soils can be influenced powerfully by rock type or the lie of the land, particularly as this affects drainage. On a Javanese mountain, a black soil can form on one slope and a red oxisol on another, the different events being directed by the strike of the strata, which alters the pace of the flow of water. Rainfall and heat are the same on both sides of the mountain divide and the vegetation all belongs to the same biome, but the running of the water results in different profiles on opposite sides of the ridge (Figure 20.8). These patterns of local shuffling of profiles are common.

When soil surveyors do their own mapping they have little use for the great soil groups. Not only are these terms too vague, but their scale is too large to be useful. National Soil Surveys derive more particular descriptions of soils that seem to be useful in local circumstances. These classifications are as concerned with fine structure, texture, color, acidity, mineral composition, and drainage as they are with the genetic properties of the soil profile. One of the most ambitious of these classifications is that of the U.S. Soil Survey Manual (1975). This creates an elaborate, and formal, procedure for describing soil appearance and properties, complete with a very complex terminology based on Greek and Roman roots (Tables 20.1 and 20.3).

But even when using a system so complex as that of the U.S. Soil Survey the practical worker still needs air photographs. Either plants are mapped and called "soil" or, if land is bare from ploughing, the shadows on the land are used to tell the tale. Describing a soil is one thing (often very difficult) and mapping the soil is another (often more manageable). The local units of soil mapping are empirical. We talk of "soil series"

Figure 20.7 Podzol under tundra.
Soil trench on a gravel beach ridge overlooking Imuruk Lake, Seward Peninsula, Alaska. The site is well beyond the tree line and the bottom of the soil pit is ice (permafrost). But the gravel lets the ice melt deeply enough in summer for water to move sufficiently to leach the surface. The whitish "A_2" horizon of a podzol can be seen clearly just under the organic mat. The plants are mostly cotton grass (*Eriophorum vaginatum*) and the dwarf birch (*Betula nana*).

Figure 20.8 Influence of slope on tropical soil formation. On north and south faces of a Javanese ridge different soils form, black (morgalitic) on one side and red (lateritic) on the other. Rainfall appears to be the same on both sides but the inclined strata of the underlying rock make the pattern of drainage different on the two sides. (From Mohr and van Baren, 1954.)

Table 20.3
Subdivisions of an Order by U.S. Soil Survey Manual

Notice that every subdivision of Alfisol *contains the suffix* -alf. *Notice also that both climate and local physiographic circumstances (water-logging) can be used to erect a subdivision. The possible permutations of a complete classification are therefore very large.*

Order Alfisol (= podzol)
Aqualf = with gleying (water-logging)
Boralf = in cold climate
Udalf = in humid climate
Ustalf = in subhumid climate
Xeralf = in subarid climate

or "catenas" to describe the purely local result as the genetic processes leading to a soil profile interact with local bedrock and drainage to produce the soil with which a real ecosystem must cope.

SOIL SERIES, CATENA, AND PEDON

Practical soil surveyors map on the scale of farm fields or parts of counties. Usually only one of the great soil groups of the zonal classification will be present because all the soils of an area small enough to be mapped on this scale are likely to have had similar histories under similar climates. Yet the units "SOIL SERIES" that appear in a soil survey are real units. They can be mapped with reasonable precision and are usefully different. They are the expression in soil of patterns in the underlying rock.

A SOIL SERIES is a GROUP OF SOILS DEVELOPED FROM THE SAME KIND OF PARENT MATERIAL, BY THE SAME GENETIC COMBINATION OF PROCESSES, AND WHOSE HORIZONS ARE QUITE SIMILAR IN THEIR ARRANGEMENT AND GENERAL CHARACTERISTICS (Buckman and Brady, 1969).

Practical soil surveyors are soil taxonomists. When they describe their *soil series* they look at soil and handle it as a biologist examines a specimen. An auger brings up samples from the different horizons, a rubbing between the fingers gives the "texture" (proportions of sand, silt, and clay), a glance tells the color; the like has been seen before, it is given a *series* name. This works because, naturally enough, outcrops of the same geological formation in the same neighborhood yield the same physical and chemical structures in the soil.

When soil taxonomy moves from classifying *great soil groups* to describing *soil series* it has taken a qualitative as well as a quantitative change. The two classifications are of different kinds; the one of a process that stamps the pattern of climate on the land whatever the underlying rock may be, and the other the expression of rock or topography triumphing over climate.

An alternative unit of soil taxonomy revealed by the mapping process is the CATENA. This unit was developed from the observation that local topography can alter many fundamental qualities of the soil, even though the soil profile is built from the same regolith from the same parent rock. In particular, poorly drained bottom land will develop a *soil series* that is distinctly different from the *soil series* developed upslope on the same outcrop.

The concept of the CATENA of related soil series was developed as a practical mapping unit to be used on the ancient peneplains of East Africa, where undulating topography repeatedly rang the changes of a sequence of a few *soil series* (Milne, 1935a, 1935b), but it has been found to have very general utility. Where catenas are recognized, the set of associated soil series of each will blend together, as the drainage blends down a slope from good to bad.

It is obvious that fine details of soils will vary within any soil or catena; these units, therefore, will always be somewhat arbitrary. We rely on the training of the soil surveyors to yield useful assignments of a soil to one series or another. This leaves to an individual worker the task of deciding what to describe. A soil surveyor is forced to share with plant community analysts the difficulty of deciding which is the best piece of landscape worthy of detailed description. "Where do you dig your hole?" and "How large should the hole be?" are questions akin to "Where do we place our vegetation plot and how large an area should it cover?" (see Chapter 15). The compilers of the U.S. Soil Survey Manual make a useful contribution towards the resolution of this difficulty with their concept of the PEDON.

The *pedon* is a sample piece of the local soil. The soil surveyor decides what a typical specimen of the soil being mapped is, then describes this typical specimen. The description must include the soil profile, plus a number of physical and chemical measurements on all the horizons, so that the actual sample of a type soil has to be three dimensional: a volume of ground rather than an area. The surface area of this volume must be large enough to include the minor irregularities that occur in all soils. The procedure is entirely empirical and based on the judgment of the soil surveyor. Practical experience has led to the following definition of the resulting unit: A PEDON IS A SMALL VOLUME OF SOIL LARGE ENOUGH FOR THE STUDY OF HORIZONS AND THEIR INTERRELATIONSHIPS WITHIN THE PROFILE AND HAVING A ROUGHLY CIRCULAR LATERAL CROSS SECTION OF BETWEEN 1 m AND 10 m (Simonson, 1962; Anon., 1960).

We get from the "pedon" to the "soil series" by means of the POLYPEDON defined as A GROUP OF PEDONS CONTIGUOUS WITHIN THE SOIL CONTINUUM AND HAVING A RANGE IN CHARACTERISTICS WITHIN THE LIMITS OF A SINGLE SOIL SERIES.

Ecologists will sympathize with the difficulty of defining these various units when they reflect on their own difficulties of defining units in the endless continua of plant communities (the units *association, sociation,* and *stand* come to mind, see Chapter 15). When ecologists publish data on a habitat, it should be standard practice to name the soil series that sets its limits. If their study has any business with the ground underfoot, they would do well to describe the soil profile in a well-chosen "pedon" as well.

PROBLEMS OF SOIL TAXONOMY

Whenever a scholar classifies, whatever the subject of interest, the aim is to group the similar into categories; and then to group the categories into higher categories so that a hierarchical order results. In biology, of course, this exercise led to major discoveries, the processes of evolution and natural selection among them. But efforts at the taxonomy of soils have not resulted in any comparable discoveries.

The most ambitious attempt to classify soils is that of the Comprehensive System of the U.S. Soil Survey. This system was devised with the deliberate intention of finding natural order in the soils of the world (Anon., 1960; Simonson, 1962). In a passage written in an apparent spirit of advocacy in a standard soil text (Buckman and Brady, 1969) we find this system compared directly with the Linnaean system as in Table 20.4. The table compares the formal classification of the Miami silt loam, a local soil with a podzolic profile, to the standard classification of sweet clover. Reference to Table 20.3 helps clarify the soil taxonomy. We observe the profile of a podzol *(alfisol),* and it has the deep profile that we expect in the local humid climate, making it an *udalf.* Next there is the prefix *"hapl,"* which separates a *hapludalf* from an *udalf,* and this prefix simply means that the diagnostic horizon is

Table 20.4
Linnaean and Soil Taxonomic Systems Compared
The soil taxonomy is that of the comprehensive system of the U.S. Soil Survey. This system is NOT comparable to the Linnaean taxonomy of species because different kinds of criteria are used at different stages in the hierarchy.

Plant Classification	Soil Classification	
Division: Pterophyta	Order: Alfisol	Climate
Class: Angiospermae	Suborder: Udalf	
Subclass: Dicotyledoneae	Great Group: Hapludalf	
Order: Rosales	Subgroup: Typic Hapludalf	
Family: Leguminosae	Family: Fine loamy mixed mesic	Parent material
Genus: *Trifolium*	Series: Miami	
Species: *repens*	Type: Miami silt loam	

minimal. So we have the sort of profile that the podzolic process yields in a wet place: podzolic, but without the beautiful array of horizons expected in the cooler clime of a spruce forest. The classification so far speaks to the effects of climate on the surface regolith, but we have not gotten to the local soil, the *soil series,* yet. When we do we must abandon classifying the effects of climate to a large extent and classify by texture, as in the lower half of Table 20.4.

Different sets of criteria must be used by a soil taxonomist to identify the soil series of a neighborhood (texture, mineralogy, the effects of drainage) from those used to identify the effects of climate on a soil profile. Ecologists will notice how this property of a classification of world soils is similar to properties of a classification of world vegetation. Plant *formations* reflect climate, and can be classified by the shapes that climate decrees, just as soil *orders* or *great groups* can be classified by the results that climate produces. But quite different criteria are needed to order individuals and species of plants, just as special criteria are needed to order individual local soils. Soil classifications are like ecological classifications rather than true Linnaean taxonomy (Figure 20.9).

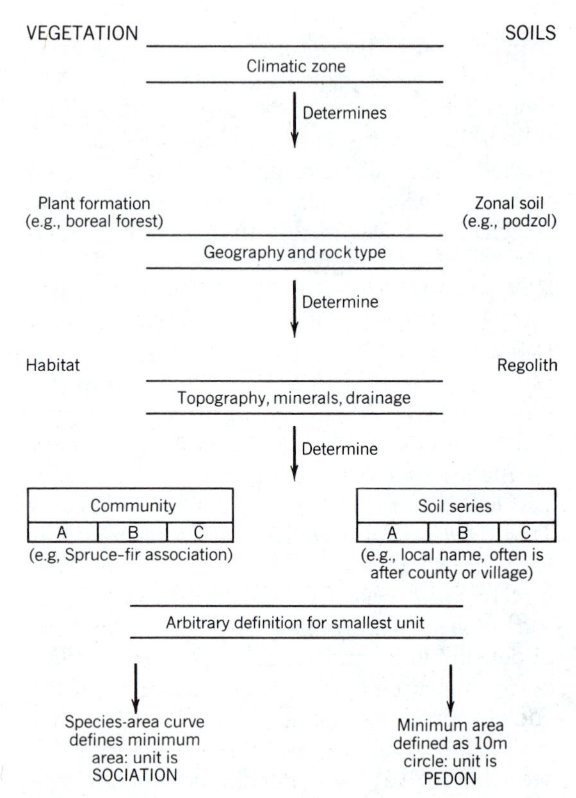

Figure 20.9 Soil and vegetation taxonomy compared. Making a hierarchical classification of soils involves decisions comparable to making a classification of plant communities.

The similarity of the classification of the Miami soil series in Table 20.4 to the taxonomy of sweet clover is superficial. All the taxa of higher order than species in the Linnaean system have the logical support that they reflect the workings of the same processes that separate the species themselves. Genera, families, and orders are somewhat arbitrary groups, but all can be defended as reflecting ancient splits in species populations comparable to the splits that make modern species. Species, families, and orders all originate in acts of speciation. The *pedologist's* families, groups, and orders have no like relationship to their *soil series.*

COLORS IN A LANDSCAPE: SIMPLE RESULT OF COMPLEX PROCESS

The color of a soil is one of the most notable properties of a landscape. Whether we live in places of red dust, brown, or gray depends on the zonal soil of our region. There are local exceptions where the color of peculiar rock is stamped on a countryside, as in the ancient county of Rutland *(redland)* in England, but usually the color of mud and dust is the color of a regional soil.

It is at least of aesthetic interest to know why soil formation over a whole climatic region leads to a characteristic color on the ground, but soil processes are immensely complicated and it is not easy to give satisfying answers. This can be seen in the different chemistries of laterites and podzols; what I like to call the problem of the red and the gray.

The north is gray or brown because the soils are podzols (spodosols); the tropics are red because soils are laterites *(oxisols)*. The gray of a podzol is the gray of silica minerals that have been washed clean of iron and aluminum compounds. We may say, therefore, that north temperate lands typical of the European West are tinted towards the gray or gray-brown because washed silicates collect at the surface of the ground. Extra pigment is supplied to this silicate paint by organic matter and, of course, colors brought up from below by ploughs.

The wet tropics are red because silicate minerals, far from being conserved in the soil, are washed away. Oxisols are made of aluminum and iron oxides or complexes of the two. These minerals are usually referred to as *sesquioxides* in conversations between pedologists, being typically of the forms Al_2O_3 and Fe_2O_3. Their color is red-brown or actually red. Where the base pigment of the north is grayish silicate, the base pigment of the tropics is red sesquioxide. And since humus is removed from tropical soils by voracious decomposers, there is little organic pigment to blend into the base color. So tropical lands are colored red.

A simple answer to why temperate and tropical lands are of different colors, therefore, is that the local soils collect different minerals. But then we must ask why these different minerals collect. The question actually resolves into "Why are sesquioxides removed from the surface in cold regions of podzols whereas silicates are removed from warm regions of oxisols?"

Attempts have been made to suggest that soil pH decides whether silica or sesquioxides are removed. The appeal of pH is that silicates are soluble at high pH (pH 8+) but insoluble at low pH. Podzolic soils tend to have lower pH than oxisols (pH 3 to 5 versus pH 5 to 6), the difference being largely due to the accumulation of organic acids in podzols, together with carbonic acid of soil respiration trapped in the surface layers. But in fact the pH of both soil types remains in the range in which silica is largely insoluble making the pH hypothesis less than convincing. A more likely explanation is that sesquioxides are removed from the surface horizons of podzolic soils in organic complexes. The absence of accumulations of humic substances in the tropical oxisols prevents this removal, allowing sesquioxides to accumulate.[3]

If the organic content of the soil is indeed the arbiter of sesquioxide transport, and hence of regional soil color, then ecologists have the satisfaction of knowing that the very color of the ground underfoot depends on living processes in ecosystems.

---

[3]This explanation was suggested by F. Ugolini.

PREVIEW TO CHAPTER 21

Lakes are particularly useful subjects for ecological study because they house communities that are unusually well defined in space, and whose fortunes are strongly influenced by fairly easily understood physical events. Most lake basins are made by glaciers, volcanism, solution of limestones, or by rivers. Lakes become stratified with an upper warm layer, the *epilimnion*, floating over a bottom layer of cold water, the *hypolimnion*. In lakes of low productivity (*oligotrophic lakes*) this floating of warm water over cold has few dramatic consequences, but if a lake is very fertile (*eutrophic lake*) the concentration of plant life in the warm surface waters may be so great as to form an effective canopy shutting out light from bottom waters. The cold, dark bottom water is then cut off from the oxygen of the air above and denied the oxygen resource that would have been provided by photosynthesizing plants. A rain of dead matter comes down from the life zone above and decomposes. Biological oxygen demand (BOD) may use up the oxygen reserve making the bottom waters anoxic. The epilimnion of a stratified lake is mixed by the wind, which generates a system of currents. Prominent among current systems are *Langmuir circulations*, which result as surface movements drag water up from below, a process that results in a series of helical circulation cells. Stable systems of stratification also occur when the bottom water contains sufficient dissolved ions for it to be significantly denser than the surface water. These lakes are called *meromictic*. Solar ponds, used as energy collecting devices, are artificial meromictic lakes. The nutrient status of a lake is controlled largely by inputs from the lake watershed, principally because lakes do not have many long-lived plants to regulate the nutrient stocks as do terrestrial ecosystems. This sensitivity to watershed input is the reason that lakes change their state so rapidly in response to pollution. As lakes fill with mud the volume of the hypolimnion is reduced, requiring that the oxygen reserves of the deep water are reduced also. This observation led to the spurious claim that polluting a lake is actually artificial aging. The drifting algae of lakes (*phytoplankton*) belong to a number of different divisions of the plant kingdom, having fundamentally different systems of biochemistry. Drifting animals (*zooplankton*) are taxonomically less diverse, coming from few taxa. Crustaceans and insects, together with rotifers in the phylum Aschelminthes, are the most prominent animals. Concentrations of nutrients in lake waters are generally very low. Carbon is almost always not limiting, and phosphorus is generally so scarce that enrichment always produces more algal growth. Many of the properties of lake life result from the fact that plants of the open water are microscopic. No general theory to explain this small size exists but the size may be an adaptation to a drifting habitat. Most algae are denser than water and sink slowly according to Stoke's law. This slow sinking may have selective advantage in that it provides cheap movement through the water for the collection of nutrients. The algae and smaller zooplankton generate laminar flow as they move through water, their size and velocity being such as to yield *low Reynolds numbers*. Movement under these conditions constrains the way in which herbivores can feed, and results in systems of filter feeding. A major theoretical question posed by planktonic life is the coexistence of numerous species in a place where physical separation of niches seems difficult (*the paradox of the plankton*). Another planktonic habit of wide theoretical interest is *cyclomorphosis*, the change of shape of plankters with seasonal changes in water temperature. These changes of shape seem best to be explained as providing growth in ways that hide the increased bulk with transparent structures so that the growing animal is overlooked by fish that hunt by sight. Avoidance of predation by fish is also the most prominent of a series of explanations for vertical diurnal migrations of the plankton, since these journeys to deep water by day put the animals in the dark where hunting fish cannot find them.

A general thrust of recent studies of the biota of lakes has been that predation, both by fish and by invertebrate predators, is of very major importance in establishing the structure of lake communities.

CHAPTER 21

LAKE ECOSYSTEMS

The study of lakes, LIMNOLOGY,[1] has had a special importance to ecologists from early in the development of our subject. Lakes are convenient ecosystems in that they are bounded by distinct edges. Inputs and outputs are clearly defined (Figure 21.1). The primary producers are tiny phytoplankton sifting through a void of water. This is one system where animals are not hidden by the forest, but can be caught at will with nets of appropriate sizes. Lifetimes are short, making experiments spanning several generations possible. Nutrients are held in solution, where they are measured easily. Even past events can be studied from fossil traces in the mud.

But life in lakes is constrained by physical circumstances. Water has curious thermal properties that impose seasonal changes on the habitat, stratifying the habitat into discrete subhabitats. And water has one almost bewildering property in that it presents obstacles to movement of the very small that can best be likened to the drag of molasses on a human-sized swimmer. This property comes from life at low Reynolds number. Understanding life in lakes thus has to follow mastery of the special, even slightly bizarre, physical properties of water filling a hole in the ground under the sun.

[1]LIMNOLOGY comes from the Greek *limnos*, "a lake," and *logos*, and means literally "the science of lakes."

ORIGIN OF LAKES

A lake is a hole in the ground filled with water. Water is common enough outside deserts, and the distribution of lakes is largely set by the distribution of suitable holes. But holes are not often made on the surface of the earth; most geological processes work to fill holes in. Even a lake itself may be viewed as a process tending to fill holes with sediment, meaning that lakes are not only unusual objects on the earth but also that they do not last long. With a very few special exceptions, lakes tend to be young in geological or evolutionary terms.

Most lake basins have been made by glaciers, volcanos, or wandering rivers. Less common are lakes occupying solution basins. Less common still are lake basins made by crustal movements, wind, and even meteorites.

Lakes of Glacial Origin

Most characteristic are KETTLE LAKES, which occupy holes left when large blocks of glacial ice melted out among the debris after the ice sheets retreated (Figure 21.2). Kettles tend to be deep, generally holding more than 10 m of water over 10 m of sediment. The thousands of lakes in the U.S. Midwest are mostly kettles, or similar depressions in the old glacial moraines.

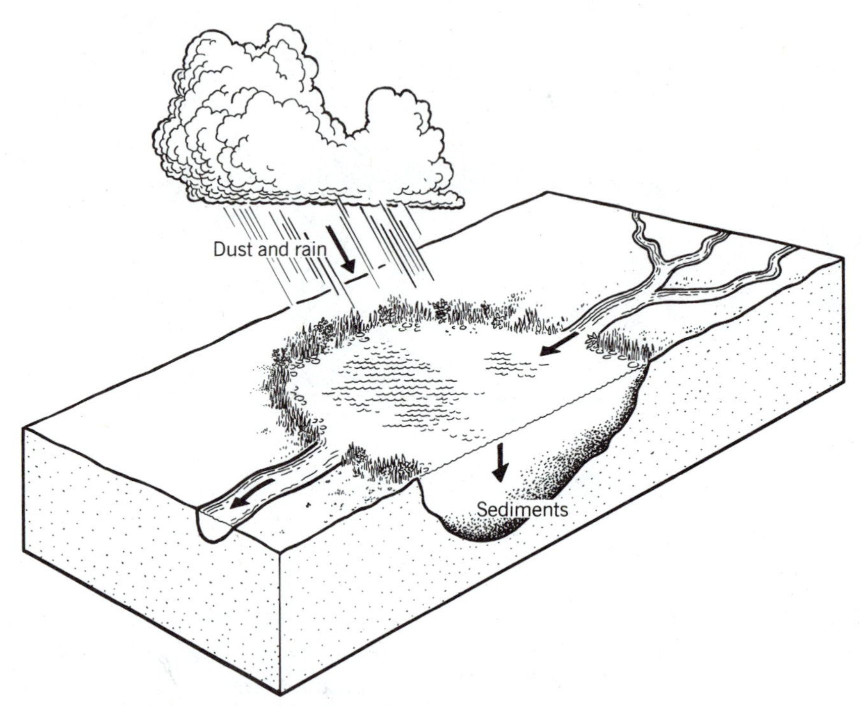

Figure 21.1 A lake as a flowthrough system.
Inputs of dissolved substances from the atmosphere or the watershed will tend to be balanced by outputs in drainage streams and to bottom mud.

Figure 21.2 A kettle lake in arctic Alaska.
Toolik Lake lies in the northern foothills of the Brooks Range. It was formed when mountain glaciers retreated at the end of the ice age and left behind ice buried in bulldozed rock and soil. The ice melted, leaving behind an irregular hole 1000 m across and 25 m deep at its deepest point. A radiocarbon date on the sediments shows that the basin first held water between 14,000 and 12,000 years ago.

Larger glacial lakes occupy the dammed-up valleys of the glaciers themselves, the dam usually being the debris of the old end moraine. The Finger Lakes of New York State, and the lakes of the English Lake District, like Windemere, are of this kind; so is Loch Lomond and other famous lochs. These lakes are long, narrow, and deep, some 70 m of water in the middle of Lake Windemere, for instance. These long, narrow basins are shaped like Norwegian fjords, which are also glacial valleys but flooded by the sea, and these lakes are accordingly called FJORD LAKES. They can be formed by mountain glaciers as well as by ice sheets, even on the equator if the mountains are high enough (Figure 21.3).

Volcanic Lakes

Volcanos may dam valleys by blocking them with lava flows, or the irregular surfaces of large sheets of lava may hold small lakes. But the more interesting volcanic lakes are those occupying craters, both collapse and explosion craters.

The classical large crater at the summit of a volcano is a collapse crater and is called a CALDERA. The cavity is formed at the end of an eruptive episode when the roof falls into the old magnum chamber out of which the lava poured. Such CALDERA LAKES are not common, but may be large, spectacular, and deep when they are found. Some, like Galapagos Fernandina (Figure 21.4), hold water that came partly from the volcano itself (*primary volcanic water*) and may be rich in magnesium sulfate (Table 21.6).

More common than caldera lakes are those occupying explosion craters. Typically volcanic explosions occur around the rims of calderas, or in flatter volcanic flows. Explosions produce what look like shell holes of large, or even gigantic, size. The craters are sometimes called MAARS and the lakes MAAR LAKES (Figure 21.5). Maar lakes are very likely to be closed in that there is no outlet, with drainage always inward down the old crater rim. Closed lakes like this are called ENDORHEIC.

Lakes in Solution Basins

Where surface rock is soluble, or partly so, water percolating in cracks may etch its own basin. A common sequence of events is for percolating

Figure 21.3 A fjord lake in the Andes.
Lake San Marcos lies close to the equator in Ecuador at an elevation of 4000 m. The typical shape of the glacial valley can be seen clearly. There is a flat bottom under 30 m of water. A glacier still occupies the valley higher up.

Figure 21.4 Fernandina crater lake before the eruption of 1968.
This is the caldera of the volcano that forms the entire Galapagos island of Fernandina (Narborough). The caldera is about 9 km across and nearly 1000 m deep. The volcano is still active and the lake was violently disturbed by the major eruption of 1968. Apparently no really ancient caldera lakes are known, perhaps because they are usually drained in the later stages of volcanism.

Figure 21.5 Maar lakes from the arctic and the equator.
The upper panorama is of Cagaloq Lake of St. Paul Island in the Bering Sea, the lower of Lake El Junco from the Galapagos island of San Cristobal (Chatham). Both are shallow and hold essentially rain water. Cagaloq is 20,000 years old and El Junco more than 45,000 years old.

water to form underground caverns, the roofs of which then collapse, both damming the percolating stream and forming a surface hollow that fills with water. Many such lakes exist in the Karst region of Yugoslavia and so solution lakes are known to limnologists as KARST LAKES. These lakes may be very numerous in limestone country, particularly where limestone is sandwiched between impermeable bedrock and a surface mantle of sands or other unconsolidated material. This is the pattern in Florida, where the characteristic and abundant round lakes and ponds are all *karst lakes*.

Solution of a different kind is also responsible for immense swarms of shallow lakes in the arctic, most particularly on the Alaskan coastal plain (Figure 21.6). The basins are dissolved, or rather melted, out of the permanently frozen ground. In this region of PERMAFROST the surface rock has a large content of permanent ice, perhaps making up 80 to 90% of the total volume. The surface ice is prevented from thawing in summer by an insulating layer of tundra vegetation but, if this insulation is damaged, a small puddle or pond develops. Moving water, including waves when the pond is large enough, transfers heat to the sides and the lake grows. Limnologists see similarities in this process to the solution of *karst* basins and so call these arctic melt-holes THERMOKARST LAKES. They remain shallow (1 to 2 m) and close over again by encroaching vegetation within a few thousand years.

Figure 21.6 Thermokarst lakes on the Alaska coastal plain. The basins have been melted out of the ice of the frozen ground (permafrost) by water circulated by wind. Wind is also thought to be responsible for shaping and orienting the basins.

Tectonic Lakes

The deepest and most ancient of the world's lakes were formed by crustal movements. Lake Baikal in the Soviet Union and Lake Tanganyika in East Africa approach the size of inland seas, are of Pliocene age or older, and are more than 1000 m deep. They occupy rift valleys, the bottoms of which have fallen as a series of GRABENS. Smaller lakes also occupy rifts and grabens in various parts of the world. Characteristically they have steep sides and flat bottoms, like Lake Tahoe in the Sierra Nevada. Small (1 km) Lake Yambo in the Ecuadorian Andes (Figure 21.7) occupies such a graben, is filled by ground water, and gives limnologists the remarkable find of a freshwater lake in the middle of a desert (Stenitz-Kannan *et al.*, 1983).

Lakes Formed by Rivers

Where great rivers meander across plains they leave behind as lakes portions of abandoned channels. The lakes are long, narrow, and shallow, often being in the most curved portion of an old meander and thus bow-shaped, a property that leads to their being called OXBOW LAKES. Towards the deltas of great rivers like the Mississippi and Amazon these lakes are abundant, but the lakes are ephemeral since they last only until the next passage of the river picks them up again. They have biota relict of the river system that gave them birth, variously modified depending on the time since isolation. Variants on these true *oxbows* are lakes that are flooded seasonally by the parent river, a type prominent in the lower reaches of the Amazon; these, called *varzeas*, are parts of the great river when it spreads in the wet season but isolated lakes for a few months in the dry season each year.

Figure 21.7 Graben lake in the Andean desert.
Lake Yambo lies in the Interandean Plateau at 2000 m elevation, close to the equator. The flat portion of the bottom now lies under 24 m of water but also holds 8 m of sediment.

THERMAL STRATA IN TEMPERATE LAKES

Water in lakes frequently becomes stratified, with surface warm water floating over colder water in the depths of the lake. Where the two layers meet temperature changes rapidly, which is called the THERMOCLINE. A lake so stratified is effectively divided into two separate compartments, an upper lake, the EPILIMNION, and a lower lake, the HYPOLIMNION. The layer between the two across which the *thermocline* runs is technically called the METALIMNION, although many limnologists use the word "thermocline" to refer both to the water mass (metalimnion) and to the temperature gradient that crosses that mass (Figure 21.8).

This thermal stratification comes about, of course, because the lake surface is heated preferentially by the sun and because warm water expands. Once established, a system of warm water floating over cold water is stable and will require considerable force to overturn it. Surface currents started by gentle winds will not penetrate the thermocline, but will mix the epilimnion while leaving the hypolimnion undisturbed. Strong winds, perhaps coupled with surface cooling, are needed to force currents of warm surface waters down into the depths before the lake can be completely mixed.

In winter, lakes are stratified differently. Because water is at its densest at 4°C (Table 21.1) and freezes at 0°C, the deep water, at 4°C, may be the warmest with a temperature gradient from there to the surface.

In seasonal climates with cold winters and hot summers lakes have corresponding seasonal histories of stratification. In winter they are stratified with ice floating over warmer water. This episode ends when the ice melts and strong winds completely mix the waters of the lake, the SPRING OVERTURN. The lake then goes into its second period of stratification with a warm *epilimnion* floating over a cold *hypolimnion* and a thermocline between. Surface cooling and strong winds at the end of summer mix the water once

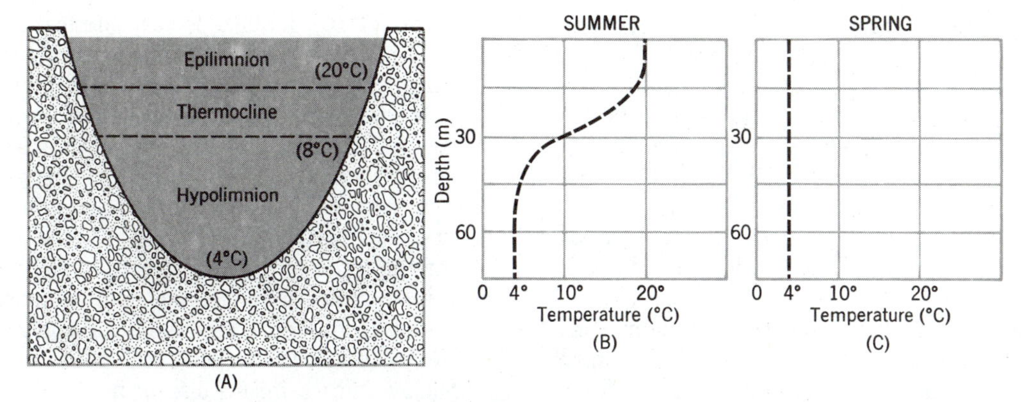

Figure 21.8 Temperature stratification of a lake of temperate latitudes in summer. Figure *A* is a diagrammatic cross section of a stratified lake. A warm epilimnion floats over the cold water of the hypolimnion below, and the two are separated by a layer, the thermocline, of rapid temperature change. Figure *B* is a temperature profile of the very deep Seneca Lake in New York as it appears in August, drawn to the same scale as Figure *A*. The practical effect of such stratification is to isolate completely the bottom water from the air until cooling and the winds of the fall effect a mixing of the lake once more. In the following spring there will be a time when the lake water has a temperature of 4°C at all depths, as in Figure *C*. (Data for Seneca Lake from Birge and Juday in Ruttner, 1963.)

Table 21.1
Density of Water as a Function of Temperature
Water is densest at 4°C. The steepest temperature gradient is between cold water and ice but the gradient is also relatively high at the warmer temperatures like those encountered in tropical lakes. (Data from Hutchinson, 1957.)

	°C	Density
Ice	0	0.9168
Water	0	0.9999
Water	4	1.0000
Water	7	0.9999
Water	10	0.9997
Water	15	0.9991
Water	20	0.9982
Water	25	0.9971
Water	30	0.9957

more, the FALL OVERTURN. This time of mixing continues until the first layer of ice forms, when the annual cycle is complete.

Very large temperate lakes like Lake Erie can be mixed by storm force winds even in summer, so these lakes sometimes alternate periods of stratification and *mixis*. But most temperate lakes remain stratified throughout the summer.

Only the epilimnion can receive oxygen by direct infusion from the atmosphere, the hypolimnion being quite cut off. Moreover the epilimnion also receives most of the oxygen available from photosynthesis within the lake because only the surface water is well lighted. Only when lakes are so transparent that light penetrates to the depths is there significant photosynthesis in the hypolimnion. In other lakes, including all the more fertile ones, photosynthesis is restricted to the epilimnion. A consequence of thermal stratification, therefore, is that the epilimnion receives oxygen throughout the summer but the hypolimnion does not.

The demand for oxygen continues in the hypolimnion, even though supplies are cut off, because animals living there must respire. Even more important is the respiration of decomposers, because all dead matter from the more productive surface regions of the epilimnion falls down into the hypolimnion where it is food for decomposers. The BIOLOGICAL OXYGEN DEMAND

(BOD) from all this respiration in the hypolimnion must be met by oxygen already held in the water before the lake stratifies. If the lake is so fertile that the amount of dead matter to be decomposed is large the BOD can exceed the oxygen reserve so that the hypolimnion becomes entirely anoxic in late summer. Then not only does a warm upper lake float over a cold lower lake, but a lake with oxygen floats over a lake without.

The nutrient status of surface and bottom water also changes during the summer. The epilimnion loses nutrients as these are taken up by phytoplankton and then lost to the bottom in dead matter falling into the hypolimnion. The hypolimnion gains these nutrients as decomposers release them from the dead matter. If BOD is so great that the hypolimnion is completely anoxic, this regeneration of nutrients at the bottom is expedited as the resulting reducing redox potentials let nutrients like phosphorus and iron redissolve more readily. As a period of stratification is prolonged, therefore, the epilimnion becomes depleted of nutrients whereas the hypolimnion is enriched.

The consequences of thermal stratification depend on the fertility of the lake water (Figures 21.9 and 21.10). If a lake is very infertile, some photosynthesis is carried on in the depths and little dead matter falls to the bottom so that the BOD is small. The hypolimnion never becomes anoxic, nor is it heavily enriched with nutrients. Fish like trout thrive in these lakes, spending much time in the cold water of the hypolimnion and making only brief foraging excursions to the warm epilimnion for food. If a lake is very fertile, however, the hypolimnion becomes both anoxic and charged with nutrients. Fish like trout would suffocate and only surface-living fish could survive. At fall overturn a fresh pulse of nutrients from the depths is injected into the surface water causing a short bloom of production before winter sets in.

The annual cycle of events in lakes in temperate seasonal climates may be summarized as follows.

(i) Winter Water evenly mixed and tending to the same temperature and chemical composition at all depths. Deep water never colder than 4°C, the temperature at which water has the greatest density. If the lake freezes, there will be a temperature gradient downwards from 0°C at the bottom of the ice to the maximum temperature

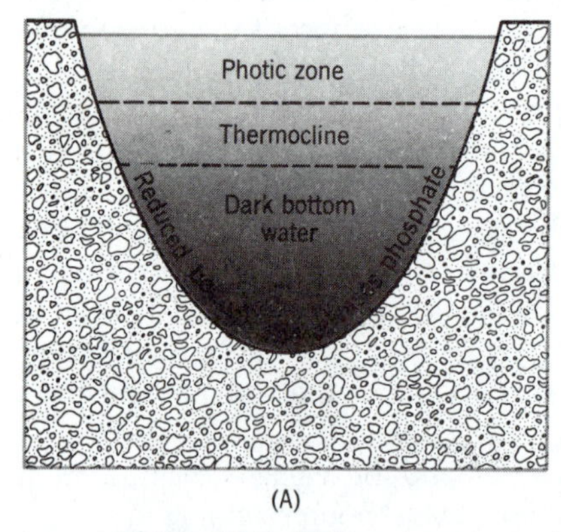

(A)

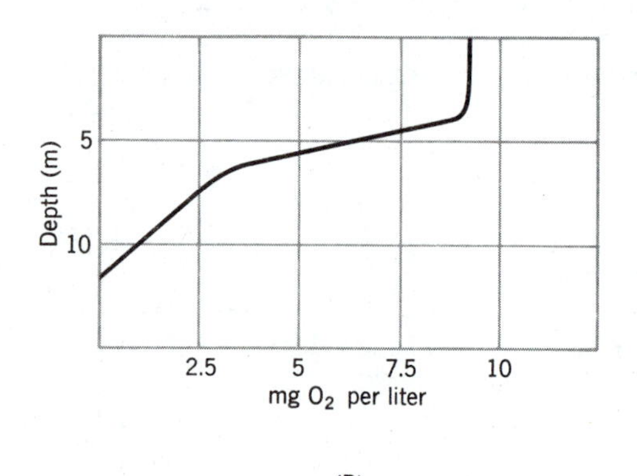

(B)

Figure 21.9 Light and oxygen profiles of a stratified fertile (eutrophic) lake in summer. (*A*) Section through the lake showing that light has been almost completely absorbed by the plankton of the top few meters so that too little light penetrates to the thermocline and beyond to support photosynthesis. But there is a rain of corpses into the deep water, whose decomposition requires oxygen. Since the deep water is cut off from the air until fall overturn, there develops an oxygen deficit in the deep water, and the bottom mud is reduced. An oxygen profile typical for such a lake is given in (*B*).

of 4°C. The nutrient concentration of the water will be at a maximum. Oxygen concentration will remain high but can be depleted under the ice in shallow lakes. Sediment at the mud/water interface usually will be oxidized.

(ii) Late Spring–Early Summer Lake stratifies and a thermocline separates a warm epilimnion from a cold hypolimnion. There is a bloom of phytoplankton in the surface water. Nutrient concentrations begin to fall in the epilimnion and oxygen concentrations begin to fall in the hypolimnion. Sediment at the mud/water interface is still oxidized.

(iii) Late Summer Maximum thickness of epilimnion and maximum temperature difference between epilimnion and hypolimnion. Epilimnion may be depleted of nutrients and hypolimnion may be depleted of oxygen. In very fertile lakes the hypolimnion may actually be anoxic and the sediments of the mud/water interface may be chemically reduced. Waters of the hypolimnion may be nutrient-rich, both from solution of particles settling down from the surface water and from solution of nutrients from reduced bottom mud.

(iv) Fall Overturn Evaporative, convective, and radiant cooling of the surface water reduces the stability of the stratified system. Strong autumn winds then mix the lake waters, imposing uniform temperatures and chemical composition throughout the water column. Bottom mud and water are resupplied with oxygen. Surface water is resupplied with nutrients, which may result in a fall bloom of phytoplankton at the surface. The lake continues to cool towards its winter condition.

EUTROPHY AND OLIGOTROPHY

Fertile lakes, as we have seen, tend to acquire an OXYGEN DEFICIT in the deep waters during the summer, whereas infertile lakes do not. Nutrient cycles also are influenced by the oxygen status of the deep water. Because of these differences, the German limnologist August Thienemann (Hutchinson, 1957) set up a classification of lakes on the basis of their fertility, separating them into EUTROPHIC and OLIGOTROPHIC lakes.[2] The essential qualities of each kind of lake may be summarized as follows:

[2]EUTROPHIC and OLIGOTROPHIC: Greek for good nursing and few nursing (trophos being the word for suckle, hence feed).

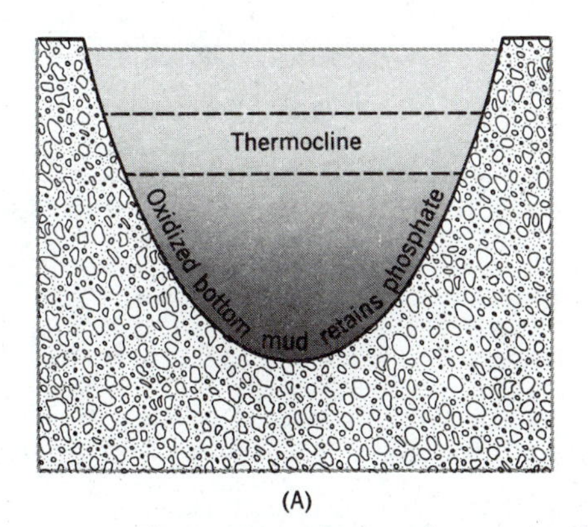

(A)

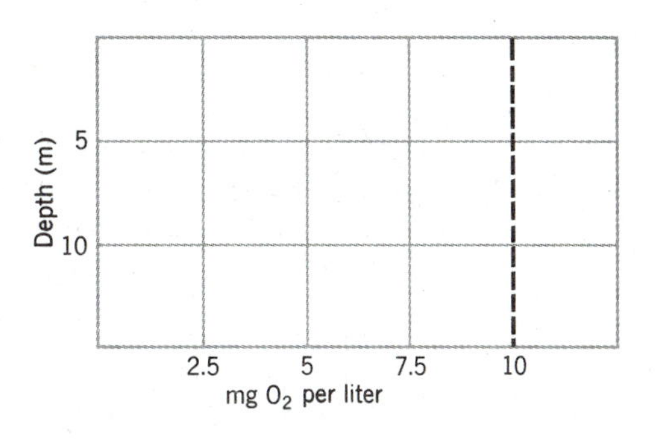

(B)

Figure 21.10 Light and oxygen profiles of a stratified infertile (oligotrophic) lake in summer. An infertile lake may stratify in summer, with consequent isolation of the bottom water from the air, but there is so little plankton floating in the top water that light penetrates deep into the lake, permitting photosynthesis and oxygen generation even in the hypolimnion. Also there is less oxygen demand in the deep water, since there is less detritus coming from above to be decomposed. (*A*) is a cross section of such a lake; (*B*) is a hypothetical oxygen profile.

Oligotrophic Lakes Concentration of dissolved nutrients is low; water appears transparent or blue; algal blooms at the surface not visible; productivity low; deep water retains oxygen at all seasons; sediment at mud/water interface oxidized at all seasons; accumulation of nutrients in the hypolimnion during summer low; injection of nutrients to surface water at fall overturn low; some plant production throughout the water column and on the mud surface (Figure 21.10).

Eutrophic Lakes Concentration of dissolved nutrients is high; water is turbid, opaque, green or brown; dense algal blooms in the surface water with the possibility of floating mats of algae; dense shade cast by plankton of the surface water prevents photosynthesis in the lower parts of the water column; productivity is high; community respiration is high and oxygen demand is high; deep waters may become anoxic in summer; sediments at mud/water interface may be reduced in summer and settling nutrients may be redissolved; hypolimnion is enriched with nutrients in summer; there is a strong fertilizing effect at fall overturn when nutrient-rich water is brought to the surface (Figure 21.9).

It will be apparent that the terms *oligotrophy* and *eutrophy* are more relative terms than they are an absolute classification. Limnologists sometimes describe the halfway condition as MESOTROPHIC.

Whether a lake is oligotrophic or eutrophic depends on geography and climate. Nutrient status of the water must depend on nutrient input and nutrient output (Figure 21.1). All the syndromes of properties described by these terms, therefore, depend on events outside the lake basin.

LANGMUIR CIRCULATION AND THE DESCENT OF THE THERMOCLINE

Within the epilimnion there is no change of temperature with depth (Figure 21.8), showing that mixing currents must penetrate from the surface to the thermocline. Apparently these currents determine the depth of the thermocline and they force it deeper as a season progresses. An important generator of these currents is LANGMUIR CIRCULATION, which was first discovered in the sea. Debris, particularly the torn-up remains of seaweeds from rocky shores, floats in rows on the sea surface, the rows being parallel to the direction of the wind. Langmuir (1938) reasoned that water currents must be moving at right angles to the wind and nudging the debris into rows, and he imagined a pattern of circulation that would provide for this. The wind would tend to push surface water along in the direction in which it was going, but this moving water must be replaced. Part of the replacement water would come in from behind, but some would be expected to come in from *below*. It must thus happen that a horizontal current will be accompanied by a vertical, ascending current. And the water moved upwards in the vertical current must be replaced by water that sinks, meaning that a system of up and down circular cells must be associated with the horizontal movement of water. When such a system develops in the three dimensions of a body of water, a series of helical circulation cells results (Figure 21.11). It is this pattern of circulation cells that we call Langmuir circulation.

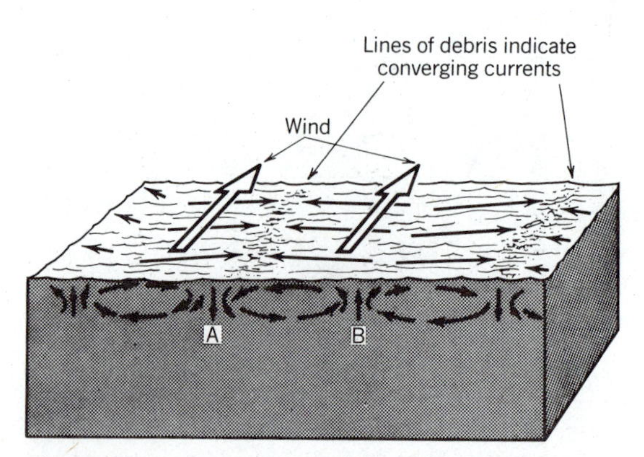

Figure 21.11 Langmuir circulation cells.
Turbulence in the surface water, which is driven by the wind, apparently becomes organized by drag and wave effects into helical cells; where outward-flowing currents converge, debris collects. The descending and ascending currents are probably important in mixing the epilimnion of a lake and in fixing the depth of the thermocline.

The system of helical cells that Langmuir postulated would explain the lines of seaweed, because these would collect where the horizontal water vectors met. This hypothesis was testable because it predicted the current systems generated in all open water. Langmuir tested the hypothesis in a large lake. He examined the current systems directly, using dies or neutral density floats to plot the paths of currents and their velocities. Not only was the postulated pattern of currents shown to exist but the velocities of the ascending and descending currents were found to be considerable. Downward currents where the cells merge (Figure 21.11A) were as high as 4 cm per second and the corresponding upward currents where the cells parted (Figure 21.11B) were as high as 1.5 cm per second.

The downward currents were found by Langmuir to extend to depths of up to 7 m. Here then is a mechanism that will account for the mixing of the waters of the epilimnion. When the wind blows strongly, Langmuir circulation cells are set up in the surface waters of lakes and they mix the warmed surface water with a thin layer lying immediately underneath. Because the underlying water is denser, the mixing process is resisted. But as the season progresses a repeated pressure is exerted on the metalimnion by the descending currents of the Langmuir cells and the thermocline is driven down (Figure 21.12).

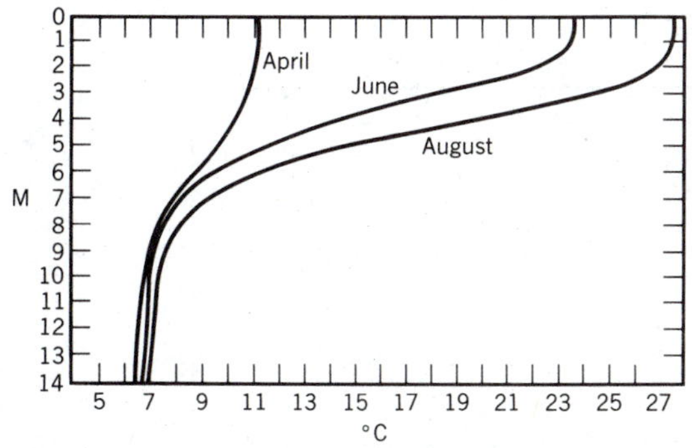

Figure 21.12 Development of the thermocline in a temperate lake.
Data are from Linsley Pond, a kettle lake in Connecticut. As the summer progresses the temperature stratification intensifies. (From Hutchinson, 1957.)

In tropical lakes, the range of temperature from top to bottom of the columns of water is not great but the position of a thermocline can still be detected. The reason for this lies in the fact that the density of water changes more rapidly at higher temperatures (Table 21.1). The loss in density of water between 25 and 30°C is more than twice the loss between 10 and 15°C, so that a vertical difference of one or two degrees is sufficient to stratify the lake. In places like the Amazon forest, where strong winds rarely reach a lake, thermal stratification may last for long periods. The hypolimnion is nearly always without oxygen. Even these lakes, however, may be overturned by exceptional storms, or because they have been flooded by cold rain or ground water. Accordingly they properly are called OLIGOMICTIC lakes (few mixing). They may be stratified for weeks or months before there is a short-lived episode of mixing.

CHEMICAL STRATIFICATION: MEROMIXIS AND A THERMAL POND

In simple thermal stratification, the bottom water is denser because it is colder. But water can also be dense because it is a strong solution of ions. Dense solutions of solutes can collect under fresher water in various ways, resulting in a stable system in which the lake is permanently stratified. This condition is called MEROMIXIS. The dense bottom water is the MONIMOLIMNION, the floating upper water is the MIXOLIMNION, and there is a CHEMOCLINE between the two (Figure 21.13).

Because the deep water of a chemically stratified lake is never exposed to the atmosphere it is possible for the depths to be heated by the sun until the bottom water is far hotter than the surface. This is the principle behind the solar pond.

To construct a solar pond, all that is needed is a tank of swimming pool proportions, the bottom of which is painted black. Strong saline solution fills the bottom half of the tank and a layer of fresh water is run onto the top. The salt *monimolimnion* then heats up and heat is taken from it by a system of immersed pipes, effectively a

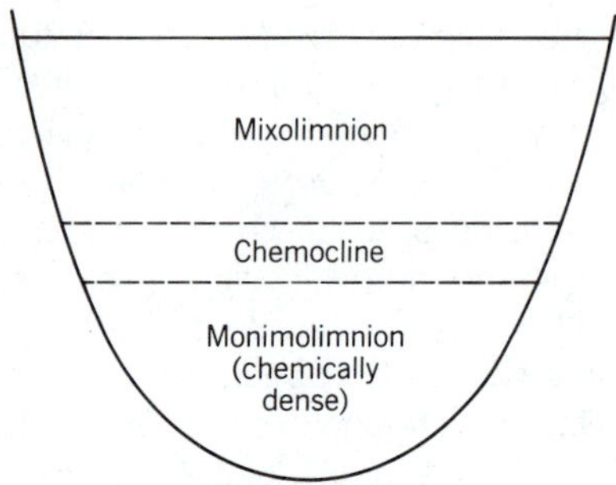

Figure 21.13 Strata in a meromictic lake.
If a dense layer forms deep in a lake (e.g., seawater injected under fresh water) the lake may be permanently stratified with only the upper lake (mixolimnion) being subjected to overturn.

radiator in reverse. The pond lasts until slow diffusion across the *chemocline* breaks down the density gradient sufficiently for the energy of wind to overcome the stratification.

Figure 21.14 shows the temperature and chloride profiles of natural solar ponds in Nevada and Japan. In Big Soda Lake the temperature of the monimolimnion is higher than the temperature across the chemocline, but lower than that of the surface water. Lake Sinmyo is a coastal lake with seawater in the bottom; a close analog to an artificial thermal pond.

It has been noted that very eutrophic lakes that are thermally stratified collect dissolved nutrients in the bottom water every summer. These nutrients, of course, act to increase the density of the bottom water. In most cases this extra density is not enough to prevent overturn in the autumn when the epilimnion cools. But if thermal stratification is prolonged, there remains the possibility that the concentration of nutrients in the bottom water can increase until the lake becomes chemically meromictic. The old hypolimnion then becomes a nutrient-rich monimolimnion and the lake is permanently stratified. Hutchinson (1957) calls this condition BIOGENIC MEROMIXIS.

THE DETERMINANTS OF EUTROPHY OR OLIGOTROPHY

The fertility of a lake is a function of the whole watershed or regional ecosystem of which the lake is a part. This is because the rate of input of nutrients to the lake is of overriding importance in determining its trophic state. The usual concentration of nutrients in the water is, of course, also a function of the rate at which they are exported via the outlet or buried in the mud, but these processes are not nearly so flexible as the inputs. If the inputs from the watershed change little. the fertility of a lake will be constant.

Lake fertility responds very quickly to nutrient enrichment by human activity, whether as sew-

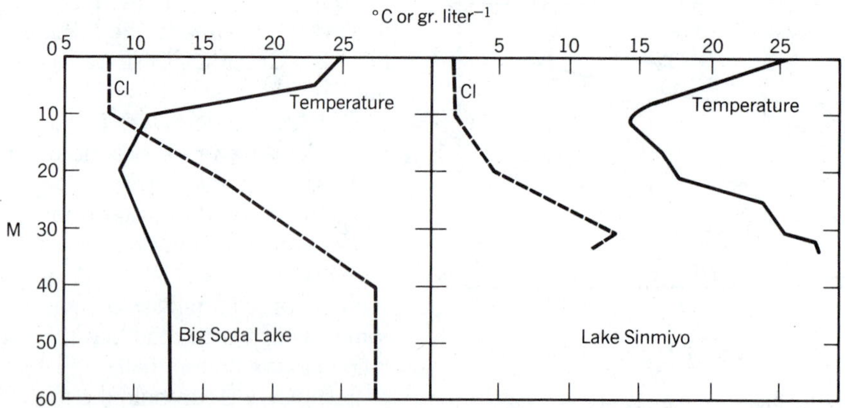

Figure 21.14 Temperature and chloride profiles of meromictic lakes. (From Hutchinson, 1957.)

age, phosphates from detergents, or agricultural run-off. This is because these processes effectively change the watershed inputs. From the point of view of the lake these are processes in the drainage basin; they change the inputs without there being commensurate changes in the outputs. As long as the nutrient enrichment is maintained, these *polluted* lakes are more eutrophic than they once were. Pollution, therefore, leads to eutrophication. But the new watershed regime of pollution must be maintained or the lakes will tend to revert to their former condition. The principal response to nutrient enrichment is an increase in the rate at which nutrients are sequestered in the mud. What happens is that biological production increases with nutrient concentration in the epilimnion. The rain of dead matter to the mud is thus increased, and the burial of nutrients in the mud increases in proportion. Nutrients are also stored in an increased biomass of rooted aquatic plants. Thus there are processes within a lake that regulate nutrient outputs. These act to damp small-scale fluctuations of incoming nutrients. But the main regulator of the nutrient state of a lake remains the control of nutrient inputs within the watershed as a whole.

Monterosi and the Roman Empire

Studies of the sediments of an ancient lake near the city of Rome give us a nice case history of the trophic stability of a lake being regulated by the watershed. Lago di Monterosi is a maar lake formed by a volcanic explosion about 26,000 years ago (Hutchinson, 1970). Drill cores of the lake sediments have been used to investigate the history of the lake (Chapter 22).

For the first 24,000 years chemistry and fossils of Monterosi mud suggest very few changes in lake trophy. The lake was moderately oligotrophic and remained so. This is really remarkable because those 24,000 years saw enormous changes in the surrounding land. For the first 10,000 of these years an ice age climate prevailed in Italy and the glaciers were not far away. Then came the climatic catastrophe of the lateglacial, a 4000-year period of irregular and drastic climatic change, followed by the familiar climate of the Holocene for the next 8000 years. Pollen analysis (Chapter 22) showed that local vegetation changed from sagebrush steppe to grassland and then to the Italian forests of prehistoric times. But the life of the lake went on, supported by the old nutrient steady state, without much change.

It took the Roman Empire to alter the trophic status of Lago di Monterosi. The Romans built a road, the Via Cassia, and this piece of engineering changed the direction in which groundwater flowed. Ever afterwards water reaching the lake percolated through volcanic debris and limestone that had been by-passed for the first 24,000 years of the lake's history. The lake water hardened (more calcium carbonate) and the lake became more eutrophic, complete with summer episodes when the hypolimnion became anoxic. This condition has persisted since the Via Cassia was built in 200 B.C. until the present day (Figure 21.15). But notice that the Romans only manipulated drainage in the watershed, though inadvertently. This is not a history of pollution. The Romans made the lake part of a new ecosystem drainage and a new steady-state supply of nutrients was established in the lake.

Clean-up at Lake Washington

A modern history of human interference with a watershed shows how very quickly a lake can respond to changes in inputs. Lake Washington, the large lake beside which the city of Seattle is built, has been first "polluted" and then "cleaned up" over a span of less than forty years.

Lake Washington is several kilometers long by a kilometer or so wide along its rambling length, and it stratifies in summer. The original mesotrophic state was set by the forested soils of its watershed. But then Seattle vented its sewers into the lake and expanded rapidly. By 1963 the concentration of phosphate and nitrate in the open water had multiplied many times (Figure 21.16). Lake Washington was now strongly eutrophic. Dense blooms of phytoplankton grew in the lake and a scum of blue-green algae floated on the surface. The hypolimnion remained anoxic all summer.

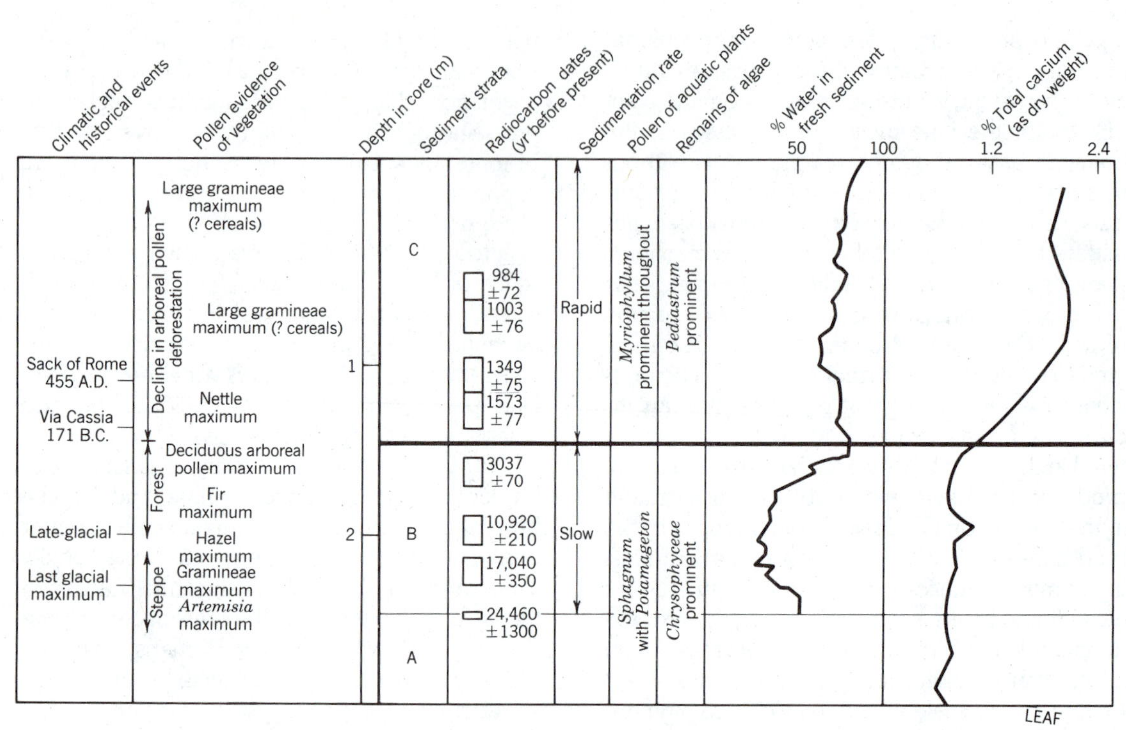

Figure 21.15 Reconstruction of history of Lago di Monterosi from sediments. Three strata, A, B, and C, are recognized in the sediment cores. All the data show that the B–C boundary (heavy type) marks a profound change in the state of the lake ecosystem. Sedimentation rate, aquatic plants, water content, and sediment chemistry changed in a way that reflects the establishment of a fresh steady state. Radiocarbon dating shows that this change was synchronous with the building of the via Cassia by the Roman Republic. (Redrawn diagrammatically from figures in Hutchinson, 1970.)

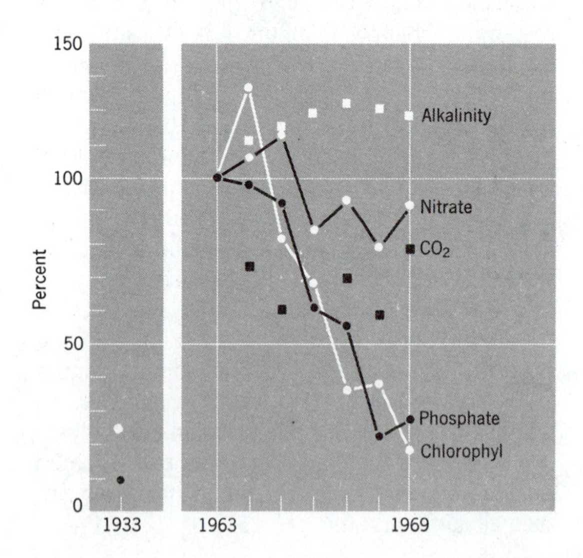

Figure 21.16 The cleaning of Lake Washington. These were crucial measurements on the waters of Lake Washington over the period of its recovery from cultural eutrophy. At the left are measurements of phosphate and chlorophyll in the water before the massive pollution, fortunately recorded in 1933. Between 1965 and 1969 most of the sewage discharge into the lake was diverted. The phosphate content of the water plunged, presumably as the phosphate originally in the polluted lake was buried in the mud. The decline of chlorophyll records the decline of the algae. The record clearly indicates phosphate as the primary pollutant, for the lake lost the symptoms of pollution while nitrates and other potential fertilizers remained in large amounts. (From Edmondson, 1970.)

Eutrophic lakes were getting bad press in the 1960s; people objected particularly to the floating algae and the replacement of fish needing plenty of oxygen in the hypolimnion (most sport fish are of this kind) with other fish that could get along with little. So Seattle built a new sewage system that dumped the remains into the sea instead of the lake. In just six years the concentration of phosphates and nitrates of the open water had fallen back to the 1933 levels (Figure 21.16). Oxygen once more persisted in the hypolimnion all summer, and the characteristics of eutrophy to which people objected vanished (Edmondson and Lehman, 1981).

The important point of the Lake Washington saga is that these dramatic changes were brought about by manipulating nutrient inputs alone. Even a comparatively large lake like Lake Washington is sensitive to changes in nutrient flux from the watershed. Buffering powers of the lake system are shown to be modest. The pollution episode was an experiment in watershed management and it demonstrated that lake trophy is a watershed property. Whether a lake is oligotrophic, mesotrophic, or eutrophic depends on local rock, local climate, local soils, and local biological activity.

LAKE AGING: EUTROPHY AND THE OBLITERATION OF THE HYPOLIMNION

Ancient lakes fill with mud. The process may be slow, but it is very steady—typical kettles from the last glaciation, for instance, are now about half full. This is the true process of lake aging—filling with mud. But this infilling sometimes alters the trophic status of the lake, because it shrinks the hypolimnion while having little effect on the epilimnion.

Figure 21.17 shows how sediment gradually can obliterate the hypolimnion of even a deep lake. Suppose that the lake is oligotrophic with a small biological oxygen demand (BOD) in the deep water and an original hypolimnion that retains oxygen throughout the summer; as accumulating sediment replaces the old hypolimnion the BOD will not change but the oxygen reserve by which this BOD can be met will lessen progressively. When the hypolimnion is not more than a thin layer between mud and thermocline (Figure 21.17*B*) the oxygen reserve held by this small volume of water may be so small that the BOD of even an oligotrophic lake cannot be met. The hypolimnion then becomes anoxic, and an

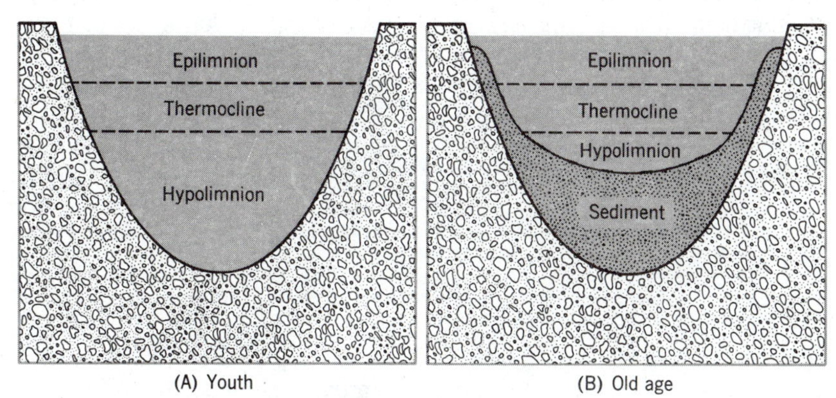

Figure 21.17 The effect of aging on an infertile lake.
The young lake at the left (*A*) is infertile (oligotrophic), retaining oxygen in its deep water all summer. The drawing at right (*B*) is the same lake when infilling of the basin is far advanced. The thermocline is in the same place, and the epilimnion is scarcely altered. But the hypolimnion is a vestige of its former size. The oxygen reserve of the reduced hypolimnion may be too small to support as much decomposition as before, and the bottom becomes anoxic. As defined by the oxygen content of the bottom water, the oligotrophic lake has become eutrophic by aging.

oligotrophic lake has taken on some of the properties of eutrophy.

An anoxic hypolimnion, of course, also means a reduced redox potential in the mud with consequent release of nutrients to solution. This means that shrinking of the hypolimnion by mud can in the end actually fertilize the lake. The lake becomes more eutrophic in its old age.

When this process was first pointed out by Deevey (1955) it was an interesting theoretical curiosity, nothing more. But it had very powerful repercussions in limnology and beyond when others extrapolated from this observation to a quite unjustified inference—that pollution was artificial aging! This argument said that since lakes were more fertile in their old age, therefore making lakes more fertile must make them old. Students of philosophy will recognize the classic intellectual fallacy in this argument. But "artificial aging" became an environmentalist slogan. Eventually it led to the absurdity of a news magazine with the headline "Who Killed Lake Erie?"

What had happened to Lake Erie was that the bottom water of the shallow west end of the lake became anoxic during short periods in summer. This eutrophication was a direct result of pollution, particularly of agricultural run-off and phosphatic detergents, which increased the productivity of the west basin. Far from dying, the lake was highly fertile and full of life. The life was undesirable life, which was why the public became interested in stopping the pollution. High BOD caused fish kills by removing oxygen, and dead fish are common in highly eutrophic lakes. But talk of the lake itself dying is rubbish.

THE VARIETY OF LAKE LIFE

Life is lived in lakes both in the open water, whether drifting or swimming, and attached to the many surfaces of a lake basin and its water. Drifting life of the open water is called collectively the PLANKTON, the animals being the ZOOPLANKTON and the plants the PHYTOPLANKTON. Many or most of the plants are very tiny indeed, so tiny that they passed through the fine nets used in early lake and ocean work and were thus often overlooked. This led to their being given a special name, the NANNOPLANKTON (dwarf plankton), but this is not a rigorously defined term.

Nearly all the phytoplankton are microscopic, the smallest of them being less than 10 microns ($1\ \mu = 10^{-3}$ mm). Some are flagellated, and so motile, some (the diatoms) are encased in silica valves, and some live as colonies, as filaments, in spherical masses of cells, or as flat plates. The varieties of their shapes are as elegant as their taxonomy is complex. But probably the most interesting of their general properties is that they come from several ancient plant lineages; from different divisions of the plant kingdom. Although, therefore, they are all alike in being tiny, drifting, primary producers, they offer chemical systems and primary metabolic pathways as fundamentally different as any in nature. The principal groups are listed in Table 21.2.

Of these very diverse forms of plant life, the blue-green algae and the diatoms are of particular interest to ecologists. The blue-greens form the floating mats in polluted lakes, which are held up by gas bubbles. Their chemistry is such that few herbivores eat them at all readily, so that the mats persist instead of providing forage. Moreover they are able to fix atmospheric nitrogen so that lakes polluted with phosphates but not nitrates support dense populations. Diatoms are interesting because of their prominence in many lakes, both in number of species and in total biomass. Their silica skeletons make dissolved silicon a limiting factor in some lakes. Their taxonomy has been developed in great detail based on structures in their silica skeletons that can be defined and measured. Their skeletons are preserved in lakes where the pH is acid, so that the histories of populations and diversities can be reconstructed.

The ZOOPLANKTON includes fewer phyla than the phytoplankton does divisions. Protozoa are always present. The phylum Aschelminthes of pseudocoelomate worms is represented by many classes, but of overwhelming importance is the one class Rotatoria, the ROTIFERS. These are microscopic metazoans propelled by cilia. They are among the most abundant herbivores feeding on the phytoplankton and are probably the most important primary consumers in many lakes. A few are predators on other rotifers.

The other animals of major importance in the zooplankton nearly all belong to the arthropod classes Crustacea, Arachnida, and Insecta. All the familiar cladocerans (water fleas; *Daphnia, Bosmina, Chydorus*) and copepods are crustaceans. Water mites (arachnids) look superficially similar as small orange-red swimming things but they exist in many species, though their taxonomy is difficult and poorly known. Most water mites are inedible to limnic predators and they may at times achieve high densities.

The most abundant insects are diptera larvae, particularly midge larvae like chironomids and the active predator *Chaoborus*. An understanding of the dynamics of a lake system is not possible without realizing that insects may be extremely important as both primary and secondary consumers. The familiar flying forms outside lakes are best thought of as mere phases of egg transport for the next generation of lake animals. These planktonic communities support fish, whose grazing has profound effects on the planktonic communities. Larger lake animals like crocodilians are often based on land food chains as much as on the production of the lake itself.

Table 21.2
Principal Groups of Algae in the Plankton of Lakes

Blue-Green Algae (Cyanophyta or Myxophyceae)	Primitive. Like bacteria they are prokaryotic. Many are filamentous. Some float with gas bubbles and form floating mats. Many are nitrogen fixers. Most are not eaten by the majority of lake herbivores.
Green Algae (Chlorophyta)	Very diverse group, mostly fresh-water. Particularly abundant are desmids. (*Spirogyra, Volvox, Chlamydomonas*)
Yellow-Green Algae (Xanthophyceae)	Characterized by special arrays of photosynthetic pigments in which yellow carotenoids predominate over green chlorophylls. (*Chlorobotrys, Gloeobotrys*)
Golden-Brown Algae (Chrysophyceae)	Another group characterized by photosynthetic pigments, particularly xanthophyll carotenoids. (*Dinobryon*)
Diatoms (Bacillariophyceae)	Possess rigid silica tests in the form of two valves.
Cryptomonads (Cryptophyceae)	Possess a specialized array of pigments that make them biochemically distinct. (*Cryptomonas*)
Dinoflagellates (Dinophyceae)	Single-celled, flagellated algae, the characteristic shape of most having a groove running around the "waist" of the cell in which the flagellae function. (*Peridinium*)
Euglenoids (Euglenophyceae)	Single-celled, flagellated, and without a cell wall. (*Euglena*)

PHYSICAL ATTACHMENT ZONES WITHIN LAKES

Lake basins provide four different kinds of physical surface to which living things can attach themselves: mud of the lighted shallows, mud of the dark bottom, the sides of plants anchored in shallow water, and the thin surface tension film of the water itself (Figure 21.18). Inhabiting each of these places requires different adaptations and imposes different physical constraints.

Life attached to the bottom or moving in the bottom mud is called BENTHOS (Greek for the depths). In lighted shallows green plants, both angiosperms and algae, make thick stands and are properly called part of the benthos. Algae growing this way in the sea are usually called "benthic" but it is more usual to refer to the anchored plants of the lake benthos as "*rooted aquatics*" or "*aquatic macrophytes*." The part of the bottom on which they live is called the LITTORAL ZONE (Figure 21.18). Limnologists, however, think more of burrowing benthic animals like the larvae of chironomid midges or tubificid worms when talking of lake benthos.

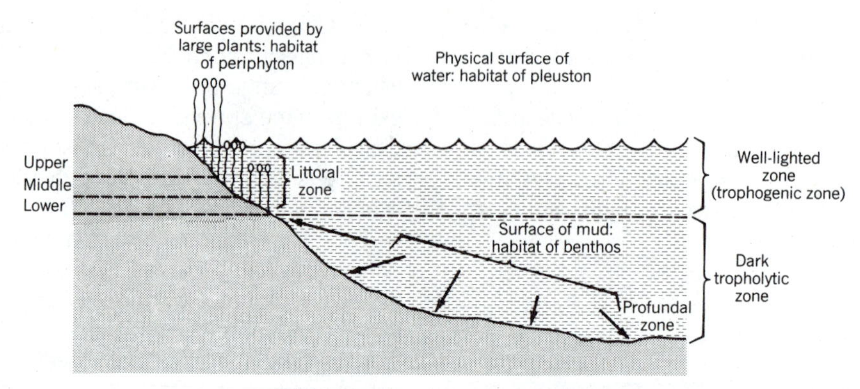

Figure 21.18 Life zones in a lake.

Small animals and some plants attached to, or closely associated with, the fixed plants of the littoral zones are called the PERIPHYTON and those attached to the surface film are the PLEUSTON.[3]

LAKE CHEMISTRY AND THE LIMITS TO PRODUCTION

Primary production in lakes varies; by geography, by latitude, from lake to lake, and from season to season or from day to day (Table 21.3). Factors controlling production include temperature and whether a lake remains stratified for long periods. But more general limits to production are set by the generally low concentrations of a few essential nutrients in lake water. The living part of lake ecosystems has little control over these essential nutrients, unlike terrestrial ecosystems like that at Hubbard Brook (Chapter 16), where nutrients are conserved and cycled within the large standing crop of biomass.

The most abundant cations of lake water are sodium, potassium, magnesium, and calcium; the most abundant anions are carbonate, sulfate, and chloride. These are present in remarkably con-

[3]PERIPHYTON: Greek, around plants. PLEUSTON: Greek *plein,* to sail or float.

Table 21.3
Productivity of Lakes of Varying Fertility
(Data collected by Wetzel, 1975.)

	Primary Production (mg C/m^{-2}/d)	Phytoplankton Biomass (mg C/m^{-3})	Prominent Phytoplankton	Total P (μg/l)	Total N (μg/l)
Oligotrophic	<50–300	<50–100	Chrysophyceae Cryptophyceae Dinophyceae Bacillariophyceae	<1–10	<1–600
Mesotrophic	250–1000	100–300		10–30	500–1100
Eutrophic	600–8000	>300	Bacillariophyceae Cyanophyta Chlorophyta Euglenophyta	30–>5000	500–>15,000

Table 21.4
Mean Equivalent Proportions of Important Ions in Natural Waters
(From Hutchinson, 1957.)

	Mean Igneous Source Material	Mean Sedimentary Source Material	Water from Igneous Rock	Wisconsin Soft Waters	Uppland	Central Europe	Mean River	N. German Soft Waters
Na^+	20.1	4.8	30.6	10.9	13.6	4.5	15.7	43.0
K^+	9.8	8.0	6.9	4.8	2.2	1.9	3.4	6.7
Mg^{++}	33.1	34.0	14.2	37.7	16.9	25.4	17.4	14.3
Ca^{++}	37.1	53.2	48.3	46.9	67.3	68.2	63.5	36.0
CO_3^{--}	—	93.8	73.3	69.6	74.3	85.4	73.9	42.4
SO_4^{--}	—	6.2	14.1	20.5	16.2	10.8	16.0	14.1
Cl^-	—	0(?)	12.6	9.9	9.5	3.9	10.1	43.5

sistent proportions in surface fresh water of different origin and in different parts of the world (Table 21.4). Typically their dissolved masses are much greater than those of all other ions. These most abundant ions are the least likely to be limiting though their supply, particularly that of sulfur, can influence productivity. The essential nutrients phosphorus and nitrogen are less abundant and their concentrations in surface waters correlate with various measures of lake productivity (Table 21.5). In addition, other elements found in living things, like molybdenum, iron, cobalt, gallium, and silicon (for diatoms), are far from prominent in analyses of lake waters. These data allow a general hypothesis of lake productivity being set by concentrations of these nutrients (Tables 21.6 and 21.7).

A general test of the hypothesis comes from the phenomenon of seasonal changes of productivity in temperate lakes. In these lakes algae bloom in spring and in autumn, with a time of lowered productivity in high summer (Figures 21.19 and 21.20). Seasonal changes in nutrients apparently explain these histories as follows. Productivity is low in winter due to short days and low temperatures. Algal growth accelerates

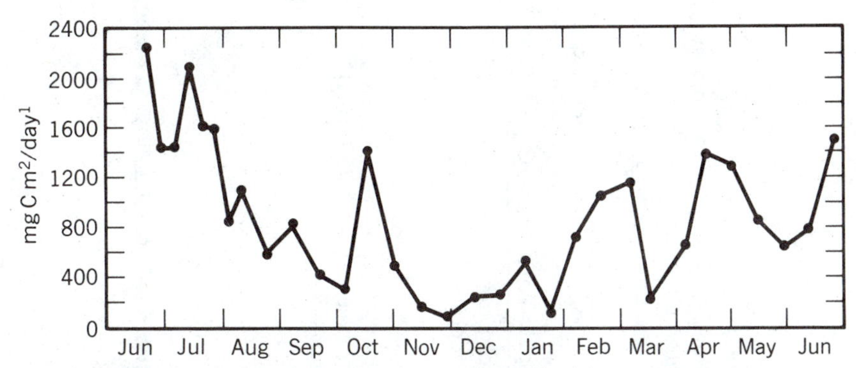

Figure 21.19 Seasonal productivity in a seasonal eutrophic lake.
Data are for the very eutrophic Wintergreen Lake in Michigan. The October peak of productivity coincides with the fall overturn when nutrients regenerated in the hypolimnion are brought to the surface. (From Wetzel, 1975.)

Table 21.5
Relationship between Nitrogen, Phosphorus, and Productivity of Lakes
(Modified from Likens, 1975.)

Trophic Type	Mean Primary Productivity (mg C/m^{-2} day)	Phytoplankton Biomass (mg C/m^{-3})	Chlorophyll *a* (mg/m^{-3})	Dominant Phytoplankton	Total Organic Carbon (mg/1)	Total P (μg/1)	Total N (μg/1)
Very oligotrophic	<50	<50	0.01–0.5			<1–5	<1–250
Oligotrophic	50–300	20–100	0.3–3	Chrysophyceae	<1–3		
				Cryptophyceae			
Moderately oligotrophic				Dinophyceae		5–10	250–600
				Bacillariophyceae			
Mesotrophic	250–1000	100–300	2–15		<1–5		
Moderately eutrophic						10–30	500–1,100
Eutrophic	>1,000	>300	10–500	Bacillariophyceae	5–30		
				Cyanophyceae			
Hypereutrophic				Chlorophyceae		30–> 5,000	500–> 15,000
				Euglenophyceae			

Table 21.6

Chemistry of Lake Waters

The suite of lakes from the Andes, the Amazon, and the Galapagos Islands of Ecuador shows the general similarity of fresh waters. Data are from lake surfaces; deep water below the thermocline often has very different chemistry. Notice the extremely low concentrations of phosphate. Data for a Galapagos lake (Genovesa) filled with evaporated seawater and from two calderas holding primary volcanic water are given for comparison. All data are in parts per million (ppm). — means no measurement made; Tr means trace. (From Steinitz Kannan et al., 1983a, 1983b; Colinvaux, 1968; and Steinitz Kannan, unpublished.)

	pH	Hardness ($CaCO_3$) mg/e	Ca	Mg	PO_4	SiO_2	NO_3	SO_4	Cl	F	Na	K	Zn	Cu
ANDES														
Cuicocha	8	226	5.0	25	0.01	100	7	26	73	—	—	—	—	—
San Pablo	8	94	1.6	13	0.16	150	35	10	5	—	30	6	0.32	0.011
Yaguarcocha	8	170	2.7	25	0.13	46	21	18	20	—	103	55	0.04	0.002
Conru	7	32	0.6	4	0.05	180	20	9	5	—	15	13	0.03	0.005
Yambo	9	636	2.0	150	0.27	110	131	160	112	1.4	263	56	0.05	0.005
Caricocha	6	16	0.3	2	0.02	15	28	7	5	—	—	—	—	—
San Marcos	7	43	0.0	1	—	—	1	1	<5	0.1	3	1	0.05	—
AMAZON														
Agrio	6	—	7.0	1	(0.09)	—	—	<10	(3)	0.1	(3)	3	—	—
Sta. Cecilia	5	—	9.0	1	—	—	—	—	(3)	0.1	<5	3	—	—
GALAPAGOS														
El Junco	6	2	0.0	1	0.04	1	18	5	7	—	67	6	—	—
Tortoise	6	80	7.3	Tr	0.02	8	17	—	70	—	—	—	—	—
VOLCANIC														
Quilotoa	8	4,500	300	720	0.44	130	118	2,560	4,160	—	—	—	—	—
Fernandina	7	612	156	54	0.09	44	3	1,990	240	5.4	856	33	—	—
BRINE														
Genovesa	7	16,900	4,390	1,440	0.05	24	—	2,780	44,000	—	22,000	602	—	—

Table 21.7
Nutrients Required by Algae
It is considered that most phytoplanktonic algae will grow in water provided with these 17 nutrients plus the three vitamins, though many are needed only in very low concentrations. (From Hutchinson, 1967.)

Carbon	C	Manganese	Mn
Nitrogen	N	Zinc	Zn
Phosphorus	P	Copper	Cu
Sulfur	S	Boron	B
Potassium	K	Molybdenum	Mo
Magnesium	Mg	Cobalt	Co
Silicon	Si	Vanadium	V
Sodium	Na	Vitamins	Thiamin
Calcium	Ca		Cyanocobalamin (B_{12})
Iron	Fe		Biotin

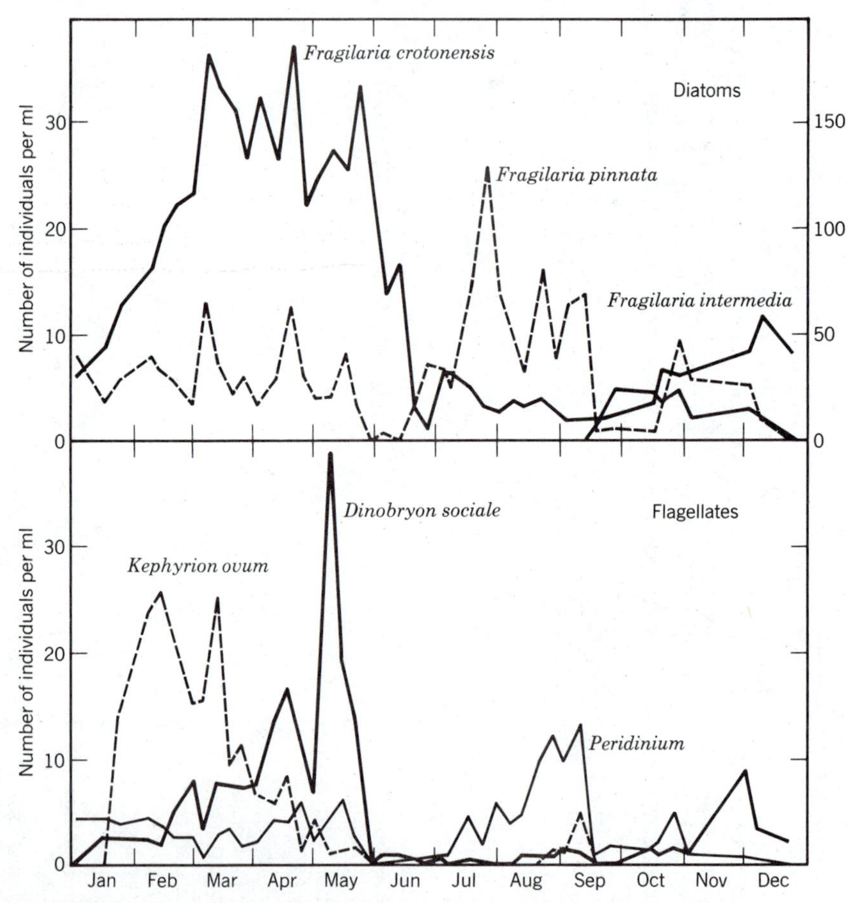

Figure 21.20 Seasonal populations of algae in Lake Tahoe.
Lake Tahoe is an oligotrophic lake, deep, with blue water. Seasonal blooms of algae are not evident to casual inspection, but they are present. (From Goldman, 1975.)

in the spring with the warming that precedes summer, causing the first bloom. But the lake then stratifies and the process of exporting nutrients to the hypolimnion via precipitating organic matter proceeds to lower nutrient concentrations in the epilimnion. Productivity then falls during the summer, despite the fact that this is the warmest time of the year with long days. The tiny, short-lived algae of the spring bloom are not replaced and the standing crop is lowered, ending the bloom. At fall overturn nutrients are brought up from the hypolimnion causing the fall bloom.

That nutrients are limiting is also suggested by a phenomenon known to limnologists as LUXURY CONSUMPTION (Lund, 1965). Algae will take up excess nutrients, particularly phosphorus, as fast as they are added to their water and apparently in excess of their immediate needs. They are then able to draw on their body stores for growth later when low concentrations in the water might be expected to be limiting. This behavior can be understood as an adaptation to restricted or uncertain supplies of essential nutrients.

Direct tests of the nutrient limiting hypothesis come from enrichment experiments, particularly the many demonstrations that enriching lakes with phosphatic fertilizer or nitrogenous wastes increases productivity greatly. The extra production, however, does not last because the added nutrients find their way to the lake mud as described earlier. Some nutrients are held for awhile in the rooted aquatic plants, but these generally are short-lived and the holding of nutrients by them is temporary. The whole lake ecosystem does not conserve or regenerate nutrients in biomass as do most terrestrial ecosystems so that lake productivity remains under control of nutrient input from the watershed.

With lake chemistry being so variable it must happen that different nutrients become of major importance in limiting productivity. However, the predominating importance of phosphorus and nitrogen seems clear enough. Enriching with phosphorus alone is sufficient to make a lake eutrophic, since this is sufficient to support blue-green algae, which can supply themselves with combined nitrogen by fixation of atmospheric nitrogen. This means that they can compete strongly for phosphorus with other algae that must also compete for nitrogen. Heavy fertilizing with nitrogen as well as phosphorus takes away this advantage and other algae may benefit at the expense of the blue-greens. The suggestion has been heard from limnologists that the way to rid a lake of algal scum resulting from excess phosphates is to increase the loading of sewage (rich in nitrogen), thus tipping algal competition away from the floating nitrogen fixers. The lake would still be pea-green productive, however.

The hypothesis that lake productivity might also sometimes be carbon limited has recently been examined by limnologists and essentially dismissed. Photosynthesis of terrestrial plants does, of course, have an upper limit set by the flux of carbon dioxide, but aquatic plants receive their carbon dioxide in solution from the carbonate–bicarbonate system. This should provide a larger flux of carbon than is available to terrestrial plants and limnologists have never thought it likely that carbon could be limiting to photosynthesis in lakes. However, the hypothesis of carbon limitation became important in studies of pollution because of the possibility that the carbon in sewage and other wastes could be as important to eutrification as the accompanying enrichment of phosphorus and nitrogen. In support of the hypothesis was the observation that typical pollution causes a bacterial bloom well before an algal bloom. The hypothesis was that the bacteria broke down organic matter to release carbon to solution as CO_2 and that this flux of CO_2 was a prime sponsor of the algal bloom.

Field data do not support the hypothesis that carbon can be limiting (Likens, 1975). The various dissolved forms of carbon dioxide and carbonates maintain near equilibrium with the atmosphere and are always present in sufficient concentration to replenish carbon drained by photosynthesis. In eutrophic lakes in bright sunlight, photosynthesis can remove carbon more rapidly than it is replaced, but only briefly, after which community respiration restores CO_2 tensions. The fact that algae respond to pollution more slowly than bacteria is probably due to slower growth and to bacterial release of nutrients other than carbon. It seems certain that carbon in pollutants is not important to productivity (Hobbie, 1975; Wetzel, 1975).

LIFE AT LOW REYNOLDS NUMBER

Perhaps the most outstanding characteristic of the life of aquatic systems, both fresh-water and marine, is the small size of the plants. Only benthic plants of shores or shallows are large; all those of the open water are microscopic. A direct consequence is that grazing animals are small to microscopic also. All these small animals, and many of the plants, must swim or else make the water move past them. But water has what seem to be bizarre properties when moved by things that are very small—it passes by with laminar flow, offering the sort of resistance a person should encounter if moving through molasses. This property of water that relates the way it flows to the size and speed of objects moving through it is defined by a function known as Reynolds number. Because life in the open water is lived at low Reynolds number it has its own rules of movement, of hunt, and of escape.

On the Adaptive Significance of Small Size of Phytoplankton

There appears to be little modern literature on the selection that has pressed small size on plants of the open water, although the advantages of smallness do not seem to be self-evident. What does seem obvious are the disadvantages of small size. The following list describes some obvious advantages of *large* size that are *foregone* by the phytoplankton. Large plants:

1. Should be at a competitive advantage against small plants.

2. Can store large reserves of nutrients (cf. rain forest trees).

3. Can use reproductive strategies based on stored energy and mass release of propagules at favorable times.

These advantages seem sufficient to ensure that large plants exist in all terrestrial habitats and in all benthic habitats of lakes or oceans where plants can anchor. Only in open water are the large plants absent. There the phytoplankters are not only denied all those advantages of bigness but they also pay a high cost of smallness in the loss of a significant part of net photosynthetic production by excretion through the cell walls; an apparent consequence of a relatively huge surface area and thin cell wall.

It has been argued in the older oceanographic literature that small size, by providing a large surface area in contact with the water for each cell, promotes nutrient uptake and prevents sinking (Hardy, 1956). Both of these functions can be met as well by adaptations to large shapes, by use of floats (as in the seaweed *Sargassum* or by the floating systems of pond weeds like *Lemna*), and by adopting convoluted edges.

We are left with the question of why nearly all plants of open water are microscopic. It may be that the selective advantage of small size must be that it minimizes loss by drifting and promotes rapid recolonization of water newly brought to the surface trophogenic zone. Large floating plants would be removed from open water of lakes by wind, or in the sea would be swept clear of the oceanic environment to which they were adapted. Smaller plants could circulate with water currents, repeatedly returning to the place of origin and recolonizing it. This is what one might call the ANTI-DRIFT HYPOTHESIS for the small size of the phytoplankton.

Whether the anti-drift hypothesis turns out to be correct or not, the fact that phytoplankton and many of their predators (herbivores) are tiny has profound consequences. One of these is the structure imposed on lake communities by the "sink or float" adaptations of algae.

The Mechanics of Sinking

Most algae are nonmotile, or nearly so, and they are denser than water—typically between 1.01 and 1.03 times the actual density of the water solution in which they live (Hutchinson, 1967). A few algae have neutral or positive buoyancy by possessing oil or air flotation devices or gelatinous sheaths, but these are in the minority. Most are dense and must sink. Since this is the norm, and since there appears to have been ample time for natural selection to have favored floating forms if this habit led to greater fitness, it must be assumed that it is advantageous to sink.

The most obvious advantage to sinking is that this is a cost-effective way of moving through the medium in quest of nutrients. A perfectly stationary alga permanently in one micropacket of water would soon sweep this clean of all dissolved nutrients and be unable to obtain fresh supplies other than by the fatally slow rate of ionic diffusion.

The obvious disadvantage to sinking is that the algae might be carried below the lighted zone to a dark death at the bottom of the lake. But this fate is mitigated by turbulence within the water column that tends to return algae to the surface. There should inevitably be a continued reduction of a static algal population by sinking, of course, however turbulent the water, but real algae reproduce. The active algal population of the lighted water gains from reproduction even as it loses from the net excess of sinking loss over turbulent return. Each individual alga apparently stands to gain more fitness by slowly sinking, and thus winning nutrients needed for photosynthesis and reproduction, than it stands to lose by falling out of the productive region. There are of course other selection pressures at work, like the effect of sinking on predation, and these are probably responsible for such developments as swimming or floating in other algae. But a strong case can be made that sinking generally is an adaptive response to the needs of nutrient uptake.

Objects of the size of typical algae (less than 0.5 mm) sink according to a classical formulation of hydrodynamic theory known as STOKES' LAW, which says that spherical objects fall through fluids as a function of the square of their radii:

$$V = \frac{2gr^2 (\rho_1 - \rho_2)}{9\mu}$$

where

V = velocity (cm per sec)

g = acceleration due to gravity (cm per sec^2)

r = radius of sphere

ρ_1 and ρ_2 = density of the falling object and of the medium (g per cm^3), respectively

μ = viscosity of the medium (poises, or dyne-sec per cm^2).

Calculations based on this relationship show that sinking velocities for typical algae are likely to be very slow, on the order of hours per millimeter (Table 21.8).

Hydromechanics also tells us that sinking speeds can be modified as a function of shape, in particular long, rod-like objects fall more slowly than spheres. This may be the reason for the prevalence of filamentous forms among algae. Protuberances and fan-like structures also lower sinking speeds, and these may be particularly important in the smaller zooplankton that otherwise would have faster settling velocities because of their larger size. It is possible, therefore, to put together a convincing general theory that explains the shapes of many planktonic organisms as being partly devices to retard or promote sinking velocities (Hutchinson, 1967).

Some phytoplankton, however, are self-motile, and most of the animals of the plankton move. Other animals, the filter feeders, move algae in streams of water that they generate. Because these motions are more rapid than sinking, these activities come up against the hydromechanics of laminar flow, imposing on life the peculiar constraints of operating at low Reynolds number.

Reynolds Number and Its Consequences

The water a human knows is turbulent. Push it with your hand and the water parts, filling in behind the moving hand with spinning eddies. Swimmers thrust water aside, pressing back the eddies and hurling themselves forward as the

Table 21.8
Sinking Speeds of Spheres in the Size Range of Phytoplankton
The velocities (mm/sec) are calculated on the assumption that Stokes' Law applies. The dashed line to the right denotes the domain where size and velocity are such that Reynolds numbers are less than 0.6, where Stokes' Law no longer holds.

	Density of Sphere Less Density of Water						
Diameter	**0.0001**	**0.001**	**0.002**	**0.005**	**0.01**	**0.02**	**0.05**
0.01 mm	0.00000545	0.0000545	0.000109	0.000273	0.000545	0.00109	0.00273
0.10 mm	0.000545	0.00545	0.0109	0.0273	0.0545	0.109	0.273
0.50 mm	0.0136	0.136	0.273	0.681	1.36		
1.00 mm	0.0545	0.545			*Re* > 0.6		

water pours in behind. A fish does it better, causing less commotion but still riding on the turbulence its motion creates. But to the small and the slow, water has very different properties. For them flow is laminar; the water glides by as an intact layer, smoothly, without the rapid circular displacements of turbulent flow. One of the strangest properties of this laminar flow is the reversibility of position: the moving particle or water mass can be moved back the way it came and the whole system is replaced in its original position with each parcel of water back where it was. If you were to stir a mixture under conditions of laminar flow, you could unstir it by reversing the motion exactly. Many of the smaller motile inhabitants of a lake actually live under these bizarre circumstances.

Whether flow is turbulent or laminar is a function of size, velocity, density, and viscosity. The formal relationship is called in hydromechanics REYNOLDS NUMBER (*Re*) and is given as:

$$Re = \frac{lV\rho}{\mu}$$

where

l = length of moving object (cm)
V = velocity (cm per sec)
ρ = density of liquid (g cm^3)
μ = viscosity of liquid (poises, or dyne-sec per cm^2).

At high Reynolds numbers flow is turbulent, and at very low numbers it is laminar. The dividing line is a Reynolds number of about 1.0. From the Reynolds number equation it is obvious that all large animals will have large Reynolds numbers, because length is in the numerator. Humans swimming in water have Reynolds numbers of more than a million and small fish have numbers in the thousands (Table 21.9). Small cladocerans like *Bosmina* or *Daphnia* live close to the dividing line, but the motile algae live well below.

One of the obvious manifestations of life at low Reynolds numbers is the way that flagellates swim: the undulating flagellum goes first and the *Chlamydomonas* or *Euglena* is dragged along behind. Larger swimmers have their propelling

Table 21.9
Reynolds Numbers of Animals in Water
Below Reynolds numbers of 1.0 flow tends to be laminar. (Data of R. Zaret, 1980.)

Animal	Length	Velocity	Reynolds Number
Human	1.5 m	1 m/s	1.5×10^6
Fish	5 cm	1 cm/s	5×10^3
Cladoceran	—	Sinking	0.1
Cladoceran	—	Swimming	3
Copepod	—	Swimming	500

structures behind them, driving back the turbulence they make.

Zooplankters as a class span the Reynolds number divide at 1.0. The herbivores among them must catch phytoplankters. One-on-one encounters by simple reaching for the algal prey would be difficult in conditions of laminar flow, since the target is pushed away in an intact packet of water. The response of these zooplankters is to generate with cilia flows of water that bring the prey to the predator. The phytoplankters come smoothly as part of the laminar flow of water, thump against sensitive parts of the zooplankters' feeding apparatus, and are seized. In high-speed photographs by J. R. Strickler (for example, in Alcaraz *et al.*, 1980) and Gilbert (1980) this process has been witnessed directly. A rotifer, for instance, drawing a long filamentous diatom colony into reach of its mandibles can be seen to crunch the filament in sections like brittle spaghetti, the parts being swept down its open mouth.

This procedure actually gives filter feeders some selectivity over what they eat because the photographs show some algal cells being rejected whereas others are seized. Presumably chemical cues are used to choose food, although this is not certain. These observations do suggest that "filter feeding" animals do not act in quite the indiscriminate way suggested in the theoretical predatory model of the type 1 functional response (Chapter 10).

Hunting in conditions of low Reynolds number brings special difficulties. Detection of prey is a major problem, though it has been suggested that laminar passage of water transmits pressures from nearby objects that can be detected (Zaret, 1980). Seizing prey of any bulk also is difficult and the high-speed photography shows that attacks can fail if the aggressor is unable to retain a grasp on the prey.

A special pattern of activity in some zooplankters comes about because they can work on both sides of the Reynolds number divide. In hop-and-sink behavior the animal launches itself with a stroke of appendages in a turbulent flow jump, then sinks quietly in what may be near laminar flow conditions (Figure 21.21). This pattern possibly serves both the search for prey in the sinking episodes and as an avoidance reaction to

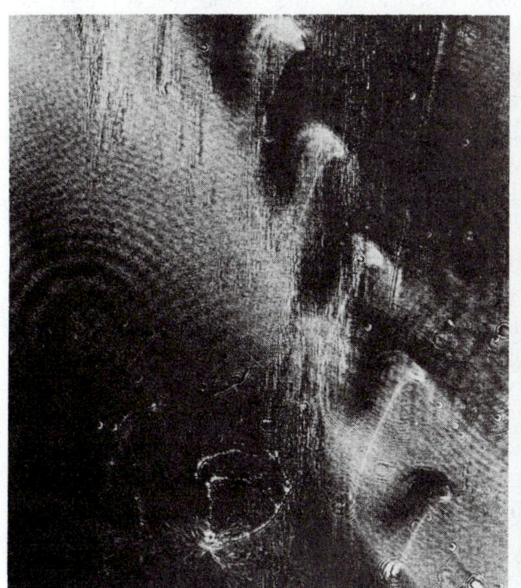

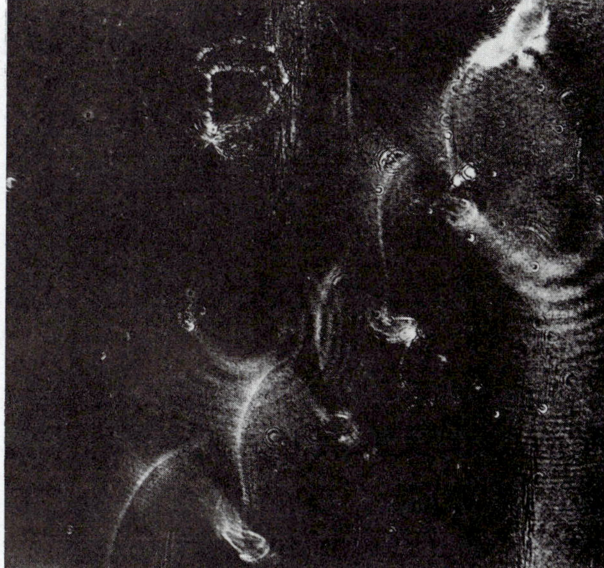

Figure 21.21 Hop-and-sink swimming behavior of a copepod. Two views of *Cyclops scutifer* swimming by time-lapse Schlieren microphotography described by Strickler (1977). (From Kerfoot *et al.*, 1980.)

larger predators. An elaboration of the technique may be shown by predaceous insect larvae like *Chaoborus*, which hang suspended in the water, as if in ambush, and are able to cross the Reynolds number divide for a rush at prey with the speed of an animal generating turbulent flow. These many possibilities are explored in the volume edited by Kerfoot (1980).

THE PARADOX OF THE PLANKTON

Planktonic life in the epilimnion provides one major puzzle, the answer to which is still not clear. It is hard to understand how there can be so many species of phytoplankter coexisting in the open water. Planktonic plants all acquire energy in the same way, by photosynthesis; they all require a similar set of dissolved nutrients, such as the limiting phosphorus; and they live stirred up together in a transparent medium that they can hardly divide up into private spaces. Why, therefore, do most of them not suffer competitive exclusion? On the face of it one would expect phytoplankters to be engaged in as fierce a competition as the paramecia in Gause's centrifuge tubes when he performed the experiments that led to the exclusion principle (Chapter 6). And yet we find scores to hundreds of species of phytoplankton coexisting in the open water of a lake. Hutchinson (1961) called attention to this remarkable feat of coexistence, calling it the PARADOX OF THE PLANKTON.

Several alternative hypotheses have been put forward to explain this paradox, though none has been tested to complete satisfaction.

Hypothesis 1: Competitive equilibrium is never attained Hutchinson (1961) suggested that in temperate lakes there was never time for a competition to eliminate species before the next seasonal change in the lake. Rather than competitive exclusion, there was seasonal succession, as different species became abundant in turn though none became extinct. What we know of seasonal changes in lakes (Figure 21.19) is compatible with this view. Indeed, many phytoplankton species are provided with resistant cysts or spores that can lie dormant on the mud surface to pass an unfavorable season. These resting times might also serve to get species through the worst competitive episodes. One appealing aspect of this hypothesis is that it makes at least one broad prediction that is potentially falsifiable: that lakes in less seasonal places like the lowland tropics should have fewer species than temperate lakes. This is because phytoplankters at the equator should get closer to competitive equilibria, with consequent increased chance of extinction. Really good comparable data of plankton diversity in tropical and temperate lakes to test this prediction are not available, but there are indications that the species list may be somewhat less in equatorial lakes.

Hypothesis 2: Niches of phytoplankton are so similar that competition works very slowly Exclusion happens only if individuals are sufficiently different for selection of one over the other to be possible. All the members of a single population, for instance, coexist. Riley (1963) suggested that phytoplankton niches have so converged in their long histories (perhaps 3×10^9 years) that competitive exclusion works only slowly. Apparently no suitable test for this hypothesis has yet been devised.

Hypothesis 3: Some species coexist in cooperation If neighbors are symbiotic or commensal, there obviously would be no exclusion. The few data available suggest, in fact, that there is little such cooperation in the algae (Hutchinson, 1967).

Hypothesis 4: Adequate niche differentiation is possible, in spite of appearances Although stirred together, the algae might have different nutrient requirements. One model assumes different nutrient limits, either as a single nutrient or nutrient sets, for each alga, and allows a large enough number of permutations of the known limiting nutrients to account for the observed diversity (Peterson, 1975).

Hypothesis 5: Competitive exclusion does take place in the open water but lost populations are continually replaced by immigration from refuges at the peripheral mud This hypothesis arose when evidence was found that equilibrium conditions probably did occur in tropical lakes as predicted by hypothesis 1, but that phytoplankton species were still so numerous that the paradox of the plankton remained (Colinvaux and Steinitz Kannan, 1980). It was suggested that competitive exclusion would be much less likely at the lake periphery where physical separation of niches on mud and among rooted plants was more possible. The periphery also would continually collect species by long-distance immigration in the wind. From this refuge of the peripheral mud populations would colonize the open water as fast as other populations went extinct there by competitive exclusion (Figure 21.22).

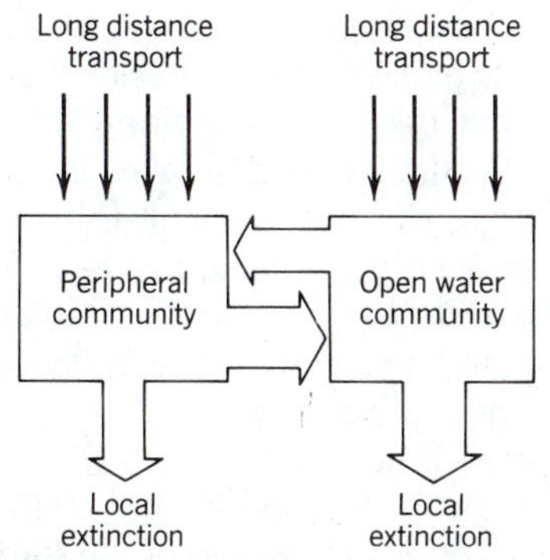

Figure 21.22 Mechanism for maintaining high species richness in equilibrium phytoplankton community.
The model postulates extinction by competitive exclusion in the open water balanced by immigration from populations in mud refuges at the periphery and by long-distance transport. (From Colinvaux and Steinitz Kannan, 1981.)

Since all these proposed mechanisms are plausible, it is likely that all operate to some extent. The result is one of the remarkable properties of lake systems: open water crowded with great numbers of species. A similar pattern is found in the sea. Yet the consequences for community structures in lakes do not include a large array of herbivorous animals. Perhaps this is related to the fact that any herbivore seeking those phytoplankters must enter a low Reynolds number world. Filter feeding, rather than individual hunting, is the rule for these animals. Even though we now know that some of these "filter feeding" animals can discriminate between algae, it is probably still true to say that they are not specialized enough for many kinds of herbivore to coexist, even when the algal food comes in many species.

CYCLOMORPHOSIS: WHEN SHAPE CHANGES IN THE EPILIMNION

An oddity of lake life is that the shapes of some planktonic organisms change with the seasons. A species may have one shape in the spring, but members of the same population a few generations later in high summer will carry structures that make them look quite different. This phenomenon is called CYCLOMORPHOSIS. It is most noticeable in zooplankton, both in crustaceans like *Daphnia* (Figure 21.23), *Bosmina, Ceriodaphnia,* and *Chydorus* and in rotifers (phylum Aschelminthes). That species in differ-

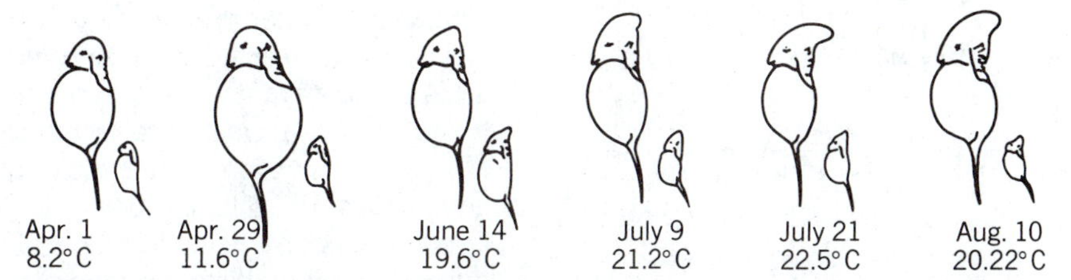

Figure 21.23 Cyclomorphosis in *Daphnia retrocurva.*
The drawings are of young and adults at each date, drawn to scale. The changing shape of the helmet is probably a stratagem to increase size by storing reserve calories in a transparent structure, thus minimizing the target offered to predators that hunt by sight. (From Brooks, 1946.)

ent phyla show essentially the same phenomenon argues for a fundamental adaptation to lake life. Even many plants, notably the dinoflagellate *Ceratium* (Figure 21.24) and some diatoms, show something similar.

The prevailing explanation of cyclomorphosis in *Ceratium* is that successive generations of the alga are shaped so that their sinking speeds are adapted to the changing viscosity of the water (Hutchinson, 1967). As the epilimnion warms in summer its viscosity falls. The extra spines (Figure 21.24) act as brakes. Certainly the possession of extra spines can be correlated with water temperature and can be induced by growing cultures in water of appropriate temperature. Other possibilities must remain; perhaps that extra spines interfere with filter feeding animals when these are at their most abundant, or that there is some other advantage as yet unknown.

Cyclomorphosis seems much less likely to control sinking speeds in animals, however, because their relatively large size makes any change in speeding negligible. The most convincing explanation for zooplanktonic cyclomorphosis was offered by Brooks (1965), who suggested that the new shapes reduced fish predation in summer. *Daphnia* and other planktonic animals are the quarry of fish that hunt by sight. The *Daphnia* are poised in a glassy void, moving for the most part sluggishly due to low Reynolds number. The fish swiftly ride their turbulent flow and are able to take any plankter they see. *Daphnia* and the rest are largely transparent, yet they must be partly visible. The larger they are the more visible a target they will make for a fish, and summer is a time of rapid growth, or else of the rapid storage of food reserves to be used in reproduction. Brooks put forward the hypothesis that "helmets" and other protuberances of cyclomorphic crustaceans were transparent storage structures that let the animal grow without increasing the visual target for a hunting fish.

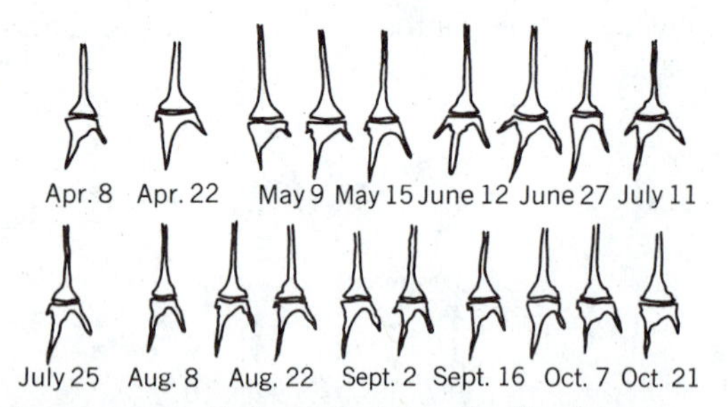

Figure 21.24 Cyclomorphosis in a dinoflagellate. *Ceratium hirundinella* generations living in warmer water (June to July) tend to be made up of individuals with more spines. (From Hutchinson, 1967.)

Brooks' predation hypothesis conforms to modern knowledge of the potency of fish predation on zooplankters. Small fish can prey so heavily on large zooplankters that these can be completely removed from a lake, or nearly so, leaving the water to smaller species that escape fish. This first became clear when Brooks and Dodson (1965) showed that lakes in Connecticut with populations of a small fish of marine origin, the alewife *(Pomolobus pseudoharengus)*, were without larger species of cladocera, whereas similar lakes with no alewives had the large species in abundance.

Since then it has been shown that planktiverous fish react to the presence of plankton depending on the apparent size of the target. The apparent size of a plankter is a function of its actual size and its distance away, leading to the concept of REACTIVE DISTANCE. Fish can see, and strike at, larger plankters from farther away (O'Brien, 1979). This work has in fact shown that hunting fish operate an accurate *optimal foraging strategy* (Chapter 5) and are able to concentrate on larger prey as they become numerous within the reactive distance (Werner and Hall, 1974). The advantage to a cyclomorphic plankter of faking a small size is readily apparent. Moreover, there are sometimes equally strong advantages of actually being large. The increased bulk of a helmeted form of *Daphnia* reduces its vulnerability to attack by the predaceous copepod *Heterocope* in arctic lakes (O'Brien *et al.*, 1979).

It is now, therefore, a reasonable conclusion that cyclomorphosis in zooplankters adapts the animals to minimize predation. The cyclomorphic forms allow bulk to be increased without also increasing the visual target. The new shape and size is the best mix to foil both invertebrate predators that hunt the small and vertebrate predators that hunt the big. The strength of this hypothesis must increase the plausibility that predation is involved in selection for cyclomorphosis in algae also. Perhaps extra spines on a *Ceratium* increase the chances of rejection when it thumps against the mouth parts of a filter feeding herbivore.

VERTICAL DIURNAL MIGRATIONS

A ubiquitous property of the zooplankton is the habit of swimming up and down on a day and night cycle. The animals are found near the surface at night but in deeper water by day. When the small size of the animals is considered, some of the distances covered are remarkable; a vertical journey of about 10 m for an animal 1 mm or so long, for instance (Table 21.10). Most of the larger limnic zooplankters undertake these daily journeys. Smaller ones like the rotifers migrate also, but not with such consistency and they may be out of phase with the large plankton (Wetzel, 1975). Some movements in flagellated algae have been found also but these are not so striking as the vertical migrations of large plankton. Even greater are vertical migrations in the sea, where the larger members of the plankton may make daily descents of up to several hundred meters, when their massed presence shows up on sonar traces as the DEEP SCATTERING LAYER.

A diurnal cycle obviously seems to be linked to light, and this has been well documented (Hutchinson, 1967). Light is the trigger for the movements, the animals rising in darkness but letting themselves sink in proportion to rising light intensity. When flagellated algae move, it is reasonable to suspect that they may be positioning themselves at light intensities optimum for photosynthesis, but this explanation will not do for

Table 21.10
Vertical Migration of Zooplankton
Data are maxima that have been measured. The larger amplitudes and velocities are for the larger animals, there being a general correlation between range of migration and size. (Approximated from data compiled from various sources by Wetzel, 1975.)

	Maximum Amplitude (m)	Maximum Velocity (m/hr)	
		Ascent	Descent
Cladocera			
Daphnia retrocurva	24	10.6	5.0
Daphnia schoedleri	2	1.4	1.4
Bosmina longirostris	20	19.0	12.0
Leptodora kindtii	9	4.2	2.0
Copepoda			
Limnocalanus macrus	24	18.0	9.8
Amphipoda			
Pontoporeia affinis	40	11.7	13.9

the animals. There are three general hypotheses of the selective advantage the behavior may confer on zooplankton.

Hypothesis 1: Predator avoidance The dangers from fish that hunt by sight ought to be reduced if the plankton spend the day in dark, deeper water. The animals rise at night to feed at the surface when fish cannot see them. Hunting zooplankters can feed at night since they do not depend on light for prey detection.

Hypothesis 2: Metabolism at low temperature This hypothesis rests on the undoubted fact that there are advantages in "feeding warm and digesting cool." The hunting efficiency of ectotherms increases with temperature as they become more active (Chapter 5). On the other hand their metabolic costs go down with temperature, so it is an advantage to remain in colder water when not actively feeding. Thus the hypothesis states that vertical migration takes advantage of the fact that lakes (and the oceans) stratify (McLaren, 1963). This hypothesis may be coupled with the suggestion that algae are more nutritious at night, since this is when they synthesize proteins. So: eat quality food at night at the warm surface, and digest by day in the cool of the depths.

Hypothesis 3: Niche separation and the food search A pattern of regular movement should be useful in the food search, and it may also serve in separation of niches (see discussion of the paradox of the plankton). In a transparent mass of water, ordination cues are rare, the most obvious being light from above and gravity from below. Gravity may be less useful in orientation for low-density animals than light. Hence many regular movements will be constrained to the vertical using light intensity as the ordinator.

Recent studies provide convincing evidence that the mechanisms of all three hypotheses can be working in some circumstances. In Lake Gatun in Panama a common planktivorous fish can be shown to forage voraciously on the copepod *Diaptomus* in the laboratory but the stomachs of wild-caught fish hold few *Diaptomus*. It was shown that vertical migration of *Diaptomus* kept it away from the foraging fish in the lake, thus demonstrating the selective advantage directly (Zaret and Suffern, 1976).

In another tropical lake, Lanao in the Philippines, *Chaoborus* larvae are restricted to parts of the lake where deep water lets them descend more than 30 m to where fish cannot get them in daytime (Lewis, 1979). On the other hand water mites in a lake in Quebec go through a pronounced vertical migration although it is known that they are distasteful to fish, and thus almost never eaten. Both energetic efficiency and pursuit of food seem plausible advantages to these water mites (Riessen, 1981).

One very elegant demonstration showed how the habit can serve as a food search: A population of marine copepods went through the classic diurnal migrations until a layer of algae was injected into the water, after which they tended to remain around the algal layer (Bohrer, 1980). Finally, studies in Toolik Lake of northern Alaska, where there is permanent daylight in the summer months, found no vertical migration, but the animals positioning themselves according to temperature and environmental cues (Buchanan and Haney, 1980).

It therefore seems certain that the selective advantages of vertical migration can be various. Avoidance of predation by fish is probably the most all-pervading advantage but the others must all operate to varying degrees. The habit essentially is a property of the system in which the animals live—transparent, thermally stratified, and intermittently lighted from above. The simplest of responses to these physical circumstances involves swimming up and down. Such swimming has little energetic cost associated with it, since passive sinking and active ascent would use no more energy than swimming to stay in one position (G.A. Riley in Hutchinson, 1967). A number of adaptive advantages can come from this simple habit, which is then fixed and preserved by natural selection.

CONCLUDING NOTE

Contemporary studies emphasize the role of predation in setting structure to lake life (Lewis, 1979; O'Brien, 1979; Zaret, 1980; Kerfoot, 1980). Predation by fish is, as we have seen, thought to be most important in generating cyclomorphosis and vertical migrations. Ever since the seminal paper by Brooks and Dodson (1965) it has been clear that the species of zooplankton living in a lake, and their relative abundance, have been very largely influenced by the fish populations present. But even strong predation by fish leaves different species of zooplankton coexisting. The relationships between these might be set by competition, but it is also likely that invertebrate predators, particularly insect larvae, are very important in the lives of the smaller planktonic herbivores. We have a general picture of the whole structure of planktonic life being heavily dependent on predation.

But the conditions in which this predation is so important are set by the physical properties of lake systems. Water is transparent, stratified, and fluid. Plants respond to fluidity with microscopic size. There is little storage of biomass in this system, with consequent little storage of nutrients. There is a very strong vertical component to the system's physical structure. These are the facts that make food chains start with tiny plants and that make herbivores unable to escape predation by huge size as on land. If predators control lake communities, it is the physical circumstances of lakes that make this possible.

PART FOUR

COMMUNITY BUILDING

PREVIEW TO CHAPTER 22

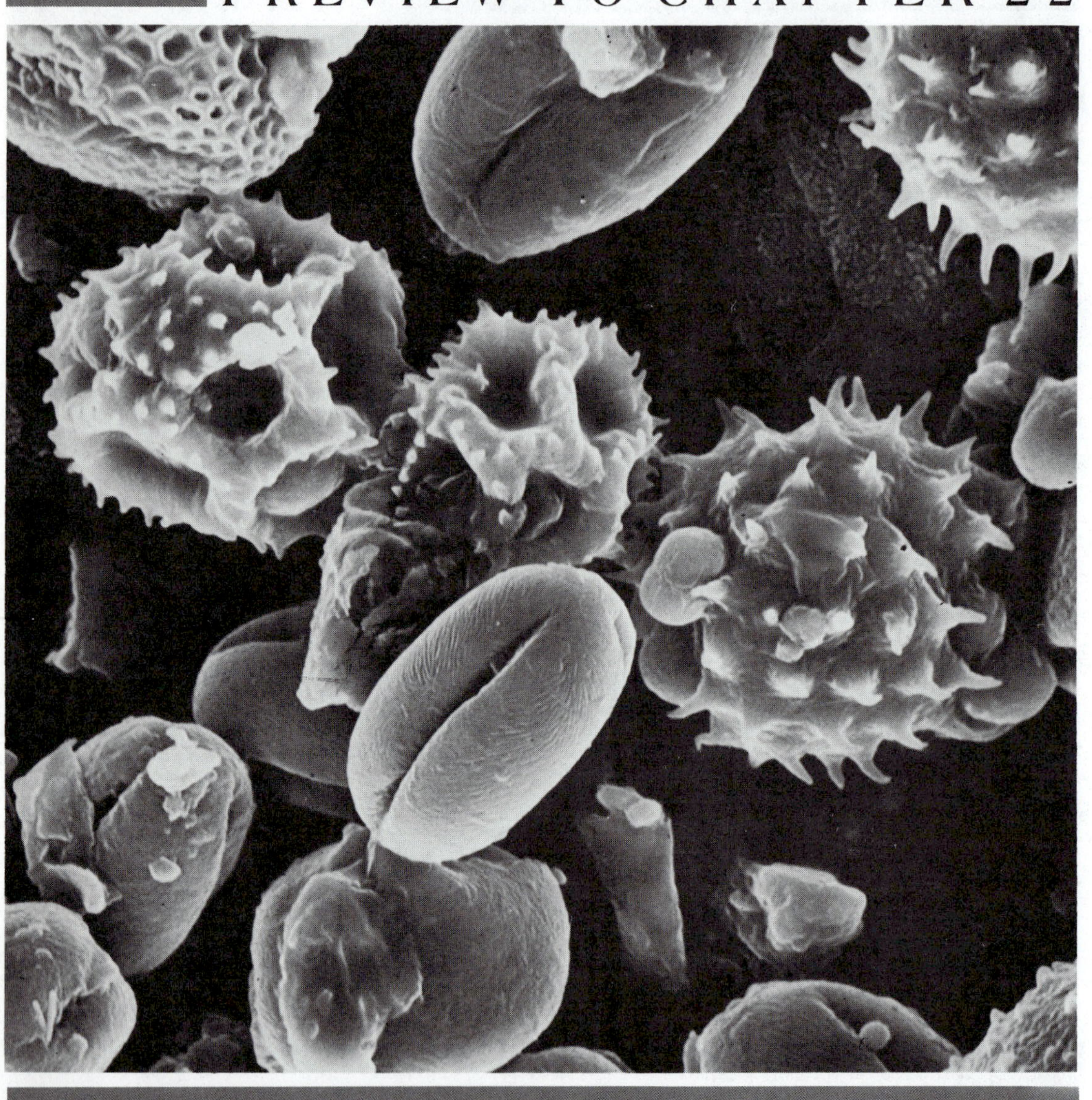

Ecologists use the fossil record to test hypotheses. We know that the distribution and abundance of life on earth has developed gradually from migrations and competitions that may have taken thousands of generations to run their course. We make hypotheses about these past events to explain modern patterns of life, but we must go to the fossil record to test the hypotheses. The ecological approach to paleoecology is thus different from the classical approach of geology. Geologists use fossils to reconstruct the past, but ecologists use the record of the past to reconstruct lives and population histories of organisms. Most useful have been the quantitative records of microfossils preserved in sediments of lakes, bogs, or the sea because with these we are able to follow the histories of ecosystems, both changing species populations and environmental change. Pollen grains preserved in lakes and bogs provide histories of plant communities and we can measure both the proportions of different species represented in the pollen collection and the influx of individual pollen types, thus obtaining estimates of changing size of plant populations. Applying this technique to the American Midwest has been decisive in showing that plant associations are temporary groupings that have changed continuously since the last ice age. An earlier tradition in pollen analysis is the reconstruction of past climates using assemblies of species as rough indicators of climate, an approach that has recently been refined by multivariate statistical techniques that derive transfer functions relating pollen percentages to temperature and precipitation. Pollen analysis has value for archaeology too, as when a Europe-wide decline of elm trees, possibly from elm disease, is demonstrated at a time when stone-age peoples were expanding agriculture, or in defining the living conditions of the aboriginal American populations who lived on the old Bering land bridge. The best maps of the climate of the ice age have been made by applying the transfer function technique to assemblies of planktonic foraminifera preserved in deep sea cores. These fossils also yield a chronology of the Pleistocene because concentrations of the isotope ^{18}O in their skeletons record changes in ocean chemistry consequent on water locked up in ice sheets. A more general method of dating is measuring the rate of decay of ^{14}C in dead organic matter, a method that usually gives ages only to 35,000 years ago. Various other isotopes, including those of uranium and cesium, also are used as clocks to date the past. More complete calendars are now available for the last 7000 to 9000 years in both America and Europe from tree rings. Analyses of individual tree rings yield detailed environmental data, as when the ^{18}O content is used as a thermometer, or data about individual or population growth are deduced from ring widths. Another potential thermometer of the past is the ratio of C4 to C3 grasses deduced from carbonized leaf fragments in lake sediments. But the population and community histories recorded by microfossils remain of most interest to ecologists. The evidence of carabid beetle fragments in bogs, for instance, shows that these insects have undergone almost no morphological or physiological evolution for several hundred thousand years. The diatom record in lake sediments can be used as a direct measure of the changing composition of the phytoplankton, letting predictions of theories of community building be tested directly. Fossils of zooplankton in lakes, coupled with evidence of mud chemistry, have been used to test the hypothesis that lake ecosystems are self-regulating and controlled by the biota, showing in fact that the biological portion of lake ecosystems is almost totally determined by physical inputs from outside the lake.

CHAPTER 22

PALEOECOLOGY: THE FOSSIL RECORD OF ECOLOGICAL PROCESS

PALEOECOLOGY means literally "the ecology of the past." There are subtly different kinds of paleoecology, however. Geologists use fossils as TIME-STRATIGRAPHIC INDICATORS and as tools for reconstructing past environments. Standard divisions of geologic time (Figure 22.1) are still based on fossil assemblages. But, more than this, geologists use their fossils to infer such things as temperature, habitat, and climate of the past epochs they name. They make use of the PRINCIPLE OF UNIFORMITY, which says that PROCESSES IN THE PAST WERE THE SAME AS PROCESSES ON THE CONTEMPORARY EARTH, a principle paraphrased as "the present is the clue to the past." Applying this principle to the study of fossils is to say that the nearest living equivalent to a fossil form lives in such and such an environment, therefore, the same environment pertained in the time of the fossil. Much knowledge has come from this approach and the method can be given a formal definition: PALEOECOLOGY (geological usage) IS THE RECONSTRUCTION OF PAST ENVIRONMENTS FROM THE EVIDENCE OF FOSSILS.

The alternative approach is to use independent evidence of past environments to reconstruct the living conditions and life of a fossil, yielding a quite different definition: PALEOECOLOGY (biological usage) IS THE RECONSTRUCTION OF LIFE HISTORIES, LIFE STRATEGY, AND NICHE OF EXTINCT SPECIES BY COMBINING PHYSICAL AND CHEMICAL EVIDENCE OF PAST ENVIRONMENTS WITH THE LIFE FORM DATA OF THE FOSSILS.

But the fossil record can also be used to test hypotheses of population or community ecology. All our ideas of community building or population size are about processes that produce their effects over very long spans of time. We cannot run experiments of such duration, but we can look at the fossil record to see what has happened in the past. This is the process that Edward Deevey (1969) has called "Coaxing history to conduct experiments." The method is particularly useful when applied to the recent past, the Holocene and later Pleistocene, back to a few tens of thousands of years ago. We can trace the

Subdivisions Derived from Strata				Important Fossils	Age $\times 10^6$ yr
	Systems	Series	Stages		
PALEOZOIC	Quaternary	(Recent) Pleistocene	More than twenty widely recognized	First humans	2.5
	Tertiary	Pliocene		Elephants, horses, and large carnivores.	13
		Miocene		Mammals diversify	25
		Oligocene		Grasses and grazing animals become abundant.	36
		Eocene		Primitive horses appear	58
		Paleocene		Mammals prominent for first time	
					63
MESOZOIC	Cretaceous		About thirty widely recognized	Dinosaurs become extinct; flowering plants appear	
					135
	Jurassic			Dinosaurs reach climax	
				Birds appear	
					180
	Triassic			Primitive mammals appear; conifers and cycads become abundant	
				Dinosaurs appear	
					230
CENOZOIC	Permian		Many recognized	Many reptiles; conifers develop	
					280
	Pennsylvanian (Upper Carboniferous)			Primitive reptiles appear; insects become abundant	
				Coal-forming forests widespread	310
	Mississippian (Lower Carboniferous)			Fishes diversify	
					340
	Devonian			Amphibians, first known land vertebrates, appear	
				Forests appear	
					400
	Silurian			Land plants and animals first recorded	
					430
	Ordovician			Primitive fishes, first known vertebrates, appear	
					500
	Cambrian			Marine invertebrate faunas with hard skeletons	
					570
PRECAMBRIAN Complex assemblages of rocks, largely metamorphosed				Procaryotes long present; first metazoa with soft skeletons	

Figure 22.1 (opposite page) Geologic time scale.
The stratigraphy was erected using fossils as time-stratigraphic markers. Dating in years came later by radiometric methods. (Modified from Longwell *et al.*, 1969.)

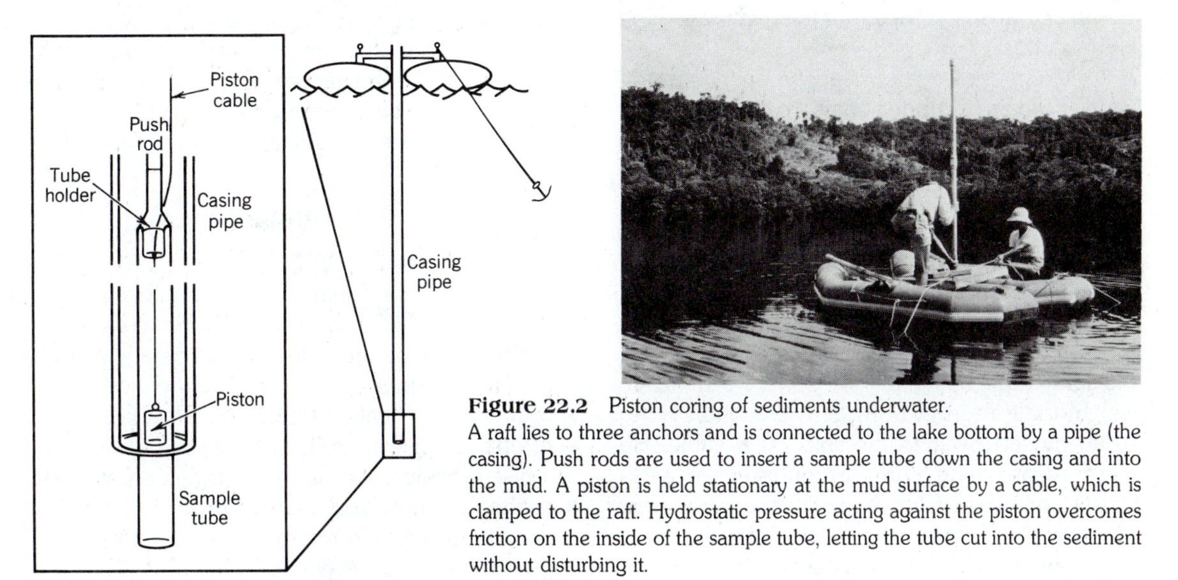

Figure 22.2 Piston coring of sediments underwater.
A raft lies to three anchors and is connected to the lake bottom by a pipe (the casing). Push rods are used to insert a sample tube down the casing and into the mud. A piston is held stationary at the mud surface by a cable, which is clamped to the raft. Hydrostatic pressure acting against the piston overcomes friction on the inside of the sample tube, letting the tube cut into the sediment without disturbing it.

processes that have led to distribution and abundance on the contemporary earth. Therefore: PALEOECOLOGY (ecological usage) IS THE USE OF THE FOSSIL RECORD TO RECONSTRUCT THE HISTORIES OF POPULATIONS OR COMMUNITIES IN ORDER TO TEST CONTEMPORARY ECOLOGICAL HYPOTHESES.

Reconstructing population or community histories requires quantitative samples. Fresh-water or bog deposits on land yield records of plant pollen, diatoms, remains of zooplankton, and sometimes beetle fragments. Some of these microfossils may be found in concentrations of 10^3 to 10^5 fossils per milliliter of sediment. Similar fossils are found in the sea, though the most useful marine microfossils are the tests of calcareous foraminifera. A sample of a hundred thousand individual diatoms or pollen grains, or some lesser number of foraminifera, from mud can be a powerful tool for reconstructing a past community. Ecologists, therefore, have learned to sample mud under water.

The mud of a lake or ocean must be sampled as a CORE, a long, continuous section through the sediments. Moreover, the sediments must not be disturbed in the process. This is done with piston samplers, using principles first introduced into oceanography by Kullenberg (1955) and then adapted for use in lakes by Livingstone (1955). A PISTON SAMPLER works by using the hydrostatic head of water against a stationary piston to overcome friction on the inside of a sample tube as this is pushed into the sediment (Figure 22.2).

The difficulty in coring mud is not penetration, which is usually easy. Much lake sediment, for instance, can be penetrated by a good push. A 5-kg hammer sometimes helps but nothing like a rotary drill is needed. The real problem is that mud will be prevented from entering the sample tube by friction on the inside. A good shove on an open tube merely results in bunging up the end. The captive piston solves this.

In lake work with a Livingstone sampler, the piston is held by clamping its cable to the raft (Figure 22.2). The tube is then pushed down around it. What happens can be visualized by thinking of a hand-held bicycle pump, which is used by pushing the piston and holding the pump tube stationary. In piston sampling you hold the piston and push the tube, but the relative motion is the same.

In lake work it is necessary to anchor a raft, usually a pair of rubber boats, to three anchors in a "Y" so that the raft cannot move, and to lower a pipe (the casing) through the water to the mud. The piston sampler is then used inside this casing, hauling up the sediment a meter at a time, then going down the old hole for the next meter, and so on. In ocean work it is not possible to work with rods and casing and so a Kullenberg typically takes 10 m cores at one bite, the long tube being driven down by a heavy weight (half a ton of iron), and the whole thing being hauled back to the ship on a cable. By these means long cores of undisturbed sediments are raised from both lakes and oceans. Microfossils can then be extracted at intervals to work out histories of populations and communities.

THE POLLEN TOOL

Over any community in spring and summer there hangs a cloud of pollen undetected, save by sufferers from hay fever, yet composed of an almost unthinkably vast multitude of tiny, drifting grains, gently settling as a pollen rain. Pollen grains from different plants are easily identified to genus with a light microscope (Figures 22.3 and 22.4), although not to species. Pollen that settles in waterlogged places is preserved, for the outer coat of pollen grains is made of one of the most resistant materials produced by living things. The pollen contents quickly rot but the outer shell may persist in the sediments of a bog or lake for thousands of years, and hundreds of thousands of years if the sediments remain wet and anoxic.

It is a simple, although time-consuming, matter to sort a few thousand pollen grains from sediments, destroying the organic matrix with sodium hydroxide and a sulfuric acid–acetic anhydride mixture, and dissolving minerals with hydrofluoric acid (M. Davis, 1963; Faegri and Iversen, 1964; Erdtman, 1969; Moore and Webb, 1978; Birks and Birks, 1980). The pollen extract is mounted on a microscope slide, after which it is a comparatively simple task to traverse the slide at a magnification of 400 diameters and to identify and tabulate every pollen grain until enough have been counted to calculate the percentage composition of that ancient pollen cloud.

The Pollen Percentage Diagram

Figure 22.5 is a POLLEN PERCENTAGE DIAGRAM from the sediments of a small kettle lake in Vermont (M. Davis, 1965). Depth in sediment is plotted on the ordinate and percentage of each pollen type on the abscissa. Each horizontal line of histograms (called POLLEN SPECTRA) represents a single analysis of the past vegetation at a discrete interval of time, giving a complete list of the common taxa that make up most of the pollen rain with an importance value (% pollen present) for each. A glance at Figure 22.5 shows that vegetation has changed radically around the lake. In the early days there was much spruce and pine pollen but little beech and hemlock. Since the lake was made by glacial ice, it is obvious that in the spruce–pine episode we are looking at a trace of the landscape of Vermont in early postglacial time, and that much was to change before final establishment of the forest that the first New Englanders were to see.

Figure 22.3 (opposite page) Pollen of European tree genera drawn to a common scale.
The scale is in microns, which means that the 100 divisions shown represent one-tenth of a millimeter. The largest pollen grain in the figure, the pine, would be just visible to the naked eye as a minute speck. Notice how distinctive are the grains of different genera. Even when built on a common plan, as are the 3-pored grains of birch and hazel, there are distinctive differences (between these two the shape of the pores) with which the analyst quickly becomes familiar. But it is seldom possible to distinguish between species within a genus. Alder grains, for instance, may have 4 or 5 pores (as shown) or even 6 or 7, yet there seems to be no correlation between the number of pores and the different species of alder. Sometimes you are lucky, however, and two species of lime can be told apart by the shape of their pores, as shown. (Redrawn from Godwin, 1956.)

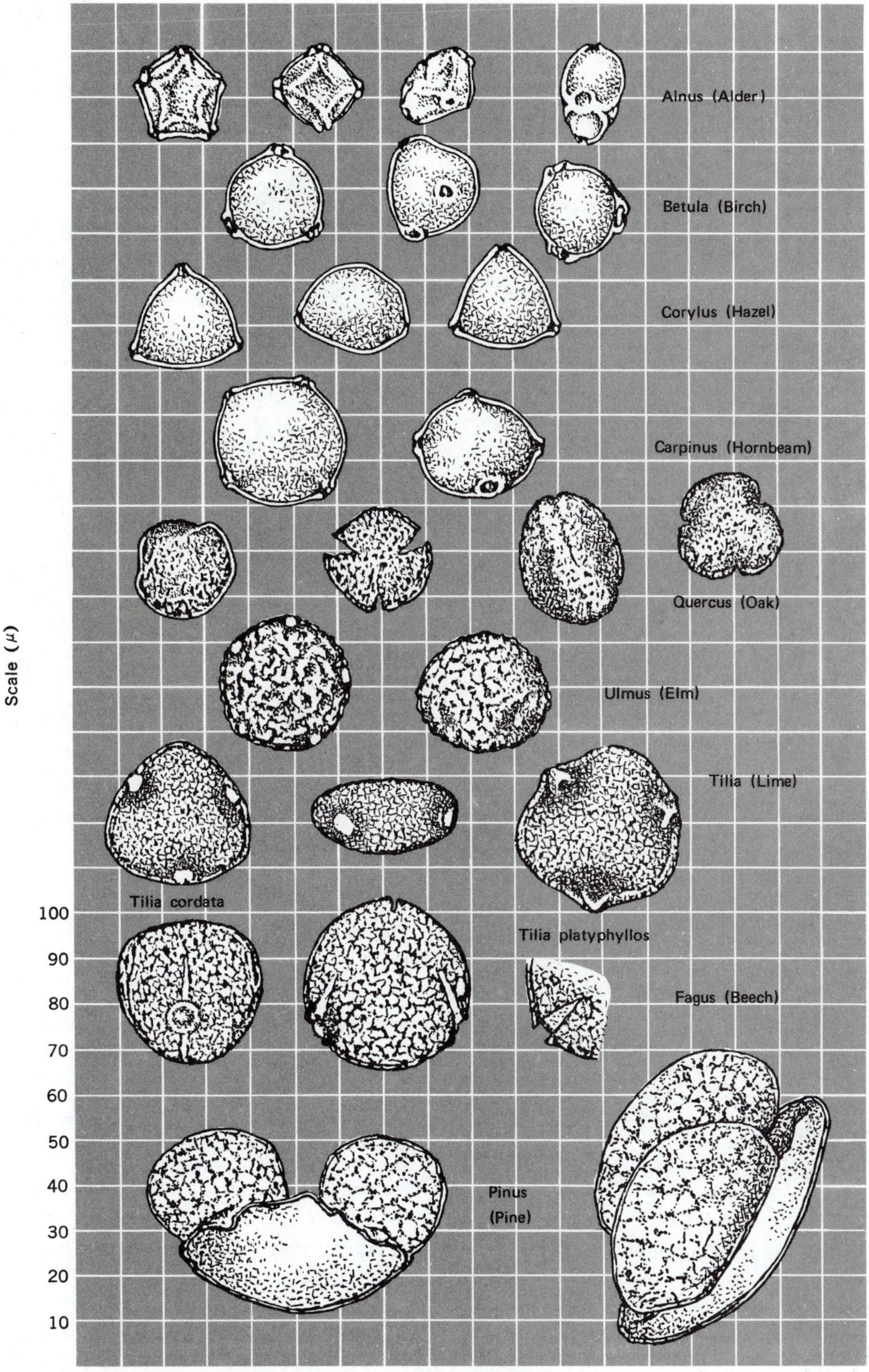

Alnus (Alder)

Betula (Birch)

Corylus (Hazel)

Carpinus (Hornbeam)

Quercus (Oak)

Ulmus (Elm)

Tilia (Lime)

Tilia cordata

Tilia platyphyllos

Fagus (Beech)

Pinus (Pine)

Scale (μ)

100 90 80 70 60 50 40 30 20 10

0 10 20 30 40 50 60 70 80 90 100

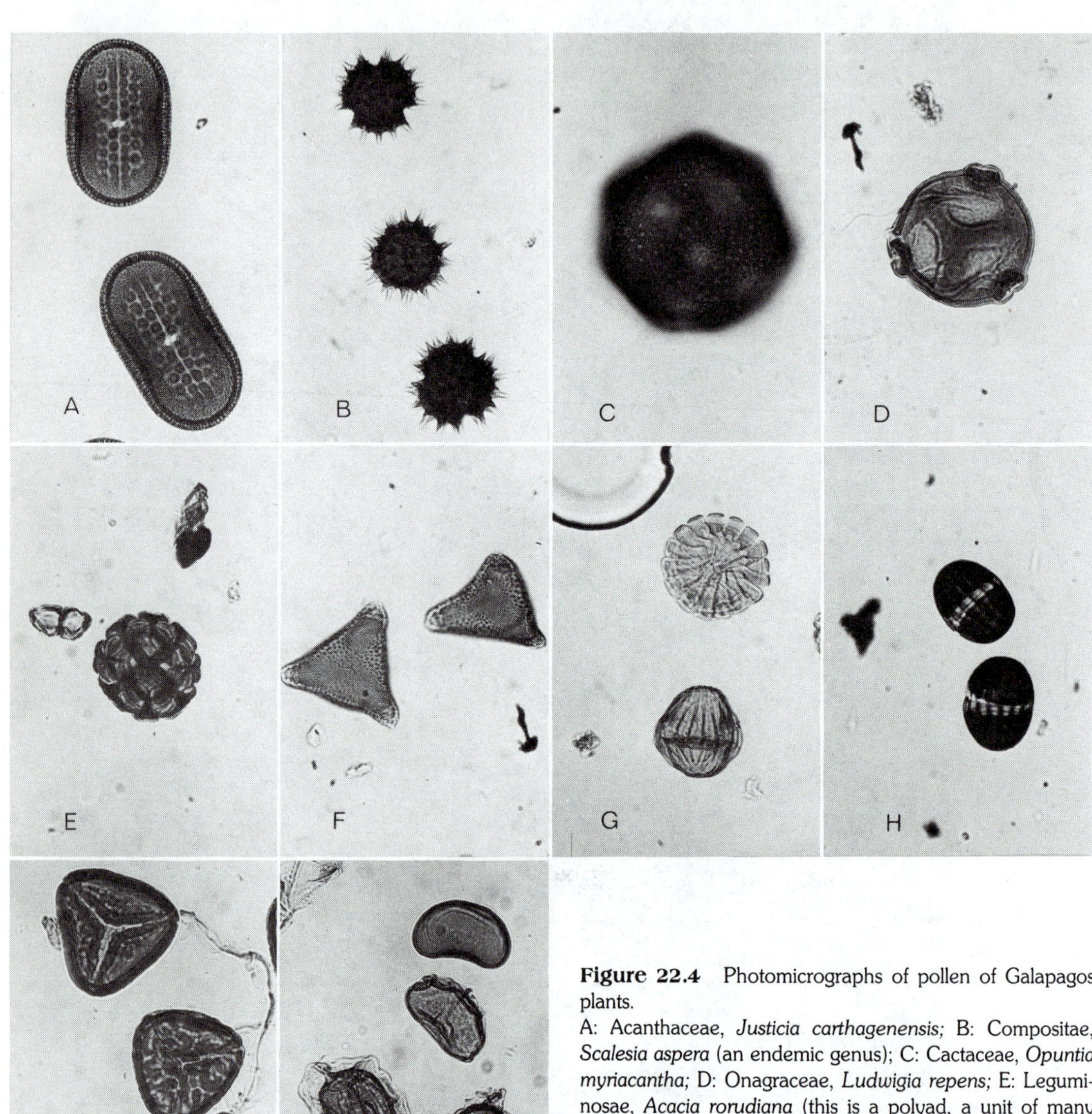

Figure 22.4 Photomicrographs of pollen of Galapagos plants.
A: Acanthaceae, *Justicia carthagenensis;* B: Compositae, *Scalesia aspera* (an endemic genus); C: Cactaceae, *Opuntia myriacantha;* D: Onagraceae, *Ludwigia repens;* E: Leguminosae, *Acacia rorudiana* (this is a polyad, a unit of many individual pollen grains); F: Sapindaceae, *Cardiospermum galapageium;* G: Lentibulariaceae, *Utricularia foliosa.* H: Polygalaceae, *Monnina chanduyensis;* I: Polypodiaceae, *Anogramma leptophylla* (a trilete fern spore); J: Polypodiaceae, *Asplenium serratum* (a monolete fern spore with ornate outer coat that easily separates from the smooth spore).

To control the large amount of data in a pollen diagram, and to help interpret the record, pollen analysts divide the diagram into POLLEN ZONES (zones A–C in Figure 22.5). This can be done by a multivariate analysis that identifies regions of comparative homogeneity but is usually done by eye.

Most of the thousands of pollen diagrams that have been drawn in the sixty years of palynology have been *pollen percentage diagrams* like this,

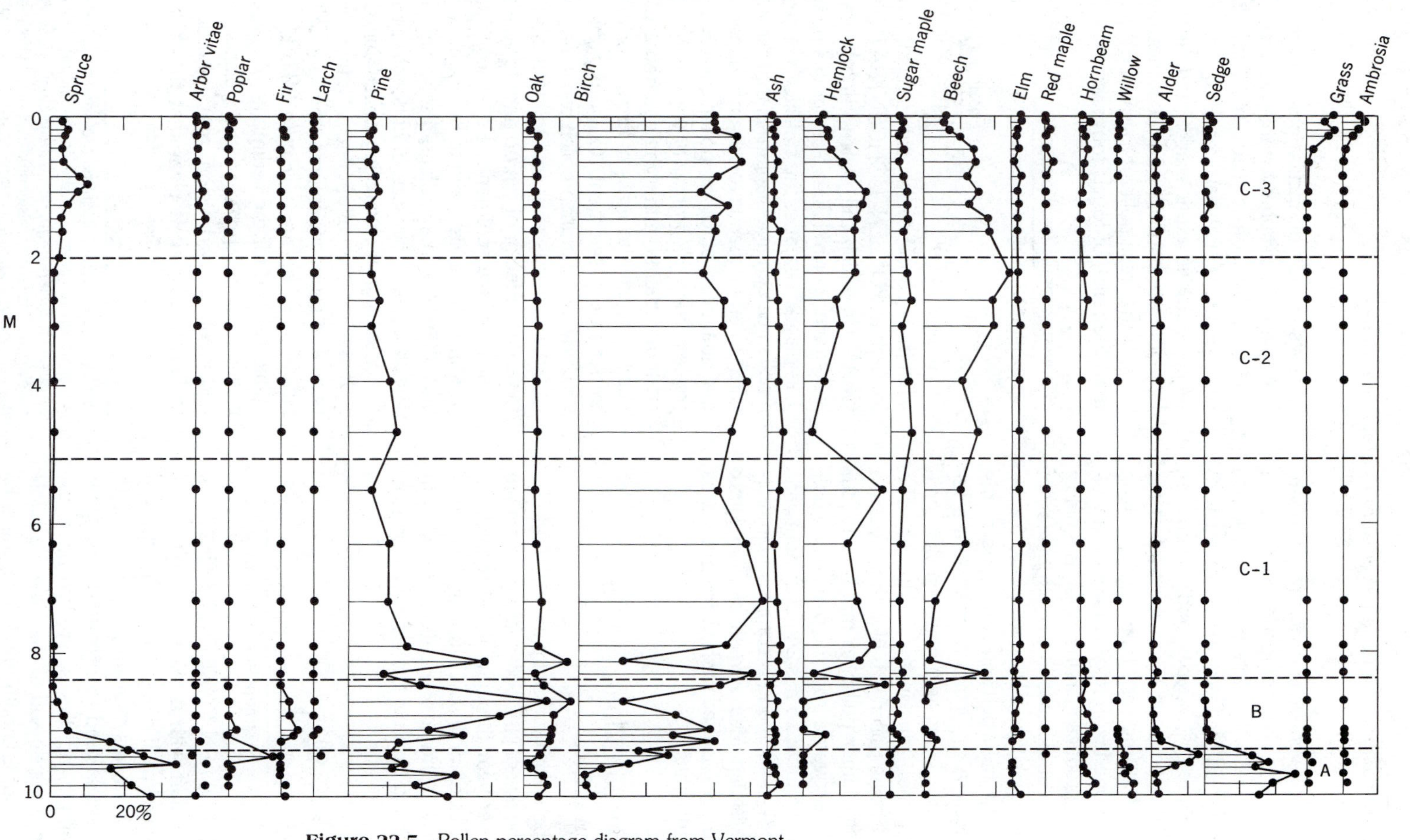

Figure 22.5 Pollen percentage diagram from Vermont.
The pollen zones A–C are the standard zones for New England and cannot be correlated with the European sequence (Deevey, 1939). Percentages are calculated as percent total terrestrial pollen. (From M. Davis, 1965.)

but they have one very great disadvantage—there is no effective way of deciding if a small change in the percentage of a single pollen taxon has any statistical validity. There are as many variables as there are pollen types, and a change in real numbers of any one taxon must affect the percentages of all the rest. Conversely, a high percentage of one taxon might represent a lowering of pollen production in the forest as a whole rather than a real increase in the population of the taxon in question. We now know, for instance, that the high percentage of pine at the bottom of zone A in Figure 22.5 is more a reflection of there being few heavy pollen producers other than pine in the landscape. The percentage of pine pollen was high then because there was little other pollen to mix with it. But this was not discovered from the percentage pollen diagram. Figure 22.5 has 24 dependent variables (the pollen taxa) and is completely intractable for a statistician. Measurement of *pollen influx* (Figure 22.6) was needed to show the true status of the ancient pines.

Pollen Concentration and Pollen Influx

The truth about the Vermont pine was found by calculating the flux of pine pollen falling per square centimeter per year throughout the record. To calculate POLLEN INFLUX like this it is necessary to measure the concentration of pollen in the sediment and to know the sedimentation rate. Then

$$\text{Pollen influx (grains cm}^{-2}\text{ yr}^{-1}) = \frac{\text{Pollen concentration (grains ml}^{-3})}{\text{Sedimentation rate (cm yr}^{-1})}$$

To measure POLLEN CONCENTRATION it is necessary to start with a measured volume of fresh, wet sediment. It is possible to extract every pollen grain from this measured volume, make a small subsample, and then count every grain in the subsample, but this is very laborious. Instead, pollen analysts add to the measured volume a "spike" of a known number of grains of exotic "marker" pollen. Temperate zone pollen analysts use the distinctive pollen of the Australian genus *Eucalyptus* for this purpose. Then a pollen preparation is made from the "spiked" sample in the usual way, resulting in a slide containing both the exotic marker and the fossil pollen. Both are counted and the concentration of each of the fossil pollen types can be calculated from the ratios of fossils to marker as follows:

$$\frac{\text{Concentration of fossil pollen}}{\text{Concentration of marker pollen}} = \frac{\text{Fossil pollen counted}}{\text{Marker pollen counted}}$$

With these concentration data alone it is possible to plot a POLLEN CONCENTRATION DIAGRAM[1] in which the pollen importance values are "grains per milliliter" or "grains per gram." This diagram escapes the statistical constraints of a pollen percentage diagram in that the numbers given to any pollen taxon are not dependent on the quantities of other pollen present. *Pollen concentration diagrams,* however, introduce a fresh uncertainty in that they are ratios of pollen number relative to mass of organic or inorganic matter in the sediment. Obviously the rate at which debris other than pollen can collect is highly variable and it is easy for pollen concentration to vary by a factor of ten up and down a section of peat or lake sediments. Pollen concentration diagrams, therefore, tend to be no more informative than percentage diagrams, perhaps less so. The value of a measure of concentration lies in the possibility of calculating pollen influx.

If the sedimentation rate is known from all parts of a sediment column, then the influx calculation can be made. Figure 22.6 is a pollen influx diagram for the sediments of glacial lakes in New England, showing that populations of pines had a complicated history that was not evident in Figure 22.5. It is clear that the percentage

[1]Pollen concentration diagrams are sometimes called "absolute pollen diagrams" or the concentration is referred to as *absolute pollen frequency* (APF). Some pollen analysts believe that this term is misleading and should be abandoned (Maher, 1972; Colinvaux, 1978) but the term still appears in the literature (Birks and Birks, 1980).

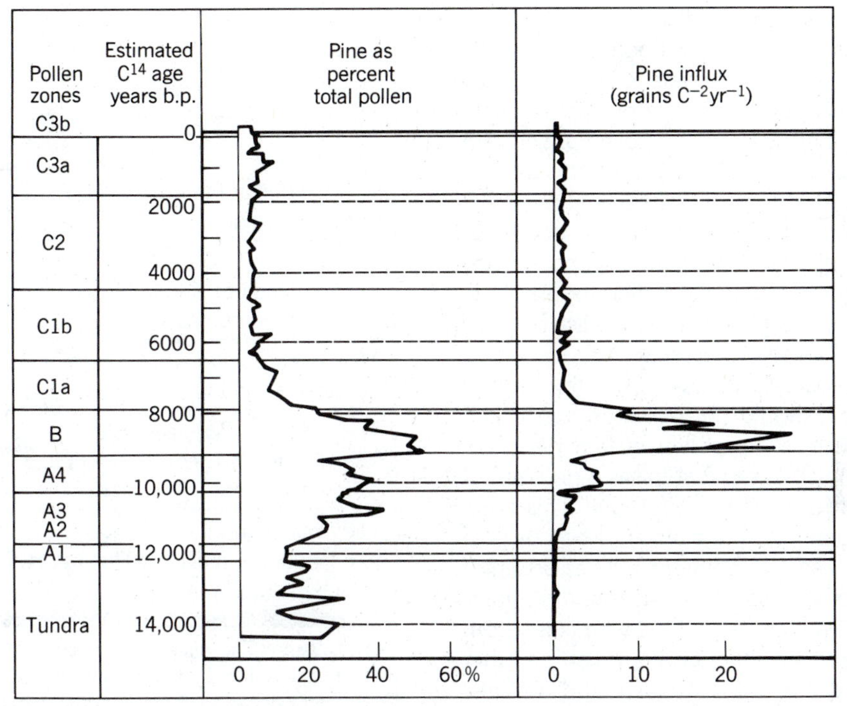

Figure 22.6 Pine pollen percentage and influx in New England.
Calculation of pine influx shows that there were no pine trees in the tundra episode at the bottom of the diagram although pine percentages were high. A true population event in the pines happened later but lasted less than 1000 years. (Data of M. Davis, 1969, from Roger's Lake.)

diagram gives a misleading account of the history of pine (M. Davis, 1969).

Special Problems with Pollen Data

A *pollen spectrum* is a very incomplete species list of a plant community because some plants produce much more pollen than others. Plants are pollinated either by wind (ANEMOPHILY) or by animals, principally insects (ENTOMOPHILY), though hummingbirds and bats serve for orchids and other tropical flowers. Wind pollination means the dispersal of very large amounts of pollen; insect pollination of very little. Furthermore, pollen dispersed by insects is not loosed into physical circulation, except on an insect corpse, so that the pollen analyst rarely finds the pollen of entomophilous plants. A pollen spectrum, therefore, tends to list mostly wind-pollinated plants.

In temperate lattitudes most forest trees are wind pollinated and the pollen spectra in these places are mostly of trees (Figure 22.3). In the Arctic, and on prairies, wind-pollinated grasses and sedges are well represented in pollen spectra, though sometimes much tree pollen blows over open country from distant forests and this will show up in the pollen rain. In the wet tropics most of the trees join the herbs in using animals for pollination. Pollen influx in the tropics thus tends to be very low, but standard pollen counts include more species because the rare kinds are not drowned in a rain of copious wind-dispersed pollen. Tropical pollen diagrams thus have special problems of interpretation (Livingstone, 1971; Salgado-Labouriau, 1980).

It is usually possible to identify pollen only to genus by light microscopy (Figures 22.3 and 22.4). Sometimes it is possible to do better with an electron microscope but the practical difficulties of surveying thousands of pollen grains with this instrument means that we do not often try. So a *pollen percentage diagram* like Figure 22.5 lists mostly generic names only. For some kinds of plants, most frustratingly grasses and sedges, it is possible only to identify pollen to family, so that the taxa in a pollen analyst's list become "Gramineae" and "Cyperaceae."

If a pollen sample is taken from the ground under a tree it should not be surprising to find the sample rich in the pollen of that particular tree. Pollen analysts say that this is a problem of LOCAL OVERREPRESENTATION. This is always a problem with palynology of bog or peat deposits because the sediment collects pollen of bog plants as well as of the surrounding community. The problem is avoided with cores from lakes of modest size, though there can be other problems: pollen is swept by waves and currents, and so may be concentrated by size in different parts of the lake, or it may be mixed down into the mud away from its original layer by burrowing animals (R. Davis, 1967). Though these various problems can lay traps for the unwary, they can be avoided or allowed for with care (Birks and Birks, 1980).

CLASSICAL PALYNOLOGY: THE EUROPEAN SEQUENCE

Pollen analysis was invented early in this century to test hypotheses about climatic change in northern Europe. The hypotheses came from studies on the gross stratigraphy of bogs and suggested that there had been a series of widespread climatic episodes, some six to ten in all, since the end of the last ice age. The proposed sequence of climatic events is called the BLYTT–SERNANDER SEQUENCE (Figure 22.7). The extent to which this climatic reconstruction is correct is central not only to the history of European climate and vegetation but also to the archaeology of Europe.

Climatic Periods
The Subatlantic Period Climate humid, and, especially at the beginning, cold.
The Subboreal Period. Climate dry and warm, much as in central Russia
The Atlantic Period. Climate maritime and mild, probably with warm and long autumns.
The Boreal Period. Climate dry and warm.
The Subarctic Periods of Blytt. The climatic conditions more or less undetermined.
The Arctic Period. In Scandinavia a climate like that of South Greenland.

Figure 22.7 The climatic sequence constructed by Blytt and Sernander from bog profiles in southern Scandinavia. Tree stumps define the Boreal and Subboreal periods, and leaves of arctic plants like *Dryas* define the Arctic period. The Atlantic and Subatlantic periods have peat of relatively moist and warm times. (From Sernander, 1908.)

The Swedish bogs may be more than 5 m thick. Where peat cutters have been working it is possible to observe a cross section of peat, where some remarkable layering becomes obvious. Most striking are layers of tree stumps, erect and in place, showing that trees once grew on the bog. These traces of ancient woodlands might, of course, mean no more than that a particular bog had been drained by some local geographical event and later flooded once more, drowning the trees. Geographical changes of this kind occur all the time, as streams cut out valleys and landslides block them again. But the pattern seemed common to many Swedish bogs, suggesting that changes in climate were responsible for drying and wetting bogs of a large area (Sernander, 1908; West, 1968; Flint, 1971).

Under the bottom layer of stumps was marshy peat, but included among the leaves and seeds of sedges were leaves of the little arctic plant

Dryas, a small, white herbaceous rose with a yellow center, a plant exclusively of the arctic tundra. This bottom layer thus represented vegetation of what could be called an ARCTIC PERIOD. Higher up was the first line of tree stumps, commonly birch, but sometimes other trees too. It seemed certain that a bog must have been partially drained to allow trees to root, suggesting climatic drought, the BOREAL PERIOD. But there is peat between the *Dryas*-bearing bottom mud and the first line of stumps, a gradation probably, but one that could be used as a stratigraphic unit—the SUBARCTIC PERIOD. Above this was more wet peat overlying the tree stumps, doubtless due to flooding and a wetter climate, the ATLANTIC PERIOD. Then there was another line of stumps, commonly pine tree stumps, thus another dry time, called the SUBBOREAL PERIOD. Next was the modern peat that continues to the treeless top of the bogs, which represents the modern climatic epoch, a wet time, the SUBATLANTIC PERIOD. In this way (Figure 22.7) Blytt and Sernander used plant fossils to infer climate and then to separate postglacial time into a series of discrete climatic periods.

From the first moment this scheme was put forward, there was room for doubt. What did we really have other than a record of water levels in bogs? When the bogs dried a little, trees grew on the tops of them; when they flooded again, the trees were drowned. A climatic explanation certainly seemed necessary to synchronize the water levels in bogs all over Scandinavia, but there was nothing in this to show that climate was divided into discrete epochs. All we really had was evidence that the bogs had dried out twice, slowly each time, and enough each time for trees to flourish for a generation or two. There was quite a leap from this to the impression of climatic succession given by Figure 22.7. Pollen analysis was invented to discover what was happening outside the bog itself (von Post, 1916).

Sweden is a fine country for pollen analysis because it is vegetated with but few species of trees, all wind-pollinated. Von Post found that zones in his percentage pollen diagrams matched the periods of the Blytt–Sernander sequence. Since then hundreds of percentage pollen diagrams show that this sequence can be identified all over northern Europe, not just in Sweden. Furthermore, extra pollen zones allow the sequence to be refined. In all, nine (ten granting modern times a thin zone at the top) pollen zones are identified since the retreat of the continental glacier. They are known by Roman numerals, counting from the bottom up.

Figure 22.8 shows a generalized percentage pollen diagram of zones IV through IX from southern Norway, and Figure 22.9 shows the bottom four zones from Denmark. The main additions to Blytt–Sernander by palynology are in the *Arctic Period*. Pollen diagrams showed that there was an extrabrief time of trees, called the ALLERØD OSCILLATION, that left no record in stumps. The episodes of tundra on either side of this forested period are the OLDER DRYAS and the YOUNGER DRYAS. The full sequence is as follows:

IX Subatlantic
VIII Subboreal
VII Atlantic (later)
VI Atlantic (early)
V Boreal
IV Preboreal
III Younger Dryas }
II Allerød } Late glacial, the original Arctic Period
I Older Dryas }

The hypothesis of a sequence of large-scale climatic changes since the last ice age, therefore, could not be falsified by pollen analysis and became established as a working hypothesis (Flint, 1971). This early success of palynology established the pollen tool as of use in paleoecology according to the geological definition. Traces of fossil communities had been used to infer past environments. The march of logic had gone:

1. from pollen spectra describe a plant *formation*.

2. from this plant *formation* describe a climate.

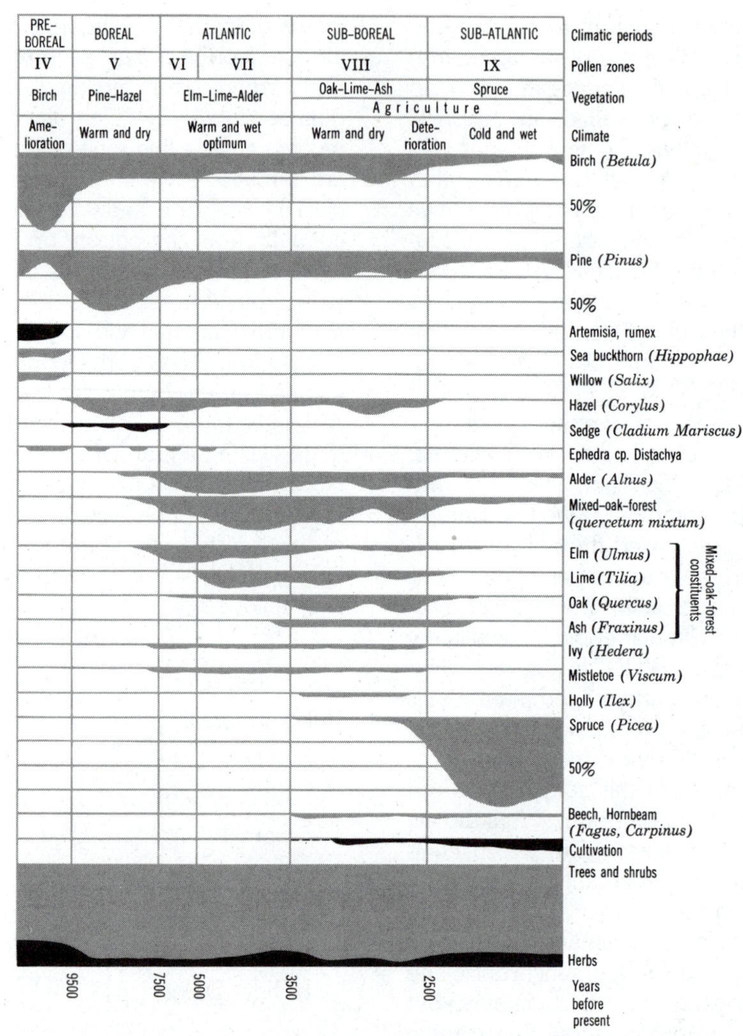

Figure 22.8 Schematic pollen diagram from bogs of southern Norway showing the correlation between Blytt–Sernander climatic periods defined by bog stratigraphy and zones of pollen spectra.
The pollen are shown as percentage total arboreal (tree) pollen, and each division represents 25%. The few herbs in the total count (non-arboreal pollen or NAP) are shown in darker shading. It is common practice to show the relationship of herb to tree pollen in a separate part of the graph, as is done here at the right of the diagram.
(After Hafsten, redrawn from Faegri and Iversen, 1964.)

At the same time palynology came to be useful as a tool for dating. Over the first forty years of palynology there was no radiocarbon dating; indeed no method at all of putting an accurate age on anything older than the invention of writing. But a pollen zone could be a time-stratigraphic marker, just as geologists used older fossils to organize their stratigraphy (Figure 22.1). Up to about 1950, therefore, European pollen analysts often were called in to "date" an archaeological dig or a late glacial event—by collecting pollen from surrounding mud and matching it to one of the nine pollen zones. This did not give ages in years but assigned a dig to its relative position in the archaeological sequence.

In these early circumstances the use of pollen to test ecological hypotheses themselves was obviously limited. Also, without independent dating, calculations of pollen influx were impossible. More seriously, the habit of using assumptions about pollen to construct climates became so established that reversing the logic was not easy.

DEVELOPMENTAL HISTORY OF EASTERN AMERICAN VEGETATION

A second major success story of palynology has been the test of the hypothesis that plant communities are evolved entities whose organization persists through time. This view was held by some phytosociologists when trying to classify plant *associations* into higher natural groupings (Chapter 15). The hypothesis predicts that plant communities of the past would have similar species lists, and similar patterns of relative abundance, to modern plant communities. Reconstructions of ancient plant communities from pollen, if sufficiently complete, would serve to test this hypothesis.

The eastern American forests are very rich in species, only the forests of China being of comparable diversity in temperate regions. Individual taxa in these forests are of great antiquity, for we find fossils of their leaves and fruits in the Miocene and Pliocene beds of Europe and Asia as well as North America.

These complex forests seem to have survived the ice ages in America because there has always been room for their plants to live on unglaciated land, that is, south of the Laurentide ice sheet and away from the mountains (Figure 22.10). The northern parts of the forest's present home might have been taken over by ice but the rest remained.

We can then ask a simple question of pollen analysis. What happened to the rich deciduous forests of eastern North America during the last ice age? Was it always present? Did it migrate? Or did this complex forest not exist at all in those days so that modern associations are merely the product of the recent mixing together of ice age survivors?

These three possibilities can be formally stated as alternative hypotheses to be tested by pollen analysis as follows:

Hypothesis 1 That vegetation was essentially unaffected by the glacial events.

Hypothesis 2 That vegetation migrated south as complete associations so that the latitudinal zoning of modern America was essentially intact but displaced to the south.

Hypothesis 3 That plants responded to changing climate as individuals, so that the different climates of the last ice age resulted in plants being associated in combinations not found in modern America.

Hypothesis 1: Vegetation Unaffected

This hypothesis was put forward and defended vigorously by a school of botanists (Braun, 1950, 1955). They concluded from studies of the vegetation itself that the unique formations of eastern America, with their many species and their ancestries far into the Tertiary, were too ancient and too complex to have been shifted around as belts when the ice sheets came and went. Braun postulated that the glaciers bulldozed their way into the edge of the forest, causing local disturbances beyond the reach of the ice, but no more.

Closeness between advancing ice and rich vegetation is not without precedent elsewhere, since it can be seen in places such as New Zealand, where tropical plants like tree ferns grow within yards of an advancing valley glacier. The hypothesis, therefore, was plausible, but any long pollen record from a critical region should test it (Deevey, 1949; Whitehead, 1967). Such records are now available from, among other places, North Carolina, due to the work of Frey (1953)

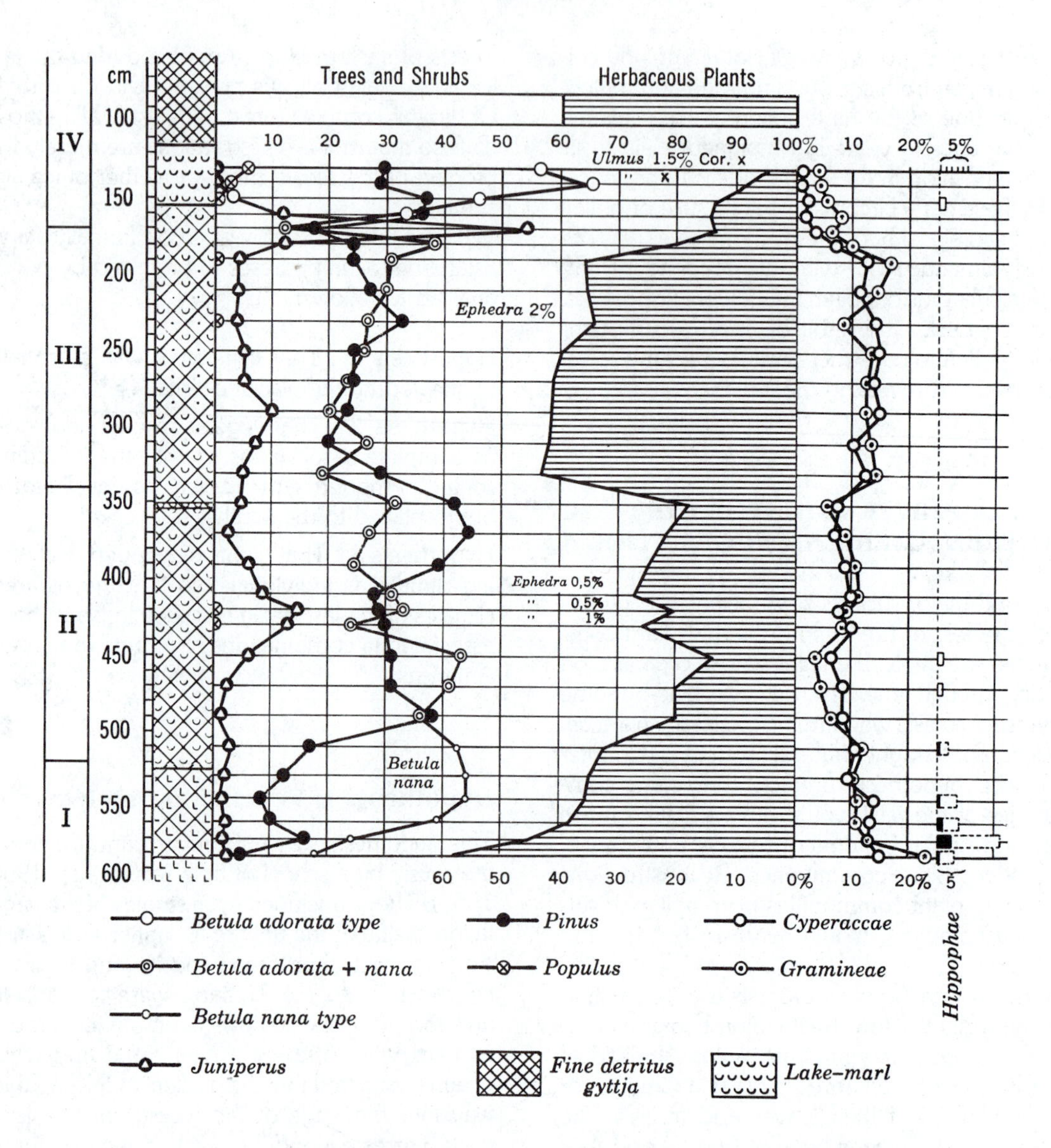

Figure 22.9 Composite lateglacial pollen diagram from Denmark showing the earlier periods added to the Blytt–Sernander sequence.
Zone I is the Older Dryas, zone II the Allerød, and zone III the Younger Dryas. The Allerød mild period shows as a blip in the juniper curve together with a loss of pollen of tundra plants such as dwarf birch, *Betula nana.* The section ends with zone IV, the Preboreal period in the Blytt–Sernander scheme and the base of Figure 22.8. The composite diagram showing the general relationship between trees and shrubs is shown to the left of this figure and it reveals that herb pollen was more important than in the later zones of Figure 22.8. These were times just after the glaciers had gone when trees were beginning to colonize the tundra. The right-hand part of the diagram shows herbs only except for willow *(Salix)*, which is assumed to be a creeping tundra plant and not a heavy producer of pollen. Most of the herbs listed do not have familiar English names but Cyperaceae are sedges, Gramineae grasses, *Rumex* docks, and *Urtica* stinging nettles. (Redrawn from Faegri and Iversen, 1964.)

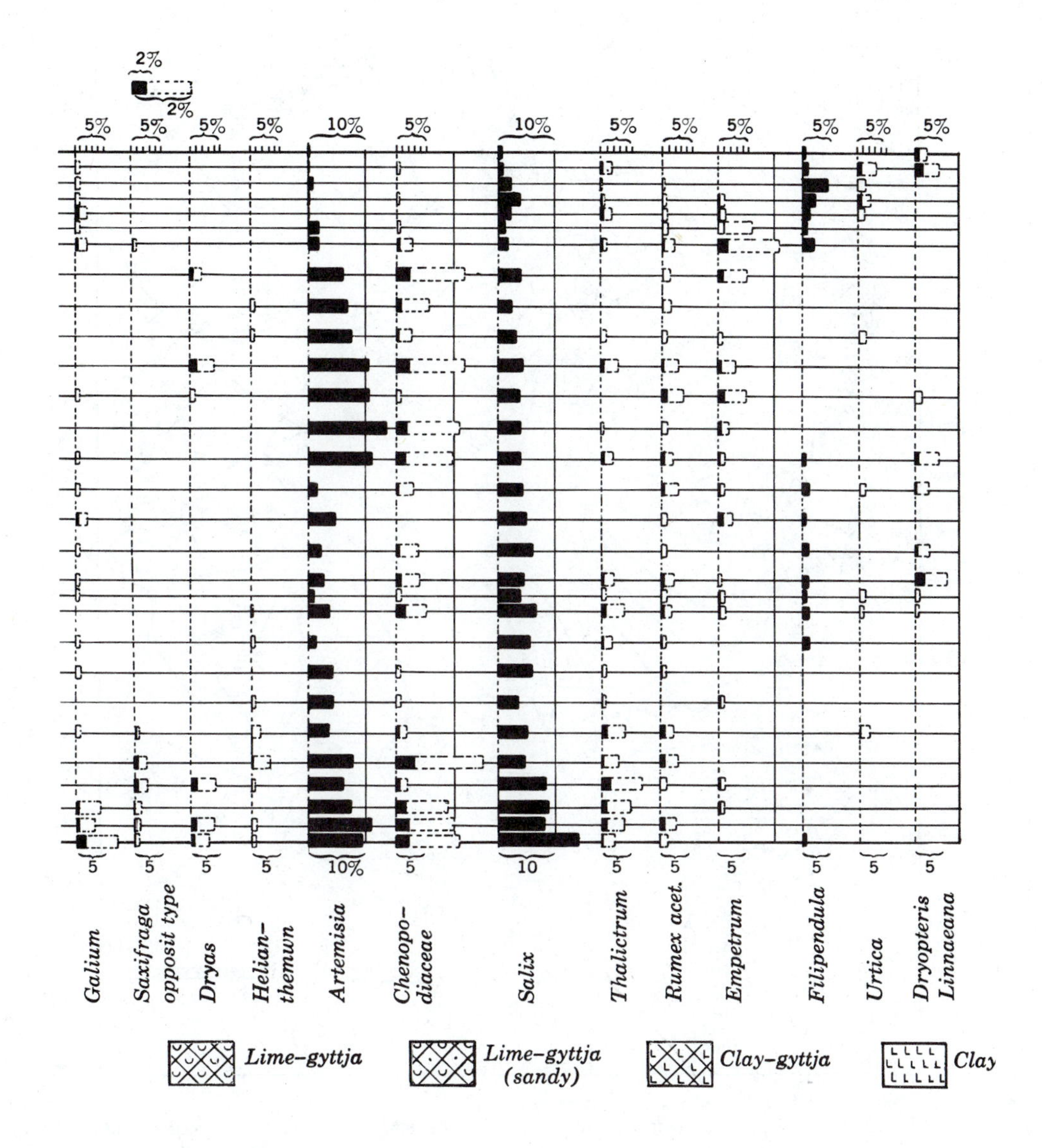

and Whitehead (1964, 1965). Figure 22.11, a composite pollen diagram of Frey's work, shows the completeness with which pollen analysis disposes of the hypothesis. Radiocarbon dating shows that Frey's history spans from sometime during the last ice age on into postglacial time. It is not necessary to attempt an actual reconstruction of vegetation to see that the changes must have been very great during the time covered, even when so far removed from the ice as North Carolina always was. The hypothesis that plant associations beyond direct reach of the ice were unaffected accordingly fails.

Hypothesis 2: Latitudinal Migrations of Associations

To test this hypothesis all a pollen analyst need do is bore lakes and bogs in all the regions to which the bands could have moved, extract pollen spectra of ice age times, and compare them with surface spectra from farther north.

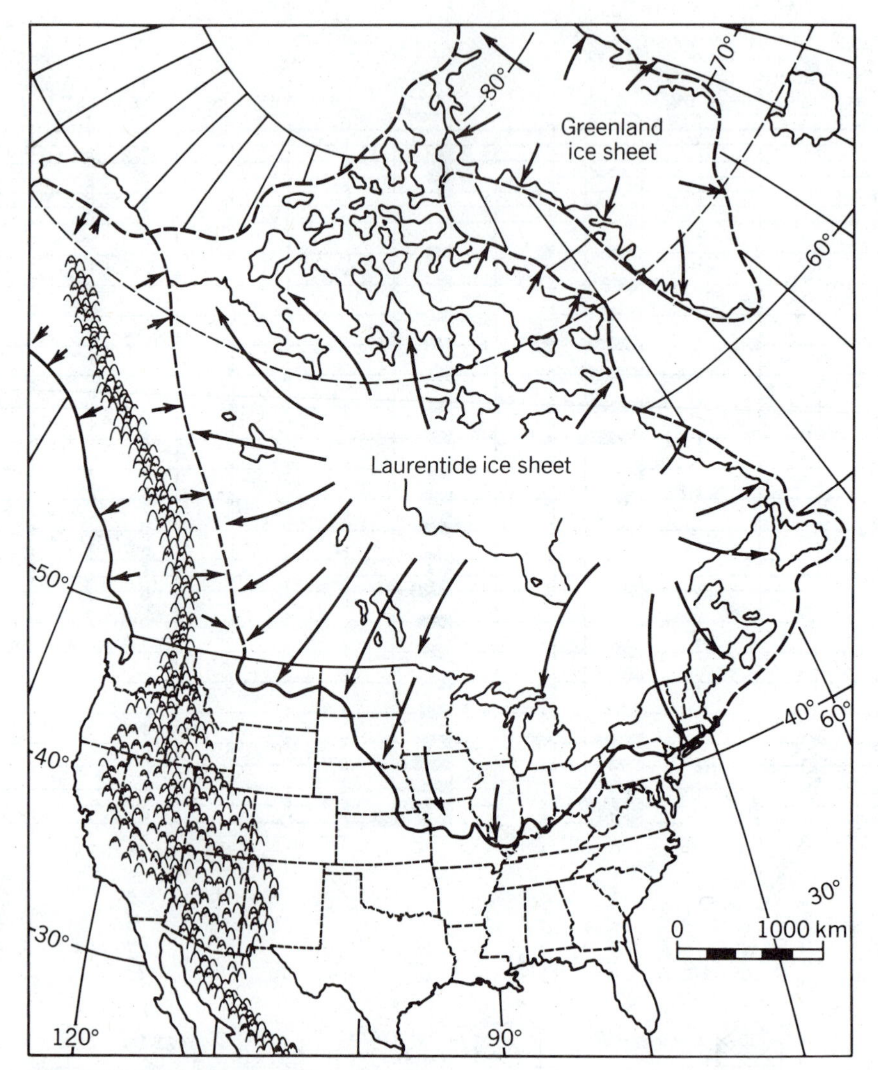

Figure 22.10 Ice sheets of North America.
The Laurentide ice sheet advanced parallel to the mountains leaving ice-free land to the south. (From Longwell *et al.*, 1969.)

Figure 22.12 is a pollen percentage diagram from the Dismal Swamp of Georgia. The earliest pollen spectra in this diagram date from just after the end of the last glaciation, when the continental ice was in full retreat. Here we find pollen spectra of spruce and birch, as well as herb indicators of northern vegetation like *Thalictrum*. In later times, these pollen spectra are replaced by others that suggest temperate deciduous forest until, at the top of the diagram, we find the surface spectrum corresponding to the blackgum, sweetgum, and hickory woods that now occur in the modern Dismal Swamp.

The first glance at this Dismal Swamp diagram seems to support the hypothesis of migrating vegetation belts. Once there was boreal

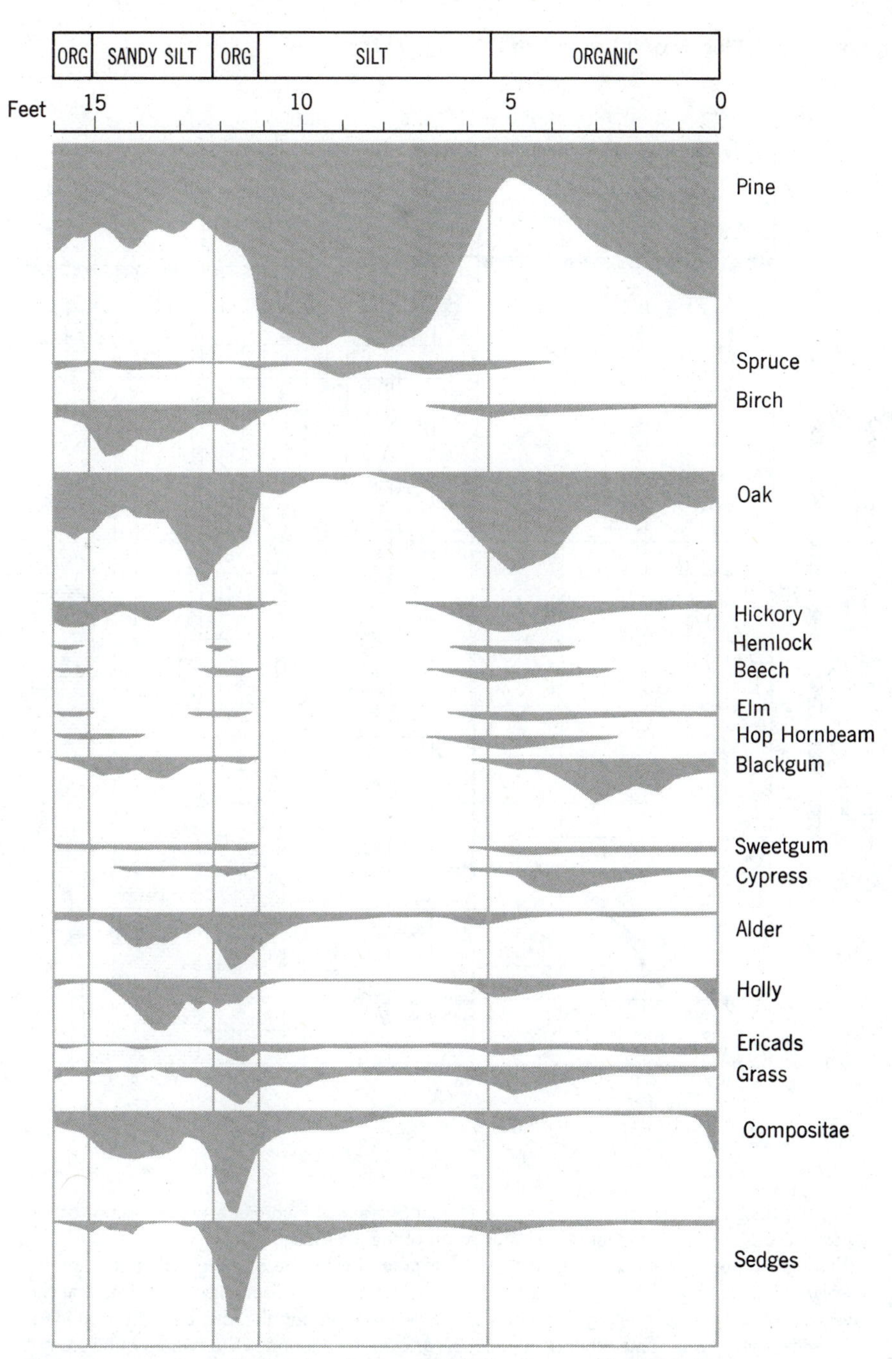

Figure 22.11 Generalized pollen diagram for Bladen County, North Carolina. Radiocarbon dating shows that much of this diagram was contemporary with glacial time, and that the top few feet represent the beginning of the postglacial period. Even without being able to deduce the exact vegetation represented by the various pollen spectra, it is obvious that there have been marked changes in the local vegetation over the time span covered. This sort of evidence seems incompatible with the Braun hypothesis that eastern American vegetation in ice age times was much as we see it now. (From Frey, 1953.)

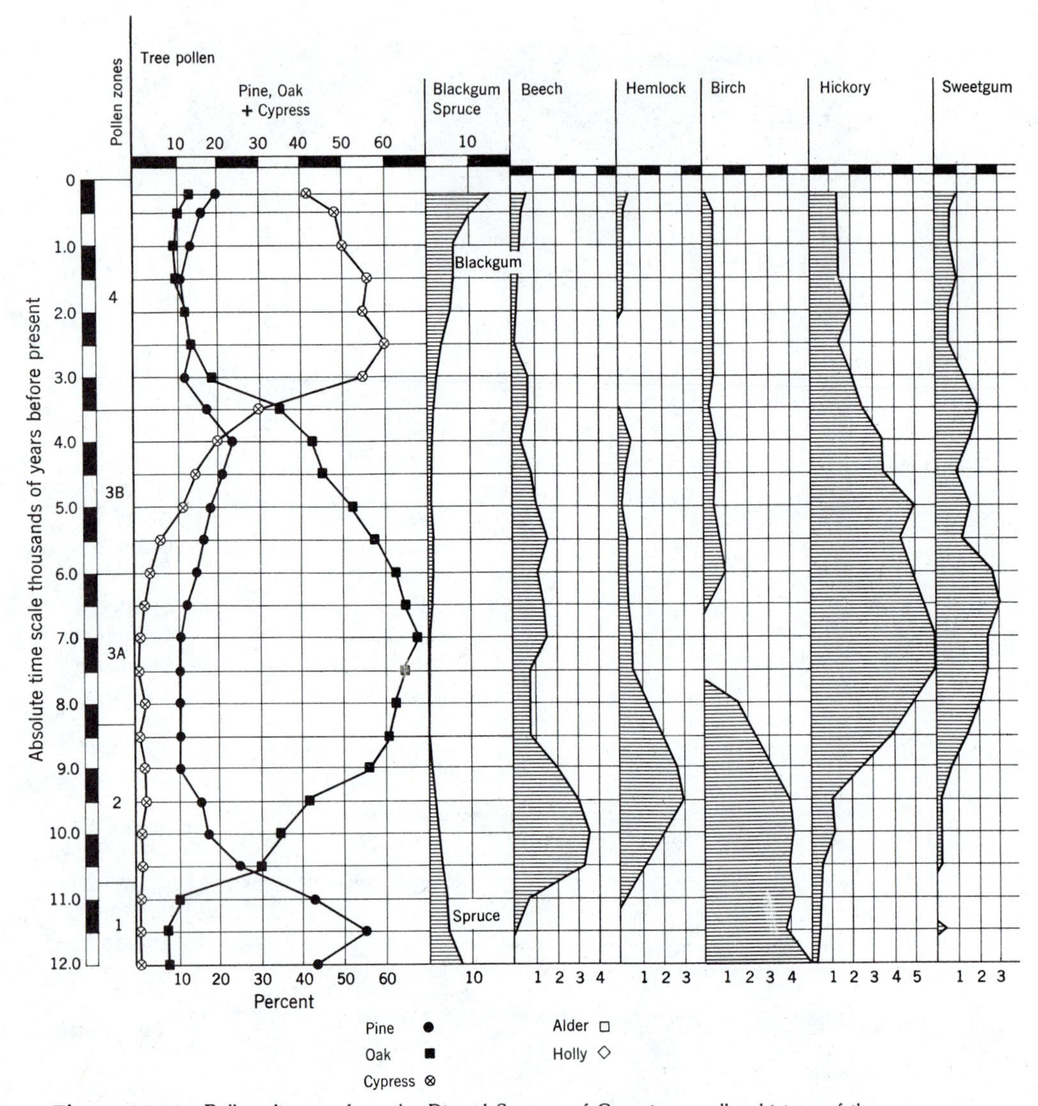

Figure 22.12 Pollen diagram from the Dismal Swamp of Georgia, a pollen history of the postglacial period at a site that was always south of the ice sheets.
Although spruce appears at the bottom of the diagram the spectra of these early times are not really comparable with spectra further north (Figure 22.5). The comparison suggests that the vegetation of Georgia in late glacial times was different from any that can be seen in modern America. Pollen from some other plants such as grasses were found in the swamp but were not included in this count. (From Whitehead, 1965.)

forest there; now there are hardwoods. This crude pattern can be seen in pollen diagrams from all over middle latitudes in the United States and it was one of the props of the original hypothesis of migration (see reviews by Deevey, 1949; Martin, 1958; Whitehead, 1965, 1967).

But closer examination of the pollen spectra of the ancient Dismal Swamp leads to different

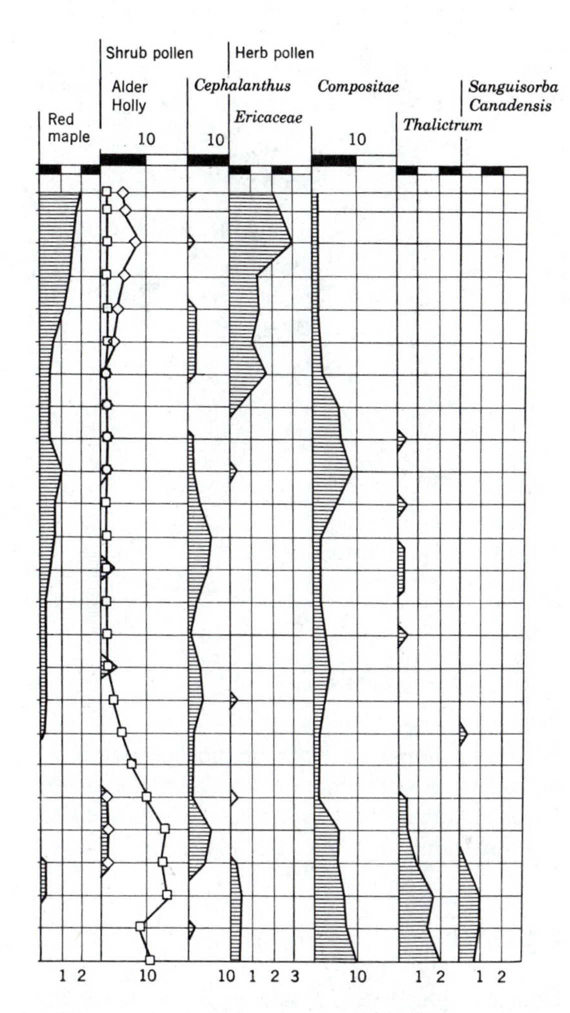

conclusions. The spectra are in fact not quite like those of the present northeast. This can be seen by comparing them with the nearest equivalent spectra in the Vermont pollen diagram (Figure 22.5). At no time in the postglacial history of Vermont were the pollen spectra quite like those of the "boreal period" of Dismal Swamp. Many similar comparisons have been made with the same mix of results (M. Davis, 1982). Trees now characteristic of the northeast grew far to the south in glacial times, but we cannot recognize whole communities of the modern north down there.

The pollen also allows, however, a more subtle attack on the problem. M. Davis (1982) used the data from all published pollen diagrams of the eastern United States to plot the migration paths of principal trees since the time of the last glaciation. She plotted the earliest appearance of pollen or macrofossils (leaves, seeds, *et cetera*) in sediments at each site and connected up *isopleths* of equal age. Figure 22.13 shows just two of her species maps. Few of the species show any synchroneity of movement, and they certainly did not move as multi-species belts.

Hypothesis 3: Modern Associations Absent in the Past

Part of the test of this hypothesis has been discussed above: differential migration throughout the Holocene shows that the associations of these times were not permanent. Some of these data are also given in Chapter 23. But there are also enough pollen diagrams, both percentage and influx, from the full glacial period between 23,000 and 16,000 years ago to examine the composition of those ancient associations directly (Watts, 1970, 1980; Watts and Stuiver, 1980; Delcourt, 1979; Davis, 1982). These data show that the trees of the ice age forests lived in different combinations from those of the present day. They did, of course, tend to match up in formations of trees of similar design, but even the formations merged differently from modern formations so that conifer and deciduous forms appear in novel relationships. The communities would not have been completely novel to a modern botanist but neither would they have matched closely the typical modern *associations* like our oak–hickory or beech–maple forests. Pollen data, therefore, do not falsify the third hypothesis. We conclude that the trees of the eastern United States respond to climatic change as individuals and that communities dissolve with climatic change to be replaced with new communities.

USE OF POLLEN FOR CLIMATIC OR SPECIES POPULATION HISTORIES RECONCILED

There is no real contradiction between using pollen data as a tool for climatic reconstructions, as

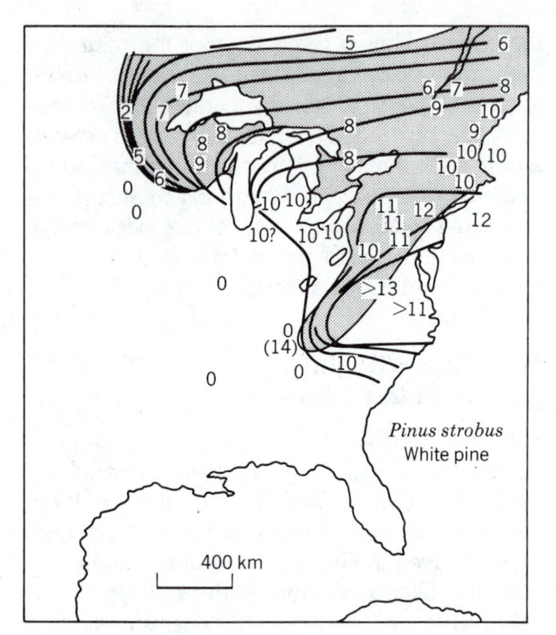

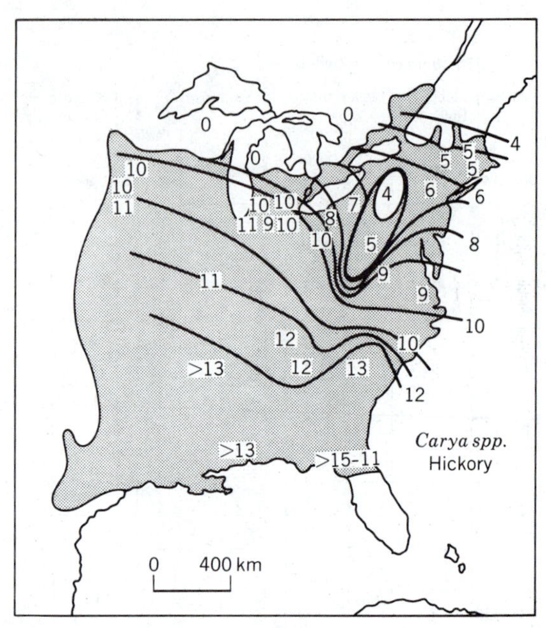

Figure 22.13 Postglacial migration of pine and hickory in the eastern United States. The lines (ISOPLETHS) show the edge of distributions at thousand-year intervals as reconstructed from pollen data. Shaded areas are modern distributions. (From M. Davis, 1982.)

they were first used in Europe, and the conclusions of the latest studies of eastern America that species responded independently to climate. Individual species do respond to changes in temperature and precipitation with altered distributions and this fact can be used to make detailed studies of past climates, particularly by the use of *transfer functions* (described below). It was actually these individual responses that were being used in the classic European work. The European forests had few species, so that the possible reassortments were minimal. Life form is related to temperature and precipitation (Chapters 5 and 14) and records of life form do give good indications of climate. For instance, showing with pollen data that land was covered with treeless tundra, as in early and later *Dryas* times, allows very definite climatic conclusions. But where vegetation is more complex it may require careful separation of the fates of climatically sensitive species to infer climates. Pollen continues to be useful for paleoecology in both the geologist's sense (environmental reconstructions) and in the ecologist's sense (examining the history of species populations). Perhaps the most potent demonstration of pollen data in the latter role is the conclusion that there are no such things as fixed or permanent associations of plants.

DATING

Dating in paleoecology can have the same dichotomy of opposed objectives as the rest of the subject. In dating, this conflict takes the form of "Shall the fossils be used for dating or is there an independent way of dating the fossils?" In classical European palynology, pollen zones were used as a rough system of dating the events of the Holocene, as we have seen. Current practice in North America is to use pollen for dating in one very special instance—the dating of first European settlement. FIRST SETTLEMENT appears as an abrupt rise of ragweed pollen *(Ambrosia)* from a trace to 10% or more of total pollen. Since the date of first settlement can be found from historical records, the AMBROSIA RISE can date a layer near the top of lake sediments to the year. This can be very valuable to a paleoecologist needing time scales over the last two centuries. But for most purposes in paleoecology and paleoclimatology we want a date that is independent of the fossils to be used. Radiocarbon is the mainstay within its range, but there are other isotopic and comparative techniques available as well.

Radiocarbon Dating

The isotope ^{14}C is synthesized in the upper atmosphere out of ^{14}N. The energy source is cosmic radiation, which induces a flux of neutrons in the upper air. A neutron striking a nitrogen atom is the primary synthesizer of ^{14}C:

$$^{14}N + n = {}^{14}C + H^-$$

The radiocarbon so formed decays spontaneously back to nitrogen, emitting a beta particle in the process:

$$^{14}C = {}^{14}N + \beta^-$$

Beta decay of ^{14}C proceeds at a constant rate, giving a HALF-LIFE of 5730 years. If cosmic bombardment of the earth was really constant, then the atmosphere would have a constant composition of ^{14}C depending on the rate of synthesis due to cosmic bombardment and the rate of beta decay. It is now known that the intensity of cosmic bombardment has fluctuated somewhat over the last ten thousand years, but not by very much. For many dating purposes we can assume constant bombardment and the consequent constant concentration of ^{14}C in the atmosphere. This ^{14}C is oxidized and remains in the air as $^{14}CO_2$, in which form it is incorporated by plants in photosynthesis. All living things, therefore, contain ^{14}C in their tissues, the concentration being a function of the concentration of $^{14}CO_2$ in the contemporary air.

Radiocarbon dating is merely an estimation of how long organic matter has been dead by a proportional measure of how much ^{14}C remains undecayed. We know the concentration of ^{14}C in the living tissue, the half-life of ^{14}C, and the residual ^{14}C in the corpse, so it is a simple matter to calculate the time of death. But there is a big catch. Until recently it was not possible to measure residual ^{14}C directly, because the concentration is so low. What can be measured with comparative ease is the beta decay rate. Instead of measuring all the ^{14}C atoms in a sample, therefore, we have been reduced to measuring that tiny fraction of atoms that are actually emitting their beta particles in our laboratory.

The standard method of radiocarbon dating is to burn the sample to convert all the carbon into gas, which is collected. Some laboratories insert the resulting $^{12}CO_2$–$^{14}CO_2$ mixture directly into their counting chambers; others convert the carbon to methane. The actual counting operation is to place the flask of gas within a ring of Geiger counters, thus to record the rate of beta emission. This rate is, of course, a function of the ^{14}C remaining in the carbon mixture and allows the age calculation to be made.

The effective range of this standard method is about 35,000 years. For samples older than this there is so little radiation left that it cannot be measured accurately. A few dates back to 60,000 years have been reported, though special enrichment techniques are needed and these ambitious datings have remained few (Stuiver *et al.*, 1979). Use of Van de Graff accelerators in tandem, or a cyclotron, as extremely sensitive mass spectrometers for the measurement of ^{14}C directly may eventually allow dating of much older samples (Muller, 1977; Bennett, 1979). The first accelerators built for dating are now (1984) in service. As yet they can be used to date very small samples but not to date very old ones.

Other Methods of Dating

In principle, any system of radioactive decay can be used as a clock, always providing that the INITIAL CONDITIONS at the time the clock was started are known. For ^{14}C the assumed initial condition was an atmosphere in equilibrium for ^{14}C with a constant cosmic bombardment. From this assumption it follows that ancient plants held ^{14}C in the same proportion as ^{14}C in living plants. With other dating systems the biggest difficulty usually is in knowing the initial conditions when the clock was set.

Several methods have been devised using the URANIUM DAUGHTER SERIES. The world oceans are so large that the concentration of uranium in seawater is either constant or else changes negligibly within a few hundred thousand years. ^{238}U decays to a series of daughter nuclides, of which thorium (^{230}Th) has a very long half-life (8×10^4 years). But thorium is precipitated in sediments so that the sea is swept clear of thorium. Any crystalline process that traps uranium from seawater, therefore, carries within it a private clock. The *initial conditions* are an input of ^{238}U and no thorium. Provided the crystal is properly isolated from its environment it begins to collect ^{230}Th from the decay of its own ^{238}U. To date such a crystal all we need do is measure the uranium/thorium ratio. This method has been applied to calcite in oceanic sediments, most notably in the carbonates of coral reefs. It gives ages in the range 5000 to 350,000 years, and is most useful in the 100,000 to 300,000 bracket that cannot be reached by radiocarbon (Broecker and Thurber, 1965; Hedges, 1979).

Uranium daughters can also be used for dating bones. Living bone is essentially without uranium but dead bones trap uranium from groundwater. Bones serve for a uranium trap, however, only as long as they contain organic components as well as the bone mineral, apatite. The organic matter in buried bones quickly rots, thus leaving only a short window of time in which uranium is collected from solution. This sets the clock, and the *initial conditions* are dead bones with uranium but no daughter products. As with oceanic calcite crystals the $^{238}U/^{230}Th$ ratio can be measured to yield a date. A particular hazard with bone dating, however, is that ^{230}Th can be deposited into the sample from outside. Uranium provides an alternative clock, though with a shorter range, which can be used as a check against this. In addition to ^{238}U there is the isotope ^{235}U in the bone samples, and this decays to protactinium (^{231}Pa). If the $^{235}U/^{231}Pa$ and $^{238}U/^{230}Th$ clocks give the same age, there has been no contamination. This method has been used recently to demonstrate that human remains in California, which were once thought to be so ancient as to be puzzling, are in fact no more than 11,000 years old like other traces of early man in the Americas (Bischoff and Rosenbauer, 1981.)

Young sediments formed over the last few decades can be dated by isotopic clocks also. The radionuclide cesium-137 (^{137}Ce) was produced by bomb-testing in the atmosphere and reached its highest concentration in lake sediments around 1963 (Pennington *et al.*, 1976). With a half-life of only 8 hours its usefulness should soon end. The lead-210 method remains, however. ^{210}Pb is in the uranium daughter series, the immediate product of the gas radon-222, and has a half-life of only 22 years. Since ^{210}Pb reaches lake sediments both from atmospheric fallout and from uranium series decay within the sediments, care has to be used in determining the *initial conditions* (Brugam, 1978). Dates back to 150 years are possible and the method has value in studies of human impact on lake systems.

Other dating methods continue to be invented. For example, amino acids in bone collagen change spontaneously from L-isomer to D-isomer depending on temperature.

TWO PROBLEMS IN ANTHROPOLOGY

Landnam and Elm Decline

Within the synchrony of the European pollen zones is a smaller synchronous event, the ELM DECLINE. This is a sharp drop in the percentage of elm pollen from over 10% to about 5% or less. The decline happened about 5000 radiocarbon years ago and was completed within 300 years. Elm, although wind pollinated, is not a heavy pollen producer and the previous pollen percentage of more than 10% means a large proportion of elm trees in the European forests. There was also oak, birch, lime (an Old World linden), ash, and alder in the forests. Then something happened to the elm population that affected all the forests of northern Europe within about 300 years (Figure 22.14). The decline of elm pollen typically is accompanied by increases

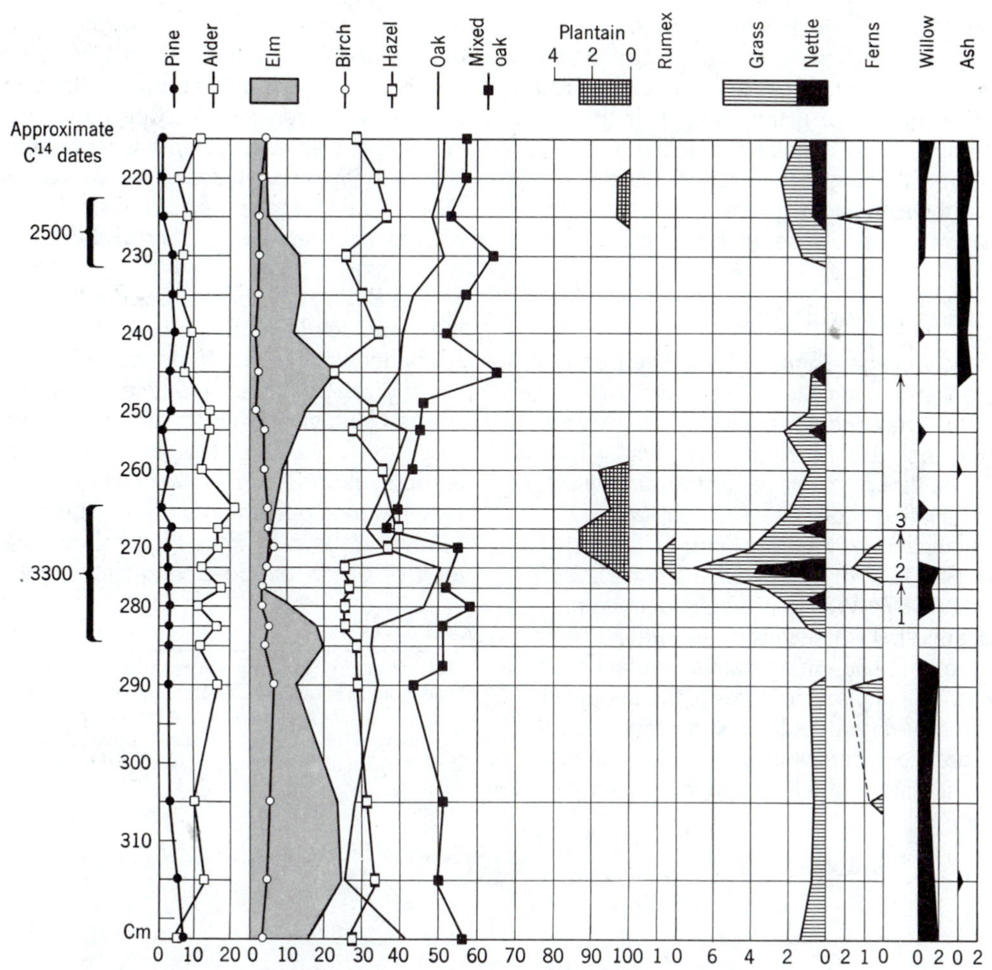

Figure 22.14 Pollen percentage diagram from Ireland showing an elm decline. The elm decline is accompanied with an increase in weed pollen that might suggest agriculture. Rival hypotheses to account for this are disease of elm trees or early shifting agriculture, a so-called *landnam* event. (From Smith and Willis, 1961.)

in the pollen of weeds like plantains, docks, and nettles. With this discovery, we are presented with evidence of an ancient population event that needs explaining. There are two hypotheses in vogue: human activities and an outbreak of Dutch elm disease.

Iversen (1949) first postulated that early agriculturalists learned to cut down trees in patches, burn the stumps, and grow a cereal crop in the ashes. He called it LANDNAM, meaning a "land-taking." This explained the decline of elms and the rise of weeds. Pollen of other trees remained from nearby forests. Later versions of the hypothesis suggest that elm was singled out over other trees for human attack because its leaves make good forage for animals. People cut the branches down for forage year after year, thus killing the trees (Garbett, 1981).

The data, however, can be interpreted just as well as representing an outbreak of Dutch elm disease, suggesting that the people of 5000 years ago saw their elm trees die just as we of this generation have seen them go. The weed pollen might represent weeds invading open spaces left by dead elms, or early agriculturalists might have taken advantage of the spaces left by dead elms to farm there. The true cause of the elm decline remains a matter of active debate. If the elm disease hypothesis turns out to be the true one, it may show that forest trees suffer recurrent epidemics thousands of years apart—revealing population histories of some theoretical interest.

The Bering Land Bridge

Pollen percentage diagrams have been used in the geological way of inferring environments to reconstruct the conditions in which the ancestors of aboriginal American peoples lived. These people are thought to have lived around what is now the Bering Strait between Alaska and Siberia, but which was then dry land: THE BERING LAND BRIDGE or BERINGIA (Figure 22.15, Hopkins, 1967). We do not have definite proof that all ancestral Americans came from *Beringia,* but there is strong circumstantial evidence that they did. The physical affinities of the people are with present-day populations of central Asia. There are no records of man in the Americas before the time of the last ice age, though there are plenty in the Old World, showing that people did invade the Americas from elsewhere. If a Bering Strait crossing was not made, then one of the great oceans had to be crossed by migrating peoples at an early date, something that is inherently very improbable. Moreover, the Bering land bridge offered a terrestrial homeland halfway between the New and Old Worlds at the critical time.

The *Bering land bridge* was dry land because the world sea level was lowered in the ice age. A continental glaciation stores water in glacial ice, the total volume thus held in the great ice sheets being very large (Figure 22.10). As a result, world sea level falls by a corresponding amount: about 100 m at the ocean surface being required to supply the required volume of water for the ice sheets.[2] Figure 22.15 shows the huge area of ocean beside Bering Strait that is less

[2]This process of lowering sea level by removing water to build glaciers is called EUSTASY and is distinguished from ISOSTASY, which is the apparent movement of sea level by the raising or lowering of a coastline through relative depression or raising of the land because of the weight of ice it supports.

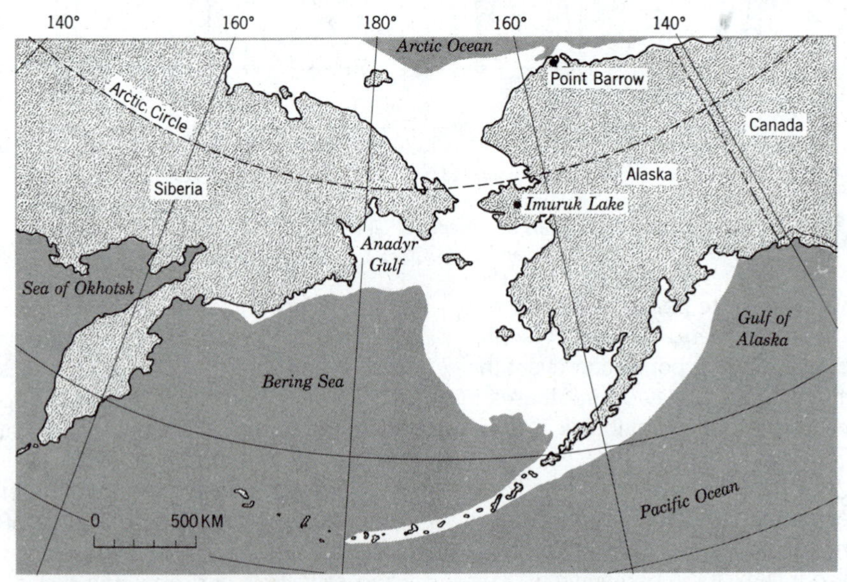

Figure 22.15 Map of Bering Sea region at the time of the last glacial maximum, showing the coasts of the Bering land bridge.
This was the land exposed when sea level fell by 100 m. (Redrawn from Hopkins, 1967.)

than 100 m deep at present and so was drained at the last glacial maximum. This drained land of Beringia had no glaciers on it, thus being cleared land available to people. But this raises the issue of what that land was like. Putting the question differently, what were the living conditions in so arctic a place in a time of world glaciation? Pollen diagrams from ancient lakes in the region have answered this.

Figure 22.16 shows the top 50,000 years or so of a pollen diagram from Imuruk Lake, near Bering Strait (see Figures 22.17 and 22.15; Colinvaux, 1964). Imuruk Lake is surrounded by tundra, the nearest spruce tree being 50 miles away, and yet there is spruce and alder pollen at the top of the diagram. This is explained because enough spruce pollen to make up 10 or 15% in a tundra pollen diagram can be blown many miles beyond the tree line. The pollen diagram ends with a time a few thousand years ago (the top sediments have been removed by wave action) when spruce and alder were near enough for their pollen to have been blown from the tree line to the lake. Earlier in lateglacial time, zone K, spruce and alder pollen must have failed to reach Imuruk, suggesting that the tree line was still far away. The birches, which show as a maximum in this zone, were tundra birches, dwarfs living among sedge tussocks, not trees. They are common around Imuruk Lake now and the apparent surge of their pollen in zone K time was because of the withdrawal of spruce and alder pollen from the pollen sum. This is an artifact of the pollen percentage method. But back in the time of the land bridge, zone J, the dwarf birches had waned, and the pollen sum was largely made up of grasses, sedges, and some of the low sage plants of the arctic, *Artemisia*. This vegetation was certainly a tundra, probably devoid even of the more shrubby tundra plants. It is now known that similar vegetation existed at all latitudes of the Bering land bridge, from the southern land bridge coast where it met the Pacific Ocean, to the edge of the Arctic Ocean to the north, and eastward into the Yukon and northern Canada (Colinvaux, 1964b, 1965, 1981; Cwynar and Ritchie, 1980; Ager, 1982). But it has proved difficult to decide what the vegetation was like in detail since no modern tundras yield pollen spectra that are exactly like it. Pollen influx was

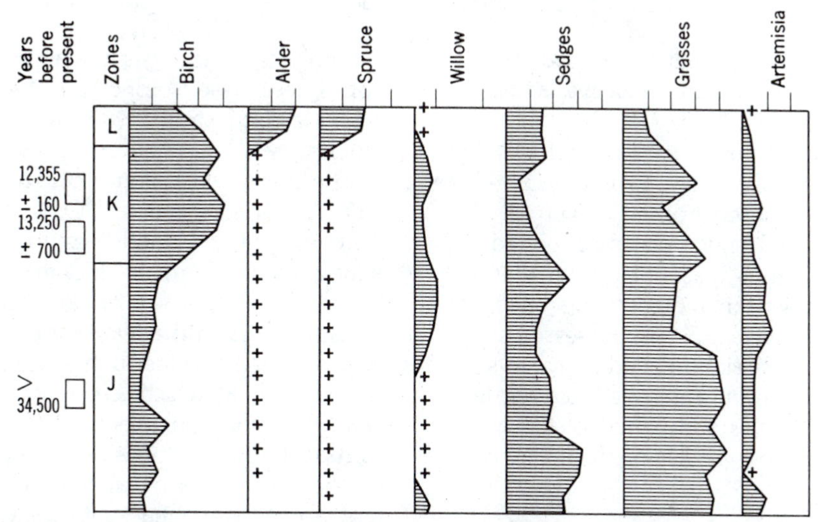

Figure 22.16 Pollen percentage diagram from the Bering land bridge. Data are from the top section of a core from Imuruk Lake (Figures 22.15 and 22.17). The three-zone history shows first an herb tundra, then a population increase in dwarf birches, and finally the approach of a distant tree line. The Bering land bridge existed only during the earliest, most extremely arctic period (zone J). (From Colinvaux, 1964a.)

Figure 22.17 Imuruk Lake on Seward Peninsula, Alaska.
The shrub tundra in the foreground consists of a turf of sedges, heaths, dwarf birches, and prostrate willows. The tallest plants in the photograph are willows, which grow best along drainage channels and, as here, at the side of the lake. White dots in the left foreground in front of the reindeer are the fluffy seed heads of cotton grasses. Pollen records from the lake mud suggest that this tundra is the least arctic during the lake's history of tens of thousands of years.

apparently lower than that from modern tundras, suggesting a very unproductive landscape (Cwynar and Ritchie, 1980).

We know that a tundra such as that revealed by the pollen history of the Bering land bridge meant a cold arctic environment. But we also know that the land bridge plains supported big game. We may thus reasonably infer that the ancestors of American Indians were big-game hunters of those ancient plains, possessed of the skills and social cohesion needed to live in such a place. Instead of being half-naked savages, they had boots and gloves and parkas of fur, as do the modern Eskimos who live on the arctic coast.

THE USES OF OXYGEN ISOTOPES

The isotopes of oxygen have been used to record temperatures of the past directly, or to supply chronologies of ice ages in marine sediments. In both roles, these isotopes provide some of the most elegant instruments in the paleoecological tool kit.

There are two isotopes of oxygen, ^{16}O and ^{18}O. The heavy isotope, ^{18}O, is stable, just like the commoner ^{16}O, and does not decay. Its usefulness comes from the fact that it enters into chemical reactions as a function of temperature. Thus the ratio of the two isotopes can, in principle, be used as a thermometer that records the temperature at which a chemical synthesis or change of state took place.

As with using isotopes for dating, it is essential to choose materials that have remained unaltered since the original synthesis, which requirement has directed much ^{18}O work to marine carbonates, particularly the mineral calcite. Over temperatures of interest to biologists the changes in concentration of ^{18}O in carbonates are ex-

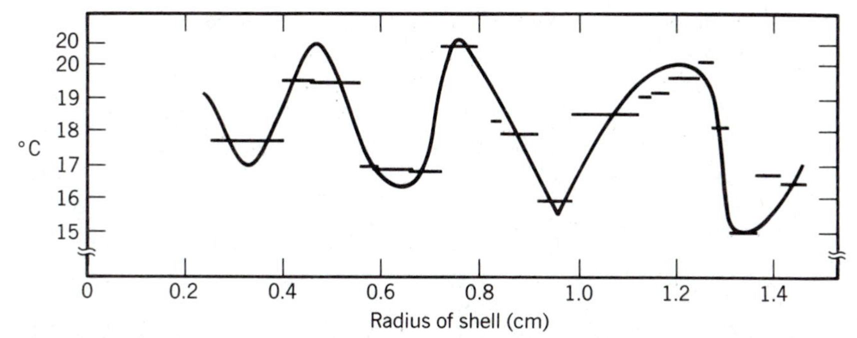

Figure 22.18 Seasonal temperatures of the environment of a Jurassic belemnite. Temperatures computed from $\Delta^{18}O$ in successive growth rings in the shell. Horizontal lines show thickness of sample used. (From Urey *et al.*, 1951.)

tremely tiny. A change of 0.5°C, for instance, results in a change in the abundance of ^{18}O of only 1 part in 25 million. Extremely sensitive mass spectrometers are required to detect accurately such changes, although several laboratories now achieve these results almost as a matter of routine.

The first demonstration of the possibilities of ^{18}O for ecology was made by Urey *et al.* (1951), who examined the shell of a belemnite from the Jurassic. Belemnites were free-swimming molluscs related to squids, but they had rather massive external shells of calcite. Sections of these shells reveal a pattern of concentric rings, almost certainly growth rings. Urey and his colleagues made a slice through a belemnite shell and assayed the $^{16}O/^{18}O$ ratio for each of the broader rings (Figure 22.18). They put their conclusions as follows:

"This Jurassic belemnite records three summers and four winters after its youth, which was recorded by too small amounts of carbonate for investigation by our present methods; warmer water in its youth than its old age, death in the spring, and an age of about four years."

This is a post-mortem on an animal that had been dead for more than a hundred million years.

This demonstration had, of course, also revealed the temperature of that part of the Jurassic ocean; a mean temperature of 17.6°C with a seasonal variation of about 6°C. Obviously there are powerful conclusions that can be drawn about any marine environment that left suitable calcite fossils as we describe temperature and seasonality in advance of collecting community and population data.

Foraminifera and Ocean Dilution

The ^{18}O method was next applied to foraminifera tests. These calcite skeletons of planktonic and benthic animals are found in large numbers in piston cores from ocean sediments. The planktonic species that live near the sea surface can be separated from the benthic forms. It seemed that oxygen isotope measurements from foraminifera up and down deep sea cores would yield a history of the temperature of the oceans, perhaps water mass by water mass. This was a particularly exciting prospect for those interested in working out changes in the world climate with the coming and going of ice ages, and the possibilities were pursued vigorously, particularly by Emiliani (1966). Oxygen isotope studies did indeed prove immensely fruitful to climatic historians, but not quite in the way that was expected. Foraminifera turned out to record ocean dilution more strongly than ocean temperature.

Among the changes of state that affect the $^{16}O/^{18}O$ ratio are evaporation and freezing of water. Glacial ice has a different ratio from that of water at ocean temperatures. When water was withdrawn from the oceans to make continental ice sheets the oxygen isotope ratio of the surface of the oceans was changed. When the glaciers melted, the ratio in the surface waters changed again. Thus it turned out that the $^{16}O/^{18}O$ ratio of surface ocean, as recorded in tests of planktonic foraminifera, was a sensitive measure of the relative glacial state of the earth. This direct record of glaciation actually drowned out the indirect record of temperature (Shackleton, 1968).

Figure 22.19 is a curve of the relative depletion of ^{18}O in sediments from the deep sea. Ages have been put against each of the isotopic events by extrapolating from radiocarbon and uranium daughter series estimations. Some of the dating is as yet tenuous, and attempts are now being made to improve the resolution of this chronology. It is likely also that some of the relative changes of oxygen isotopes record temperature as well as dilution effects, and work on these possibilities proceeds. But this oxygen isotope curve as it stands yields the most convincing evidence of the magnitude and chronology of climatic change in the later ice ages that we have. It has been used to test predictions of the astronomical theory of ice ages (Hays *et al.*, 1976).

THE POSSIBILITIES OF A C4 THERMOMETER

It may be that a land paleothermometer can be found that has an accuracy comparable to that of oxygen isotopes in marine calcites. The possibility relies on the fact that different photosynthetic pathways tend to be used by plants at different temperatures (Chapter 3). The C4 pathway is predominant in hot, dry places and the C3 pathway in cooler, moister places. Livingstone and Clayton (1980) suggest that we use plant microfossils from lakes to infer the photosynthetic pathways of local plants, and hence the local temperature.

It seems particularly reasonable to apply the method to grasses, since whole genera or subfamilies within the grass family Gramineae use either one pathway or the other, but not both. Livingstone and Clayton show that in East Africa the importance of these grass genera in the local vegetation depends on altitude and temperature. The C4 grasses are in the hot lowlands and the C3 grasses in the cool highlands. They also show that this distribution is indeed a result of temperature and not some other function of altitude. Using this fact to reconstruct past temperatures then becomes a matter of determining which genera predominated at any chosen level in a lake sediment core. Unfortunately this cannot be done by pollen analysis, since all grass pollen looks alike. But the East African sediments have many charred grass fragments that were blown into the lakes as dust from old fires and these fragments of tissue often can be identified to genus. The paleoecologist must hunt for

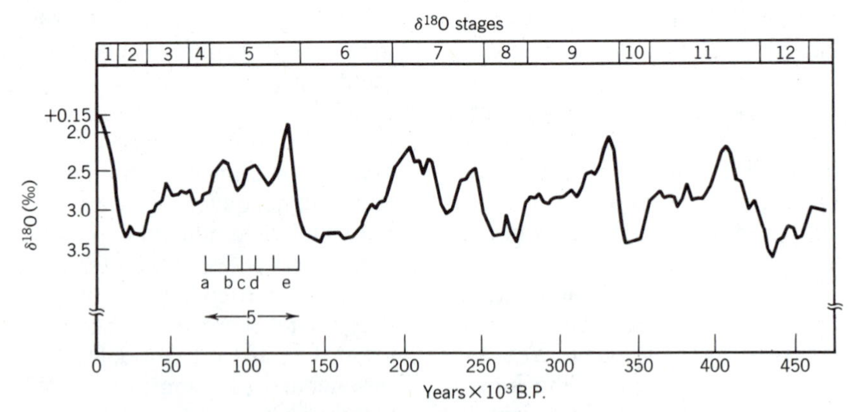

Figure 22.19 $\Delta^{18}O$ record from deep sea cores.
Low $\Delta^{18}O$ represents interglacials like the present and high $\Delta^{18}O$ glacial times when ice sheets differentially stored ^{16}O. The last interglacial is divided into stages 5a–5e. The chronology is derived by extrapolation from various radiometric dates. (Data of Hays *et al.*, 1976.)

microscopic, charred fragments of grass cuticle to use this thermometer, and also must learn to recognize grass genera from cuticular fragments. The possibilities are interesting, though the method is still being explored (1983).

TRANSFER FUNCTIONS APPLIED TO FORAMINIFERAL ASSEMBLAGES

Maps can be made of the distribution of planktonic foraminifera in the oceans. The result is something like a vegetation map in that broad groupings of foraminifera can be found in different oceans in the way that broad formations occupy areas on land. The most plausible hypothesis to explain this pattern is that it reflects water temperature. This should mean that a species list of fossil foraminifera from a deep sea core holds a record of the temperature of the sea surface.

Imbrie and Kipp (1971) showed how this temperature record might be read using the technique of TRANSFER FUNCTIONS. The method depends on finding best-fit correlations between the distribution of modern foraminifera and isotherms of the sea surface. Species lists of foraminifera from an ocean can be quite large, with perhaps 4 to 600 species being present. Whole species lists are examined, species by species, in the hunt for those whose distributions do coincide with measures of temperature. Many species do not show clear correlations and are accordingly rejected. In the end a grouping of species is left, the maps of which are good fits to maps of water temperature. From this list are calculated the transfer functions that derive water temperature from community composition.

These transfer functions are then applied to species lists of fossil foraminifera from the cores, again rejecting the species that are known to fail to meet the requirements of the model. The result is a measure of the surface water temperature at the core site at the time the foraminifera lived. The method has been applied to core material with radiocarbon ages of 18,000 years from all the world oceans to yield a map of surface temperature at the last glacial maximum (Figure 22.20). This is the map on which models of global

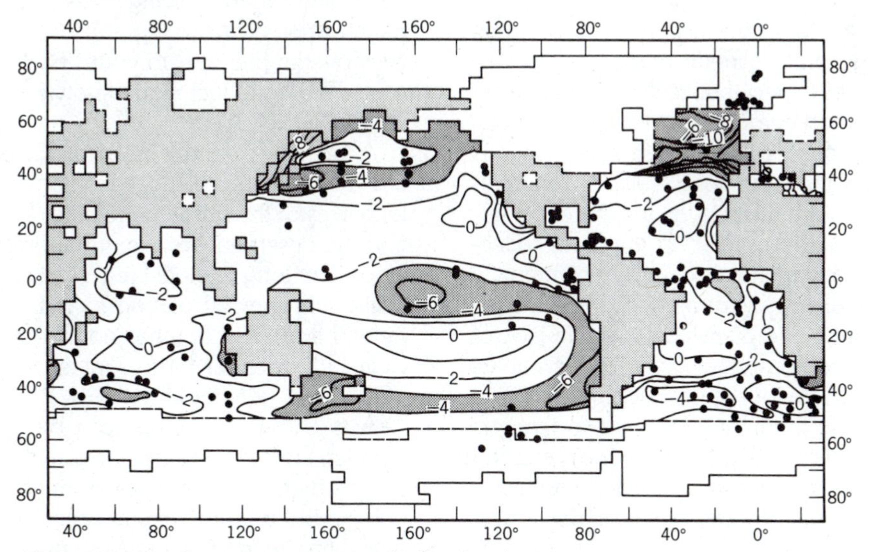

Figure 22.20 Ocean temperatures at the last glacial maximum.
Isotherms show temperatures as they differed from the present day. The data were reconstructed from species lists of foraminifera from deep sea cores using transfer functions correlating modern populations with temperature at the sea surface. Data sites (deep sea cores) are dots. Coldest water shaded. (From CLIMAP, 1976.)

climate of the last ice age are now based (CLIMAP, 1976).

It is important to notice that this transfer function method of Imbrie and Kipp makes no claims about communities of animals. There is no suggestion that discrete communities of foraminifera exist in nature; nor that communities have boundaries set by temperature; nor even that temperature is an overriding limiting factor for individual species. All that is claimed is that many species are affected by temperature, a proposition that no ecologist is likely to resist. The investigators put together artificial groupings by their computer selection program, seeking to amplify the temperature signal hidden in a species list. Then they sort out analogs of these artificial groupings in the fossil assemblages. The maps of ancient oceans that they have produced are consistent with other data we have about the ice age earth, showing that their assumptions were sound.

TRANSFER FUNCTIONS APPLIED TO POLLEN DATA

In principle, the transfer function method can be applied to generate climate maps from pollen data as well, but special difficulties threaten the precision that can be obtained. The most serious is that the species list of a pollen assemblage is much smaller than the species list of foraminifera: all the foraminiferans in the parent community may be represented in the fossil collection but not many members of a plant community turn up in a pollen spectrum. A second difficulty is that perennial one of palynology, the lack of precision in taxonomy that gives us the names of genera only: obviously a genus can have species with widely different tolerances to temperature. And a third difficulty is that most existing pollen data are in percentages rather than as influx. All surface pollen data are calculated as percentages.

Nevertheless there are some striking demonstrations of the possibilities inherent in the method (Webb and Bryson, 1971). The method begins by scanning surface samples and the fossil pollen spectra from the region to be worked in a subjective quest for pollen taxa that look as if they might have fairly reliable correlations with climatic variables. All taxa that are scarce in pollen diagrams, or that appear at the surface as a result of human activity, or that are common in fossil assemblages for unknown reasons, are excluded from further consideration. In addition, a further list of taxa is excluded on the grounds of botanical intuition: the belief that the parent plants are not closely limited by climatic variables. What is left is a set of artificial groups of species. Transfer functions are then calculated between these lists and climatic variables such as temperature and moisture, using multivariate methods as Imbrie and Kipp did for the foraminifera. Figure 22.21 shows the kinds of results that can be obtained.

THE USES OF TREE RINGS

The most familiar use of tree rings is as a clock to date a living tree: in temperate latitudes the rings are annual and it is a simple matter to count back to year one. But where long-dead trees are available alongside living trees it is possible to match rings of the living and dead where their lives overlapped and to count back far into the past. This is the technique known as DENDROCHRONOLOGY.

A tree-ring calendar in Arizona now spans the last 9000 years, taking advantage of the very long lives of the bristle-cone pine *(Pinus aristata)* and the desert environment that preserves dead wood for long periods (La Marche, 1974). A similar chronology of bog oaks from Northern Ireland and Germany has now been completed to yield a calendar for Europe spanning the last 7000 years (Pitcher *et al.*, 1984).

A tree-ring chronology can be used to date wooden artifacts or ruins if rings in the specimens can be matched to a block of rings in the chronology, but a more general use in the service of dating has been to calibrate the radiocarbon time scale. Wood from a single year or small run of years can be taken and its ^{14}C content determined in the usual way. By this means it has been possible to discover periods when atmos-

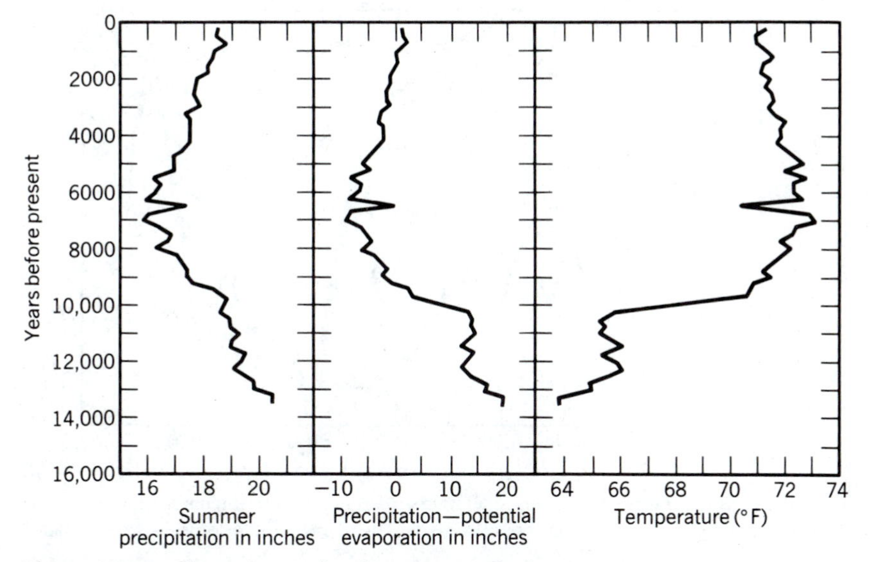

Figure 22.21 Climate from pollen data using transfer functions.
Transfer functions correlating percentage total pollen of selected taxa are correlated with modern climatic variables and then applied to pollen percentages from cores from Kirchner Marsh in Minnesota. (From Webb and Bryson, 1971.)

pheric concentration of ^{14}C was not at the assumed constant level because of temporary changes in the cosmic ray flux. At one point in postglacial time a carbon date can be as much as a thousand years in error for this reason. However, because of calibration of ^{14}C against tree rings, this can be allowed for in assigning the age of a specimen over these spans of time (Figure 22.22).

Perhaps more exciting for paleoecology, however, is the opportunity these long series of tree rings give for their own environmental records. A tree ring is a sample of living material from a known past year. The $^{16}O/^{18}O$ ratio in the ancient wood has been investigated for its possibility as a direct thermometer recording past air temperatures (Libby *et al.*, 1976; Gray and Thompson, 1977). Figure 22.23 shows how well temperatures reconstructed in this way from oak trees in Germany compare with the historical record of temperature from England over the last 600 years.

Isotopes of hydrogen in wood have also been investigated for climatic traces (Wilson and Grinstead, 1975). Use of the isotope series of carbon from tree rings, particularly ^{13}C, was discussed

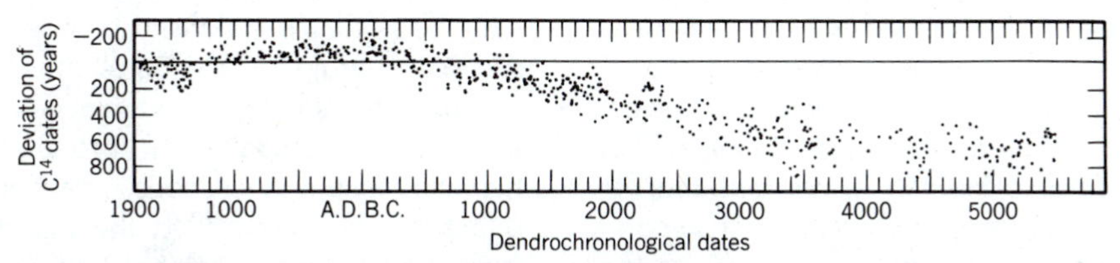

Figure 22.22 Calibration of radiocarbon with tree rings.
Data points are the departures of radiocarbon ages from dendrochronological ages of wood samples. (From Ralph and Klein, 1979.)

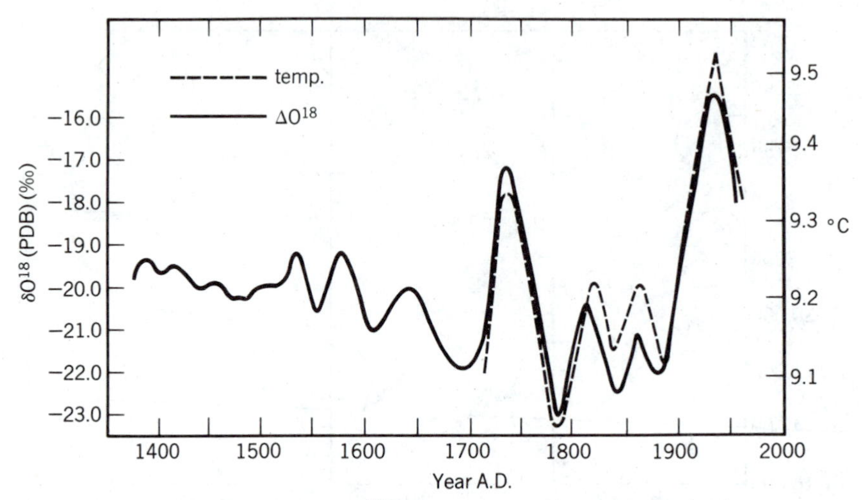

Figure 22.23 Correlation between $\Delta^{18}O$ in tree rings and historical records of temperature. The ^{18}O was measured on oak trees from Germany. The temperature records are from England. (From Libby *et al.*, 1976.)

in Chapter 18 as it was used for computing concentrations of carbon dioxide in air from before the industrial revolution.

All of these investigations encounter complications because of the properties of wood: the results are affected by using cellulose as opposed to lignin, by using old growth as opposed to young growth, by the subsequent history of the tree, and even by the tree's health. Each set of conclusions seems to be challenged or debated in the literature, but there is no doubt that profound and detailed knowledge of the past will be gathered by using rings of long-dead trees as time capsules of the environments in which they lived.

Yet so far the most thorough knowledge of past climates from tree rings has come from applying transfer functions that relate simple tree-ring widths to climatic variables (Fritts, 1976). It has long been known that a tree will put on a wider ring in a "good" year than in a "bad" year, showing that there is a correlation between ring width and temperature or precipitation. The individual record, of course, is confounded by both the health of a tree and accident. The contribution of Fritts (1976) and the Tree-ring Laboratory at Tucson was to quantify the relationship between ring widths of many trees at each of many sites and a series of environmental measurements. Multivariate analysis of these data sets yields correlation coefficients between ring widths and specific environmental variables that can then be used as transfer functions to apply to sets of tree rings from the past. Figure 22.24 shows a specimen of these results, where climate is determined in terms of departures from mean width by rings of different periods.

An alternative use of tree rings is for what might be called "very recent paleoecology": trees can be treated experimentally over a span of a few years and then the record in their rings used to compare their growth before and after the experimental manipulation. A particularly nice example of this approach comes from Australia with experiments on wild *Eucalyptus* populations (Morrow and LaMarche, 1978).

A remarkable characteristic of Australian ecosystems is that the density of insect herbivores seems to be higher than in all other parts of the world that have been examined. Morrow and LaMarche investigated the probability that this herbivore load had a serious impact on the growth of trees. They controlled the insects on some trees, and on some isolated branches of others, with insecticides for three years. Then they compared growth before treatment, and growth of

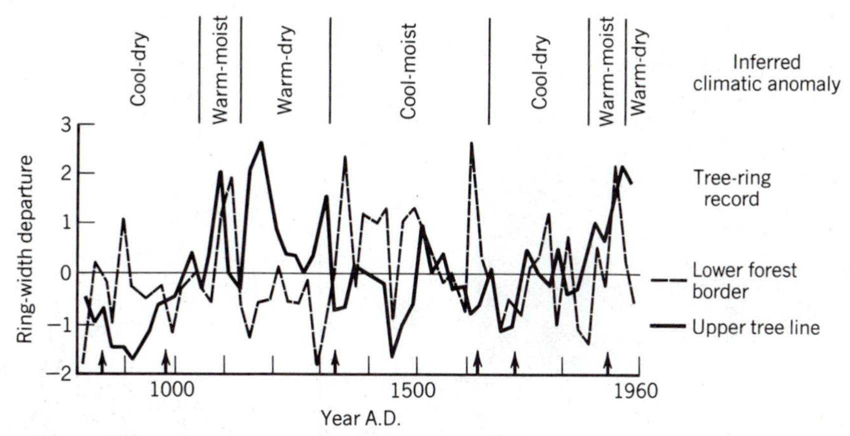

Figure 22.24 Climate inferred from widths of tree rings.
Departure from mean width of rings is correlated by transfer functions to the various climatic parameters and these transfer functions are then applied to past tree rings to yield a climatic history. (From La Marche, 1974.)

untreated trees, with growth in the treated trees by looking at the width of tree rings. Figure 22.25 demonstrates the correctness of their prediction that removing herbivores would increase growth.

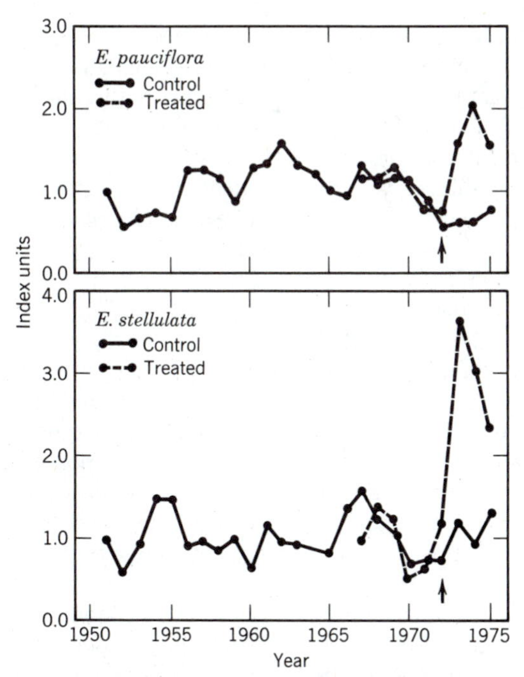

Figure 22.25 Tree-ring data used in experimental test of effects of herbivory.
Data are measures of ring width put down by *Eucalyptus* trees with and without treatment with insecticides. (From Morrow and La Marche, 1978.)

THE USE OF BEETLES

Beetle fragments are common constituents of sediments, though it takes large samples of mud to get a good collection. A 4-kg bag of mud from a Pleistocene deposit, for instance, may yield fragments of 2000 individual beetles. Beetle analysis cannot be applied to typical lake cores raised with a Livingstone sampler, because these yield only grams per stratum, rather than kilograms. But where there are quarries, road cuts, or bluffs through old deposits it is possible to collect large samples of old beetle communities without much difficulty. Typically the mud is sieved to remove large particles and then dispersed, and the beetle remains float out. Beetle analysis was invented by G. R. Coope (1959) and has since been developod by him and others into a major tool of paleoecology (Coope, 1977, 1979; Morgan, 1973).

Perhaps the most striking novelty that Coope introduced was the idea that a beetle fragment

Figure 22.26 Beetle fragments washed from mud. Fragments like these can be identified to species by a worker who "knows the vocabulary." When fragments can be concentrated from about a kilogram of mud or peat, the species composition of ancient beetle communities can be reconstructed.

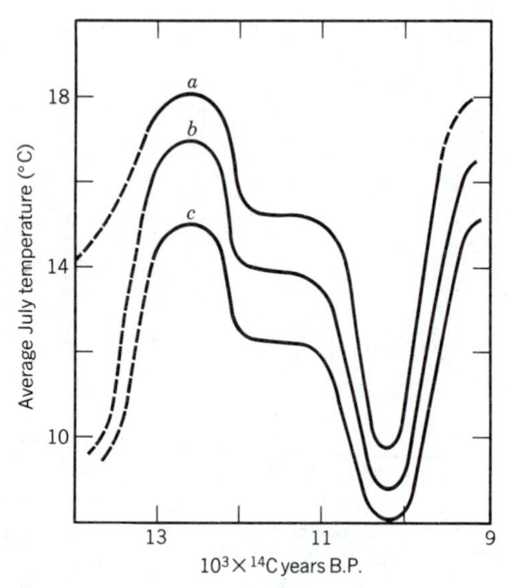

Figure 22.27 Climatic reconstruction of ice age England from fossil beetles.
Primary data are species lists of co-occurring beetles identified to species from subfossil fragments. July temperatures are inferred from present-day distributions. *(a)* southern, *(b)* central, and *(c)* northern England. (From Coope, 1977.)

could be identified to species. The remains consist of *elytra* (wing cases), the thorax, which is often in one piece, heads, fragments of all of these, and many disarticulated legs. Nothing can be done with the legs, but bits of elytron, head, or thorax usually contain enough ornamentation to be identified to species. The thing that makes this possible is, in fact, this ornamentation (Figure 22.26), which is often highly characteristic of a species. Beetle parts are made of chitin, which preserves very well, so that all this fine structure is maintained.

Beetle faunas are very species rich, so that the first essential for beetle paleoecology is a thorough taxonomic knowledge of local taxa. Even the depauperate fauna of Britain, where Coope started his studies, has over 3000 species of named beetles. Elsewhere the lists of possible species are many times larger. To know a fauna so well that the species can be identified from fragments obviously implies much learning, and fossil beetle specialists will only work where, in Coope's words, they "know the vocabulary." Because of this need for special knowledge, beetle studies so far are restricted to Britain and related areas of northern Europe, to parts of the USSR and to a few regions of North America. For communities in these northern latitudes striking conclusions about both community ecology and climatic history have been made (Figure 22.27).

Ecology of Beetle Communities

A first major finding is that there has been no morphological change in the beetle species examined for at least 100,000 years. The chitin fragments over this time span, and probably for much longer, are identical to modern fragments. The youngest material in which any significant change has been found is from the Beaufort Formation of the Canadian Arctic, which is of late Miocene age and in excess of 10 million years old (Matthews, 1976). It is quite certain, therefore, that arctic and temperate beetle faunas have undergone almost no morphological evolution throughout the very large climatic changes imposed on their habitats by the ice ages.

A second discovery is that the beetles nearly always appear in familiar assemblages, with no more than 5% of species being what might be called "outsiders." This conclusion almost seems to oppose the conclusions of palynologists that, for plants at least, there are no permanent communities. Plant communities throughout the Pleistocene continually dissolve as individual species populations migrate with shifting climate at different rates. The beetle arrays from past times, however, can nearly always be matched with modern species lists quite closely. The explanation seems to be that the beetles of these strongly seasonal climates are more susceptible to physical constraints of climate than to competition.

Beetles are ectotherms (Chapter 5). In seasonal climates their population maxima in summer are very sensitive to weather and tend to be set in density-independent ways (Chapter 8). These considerations allow the hypothesis that beetle species of a strongly seasonal place have their ranges set by the physical properties of the place rather than by the activities of the other beetles (competition) or other animals. This hypothesis predicts that beetles with similar physical requirements will be found together, which is what the data show.

This conclusion of stability of beetle associations needs to be viewed with some caution. The studies concentrate on species that are carnivores or scavengers, animals that are the least likely to be restricted by food supplies. Many herbivorous beetles are specialist feeders and should go where the food plant goes, but there are few of these in fossil beetle lists. Also beetles are able to migrate rapidly and to colonize rapidly, so that a sample of beetles from many summers collected together is very likely to include a goodly number of any species able to reproduce for a time at local temperatures and humidities. These "communities" of beetles, therefore, are not to be compared with communities of plants, which are likely to be of slowly migrating, strongly competing individuals tending to exist in equilibrium populations.

Yet this discovery that carniverous or scavenging beetles of seasonal places always turn up in familiar array tells us more of the slow rates of evolution in these organisms. Their physiological tolerances apparently remain as stable as their morphology, since ancient populations congregate with the same kinds as do modern populations. Evolution of both shape and function in these animals is extremely slow.

These highly conservative animals have kept pace with strongly fluctuating environments by migration, not by speciation. A tundra species in an ice age, for instance, persists through an interglacial on mountain tops without change. With the next ice age the range spreads and populations from far mountain tops are mixed. But the process of recurrent isolation and mixing does not result in new species. This is an important finding because of arguments advanced in other groups that evolution is speeded-up by periods of isolation in refugia. This is an essential part of arguments about allopatric speciation models (Chapter 7), and becomes important in trying to explain high diversity in some tropical areas (Chapter 26). Evidently the length of isolation is important, and an interglacial or an ice age is not long enough for beetles.

PALEOLIMNOLOGY

Lake mud holds traces of the lake's own inhabitants, as might be expected. Most lake biota are small (Chapter 21) and the fossils are even smaller. One milliliter of lake mud may hold many thousand claws from cladocera, together with diatom remains and algal cells even more abundantly than pollen grains. All these remains are of select subgroups of the lake's inhabitants, because not all preserve in the lake mud.

Copepods and rotifers generally leave no traces. Insect larvae, particularly midge larvae (Chironomidae), leave pieces of integument and their jaws (Stahl, 1959). These have been of only moderate use because the taxonomy of the modern forms is difficult and the taxonomy of the remains is even more difficult. There is nothing in lake mud that can be identified with the clarity of beetle fragments, but cladocerans of the family Chydoridae do leave traces that can be identified to genus and sometimes to species (Frey, 1960).

A chydorid analysis proceeds in much the same way as pollen analysis, namely, extraction from

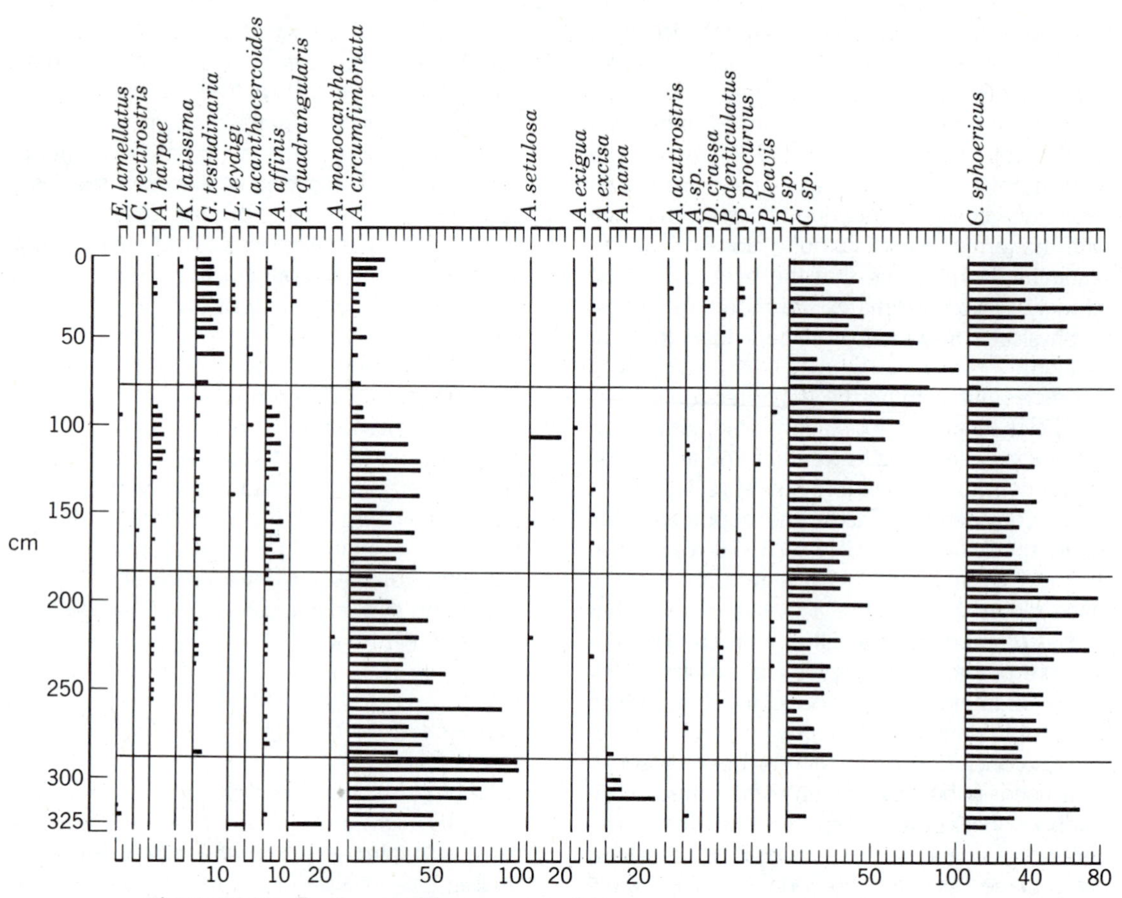

Figure 22.28 Fossil zooplankton percentage diagram.
Data are remains of cladocerans of the family Chydoridae from sediments of a lake in Wyoming. The record spans all of postglacial times. (From DeCosta, 1968.)

unit volume of mud, counting measured subsamples, and calculating fossil influx if possible, otherwise expressing results in a percentage diagram. Every recognizable fragment is scored as an individual. A sample study is that by DeCosta (1968) of a small mountain lake in Wyoming, Whimpy Lake. The record spans from near the end of the last glaciation until the present—a typical postglacial sedimentary sequence, the pollen of which shows the usual range of postglacial vegetation change. A total of twenty-five chydorid species were found in the sediment column, though the mean number at any one level was only five. Evidently some species come and go, at least in abundance sufficient for them to turn up in the counts. But just three species are the common ones throughout the record, though which of these three is most important itself changes (Figure 22.28). The most striking change is early in the history of the lake when *Alona circumfibriata* ceases to be superabundant. Interestingly, this change happens before there is any change in the pollen diagram. Apparently chydorids in a lake respond to the physical environment (probably temperature) before the plants can complete their response.

Probably the most complete record of a portion of a lake's biota is given by diatoms. These are the best fossils of all for taxonomists, since the complete silica skeletons collect in the mud.

They can be extracted and cleaned with nitric acid or strong oxidizing agents like 30% hydrogen peroxide, yielding a preparation of fossils that looks no different from a preparation of a modern plankton sample. Since diatoms are the most abundant taxa in freshwater phytoplankton, a series of extracts from the mud of a sediment core yields a history representative of much of the phytoplankton (Bradbury, 1971; Evans, 1970; Stockner, 1972). Diatom diagrams look much like pollen diagrams, though they tend to be complicated by having many more species included—the whole flora instead of just a small portion as with pollen.

More can be learned from the mud itself, rather than just using its fossils. The organic content is easily measured by weighing an oven-dry sample and burning it before weighing it again. The second weight is less by reason of the organic matter that was lost as CO_2 and the LOSS ON IGNITION is a crude measure of the total organic content. The usual procedure is to put the sample in a tared crucible in a muffle furnace at 500°C for three hours to do the burning. In some circumstances the organic content of mud is a clue to past productivity, since the rate at which organic matter descends to the bottom will be a function of productivity. Other clues to past production may be the breakdown products of chlorophyll in mud (Vallentyne, 1956; Gorham *et al.*, 1974).

Chemistry and mineralogy of sediments can be used to reconstruct much of the physical state of the ecosystem. The concentration of solutes dissolved in the water of a lake is a function of burial of nutrients in the mud, as well as of the rate of input from the watershed (Chapter 21). It follows that sediment chemistry allows inferences about past nutrient states. The mineral species in mud tell of the rate of weathering and the source of the mineral fraction.

The record in mud, therefore, can be used to reconstruct the history of a lake ecosystem and its external environment on a scale that is not possible for any other ecosystem type. Pollen and macrofossils give the surrounding vegetation. Minerals and chemistry tell of watershed inputs; sometimes ash layers tell of volcanic eruptions. Oxidized layers tell of low water or drought. All these can be dated. Then *loss on ignition* and pigments may tell of productive state, and fossils of diatoms, chydorids, and other biota tell of the response of the lake's inhabitants.

One of these studies was described in Chapter 21, that of Lago de Monterossi, where pollen gave vegetation and various sedimentary studies gave the nutrient state of the lake, the result being a demonstration of how a lake is controlled by its watershed (Hutchinson, 1970). This lake story crossed the history of ancient Rome. Equally elaborate studies of lakes have shown the environmental impact of Mayan populations in Central America (Cowgill *et al.*, 1966; Deevey *et al.*, 1979), pollution in the American Midwest (Birks *et al.*, 1976), the environmental history of southern Venezuela (Bradbury *et al.*, 1981), and the history of ancient Lake Biwa in Japan (Horie *et al.*, 1972–1979). Each paper has many authors. They are team studies because they take many workers, and they would take a chapter each to describe, but the following two studies illustrate some of the possibilities.

9000 Years of the East African Rift

In the highlands above Lake Victoria of the East African Rift is a line of old volcanic craters holding lakes: Baringo, Nakuru, Naivasha, and others. Ringed around them to the east, in an arc more than 500 km long, is a line of mountains with legendary names, including the Aberdares and Mounts Elgon, Kenya, and Kilimanjaro. The plateau on which the lakes lie is a place of savanna and grassland, 1800 m (6000 feet) high, supporting an ungulate fauna with its complex array of predators. The climate is seasonal, but generally dry, and all the lakes are endorheic. Some, like Nakuru, are shallow and saline, bright green with phytoplankton, and support immense flocks of flamingos, a bird that feeds by filtering the plankton out of algal soups that are shallow enough to wade in. But ancient strandlines raised far from the present shores tell of past times when the lakes were bigger, suggesting wetter climates, the so-called PLUVIALS of East Africa.

Lake Naivasha is fresher than the rest and

deeper, although still closed. Apart from nearby Lake Victoria (Kendall, 1969), Naivasha suggested the best chance of an ancient history and was chosen for coring (Richardson and Richardson, 1972). A core 28 m long was raised from under 14 m of water with a Livingstone sampler from a raft of rubber boats. It is certain that this core did not penetrate all the sediment body and its length represents merely the limits to drilling by hand. It turns out to span only 9000 radiocarbon years, yielding a mean sedimentation rate of 0.33 cm yr^{-1}, the highest sedimentation rate known for almost purely organic sediments. Other high sedimentation rates come when rivers or glaciers tip silt and clay into lakes, but for limnic autochthonous materials this centimeter every three years is a record. It seems reasonable to relate this high sedimentation rate to the very high productivity of the lake.

Figure 22.29 shows data for moisture, loss on ignition, sodium, and potassium for the sediments. The layer about 3000 B.P. with low water and organic matter was sandy and brightly colored, making it quite certain that this was a time when the lake had dried completely. For the rest, the data show a history of lake chemistry and productivity in three stages: a long, earlier time of fresher water, concentration of the lake solutes, and a later freshening episode. The freshening episode can be seen particularly in the potassium curve, but it is also visible in data for calcium, strontium, and magnesium (Richardson and Richardson, 1972). Evidently we have a history of changing lake chemistry set by changes in water volume, inputs, outputs, or all three.

The Richardsons made a diatom analysis of the whole core, finding that diatom communities changed in close conformity to the chemistry of the sediments, a good indication that sediment chemistry was measuring nutrient state of the open water. Throughout the early fresher-water stage the diatom community was very stable, the list being always the same and the most common species being the same. In the upper half of the core there was a complete change, both of taxa present and of the abundant species. Finally, the upper diatoms tended to be species known to favor shallow water or to be benthic. This history

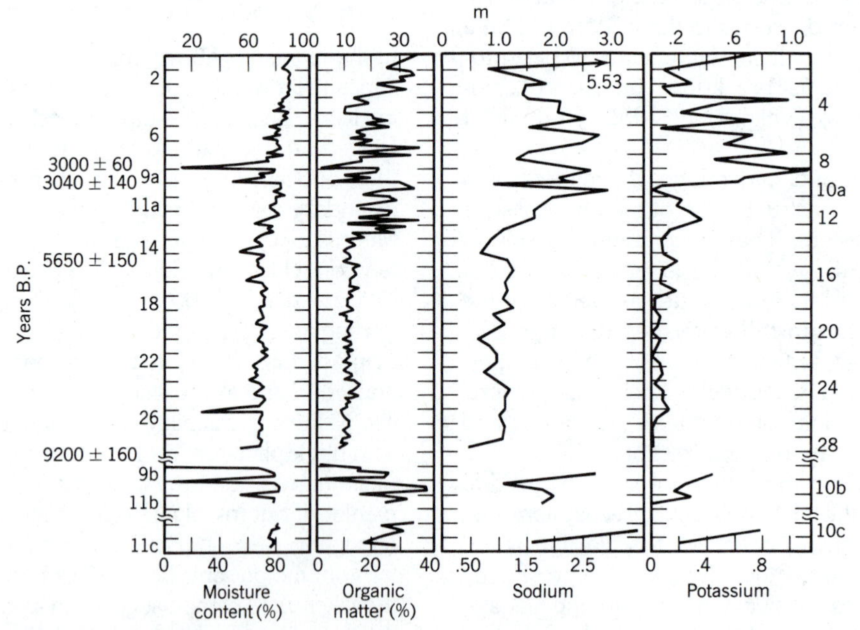

Figure 22.29 Chemistry of Lake Naivasha sediments.
The data record changes in past water chemistry that were functions of changed rainfall. (From Richardson and Richardson, 1972.)

is in complete accord with a lake, once deeper and stable, that first shrank, then dried, then oscillated in level more widely than in the earlier time.

This record can now be read in conjunction with pollen histories from Victoria and Naivasha (Kendall, 1969; Livingstone, 1975) and from lakes on Mount Kenya (Coetzee, 1967) and shows the pattern of environmental change in East Africa over the Holocene. Lake Naivasha did respond to changes in precipitation, with changed chemistry and changed biota, in step with changes in the surrounding vegetation. Even at this equatorial place, climatic change has been marked and none of the ecosystems are immune from this environmental instability, not even the biota insulated in a tropical lake.

Falsifying the Hypothesis of Lakes as Self-developing Microcosms

Linsley Pond in Connecticut is a small kettle lake that was the site of some of the first modern studies in paleolimnology. Deevey (1939) had cored the sediments with a peat borer from anchored rowboats before piston samplers were even invented. His pollen diagram from Linsley Pond (Figure 22.30) was one of the pioneer dia-

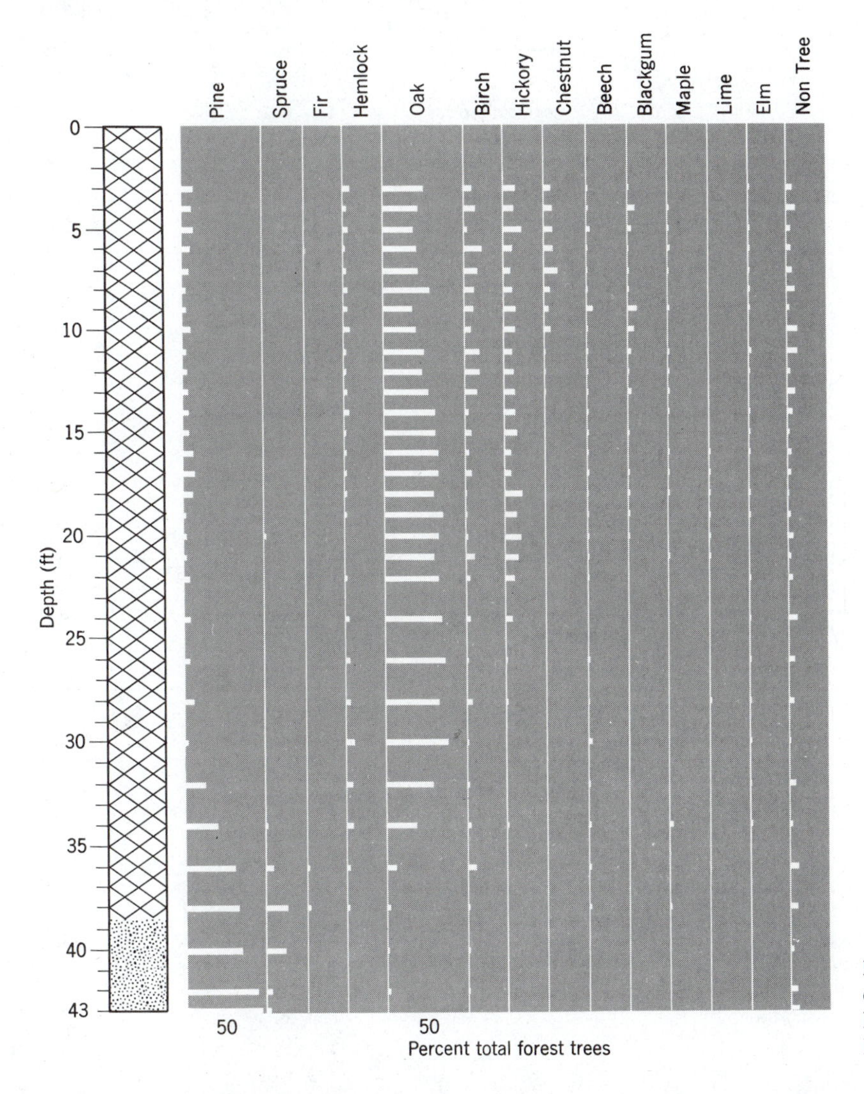

Figure 22.30 Pollen percentage diagram from Linsley Pond, Connecticut (From Deevey, 1939).

grams from New England and was decisive in showing that the European pollen sequence could not be extrapolated across the Atlantic. But his *loss on ignition* data seemed to raise matters at least as interesting.

Figure 22.31 shows the loss on ignition curve for Linsley Pond sediments with depth in core on the horizontal axis. If this curve is a crude measure of lake productivity it seems to suggest some interesting things about the productive history of the whole ecosystem. At first, when the lake was in a raw hole left by melting ice, production was very low, but then it rapidly increased. At a depth of about 30 feet down the core productivity had apparently leveled off, after which it remained constant for most of the history of the lake. This "growth in productivity" looks very much like the growth in numbers of

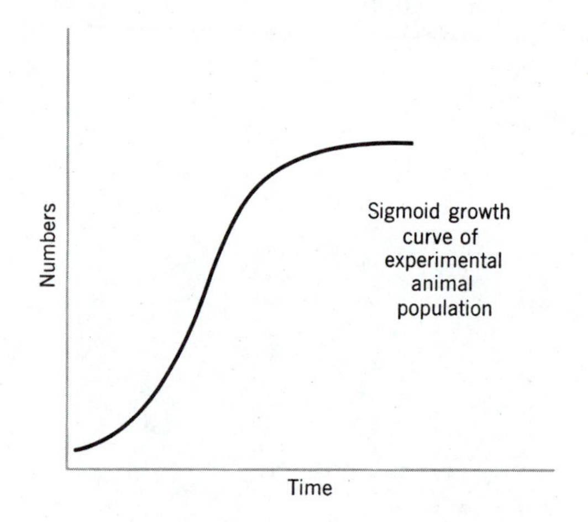

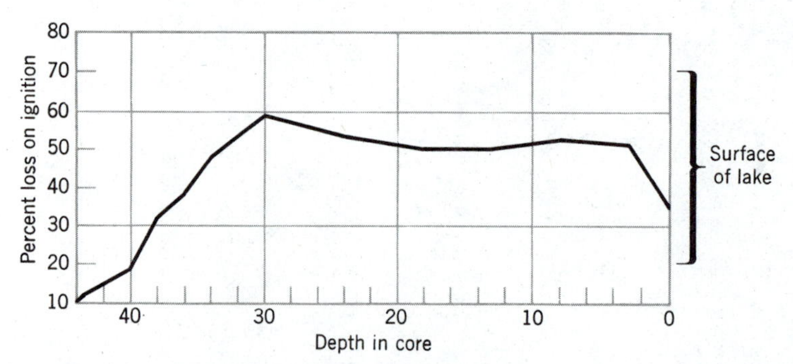

Figure 22.31 Sigmoid growth in Linsley Pond.
The accumulation of organic matter in Linsley Pond was once compared to a sigmoid growth curve of an animal population. A good estimate of the organic content of sediments is given by weighing a dried sample, burning it, and then reweighing. The percentage *loss on ignition* that can then be calculated is a useful measure of the organic content. The increase of organic matter in the early sediments of Linsley Pond was sigmoid, imitating the shape of a population growth curve. This allowed speculations that the whole lake ecosystem somehow grew to productive maturity like a population. Later work showed that the sigmoid pattern merely reflected decreased erosion in the drainage basin, which lessened the input of inorganic sediment to the lake. (From data of Livingstone, 1957).

a population when experimental animals are introduced into an empty container (Chapter 6).

These data of Deevey's let limnologists speculate that growth in production of a lake was logistic. They suggested that a whole ecosystem went through a growth history that was in some remote fashion comparable to the growth of the individual populations that made it up. This theme of communities growing through time until a stable state was achieved was widely discussed in the 1930s and 1940s, usually being thought of as being brought about by ecological succession (Chapter 23). The Linsley Pond data seemed to encourage this view, suggesting that the whole ecosystem matured to a stable state of maximum productivity through logistic growth.

These speculations can be formally stated as the *hypothesis that a lake is a self-regulating microcosm in which adjustments are made through biological process to maintain maximum productivity.* It will be evident that this hypothesis is contrary to modern ideas of how natural selection fashions life in ecosystems, where individuals are expected to act selfishly without regard to the common good. They should maximize individual productivity, which need not at all result in maximal community productivity. Furthermore, our modern knowledge of lakes shows that nutrient state, and hence productivity, is set by events within the watershed beyond the control of the lake's biota. But these things were not so clear forty years ago and the hypothesis of self-regulation was an attractive one.

The hypothesis was disproved when some of the earliest influx measurements were applied to the loss on ignition data from Linsley Pond. Livingstone (1957) recored the sediments with his piston sampler and set about calculating sedimentation rates. Radiocarbon was just becoming available as a dating tool, but Livingstone was able to use something better over the region of proposed sigmoid growth—laminae in the mud that could be shown to be annular bands and so true VARVES. Rate of sedimentation could be calculated from the numbers and thicknesses of these bands. Loss on ignition was then measured on unit volume of mud and the influx of organic matter to the sediment per year was calculated. All trace of the sigmoid history then vanished.

The influx data showed that dilution by inorganic sediment changed in the history of the lake while the organic influx remained constant. When the lake was first formed, input of minerals from the bare landscape was high, but this leveled off as the landscape was stabilized by vegetation. The long equilibrium period of Figure 22.31 represented most of Holocene time when a complete vegetation cover controlled the watershed in the way that we now know so well from the Hubbard Brook studies (Chapter 17). Both allochthonous input of minerals and autochthonous input of organic matter were stable throughout this time, even though there were considerable shifts in the make up of the vegetation (Figure 22.30). This control of the watershed by the vegetation was only broken when the Yankees arrived with their axes and plows, which promptly increased the rate of mineral sedimentation, thus causing the drop in percentage loss on ignition shown at the right of Figure 22.31.

PREVIEW TO CHAPTER 23

In ecological successions plant species replace each other in sequences that seem to be roughly predictable. Often successions are accompanied by changes in habitats leading to greater productivity. Succession ends with a climax community characteristic for the region or habitat type. In familiar old field successions, the sequence of *seral* communities is often well known and predictable. These phenomena can be explained as the result of the coexistence in any landscape of plants with different strategies and life histories. In the simplest explanation the early or pioneer communities are opportunistic *r*-strategists that are eventually replaced by equilibrium species that have been increasingly *K*-selected. Refinement of this analysis suggests three main classes of plant strategies: ruderal, competitor, and stress-tolerator. These general strategies are suited for invasion in sequence, first into unoccupied habitat, then to invasion against competition, and finally to persistence in a community where competition is so strong that individuals suffer physical stress. Species replacements in forest succession are particularly well understood, notably as shade-tolerant monolayer designs replace light-adapted multilayers. Forest succession has been modeled successfully by computer simulation and as Markov chains.

Physical development of ecosystems is particularly noticeable in primary successions. This can be *autogenic,* when community action changes the physical habitat, or *allogenic,* when changes are imposed from outside. Autogenic ecosystem development is illustrated by primary succession on Alaskan glacial deposits. Whether this succession develops to forest or to a peat bog, however, depends on drainage. Successions on sand dunes are autogenic in their early stages but are unlikely to progress beyond simple communities without allogenic changes such as raising of water tables. Hydrarch successions on wetlands typically end in systems of swamp communities alternating with open water unless the water table is lowered by physical drainage. Ecosystem maturity can be defined as the state with maximum biomass in which production is balanced by respiration. Attempts have been made to describe succession in terms of energetics, including examining the possibility that increasing complexity and biomass in climax communities of some successions implies the maximizing of physical organization or order. The paradox of community organization revealed by succession, however, appears to be more amenable to explanation from persistent invasions by individuals of species with different life histories.

CHAPTER 23

ECOLOGICAL SUCCESSION: COMMUNITY BUILDING IN ECOLOGICAL TIME

The more permanent of complex communities and ecosystems develop only after a prolonged series of communities occupies a site, usually with an accompanying series of physical modifications of the habitat. The series of communities taking part in this process is called an ECOLOGICAL SUCCESSION. Succession may be looked upon as a process of total ECOSYSTEM DEVELOPMENT, so that a formal definition can be that ECOLOGICAL SUCCESSION IS THE GRADUAL CHANGE THAT OCCURS IN AN ECOSYSTEM OF A GIVEN AREA OF THE EARTH'S SURFACE ON WHICH POPULATIONS SUCCEED EACH OTHER.[1]

The most familiar ecological successions are those on abandoned farmland, like the succession from weeds to forest of temperate latitudes described in Chapter 1 (Figure 1.1). A series of plant communities occupies the fields in roughly predictable order while the habitat progressively changes. Billings (1938) reconstructed one such succession in the North Carolina Piedmont from observations of recently abandoned fields and from a series of farms abandoned at different times over most of the previous 150 years (Table 23.1). As the herbal communities of the old fields succeed each other, and give way to pines, the organic layer of the soil deepens (Figure 23.1). The water retaining capacity of the soil increases (Figure 23.2). And the progressive changes continue in the plant populations even after the changes in soil grow slower, as is illustrated by the relative abundance of pine and oak seedlings (Figure 23.3).

Comparable phenomena are found in other ecosystems. In lakes or the oceans, for instance, seasonal successions of short-lived algal communities follow each other from spring to autumn year after year. And newly formed marshlands or sand dunes are colonized by regular successions of communities even as the habitats change, most strikingly when moist soil develops on dry dunes.

The forcing-function of ecosystem change often seem to be provided by vegetation, as the biomass of developing plant communities both changes the physical state of the ecosystem (Chapter 17) and provides the habitat for animals or decomposers. But animal communities also change in succession. They may modify the habitat for plants and replace other animal communities in a colonizing dynamic of their own.

[1]Adapated from Tansley (1920), who referred to just vegetation rather than the whole ecosystem. Tansley (1935) only invented the word "ecosystem" fifteen years later.

Table 23.1
Old Field Succession on the Piedmont of North Carolina
(From data of Billings, 1938.)

Years Since Abandonment	Most Abundant (Dominant) Species	Other Common Plants
0(fall)	Crabgrass	
1	Horseweed (*Erigeron*)	Ragweed
2	Aster	Ragweed
3	Broomsedge	
5–15	Short-leaf pine	Loblolly pine
50–150	Oaks and other hardwoods	Hickory

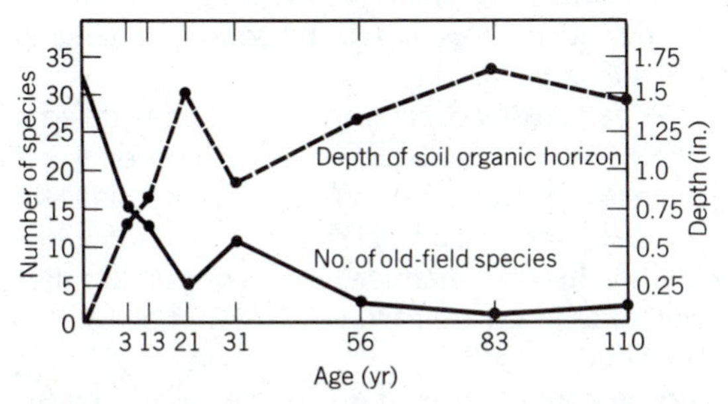

Figure 23.1 Changing depth of soil and number of herb species in old fields in the piedmont of North Carolina. The organic horizon grows progressively deeper as the number of colonizing herb species falls. It happens that pines, arriving between years 5 and 15, grow after the organic layer is about as thick as it will become. (From Billings, 1938.)

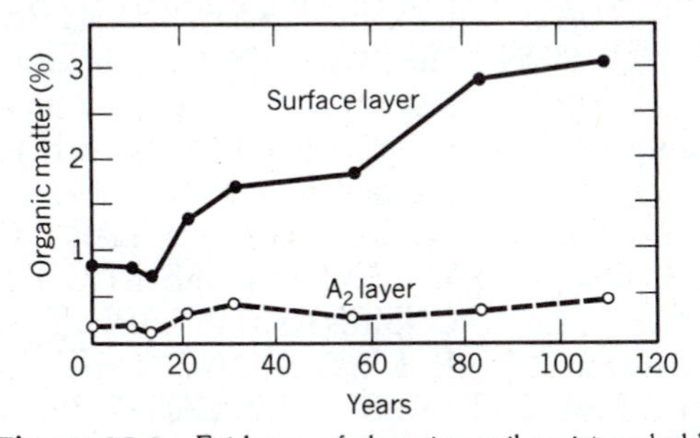

Figure 23.2 Evidence of changing soil moisture holding capacity in old fields in the piedmont of North Carolina. The capacity of soil to retain water is a function of the organic content. The thickness of the elluviated "A_2" horizon is a record of relative leaching (Chapter 20). These two curves suggest that the soil capacity for moisture increased continuously for a hundred years of succession as the rate of leaching decreased. (From Billings, 1938.)

These facts of ecological succession have attracted ecologists principally because they suggest ordered development. Succession has the appearance of a cooperative enterprise. Plants and animals come and go in sequence, each apparently making its contribution to ecosystem change and then departing. The final result is a perfected complexity that persists indefinitely.

Ecologists use several approaches to study processes of community building. One is to model the changing rates of colonization and extinction when communities are built on isolated raw habitats, the approach known as *island biogeography* (Chapter 24). A particular aim of this approach is to understand the conditions setting

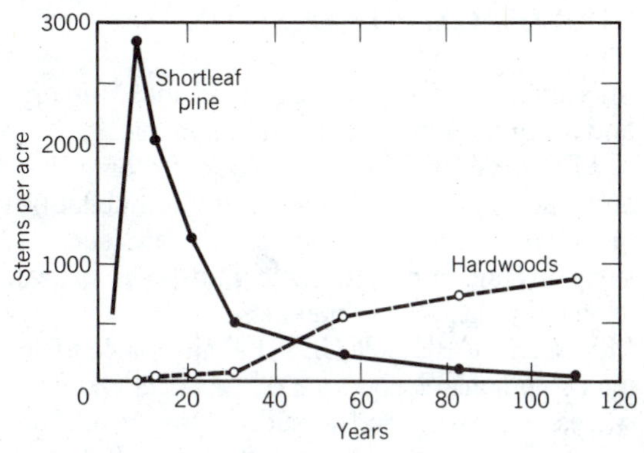

Figure 23.3 Relative abundance of pine and oak seedlings in old fields in the piedmont of North Carolina. (From Billings, 1938.)

the equilibrium number of species eventually reached. Other approaches are evolutionary, asking how accommodations between species in complex communities come about. This approach addresses matters of mutualism, mimicry, and the relationship between predators and their prey that allow the coexistence of many species (Chapter 25).

But the process of building plant communities in relatively short spans of time through ecological succession is yet a separate phenomenon requiring its own explanations. Communities are built not in one quick fitting together of coevolved species but from pools of local species through the more or less regular replacements of succession. The endpoint is the complex community dominated by a few characteristic species that typically arrive late in the succession. There must be reasons for these ordered replacements. In the quest for these reasons we have always had the knowledge that habitat changes proceed alongside the sequence of species replacement. Would the species replacements occur without the habitat changes, or is succession a process of ecosystem building in which the habitat changes caused by the early communities trigger invasions by plants of the later communities?

SUMMARY OF TERMS AND CONCEPTS OF ECOLOGICAL SUCCESSION

Succession traditionally is divided into *primary* and *secondary; autogenic* and *allogenic.*

PRIMARY SUCCESSIONS colonize bare sites and lead to the first occupation of the habitat by the climax community. Examples are successions on sand dunes, volcanic mud-flows, glacial till, filled-in lakes, and marshes.

SECONDARY SUCCESSIONS replace a climax community following a disturbance. Old field successions are secondary successions, as are the successions that replace forest in gaps after hurricanes or fires.

AUTOGENIC SUCCESSION is succession directed from within the ecosystem itself. The term particularly refers to habitat changes brought about by the biota, as when soil is built and nutrients collected. If these enrichments of soil promote the next community replacement then the succession is *autogenic.*

ALLOGENIC SUCCESSION is succession driven by forces outside the ecosystem, as when a progressive fall in a marsh water table due to stream down-cutting leads to a succession of plant communities suited to progressively drier habitats. The communities themselves may have had no influence on the critical habitat changes. The conceptual differences between *autogenic* and *allogenic* successions are significant. In the former, plants and animals are the genesis of change; in the latter, they merely respond to changing climate or geography.

A particular, recognized succession is called a SERE. A HYDROSERE is a successional sequence on wet land. Wetland successions in general are called HYDRARCH SUCCESSIONS. In a like manner successions on dry sites like sand dunes are called XERARCH SUCCESSIONS.

The communities of a *sere* are called SERAL COMMUNITIES. The earliest *seral community* on a site is the PIONEER COMMUNITY. The succession is said to end with the CLIMAX, which is the COMMUNITY WITH WHICH A SUCCESSION ENDS AND IN WHICH SPECIES PERPETUATE THEMSELVES THROUGH REPRODUCTION (Mueller-Dombois and Erenberg, 1974). Since slow changes continue in a "climax," even when climate is constant (Chapter 22), an alternative definition may be that the CLIMAX IS A COMMUNITY SIMILAR TO CONTEMPORARY COMMUNITIES IN COMPARABLE ENVIRONMENTS THAT ARE FREE FROM PHYSICAL DISTURBANCE AND IN WHICH IMPORTANT SPECIES CAN PERSIST FOR MANY GENERATIONS.

The composition of the climax community depends on the physical properties of the habitat and of the local climate. The climax for each local habitat type can be recognized by its characteristic dominant plants, like the beeches and maples of beech–maple forest of the American

Midwest or the oaks of English woodlands (Chapter 15).

But this typical result of a local succession or *sere* is often prevented by repeated disturbance. The succession can be *arrested* by events like repeated fires or heavy grazing, leading to a quite different climax community. This community is called a PROCLIMAX (Grime, 1979). The species important to the *proclimax* will be quite different from those that would be important if the succession had been left to proceed without the impositions of the disturbance. In heavily grazed pastures, for instance, unpalatable grasses or forbs may be common, or even dominant, though they might quickly disappear if the grazing pressure were removed. Perhaps the most extreme example of a proclimax is an agricultural field where repeated ploughing produces a community of opportunistic herbs.

This listing of successional phenomena emphasizes plants. Parallel replacements may proceed among the animals, and for the maintenance of a *proclimax* animal consumers can be decisive. But the more readily observed replacements are among the plants. Most studies of ecological succession, therefore, concentrate on changes in plant populations, and the rest of this chapter will largely be concerned with plant successions.

Two sets of successional phenomena require explanation: the population replacements themselves, and the parallel modifications of habitats that are particularly notable in some successions. We begin by considering the population replacements, asking to what extent these can be explained without reference to changes in the physical habitat.

SUCCESSION AS A CONSEQUENCE OF LIFE HISTORY PHENOMENA

Many of the population replacements of a typical sere can be understood as a necessary consequence of opportunist and equilibrium species living together in the same area. This is the *r* and *K* hypothesis of succession (Chapter 11; Colinvaux, 1973; Drury and Nisbet, 1973).

The essence of this hypothesis is that both opportunist and equilibrium species are able to coexist within an area of country that might be called *reasonable dispersal distance.* No evolution of fresh taxa is required for ecological succession because this is a comparatively short-term event. It happens in *ecological time* rather than in evolutionary time. We can discount from a hypothesis of causes both invasions from remote places and the production of fresh actors by speciation. Both are possible, of course, but their appearance cannot affect our general explanation. The familiar seres we see are compounded from the life histories of species already existing in the neighborhood.

The hypothesis predicts that the plants of the pioneer seral community will be extreme *r*-strategists, adapted for dispersal and rapid growth but not for sustained existence at a species equilibrium. They should be highly fecund, short-lived, and without elaborate storage organs and the more costly predator defenses. If these plants exist in a neighborhood, it can be predicted that they will occupy vacant habitat first. This prediction is upheld by the plants of pioneer communities, which are the typical annual weeds of an old field.

These *r*-strategy life history phenomena can be present in a neighborhood even though surrounding land is occupied by equilibrium plants of a climax forest. This is because the *r*-strategy adapts individuals to taking advantage of small-scale disturbances that often occur—the bare ground of a flooded stream bank, spoil from a rodent burrow, or ground torn up by a fallen tree. Another adaptation that is common among these plants is the C4 pathway for photosynthesis, which evidently adapts them to the hot, desertic properties of bare soil (Chapter 3). This suite of *r*-strategy and bareground adaptations preadapts individuals to colonize larger vacant habitats when these occur.

But the hypothesis predicts that the pioneers will certainly be displaced as *K*-strategists invade the site from surrounding complex communities. This is what we observe, most notably in seres that lead to a forest climax. The progressive invasions of perennial herbs, shrubs, and trees are by species whose life history phenomena are of progressively longer life, with all the diversion of resources from reproduction to storage and defense that these life histories demand.

Of particular appeal is the way this hypothesis predicts the orderly and repeatable properties of old field successions. The sere will proceed at a rate that is a function of the number of species strategies available locally, that is to say of the local species pool (Figure 23.4). As a result, a good naturalist can always tell to within a year or two when a given old field was first abandoned by farmers from looking at the seral stage then occupying the field.

Continued Replacements in Forest Seral Stages

The hypothesis that succession is the predictable replacement of *r*-strategists by *K*-strategists is readily convincing for all seral stages up to forest. But when the whole habitat is covered by large trees, all following a *K*-strategy of sorts, then continued successional change might not be so self-evident. And yet in classic old field successions to forest a long succession of tree populations continues between the first establishment of forest and the final slow change of climax.

In forest successions, tree populations are replaced in a regular sequence familiar to foresters, who rank trees according to their shade tolerance. These replacement series of trees are discussed in Chapter 5. They have been shown to reflect tree geometry, which is different for optimum growth in light and in shade. The first trees that replace the shrubs and herbs of an old field are of multilayer design. These trees cannot grow in the dense shade of their parents because their leaf geometry does not then yield a sufficient excess of net photosynthesis over respiration. Shade-adapted trees with fewer effective layers of leaves, therefore, invade, and Horn (1971) was able to show that trees of a successional sequence were equipped with an ever-diminishing number of effective layers until the climax trees were close to being monolayers (Table 3.4). Even in forests, therefore, a variety of strategies results in succession.

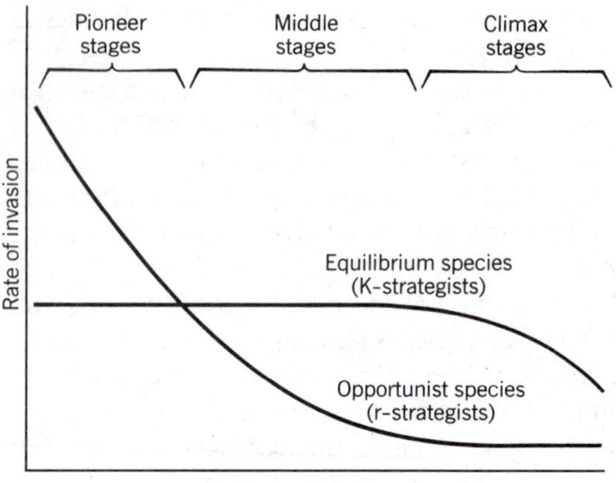

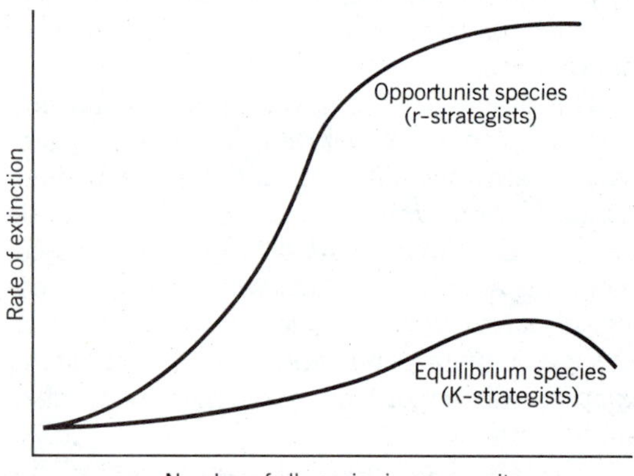

Figure 23.4 Model of succession as a function of rate of invasion.
Opportunist species invade early and rapidly *(top)* but their extinction by competition is also early and rapid *(bottom)*. The final number of species at climax will depend on the rate of continued invasion (which may proceed for thousands of years) and the rate of local extinction, which eventually should be very slow.

A Three-strategy Model for Plants in Succession

The conditions late in forest succession that lead to the prevalence of monolayer designs involve special stresses. The monolayer itself is an adaptation to stress in that it adapts seedlings to grow in dense shade. Other stresses in old established forests must be present from the massed demands of so much biomass—nutrients and water at least can be expected to be in short supply at times. The point here is not just that these scarce resources have to be competed for but that they are absolutely scarce in relation to the demands made upon them by the biomass. Competition in these mature communities is of a different order to competition in earlier seral stages where total community demand is less.

Grime (1979) has suggested that a series of three base strategies should be recognized for plants instead of just the opportunist and equilibrium strategies of r and K. These three strategies would be that of colonizer, competitor, and stress competitor, called by Grime RUDERAL, COMPETITIVE, AND STRESS-TOLERANT STRATEGIES or R, C, and S STRATEGIES.

RUDERALS are plants suited to frequent disturbance but not to competition. This concept is close to that of the *fugitive species* (Chapter 11) since powers of escape to uncontested habitat are essential to the strategy. COMPETITORS are adapted to low disturbance and low stress, in which conditions they maintain dense equilibrium populations as predicted for K-selected species. But STRESS-TOLERATORS are plants suited to low disturbance with high stress. In this *stress-tolerator* strategy the resources allocated to competition are constrained by resources allocated to surviving stress. Thus the model has three patterns of resource allocation:

1. Most resources to fecundity and dispersal = *ruderals* (R species).

2. Most resources to competition = *competitors* (C species).

3. Resources divided between stress resistance and competition = *stress-tolerators* (S species).

An R–C–S continuum fits the facts of succession more closely than the simpler r and K continuum (Figure 23.5). The particular value of the model is that it predicts continual replacements even after dense populations of large plants are attained.

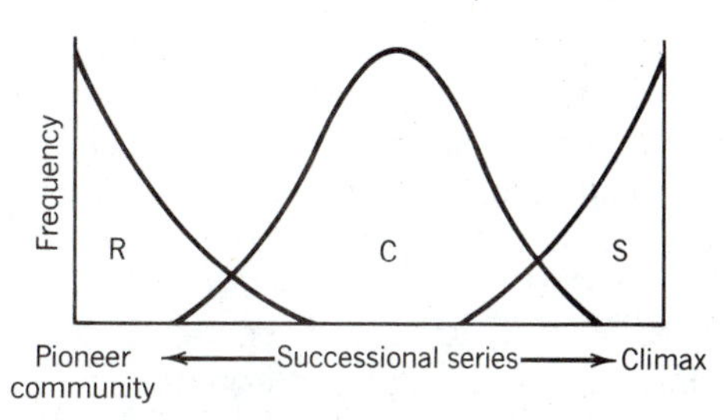

Figure 23.5 Three-strategy model of succession. Ruderals (R) are pioneer herbs that are replaced by competitors (C). At the climax stress-tolerators (S) invade. For details see text. (From Grime, 1979.)

Regeneration and Growth Strategies in Succession

Life histories of plants depend on dispersal and germination of seeds. Wind, bird, or mammal dispersal of seeds are obvious alternate strategies that require seeds of particular size or number, and result in different dispersal distances. At least as important, however, are alternate strategies for germination once the seed comes to rest on soil. Germination can be immediate or it can be delayed, perhaps for years, until some necessary environmental cue is received. A result of delayed germination is that SEED BANKS collect so that a soil is charged with viable seeds waiting for an environmental cue.

The most familiar manifestation of seed banks is in agriculture, where ploughing always results in the growth of weeds. The weed seeds may have germinated when brought to the surface and exposed to light, or germination may have been triggered even for deeply buried seeds by changes of temperature caused by breaking of the soil surface (Thompson and Grime, 1978). This *seed bank strategy* of the pioneer plants presumably is an adaptation that lets the pioneers colonize gaps that form by accident in forests or other vegetation or that are left when a large tree dies. The seed bank strategy is also used by species that colonize old burns in the boreal forest, when the heat of fire is the germination trigger. Jack pine, *Pinus banksiana,* follows this strategy.

An alternative to a seed bank used by some species is a SEEDLING BANK. Under closed forest, seeds germinate on the forest floor and the resulting seedlings persist for a long time, perhaps for years. In the dim light, photosynthesis may be barely sufficient to balance respiration, but is enough to keep the seedling alive. It is expected that seedlings using this strategy will have well-developed chemical defenses against herbivores. The gain from this strategy is a head start in the race for the canopy when a gap appears from the removal of a shading tree.

Many plants regenerate asexually as vegetative clones (Chapter 12). This strategy seems to work particularly well in dense carpets of vegetation like arctic tundras, where sites for seed germination are extremely rare. Herbs of forest floors also sometimes use this strategy as do trees like aspens.

Rates of growth and shapes of growth can be varied, allowing other strategies. Within forests, the members of a seedling bank can grow as rapidly as possible in the dim light, becoming tall and thin, or they can grow broad and slowly, perhaps slowly increasing energy reserves for a sudden spurt later. The possible variants of growth and regeneration strategies are thus many and are not yet fully explored (Grime, 1979; Bazzaz and Pickett, 1980). The variants are mostly constrained by the needs of plants to regenerate in a vegetated landscape when all light-receiving surfaces are occupied most of the time. But the existence of an array of regeneration strategies in all vegetations makes ecological succession certain whenever a clearing is made.

Simulation and Markovian Models of Succession

Successful computer simulations of some temperate forest successions have been made using data for invasion, growth, and death of individual trees. One such program, called JABOWA, was developed as part of the Hubbard Brook study (Chapter 17; Botkin, *et al.*, 1972). Each individual tree in the simulation was assumed to grow, and to have an expectation of life, like those measured for Hubbard Brook trees of various ages. Empirical equations relating growth to size, temperature, precipitation, and site defined a GROW SUBROUTINE. Known expectations of life defined a KILL SUBROUTINE. Fresh invasions were assumed to be random, but proportional to presence of saplings in existing stands, yielding a BIRTH SUBROUTINE. All interactions of one plant on another were assumed to reflect competition for light, modified by the known adaptation of each species to relative shade. These interactions were described by equations in the grow subroutine. Simulations then proceed in the order described in Figure 23.6.

Runs with JABOWA, started with early succession data, have correctly described the course of forest succession at Hubbard Brook. They then give a most realistic impression of what climax ought to be like. Figure 23.7 shows what should

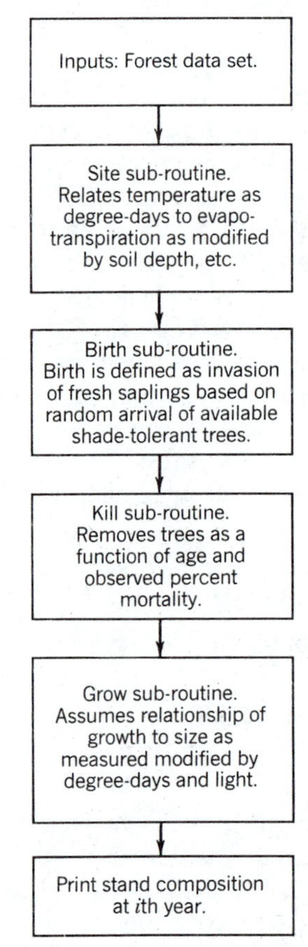

Figure 23.6 Partial flowchart of JABOWA program. (Much modified and simplified from Botkin *et al.*, 1972.)

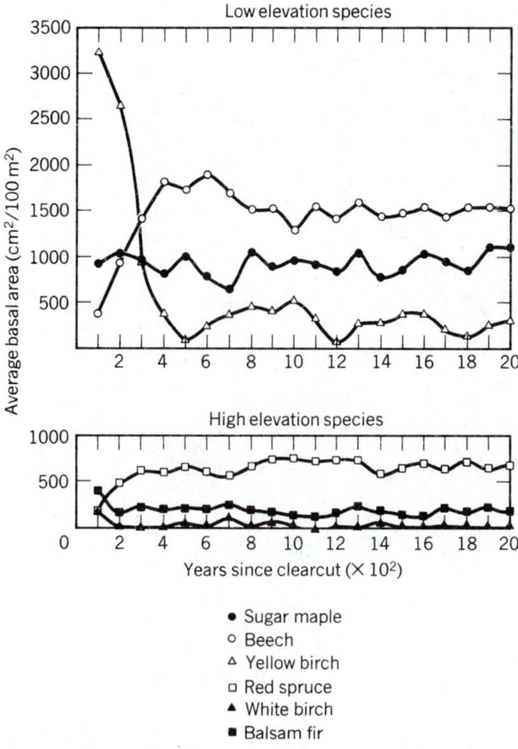

Figure 23.7 JABOWA simulation of 2000 years of forest growth.
The run begins with a clear-cut and describes the expected proportions of six trees in future forests at an elevation of 610 m. (From Botkin *et al.*, 1972.)

happen from an early forest seral stage throughout 2000 years of history in a constant climate, for six species whose ranges overlap at an elevation of 610 m. After the initial successional events characterized by the decline of yellow birch, the community settles down to a dynamic system in which all species persist though their populations oscillate, apparently at random. New Hampshire forest stands at this elevation that have been declared to be climax by field naturalists are closely comparable with the outcome of this simulation.

Simpler assumptions than those of a simulation model let succession be modeled as a Markovian process. This method uses a statistical procedure in which chains of random events are constrained by current states but no other. Markov chains of stochastic events eventually settle down into final stationary distributions quite comparable to the relationships of tree populations at climax (Kemeny and Snell, 1960).

Horn (1974, 1976) has applied the mathematics of Markov chains to forest data to predict future states. The initial data set is an estimate of the probability that any one tree in a forest will be replaced by one of its own kind or one of another kind. These data most easily could be derived by mapping every tree in a forest and then remapping the same forest at some future year, say, 50 years off. A more practicable way to get usable data is from estimates of replacement probabilities made from measures of saplings actually under trees in the forest, supplemented by measures of relative shade tolerance.

When a matrix of probabilities of individual replacement for all species of a forest is known, fairly straightforward calculations allow extrapolations to the future composition of a forest. The only data going into these calculations, therefore, are present composition of a forest and measured probabilities of individual replacements. The method has been applied to early succession data in New Jersey forests and predicts the composition of future climax communities to be closely similar to that existing in a virgin New Jersey stand now (Horn, 1975).

Succession in Simple Forests

In simple forests, trees need less of the *K* or stress-tolerator (S) strategies, since few competitors exist. Both remote islands and forests near tree line may have a single species of forest tree in some communities, the strategies of which reflect this lack of competition.

The moist uplands of the Galapagos island of Santa Cruz is occupied by a cloud forest (Figures 14.12 and 14.32). The upper part of this forest is almost a pure stand of one species of tree, *Scalesia pedunculata,* with an understory of small *S. pedunculata* saplings, ferns, and a few other bushes (Figure 23.8). Hamman (1979) showed that the forest was replacing itself, though the trees appear to be *r*-strategists.

Scalesia pedunculata belongs to a genus of

Figure 23.8 *Scalesia pedunculata* forest on Galapagos island.
The climax forest in the Santa Cruz highlands has only one canopy species belonging to an endemic composite genus. Although the strategy of the tree appears to be that of an early forest succession species, it appears to maintain itself as the climax tree indefinitely in the absence of other species adapted to invade under its canopy.

Figure 23.9 The canopy of *Scalesia pedunculata* on Santa Cruz.
The canopy does not cast dense shade, suggesting that the tree would be ill-equipped to take part in forest successions elsewhere.

the aster family (Compositae) that is endemic to the Galapagos. The trees have large aster- or daisy-like flowers and their whole structure suggests their obvious ancestry to shrubby or even herb-like recent ancestors: rapid evolution has made a small canopy tree out of unpromising material. The trees grow very rapidly and they are highly fecund, dispersing seeds widely as other composites do. Their canopies do not cast dense shade (Figure 23.9). Typical large scalesias of the canopy are 10 to 12 m high and about ten years old. They can reach 7 to 8 m in only 3 to 6 years of growth, and they begin to flower in their fifth year. This strategy actually serves them well as climax trees in the absence of all tree competition. Their own short lives ensure a constant supply of gaps that are efficiently colonized by their own wind-dispersed seeds, and the scalesia population replaces itself.

A result closely parallel to that attained by the scalesias is the maintenance of single-species stands of the birch *Betula pubescens* in northern Scandinavia over several thousand years. Moore (1979) argues that this species maintains itself in a monospecific stand because no other trees are available with the physiological adaptations needed to grow in this marginal arctic climate. The birches combine properties of ruderals and stress-tolerators in a community without tree competitors.

The scalesias and the northern birches may have a special adaptation to their role of sole climax species in that they do not cast dense shade (Figure 23.9). This means that their seedlings can survive under the shade of the canopy.

Conclusion: Succession Is a Result of Varied Strategies

The facts of species replacements in plant successions seem adequately to be explained as a consequence of the coexistence of plants with different strategies for establishment or exploiting disturbance. Of particular importance are strategies for exploiting gaps in plant cover. In forests, light for rapid growth occurs only in gaps, but the necessary gaps are produced regularly by deaths of trees. When species are limited in number, as in temperate forests, tree strategies are necessarily few also, and they may be distinctive. Despite the randomness inherent in any system of colonization, therefore, the order of succession is largely predictable in temperate vegetation.

Succession may be more varied in very complex systems. In tropical rain forests many tree species coexist, light occlusion by the canopy is nearly complete, and trees are relatively short-lived (100 to 150 years; Whitmore, 1975). Gaps from tree-fall are an essential resource for the regeneration of nearly all tree species and strategies to exploit gaps are various. Mass dispersal of small seeds, seed banks activated by light or temperature, seedling banks, slow growth in shade, or dispersal of large seeds to gaps by animal carriers are all basic strategies for which variations are possible. In these systems chance can be more important than in simpler temperate systems, making successions in rain forest gaps less predictable.

Consideration of forest gaps reveals the primary conditions leading to the general phenomenon of ecological succession. Small breaks or clearances in all vegetation are necessary for the establishment of plants once a canopy has formed. The ubiquitous array of adaptations for exploiting gaps leads to the larger-scale successions in big clearings characteristic of human activities or natural catastrophes. (Recent reviews include Waggoner and Stephens, 1970; Botkin *et al.*, 1972; Drury and Nisbet, 1973; Horn, 1974; Connel and Slatyer, 1977; Grime, 1979; Bazzaz, 1979; Bazzaz and Pickett, 1980; McIntosh, 1980; Tilman, 1982; Finley, 1984).

SUCCESSION AS ECOSYSTEM DEVELOPMENT

Alongside the population changes of many successions go changes in the physical system that can be called "ecosystem development." These changes may be minor in successions filling forest gaps but they can be large in old field succession and become striking in many primary successions. The possibilities of ecosystem development, whether *allogenic* or *autogenic,* are best illustrated with classic examples of primary successions: development of forest on fresh glacial till, the xerarch succession on sand dunes, and the hydrarch successions of wetlands.

Primary Succession on Glacial Till

Valley glaciers at Glacier Bay, Alaska, have been retreating for about 200 years and their progress has been monitored by a series of expeditions ever since 1890 (Cooper, 1939; Crocker and Major, 1955; Lawrence, 1958). They leave behind them moraines of glacial till; material in many ways comparable to a soil regolith (Chapter 19). It is of high carbonate content, because of fragments torn up from carbonate bedrock by the glaciers, and consequently is alkaline (pH 8.0 to 8.4). The material is without organic matter or combined nitrogen. Thus potential habitats near the ice front are like the empty "habitat" described as the start for a study of nutrient cycling (Figure 17.2, Chapter 17).

A range of habitats up to 200 years old in the path of the glacial retreat records ecosystem development on this till. Succession goes from a pioneer community of arctic herbs and dwarf willows, through willow scrub, then to an almost pure stand of alder bushes 10 m high at the 50-year mark. The alders are slowly invaded by sitka spruce *(Picea sitchensis)* and replaced, until a forest of conifers is present at the 120-year mark. The spruce forest, however, is progressively invaded by two species of hemlock (*Tsuga mertensiana* and *Tsuga heterophylla*) for the next 80 years, resulting in the climax spruce–hemlock forest of the region (Figure 23.10).

Profound changes in the physical habitat are brought about during this 200-year succession, starting with bare, basic, nutrient-deficient till and ending with an acidic podzol (spodosol, see Chapter 20) carpeted with needles, with ample nitrogen and other nutrients, organic matter, and a complex structure. These soil properties seem to be added bit by bit from the activities of the several seral communities.

Figure 23.11 shows changes in pH of the litter (dashed line) and the top of the mineral soil (solid line). The rapid change of pH occurs when the alders are in possession, not when pioneer plants leave bare ground for the rain to wash. The inference is inescapable: that it is the acid residues of alder leaves that are responsible for dissolving carbonates and that the removal of these carbonates is not achieved by percolating rainwater alone. Lowering of pH, therefore, is a consequence of plant growth.

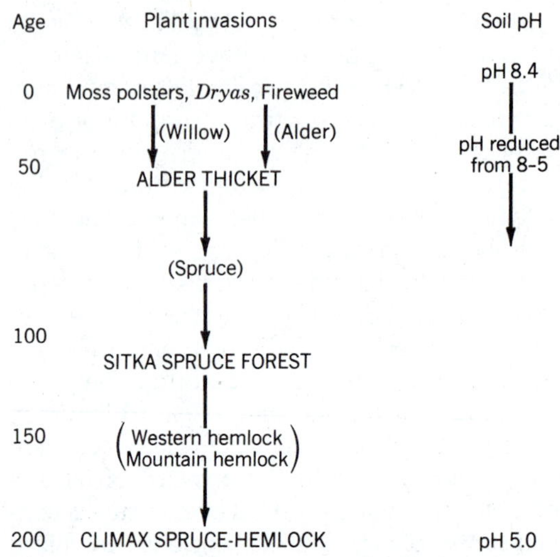

Figure 23.10 Primary successions at Glacier Bay on well-drained sites. (Age in years, data of Crocker and Major, 1955.)

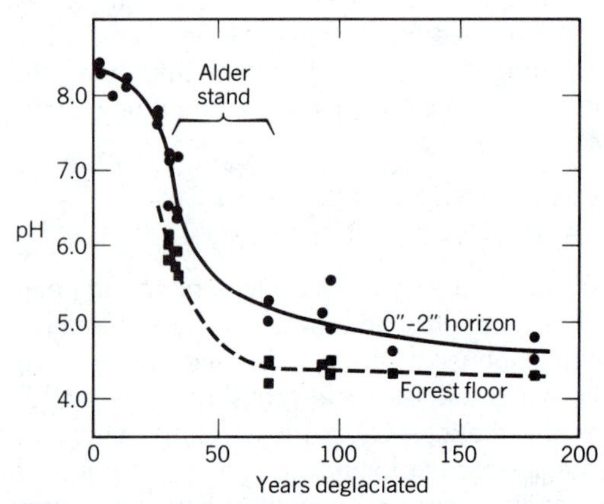

Figure 23.11 Changing pH in Glacier Bay succession. The years of rapid change are those when alder occupies the habitat. (Based on Crocker and Major, 1955.)

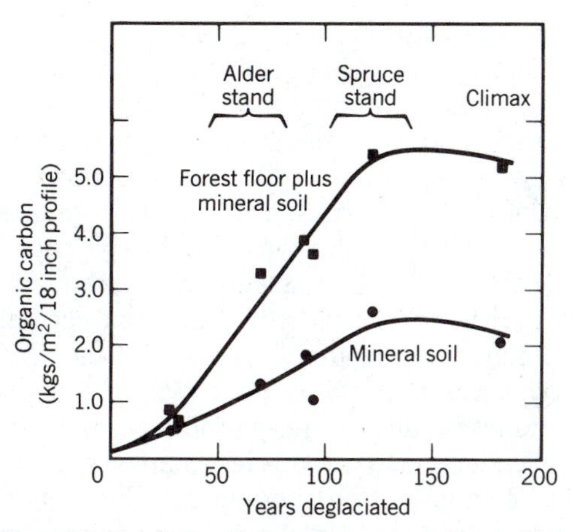

Figure 23.12 Accumulation of organic carbon in Glacier Bay succession.
Carbon continues to accumulate in the soil throughout the pioneer and alder stages before reaching what looks like a steady state under evergreen forest. (From Crocker and Major, 1955.)

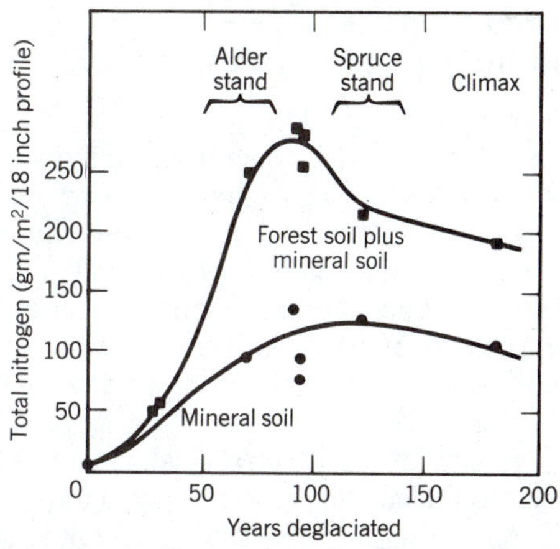

Figure 23.13 History of combined nitrogen in Glacier Bay succession.
Nitrogen is highest under the stands of alder before falling under evergreen forest. High inputs by nitrogen-fixing bacteria in alder root nodules explain this. (From Crocker and Major, 1955.)

Figure 23.12 shows that organic matter accumulates continuously and is less dependent on the kinds of plants growing than is the change in pH. Possibly a steady-state reservoir of soil organic matter is achieved in about 100 years, with fresh inputs being balanced by losses due to respiration.

Figure 23.13 shows that nitrogen rises slowly at first in the pioneer stage, but some nitrogen fixation occurs even then. Probably *Dryas* populations (an herbal rose of the tundra) are important in this since they have nitrogen-fixing bacteria like legumes. The big increase in nitrogen occurs in the alder stage, which consists of a pure stand of plants with copious nodules of nitrogen-fixing bacteria. Nitrogen falls as the alders are replaced by spruce, probably reflecting the lowered inputs.

The record from Glacier Bay shows that climax spruce–hemlock forest cannot grow on the raw habitat left by the glacier. pH must first be lowered and in all probability nitrogen and other nutrients must be added before the climax trees can invade the site. These improvements of the habitat require that the site be occupied by other plants first. These earlier plants must be adapted to high pH, must produce litter, the acid decomposition of which lowers pH, and must be associated with nitrogen fixation. A mixture of ruderals, opportunists, and stress-tolerators in the local flora have these properties. Not only are they first invaders but their development of the habitat is necessary for the invasion by *K*-strategists of the climax. This is an example of *autogenic ecosystem development.*

On marshy habitats at Glacier Bay a different sere develops. A pioneer community of marsh plants is invaded by alders in this sere also. Again pH falls, but its lowering promotes an invasion by the bog moss *Sphagnum* instead of conifers. *Sphagnum* absorbs water, making the habitat even marshier, and further lowers the pH to about pH 4. In these conditions *Sphagnum* cannot be replaced by other plants and the climax community is *Sphagnum* muskeg. On slopes of intermediate marshiness the succession may go all the way to a forest of spruce and hemlocks before *Sphagnum* hummocks begin to grow, eventually to drown the trees and create a muskeg where once was an evergreen forest. The outcome of autogenic successions at Glacier Bay, therefore, depends on drainage.

Xerarch Succession: Sand Dunes and Rocks

In all parts of the world colonists of bare sand have similar strategies, typically including vegetative reproduction by creeping stems (*stolons* or *rhizomes* usually) and drought-resistant features like sunken stomates, thick cuticles, and C4 photosynthesis. Whenever abandoned beaches or other patches of bare sand are covered with vegetation, these sand pioneers are always the first-comers. Not only do they form a pioneer seral community but they make necessary changes to the habitat by holding sand together.

XERARCH[2] SUCCESSIONS on sand dunes, therefore, always begin with a phase of autogenic succession as a pioneer community modifies the habitat. What happens next, however, depends on allogenic factors like those controlling the water table: are the sands kept moist or does drainage remain excessive? A classic example is on the sand dunes that line the southern shore of Lake Michigan.

Lake Michigan has been retreating northward ever since the end of the last glaciation 14,000 years ago, leaving behind beach sands thrown up into a long array of dunes that mark different positions of the retreating shoreline. The retreating lake has thus provided a series of habitats of different ages, rather as retreating glaciers do but giving a spread of more than 10,000 years.

Succession on these sand dunes has been discussed since Cowles (1899) first described the possibilities, but the most complete data are those of Olson (1958), who was able to age ancient dunes by radiocarbon assay and so arrange the communities that covered them in an accurate time sequence. Figure 23.14 shows the chronology revealed by this exercise. When very fresh, the dunes hold only the specialized grasses, but shortly thereafter, perhaps less than a century, a stand of poplars (the cottonwood, *Populus deltoides*) is in place. Nearby, or mingled with the poplars, are stands of pine. But dunes of all ages from 300 years to 12,000 years support woodlands of black oak *(Quercus velutina)*. Occasional small stands of poplar or pine can be found on older dunes, but in circumstances that suggest recent disturbance.

Both the poplars and the pines are early succession trees that cast shade too dense for their own seedlings. Shade-adapted oak seedlings grow underneath them and lead to a self-perpetuating community, the climax. This much seems familiar and straightforward. But the climax community remains very simple, essentially a stand of black oaks. The reason for this is that old sand dunes are too acid for all but the acid-tolerant oaks. The young dunes—where grasses, poplars, and pines grow—are alkaline with a pH above pH 8. Carbonate minerals are quickly

[2]Xerarch: literally "dry commanding" (Greek). The "arch" has the same root as in "archbishop," meaning "bishop in command."

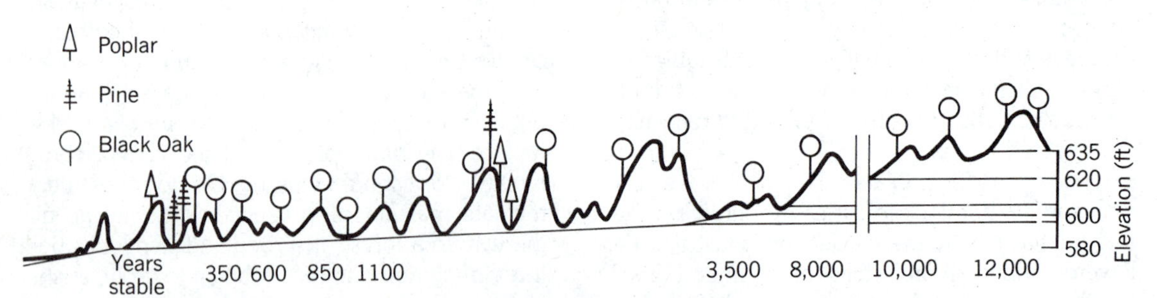

Figure 23.14 Time series across Lake Michigan sand dunes.
The lake is at the left. Dunes are dated by radiocarbon. The time series shows that succession stops with black oak woodland and does not continue to beech-maple forest as once postulated. (Transect near Gary, Indiana, from Olson, 1958.)

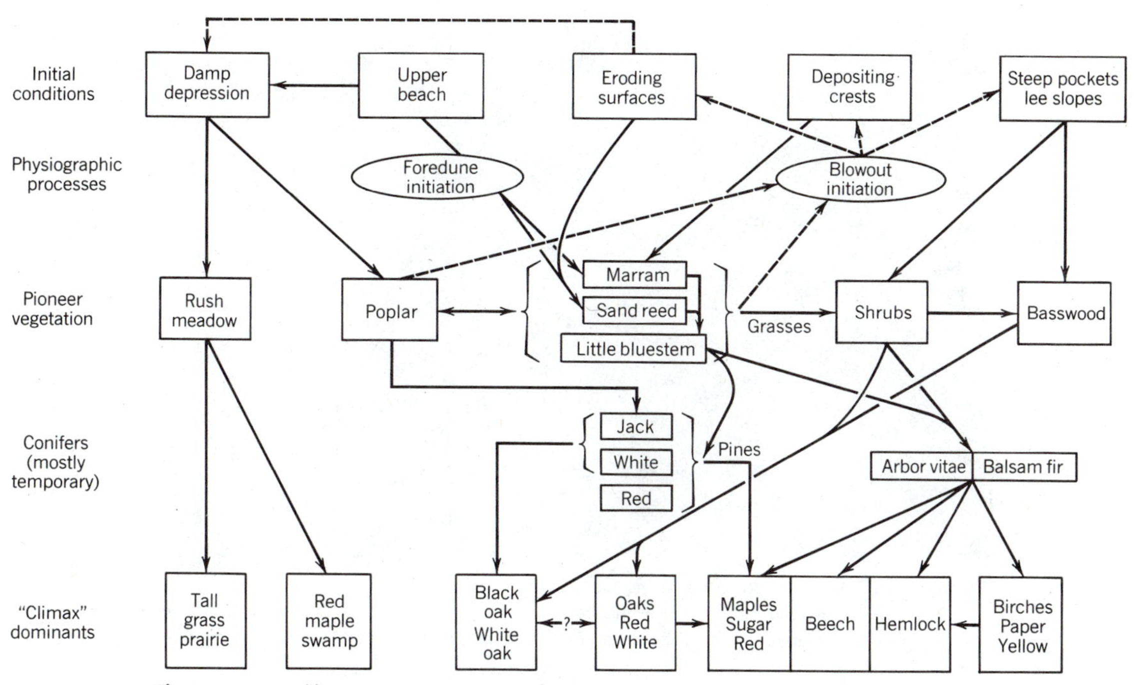

Figure 23.15 Alternative successions on Lake Michigan sand dunes. What communities develop on any part of a dune is a function of aspect and physical history more than of preceding plant populations. (From Olson, 1958.)

leached from the well-drained surface of these dunes during the first centuries of exposure and the pH has fallen to about pH 4 when the oaks are in place (Olson, 1958).

The black oak climax thus depends on sands remaining well drained. In large swales, or if the water table is close to the surface for other reasons, carbonates are not leached from the sands and quite different forest develops. On some of the most mesic sites of the region old beach sands have held beech–maple forest nearly as complex as anywhere in the Midwest. When the water table is very high the result may be a swamp, and many other alternatives are possible (Figure 23.15). Ecosystem development on old sand dunes, therefore, is primarily *allogenic*. Habitat drainage is all-powerful. Only in the pioneer stages is there a significant *autogenic* component as pioneer plants fix the shifting sands.

A comparable process of ecosystem development takes place on rocky ground that also has been called *xerarch succession*. On smooth, hard rock in exposed places only lichens have the necessary adaptations to grow (Figure 23.16). In moister places like hollows thick mosses grow. If cracks or depressions trap wind-blown organic debris, sufficient soil can collect for a community of drought-tolerant grasses and heath-like plants. In the deepest soil patches, trees with rock-hugging strategies can grow (Figure 23.17). Ecosystem development on rocks, therefore, depends on exposure and physical entrapment of soil. The process is largely *allogenic* like the developments on sand dunes. Possibly mosses and the grass–heath communities can be recognized as seral stages before sufficient soil collects for invasion of the trees. If this is so, then the development includes a xerarch primary succession running from moss to grass–heath to woodland, almost entirely allogenic but with the autogenic component of contributed organic matter. The lichens are probably outside the succession.

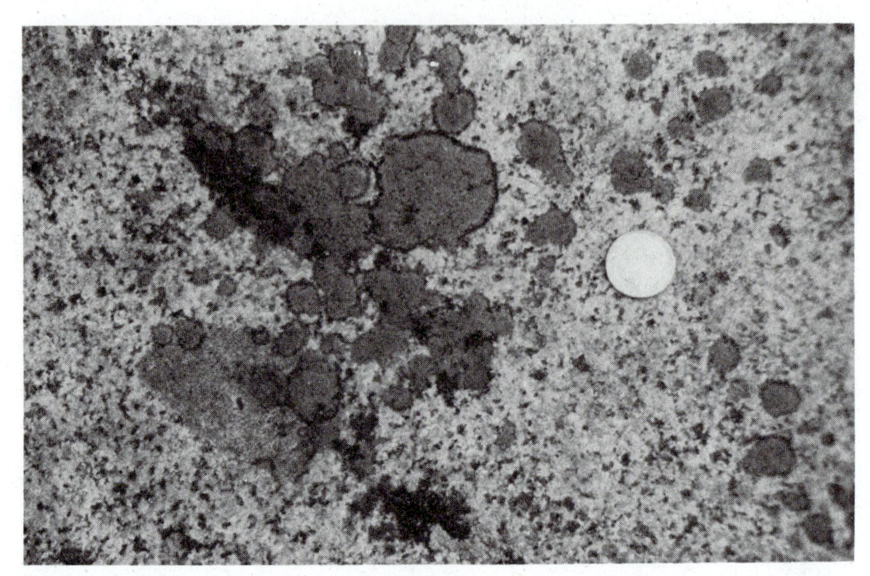

Figure 23.16 Lichens on rocks on Mount Rainier.
These lichens have been growing since the boulders were dropped by a retreating mountain glacier about 200 years ago. The lichens continue to grow in solitary state without taking part in successions.

Hydrarch Successions

Ecosystem development on wet sites depends on control of the water table. Lowering of water tables is by drainage, typically because streams cut their channels deeper. Raising of water tables follows the damming of outlet streams by landslips, volcanism, glaciation, or earth movements. On completely flat land on poorly permeable surface rock or sediments, the water table is likely to be maintained as a steady state with stream outputs balanced by stream inputs. Plants appropriate to the state of drainage grow and any successional change is allogenic.

Wet habitats have one special property that provides a directional autogenic process for HYDRARCH[3] SUCCESSIONS in that organic matter collects. Over long spans of time biomass can displace water, or even raise the ground surface above the water table. It is possible to argue, therefore, that ponds or lakes may be filled, and covered over with vegetation, as a result of an autogenic hydrarch succession. First organic mud

Figure 23.17 Tree colonizes rock.
Woodland can establish itself on rocky ground when sufficient soil collects in cracks. The trees undoubtedly are preceded by smaller colonists that grow more quickly.

[3]Hydrarch: literally "water-commanding," (Greek).

Figure 23.18 The vegetation at the edge of a pond in temperate latitudes. In the water can be seen floating leaves of water plants that are rooted in the bottom mud. In the wet mud at the edge is a line of reeds, and behind them bushes and small trees that grow in damp places. Well away from the pond, on drier ground, are trees of the local woods. This observation gave rise to the hypothesis that these successive kinds of vegetation were succeeding each other in time, as the pond filled with sediment and the encircling bands of vegetation constricted towards the center.

collects in the open water until the lake is shallow enough for rooted aquatic plants. When shallower still, cattails or other reed-like plants grow and a marsh results. In theory it might be possible for bushes to lower the water table and for forest to close over the site. This logic seems particularly appealing when looking at the vegetation on the bank of a pond or lake (Figures 23.18 and 23.19). The series of communities lining the shores looks as if it is closing in on the open water, with the local forest close behind.

High school and general biology texts often illustrate this hypothesis of ecosystem development by a hydrarch succession with a diagram like that of Figure 23.20. Whether these diagrams approximate reality depends on the kinds of ecosystems involved. Infilling of an Amazonian lake like Limoncocha (Figure 23.19) can lead to lowland swamp forest, which is the commonest type of forest in much of Amazonia where the ground is flooded frequently. In temperate latitudes, however, these successions never end in regional forest, as suggested by the diagrams, unless the water table is lowered by outside drainage. This conclusion is based on direct fossil evidence of lake infilling and community reconstructions by pollen analysis (Chapter 22).

Figure 23.21 is a reconstruction of a section of peatland in Minnesota at three stages in its history: early in postglacial time, at an intervening period, and at the present day (Heinselman, 1963, 1975). The reconstructions are based on more than fifty borings through the blanket peat bog that covers the landscape and through the

Figure 23.19 Plant communities at the edge of an Amazonian lake.
The lake, Limoncocha, occupies an abandoned river channel in the Oriente Province of Ecuador, and is shallow with gently sloping sides. Wide bands of different plant populations separate the open water from the tropical rain forest on the banks. Each band represents a community that could be considered to take part in a hydrarch succession.

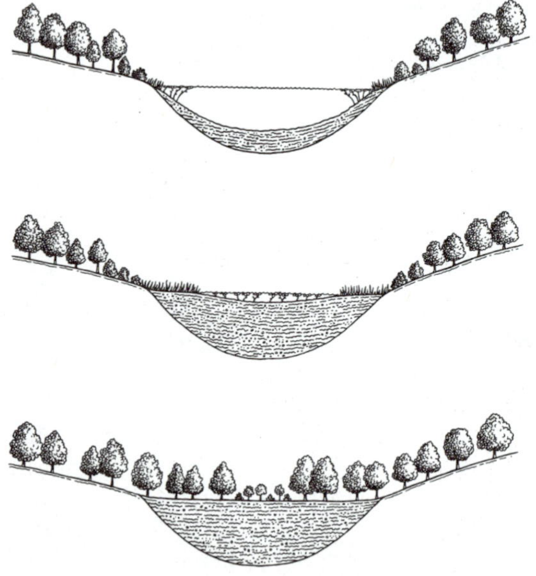

Figure 23.20 Conventional hydrarch succession diagram: probably wrong.
Floating mats and peaty sediments are supposed to displace water. The concentric rings of vegetation constrict until forest grows at the climax where once was water. The hypothesis pictured here, and in many diagrams like this, probably is wrong. The most likely outcome of filling a bog lake with peat is a permanent peat bog. Compare the real data of Figure 23.21.

bottom of Myrtle Lake itself. The data show that Myrtle Lake has never been filled but has risen with the growth of blanket peat, keeping pace with the deposition of peat on the lake bottom. Logs in the peat are certain evidence that spruce trees had invaded the bog in the past, but they died out, probably drowned by the growing *Sphagnum* bog. The climax is a pattern of peat bogs and persistent ponds that will remain, under the present climate, unless the land is physically drained.

Large areas of boreal and taiga biomes have boggy areas like that around Myrtle Lake, particularly the large muskeg areas of northern Canada south of the tundra. From the air these regions can be seen to be studded with ponds, all ringed around with floating *Sphagnum* or other marsh plants, and all in a countryside dotted with black spruce *(Picea mariana)* or other trees of wetlands. Some of these lakes perhaps persist indefinitely like Myrtle Lake; others perhaps do close over with floating peat. As water tables fluctuate with climatic change black spruce and *Sphagnum* bog then alternate indefinitely.

A more general paleoecological test of the hypothesis of hydrarch succession in temperate lat-

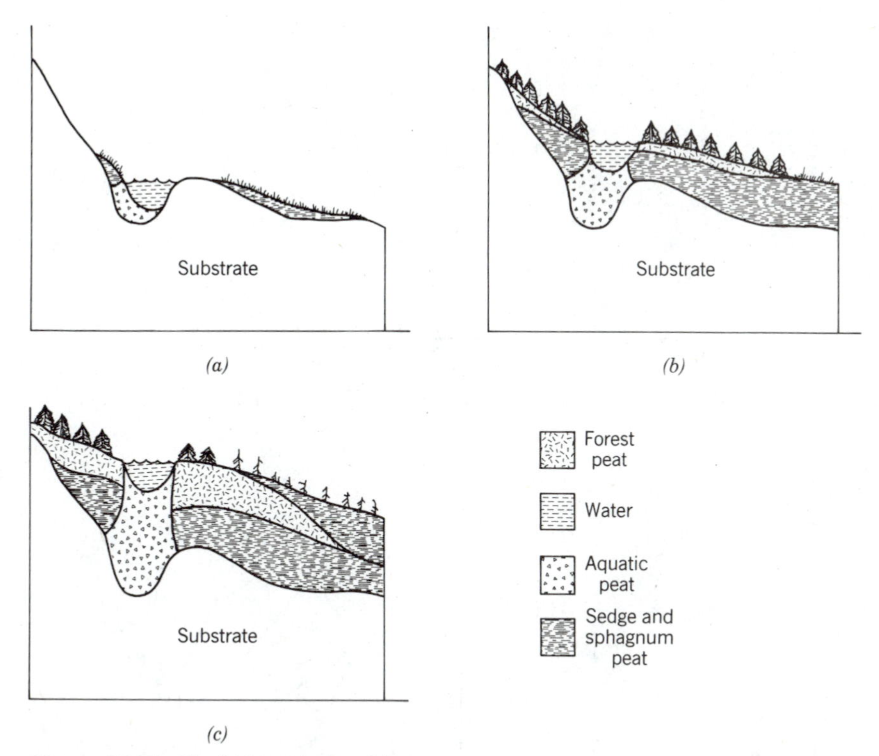

Figure 23.21 The history of a bog lake.
The three cross sections show the land around Myrtle Lake, Minnesota, in the early postglacial, several thousand years later, and at the present day. Growth of peat did not eliminate the lake. There was no succession to forest, merely a short-lived invasion of trees. This true history of a bog lake was quite unlike the postulated history of Figure 23.20. (From Heinselman, 1963.)

itudes uses pollen analysis to reconstruct the histories of numerous hydroseres. Walker (1970) identified twelve plant communities of the kind that would be expected to take part in any postulated hydrarch succession in Britain. The communities ranged from pure aquatic plants to reedswamps, wet bushlands, and *Sphagnum* bogs, and could be ranked in order of expected appearance on a continuum from wet to dry.

It would not be expected that all these communities would appear in succession at any one site. For the hypothesis of an autogenic hydrarch succession to stand, however, some subset of these communities should appear *in the right order*. To test if this was so, Walker searched the literature for stratigraphic histories where the pollen record was sufficiently unequivocal for it to be possible to describe the order in which plant communities actually had appeared. He found a total of 159 measured directions of transition between pairs of communities.

Walker's results for sets of small and large lakes are given in Figure 23.22. The data are plotted as frequencies (simple numbers of occurrence) of transition between pairs in the left-hand side of each diagram with quasi-vector diagrams showing directions of change to the right. Some community replacements are more likely than others, and there is a general tendency for "drier" communities to replace "wetter" communities. But there is a large amount of apparent randomness about the sequence of events, with many replacements actually going in the "wrong" direction (that is, from dry to wet).

Where the pollen histories were long, the concluding community of the hydrosere was always

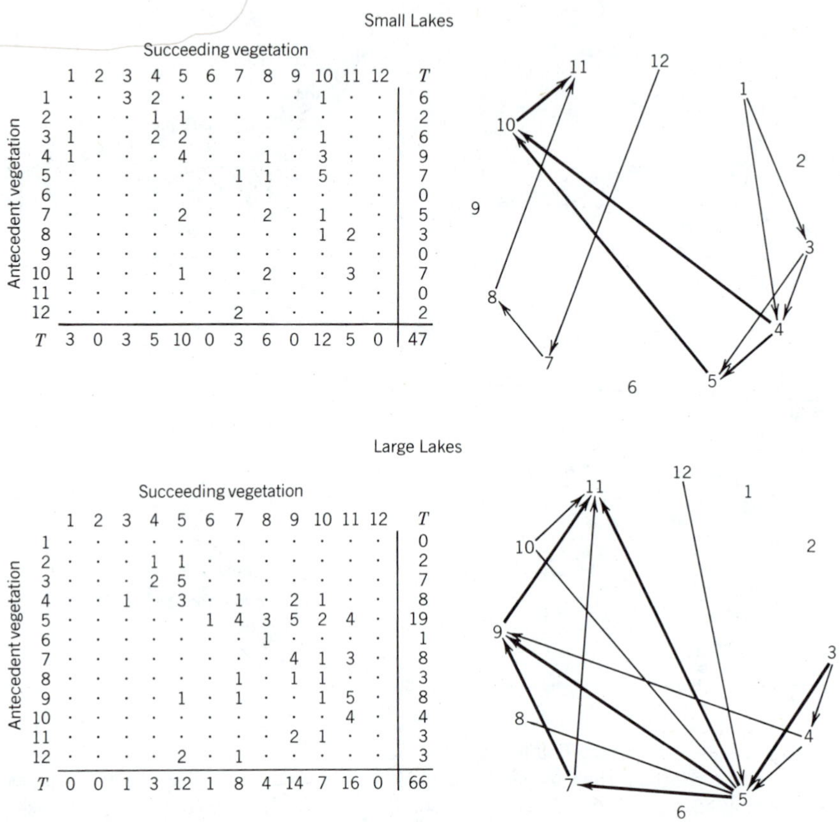

Small Lakes

Antecedent vegetation \ Succeeding vegetation	1	2	3	4	5	6	7	8	9	10	11	12	T
1	·	·	3	2	·	·	·	·	·	1	·	·	6
2	·	·	·	1	1	·	·	·	·	·	·	·	2
3	1	·	·	2	2	·	·	·	·	1	·	·	6
4	1	·	·	·	4	·	·	1	·	3	·	·	9
5	·	·	·	·	·	·	1	1	·	5	·	·	7
6	·	·	·	·	·	·	·	·	·	·	·	·	0
7	·	·	·	·	2	·	·	2	·	1	·	·	5
8	·	·	·	·	·	·	·	·	·	1	2	·	3
9	·	·	·	·	·	·	·	·	·	·	·	·	0
10	1	·	·	·	1	·	·	2	·	·	3	·	7
11	·	·	·	·	·	·	·	·	·	·	·	·	0
12	·	·	·	·	·	·	2	·	·	·	·	·	2
T	3	0	3	5	10	0	3	6	0	12	5	0	47

Large Lakes

Antecedent vegetation \ Succeeding vegetation	1	2	3	4	5	6	7	8	9	10	11	12	T
1	·	·	·	·	·	·	·	·	·	·	·	·	0
2	·	·	·	1	1	·	·	·	·	·	·	·	2
3	·	·	·	2	5	·	·	·	·	·	·	·	7
4	·	·	1	·	3	·	1	·	2	1	·	·	8
5	·	·	·	·	·	1	4	3	5	2	4	·	19
6	·	·	·	·	·	·	·	1	·	·	·	·	1
7	·	·	·	·	·	·	·	·	4	1	3	·	8
8	·	·	·	·	·	·	1	·	1	1	·	·	3
9	·	·	·	·	1	·	1	·	·	1	5	·	8
10	·	·	·	·	·	·	·	·	·	·	4	·	4
11	·	·	·	·	·	·	·	·	2	1	·	·	3
12	·	·	·	·	2	·	1	·	·	·	·	·	3
T	0	0	1	3	12	1	8	4	14	7	16	0	66

Figure 23.22 Paleoecological test of the hydrarch succession hypothesis. Twelve potential seral communities are ranked from wet (1) to dry (12). Pollen and macrofossil evidence shows the order in which pairs of communities succeed one another. Almost any order was possible, though there was a slight bias towards the direction of increasing dryness. Diagrams at the right show the most common replacements of one community by another. (From Walker, 1970.)

a peat bog. Even if the pollen showed that a seral community of bushland or swamp woodland occupied the site at some time, this was always replaced in succession by a peat bog that drowned out the trees. That this should happen is understandable from the known properties of *Sphagnum* peat communities, because they are strongly nutrient limited as the nutrient supply of a habitat becomes locked in organic matter. Any lowering of a water table accompanied by an invasion of bushes leads to decomposition with consequent release of nutrients. This promotes bog growth with consequent drowning of the tree or bush community.

Walker (1970) concluded that the climax of hydrarch succession in Britain was a peat bog. Wetlands never would be drained by hydrarch successions. The actual sequence of seral stages in a hydrosere was a function of water table changes due to stream down-cutting or damming. These results are like those from Minnesota marshes and the Glacier Bay succession on wet sites described above, and other paleoecological studies from temperate wetlands. Figure 23.23 is a better description of the history of a pond than conventional figure 23.20. Autogenic hydrarch successions cannot replace swamp vegetation with communities typical of better-drained sites. Only physical drainage can lead to this.

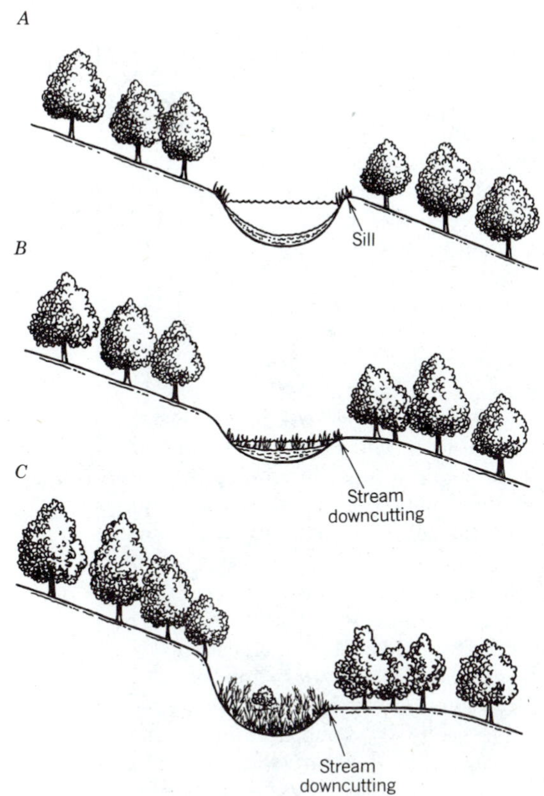

Figure 23.23 History of a typical pond.
Small ponds and lakes are drained by stream down-cutting even as sediments collect. Seldom, if at all, are they "reclaimed" by surrounding vegetation. When stream down-cutting lowers the water table sufficiently a permanent marsh is established (compare Figure 23.20).

HYPOTHESES OF ORGANIZATION THROUGH SUCCESSION

The idea that ecological succession is an organizing process has been important to ecology from the beginning. The successions most obvious to scholars in universities of the north temperate belt are old field successions, or primary successions on fresh habitat like the Michigan sand dunes. These successions are, in the earlier stages that can be watched in a human lifetime, particularly predictable. They always involve physical improvements to the habitat by the seral communities, and they end in climax communities like forest, noted for size, biomass, and complexity. It is a reasonable use of the English language to say that the climax community is built or organized through the process of succession.

Some ecologists argued that whole communities or ecosystems were self-organizing entities of plants and animals so completely coevolved as to be self-perpetuating. A principal spokesman for this point of view was F. E. Clements (1916, 1936). Clements' experience was with prairie successions during a pioneer plant survey of Nebraska in the 1890s, where he saw the old buffalo trails being closed over by secondary succession that took about 50 years to replace a climax community of prairie grasses. Later he worked in the forested east and saw old field successions, where replacement of the old forest climax on cleared fields was even slower, evidently requiring a century or more to complete. He was influenced by Cowles' (1899) original work on the sand dunes of Lake Michigan, which postulated (incorrectly) that succession on the dunes could proceed almost everywhere to the regional climax of the beech–maple forest if given time enough.

Clements was particularly struck by two apparent common denominators in these examples: complex communities always were built by systems of replacements during which habitats changed, and an ultimate limit to community building seemed to be set by climate. Community building was to him a process of growth comparable to the growth of individual plants. A community was made up of plants that had different properties to serve the community, something made clear in succession when seral communities spread over such physical damage to a community as a farmer's field—closing the wound as it were and repairing the habitat for the climax to live there once more. Moreover, these plants of seral communities served to make the climax community grow by occupying fresh sites like sand dunes. In Clements' thinking the community grows like an organism (Figure 23.24).

The Clementsian view led to attractive systems for classifying plant communities. In every climatic region there was a single climax plant community, the CLIMAX FORMATION long recognized by plant geographers (Chapter 14).

I. CONCEPT AND CAUSES OF SUCCESSION.

The formation an organism.—The developmental study of vegetation necessarily rests upon the assumption that the unit or climax formation is an organic entity (Research Methods, 199). As an organism the formation arises, grows, matures, and dies. Its response to the habitat is shown in processes or functions and in structures which are the record as well as the result of these functions. Furthermore, each climax formation is able to reproduce itself, repeating with essential fidelity the stages of its development. The life-history of a formation is a complex but definite process, comparable in its chief features with the life-history of an individual plant.

Universal occurrence of succession.—Succession is the universal process of formation development. It has occurred again and again in the history of every climax formation, and must recur whenever proper conditions arise. No climax area lacks frequent evidence of succession, and the greater number present it in bewildering abundance. The evidence is most obvious in active physiographic areas, dunes, strands, lakes, flood-plains, bad lands, etc., and in areas disturbed by man. But the most stable association is never in complete equilibrium, nor is it free from disturbed areas in which secondary succession is evident. An outcrop of rock, a projecting boulder, a change in soil or in exposure, an increase or decrease in the water-content or the light intensity, a rabbit-burrow, an ant-heap, the furrow of a plow, or the tracks worn by wheels, all these and many others initiate successions, often short and minute, but always significant. Even where the final community seems most homogeneous and its factors uniform, quantitative study by quadrat and instrument reveals a swing of population and a variation in the controlling factors. Invisible as these are to the ordinary observer, they are often very considerable, and in all cases are essentially materials for the study of succession. In consequence, a floristic or physiognomic study of an association, especially in a restricted area, can furnish no trustworthy conclusions as to the prevalence of succession. The latter can be determined only by investigation which is intensive in method and extensive in scope.

Viewpoints of succession.—A complete understanding of succession is possible only from the consideration of various viewpoints. Its most striking feature lies in the movement of populations, the waves of invasion, which rise and fall through the habitat from initiation to climax. These are marked by a corresponding progression of vegetation forms or phyads, from lichens and mosses to the final trees. On the physical side, the fundamental view is that which deals with the forces which initiate succession and the reactions which maintain it. This leads to the consideration of the responsive processes or functions which characterize the development, and the resulting structures, communities, zones, alternes, and layers. Finally, all of these viewpoints are summed up in that which regards succession as the growth or development

3

Figure 23.24 Facsimile of the first page of F. E. Clements' book on "Succession" (1916). His claim that the study of succession "necessarily rests . . . etc." though now seen to be false, was for long taken seriously.

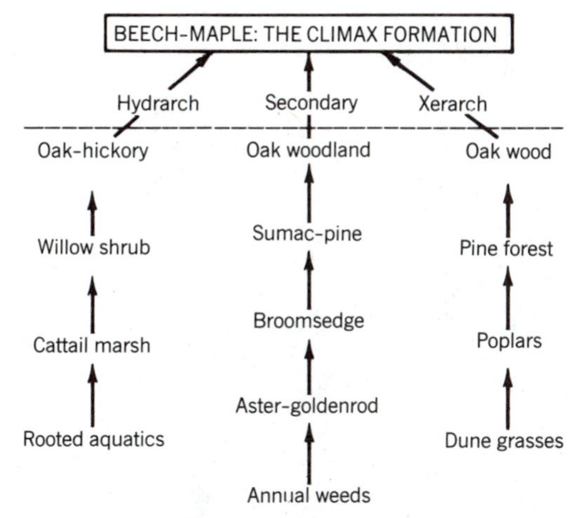

Figure 23.25 The kind of community taxonomy proposed by Clements.
All associations are supposed to be related to the climax formations by their roles as seral stages in successions.

All other communities found in the region were related to the *climax formation* as various stages of its development (Figure 23.25). For the more ephemeral communities of the early successions this classification was easy and convenient. But the system was not so tidy for communities that would never change under existing circumstances into the climax formation. The black oak communities of the Michigan dunes, for instance, show no signs of giving place to beech–maple forest after 12,000 years (see page 596). Clements suggested that all such communities were arrested climax. The black oaks would be a *preclimax*. A similar end to succession on wetland would be *postclimax*. Grazing animals or fire led to a *proclimax* (page 587) or *disclimax*. In this way all known plant communities of a geographic region were related, either as seral components of the climax or as some variant of the climax held back from full expression by some local factor of soil, drainage, or disturbance.

Essential to this point of view is the idea that a community is a SUPERORGANISM, an entity of many species that has EMERGENT PROPERTIES of its own. Realizing that his *superorganism* drew some of its properties from animals as well as plants, Clements coined the word BIOME to replace the earlier *climax formation* for his ultimate community unit.

Clements' vision of a superorganism attracted philosophers. Smuts (1926), the philosopher prime minister of South Africa, acknowledged the ecological writings of Clements as the basis of his theories of "Holism." The universe was made up of series of "holes," each more than the sum of its parts, an idea that had some vogue. Many teachers used the Clementsian world view, since this gave ecology a special place among the biological sciences (Phillips, 1934, 1935). Individuals were but part of the whole community, so superior learning was to be found in studying the whole community. The key process to study was, of course, ecological succession.

Clements' work is still important because it lies at the root of many of the political or social movements that take their names from ecology in the present day. Whenever activists accuse their political or exploiter adversaries of "ecocide" they invoke Clements' teachings. They borrow from him the idea that the ecosystem of the climax is an organism, saying that therefore it can be killed.

The modern view is that succession is an inevitable consequence of the coexistence of plants with different strategies, as we have seen, although some writers complain that the habitat changes (facilitation) in some successions are ignored (Finegan, 1984). The idea of a superorganism has no scientific basis and is unnecessary.[4] Plants, like all products of natural selection, are individualists. This essential truth was argued strongly even in Clements' day, most notably by Gleason (1917, 1926) (Figure 23.26). But the final triumph of Gleason's *individualistic hypothesis of succession* came only with the concept of species strategies in the 1960s.

[4]Superorganism: recently the term has been used in a somewhat different context as the product of selection for community properties through a system of structured demes (Chapter 25).

Where one or both of the primary causes changes abruptly, sharply delimited areas of vegetation ensue. Since such a condition is of common occurrence, the distinctness of associations is in many regions obvious, and has led first to the recognition of communities and later to their common acceptance as vegetational units. Where the variation of the causes is gradual, the apparent distinctness of associations is lost. The continuation in time of these primary causes unchanged produces associational stability, and the alteration of either or both leads to succession. If the nature and sequence of these changes are identical for all the associations of one general type (although they need not be synchronous), similar successions ensue, producing successional series. Climax vegetation represents a stage at which effective changes have ceased, although their resumption at any future time may again initiate a new series of successions.

In conclusion, it may be said that every species of plant is a law unto itself, the distribution of which in space depends upon its individual peculiarities of migration and environmental requirements. Its disseminules migrate everywhere, and grow wherever they find favorable conditions. The species disappears from areas where the environment is no longer endurable. It grows in company with any other species of similar environmental requirements, irrespective of their normal associational affiliations. The behavior of the plant offers in itself no reason at all for the segregation of definite communities. Plant associations, the most conspicuous illustration of the space relation of plants, depend solely on the coincidence of environmental selection and migration over an area of recognizable extent and usually for a time of considerable duration. A rigid definition of the scope or extent of the association is impossible, and a logical classification of associations into larger groups, or into successional series, has not yet been achieved.

The writer expresses his thanks to Dr. W. S. Cooper, Dr. Frank C. Gates, Major Barrington Moore, Mr. Norman Taylor, and Dr. A. G. Vestal for kindly criticism and suggestion during the preparation of this paper.

Figure 23.26 Gleason's reply to Clements.
The second paragraph is a succinct statement of the outlines of what would be established fifty years later. (From Gleason, 1926.)

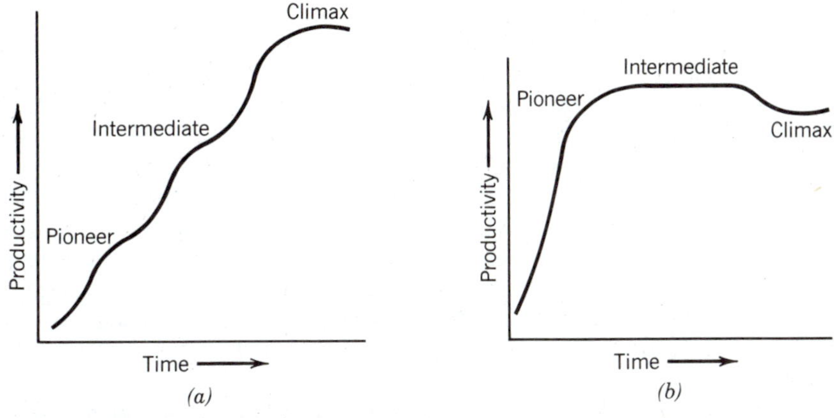

Figure 23.27 Productivity during succession.
(a) is a statement of the Lindeman (1942) hypothesis that productivity increases throughout succession. *(b)* is what is thought actually to happen in many successions, forty years of measurement later.

The Hypothesis That Successions are Regulated by Community Energetics

By many measures, succession leads to complexity and order. This appearance of order is part of the second paradox discussed in the introduction. We reject a superorganismic theory of succession, concluding that all the phenomena can result from individual self-interest. But the paradox of apparent order remains. A number of models have been proposed over the years that seek to describe this apparent ordering in terms of community energetics.

Lindeman (1942) postulated that successions should proceed towards maximizing the efficiency of energy conversion. The pioneer communities would be poorly efficient at transforming solar energy on a field scale so that ecosystem productivity was low. With succession, productivity would progressively increase, reaching maximum values in the complex vegetation of the climax (Figure 23.27*a*). Maximum order would then be supported by maximum energy supplies. This was a good hypothesis because it was testable: it predicted that seral communities could be ranked by community productivity, the most productive being the climax.

Measures of production made over the last forty years are adequate to refute the Lindeman hypothesis (Chapter 16). Early seral stages of secondary successions are highly productive, contrary to the hypothesis, being perhaps the most productive communities of all. Moreover some intriguing data suggest that productivity actually may be lowest in the climax in some successions, the exact opposite of what Lindeman predicted.

Figure 23.28 shows the results of productivity measures from Southeast Asia by the plantation-stand method (Chapter 16). Both gross and net production are found to fall in the mature plantations. What we now know about tree strategies

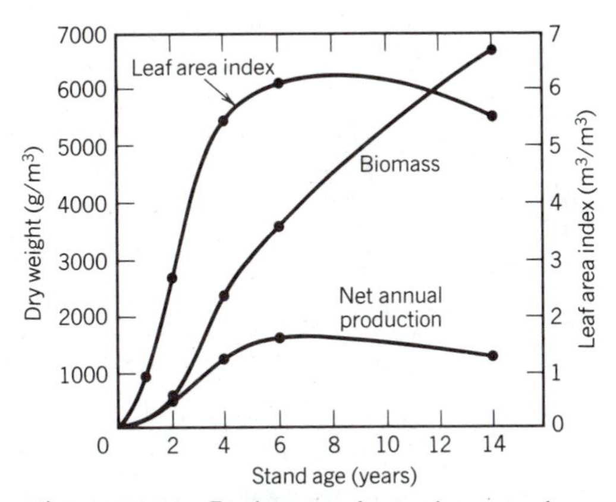

Figure 23.28 Productivity of aging forest stands. Measurements are by the plantation series method applied to tropical forest plantations in Southeast Asia. Productivity falls in aging stands. (From Kira and Shidei, 1967.)

in climax forest suggests that gross production should be low in natural climax forests also. Canopy trees of the climax are monolayers adapted to peak efficiency in shade and cannot be expected to collect carbon dioxide as rapidly as the multilayer early succession trees that they replace (Chapter 3). Moreover it is likely that climax forest trees often are under nutrient or water stresses not experienced earlier in succession, perhaps imposing further limits on production (Grime, 1979). The likely actual change of productivity with secondary succession is given in Figure 23.27*b*.

Nutrient stress in the climax also lowers productivity in aquatic microcosms (Figure 23.29). These data describe the history of production measured by the pH method (Chapter 16) in beakers of water containing planktonic and benthic algae and supplied with artificial light. The fall in productivity when succession ends is striking, but an inspection of the microcosms explains this easily enough. The sides of the glass beakers at "climax" are covered with films and festoons of moribund benthic algae, like the walls of all neglected aquaria. In this moribund biomass is locked the better part of the nutrient supply originally in the water. Productivity falls because of nutrient deficiency (Cooke, 1967).

Thus no general trend towards greater efficiency in the climax exists. In some successions, at least, the opposite is the truth with the climax being the least productive stage. This conclusion is quite in keeping with the individualistic hypothesis of succession, suggesting that the anarchy of self-interest results in an eventual community with most individuals existing under stress and total community function impeded by the results of selfishness.

Odum (1969) has interpreted the data on ecosystem productivity to mean that, although production does not increase towards the climax, total biomass does. Both in the forest stand and microcosm data, biomass was highest in the poorly productive climax (Figures 23.28 and 23.29). The same must be true of many other ecosystems, including all forests, prairies, and other terrestrial systems where large size and organic soils characterize the climax. In these systems biomass collects until further accretion is balanced by losses from community respiration

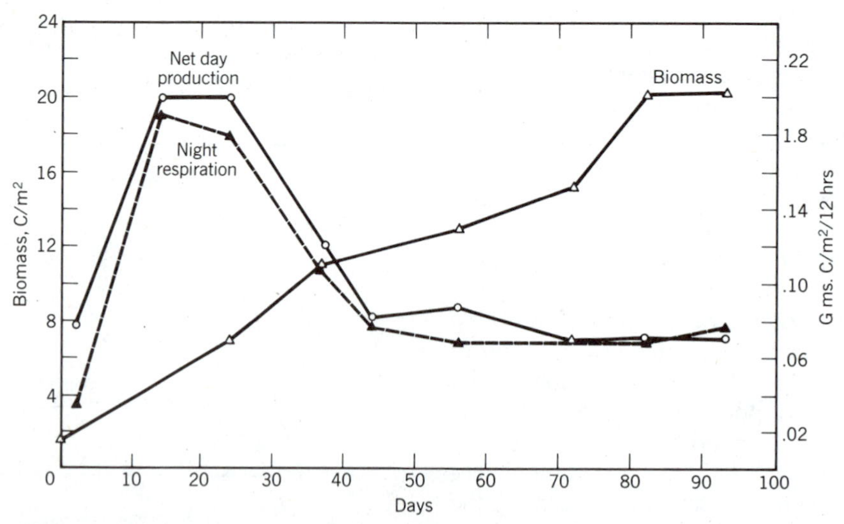

Figure 23.29 Productivity and biomass in developing microcosms.
The fresh-water microcosms in 300-ml beakers were closed but provided with an energy source. Productivity and respiration both fell at the climax, even as biomass rose. Low climax productivity is caused by nutrient shortage as these are locked up in dead organic matter. (From Cooke, 1967.)

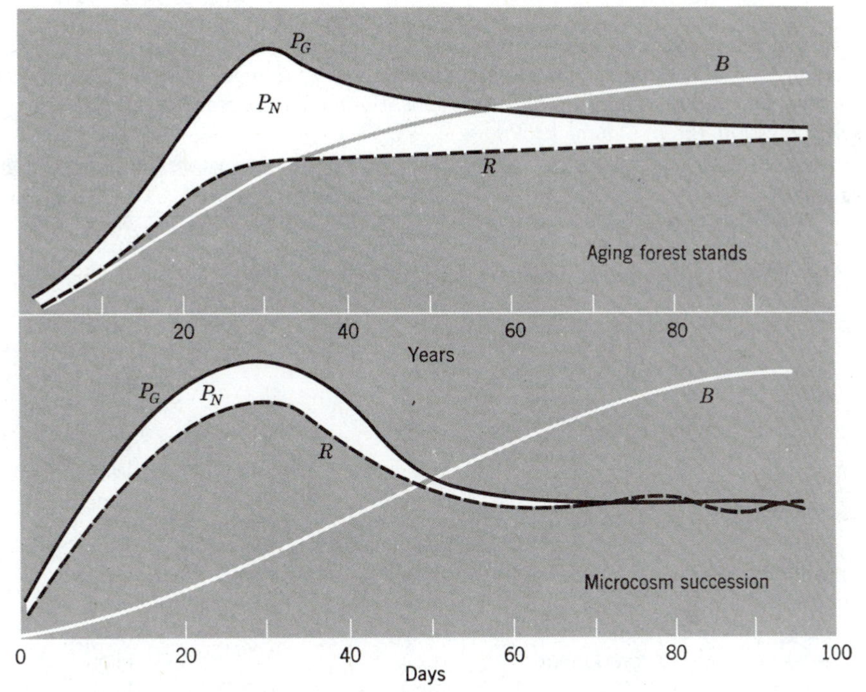

Figure 23.30 The increase in biomass during different successions. The upper diagram is compiled from data from forest stands of different ages. The lower diagram results from measurements of replicated laboratory microcosms, essentially of algae in water. P_G is gross production, P_N is net production, R is total community respiration, and B is total biomass. The data suggest that as systems age they accumulate biomass (including detritus) until respiration balances production, when they may be called mature. Notice that gross productivity falls at maturity in both systems. (Redrawn from Odum, 1969.)

(Figure 23.30). The climax community then exists in a mature ecosystem when A MATURE ECOSYSTEM is one in which COMMUNITY RESPIRATION EQUALS GROSS COMMUNITY PRODUCTION. To some ecologists this may seem a definition that lacks interest, for there is no obvious reason why species replacements and other interesting successional changes should be influenced by the attainment of equilibrium between production and decomposition. The largest steady-state supply of moribund organic matter may not be the best indicator of community maturity.

The Hypothesis of Control by the Management of Information

In classical thermodynamics the relative order of a system is defined in terms of free energy within that system. A highly ordered system, where events are non-random, has low *entropy* but high *enthalpy*. Chemical reactions that are endothermic, like the production of reduced carbon compounds in photosynthesis, increase the enthalpy of the system. In the process, the system loses randomness; which is to say that entropy falls. The system now has less free energy but is more highly ordered.

If, therefore, we ignore the fact that much biomass is merely moribund organic matter and think of it only as energy, arguments of classical thermodynamics can be mustered to show that a large deposit of biomass represents order. Phenomena of succession other than increasing biomass also can be listed that represent increasing order (Table 23.2). Most notable are increases in all measures of diverseness: more biochemical diversity, more spatial variety, more layering, and

Table 23.2
Odum's Summary of the Consequences of Succession
Some of the statements in this summary are open to objection (e.g., changes in "information") and these are discussed in the text. Others do not apply to all successions. But the table does give an overview of the general consequences of succession. (From Odum, 1969.)

Ecosystem Attributes	Developmental Stages	Mature Stages
	Community energetics	
1. Gross production/community respiration (*P*/*R* ratio)	Greater or less than 1	Approaches 1
2. Gross production/standing crop biomass (*P*/*B* ratio)	High	Low
3. Biomass supported/unit energy flow (*B*/*E* ratio)	Low	High
4. Net community production (yield)	High	Low
5. Food chains	Linear, predominantly grazing	Weblike, predominantly detritus
	Community structure	
6. Total organic matter	Small	Large
7. Inorganic nutrients	Extrabiotic	Intrabiotic
8. Species diversity—variety component	Low	High
9. Species diversity—equitability component	Low	High
10. Biochemical diversity	Low	High
11. Stratification and spatial heterogeneity (pattern diversity)	Poorly organized	Well organized
	Life history	
12. Niche specialization	Broad	Narrow
13. Size of organism	Small	Large
14. Life cycles	Short, simple	Long, complex
	Nutrient cycling	
15. Mineral cycles	Open	Closed
16. Nutrient exchange rate, between organisms and environment	Rapid	Slow
17. Role of detritus in nutrient regeneration	Unimportant	Important
	Selection pressure	
18. Growth form	For rapid growth ("*r*-selection")	For feedback control ("*K*-selection")
19. Production	Quantity	Quality
	Overall homeostasis	
20. Internal symbiosis	Undeveloped	Developed
21. Nutrient conservation	Poor	Good
22. Stability (resistance to external perturbations)	Poor	Good
23. Entropy	High	Low
24. Information	Low	High

more species as the succession approaches climax.

It happens that similar equations are used in physics to describe *information, energy,* and *entropy* in physical systems. The three concepts converge. A development of the 1950s and 1960s in ecology was the use of the mathematics of *information theory* for a combined measure of relative abundance and species richness in communities, a statistic with so many difficulties that it is now going out of fashion (Chapter 26). The information theory measure of species diversity is high for climax communities.

Margalef (1958, 1963, 1968), among others (Odum, 1969, 1971), pointed out that the information theory measure of diversity was high when biomass, as a measure of information, also was high. This allows the hypothesis that succession proceeds to maximize information or order. Margalef (1963) suggested that rapid accumulation of organic matter early in a succession (high production–low respiration; Figure 23.28) represented collecting information that would be passed on to later seral stages and used to create order. Control of erosion by organic soil in old ecosystems is, using this logic, order imposed by earlier succession stages. Since organic matter often is exported from ecosystems, communities also can be said to impose order on their surroundings. Margalef's example was a phytoplankton community that exported dead organic matter to the bottom of a lake or sea (Chapter 21). The plankton community, in a sense, "organizes" the benthos.

Intellectual exercises like these please ecologists in their debates. The approach describes the relationships between communities in ways that might make possible analysis and modeling. It is probably true to say, however, that contemporary ecologists find that studying the strategies of coevolved species gives more understanding of community process than do models based on analogies between biomass or structure with such physical concepts as enthalpy and information.

PREVIEW TO CHAPTER 24

The equilibrium theory of island biogeography states that the number of species found on an island is a function of the rate of immigration and the rate of extinction of species on the island. Both immigration and extinction are thought to be continuous so that there is almost constant turnover of species on an island, with fresh invaders taking the place of those that have gone extinct. The actual number of species on the island remains roughly constant after a while and this number is only a small fraction of the number that could live there. It is theoretically possible for an equilibrium number to be established without there being interaction between the island inhabitants as invasions and extinctions proceed by chance alone. The theory predicts, however, that species eventually should interact on a crowded island, making invasion more difficult and extinction more likely. This will result in a somewhat different interactive equilibrium. The essential predictions of the theory have been tested both experimentally and by using correlation data. Tiny mangrove islands have been shown to hold but a small fraction of their potential invaders as predicted. When all the arthropods on one of these mangrove islets are killed, there is a rapid reinvasion ending with a species equilibrium at about the original number, though the actual species involved may be different. The correlation tests rely on more detailed predictions of the theory, particularly that large islands should hold more species than small islands and remote islands should hold fewer species than near islands. Studies of species–area relationships of birds on islands cannot falsify these predictions. Studies of plants of oceanic islands lead to different conclusions, however, probably because plants exist in denser populations and because closed vegetation is difficult to invade. The theory has been applied to many other isolated habitats, including beakers of water, closed basin lakes, agricultural fields, and individual plants. Like succession theory, equilibrium island biogeography explains ways in which communities are built from a pool of species able to invade a site, but in the special circumstance that most equilibrium species or strong *K*-strategists are excluded.

CHAPTER 24

ISLAND BIOGEOGRAPHY: THE IMMIGRATION–EXTINCTION EQUILIBRIUM

ISLAND BIOGEOGRAPHY is the study of how the number of species living on an isolated space can be set as a balance between immigration and extinction. The theory was developed originally to predict the numbers of species that would be found coexisting on oceanic islands, but it has been found to be applicable to virtually all isolated habitats that must be invaded across obstacles.

Consider a totally new oceanic island created by a volcanic explosion. It is without life and can receive terrestrial animals only by long-distance transport across the sea. Colonists begin to arrive, and observations of actual young islands like Surtsey, the young island that appeared off Iceland a few years ago, show that these first colonists are prompt (Fridriksson, 1975). Some colonists establish themselves and live on the island, but this makes no difference to the rate of immigration at first, because the island is underpopulated for some time so that there is no competitive bar to the establishment of newcomers. Arrival of colonists continues at a constant rate though the number of new species among them falls progressively.

But a small island, particularly a young one, may be an uncertain place in which to live, suggesting that some of the first-comers who became established die out. Smallness of populations also makes extinction likely. Since the island is still largely empty habitat, this chance of extinction will not depend on the presence of other animals; it will depend only on physical adversities. The rate of extinction, therefore, can be treated as constant in this early phase of colonization, just like the rate of immigration.

Figure 24.1*A* illustrates this early phase of community building on an island. Rates of immigration and extinction are both linear because neither is dependent on population density. As the number of species living on the island rises, the rate of species arrival falls as a simple function of the number already there. The rate of extinction rises as a simple function of the number of species at risk. The model suggests that an equilibrium number, S_1, will be established when the rate of immigration is balanced by the rate of extinction. This equilibrium does not depend on any interactions between populations on the island for its effect. The assumption is that

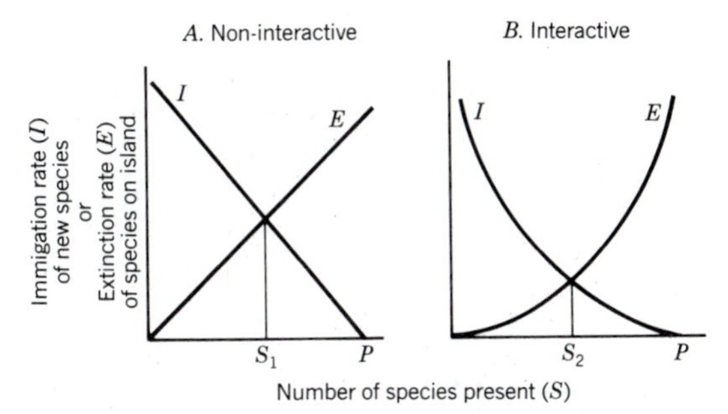

Figure 24.1 General hypothesis for the regulation of species number on islands.
When an island is lightly populated, so that most populations are not under density-dependent control, the rates of immigration and extinction are linear (non-interactive model, *A*). In a well-populated island (interactive model, *B*), the rates are exponential. The equilibrium numbers, S_1, S_2 are unlikely to be equal. *P* is the total species pool from which the colonists must come. (Modified from Wilson, 1969.)

there is no competition and no predation. The result is a NON-INTERACTIVE SPECIES EQUILIBRIUM.

The non-interactive species equilibrium, S_1, should persist only until populations on the island become large enough for competition and predation to become important facts of life. After that successful immigration requires both arrival and establishment, which depends on resistance by species already present. Both immigration and extinction then become density dependent, and both rate curves become exponential (Figure 24.1*B*). Where these curves intersect there is a new equilibrium number of species, S_2. This number is very much dependent on interactions between species on the crowded island and is an INTERACTIVE SPECIES EQUILIBRIUM.

This model for community building on islands was first put forward independently by Preston (1962) and MacArthur and Wilson (1963) and was later developed into a formal *Theory of Island Biogeography* by MacArthur and Wilson (1967) (Simberloff, 1974; Diamond and May, 1976). A first prediction of the hypothesis is that there can be two kinds of species equilibrium: *interactive* and *non-interactive*. In the simplest island model, the non-interactive equilibrium, S_1, has only a transitory existence before increasing population densities drive the species list to S_2. But if environmental circumstance should prevent the establishment of population equilibria, then the non-interactive equilibrium should persist indefinitely. Essentially this is the claim of those who argue for density-independent control of populations in stressed habitats (Chapter 8). There can be an equilibrium, and hence constant, number of species whether populations are controlled by density-dependent factors or not.

Satisfying though this preliminary property of the model may be, the importance of the hypothesis depends on its yield of more testable predictions. Many such predictions arise from the fact that the interactive species equilibrium, S_2, must be a function of both *island remoteness* and *island size*.

Figure 24.2 shows the predictions of the model for islands near and far and islands large and small. Distant islands will receive immigrants at slower rates than islands near the mainland, simply as a function of the distance to be covered by potential colonists. The extinction rate, however, is not affected by distance. The hypothesis predicts, therefore, that distant islands will hold fewer species at equilibrium than will near islands.

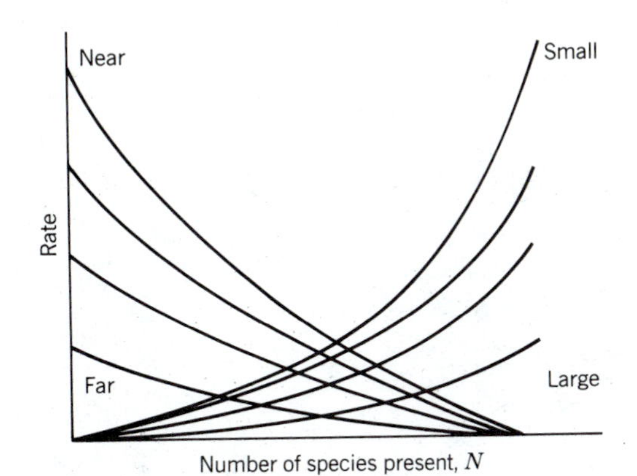

Figure 24.2 The effects of island size and distance on equilibrium number of species.
An increase in distance (near to far) lowers the immigration curve, whereas an increase in island areas (small to large) lowers the extinction curve. (From MacArthur and Wilson, 1963.)

The size of an island, as opposed to remoteness, affects both immigration rate and extinction rate. Immigration goes up with size because the target is larger, and extinction goes down with size, because there should be more opportunity to escape competition or predators on a large island. Large populations on large islands also face less chance of random extinction. The equilibrium number, therefore, will be larger on large islands.

There have been two alternative approaches to testing these predictions of the relative number of species at an interactive equilibrium: experimental manipulation of very small islands and by comparing the species lists recorded from archipelagos. These tests have generally supported the hypothesis.

DEFAUNATION EXPERIMENTS ON MANGROVE ISLANDS

The red mangrove (*Rhizophora mangle;* Figure 24.3) grows in shallow water in suitable sites at the edge of tropical seas. Clumps of the trees effectively form small islands cut off from mainland swamp forest by reaches of water varying from a few meters to up to half a kilometer across.

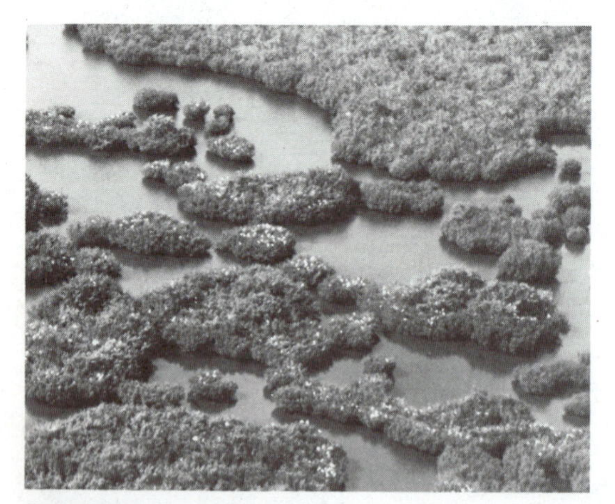

Figure 24.3 Mangrove islets.
Although nearly a thousand species are known to be able to live on islets, the equilibrium numbers seem to be between 10 and 40 species per islet.

Islets can be chosen that are discretely separate and from about 10 to 20 m across. About a thousand species of terrestrial arthropods have been collected from these islets in Florida, showing that the species pool from which the islets are colonized is $P = 1000$. Yet the actual number of species found on any one islet is usually less than forty.

Wilson and Simberloff proposed that this discrepancy between 1000 possible species ($P = 1000$) and less than 40 actual ($S < 40$) was due to the establishment of equilibria between immigration and extinction at a small fraction of the total pool. General support for this hypothesis was available at the start of the investigation because the species lists were different on different islets. Wilson and Simberloff chose six islets 11 to 18 m in diameter and 2 to 500 m from shore and killed all the arthropods on them. They did this by enclosing the entire islet in a plastic tent and injecting a massive dose of insecticide. They then had six islands that were completely "defaunated" and could watch the process of recolonization directly, making complete censi of the islets every few weeks (Wilson and Simberloff, 1969; Wilson, 1969; Simberloff and Wilson, 1969, 1970).

Results for four of the islets are given in Figure 24.4. Before the animals were killed the number of species present appeared to be a function of distance, as predicted, with the least in the distant islet E_1 and the most in the near E_2. After two years of recolonization these approximate numbers had been reestablished on each islet. As predicted by the hypothesis, however, the actual species lists were not the same; it was only the total number that was restored. Moreover, the periodic surveys clearly demonstrated a frequent replacement of species as predicted by the hypothesis.

Of particular interest in these studies is the question of whether the demonstrated species equilibrium was *interactive* or *non-interactive:* Was it competition and predation that set the species numbers so low, or was this merely the result of random extinctions in small populations? Two lines of argument suggest that both processes are involved at the same time.

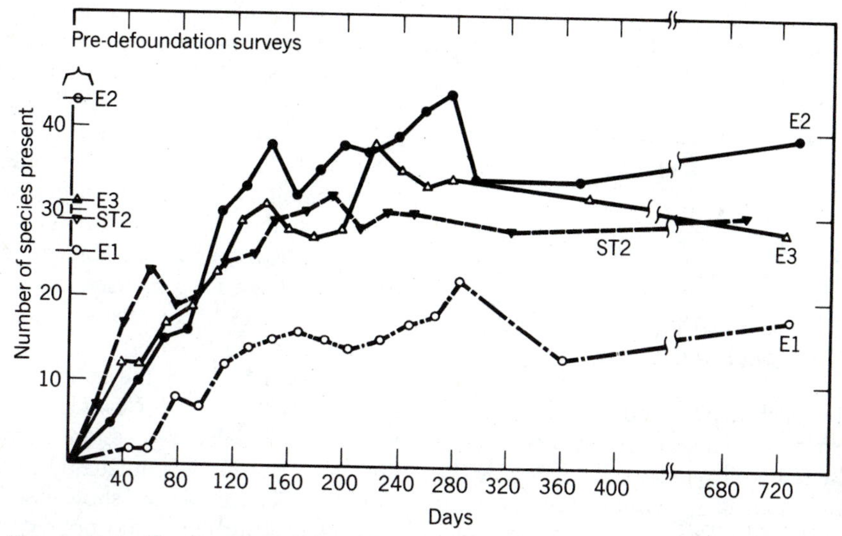

Figure 24.4 Recolonization of defaunated mangrove islets.
The species numbers before defaunation for each islet are shown at the left. E_1 is the nearest island, E_2 the farthest, and the other two are intermediate. (From Simberloff and Wilson, 1970.)

Simberloff and Wilson calculated turnover rates and were able to show that these were in the order expected from the non-interactive model. For the non-interactive model (Figure 24.1*A*)

$$\frac{dS_1}{dt} = \lambda_A(P - S_1) - \mu_A S_1 \qquad 23.1$$

where

λ_A = average immigration rate per species
μ_A = average extinction rate per species.

For the special case of equilibrium, and with a sufficient lapse of time, it can be shown that non-interactive equilibrium predicts that turnover rates must be within a narrow range (Wilson, 1969). The data from the islets do yield turnover rates within this range, showing that the results could be due to completely non-interactive processes.

However, an assessment of these same data show that separate equilibria are maintained by arthropods in each of a few major categories of habit: detritus feeders, herbivores, wood-borers, *et cetera* (Heatwole and Levins, 1973). Species equilibria were maintained in each of these broad-niche classes independently of the rest. Clearly there is a trophic structure on the island into which immigrants must fit.

One datum that is particularly suggestive of interaction is that the colonization curves "overshoot" the eventual equilibrium number before settling down to the equilibrium that existed before defaunation (Figure 24.4).

EVIDENCE OF BIRDS FROM ARCHIPELAGOS

The celebrated explosion of the volcanic island of Krakatoa in 1883 was used as an opportunity to study recolonization of an island (Dammermann, 1948). This explosion appears to have sterilized the island completely, since those parts not destroyed in the blast were overwhelmed with more than 30 m of hot ash and pumice. The island had to be recolonized from across the sea. Field data suggest that an equilibrium had been reached within 30 years (Table 24.1). The equilibrium number of about 30 species is what is predicted by empirical extrapolation from the data of other islands of similar size in the region.

Table 24.1
Recolonization of Krakatoa by Birds
(Data of Dammermann in MacArthur and Wilson, 1967.)

Census Year	Present	Extinct
1883	0	—
1908	13	—
1921	31	2
1934	30	5
Extinction rate 0.2–0.4 species yr^{-1}		

The expectation is shown in Figure 24.5, which shows the correlation of species number with island area in the region; Krakatoa with 30 species of birds would find its place in this figure. MacArthur and Wilson (1967) also calculated the expected extinction rate, assuming the known species pool and curves for immigration and extinction to be of uniform shape, and showed that the actual turnover of 0.2 to 0.4 species per year (Table 24.1) was as expected.

Figure 24.5 shows how the number of species of birds on islands generally is correlated with area. Many such studies and data sets are now available, and they generally confirm that species number does correlate with area. The strongest correlations, however, are with islands of similar kind, particularly of similar relief. Comparing low islands with high islands, for instance, does not result in a good species–area correlation (see review by Simberloff, 1974). This is to be expected since high islands should have more varied habitats, and hence a greater variety of resources, than low islands. But for similar islands, and over a considerable range of size and remoteness, a strong correlation with island area and number of bird species is established as a general rule.

Figure 24.6 shows the species–area correlation for small islands off New Guinea. The straight line is a calculated regression. Dispersal in the data turn out to be particularly revealing. Diamond (1973) was able to show that these islands had been disturbed in the not too distant past. In particular many of the offshore islands had been connected to the mainland at the time of the last ice age as a result of the eustatic lowering of sea level (Chapter 22). Since these islands were then easily penetrated by immigrants they had collected more species than they would have

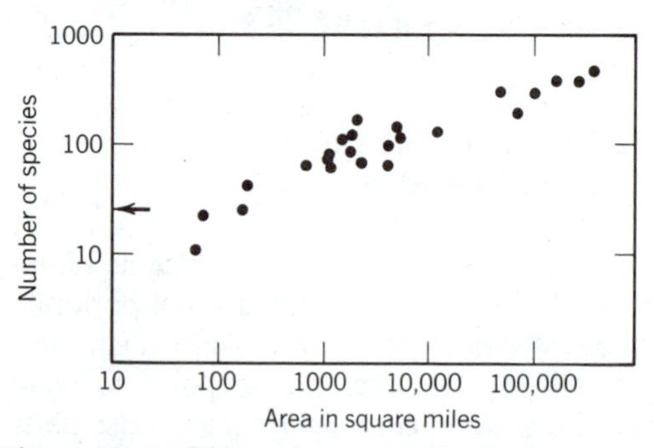

Figure 24.5 Bird numbers on pacific islands.
The islands range from Christmas Island, the smallest, up to New Guinea. The islands are set in a ring around the western Pacific Ocean, all of them being close to continental land. There is thus no effect of distance on species number. The correlation with area, however, is very strong. Krakatoa birds of the post-irruptive equilibrium would place at the left end of the correlation just off the graph. (Arrow: 30 species, 8 square miles.) (From MacArthur and Wilson, 1967.)

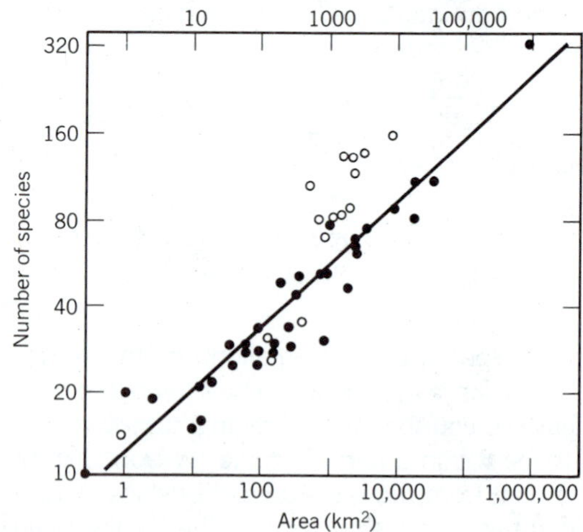

Figure 24.6 Resident bird numbers of islands offshore from New Guinea.
Species numbers and areas are plotted on logarithmic scales. Solid circles are islands thought to be at equilibrium. Open circles are of islands not yet recovered from disturbances, either having too many species from a recent land connection or too few following volcanism, *et cetera.* (From Diamond, 1973.)

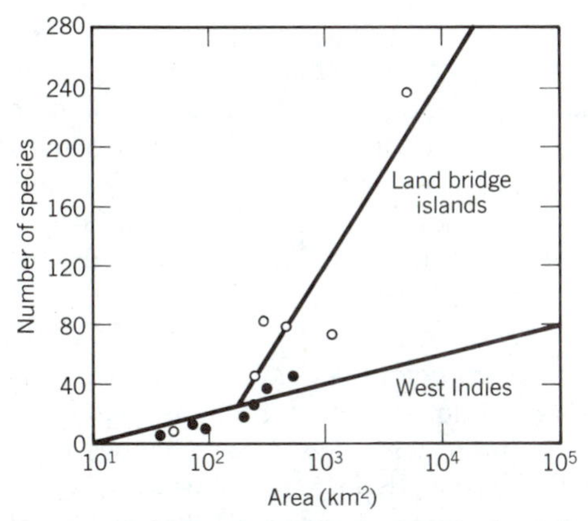

Figure 24.7 Species–area plot for land birds of Caribbean islands.
Open circles are islands that were connected to the mainland when sea level fell during the last ice age; solid circles are islands that have never been part of the mainland. The lines are computed regressions. The larger land bridge islands apparently are still oversaturated with birds from the days of easy immigration, though smaller land bridge islands have almost relaxed to equilibria appropriate for oceanic islands. (From Terborgh, 1974.)

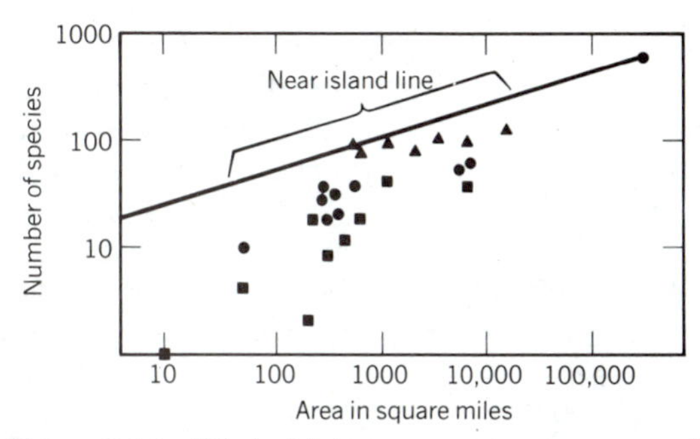

Figure 24.8 Effect of distance on species–area curves of archipelagos.
Squares denote islands more than 2000 miles from the mainland; triangles are islands less than 500 miles from mainlands; and circles are islands at middle distances. The line represents the slope for near-shore islands. The much steeper slope of the distant islands is predicted because rates of immigration are expected to be slowest for these islands. Some part, but not all, of the low species numbers on the remote islands of this particular set, however, may be accounted for by the fact that the remote islands are low-lying and so have little habitat diversity. (From MacArthur and Wilson, 1967.)

in their present island state and are now probably progressively losing species, called RELAXATION by island biogeographers.

Figure 24.7 shows comparable data from islands in the Caribbean, where the land bridge effect is particularly noticeable. Some of the larger of these old land bridge islands hold species that have never been known on islands elsewhere, but these truly continental animals are not now found on the smaller of the old land bridge islands. This strongly suggests that this part of the continental fauna already has gone extinct on the smaller islands and that the process of extinction still proceeds towards the equilibrium number (Terborgh, 1974).

These data, and others like them, show convincingly the determining role of area. Data showing the predicted effect of distance are also available. It has long been an observation of biogeographers that very remote oceanic islands have depauperate faunas. Equilibrium island biogeography predicts that this is because the immigration rate is slowed by distance, resulting in a lowered equilibrium number. However, the poverty of oceanic islands could simply be because dispersal across oceanic distances takes so long that a species equilibrium has never been reached. The theory offers a general test of the predicted effect in the predicted slope of the regression line.

Figure 24.8 shows the species–area relationship for separate archipelagos at varying distances from the mainland. Islands of the most remote archipelago are shown as squares and it is easily seen that the slope of the species–area relationship is steeper for these islands. The area effect is in fact exaggerated because small remote islands are tiny targets for dispersal at very long range and so have "too few" species.

For birds we also have actual time-series data from some islands where observations of different dates document the changing composition, but constant number, of species present. The best known of these data are from the Channel Islands off the coast of California (Diamond, 1969). The base data are from a census of 1917 and another of 1968. The numbers of species on the different islands had not changed in this interval but the species lists had. We do not know how many species both arrived and disappeared between the two census dates but the known changes show that the turnover rate was at least between 0.3 and 1.2% a year. There is argument over the precision that is possible from these data (Simberloff, 1974) but their general concurrence with the predictions of the model cannot be set aside.

OTHER ORGANISMS AND OTHER ISLANDS

For plants on oceanic islands the available data are less convincing. There is a broad correlation between plant variety and area, as shown in Figure 24.9 for plant genera and area of islands from all over the Pacific, but there is considerable scatter in the data. Partly this scatter may reflect the fact that species lists are either not complete or not based on comparable collections. In the Galapagos Islands, for instance, there is disagreement as to whether species lists from different islands reflect actual species lists or merely the intensity of collecting effort in the different regions (Johnson and Raven, 1973; Connor and Simberlof, 1978).

Pollen data from the Galapagos Islands give a very imperfect history of plant occupation but nevertheless are suggestive of a subtly different mechanism regulating the numbers of plant species. The Galapagos highlands were much drier in the last ice age than they are now (Colinvaux, 1972; Colinvaux and Schofield, 1976). When the moister cloud cover of the present day first developed about 10,000 years ago the dry highlands were colonized. Pollen data from a highland lake show that between 500 and a 1000 years passed before vegetation similar to that of the present day occupied the moist high ground but that there were few other changes for the next 9000 years (Figure 24.10). Pollen data give a very poor representation of the total species list so that many species changes could take place without mark on the pollen record. Nevertheless, the data do show a remarkable constancy in the Galapagos vegetation that is not at all suggestive of constant turnover of species. It may be that an island that is well vegetated, with a closed tree canopy, is a difficult place to invade. The common plants have large populations and are most unlikely to go extinct. Newcomers are closed out by competition so that turnover, or the attainment of a true equilibrium, is slower for plants on oceanic islands.

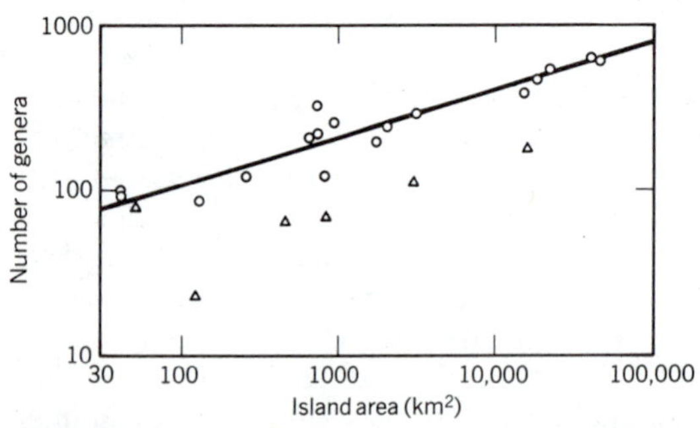

Figure 24.9 Species–area plot for plant genera on Pacific islands.
The line is a calculated regression for the less remote islands (circles). Triangles show the more remote islands. (Data of Van Balgooy, 1971, from Cox and Moore, 1980.)

At the other end of the plant size scale, microscopic algae, species equilibria may be reached rapidly. Maguire (1963) set out beakers of water along a transect running from a fresh-water pond in Texas. These beakers were islands for water life in a sea of land. The number of species of microorganisms (algae and larger protozoa) rose progressively over a two-month span as the rate of colonization fell, very much as predicted by the theory (Figure 24.11).

Lakes, provided they have no significant inlets through which floods of immigrants may be introduced, also serve as islands surrounded by land. Figure 24.12 shows species–area curves for closed lakes of Ecuador. Area is shown to be a good predictor of species number of algae, though such parameters as water chemistry and age of the lake basins were not (Colinvaux and Steinitz Kannan, 1981). Notice that the regression line for the Galapagos lakes taken separately is steeper than the regression for mainland lakes. This conforms to the requirement for dis-

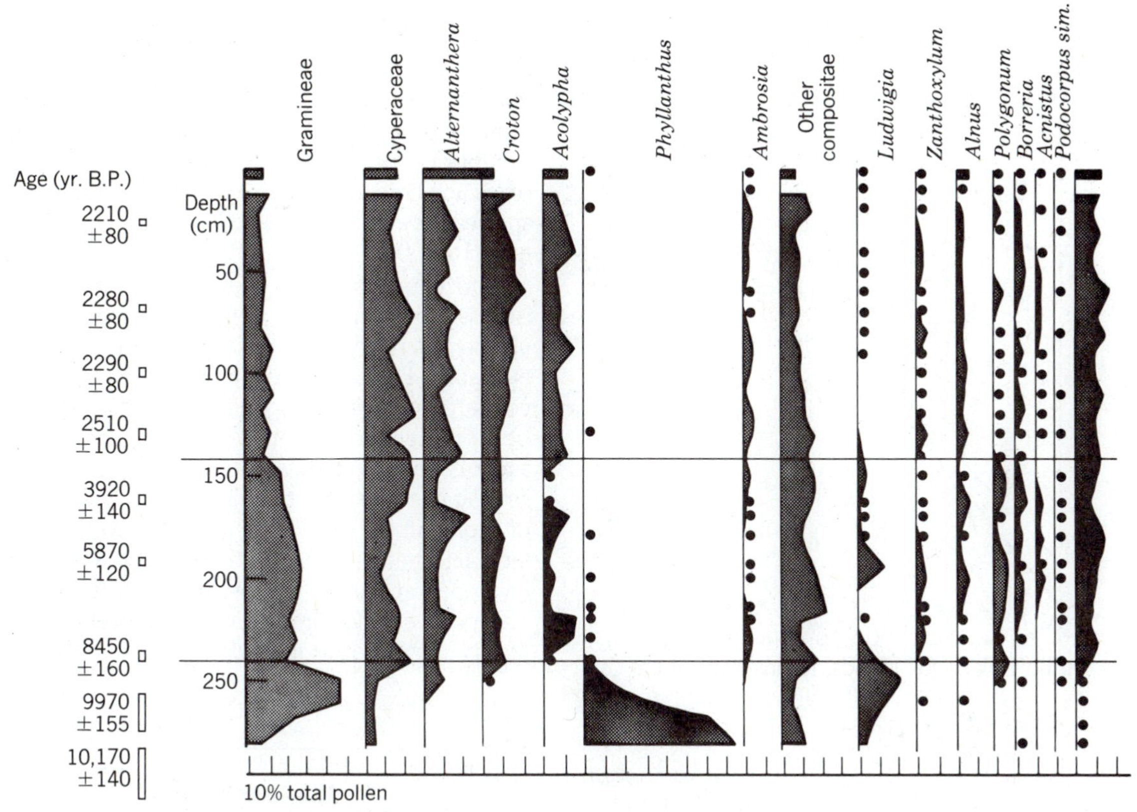

Figure 24.10 Resistance of Galapagos mesic vegetation to invasion.
Pollen percentage diagram from sediments of Lake El Junco on Isla San Cristobal (Chatham). The sediments below 3 m showed that the island was dry and without mesic forest before 10,000 years B.P. The pollen show that there were no major changes in vegetation following the initial colonization phase. (From Colinvaux and Schofield, 1976a.)

tant islands resulting from the greater inaccessibility to the species pool.

An early application of the theory was to the mammals of mountain tops, considered as islands in seas of valleys. Since the time of Darwin and Wallace, naturalists have speculated that many mountain tops were populated in the last ice age by cold-adapted animals that then could cross the intervening lowlands more easily than now. This makes their position very similar to the birds of old land bridge islands who penetrated when the ice age lowered sea level: extinction would continue although replacement by immigration stopped. Brown (1971) put forward

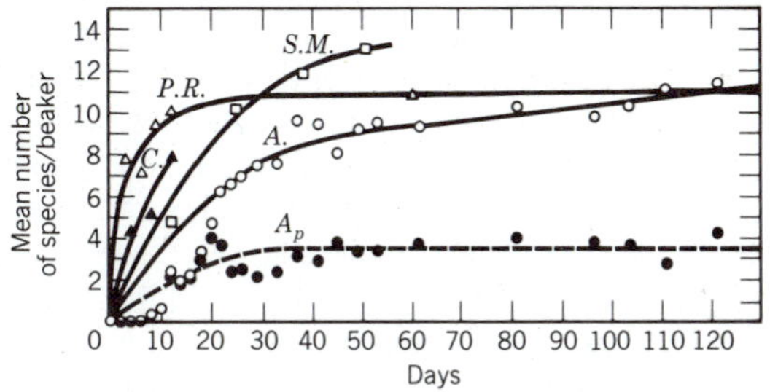

Figure 24.11 Immigration to beakers of water by plankton.
Sites are in Colorado (C), Puerto Rico (P.R.), and Texas (S.M., San Marcos; A., Austin). A_p is the curve for protozoa only at Austin. (From Maguire, 1971.)

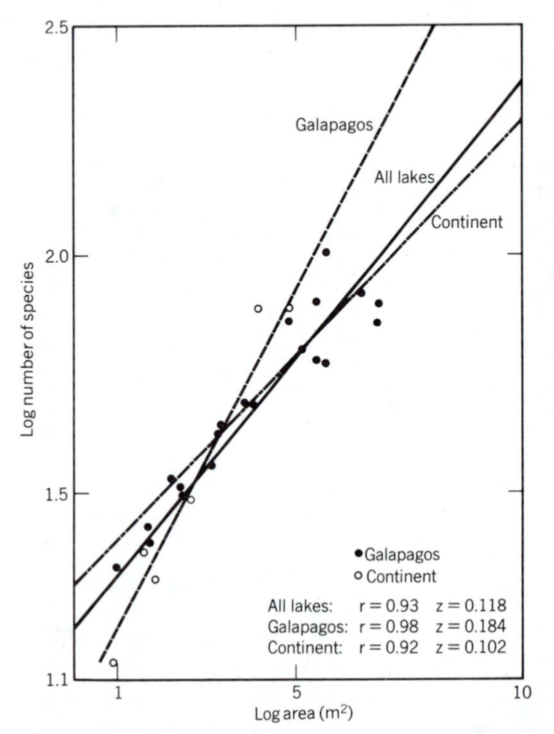

Figure 24.12 Species–area curves for algae of open water in lakes of Ecuador.
The lakes are all closed basins, or have minor outlets only. (From Colinvaux and Steinitz-Kannan, 1980, with additional lakes.)

the hypothesis that species numbers of mammals on mountain tops should have been falling since the last glacial retreat 10,000 years ago.

Brown's "islands" were small mountain ranges between the Sierra Nevada and the Rocky Mountains, the two chains that formed the "mainlands" of his study. He defined 17 of these mountain islands by arbitrary criteria of height and remoteness from each other, and then showed that number of species of mammals on the islands was significantly correlated with area (Figure 24.13). The slope of the logarithmic plot, however, was very low. This slope is a measure of remoteness of islands, or at least of the difficulty of immigration. The low slope for these mountain islands seemed, on the face of it, to suggest very easy immigration, resulting in a fast immigration rate and an unexpectedly high number of species at equilibrium. But Brown examined various factors that could make immigration easy between the various islands and could find no correlations between ease of access and number of species. His conclusion was that the mountain tops are oversaturated from invasions of ice age time and that the numbers of mammal species have not yet relaxed to equilibrium, even after 10,000 years.

Agricultural fields of monoculture crops are islands from the point of view of crop pests, yielding island systems for which many data are available. Figures 24.14 and 24.15 are species–area curves on logarithmic plots for numbers of insect pests on plantations of cacao and sugar cane. These data are perhaps particularly striking because of our intuitive belief that a field of monoculture must be vulnerable to pest attack, and yet countries with small acreage have fewer pest species than those with large acreage. The pests are not omnipotent and local extinction is shown to be the rule. Many more pests

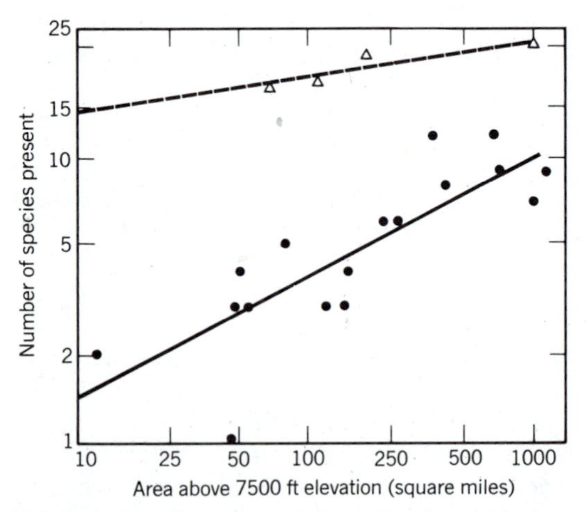

Figure 24.13 Species–area relationships for mammals on mountain tops.
The solid dots, and the solid regression line, are for 17 isolated ranges in the Great Basin. Triangles, and the dotted regression line, are comparable areas attached to the present-day Sierra Nevada. The steeper regression for the isolated mountain ranges suggest that they still retain extra species of mammals from the time when the ice age climate connected them with alpine-type environments in what are now dividing valleys with warmer climates. (From Brown, 1971.)

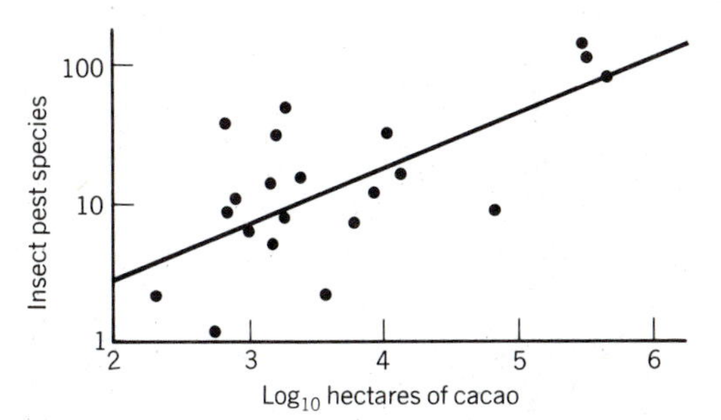

Figure 24.14 Insect pests of cacao as a function of hectares planted.
Data are from all parts of the world where cacao is grown. Data are numbers of species of insect pests recorded for each cacao-growing country and the acreage of the crop planted in that country. (From Strong, 1974.)

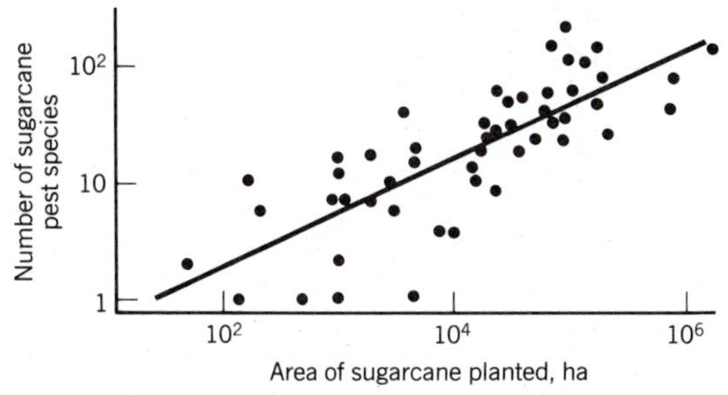

Figure 24.15 Insect pests of sugar cane as a function of hectares planted.
Data were gathered in the same way as for cacao in Figure 24.14. (From Strong, 1977.)

are available than the ones actually doing damage and more come to a farmer's notice as the island of crops increases in size.

These data from relatively large areas of uniform planting suggest that similar effects should be present for dispersed individual plants. Each plant then is an island, and a very small one. If the number of insects attacking at any one time (the *herbivore load*) goes down sharply as planted area becomes small (Figures 24.14 and 24.15), then an isolated plant should carry few pest species indeed. There should be a good chance that it holds none at all, that it should be a zero class in the random game of immigration and extinction.

Within a population of plants dispersal and clumping serve as distance measures that affect the rate of immigration of the animals using the plant for food or as a habitat. Clumping also serves as area in the island biogeographical sense. A very large number of animal species can be related to a single plant host as a result (Janzen, 1968).

EVOLUTION AS A PROCESS PROVIDING IMMIGRANTS

New species appear on islands as a result of genetic selection from pre-existing stocks as well as by long-distance transport. Effectively this evolution works as another source of immigrants. It is noteworthy that the fauna of the most remote of oceanic islands are those with a high proportion of endemics in their populations. These islands also are those with a low number of species per unit area as a result of low immigration rates. It is likely that low rates of colonization from outside the island, and the consequent low rate of extinction, increase the chances that new varieties selected within the island will be preserved.

It is well known, of course, that isolation is important to speciation; this is the essence of the allopatric speciation model. But the island biogeography approach gives an added importance to area also, since the establishment of a viable population is critically dependent on area. Recent studies of species lists of plant herbivores as a function of the areal distribution of the host species suggest that this effect has been important in deciding the numbers of species living.

Figure 24.16 is a plot of the number of species of leaf-mining lepidoptera found on each of 18 species of live oak in California. Leaf miners are small moths whose caterpillars feed within the leaves, producing a tracery of tunnels. This is a specialized way of life, but nevertheless more than a dozen species of miners may be found within the tissues of a single species of oak. Oaks are well provided with chemical defenses, so that all herbivores of oaks tend to be specialists on particular species of oak, and the leaf miners are no exception. But, as Figure 24.16 shows, the

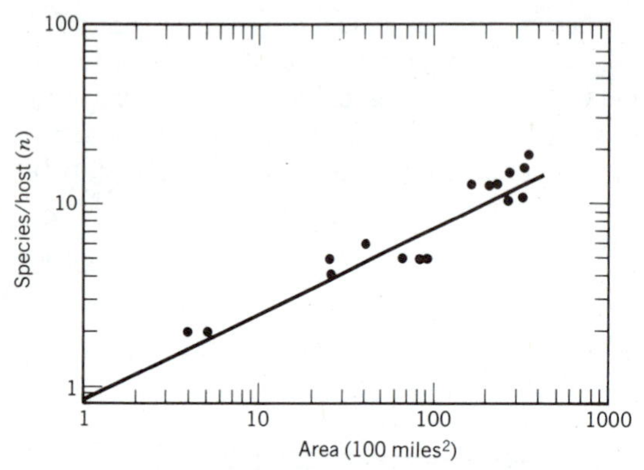

Figure 24.16 Oak trees as islands.
Data are numbers of leaf-mining lepidoptera known from each of 17 species of Californian live oak plotted against the area occupied by populations of each species of oak. The line is a calculated regression. (From Opler, 1974.)

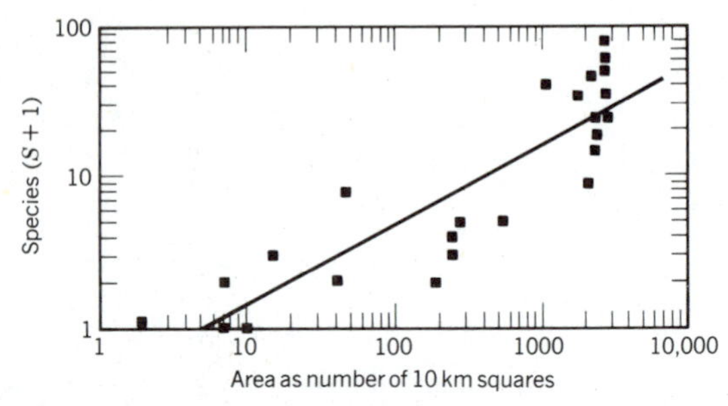

Figure 24.17 Species–area relationship between insect herbivores and their hosts in Britain.
Data are the known number of insect species found feeding on each of 27 British herbs plotted against a measure of the area of Britain in which the herb is known to exist. The area of distribution of each herb was measured as the number of 10-km squares in the British floral atlas for which the species was reported. The line is a calculated regression. (From Lawton and Schroder, 1977.)

number of species of leaf miner known for each oak species is a function of the range of the oak (Opler, 1974). These data suggest strongly that the peculiar form of immigration called "evolution" has been as sensitive to the area of the "target" as has the more orthodox form of immigration.

An even larger data set has been compiled for the British Isles by Lawton and Schroder (1977). They took advantage of the unusually complete knowledge of this flora and of the British insects to plot number of insects per host plant species versus area for all the non-arboreal flora of the island. Figure 24.17 is a sample of their findings, being the data for all terrestrial, herbaceous, dicotyledonous plants in Britain. The known species list of associated insects correlates very closely with the area of Britain occupied by the host plant.

These data are of particular interest in view of the fact that there have been many reasons for thinking that the number of species in a region, or using a host, should be a function of evolutionary or geological time (Chapter 26). They suggest that the rate of production of new immigrants *in situ* by speciation may sometimes be a function of area and isolation like the establishment of new colonists by true invasion.

PREVIEW TO CHAPTER 25

Adding species to communities where individuals interact already requires coevolution. The newcomer must share resources already committed, requiring a reallocation of resources, and this must follow whether the new arrival comes as competitor, predator, or friend. Only if a community is a loose assembly of non-interacting individuals whose numbers are kept low by density-independent constraints can this reallocation of resources be avoided. In all other circumstances a new arrival starts a process of coevolution, the records of which are seen in coadaptations and mutualisms. Ubiquitous are the coadaptations between plants and their herbivores, some of the consequences of which can be predicted by optimum defense theory. The alkaloid toxins of widely dispersed plants, for instance, appear to be adaptations to poison all but a few herbivores with specialized detoxification systems, thus largely preventing attack from herbivores based on other kinds of plants nearby. Protein-denaturing chemical defenses like tannins are costly, but cannot be detoxified and are general poisons with some effect on all herbivores—they are accordingly present in plants like climax trees that have no defense in dispersal. A general result of allelochemic plant defenses is to increase diversity of both plants and herbivores as specialist defenses produce specialist attacks, producing further specialist defenses, *et cetera.* Passion flowers and their butterfly herbivores in tropical forests are well-known examples. The effect is passed on up food chains to specialist predators on the herbivores. One system of magnifying the effect is through mimicry as prey species with chemical defenses (nasty-tasting butterflies or poisonous snakes) become models for which selection produces counterfeits out of eatable species. Mutualisms also result in reallocation of resources in time, socalled phenology. Many mutualisms pay for the cost of a new species by relieving the base population of a function for which resources had originally to be allocated, as when plants use bees for pollination or ants for defense. In pollination by insects, selection will favor those traits in plants that maximize visits from insects to the host plant species while minimizing costs of the rewards offered as nectar or for signaling. The insects will be selected for traits that maximize returns in nectar for the minimum search effort and transport time. Likewise in an ant-guard symbiosis, plants that use ants for defense against herbivores should be selected for supplying the minimum rewards in ant food or ant nest sites, while the ants are selected for ability to extract the most resources from the plant. In both these examples, extra species in the community provide plants with services (pollination and defense) from resources that the plants would have had to allocate to these functions anyway. The reallocation of resources resulting from mutualisms can be understood in part from optimization theory, but not completely because truly optimal solutions cannot be predicted for evolution. Every adaptation is a function of species history rather than of optimum design, as natural selection works as a process of the possible to modify what is already present. Recently it has been shown that limited communal functions can be preserved by a narrowly constrained form of group selection in systems of structured demes. If small groups live together for but a short while before dispersing, genes for mild community support functions can be spared the selective removal that the cost of these functions otherwise entails. But individual selection can account easily for most apparent organized properties of communities as coadaptations and mutualisms are compounded. Much community building comes about by exploiting general classes of adaptation (pollinators, ant defenses, allelochemics), many perhaps coming on the scene rather suddenly as the history of a changing taxon makes the new class of adaptation possible.

CHAPTER 25

COEVOLUTION: COMMUNITY BUILDING IN EVOLUTIONARY TIME

Life in communities goes on evolving long after successions are complete and equilibrium numbers are reached. These first replacements of colonists are soon over, finished in what might be called *ecological time*. Then comes the long process of coevolution as new species are added in *evolutionary time*.

A new member of an established community must come as competitor, predator, or friend. In whichever of these three guises, the new species must use resources that previously had been used by other species populations, even if only by decomposers. Adding a new species to a community, therefore, is a process of reallocation of resources.

The opportunities for a new species to enter an already complex community are narrowly constrained by the strategies of species already there; the newcomer must in a sense fit in. At the same time the options of the newcomer are even more narrowly constrained by the history of its parent stock. The feats of engineering or food opportunism possible for natural selection at any one time are prisoners of past history.

Bats, for instance, who are guided in flight by sonar images rather than by sight, enter communities as pollinators, fruit eaters, or insect hunters of the night. Success for a bat requires that the target foodstuffs be available at night. Bat success, therefore, is tied to plants that fruit or flower by night and to insects that fly by night (Figure 25.1). Birds, on the other hand, navigate

Figure 25.1 Winged mammal: night hunter. Equipped by ancestry with sound location as a prime sense, new species of bats enter communities as night foragers.

by sight and their success is dependent on food availability by day. Thus the options for community building are always set by the engineering of the would-be newcomer and the engineering of the others already in place.

Much of the evolution that adds species to communities is necessarily COEVOLUTION. The newcomer uses other organisms for resources, which at once makes for a selection pressure adapting the other towards maximum fitness despite being used. Bee–plant pollination systems illustrate this relationship.

Bees as Pollinators

Many plants use bees as pollinators, thus turning bees into a resource, but the plant must feed the bee with pollen or with nectar. For a plant to maximize fitness, bees must be attracted for frequent visits. Yet donating nectar and pollen to bees must divert calories and nitrogenous nutrients that might otherwise have been used for making seeds and fruits, requiring that plants attract the most bees with the least possible bait of nectar and pollen.

Plants that use bees as pollinators, therefore, have a problem in optimal resource allocation between the bee bait and reproduction. The optimization problem is compounded when a plant must compete for bees with other species of plants, perhaps forcing it to spend extra resources on bee bait. The bees have comparable problems in optimal resource allocation, needing to maximize their return in calories and nitrogenous nutrients with the minimum number of flower visits and minimal journey time. Bee–plant pollination systems, therefore, are likely to be highly coevolved.

A simple structural adaptation in flowers of the desert willow *Chilopsis linearis* results in the maximizing of bee visits (Wittham, 1977). Fine grooves run out from the main pool of nectar in the flowers; both grooves and pool are filled with nectar in the early morning. Bees that visit flowers early can maximize their retrieval of nectar by draining pool nectar only and then leaving to collect pool nectar from another flower. Later in the day when all pool nectar has been collected all flowers still retain nectar in grooves where it is held by capillary action. It pays the bees to make a second visit to collect this nectar, a groove at a time, although their returns are much less than on the earlier foraging visit. This adaptation of the plant therefore uses the bees for multiple visits, thus maximizing the chance of cross-pollination. In this coevolved system both plants and bees can be seen to maximize their returns.

A familiar habit of bees seen in gardens is the working of tall spikes of flowers like *Delphinium* or *Aconitum* from bottom to top. For the bees this is a good search pattern because the lowest flowers develop first and hold more nectar. Taking the flowers in order on a vertical climb minimizes the search effort and maximizes the return in nectar. Bees abandon their climb before they reach the top of the spike because the undeveloped terminal flowers are without nectar.

But flowers on the spikes are arranged spirally, which means that the vertically moving bee misses some flowers in its ascent. Pyke (1978) has shown how the spiral arrangement of flowers serves to exploit bees for efficient pollination. An arriving bee does not carry sufficient pollen to pollinate all the female flowers on a spike, and the spiral arrangement, which forces bees to miss some flowers, serves to present bees with no more flowers than it can pollinate in a single ascent. More subtle in the use of bees is the order in which anthers and stamens develop. The flowers are *protandrous,* meaning that each flower changes from male to female as it ages. Old flowers at the bottom of the spike, therefore, are female and receive pollen brought from outside, whereas the last flowers visited high on the spike are male, thus recharging the bee with pollen to take to another flower.

Optimization Problems

Coevolved mutualisms between bees and plants can be understood as giving optimal returns on energetic investments; the bees act as we would expect from optimal foraging theory and the plants are adapted to use the bees efficiently as, in Krebs

Figure 25.2 Selasphorus Platycercus and Campsis Radicans: coevolved.
Both humming bird and trumpet creeper allocate resources, in self-interest, for the other. For this plant a bird, not a bee, is the flying penis.

(1979) words, "a flying penis" to effect cross-fertilization (Figure 25.2). However, although we can see that the bees are well used by the plants, it is difficult to say that this use is "optimal." Perhaps a clever engineer could think of better solutions for getting service out of bees. The groove and pool nectar system, for instance, seems rather crude. The device of spiral arrangement of flowers need not necessarily be ideal. Both these arrangements obviously result from slight modifications of preexisting flower structures; doubtless they make better use of bees than earlier modifications of those same structures but this does not mean that the designs are the best possible.

Natural selection, like politics, is a process constrained by what is possible. The result always is relative extra fitness compared to the fitness granted by some alternative adaptation. We cannot expect any community to be made up of species whose arrangements are truly "optimal." Oster and Wilson (1978) make this point with an elegant analogy to economic theory. Economists of the Friedman school have argued that business firms evolve by a process of natural selection for the most efficient, which leads to the conception that the present distribution of business economic power is both ideal and the outcome of inevitable forces. This conclusion seems quite unsatisfactory since it is obvious to most people that the present distribution of economic power is neither efficient nor inevitable. Likewise with communities. Individual species evolving into them are selected to make the best possible solution to cost benefit problems though these solutions are unlikely to be truly optimal.

Direct evidence that even the most complex of communities is made up with species of less than optimal efficiency is suggested by the fact that new species can be made by competitive exclusion and character displacement, because character displacement should be impossible if the pre-existing species already were the product of ideal optimization theory. Granted historical and structural constraints, selection should make all individuals into "optimal foragers" of a sort but those constraints are real and prevent ideal solutions.

It seems, therefore, that all communities can be invaded by new species offering fresh solutions to old problems. The resulting community building in evolutionary time often must involve mutualisms, most commonly plant–animal interactions but also interactions between animals higher on the food chains. Empirical study of these interactions is now one of the most interesting and satisfying developments of modern ecology. Theory for these studies tends to be weak, since optimal solutions can be neither predicted nor expected. Instead the habits of animals and plants reveal to the astute naturalist the selection processes that have been important to their selection.

Cost benefit analysis and optimal foraging theory can serve to demonstrate the utility of particular adaptations, like the arrangements of flowers and the flying patterns of bees, and they may also suggest where more subtle adaptations are to be found. If a habit should seem less than optimal at first analysis, perhaps it has been preserved because of circumstances that appear with low frequency but which must be survived when they do occur.

Plant–Herbivore Interactions

Apart from cooperative mutualisms like those of pollination systems, much of the evolution of communities concerns endless variations on attack and defense. In a world without perfect optimizations neither perfect attacks nor perfect defenses are possible. New variants constantly are jury-rigged out of old adaptations and exposing them can give delight to ecologists. One long train of thrust and counterthrust is suggested by data from passion flowers and the caterpillars that eat them.

Passion flowers *(Passiflora)* are vines that live scattered through many kinds of tropical vegetation. A primary defense mechanism of passion flowers against herbivores seems to be alkaloid toxins, but the plants are nevertheless attacked by butterflies whose larvae can sequester or otherwise detoxify the alkaloids. Butterflies of the genus *Heliconius* in particular feed on *Passiflora* and may become specialized to feed on only one or a few species. A good hypothesis to explain this is that ancestral butterfly stock able to detoxify the alkaloids of a range of *Passiflora* species came to specialize on a single species for ecological reasons. Factors like time of vegetative growth or shade tolerance of the plant led to preferential egg laying thereon. The *Heliconius* population thus restricted to a single species of passion flower would be selected for rapid growth. This process in turn would select against unnecessary mechanisms for detoxifying alkaloids of other passion flowers. Thus the general defense of toxins in the *Passiflora* genus is overcome and a specialist *Heliconius* evolves that can eat one species of *Passiflora* with impunity.

This victory of offense sets the evolutionary stage for the construction of a new defense. *Passiflora adenopoda* has fine, hooked hairs on leaves, called *trichomes,* which give it total protection against *Heliconius* caterpillars (Gilbert, 1971). The tiny hooks tear into the pseudopodia of the caterpillars and rip holes in the body wall through which there is a fatal loss of hemolymph (Figure 25.3).

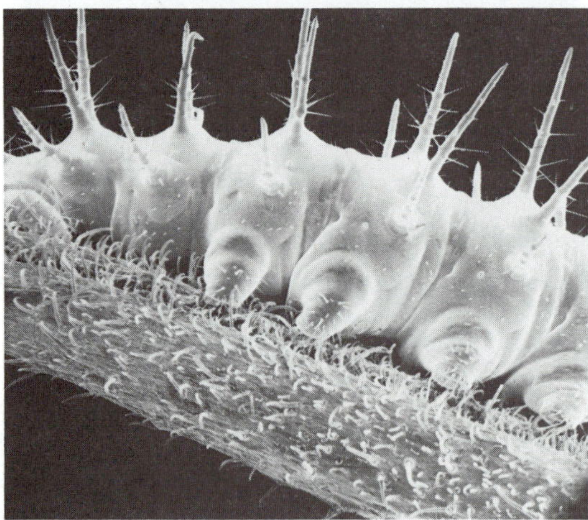

Figure 25.3 *Passiflora adenopoda:* caterpillar defense. Fine spines, called trichomes, inflict fatal wounds on caterpillars; except those selected for the communal defense of silken mats laid over the spines.

Trichomes may not be so good a general defense as toxins for passion flowers because they would not work against other kinds of herbivore, but if a specialist caterpillar immune to the toxin appears then the trichomes are apparently the most cost-effective defense against this one species. In the system studied by Gilbert the trichomes apparently give total victory for the defense. It may be, however, that this victory lasts only for a short span of evolutionary time before countermeasures are found by natural selection to unleash a fresh lethal attack. Rathcke and Poole (1975) have shown that caterpillars of butterflies of the genus *Mechanitis* cope with the dangerous barbed wire entanglements of trichomes on other plants with a communal spinning of a silken mat that lets them walk with impunity above the hooks, and thus to feed safely on the edges of leaves. The mechanism is an elegant demonstration of natural selection as the art of the possible and as an improviser. Virtually all butterfly larvae routinely spin silk as an attachment for themselves or their cocoons and the main evolutionary step necessary for this answer to the trichomes is gregarious living that makes spinning a complete surface mat on the leaf a cost-effective undertaking.

So far this silk-matting method of attack does not seem to have been applied by *Heliconius* against *Passiflora* trichomes. Possibly it never will be. But if it is, then doubtless a *Passiflora* will jury-rig a new defense in the fullness of evolutionary time, but what this would be we cannot predict. Naturalists of the future will have to discover the next stage in passion flower defenses and then rationalize the behavior with an after the fact cost–benefit analysis.

Community building in evolutionary time, therefore, is an endless process of makeshift adaptations being piled on top of one another. We can identify major themes in this process, themes that occur again and again, either because habitat properties make them likely or because the engineering or energetic designs of major classes of organisms make them necessary.

Escape or opportunity often can be manipulated with activities restricted in time; we call this PHENOLOGY. A second theme is that in all communities species interact through chemicals, either as toxins or as guides to behavior, letting us recognize a subdiscipline of ecology called ALLELOCHEMICS. Then there is the theme of dispersal, as plants, being stationary, have an endless array of behaviors that cause animals to transfer plant genetic material or disperse propagules. And the fixing of the main phyletic lines of animals and plants defines whole families of interaction, birds that largely work by day, bats that work by night, spiders that hunt with silk, or insect families whose larvae have limited mobility whereas the adults range freely. And there is one special class of possibilities, that of mimicking other species of the community for a great range of advantages. These various themes are explored in more detail below.

ALLELOCHEMICS

The massed organisms of any community give it a chemical environment. Many chemicals in the environment are used as social cues that signal between members of the same species, chemicals that are collectively called PHEROMONES. Males of some moth species can home in for several kilometers on pheromones emitted by an uncopulated female. Other organisms mark their trails chemically, territorial mammals mark their territories, and so on. But the chemical cues can be used by other animals, particularly by predators or parasites. Dogs hunt by scent. Perhaps more importantly, invertebrate predators and parasites do so also. A tick waits on branches at places marked by the facial scent glands of deer as a suitable point of embarkation (Rechar *et al.*, 1978). Bark beetles *(Ips confusus)* use terpenes released in their own excrement *(frass)* as a homing signal to take advantage of a food supply found by other beetles. But the scent in this same frass is used by predators to find the beetles (Whittaker and Feeny, 1971). Some caterpillars expel their frass explosively, so that it falls clear of the leaf on which the caterpillar is feeding to the ground underneath, a habit best explained as a defense mechanism against parasitoid hymenoptera that track prey by the scent of frass (Rhoades, 1979). Into the chemical environment of intraspecific pheromones, therefore, come selection pressures for coevolution between species.

The true ALLELOCHEMICS are compounds, the prime selective advantage for which comes from their influence on other species. Called by plant physiologists SECONDARY COMPOUNDS, these are substances like the alkaloid toxins and the protein-fixing tannins for which no obvious physiological purpose can be found. A general hypothesis of ecology is that these compounds give selective advantage either in competition with other plants (ALLELOPATHY) or as defense mechanisms in plant–herbivore interactions. The general hypothesis states that they have been manufactured at some cost in fitness to the plant for a compensating benefit in fitness.

Some examples of ALLELOPATHY are well known, like the impossibility of growing tomatoes near walnut trees in the American Midwest because of a chemical agent, called juglone, that is washed from walnut leaf surfaces to the soil (Bode, 1958). A number of examples are also known in pasture grasses, one mechanism of particular subtlety being that of the grass *Aristida oligantha* whose exuded chemicals attack not rival grasses directly but nitrogen-fixing bacteria. This tactic maintains the habitat at a low state of productivity, which in turn prevents invasion by plants competitively superior to a pioneer *A. oligantha* (Blum and Rice, 1969).

More widely studied are the coevolutionary systems between herbivores and SECONDARY PLANT METABOLITES (Rosenthal and Janzen, 1979). That these substances can act against herbivores is demonstrated in feeding trials, both with natural foods and with artificial mixtures. Chemical and physiological data also suggest that heavy costs are accepted by plants to maintain these compounds because they tend to be autotoxic and thus have to be rendered chemically harmless within the plant, as when a glycoside is combined with glucose. The general hypothesis that these compounds are maintained at a cost for herbivore defense, therefore, is plausible.

The defensive compounds work in two distinctly different ways. One class provides QUANTITATIVE DEFENSES by making food inedible. Tannins are of this class, working by being tightly bound to proteins (which is why they work for tanning leather). They make plant proteins largely indigestible to all herbivores but their effect is dosage dependent since the plant must have sufficient tannin to render most of its protein inedible.

Other quantitative chemical defenses employ simpler compounds like silica, which is deposited in plant cells as PHYTOLITHS, and the LIGNIN of wood. Lignin appears to have the special advantage in its role as a defense compound in that it does not require chemical inactivation to prevent it harming its producer. In addition the cost of its manufacture can be allocated to other budget accounts like competitive advantage through large size as well as to the defense budget.

The second class of defense compounds is the active toxins, called QUALITATIVE DEFENSES. Compounds like the alkaloids of this class interfere with the internal metabolism of animals directly and, like other poisons, are effective in low concentrations. They can, however, be detoxified by intended victims that have the appropriate chemical mechanisms. These facts lead to the many exciting sagas of coevolved systems of chemical attack and defense that have been discovered.

OPTIMAL DEFENSE THEORY states that the secondary compounds are allocated as defenses in ways that maximize inclusive fitness and that such allocation always involves costs to the plant (White, 1969; Rhoades, 1979). It will be noted that the predictive power of this theory is constrained by species histories or mechanical possibilities like all optimization theories in ecology, as discussed above. But the theory does lead to some general predictions that are testable.

First is the prediction that *less defended organisms have higher fitness in the absence of enemies*. Many data from the agricultural literature support this. Different strains of crop plants have different toxic properties and it can be shown, for instance, that non-toxic morphs of clover *(Trifolium)* grow and reproduce more vigorously than toxic morphs when tests are controlled. Tobacco *(Nicotinia)* plants especially rich in alkaloids are stunted, suggesting that energy allocated to the manufacture of nicotine has been abstracted from energy available for growth

(Whittaker and Feeny, 1971). Many pest problems of agriculture possibly result from our habit of selecting for rapid growth at the expense of chemical defenses.

Two other predictions are that *chemical defenses should be allocated to different tissues in proportion to risk* and *defenses ought to be reduced when enemies are absent if possible.* Many data can be understood in light of these predictions. Tannins are concentrated in leaves, which are prime targets for herbivore attack. Secondary compounds are transferred from leaves to seeds or other reproductive structures at the end of the growing season by annual plants, or at annual leaf-fall for perennials. In a number of commercial forestry trees secondary compound concentration depends on the history of pest outbreaks, and in fir, pine, spruce, larch, and birch, resin and phenol concentrations increase after infestation (Edmunds and Alstad, 1978). Plants stressed by drought tend to be more nutritious because they have less toxin, which would be expected if the plants under stress have fewer resources to allocate to toxin production. White (1969) suggests that some pest outbreaks in arid regions actually are brought on by the increased palatability of plants in droughts.

The theory also predicts general circumstances in which qualitative, as opposed to quantitative, chemical defenses should be used. Scattered, opportunistic, or *r*-strategy plants receive considerable protection from herbivores by reason of their separation in space. They will, however, be subject to attack from various herbivores coming from neighboring plants of other kinds. A good defensive strategy in these circumstances is to use toxic qualitative chemical defenses. Although defenses against these toxins can be evolved, the defense must be toxin-specific and is unlikely to be possessed by many of the varied herbivores finding the scattered plant. We do indeed find that herbs and other early succession plants tend to have alkaloids and other toxin defenses. It is on these plants that we find the intense coevolution systems between herbivore and host of the kind described above for *Heliconius* butterflies.

Large plants of climax communities growing closely together have no defense by spacing out and are easily found by herbivores. For these the theory predicts the quantitative defenses of tannins and the like that make the tissues inedible, because these defenses are almost impossible to overcome by biochemical means. Accordingly we find that forest trees have leaves defended by tannins (Figure 25.4). Eighty percent of woody dicotyledonous plants possess tannins but only 15% of herbs do so.

Figure 25.4 Appalachian forest: quantitative chemical defense.
Tannins make leaves scarcely edible to all herbivores, because close-packed trees of the canopy have no cheaper defense in dispersal. Uneaten leaves send a large part of the flux of net productivity to the litter and decomposers.

Large plants with quantitative defenses are said to be APPARENT and the scattered plants bearing toxins are termed UNAPPARENT. From considerations of resource availability alone, we should expect that unapparent plants would be attacked by generalist herbivores because they are hard to find. Optimal defense theory, however, shows correctly that the unapparent plants should be attacked only by specialist herbivores able to find their scattered prey, because only specialists can coevolve the detoxification systems necessary to overcome the chemical defenses of unapparent plants.

Important community and ecosystem properties follow from these general coevolutions between plant chemical defenses and herbivores. Where, for instance, quantitative, tannin-like chemicals are used, much of primary production goes to leaf litter, decomposer food chains, and maintenance of soil structure. Powerful ecosystem properties result as an indirect consequence of selection for unpalatability in plants, and the earth may be said to be green (Chapter 8) as a result of coevolution between plant and herbivore. Reviews of the extensive literature on the subject will be found in Rhoades (1979), Whittaker and Feeny (1971), Feeny (1975), Rhoades and Cates (1974), and Rosenthal and Janzen (1979).

HOW TO WIN POLLINATORS AND INFLUENCE TRAVELERS

Plants use animals in reproduction to transport both pollen and seeds. They also use wind and water for these functions but probably most plants use animals some of the time, making mutualisms between plants and animals as prevalent as the predator–prey relationships of herbivory.

To exploit animals for pollination a plant must provide both signal and reward. The signal must be appropriate to the sense systems of the animals used. Scents made from flavenoids and volatile terpenes are signals for insects for which olfaction is more important than sight, which is why flowers smell nice. Birds use visual cues, which is why bird-pollinated plants like orchids tend not to smell at all, although their shape and color may be remarkable.

The rewards offered by plants, often but by no means always nectar, vary considerably suggesting that manipulation of the rewards provides a control on the animals visiting. A high-quality reward should have not only large volume and high sugar concentration, but also contain amines and other protein derivatives. Scattered plants that require long journeys by pollinators might be expected to offer large rewards, thus making visits cost-effective for animals, and data suggest that this is so (Baker and Baker, 1975). But a rich reward of nectar in a scattered plant could be taken by visitors from nearby plants bringing the wrong pollen, a circumstance that leads to the somewhat surprising discovery that high-quality nectar typically is poisoned with toxins. Only specialized pollinators equipped with the appropriate detoxification mechanism, therefore, can win a large reward and their toxin specializations force them to serve the plant species with which they have coevolved.

Plants can also manipulate pollinators in other ways. Earlier we described the advantage of spiral flower arrangement and a groove and pool nectar system. Selection between pollinators is also arranged by flower geometry and this serves as well to control the application of pollen.

But if plants must coevolve with their pollinators, they must also sometimes coadapt with neighboring plants with whom the pollinator pool must be shared. A common result of such coadaptation is staggered times of flowering, itself one of the important mechanisms driving phenological events, particularly in tropical ecosystems. Figure 25.5 shows the times of flowering over two successive years of ten species of plants in a tropical forest all of which are pollinated by hermit hummingbirds. It seems inescapable that the plants have been selected to avoid as far as possible flowering at the same time as others in the hummingbird-using guild. Since the incidence of rain, the most important environmental cue, is very variable in this system, it seems likely that some plants of the guild pay a high cost in staggering their flowering to times that are not physiologically optimum. The value of reduced competition for hummingbirds, however, apparently makes this cost worthwhile. It is noteworthy that a temporal organization of the community is imposed in this way, as a result of individual selection that coadapts plants to flower at different times.

A similar set of relationships exists between plants and their coevolved or coadapted seed dispersers. Signals, both olfactory and visual, must

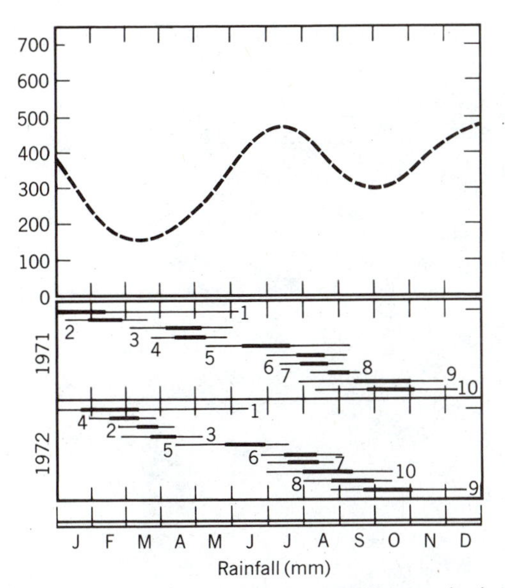

Figure 25.5 Phenology of ten species of *Miconia* bushes. Flowering times of the ten species shown for two successive years. All are pollinated by hermit hummingbirds. Flowering times are staggered, even though rainfall (top) is variable. (From Stiles, 1977.)

be accompanied by appropriate rewards, hence the soft parts of many fruits contain sugars and emit aromas. A common arrangement is for the fruit to be edible but the seed to be toxic. The edible fruit rewards visitors and the toxins in the seed are an adaptation to reduce seed predation by less welcome visitors, notably insects. A tight web of coevolution between a plant and different sets of animals, some to encourage mutualism and others to discourage predation, can result. The logic that a particularly rich reward offered by a well-dispersed plant should be laced with toxins to discourage all but specialized visitors applies equally to fruit dispersal and pollen dispersal. An example is the poisonous tree *Hippomene mancinella,* whose apple-like fruits are temptingly luscious but depressingly poisonous to humans, though presumably not to the dispersing animal for which they were made.

Plants make even more devious use of animals in their reproductive endeavors when they rely on them to deactivate seed defense mechanism implaced against insect seed predators. This happens when seeds are provided with lignified coats of extreme toughness to resist insect boring, which cannot, as a consequence, germinate unless the seed coat is abraded, usually in the gizzard of a bird. It has recently been suggested that a tree endemic to the island of Mauritius, *Calvaris major,* is now nearly extinct because its very thick seeds relied on passage through the gizzard of a dodo before they could germinate. The tree was so closely coevolved with the dodo that extinction of the bird spelled extinction for the tree (Temple, 1977).

HOW TO USE ANTS

Ants are among the most abundant of animals, small as individuals yet living in colonies of thousands or millions of individuals whose aggregate mass may range from a few grams to 20 kilograms for a giant colony of African driver ants (Wilson, 1971). They patrol large areas, often aggressively, with diets ranging from carnivory to herbivory, and they make forays into such exotica as gardening. They tend to build nests full of organic litter or food stores. They live in social systems in which morphologically discrete castes achieve divisions of labor. The behavioral responses of individual ants, however, are few and simple: ability to follow a pheromone trail, recognition of conspecifics and food, and self-sacrificing aggression if necessary.

The EUSOCIALITY, comprising reproductive division of labor, with sterile individuals serving fertile parents and cooperative care of young, is an aspect of natural history that is most attractive to evolutionary thought. Much understanding of the causes of *eusociality,* largely based on kin-selection models, has now been achieved (Wilson, 1971; Oster and Wilson, 1978). Alternatively, ecologists can think of the ants for their great importance in the trophic structure of many communities, as when leaf-cutting ants defoliate tropical rain forest trees. But the ants are equally exciting as a resource to be exploited by other organisms.

A patrol of ant sentries can provide an excellent defense for a plant against herbivore attack.

Figure 25.6 *Formica* sp. feeding: payment to the guardians.
The ants patrol the trumpet vine *Campsis radicans,* feeding from nectaries on the lower parts of petioles. This is an example of the ant guard symbiosis.

Many tropical trees (*Acacia* are particularly well known) both house and feed ants with this result. Special cavities in the branches are used by ants to build nests and the tree has extra-floral nectaries (that is, nectaries outside the flowers on stems) on which the ants feed. The ants are thus nurtured by the tree. Experimental removal of the ants with insecticides results in destructive attacks on the tree by defoliating insects, showing that the ants have the coevolved function of tree defense.

Travelers in the tropical rain forest know that it is unwise to rest a hand on any tree because of waiting ants, suggesting a general close relationship between ants and trees. The symbiosis is not restricted to trees and tropics, however, and some temperate sunflowers also attract ants with extra-floral nectaries. The trumpet creeper *(Campsis radicans)* of the American Midwest also exhibits the ANT-GUARD SYMBIOSIS with no less than four kinds of extra-floral feeding points for ants (Figure 25.6). Nectar for the ants in these trumpet creepers is quite different from the floral nectar used to attract hummingbirds and bees for pollination (Elias and Gelband, 1975).

If plants can be adapted to use ant sentries, then so can herbivores. Caterpillars of Lycaenid butterflies, among many others, are tended and herded by ants, which will carry the caterpillars from food plant to food plant. This transport by ants may be particularly valuable to Lycaenids, which are often food specialists on small or scattered herbs where the ants can take them. The caterpillars feed the ants directly for these services, giving them nectar-like secretions called honeydew. Lycaenid species seem to have coevolved with their own species of ants and their own specialist food plants. The symbioses may explain why the family Lycaenidae is one of the most species rich of butterfly families (Gilbert and Singer, 1975).

More familiar to people of the temperate zone are ants that tend aphids for honeydew rewards. These systems may not be entirely at the expense of the host plants because the ants certainly serve to protect the plants against all her-

Figure 25.7 *Calymmodesmus* sp. marching by pheromone with hosts. The millipede is symbiotic with the army ants *(Labidus praedator)* and follows their odor trails on the march.

bivores other than the chosen one. These plants are in effect rewarding the ants for their defense services with honeydew channeled through the herbivores instead of making it themselves in extra-floral nectaries. Probably various systems exist ranging from ones in which the herbivore load to the plant is small to others where the ant-enhanced herbivory is critical.

Army and driver ants also represent resources to be exploited by ingenius adaptations. They serve as lines of beaters from which other small animals, particularly arthropods, flee, leaving their shelters. Whole guilds of ant-shrike birds exist that follow army ants and feed on the insects that expose themselves. At the same time some arthropods take advantage of the army ants more directly, marching with them, following the ant pheromone trails, and escaping attack from the ants either by mimicry, as with staphylinid beetles, or perhaps by chemical means (Figure 25.7).

These various adaptations to take advantage of the ants use their fighting prowess as a resource and their simple, stereotyped behavior so that they can be duped. Another resource offered by the ants is their untidy, food-rich nests with their own reproductive factories inside. An immense array of social and nest parasites of ants is known as beetles, flies, and other organisms feed on ant litter, food reserves, and larvae, or actually induce the ant workers to feed them themselves, like insect versions of cuckoos (Wilson, 1971).

Ants represent a very special set of resources in nature that are used mutualistically or symbiotically by very many organisms. But comparable, if generally less dramatic, mutualisms are known for many organisms in all ecosystems. Use of ants serves as an example for a very general process in community building.

THE ECOLOGY OF MIMICRY

A living community is a place of myriad signals. Pheromones are emitted, songs and chatters are sounded, and every movement or shape is a signal to watching eyes or listening sonar. Every organism will both give signals and receive them, and its survival may depend on how it does its signaling.

The evolution of mimicry is the evolution of false sets of signals, and the opportunities for this in a crowded community are many. Mimicry can serve to hide the weak and eatable, to give cover to predators, and to almost any purpose for which false signals can give selective advantage. The more complex the community, the greater the opportunity for mimicry, which accordingly seems to be most prominent in ecosystems like the tropical rain forest or coral reefs.

In typical mimicry there are three actors: the model, the mimic, and the dupe. The MIMIC is a counterfeit copy that plagiarizes the MODEL. The DUPE is the enemy or victim of the mimic. Enemies and victims are sometimes called "signal receivers," but the term "dupe" seems not only more expressive but equally accurate (Pasteur, 1982).

We know disproportionately much about mimics that counterfeit light signals, since we are so dependent on light ourselves. This bias can let us imagine that mimicry is more perfect on land than in the sea when we see marvelously exact copies of butterflies (Figure 25.8) and compare them with what seems to be crude copies of some marine organisms. But the butterfly must dupe birds, which have eyes perhaps even better than ours, whereas the marine mimics might have to dupe only fish eyes, which are not very good.

The best known mimicry systems, called Batesian after the 19th-century naturalist who first discovered them in the Amazon rain forest, are defensive. The mimic is eatable, the model is poisonous, and the dupe is a bird capable of learning that the model is unpleasant to eat. Many of the Amazon systems that first attracted Bates were based on monarch butterflies of the family Danainae. The caterpillars of these butterflies are coevolved with milkweed plants that have toxic alkaloid defenses. The caterpillars store forms of the toxins as cardenolides, these being bitter-tasting heart poisons that are also powerful emetics (Brower and Brower, 1964). The cardenolides are retained in the adult butterfly, which is thus not only uneatable but also tastes so bad that birds vomit almost instantly after eating one. Birds quickly learn to avoid the danaid butterflies after a few unfortunate tries. The system represents an elegant adaptation of the butterflies to toxins in their food supply in which they not only overcome the toxins but use them for their own defense. But other butterflies can then use the same defense by being selected to be visual counterfeits of the monarchs. Many species do this.

Figure 25.8 Batesian mimicry.
The model is above, the mimic is below, the dupe is any bird that finds the model to taste nasty.

The effectiveness of defensive mimicry depends on the relative eatability of the mimic, the relative nastiness of the model, and the discriminatory powers of the dupe. Many snake species are visual mimics of the highly venomous coral snakes, *Micrurus*. Mimicry of monarch butterflies may succeed as a function of the eatability of local monarch populations, which can vary widely as optimization of local strategies sometimes results in less resources going into the storage and retention of cardenolides. Mimicry can still be strong even when the discrepancy in eatability between model and mimic is modest, as in Australian mistletoes, whose leaves are similar to the leaves of trees on which they are parasitic. The dupe in the mistletoe system is a species of opposum that can eat both the host tree and the mistletoe, but which prefers mistletoe.

Much mimicry, of course, is mere camouflage in which the model is *indifferent* to the dupe. The many imitations of plant parts by insects that serve to hide both predator and prey are of this kind. But in some offensive mimicry the model may be *agreeable* to the dupes. This happens with the saber-toothed blenny, a fish that is a mimic of a cleaner fish (Figure 25.9). Cleaner fishes have a mutualist relationship with client fishes by taking ectoparasites from their bodies and are thus allowed to approach within nibbling distance. The saber-toothed blenny is, however, a predator that bites the clients of the cleaner fish once its disguise allows it to get within nibbling distance of its dupe.

Firefly females of the genus *Photuris* are predators of the males of another firefly, *Photinus macdermotti.* Male fireflies find females with light signals, the flashes of which are species-specific. The *Photuris* females mimic the flashes of *Photinus* females, thus luring *Photinus* males to a deadly embrace. He homes on the light expecting sex but gets eaten instead.

Probaby much offensive mimicry is simple

Figure 25.9 Treacherous mimic: disguised aggressor. The saber-toothed blenny mimics a cleaner fish, is allowed close, and takes a bite of flesh instead of removing a parasite.

camouflage, as the predator lies in wait camouflaged as prey or as some immobile portion of the habitat. But issuing false signals is probably a common tactic of predators also, even down to the twitching tip of a cat's tail that may serve to distract the quarry's attention. In complex communities, misinformation is a common tactic for both defense and attack. Strong mutualistic links are forged in this way, representing another set of community properties fashioned by individual selection.

SPECIATION WITHIN A COMMUNITY

Insight on how speciation can continue within complex communities and result in coadaptation and mutualism comes from further studies in island biogeography, notably in the process known as a *taxon cycle.*

The basic island biogeography model (Chapter 24) examines only the equilibrium attained between invasions of the island from outside and local extinction, but a second source of "invasions" is the creation of new species within islands themselves. *K*-selection should be of particular importance, since *K*-strategists are less likely to reach the island from outside. We imagine the island becoming filled with *r*-selected immigrants and these being subjected to *K*-selection to produce ENDEMIC ISLAND SPECIES. It was this island scenario that first let MacArthur and Wilson (1967) introduce the concept of *r*- and *K*-selection (Chapter 11).

A TAXON CYCLE is a sequence of speciations thought likely after an island is reached by opportunist *r*-strategists. When the original *r*-strategist population penetrates the interior of an island, *K*-selection should produce a species-population able to exist at equilibrium in the equable conditions of the interior. If the island is large or physically diverse, several *K*-selected species may result. These equilibrium species of the interior are less likely to suffer local extinction than the parent stock of *r*-selected colonists that are by now restricted to marginal habitats on the island like the beach regions. Local extinctions at the margin, therefore, give opportunities for reverse invasions from the *K*-selected populations of the interior. Readapting to the physical uncertainties of marginal habitat, of course, requires *r*-selection. A complete taxon cycle then consists of *r*-selection on a distant mainland to produce colonists, *K*-selection among the colonists to produce new species for the island interiors, and then a second bout of *r*-selection producing a fresh generation of colonists from out of the island's endemic stock.

Data from remote archipelagos apparently carry the imprint of the working of this process as related endemic species are found in strings along the archipelagos (Ricklefs and Cox, 1972). The working of this process was first suggested by Wilson and Taylor (1967) when examining the fate of introduced species of ants into the islands of Melanesia.

Ants are now able to invade islands easily on the ships of commerce and the ants that travel in this way are called TRAMP SPECIES. So many tramps now travel the world on ships that the tramp species list in the Pacific (38) is nearly as large as the total list of endemic species (43). Despite this, the larger islands of Melanesia always hold more endemics than they do tramps, the ratio being about 3 to 2. This suggests that the endemics are *K*-selected and less likely to become locally extinct. The tramps invade easily, but they go extinct easily, and thus have lower equilibrium numbers (Figure 25.10).

If ships were not bringing in tramps from outside, it is reasonable to suggest an invasion pressure to the coastal regions from the interior and thus the production of potential new tramp species. Figure 25.11 shows the complete scheme put forward by Wilson and Taylor in which the final act in the taxon cycle is a tramp episode from the islands back to the mainland, from beach to beach in the opposite direction.

Figure 25.12 shows the complete postulated history within island communities themselves. After the interactive equilibrium expected by the island biogeography model is established, there comes a reassortment of species as in the late stages of an ecological succession. A relatively complex community now exists in which evolution proceeds by *K*-selection and taxon cycles

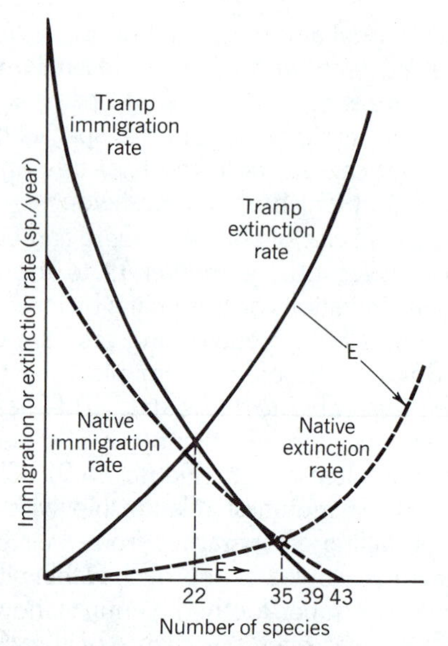

Figure 25.10 Model to explain relationship of tramp and native ant faunas on the Pacific island of Upolu, Samoa. Known species pools are for native species, $P = 43$, and tramp species, $P = 39$. Upolu now has 35 native and 22 tramp species. The model is an attempt to describe the probable increase in number of species that would result from evolution of the present tramp fauna toward the equilibrium represented by the present native fauna. (From Wilson, 1969.)

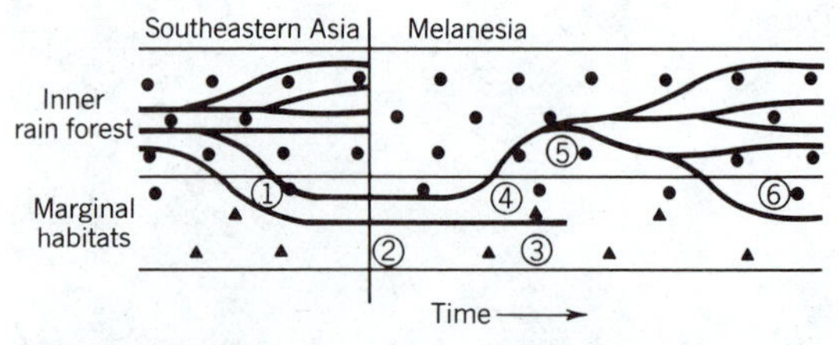

Figure 25.11 Inferred taxon cycle for Melanesian ants. The Melanesian ants seem certainly to have been derived from Asian stocks. (1) Populations adapt to marginal habitats on mainland, then cross water to islands. (2) From marginal habitat on islands, the invaders go extinct (3) or invade rain forest inland (4). Once in rain forest they may radiate into a number of rain forest species (5), and perhaps into new stocks that can recolonize the marginal habitat (6) and so be able to colonize other islands, or even the mainland. (From Wilson, 1959.)

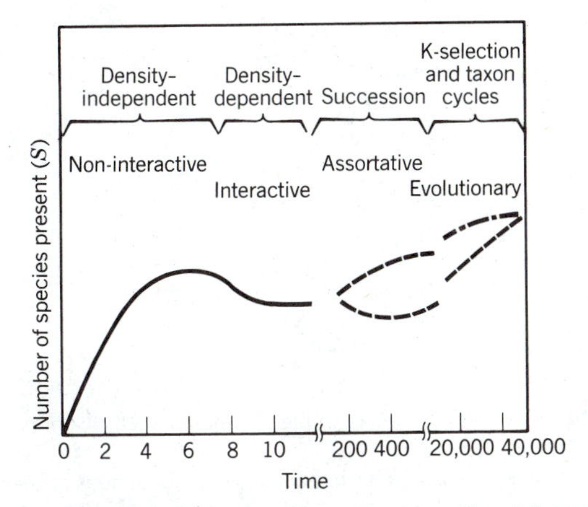

Figure 25.12 Postulated sequence of equilibria following the original invasion of an island.
The attainment of each kind of equilibrium would certainly overlap. The loss of species between the non-interactive and interactive equilibria follows the observations on the experimental mangrove islands (Figure 24.4). Two possible pathways for species number under succession and evolution take account of the fact that the attainment of an assortative equilibrium in the form of the climax in temperate vegetation actually may possibly lower species number from the interactive equilibrium of the preceding seral stage. (Modified from Wilson, 1969.)

to produce a final evolutionary equilibrium number.

Evolutionists differ over the extent to which island models can be applied to mainland communities, since the isolation of islands is so extreme. However, isolation is a matter of degree so that islands are likely to show in a simplified manner the ways in which vicariance speciation works. Island models ought, therefore, to be applicable to continental communities (Levins, 1975). Invasions may well be more rapid on mainlands, and far larger numbers of species are involved, but these may well be differences of degree only.

Particularly noticeable in the coevolutions and mutualisms of complex communities are the exploitations of a particular kind of relationship. The plant use of animals for pollination is one, and this is divided into subsets by use of insects, birds, bats, or arboreal mammals. Invasions by a new taxon class likewise opens a spate of possibilities, and the arrival of predator immunity for one species invites numerous inventions by mimicry. Many of the coexisting species in a complex community, therefore, seem to represent so many variations on a few simple themes. This observation leads some evolutionists to suggest that radiations can be rapid when a new kind of opportunity appears. Breakthrough with a toxin, or unpalatability, or an efficient pollination mechanism suddenly opens possibilities leading to rapid selection pressures.

The speed with which a response can be made is again suggested by island studies as in Hawaii, where a suite of endemic pollinating birds evolved within a hundred thousand years. One imagines the evolutionary equilibrium at the right of Figure 25.12 proceeding with a regular turnover of local extinction, reinvasions, and taxon cycles with little change in the equilibrium number within the community until some small change in morphology or habit accidentally opens fresh opportunities. A taxon speciates rapidly, new opportunities for coadaptation and mutualism result, and the long equilibrium is punctuated with an episode of rapid replacement of species. This is the essence of the PUNCTUATED EQUILIBRIUM model of evolutionary history (Eldredge and Gould, 1972).

GROUP SELECTION AND THE SUPERORGANISM REVISITED

Most ecologists find no difficulty in the idea that selection for individual self-interest can produce the most complex of communities. The intense interdependence of so many members of the more complex communities comes about through individual pairs of coadaptations and coevolutions. Mutualists use each other for individual advantage; consumers take advantage of their prey and the prey are selected for maximum avoidance of consumers. In this view, no group-organizing principle is needed to explain community properties.

And yet there remains an almost startling element of community functionalism in the activities of some species populations. This is the idea conveyed in the original use of the word "niche" by Elton, what I have called the type 1 functional niche (Chapter 2). In this sense, the niche of earthworms functions in the community in ways that improve the soil, one of the indirect consequences of which is making the habitat better for earthworms. In a like manner symbiosis with nitrogen-fixing bacteria improves the habitat for others, bark-tunneling beetles produce habitats for associated forms, and so on. From time to time ecologists worry lest all of these community functions can really be the outcome of selection for selfish self-interest or if some kind of subtle group or community selection must be involved after all. These are the questionings that I have described as the second paradox of organization in the opening chapter.

It is clear that in the most simplified of breeding or selection systems, group selection for traits that aid other individuals is not possible. Figure 25.13 shows what should happen. Of two earthworm strains, type A, a community benefactor, acts in ways that improve the environment for earthworms at some cost C, giving a gain in fitness G to itself. But type B worms in the same community have an equal gain but no commensurate cost and thus should leave more surviving offspring. Selection will remove the community benefactor's genes from the population rather rapidly. This will be the result of any community function that incurs a cost in any system with freely assortative mating.

Yet there is a single circumstance in which the A type community benefactor can be favored by selection: when living in a very small subpopulation in a generation free from intense competition. In this very small group, all individuals, both type A and the rest, will gain from the presence of type A. The group, as a whole, will leave more surviving offspring than other groups of similar size with no type A individuals. If at the end of a single generation there is an episode of dispersal, the costs of the benefaction never catch up with type A.

The key to this form of group selection is life in isolated groups without a continuous and general mixing of genes throughout the population, what D.S. Wilson (1980) calls a system of STRUCTURED DEMES. A "deme" is the geneticist's term for a panmictic population in which cross-breeding is general and random. "Structured demes" describes the fact that in many real populations potential inbreeders live much of the time in small subpopulation isolates. Wilson uses the example of a small cavity of water on a tree trunk where mosquito larvae live for a generation. All larvae in that one cavity would benefit from habitat improvement by one strain of them and then they would disperse to other cavities for the next generation. In every generation the community improvers would derive benefit from the improvement but whether the other types derived their even larger benefits would depend on whether an improver was present or not. Mathematical models leave no doubt that this mechanism would work if the limiting conditions are met.

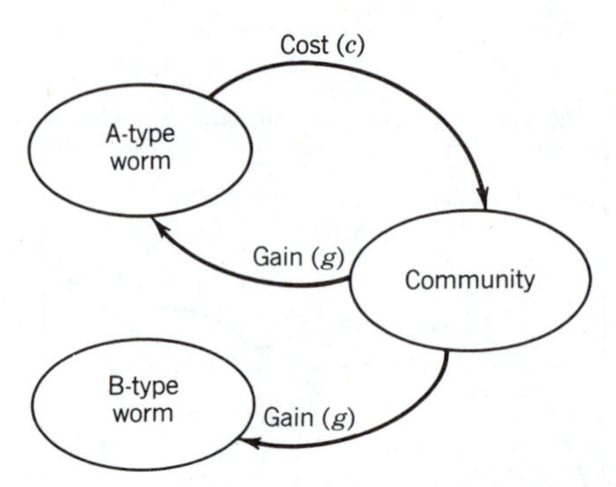

Figure 25.13 Model of indirect community effects for an individual adaptation.
The A-type must be selected against in a continuous cross-breeding system because a cost falls to the A-type but not to others. However, if the small group disperses shortly the A-type has a gain in fitness. (From D. S. Wilson, 1980.)

Wilson (1980) makes convincing arguments that this mechanism of group selection by structured demes is important in the evolution of small, tight-knit communities that disperse as whole entities. The community associated with bark-tunneling beetles is of this kind. Another is the community of mites and other organisms that travel with scarabs or other dung beetles, using the flying beetles as transport for the whole community, so called PHORESY.

For many ecologists, however, it is probably

true to say that the mechanism of group selection by structured demes is unnecessary to account for most of the functions that individuals serve in a community. Predators prey, pollinators pollinate, nitrogen fixers fix, and earthworms dig because these activities give them individual fitness and not because the community function is served thereby. Community building in evolutionary time essentially is a result of individual selection.

The community structure that results will be compounded of competition, predation, and mutualism, the relative importance of each of which probably depends on the ecosystem type. Predation is now thought to be particularly important in aquatic systems, where predators of the open water can reduce populations of many herbivores to low levels (Chapter 21). In terrestrial systems, competition may be relatively more important, though predation may have more effect on community structure even there. In the most complex systems like tropical rain forests, mutualism may be the most important structural associations of all (Strong *et al.,* 1984).

PREVIEW TO CHAPTER 26

Species are thought always to be sufficiently different to avoid competition, which suggests that a limit exists to the number of ways in which any resource can be divided. The modern attempt to explain the number of species (the *Santa Rosalia model*) began with this assumption; essentially that local resources determined how many species could be packed into a habitat. Increasingly, however, ecologists suspect that this theoretical upper limit set by competition is important only for a few groups of specialized animals like birds, whose numbers are not usually constrained by heavy predation. Most organisms live most of the time in populations kept small by predation, by weather or changes in seasons, by low productivity, or because prey have effective defenses. Studies of predation on both plants and animals show that adding predators to ecosystems can increase diversity by reducing competition in the prey populations (the *cropping principle*). Clines of diversity, like those from equator to poles, up mountains, or up from the deep sea, reflect environmental clines. Where diversity is low, environmental constraints keep populations low or so fluctuating that extinction is likely. In high-diversity places, extinction through a precarious population size is less likely. In the places of highest diversity of all, like the tropical rain forest, not only are populations seldom precarious but the environment has periodic perturbations just sufficient to prevent excessive competition from removing species by ecological dominance. *Ecological dominance* itself is a pattern of relative abundance in which a few species are more common than is allowed for in random distributions. It happens in communities where predictability of the environment and sufficient productivity allow prolonged equilibrium populations, as in the climax of temperate successions or among algae in polluted lakes. Measuring diversity is always difficult, since both total numbers of species and relative abundance vary from place to place. A simple count of *species richness* may miss the rare species, producing an estimate that is too low. This difficulty can be met by standardizing the collecting effort by ensuring that the same proportion of a relative abundance curve (the *log-normal distribution*) is collected from each habitat. Alternatives have been sought by computing indexes that combine the *richness* and *equitability* components of diversity, though confidence in these indexes is now not as high as it once was. Use of the information theory index, H', for this purpose influenced a long debate in ecology over possible relationships between diversity and population stability within ecosystems. Modern analysis suggests that any general connection between ecosystem stability and diversity is negative: the more species there are, the more populations may change. However, it now appears that ecosystem stability depends more on environmental stability. Many ecosystem changes are related directly to fluctuations in the physical environment. At the same time, the physical environment controls the processes setting species number.

CHAPTER 26

BIOSPHERIC DIVERSITY AND STABILITY

Modern diversity studies were strongly influenced by a paper by Hutchinson (1959) summarizing the then state of theory on the causes of diversity. Hutchinson wrote the paper in Palermo, Sicily, and, in a fit of scholarly whimsy, dedicated it to Palermo's patron saint so that the paper ever since has been called "the Santa Rosalia paper." Hutchinson tried to answer the question "Why are there so many different kinds of plants and animals?" by assuming that food energy set the limit. Energetics should limit complexity of food chains. Competition for energy should decide how many niches were possible and hence how many animals could coexist.

The possible number of niches is multiplied in a food chain by as many times as there are links in the chain. When a herbivore species results from natural selection, an opportunity for a carnivore niche is at once created. Appearance of a carnivore then creates the opportunity for another niche at the next trophic level, and so on. But this multiplication of niche number obviously is narrowly constrained since most food chains have only between three and six links.

Hutchinson recounted the energy flow arguments restricting the lengths of food chains (Chapter 2), noting that an animal at the fiftieth link of a food chain would have a population of 10^{-49} times the population of the first link, assuming an unrealistically high Lindeman efficiency of 20% and a modest doubling of individual size with each link in the chain.

Actually, as Hutchinson argued, food chains may be even shorter than are decreed by thermodynamics in this way, because selection works to compress the chains. Selection at the highest trophic level should always work to favor the most effective predator, which will therefore be in constant danger of exterminating its prey. If the animal at the nth link exterminates, or seriously depletes, the $(n - 1)$th link, then it must resort to cannibalism or concentrate on the $(n - 2)$th link. In this way, selection will work to shorten food chains as well as to lengthen them. The working of this process may be the explanation that food chains tend to be short in very productive systems like ocean upwellings (three links common) compared to unproductive systems like blue ocean water (five to six links). The highly productive system may allow dense populations of predators with consequent danger of over-eating prey.

The Santa Rosalia paper thus recognizes that increasing diversity through food chains is not limited only by the resource energy available. Energetics may set an upper limit to the number of links but this is unlikely to be reached. The actual number of links probably results from an equilibrium between the rate at which food chains are both shortened and lengthened by selection. Abundant resources actually may so deflect this equilibrium as to make food chains shorter than they otherwise would be.

Niche-packing

Having shown that the possibilities of multiplying diversity through food chains were strictly limited, Hutchinson (1959) then explored the possibilities for dividing energy between pairs of competing species. This was an argument of NICHE-PACKING and was the crux of the Santa Rosalia argument.

Crucial to the Lotka–Volterra–Gause model of speciation is that, when allopatric populations are brought together to live in sympatry, they can do so only if their niches are separated by some minimal distance. Hutchinson proposed that this minimum separation of niches might set a limit to the number of species. If evolution has had long enough to saturate the earth, then these finite differences would determine how many species there could be.

Studies of closely related sympatric species (Chapter 7) had shown that animals could divide resources by different behavior or through having beaks or jaws of different sizes that let them feed on food of a specific size or shape. Hutchinson (1959) reasoned that if species had evolved to the possible limits of similarity, then these limits would be reflected in dimensions of the feeding (*trophic*) apparatus. Table 26.1 gives his data for skull length of pairs of mammal species and bill lengths for pairs of birds. Mean ratio of lengths between sympatric pairs is 100:128, and the variance from this ratio is not large. Finding these rather similar ratios in size of trophic structures between pairs of sympatric birds and mammals

Table 26.1
Trophic Measurements of Closely Related Species in Allopatry and Sympatry
Measurements are length of skull of female mammals and culmen length on birds. (Modified from Hutchinson, 1959.)

Species	Measurement when Sympatric (mm)	Measurement when Allopatric (mm)	Ratio when Sympatric
Weasels			
Mustela nivalis	33.6	34.7–36.0	100:134
M. erminea	45.0	41.9	
Mice			
Apodemus sylvaticus	24.8	25.6–26.7	100:109
A. flavicollis	27.0		
Nuthatches			
Sitta tephronota	29.0	25.5–26.0	100:124
S. neumayer	23.5		
Darwin's finches			
Geospiza fortis	12.0	10.5	100:143
G. fulginosa	8.4	9.3	
Camarhynchus parvulus	7.0–7.5	7.0–8.0	100:128:162
C. psittacula	8.5–9.8	10.1–10.5	100:127
C. pallidus	11.2–12.6	10.8–11.7	
			Mean ratio 100:128

gave encouragement to the idea that they had evolved to be as similar as possible and that many communities might in fact be saturated with bird and mammal species. That other animals too might have divided resources into the smallest possible packets through specialized niches was suggested by data from insects, related species of which often seemed to exist in a series of discretely different sizes.

These arguments were applied in *Santa Rosalia* only to animal numbers. Hutchinson was careful to keep the argument to animal species only, though acknowledging that a prime cause of animal diversity was, of course, plant diversity. If every plant species had the potential of being the base of its own food chain, then a prime determinant of the number of kinds of animals was the number of kinds of plants. However, the arguments for niche separation should be directly applicable to plants also. Ways in which space under the sun is divided serve plants as niche axes equivalent to the food resources of animals, making it more difficult to find some measure of niche separation among plants to serve as length of culmen does for birds. However, even if we cannot measure niche separation so conveniently similar processes should be working for plants, though perhaps more hidden. The model had generality, therefore.

With the benefit of modern work on community evolution (Chapter 25) it is clear that the Santa Rosalia conception of numbers of species being limited by competition or lengths of food chains does not allow for all the possibilities. Species can also be packed together through the dynamics of attack, defense, cooperation, timing, or dispersal, and if populations are kept low in density-independent ways or by predation, then many more species might coexist than can be allowed for in a competition model. But the Santa Rosalia arguments persuaded a generation of ecologists that explaining the diversity of species was in theory possible. Studies of patterns of diversity became an important part of ecological research, and for these studies it became necessary to compare diversity in different habitats, an undertaking that required objective measures of community diversity, measures that have proved to be hard to devise.

THE MEASUREMENT AND DESCRIPTION OF DIVERSITY

The word "diversity" carries a clear enough meaning of its own. It implies variety. It comes from the Latin words *dis* and *vertere*, which put together can be read as "to turn away," hence stressing the difference between objects. When we talk of roads "diverging" we build on the same Latin roots. So *species diversity* means the variety of species, and the question "Why so many species?" can equally well be put as "Why this species diversity?"

But investigations into species diversity always are complicated by the fact that population size varies from place to place. Consider two simple communities, each with only one trophic level and each with a hundred individuals of four species A, B, C, and D.

Community 1 has	25 A	25 B	25 C	25 D
Community 2 has	5 A	5 B	1 C	89 D

Exactly the same species list can be constructed by diligent collecting in each of these two communities, yet they clearly are very different. And the key word in amassing the complete species list in community 2 definitely is "diligent" because that small population of species C would be overlooked in most surveys.

Two components of variableness in a species array are shown by this example: the actual number of species present, which ecologists call SPECIES RICHNESS, and differing relative abundance, which ecologists refer to with the term EQUITABILITY, defined as THE RELATIVE EVENNESS OF THE NUMERICAL IMPORTANCE OF A SPECIES IN A SAMPLE.[1]

Finding a satisfactory statistic to measure both *richness* and *equitability* at the same time has been difficult. Apparently no universally accept-

[1]Ecologists have difficulty defining equitability in a way that is applicable to all systems. Whittaker (1977) draws on Lloyd and Gelardi (1964) to offer the comprehensive but cumbersome definition of *equitability* as the *relative evenness of the importance values of adjacent species in the sequence from most important.*

able statistic exists, although several are available and have been useful in particular studies. The most widely used statistic was the Shannon–Wiener[2] INFORMATION THEORY INDEX, H', where

$$H' = -\sum p_i \log p_i$$

and p_i is the proportion of the total number of individuals in the ith species. This actually serves as a statistical measure of the probability of guessing the identity of an individual taken from a sample at random. Table 26.2 shows a range of calculations of H'.

Ecologists have come to realize that the *Shannon–Wiener* index has serious statistical shortcomings with the result that the index is now used less than it was a few years ago (Hurlbert, 1971; Goodman, 1974). The statistic has had an important place in the development of ecological ideas, however, because its use as a measure of diversity and of complexity led to important ideas about the stability of complex communities that, although now largely discounted, had a strong influence on ecological thought in the recent past (see below).

Numerous other indexes of diversity appear in the literature (Peet, 1974; Pielou, 1975). The three most encountered are these:

S = SPECIES RICHNESS

H' = SHANNON–WIENER INDEX

λ = SIMPSON'S INDEX

SPECIES RICHNESS is the simplest measure, being the number of species actually tallied by the investigator. Strangely, this apparently simple measure is not completely divorced from an estimate of relative abundance as well, because the commonness or rarity of a species will determine whether it is counted by any given sampling effort. In the examples of communities 1 and 2 given above, for instance, anything less than a very thorough sampling of community 2 would totally overlook the rare species C and report the species richness as 3 instead of 4. A superficial sampling of this community might actually report the richness as 2 or even 1, whereas any reasonable sampling of community 1 would yield the full estimate of richness as 4.

SIMPSON'S INDEX, λ, may be the safest alternative to the Shannon–Wiener index when species richness alone is considered an insufficient measure. The index

$$\lambda = \sum p_i^2$$

is the probability that any two individuals picked at random will be of the same species. The index is thus a measure of how individuals in a sample are concentrated into a few species, for which reason this is sometimes called a *dominance* index. It is, however, an inverse measure of diversity.

There is a tradition among some ecologists to distinguish between the terms *diversity* and *species richness*, arguing that "diversity" should be reserved for formal measures, like H' or Simpson's index, which combine relative abundance with richness. However, this distinction may have had its day. Measures of *species richness* alone, with control for equal sampling effort, serve to compare different communities as well as the various indexes (Grigg and Maragos, 1974; Loya, 1976) and without the possibilities for being misled inherent in some indexes (Hurlbert, 1971).

Table 26.2
Sample Calculations of the Shannon–Wiener Diversity Index, H'
Notice that to produce strikingly different values the "communities" are given extremely different relative abundances. (From Price, 1975.)

	Sp 1	Sp 2	Sp 3	H'
2 species	90	10		0.33
2 species	50	50		0.69
3 species	80	10	10	0.70
3 species	33.3	33.3	33.3	1.10

[2]The Shannon–Wiener index often is called the Shannon–Weaver index because the most accessible review of its mathematics is in a chapter by Shannon in a book by Shannon and Weaver (1949). This is incorrect, Wiener having published the statistic independently.

The advice of Connell (1978) seems sound, to use *S*, species richness, as the general measure of diversity.

GEOGRAPHICAL PATTERNS OF SPECIES DIVERSITY

Some places have more species than others. The wet tropics have more species than the temperate zone; typical land more than typical sea; the water of hot springs has few species whereas the cold water of lakes has many. These local and geographical differences in species number must be explained by any general theory of diversity.

The following is a list of widely accepted generalizations about differential diversity. Probably all the items need some qualification, because there can be exceptions. The generalizations, however, have guided much ecological thinking.

1. A cline of diversity runs from the equator to the poles, with highest diversity in tropical regions and progressively lowered diversity at higher latitudes.

2. Diversity changes with altitude on mountain sides, being lowest at high elevations.

3. Unit area of land has higher diversity than unit area of ocean.

4. Unit area of a continent has higher diversity than unit area of an island.

5. The continental latitudinal gradient is partly reversed for the planktonic populations of lakes, with lakes of temperate latitudes having higher diversity than lakes of the tropics or equator.

6. Places of extreme environmental rigor, like hot springs or polar deserts, have low diversity.

7. A cline of diversity of benthic animals in the oceans runs from low diversity in the shallows to high diversity in the abyss.

The differential diversities described in these generalizations are ecosystem properties that depend on physical conditions. That life in the open sea is divided into fewer species than life on well-watered land clearly must reflect physical constraints on life-styles in the two systems. Likewise the all-pervading cline of diversity from equator to poles must ultimately depend upon the seasonality of climate and the distribution of solar energy with latitude. Hot springs, deserts, and equatorial lakes likewise must have their species numbers set as functions of environmental regimes. Ecologists try to relate the working of various processes leading towards diversity to the differing environmental circumstances in order to explain these patterns.

INCREASED DIVERSITY THROUGH PREDATION: THE CROPPING PRINCIPLE

Heavy predation gives opportunity for increased diversity by reducing competition among the prey. The predators work by keeping prey numbers so low that resources are released for other species that are better able to escape predators than to endure competitors. Our expectation that predation should work in this way is sometimes referred to as THE CROPPING PRINCIPLE.

The essential requirement for this effect to appear is that predators be capable of reducing the population of their prey to levels significantly below what could be maintained in the absence of predators. Not all carnivorous animals do this, nor do others do it all the time (Chapter 10), but where predators of insects, plankton, or plants are highly effective, the cropping effect opens almost unlimited possibilities for diversity. This may be an important mechanism contributing to the latitudinal cline of diversity, since minimal seasons in the wet tropics allow predator populations to be maintained with little seasonal escape by the prey.

Packing in more species under predation gives a quite different meaning to the term *species packing* than is allowed for in the *Santa Rosalia* argument. Interspecific competition is no longer the prime determinant of species number—predation serves that role instead. There is, of course, always the proviso that the predation must be effective predation. With animals, or in systems, where predators cannot significantly reduce the numbers of their prey, predators might have less effect on diversity. Very likely this is so for the bird and small mammal species used by Hutchinson to test the species equilibrium hypothesis (Table 26.1). Not being under controlling predation, interspecific competition became more important for these animals.

If predation allows increased diversity in the prey trophic level, this should be followed by increased diversity among the predators also. As specialized prey species invade or evolve in response to resources released by decimation of potential competitors, so resources are created for new species of predator. The *cropping principle,* therefore, offers an almost unlimited contract with increasing diversity: more kinds of prey leading to more kinds of predator, whose predations make room for more kinds of prey, and so on.

The cropping principle has the great virtue that it leads to testable hypotheses. A formal statement of the principle is THE DIVERSITY OF PREY IN A COMMUNITY WITH AN ARRAY OF EFFECTIVE PREDATORS IS A FUNCTION OF PREDATION. This hypothesis is nicely testable by experiment:

Experiment 1 Remove predators from a community so regulated.
Predicted outcome: prey diversity decreases.

Experiment 2 Add predators to a community with little predation.
Predicted outcome: prey diversity increases.

An array of experiments of both kinds have been made, some intentional, and some the unintentional results of management practices.

Predator Removal in a Benthic Intertidal Community

On rocky substrates of intertidal or subtidal stretches of coast, population limits to both animals and plants tend to be set by space. Seaweeds live anchored to rocks against the waves and maintain productivity comparable to that of land plants of mesic habitats by extracting nutrients from the flowing water. But these plants must compete with filter feeding animals for space on the rocks (Figure 26.1) as various bivalves and barnacles feed by extracting plankton or debris from the flowing water. The flux of food reaching the rock surface typically is in superabundance, and the total number of filter feeding animals that can live is set by space, not food.

Figure 26.1 Intertidal competition.
Anemones, brown and green algae, the sea grass *Zostera,* and others compete for space on the coast of Washington and a predatory starfish crosses.

In the absence of effective predation or grazing, competition can be strong in this system. Connell (1961), for instance, showed that barnacles compete strongly for space. Intraspecific competition within single-species stands is by the simple process of crushing or prying off neighbors. Separate bands of different species like those often seen in intertidal zones (Figure 26.2) also result from competition. Larvae of *Chthamalus,* the upper species in the intertidal of Scotland, often settled and started to grow in the lower zone where *Balanus* lives but were always

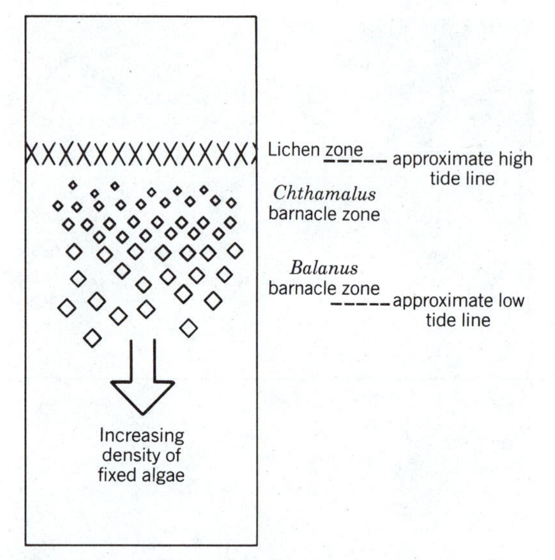

Figure 26.2 Competition between barnacles.
Barnacles squeeze competitors off the rock, a proceeding that gives the advantage to fast-growing and bigger individuals. *Balanus* individuals can always remove *Chthamalus* individuals from the lower zone but *Chthamalus* can apparently endure longer exposure to the air and a shorter feeding time. Both species push aside their own kind. Species diversity in the system is low. (Based on data in Connell, 1961.)

squeezed out by aggressively growing *Balanus*. *Chthamalus* apparently survives in the higher parts of the intertidal as a fugitive species adapted to long exposure to the air and short underwater time for feeding.

Introduction of a serious predator into this intertidal system might be expected to reduce the extreme pressures of competition, leading to a different distribution and abundance of the sedentary animals. Paine (1969) investigated these possibilities in the more complex intertidal community of the Pacific coast of Washington State. Figure 26.3 shows a component of the food webs on these shores. The top carnivore, *Pisaster,* is a starfish able to prey on all the filter feeders and herbivores that are the base of food chains in this subweb, as well as on the only other important predator, the snail *Thais*. The bivalves, acorn barnacles, and *Mitella* are anchored filter feeders. The limpets and chitons are grazing herbivores, but they too require space for anchoring themselves when passing the hours during which they are exposed to the air at low tide.

Paine removed all the *Pisaster* from a section of shore 8 m long and 2 m in vertical extent, and kept *Pisaster* away for six years through frequent visits. An adjacent area was kept as a control and both were surveyed at frequent intervals. The appearance of the control plot did not change, but there were progressive changes in the plot free from *Pisaster* predation. Within a few months a barnacle population of *Balanus glandula* grew to take up most of the space, after which the bivalves *Mytilus* and *Mitella* proceeded to displace the barnacles in their turn. After several years it is clear that the rocks are settling down to being mostly the home of a long-lived population of *Mytilus*.

The part of the rocks covered by animals also has grown at the expense of seaweeds that are being scraped off or denied space in which to grow. Perhaps even more revealing is that animals that can move, like the larger chitons, have left the area.

What was originally a 15 species system has been reduced to an 8 species system, merely by the removal of the top predator, and the changes were still proceeding when Paine reported his findings. The total effect of the experimental removal is undoubtedly even more widespread than this account allows because seaweeds, and their attendant epifauna, also have been lost. Effec-

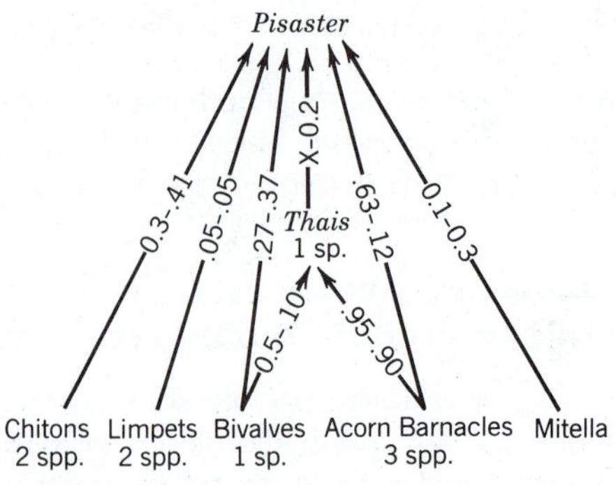

Figure 26.3 *Pisaster*-dominated subweb in Washington intertidal.
When *Pisaster* is removed from a section of reef, species diversity of the prey species falls. Numbers are calories (x = <.01). (From Paine, 1966.)

tive predation, therefore, is a prime cause of the high diversity on Washington State shores.

The Cropping Principle in Lakes

Planktivorous fish that hunt by sight can virtually remove the larger zooplankton from a lake. In the face of this predation, smaller species of plankton that are less easily seen by fish predominate. Brooks and Dodson (1965) originally suggested that the surviving smaller plankters must then suffer strong competition, though it is now believed that these small planktonic animals in fact experience heavy predation from invertebrate predators that are not limited in their hunting by the visual acuity of a vertebrate eye. What results in aquatic ecosystems is a community structure strongly dependent on predation, as is reviewed in Chapter 21.

The diversity of any trophic level in a lake might be expected, therefore, to depend on the predators at the top of the food chains. With no fish present, a few species of the larger grazing plankton ought to keep diversity low through competition. Once fish are introduced, the sway of the larger grazing plankters is ended and an increased diversity of both smaller plankters and their invertebrate predators like insect larvae should appear.

Although many of the world's lakes do have fish in them, the resulting diversity of plankton still is surprisingly modest, the number of coexisting herbivorous plankton in the open water being small. In small tropical lakes the zooplankton species list may be particularly small, perhaps one copepod, one cladoceran, and five or six rotifers. The diversity of predators may actually be higher than this diversity of prey so that other constraints must be present limiting the diversity of herbivores.

In the higher trophic levels in lakes sometimes predators can reduce diversity, instead of increasing it. As was discussed in Chapter 21, the introduction of the exotic predatory fish *Cichla* into Lake Gatun in Panama drastically reduced diversity of fishes. This was because *Cichla* was so efficient as a predator that it eliminated whole populations of native fish (Table 26.3).

Whether the addition of a top predator to an ecosystem reduces or increases diversity in the prey trophic level must be a function of predatory efficiency. The starfish *Pisaster* worked to increase diversity on Pacific reefs because its hunting reduced populations of prey species without eliminating any, but *Cichla* overdid it, as it were, thus reducing diversity. This effect of *Cichla,* however, may not be permanent. Natural selection should now be working in Lake Gatun to preserve individual prey fish equipped to escape. New species are likely to evolve, eventually casting *Cichla* into the role of diversity-inducing cropper after all. This process probably has been important in producing the swarms of species of endemic fish in ancient tropical lakes like those of the African rift.

The Cropping Principle Applied to Plants

The experience of farmers yields many records of how the intensity of cropping affects the composition of a pasture (Duffey *et al.,* 1974; Harper, 1977; Spedding, 1971). Harper (1969) used records of Welsh agricultural stations to show some consequences of sheep grazing. Sheep, like all large grazing animals, have their preferences about what they like to eat. They do not go across a field like a mowing machine, but hunt out their preferred food plants. If good pasture of palatable grasses is overgrazed, sheep make bare spots that are invaded by "weeds" unpalatable to sheep. Total plant diversity thus is increased by overgrazing.

Harper (1969) also notes that an incursion of sheep onto an old mixed pasture can have the opposite effect of reducing diversity. This will happen if the plants palatable to sheep are few and scattered, when the sheep will hunt them out and remove them from the community. Loosing sheep onto an old pasture in this way, therefore, works a little like putting *Cichla* into Lake Gatun.

A long series of experiments by Jones (1933), though not aimed at investigating diversity directly, give some quantitative data about the effects of sheep grazing. Jones took a grass sward

Table 26.3
Effect of Introduction of Predatory *Cichla ocellaris* on Fish Catches in Lake Gatun
(From Zaret and Paine, 1973.)

Family	Species	Without *Cichla* Number	With *Cichla* Number
Atherinidae	*Melaniris chagresi*	200	0
Characinidae	*Astyanax ruberrimus*	160	0
	Compsura gorgonae	120	0
	Hoplias microlepis	0	1
	Hyphessobrycon panamensis	2	0
	Pseudocheirodon affinis	7	0
	Roeboides guatemalensis	195	21
Cichlidae	*Aequidens coeruleopunctatus*	10	0
	Cichla ocellaris	0	14
	Cichlasoma maculicauda	7	36
	Neetroplus panamensis	4	0
Eleotridae	*Eleotris pisonis*	4	99
	Gobiomorus dormitor	42	10
Poeciliidae	*Gambusia nicaraguagensis*	22	0
	Poecilia mexicana	17	2
Other (25)		0	1

mostly made up of "weed" grasses of many species, divided it into plots, then experimented with different sheep-grazing regimes and fertilizer treatments. The results (Figure 26.4) show that both equitability and H' were vitally dependent on the grazing regime (Harper, 1977). Considerable changes in total species richness certainly must be hidden in those bar diagrams as well.

Also from Britain come useful data on the effects of rabbit grazing. Rabbits (*Lepus cuniculus*) are exotic introductions to Britain. They may have been there since the Norman invasion but only became abundant in the 18th century (Sheail, 1971). In a classic experiment, Tansley and Adamson (1925) fenced off some plots on old pastures on chalk in southern England, and maintained the rabbit-proof fences for six years. Their results are illustrated in Figure 26.5. The rich diversity of the chalk pasture changed into a tall stand of one grass, *Zerna* (*Bromus*) *erecta*, with a few scattered other plants. In this instance the rabbits seem to be acting as the *cropping principle* suggests. When their grazing pressure is released the available space goes to the best competitor. This tiny plot experiment of Tansley has since been "performed" over the whole of Britain when the disease myxamatosis almost eliminated the rabbits from the entire island for a number of years. The growing up of the tall grass stands became a common sight (Harper, 1977).

Like sheep, however, the effects of rabbits depends on their food preferences. The removal of rabbits from a small island off Wales by myxamatosis resulted in the discovery of 30 plant species not reported before in some 300 years of observations. Presumably invasions of the island by seeds of these plants from the mainland had previously been suppressed by the rabbits (Lancey, in Harper, 1969).

The effects of prolonged cropping were demonstrated elegantly when it was shown that pastures along the river Thames in England had been managed continuously in the same way for 900

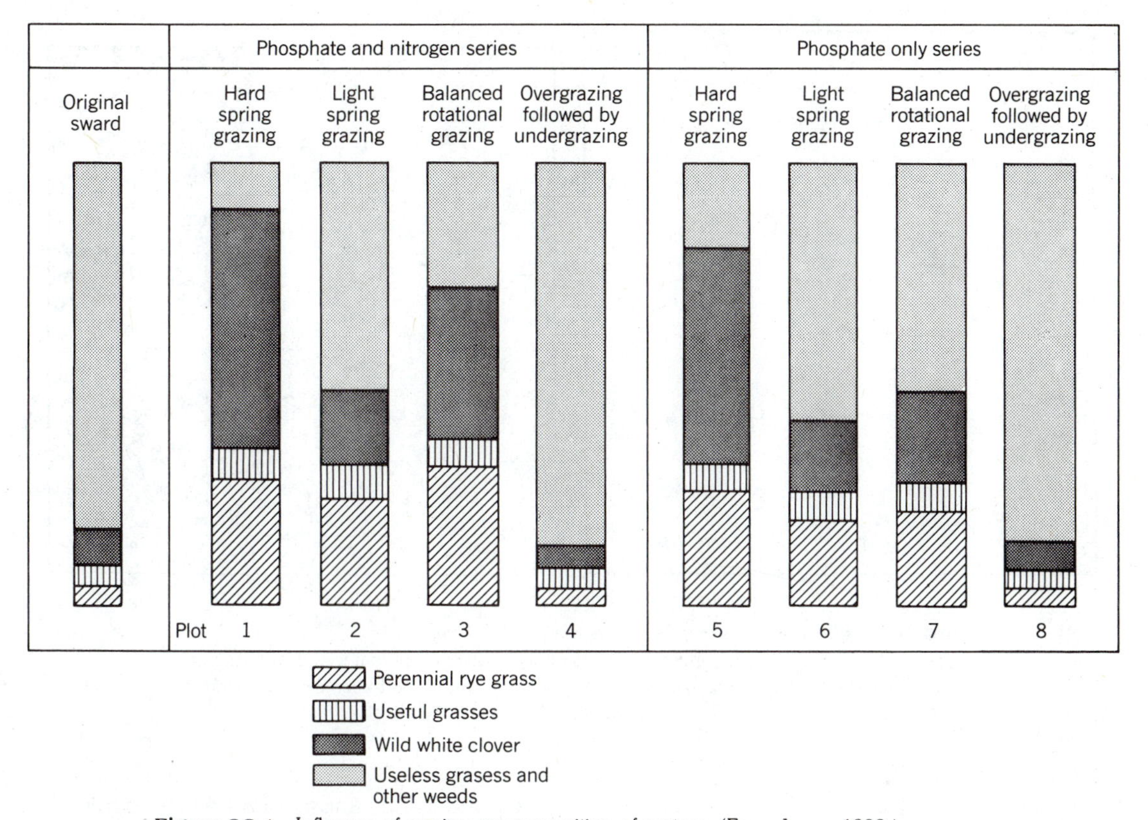

Figure 26.4 Influence of grazing on composition of pasture. (From Jones, 1933.)

years (Baker, 1937). One set of meadows had always been grazed by varying herds of cattle, horses, and geese, the only break being during three years of the British civil war in the 1600s. Adjacent meadows have been allowed to grow into hay to be cut in the summer, after which the stubble was grazed, also for 900 continuous years.

The haymakers, and the grazing livestock, loosed on these neighboring Thames meadows act as predators with different hunting techniques. Haymakers prey on tall plants, but they do so only after the plants are full grown, often after they have set seed. A consequence is that the tall grasses establish a species equilibrium and can eliminate short plants and rosette plants by competitive exclusion. Haymaking leads to a pasture of tall species (Table 26.4).

In the other Thames meadows, the grazing cattle, horses, and geese prevent tall plants from growing high, selectively eat them, and eliminate them from the community. Under this grazing pressure there is selective advantage in being a short rosette plant and it is this life form that is common in the grazed meadows (Table 26.4).

The species lists in the adjacent pastures are different after these 900 years of different predation. Of the 95 species of plants existing on both meadows in 1937 only 30 were common to pastures under both managements.

Even greater diversities should result if more kinds of grazing animals increased specialized grazing pressures. Pasture in the Middle East that is badly overgrazed by cattle, donkeys, horses, goats, sheep, and camels combined has a spe-

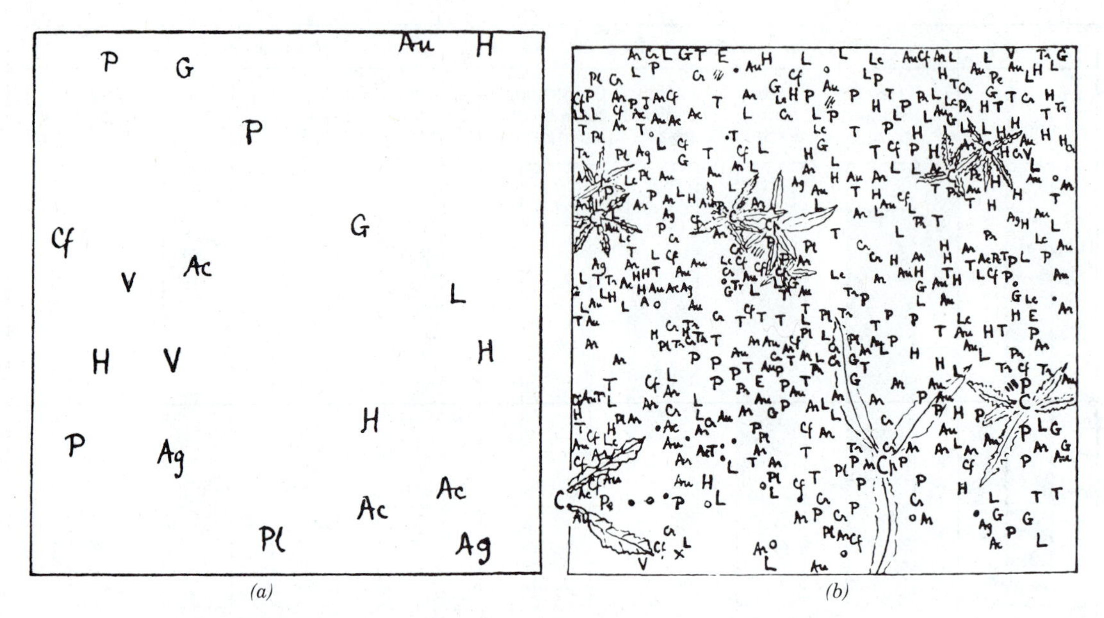

Figure 26.5 Effect of rabbit grazing on English pasture.
The two diagrams record all the principal plants in two plots, except that the plain background in plot *(a)* was a pure stand of the tall grass *Zerna erecta*. Rabbits had been excluded from plot *(a)* but had access to plot *(b)*. (From Tansley and Adamson, 1925.)

cies richness of more than a hundred herbs, although productivity is extremely low (R. H. Whittaker, personal communication).

Forest trees cannot so easily be destroyed by grazing animals when they are adults, but they are vulnerable to predation as seeds or seedlings. Janzen (1970, 1971) suggested that the area under the canopy of a tree is a particularly dangerous place for that tree's offspring because seed and seedling predators would congregate there, the more mobile of them like rodents actually using the parent tree as a "flag" to guide them to the food supply. If strong seed predation near a parent tree largely prevented regenera-

Table 26.4
Plant Diversity in Ancient English Meadows
The management of these meadows essentially has not changed for 900 years, suggesting that an interactive species equilibrium, S, is achieved. Diversity and life form depend on whether a meadow has been grazed or cut for hay. (Data of Baker, 1937.)

	S	Predominant Life Form
Total species pool	95	
Restricted to hay meadows	39	Tall
Restricted to grazed meadows	26	Short rosette
Common to both meadows	30	—

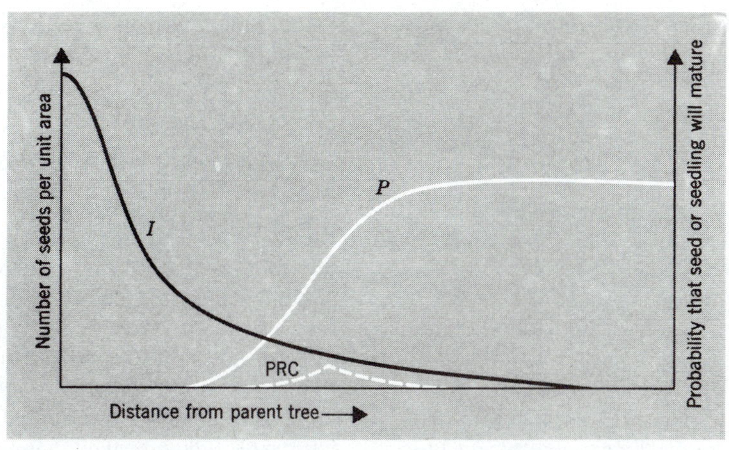

Figure 26.6 The Janzen hypothesis that tropical trees are dispersed as a result of predation on seeds and seedlings.

tion, the eventual effect would be invasion of the parental space by other species. All species under intense herbivore attack would live spaced out and diversity would be high as a consequence of cropping (Figure 26.6).

The extent to which predation on plants actually works to increase diversity depends of course on both the efficiency of the herbivore and the defenses of the plants. These will vary, both as natural selection gives temporary advantages to either attack or defense and with local environmental circumstances. Of prime importance to the working of the cropping principle is that the predators are able to maintain populations high enough to keep down the populations of their prey. The cropping principle, therefore, may be most important in places of stable climate where predators may exist at a population equilibrium. The principle may be least evident in highly fluctuating or marginal environments where the predators are *r*-strategists, opportunists, or themselves fugitives, because they are then least likely to exist at species equilibria and so to reduce the populations of their prey. Some part of the higher diversity of equatorial latitudes may be due to the minimal amount of seasonal change with consequent continuous activity of those predator populations that can keep prey numbers low.

DIVERSITY CONSEQUENCES OF STRESSFUL ENVIRONMENTS

Hot springs, salt flats, mountain tops, and the like are *rigorous* or *stressful* habitats with few species (Table 26.5A). They do, however, support some species, inviting the question "Why not more?" The answer we give is partly an island biogeographical answer, because these habitats are rare, scattered, and sometimes

Table 26.5
Diversity in "Rigorous" Habitats

Low diversity, low population densities (A)
Hot springs
Salt flats
Caves
Mountain tops
Extreme sandy deserts
Low diversity, high population densities (B)
Brackish estuaries
Abandoned fields
Polluted lakes
High diversity, variable population density (C)
Moderate deserts
Serpentine soils (with toxic ions)
Overgrazed pastures in arid regions

ephemeral. Migration to one of these places from another of its kind is difficult and slow, necessarily making the slope of the species–area curve low. At the same time, small size of the habitats (or extreme unproductivity) make population density low, so that extinction rates are high.

The island properties of these habitats are compounded by their rarity and their short existence. Hot springs are widely separated, and they cool in a few thousand years. This rarity in time and space allows little time for evolution even as it makes high extinction rates certain. A really unusual "island" like a hot spring, therefore, can only be invaded by a small portion of the *r*-strategists theoretically capable of reaching a more common habitat type of the same size.

A tentative definition of ENVIRONMENTAL RIGOR may be ENVIRONMENTAL CONDITIONS THAT ARE RARE AND WIDELY DISPERSED, OR SHORT-LIVED, OR SUBJECT TO FREQUENT CHANGES OF WIDE AMPLITUDE. When an ecosystem exists under conditions of environmental rigor so defined, low diversity is to be expected. The rigor acts either on the immigration rate, or the extinction rate, or both to produce an equilibrium species number that is low.

More familiar patches of low diversity are brackish estuaries, polluted lakes, and abandoned fields (Table 26.5B). Like hot springs, these places are islands of unusualness in a sea of normality. All three are stressed by frequent changes of state, as in a tidal estuary, or by a short existence, as forest closes over an old field or pollutants are buried in lake mud. Extinction rates, therefore, should be high and the equilibrium species number low. But these island biogeography properties of these habitats actually may be less important than the high productivity that allows some species populations to secure a competitive advantage. In a polluted lake, for instance, high nutrient loading lets a few species of algae attain very high population densities. Diversity is low in the sense that nearly all individuals belong to a few species, but it is likely that a very intensive sampling effort will discover rare individuals of many other species. Thus, when a habitat is both unusual and extremely productive, competition can keep diversity below what would attain in a place of more normal productivity.

High Diversity from Environmental Restraint of Competition

Broad regions of the earth have high diversity but to the human mind seem to be rigorous enough (Table 26.5C). The Sonoran desert, toxic soils frustrating to agriculture, and ruined pastures of arid lands all contain many species despite stress by drought or chemistry.

These habitats are *marginal* in the sense that they rank between well-watered *mesic* habitats and barren lands such as complete deserts. Typically they are supplied with rain only intermittently, or are otherwise strongly seasonal. Their high diversity can be explained because random environmental hostility serves to reduce populations rather as a grazing animal does for a pasture, giving comparable opportunities for other plants to invade.

Drought in a moderate desert like the Sonoran prevents dense cover by a few species of plants, rather as a grazing animal prevents dense occupancy of a reef by a few strong competitors. Space is not always pre-empted so that invasion by the establishment of a seedling following rain is always possible. These habitats, therefore, are easy to invade in the sense that there is little resistance by the inhabitants. Since the habitats tend to be large, although bordered by more mesic habitats from which colonists might come, moderate deserts and the like always should have a large richness of opportunists or *r*-strategy species.

In addition to the opportunists, these marginal habitats have their own biota of long-lived equilibrium species. Desert bushes and small trees live spaced out, probably reflecting competition for water and suggesting a density-dependent population equilibrium (Figure 26.7). This arrangement leaves an open canopy and bare ground that can be invaded by weeds able to complete their life cycles in the short weeks of a chance favorable season. These marginal desert communities, therefore, add a resident and unique

Figure 26.7 *Carnegiea gigantea:* spaced out in a desert.
The saguaros in Arizona are more evenly spaced than random (over-dispersed), demonstrating intraspecific interaction, probably competition for water. Theirs is an equilibrium strategy for the desert, but the bare ground between can be used by opportunists when it rains.

flora of equilibrum plants to their large number of opportunists. They win high diversity in both ways and their consequent spectacular species richness is a result. The key to high diversity in marginal habitats is that ENVIRONMENTAL HOSTILITY ACTS TO REDUCE POPULATIONS. The environment is acting as the *cropper* to "overgraze" populations.

High diversity of plants on bizarre soils like those on serpentine rocks can be explained by an analogous argument. Only generalist weeds have sufficiently broad tolerances to live there. These suffer high local extinction but they are always resupplied from surrounding habitats. In short, the unusual chemistry lets opportunists live with minimal competition (*chemistry is the cropper*) and the process of island biogeography works to set a high species equilibrium.

It is possible tentatively to conclude, therefore, that diversities in many stressed habitats are set as species equilibria, generally as variants of the theme of island biogeography:

1. The most rigorous places have properties directly comparable to true islands.

2. In some systems croppers work to prevent closed communities of equilibrium plants from regulating immigrations and extinction.

3. In marginal habitats weather, seasons, or other physical factors prevent closed communities of equilibrium plants and allows a high equilibrum number of opportunists to coexist with *K*-strategists.

DIVERSITY PATTERNS WITH INCREASING AREA

Small, uniform habitats must be shared, but relatively large patches of land may be partitioned between species. MacArthur (1965) introduced convenient terminology to describe these two conditions: WITHIN-HABITAT DIVERSITY and BETWEEN-HABITAT DIVERSITY. Earlier Whittaker (1960) had pointed out the same essential dichotomy, calling the two conditions ALPHA DIVERSITY and BETA DIVERSITY. When islands are colonized by animals with considerable resource plasticity like some colonizing birds an initial within-habitat diversity (α) results. When crowding leads to an *interactive equilibrium* some species may be confined to habitats most favorable to them, resulting in *between-habitat diversity* (β).

This formulation caters to the well-known phenomenon that sampling larger and larger areas yields a gradually increasing species list. This was clear in the early work of phytosociologists who found that species–area curves were at first steep (increasing within-habitat diversity) but then climbed more gradually (increasing between-habitat diversity) (Chapter 15). It can be argued, therefore, that the *sociation* of the Uppsala school was a community chosen to represent as pure a sample of within-habitat diversity as possible.

Cody (1975) has extended the classification for bird species diversity to allow comparison between geographic regions. POINT DIVERSITY describes the complete overlap of bird ranges over very small areas. GAMMA DIVERSITY describes the species replacements that occur over very large geographic regions. Adding these two concepts results in the following scheme:

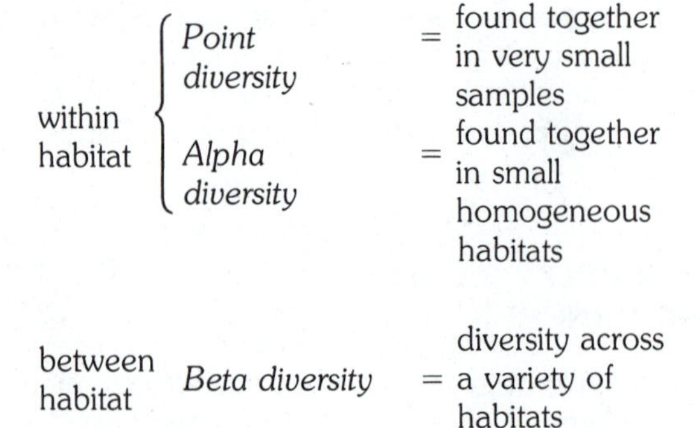

within habitat	*Point diversity*	=	found together in very small samples
	Alpha diversity	=	found together in small homogeneous habitats
between habitat	*Beta diversity*	=	diversity across a variety of habitats

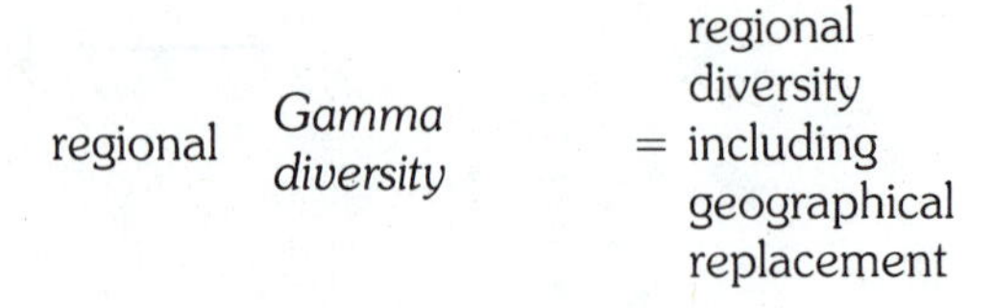

regional	*Gamma diversity*	=	regional diversity including geographical replacement

Figure 26.8 shows Cody's application of this concept to a comparison of bird faunas of California, Chile, and South Africa. The three curves are species–area curves on double logarithmic plots of the kind used in island biogeography. The mean slopes of these curves are about $z = 0.13$, which is within the range usual for continental data, and are much less steep than island curves because of the ease of immigration into parts of continents. But the slopes differ within the three regions over different increases in area.

Diversity on the three continents was most similar when measured for areas of about 10 square miles, suggesting that within-habitat (α) diversity is similar within these three southern continents. *Point diversity* (left of the curves) varies, as is confirmed by field data that show that Chilean birds have more interspecific interactions and less tolerance of overlapping territories than the others.

The three curves diverge sharply for larger areas, showing that both between-habitat (β) and regional (γ) diversities differ from continent to

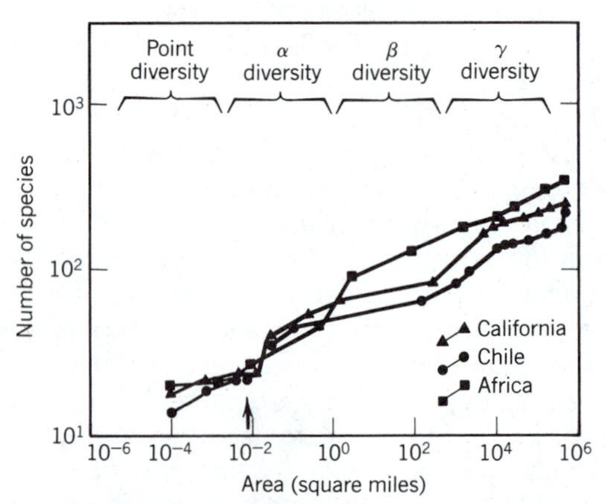

Figure 26.8 Species–area curves on three continents. Data are numbers of species of birds of broad-leaved sclerophyllous vegetation (chaparral) in California, Chile, and South Africa. (From Cody, 1975.)

continent. These may reflect different topographic diversity but may also be due to evolutionary history. In all three continents, increasing topographic diversity or climatic change increases the number of species in a collection.

THE GRAND CLINES OF DIVERSITY

Decline of diversity with altitude on a mountain side is readily demonstrated. Figure 26.9 gives data for one cline in plant diversity up high mountains in Arizona, starting from desert at about 700 m. This is near the classic site at which Merriam observed his life zone sequence (Chapter 14) and the transect crosses most of those life zones. The data show an overall decline of diversity with altitude but with a large peak at middle elevations coinciding with the rich species community of the Sonoran desert. Figure 26.10 shows a longer cline in plant diversity encountered in an ascent of the Himalayas.

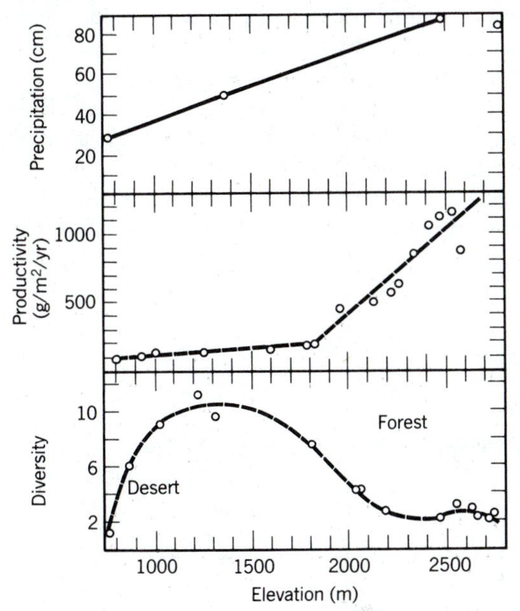

Figure 26.9 Decline of plant diversity with altitude on Arizona mountains.
Diversity is expressed as an index that relates species richness to relative productivity as well as relative abundance, but a plot of simple species richness would be of similar shape. (From Whittaker, 1977.)

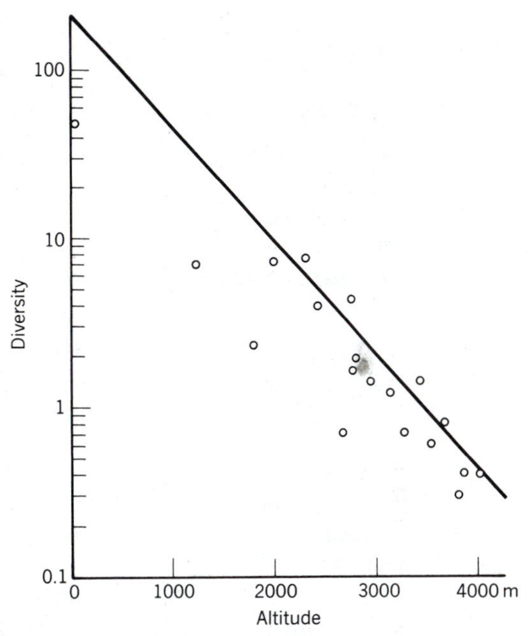

Figure 26.10 Decline of plant diversity with altitude in the Himalayas.
Diversity is expressed as an index allowing for relative size of trees based on measurements of breast height diameter (dbh). (From Yoda, 1967.)

The general proposition of a latitudinal decline of diversity from the high arctic to the equator is undoubted, and it is easy to marshall data in its support. In the state of Ohio, for instance, about 2000 species of vascular plants are known, and in the country of Ecuador about 20,000 are believed to occur.

With many animal groups, the latitudinal clines also appear self-evident. The large array of frogs and toads in the tropical rain forest can be contrasted with a fraction of that number in the U.S. Midwest, three species on the island of Britain, and none in the arctic. All the primates are tropical, reptile diversity seems to be correlated with latitude, and all naturalists suppose that the place to find lots of insects is the tropics.

But these very general statements mask local differences and many exceptions. Table 26.6 shows the result of a literature search by Price (1975) comparing species richness of beetles, ants, and dragonflies from very large areas of north

Table 26.6
Diversity of Beetles, Ants, and Dragonflies Arranged by Latitude
Data of various authors collected by Price (1975). Compare with Table 26.7.

Beetles		Ants		Dragonflies	
Labrador	169	Alaska	7	Nearctic	59
Massachusetts	2000	Iowa	73	Neotropical	135
Florida	4000	Trinidad	134		

and south latitude, which show decided, though not overwhelming, increases in the south. But Table 26.7 shows the results of actual intensive insect collecting at different latitudes, from single gardens in Sweden, Britain, Sierre Leone, and Uganda. Many more species of ichneumonid were caught in the temperate gardens, particularly in Norway, than in Africa. This is the supposed species cline reversed for ichneumonids (Rathcke and Price, 1976; Hespenheide, 1979).

Only a little ingenuity is needed to find other groups of animals that actually are more diverse at high latitudes: bears, seagulls, shorebirds, seals, and perhaps some rodent groups. These can be offset against the high numbers of frogs, bats, and monkeys of the tropics to some extent. There is also immense variation from one tropical region to another, particularly in individual groups. For instance, there are 50 to 100 species of palm tree in tropical Africa but 1100 or more in tropical South America; yet the New World island of Jamaica has as many species of orchid as the whole of Africa (Richards, 1973).

The other grand cline of diversity is in the deep sea where data suggest a cline from low diversity in shallow water to high diversity in deep water. Figure 26.11 shows data for the benthic fauna that can be sampled with a grabbing device lowered from a ship to the bottom mud. More species can be collected in this way from the deep sea floor than from continental slopes, shelves, or estuaries. Diversity is low in the well-lighted, warm, productive bottoms of shallow water; diversity is high in the dark, cold, unproductive bottoms of the deep sea (Sanders, 1968, 1969; Sanders and Hessler, 1969).

Despite numerous exceptions these three general reductions of diversity from south to north, up mountains, and down into the sea seem to be real. A number of general mechanisms to explain these ubiquitous clines have been proposed.

Table 26.7
Diversity of Ichneumonids Collected in Gardens in Temperate and Tropical Latitudes
(From Owen and Owen, 1974.)

	Species Richness	H'	Percentage Commonest Species
Temperate gardens			
Norway	758	5.5	5.5
Britain	326	4.9	3.2
Tropical gardens			
Sierra Leone	319	4.9	4.9
Uganda	293	4.5	10.1

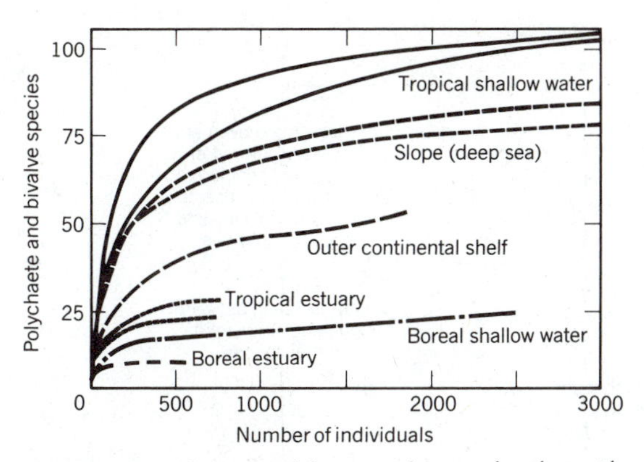

Figure 26.11 Increase of diversity of marine benthos with depth.
The deep sea mud of the abyss has more species of polychaete and bivalve than any other ocean bottom except tropical shallows. (From Sanders and Helser, 1969.)

The Productivity Hypothesis

High productivity should mean a larger total population of individuals, and hence the chance to divide the available energy between more species populations. A cline of productivity runs from equator to pole, and from lowland to mountain top (Figure 26.12). The possibilities of this cline of productivity are obvious. More energy for plants can mean dense local populations of the kind required to promote between-habitat diversity. Diverse plants and complex plants both provide fresh niches for animals, thus increasing animal diversity, and the effects are compounded with each trophic level. Connell and Orias (1964) noted that these effects would be amplified when, as at the equator, seasons are minimal because animals in a stable climate need not spend so much energy on systems to maintain their internal regulation, thus freeing more energy still for making populations and species (Figure 26.13).

This is an attractive scheme. Sharing the available energy is the logical process to set an upper limit to diversity. Perhaps for those organisms for which interspecific competition is an important regulator the argument based on productivity has merit—the greater number of bird species of the wet tropics, for instance, may reflect both the increased structure of large plants (see below) and increased, year-round productivity. At the low-diversity end of the clines, both on mountains and in the high arctic, low production or high seasonality probably do promote high extinction rates, thus lowering diversity.

But for most organisms in the more productive portions of the cline the effects of productivity do not seem to be major. Populations tend to be limited to levels well below those possible from energy considerations alone, as we have seen, so that large portions of the total energy go to decomposer chains. More damaging are the obvious exceptions to the correlation between diversity and productivity. Under high productivity, as in estuaries and polluted lakes, intense competition by dense populations can keep diversity very low, suggesting that diversity ought to be low in the tropical rain forest if productivity were the only controlling factor. In the cline of diversity on the sea floor it is the unproductive dark and abyssal depths that have the

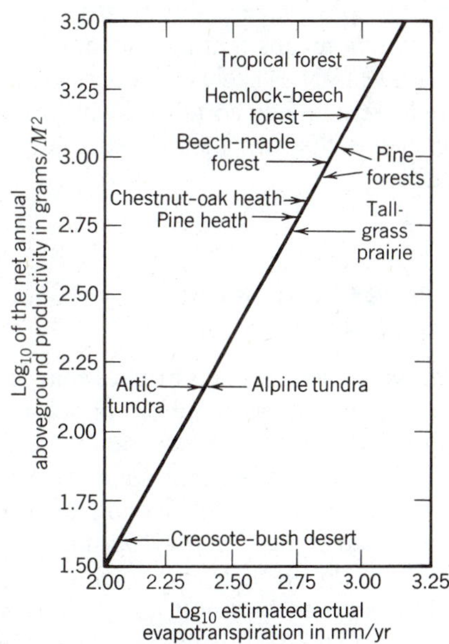

Figure 26.12 Indirect correlation of primary production with temperature.
Evapotranspiration depends on moisture and temperature. (Diagram of M. Rosenzweig from MacArthur and Connell, 1967.)

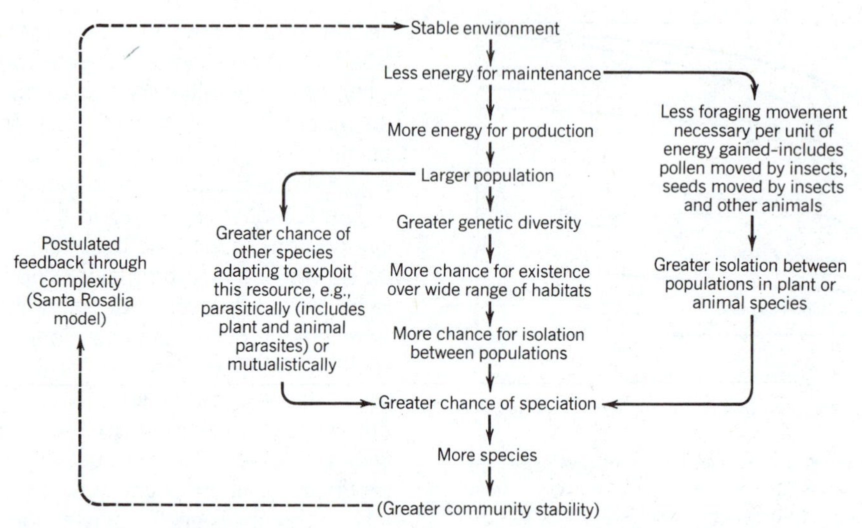

Figure 26.13 Postulated way in which diversity could be sponsored by productivity. The scheme includes a feedback from species diversity to stability, which has now been shown to be untrue. (Modified from Connell and Orias, 1964, and Price, 1975.)

high diversity, whereas the shallow productive waters have low diversity.

It cannot be said, therefore, that the correlation between productivity and diversity along a cline of latitude means that the two are causally connected. At least as likely is the possibility that both are independently correlated with seasonal climatic events that are themselves dependent on latitude.

Hypotheses of Structure and Species Saturation

Complex vegetation, as in a tropical rain forest, suggests spatial multiplication for animal niches. MacArthur (1964) and MacArthur and MacArthur (1961) compared bird species diversity with a measure of relative layering of forests, which he called *foliage height diversity*. For the biomes of the United States he obtained a good correlation between bird and foliage height diversity (Figure 26.14). A similar study in grassy, shrubby, and woody areas of southern Chile, Africa, and California by Cody (1975) found similar results (Figure 26.15). The study areas in the three continents were all within the broad-leaved sclerophyllous biome and thus should be comparable. Within-habitat diversity increased sharply along a gradient of increasing structure in each region, the most disparate data being from the South African woodlands known to have a notably impoverished bird fauna.

These data are consistent with other data suggesting that bird species diversity can be strongly influenced by interspecific competition as postulated in the Santa Rosalia model. It is intuitively reasonable that variations on a theme of flying forager are limited, suggesting that a simple release from competition provided by extra forest structure should result in more species. Even so some variations on the bird theme in the tropics clearly seem to escape structural limits. Birds like toucans and hummingbirds of the New World, or hornbills and sunbirds of the Old World, suggest ways of life in the tropics that are not available elsewhere. Lovejoy (1975) calculated that seven structural layers would be needed in a tropical rain forest, as opposed to the three or four layers elsewhere, to explain all the extra rain forest bird diversity. Seven layers of foliage for most rain forests are unlikely, but the extra birds of Lovejoy's calculations find extra niches within the smaller structure by specialized feeding that

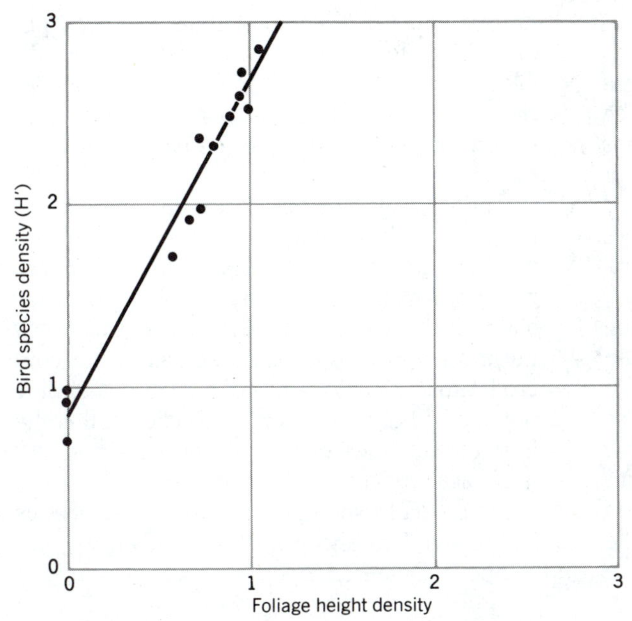

Figure 26.14 Bird diversity and foliage height diversity.
This is MacArthur's (1965) original figure. Bird diversity is plotted as the Shannon–Wiener index, H'. The graph shows so close a correlation between bird species and plant structure as to suggest that these habitats are saturated with birds. The "fit," however, is much less good if species richness is plotted instead of H'.

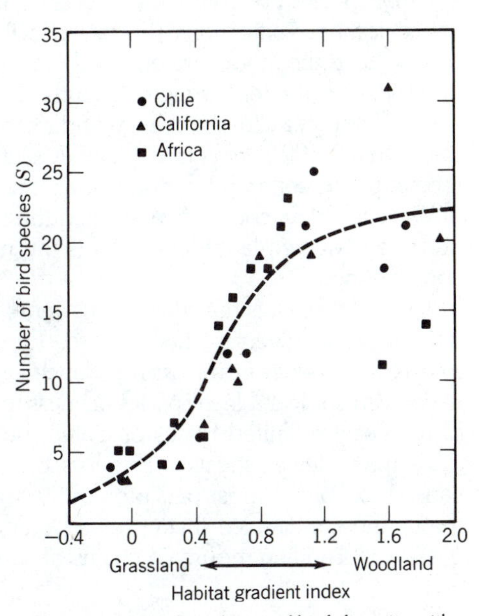

Figure 26.15 Correlation of bird diversity with habitat structure in "chaparral."
The habitat gradient index is a function of vegetation height and density. The curve is a fitted logistic curve. (From Cody, 1975.)

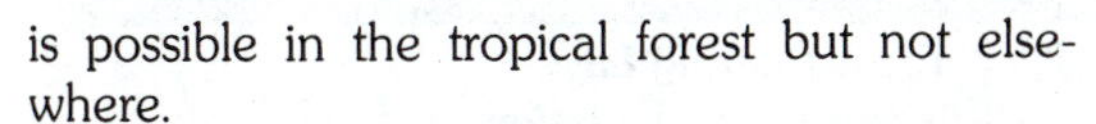
is possible in the tropical forest but not elsewhere.

It is now hard to suggest that taxa other than birds are close to saturation in most environments. Figure 26.16 shows data comparing lizards of deserts in three continents with birds of the same places. Since the relationships between species numbers in these groups are different in each continent it is certain that both groups cannot be at a saturation number. For most organisms it seems that "How many can there possibly be?" is not the right question. Usually there can be more. A better question is "Why has speciation, extinction, and history resulted in the present number?"

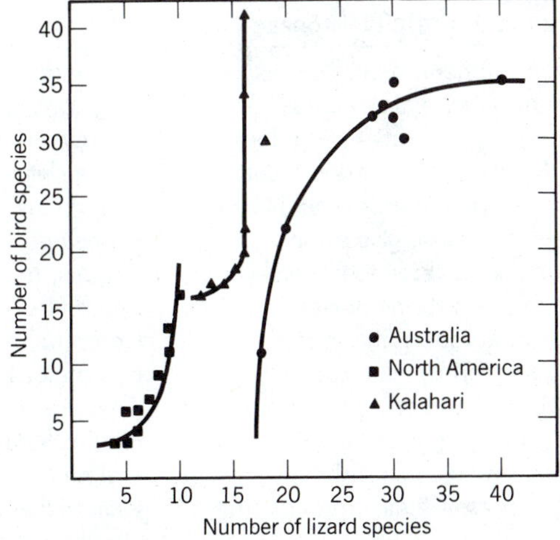

Figure 26.16 Comparison of bird and lizard diversity on deserts of three continents.
Clearly neither diversity can be predicted from the other, showing that both cannot be saturated or be at a number set by the physical environment. (From Pianka, 1971.)

The Rejuvenating Catastrophe Hypothesis

When biogeography was in its infancy, Alfred Russel Wallace (1878) put forward the hypothesis that species poverty of the high latitudes followed ice age extinctions (Fisher, 1960). Glaciers scraped the more northerly land masses bare of all species; the temperate belt south of the ice suffered massive climatic change and extinctions also. Only 10,000 years have been available for recovery. No comparable disaster, as it seemed in Wallace's day, could have afflicted the tropics and the observed latitudinal cline is the natural consequence.

Undoubtedly this history explains some of the differences in diversity that we see. European forests are known from fossil evidence once to have been comparable in species richness to those of the eastern United States or China, but now have many fewer species. This loss of species can be explained most parsimoniously because the Fennoscandian ice sheet pressed against the east–west trending mountain chains of southern Europe and Asia, denying habitat to many species of the old arcto-tertiary forest (Chapter 22). Glacial ice, and the changing climate of the glacial retreat, is also the most parsimonious explanation for the extinction of the North American megafauna (Table 26.8).

The explanatory power of this hypothesis, however, is closely limited. One powerful objection to its generality is that low latitudes were not free from major environmental changes in ice ages, of which more below. A second is that only the most northerly latitudes actually were covered with ice, leaving most of the latitudinal cline not affected directly. A third objection is that the fossil record shows similar latitudinal clines in past geological epochs without ice ages.

Figure 26.17 shows data from the Cretaceous for foraminifera and from the Permian for brachiopods. Data from deep sea cores also suggest

Table 26.8
Extinction of Large Mammals in Pleistocene North America
The megafauna of North America were much more diverse in the recent past, suggesting that the present low diversity, particularly in arctic North America, is a result of extinctions associated with the ice age, possibly including activities of early humans. (From Hibbard et al., *1959, and Martin, 1967.)*

Extinct in Early Pleistocene	Extinct during Last (Wisconsinan) Glaciation	
Rhynchotherium, mastodons	*Mammut*, American mastodons	*Mammuthus*, mammoths
Pliauchenia, extinct camels	*Megalonyx*, ground sloths	*Mylohyus*, woodland peccaries
Borophagus, bone-eating dogs	*Tanupolama*, extinct llamas	*Euceratherium*, shrub-oxen
Ischyrosmilus, saber-tooth cat	*Cuvieronius*, extinct mastodons	*Preptoceras*, shrub-oxen
Chasmaporthetes, extinct hyena	*Platygonus*, extinct peccaries	*Tetrameryx*, extinct pronghorns
Glyptotherium, glyptodons	*Camelops*, extinct camels	**Tapirus*, tapirs
Nannippus, three-toed horses	**Equus*, horses	**Tremarctos*, spectacled bears
Plesippus, zebrine horses	*Paramylodon*, ground sloths	*Bootherium*, extinct bovid
Stegomastodon, mastodons	*Capromeryx*, extinct pronghorns	*Cervalces*, extinct moose
Titanotylopus, giant camel	*Castoroides*, giant beavers	*Brachyostracon*, glyptodon
Hayoceros, extinct pronghorn	*Arctodus*, giant short-faced bears	*Boreostracon*, glyptodon
Glyptodon, glyptodons	*Nothrotherium*, small ground sloths	*Eremotherium*, giant ground sloth
Platycerabos, extinct bovid		*Neochoerus*, extinct capybara
Stockoceros, extinct pronghorns	*Chlamytherium*, giant armadillos	**Saiga*, Asian antelope
	Dinobastis, saber-tooth cat	**Bos*, yak
	Smilodon, saber-tooth cats	*Sangamona*, caribou?
	**Hydrochoerus*, capybaras	*Symbos*, woodland musk-ox

*Living in Asia and tropical America.

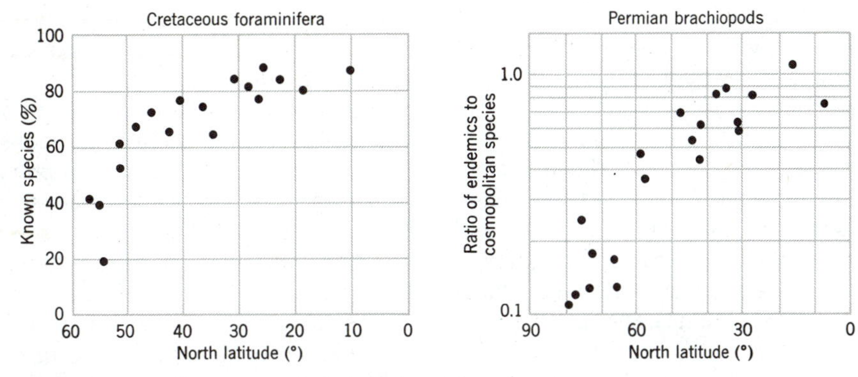

Figure 26.17 Diversity and latitude in the geological past.
(A) Planktonic foraminifera in the Northern Hemisphere. *(B)* Family diversity of Permian brachiopods. (From Stehli *et al.*, 1969).

that foraminifera diversity is a function of ocean temperature (Figure 26.18), further suggesting that the latitudinal cline in foraminifera diversity reflects temperature changing with latitude rather than catastrophic extinction in the north. Data from the land are less good but none suggest that the latitudinal cline in most terrestrial species is confined to the Pleistocene ice age. Possibly it is realistic to say that the latitudinal decline of diversity towards the poles is exaggerated during ice ages though the fundamental cause lies elsewhere.

The Time Stability Hypothesis

The TIME STABILITY HYPOTHESIS was put forward to explain the cline of diversity in deep sea benthos (Sanders, 1968, 1969). Sanders' basic data for marine benthos are given in Figure 26.11. Productive estuaries and continental shelves in most latitudes have few species of these animals, but the cold, dark, unproductive floors of the deep sea have many species. Sanders suggested that the deep sea was able to collect species because its environment was both highly predictable and stable. A deep sea floor is without seasons. It might have the same environment for the whole of the 100 million years or so between its creation in a mid-ocean ridge until its final destruction in a subduction zone (Chapter 14). In this stable environment populations

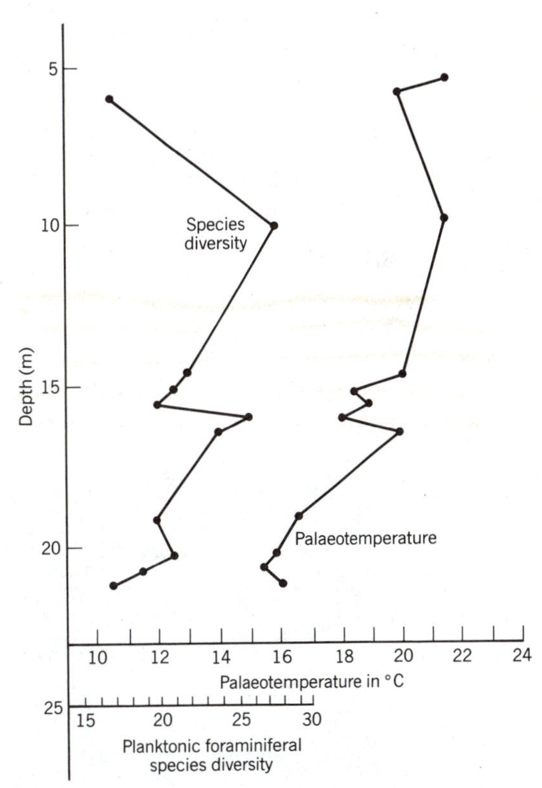

Figure 26.18 Correlation of Miocene diversity of foraminifera with water temperature.
Temperature was measured by the $\Delta^{18}O$ method. (From Jenkins and Shackleton, 1979.)

should not fluctuate, suggesting that extinction would be low. Whatever new species were produced in the deep sea would thus collect and diversity would continue to grow.

Estuaries and continental shelves, on the other hand, had frequent changes of environment, both seasonal and catastrophic. They had high extinction rates and consequently a low equilibrium number of species. Species number of any place, therefore, should be proportional to the time a habitat was in existence and its physical stability (Figure 26.19).

Other explanations for high diversity in the deep sea benthos are possible, however. The deep sea floors are of vastly greater area than the continental shelves and slopes with which the hypothesis compares them, so that any hypothesis relying on larger area can explain the data as well as this one relying on antiquity and stability. But the time stability hypothesis had wide appeal because it seemed applicable to the terrestrial cline of diversity with latitude as well. It could be said that the unpredictable, strongly seasonal, short-lived arctic environments had few species because of their youth and instability. The ancient, unchanging tropics had many species because of low extinction rates, just like the deep sea.

But data are now accumulating that show that the equatorial regions have neither the required stability nor antiquity. Most places in the wet tropics for which good data are available turn out to have seasonal climates. The seasons are not derived from temperature changes of summer and winter but on moisture changes that give wet and dry seasons.

For plants, insects, or long-lived animals the dry or the wet represent dislocations—times of changing diet, of dispersal, of preparation for breeding, of synchroneity of life-styles—just as do the seasons of the north. Tropical seasons, therefore, may carry as much chance of extinction as temperate seasons. Seasonal dangers may be less in the tropics, or they may be more; it depends probably on the type of organism involved.

The old belief that the equatorial climates were not drastically altered in the ice age is now under serious question also, both from paleoecological data and from climatic modeling (Chapter 22). The most definite data are from Africa, where long pollen records show that regions holding tropical rain forests these last few thousand years held dry woodlands or even grasslands at the time of the last glacial maximum (Livingstone, 1975; Livingstone and van der Hammen, 1978) (Figure 26.20). This is not to say that tropical rain forest climates did not exist in Africa in ice age time, because these climates certainly did exist then, but they were in different places. Most of tropical Africa experienced climatic changes from 10,000 to 20,000 years ago that were as drastic to local plant populations as those in periglacial areas farther north.

The data from equatorial South America are far less complete, though strongly suggestive of

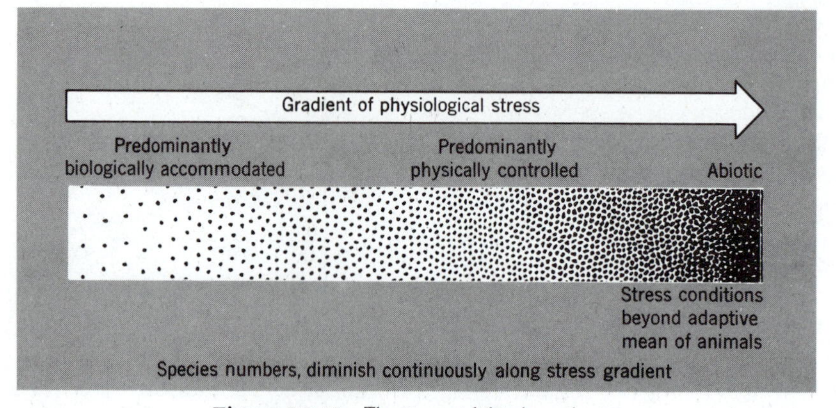

Figure 26.19 The time stability hypothesis.

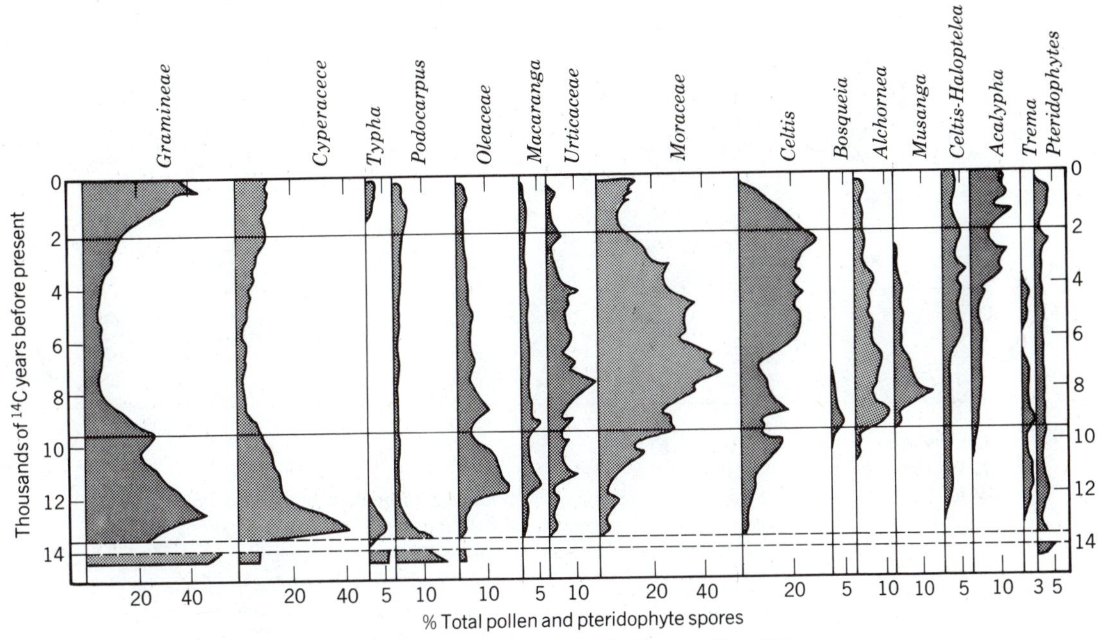

Figure 26.20 Pollen percentage diagram from Lake Victoria, East Africa. The reduction in pollen of tree taxa such as Moraceae and *Celtis* before 10,000 years ago suggests that rain forest was absent from the region in ice age time. Changes over the last few thousand years may have resulted both from human activities and from changing climate. (From Kendall, 1969.)

climatic change comparable to that of Africa. Pollen from the Andes of Colombia show wide movements of vegetation belts down the mountains in ice age time (van der Hammen and Gonzalez, 1966). An ice cap in Peru expanded. Some investigators argue that the Amazon forest was broken up into patches separated by savanna or deciduous forest (Prance, 1982).

These Amazon data are as yet very incomplete and unsatisfactory, there being, for instance, not even one radiocarbon-dated sample of glacial age from anywhere in the lowland Amazon forest obtained before 1984. Heavy reliance has been placed on modern biogeographic distributions of Amazonian butterflies and other organisms, themselves not completely known (Colinvaux, 1979; Livingstone, 1982; Figure 26.21). It seems safe to say, however, that the climate of equatorial America was significantly different in ice age times, although, as in Africa, there certainly were patches of rain forest living then as now.

Models of the global climate of the ice age, based mainly on evidence of past sea temperatures (Chapter 22), confirm that the climate of the whole earth was radically different at glacial maxima. Conditions for the detailed operation of *time stability hypothesis* on land, therefore, cannot be met.

The Latitude–Area Effect

The spherical shape of the earth requires that there be more area at low latitudes. Put simply, there is much more land in the equatorial regions than anywhere else. If a large area means a large population, and if a large population means a lowered extinction rate, then the cline with latitude may be the necessary consequence of the shape of the earth (Osman and Whitlatch, 1978).

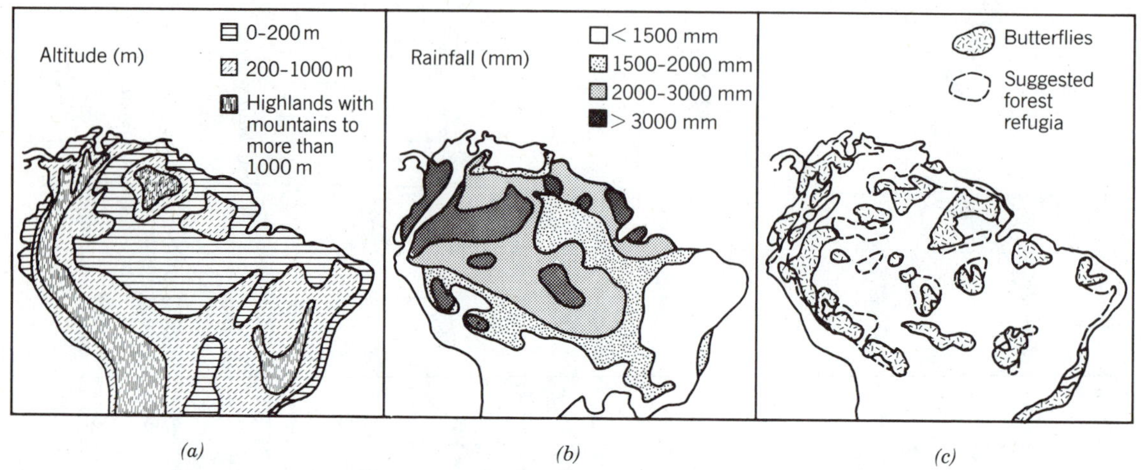

(a) *(b)* *(c)*

Figure 26.21 Topography and dispersion in the Amazon basin. *(a)* Modern topography; *(b)* modern rainfall; *(c)* distributions of modern butterfly populations (shaded) and refugia postulated from biogeographic data (dashed lines). The hypothesis that ice-age forest only existed in the indicated areas has not been tested with paleoecological data and the biogeographic data may reflect present environments. (From Colinvaux, 1979.)

This was a view put forward earlier in a different context by Darlington (1959), who marshaled evidence to show that many facts of zoogeography could be explained as a result of speciation in the tropics, followed by subsequent migration to higher latitudes (Figure 26.22). In Darlington's evolutionary view the earth is a large area of tropics, where species collect, and a peripheral temperate region, to which species migrate. This does accord with the geometry of the earth rather well.

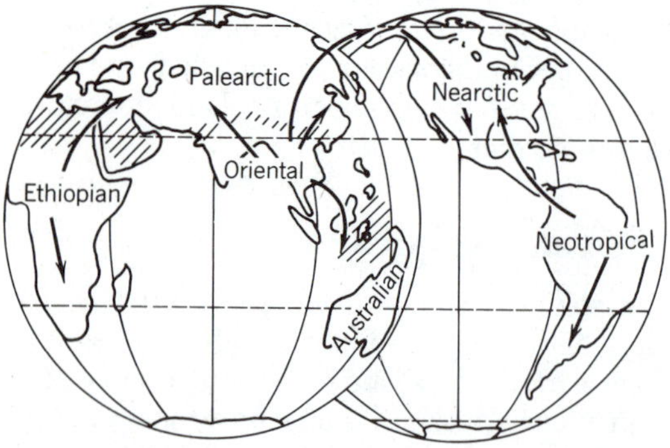

Figure 26.22 Origins of major taxa in Old World tropics. Map based on one of Darlington (1959) illustrating his hypothesis of the origin of major taxa in the Old World tropics. For zoogeographic regions see Chapter 14.

Terborgh (1973) argued that the huge expanse of equator plus tropics effectively was larger still because temperatures of surrounding regions do not change for long intervals of latitude. Equatorial regions generally have temperatures that are lower than might be expected because of heavy cloud cover. At the edges of the equatorial belt, cloudless skies make for temperatures quite as warm as those over much of the equator itself (Figure 26.23). The area occupied by equatorial species thus can be extremely large.

Large area is a property of the deep sea as well as of equatorial forest. In both, invasion of local patches by slowly dispersing species appears to offer an explanation for at least part of their remarkable species richness.

The Disturbance Hypothesis for Rain Forests

Current ecological thought sees tropical rain forests as ecosystems facing almost perpetual disturbance. This is a complete reversal of the belief held a few years ago that the wet equatorial forests were the epitome of stability. The change in

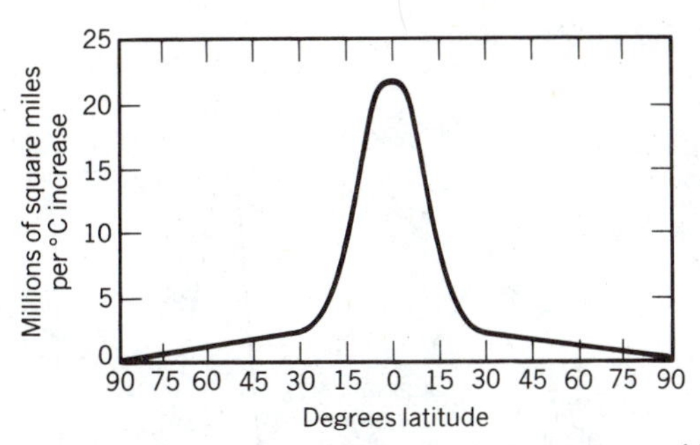

Figure 26.23 Area available at equal temperature at different latitudes. (From Terborgh, 1973.)

emphasis follows detailed studies by forest ecologists.

Tree-fall is the proximate disturbance visible in the forest: the collapse of a canopy tree, taking down with it a swath of satellites that are tied to it in a web of lianas. These patches are everywhere in the forest and the resulting gaps are the scene of a rapid succession of immigrations and extinctions. Connell (1978) showed that the places of highest tree diversity in Old World rain forests were in patches that could be traced to the extra large gaps made by shifting or swidden agriculture. Primitive farmers had amplified the gap effect, thus making its results plain. Hubbel (1979) argues from his experience of New World forests that the incidents of storm-caused gaps is sufficient to keep the whole forest in a state of perpetual change. The New World rain forests he studied appear, as a result, to be in perpetual ecological successions that never reach climax.

Physical properties of tropical rain forest ecosystems account for this instability of the trees. Warmth, water, and prolonged growing seasons make for rapid growth. Deeply weathered soils lead to tight nutrient cycles in which retrieval of nutrients from surface litter is vital, which in turn leads to shallow rooting systems. These factors combined make liana life forms conspicuous so that tall trees, shallow-rooted in wet ground, carry the load of many other plants, and these laden trees are assaulted by the frequent thunderstorms characteristic of the rain forest climate.

In the Amazon Basin, the largest of the rain forest ecosystems, the pressure of water may, in addition, leave few sites free of radical hydraulic disturbance for more than several centuries (Colinvaux *et al.*, 1985). The land is awash, and the glint of water under the trees can be seen constantly from a low-flying aircraft. The Amazon rain forest may be the most species-rich ecosystem on earth with estimates of plant diversity ranging to about 80,000 species compared to, for instance, only about 2000 species for all the varied habitat types of the state of Ohio. This huge diversity may most readily be explained because the Amazon Basin is a constantly perturbed system of very large size. Wind and weather promote diversity, as predators promote it for benthic communities of shallow seas or drought does for the Sonoran desert.

These observations may be generalized as the INTERMEDIATE DISTURBANCE HYPOTHESIS. Modest physical disturbance is sufficiently frequent to prevent interspecific competition from going to exclusion of the weaker competitor, who is repeatedly saved from extinction by environmental changes that reduce its enemy's population in time. Successions, in effect, never reach climax before being restarted. But disturbance is yet too weak to reduce populations to zero. Extinction is rare, whether from adverse environment or from competition. This pattern seems to describe conditions in much of the wet tropics nicely.

Summary of Mechanisms Leading to Diversity Clines

There is no one simple explanation for clines of diversity, only a number of mechanisms that all should work to some extent (Figure 26.24). More mechanisms certainly will be discovered, but a general outline of processes leading to clines is clear.

1. Except in rare circumstances, most regions of the earth are not saturated with species. Clines cannot be explained by answering the question "How many can there be?"

2. Local diversities are functions of local immigration (or speciation) and local extinction.

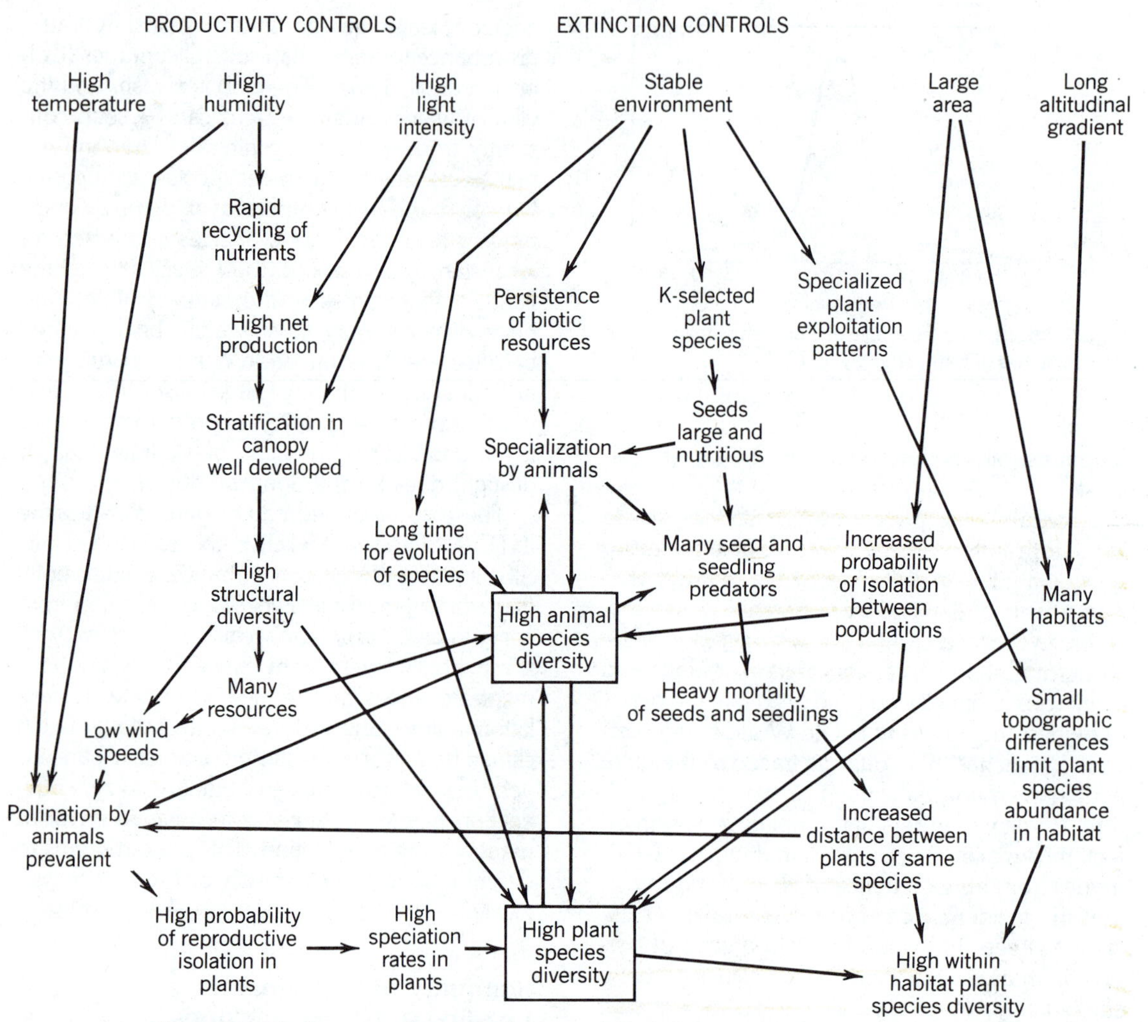

Figure 26.24 Summary of processes leading to high diversity. Modified from a scheme of Price (1975).

3. The presence of trees and structured vegetation increases diversity. Therefore the heat and water balance restrictions on tree life forms in the north (Chapter 3) account for some part of the low diversity at high latitudes.

4. The fact that the contemporary earth is passing through a series of ice ages has raised extinction rates. For some groups, notably long-lived trees and mammals, extinction rates have been increased differentially at high latitudes.

5. The antiquity and environmental stability of the deep sea floors probably contribute some part of benthic high diversity by reducing extinction rates.

6. It is possible that tropical lakes have lower plankton diversity than temperate lakes because of a form of local saturation and competitive exclusion (Chapter 22).

7. A large part of the tropical diversity may be dependent on the fact that the largest areas of the earth are those with tropical habitats.

8. Moderate physical disturbance promotes diversity, whether by seasonal drought in moderate deserts or by rain and storms in the wet tropics.

COMMONNESS, RARITY, AND DOMINANCE

The relative abundance, or equitability, component of diversity needs to be accounted for no less than species richness itself. Some animals and plants are rare; others are common. Three kinds of distributions of commonness and rarity are known. Two of these, the *log-normal* and *broken-stick* distributions, reflect random processes. The third, *ecological dominance,* results from successful ecological competition.

Preston's Log-normal Distribution

When naturalists make large collections they usually find that a few species are very common but many more are very rare. An illustration is the relative abundance of moths caught in a light trap (Dirks, 1937). Out of 56,131 specimens over 40,000 belonged to just 6 species. Most of the 349 species in the collection were rarities represented by a few individuals each.

This distribution of a few common but many rare fits a statistical pattern known as the LOG-NORMAL DISTRIBUTION and reflects underlying processes that are random. Preston (1948) revealed this distribution by arranging census data by species classes in order of descending individual abundance. Table 26.9 is a set of hypothetical data that are ranked by Preston's method. The species groups he called *octaves,* these being bounded by the series 2, 4, 8, 16, and so on. All species in Table 26.9 that had 9, 10, 11, 12, 13, 14, or 15 individuals belonged to octave D, for instance. Any collection that falls on a line is halved and shared between the octaves on either side. The procedure is well illustrated by Preston's treatment of a census of all the birds in Quaker Run Valley (Table 26.10). Few species fall in the octave at the beginning of the sequence and few species in the octave at the end. Most are in the middle octaves, so that if we plot species per octave against an octave scale we get a hump-shaped curve (Figure 26.25). This curve is the *log-normal distribution.* Log-normal distributions of commonness and rarity have been demonstrated in data from many communities, for instance, from diatoms in streams and from abundance data in vegetation as well as for birds and insects (Figure 26.26).

Log-normal distributions are themselves the product of random processes. They result when the abundance of each species population in the sample is determined at random, independently

Table 26.9
Hypothetical Census Data Ranked by Relative Abundance

Species group	A	B	C	D	E	F	G	H	I	J	K	L
Approximate specimens observed of that species	1	2	4	8	16	32	64	128	256	512	1024	2048

Table 26.10
Species Abundance of Quaker Run Valley Birds, Ranked into Octaves

Octave	<1	1 to 2	2 to 4	4 to 8	8 to 16	16 to 32	32 to 64	64 to 128	128 to 256	256 to 512	512 to 1024	1024 to 2048
Species per octave	>1	$1\frac{1}{2}$	$6\frac{1}{2}$	8	9	9	12	6	9	11	4	3

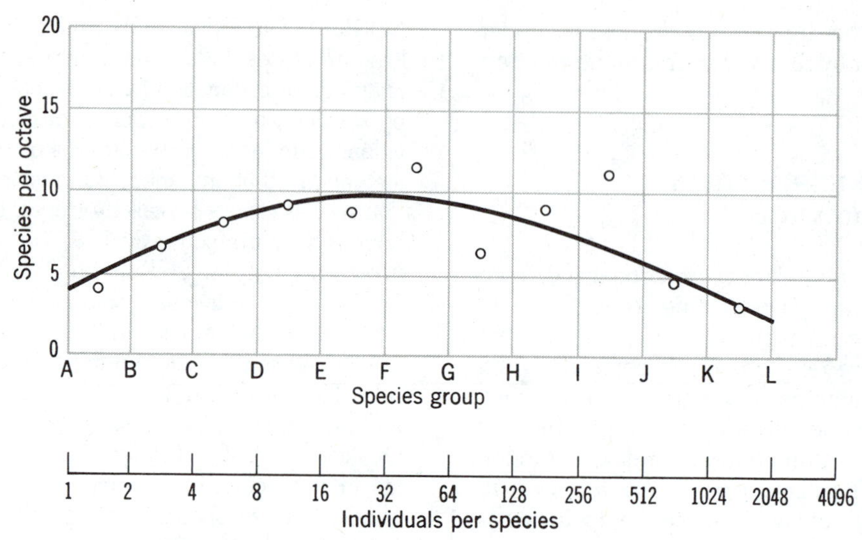

Figure 26.25 Log-normal distribution at Quaker Run Valley. Plot of the data in Table 26.10. (From Preston, 1948.)

of the other species populations (Seller, 1951). The prevalence of log-normal distributions in our data sets, therefore, shows that the relative distributions and abundance of plants and animals very often is determined by random process.

We have good reasons for expecting this randomness in community building, at least for many species. Both Margalef (1957) and MacArthur (1960) have shown that assemblies of opportunistic species must be put together by a random walk process, so that large samples of these would be expected to yield log-normal distributions. These species make up a significant part of the species list of all communities, and the major part in some. For equilibrium species, community building by random walk is not expected but a large enough sample that might include parts of more than one community probably includes random parts of many communities, which should yield a log-normal distribution for the whole sample even if distributions were different in each community.

Apart from this broad hint of the underlying randomness of much community building, the log-normal distribution is useful as an ecological tool. If only part of the distribution shows in data, it is likely that only a portion of a natural assemblage has been sampled, suggesting that more sampling is necessary (as in (*A*) in Figure 26.26; Maguire, 1971). The distribution may be particularly valuable as a test for sampling effort when comparing diversity between communities. Sampling may be standardized to yield a predetermined portion of the complete log-normal distribution at each site, thus letting species richness be used as a reliable indicator of total species diversity.

The Broken Stick

Ecologists do not believe that all the relationships between plants and animals are random, even though they find log-normal distributions. The randomness of those distributions apparently is introduced when many subdistributions are compounded: all the separate bird distributions of a large valley, the plants of many subdistributions in vegetation, or all the moths flying within sight of a light trap. On a smaller scale, relative abundance depends on competition, predation,

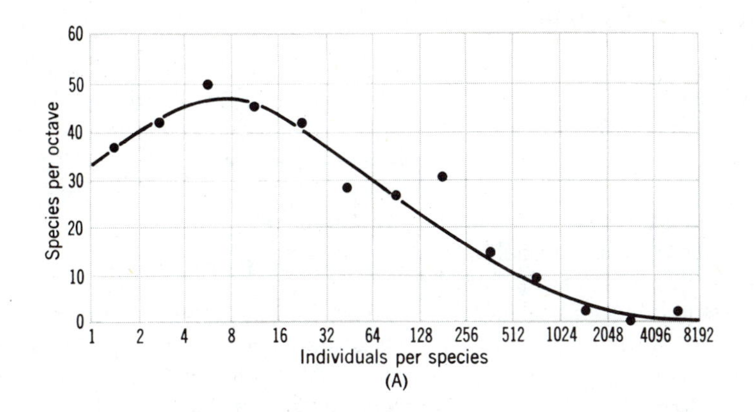

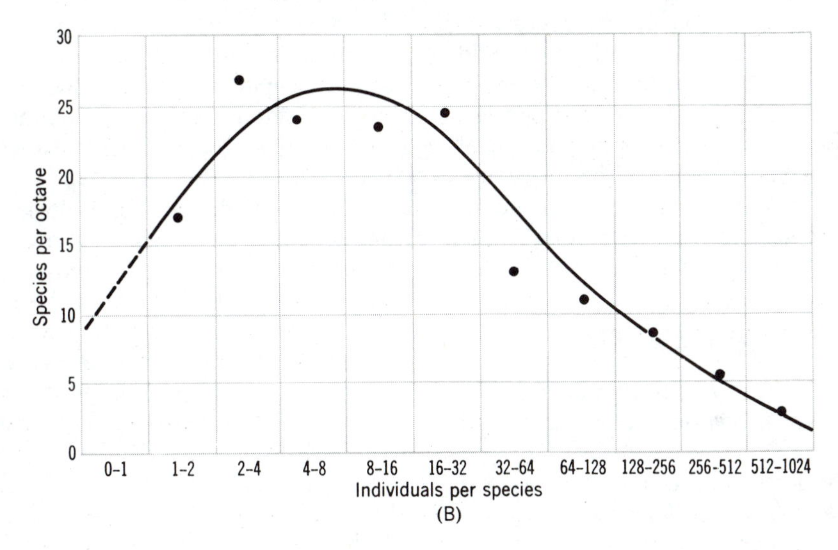

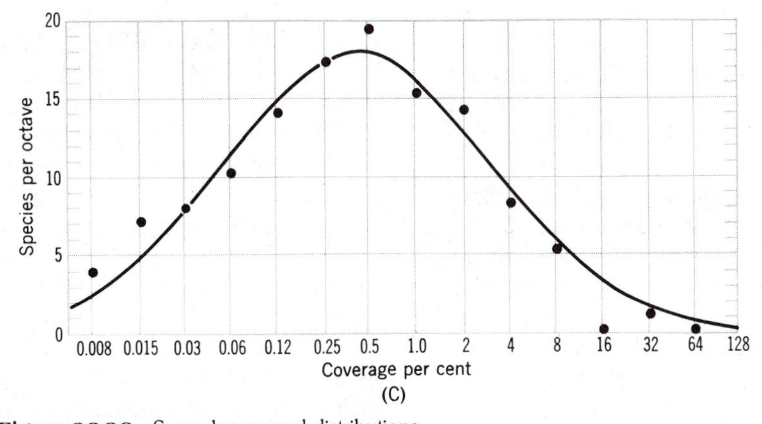

Figure 26.26 Some log-normal distributions.
(A) Dirks' data for moths that came to a light trap in Maine (from Preston, 1948). *(B)* Data for diatoms settling on a glass slide in a stream in Pennsylvania (from Patrick, 1954). *(C)* Data for plants in the Sonoran Desert (from Whittaker, 1965). The curves are all humped, but the left-hand side may apparently be missing, suggesting that rare species were too rare to be included in the sample.

and mutualism. The BROKEN STICK is an attempt to predict small-scale patterns of relative abundance when species interactions determine the pattern.

MacArthur (1957) made the basic assumption that species of a guild should divide available resources among themselves. For simplicity, he then made the second assumption that relative abundance was determined by one critical resource alone and then explored how this resource might be divided into shares, with population size being a function of the share received.

Resources divided along one niche axis could be likened to a stick. To break the stick into n lengths, to represent the shares of n species, MacArthur chose $n - 1$ points at random at which the stick would be broken. This procedure leads to the statement that the expected abundance (I_r) of the rth rarest species among n species and m individuals is given by

$$I_r = \frac{m}{n} \sum_{i=1}^{r} \left(\frac{1}{n - i + 1} \right)$$

A characteristic curve results from broken stick distributions when species importance is plotted against species rank.

Broken-stick distributions of commonness and rarity can be demonstrated for guilds of closely related organisms with synchronized life cycles that live together in small areas. These are conditions in which we expect species to interact, so that the relative abundance of all members of a guild might be expected to settle down over time in some definite relationship to other members of the guild.

Many demonstrations that broken-stick distributions exist in nature are confirmation, if any was needed, that biological interactions do influence commonness and rarity (King, 1964; Goulden, 1966; Tsukada, 1967). The broken-stick distributions, however, do not necessarily imply that MacArthur's assumption of discrete separation of resources in the broken-stick manner are involved in these distributions. It is now known that a large array of systems of apportionment lead to the distribution that we call a broken stick (Cohen, 1966, 1968). All that a broken-stick distribution implies is that some system of relationship beyond pure randomness exists. Because of this general limitation, interest in broken-stick distributions has lapsed in recent years.

Ecological Dominance

A third familiar pattern of commonness and rarity is the decidedly non-random one of ECOLOGICAL DOMINANCE. Census of a plant community showing clear dominance (Chapter 15) shows that the distribution of relative abundance is close to a geometric series. The plot on semilog paper is a straight line (Figure 26.27). In this distribution the more abundant, ecologically dominant plants have super-importance relative to the rest.

The geometric series of ecological dominance necessarily is the result of species interactions. Super-importance results from successful competition. Part of the total resources is pre-empted by the dominant plants and set aside. Part of the remaining resources goes to the subdominants, then part of the remainder to other plants, and so on until all the resource is allocated.

ECOLOGICAL DOMINANCE,[3] therefore, IS THE PRE-EMPTION OF COMMUNITY RESOURCES BY A FEW SPECIES SO THAT THESE HAVE MORE IMPORTANCE IN THE COMMUNITY THAN IS POSSIBLE IN A PURELY RANDOM PROCESS. This definition is entirely compatible with the use of "ecological dominance" in traditional plant ecology (Chapter 15), though the language is a little different.

The apparent domination of a community by a few kinds of large and omnipresent plants in this way guided the concept of the plant *association* as a unit to study. In some recent ecological work, however, DOMINANCE has come to mean THE EFFECTIVE RELATIVE ABUNDANCE OF THE MORE COMMON SPECIES. This usage probably can be traced to Whittaker (1965, 1975), who talked of the relationship of

[3]The word "dominance" is derived from the Latin *dominus* meaning "the Lord." The *ecological dominants* literally dominate the community by taking a more than randomly large share of the resources.

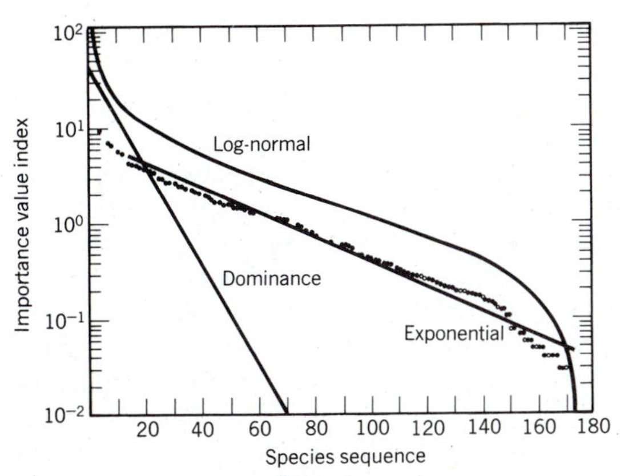

Figure 26.27 Types of abundance–diversity curves.

dominance to *diversity,* so that the plots of Figure 26.27 became *dominance–diversity curves.* The language would have been less value-laden if these had been called ABUNDANCE–DIVERSITY CURVES instead.

There are various other uses of the word "dominance" in ecology, some of them merely convenient colloquialism of the profession. The most common colloquialism is to refer to the most abundant species in a collection as the "dominant" as in "the sample was dominated by little green wicker-work people." This merely means that much green wicker-work filled the traps. Samples of anything in a universe of log-normal distributions will be "dominated" by something in this sense.

Other uses of "dominance" are in the behavioral sense of the dominant individual in a social hierarchy and in the community sense of the top predator of a food chain or, as Paine (1969) uses the word, the top predator of a subweb (page 654). These are good literal usages where the meaning is clear.

True ECOLOGICAL DOMINANCE, however, implies species interaction in general and competition in particular. By far the best known example of *ecological dominance* is in temperate vegetation, particularly in the climax communities of temperate forests. Similar dominance is achieved by a few species of algae in fertile estuaries or polluted lakes. Dominance also is possible in some simple intertidal communities with minimal predation where a few benthic organisms can out-compete the rest by pre-empting space.

Ecological dominance restricts species richness. As was noted in the discussion of diversity, the attainment of dominance by a few strong competitors in hyperfertile waters strongly reduces diversity. A reduction of diversity consequent to increasing dominance also is well known during ecological successions in temperate forests (Chapter 23). Species richness, which has increased steadily through most parts of a secondary succession, falls in the climax when most space is pre-empted by a few dominant trees.

Plant ecologists have always had difficulty identifying dominance in tropical forests when trying to describe them with the classical methods of phytosociology. In a few tropical rain forests, like those in Uganda and Nigeria, species lists are few and true dominants may exist (Hubbel, 1979; Jones, 1956). But in complex forests like those of the central Amazon, where 179 species to the hectare have been counted, identifying ecological dominants seems impossible (Prance *et al.*, 1976). Similar high diversity seems to be correlated with absence of dominance in other rain forest sites (Richards, 1956; Ashton, 1969; Whitmore, 1975). Figure 26.28 compares plots of ranked abundance against diversity for rain forests with those for temperate forests showing dominance. The tropical plots approach log-normal distributions whereas the temperate plots approach the geometric series of ecological dominance. This absence of dominance in tropical rain forests may be a prime cause of high rain forest diversity.

THE STABILITY AND COMPLEXITY OF ECOSYSTEMS

Public interest in ecology has been much concerned with whether ecosystems are stable or not. The interest, of course, is over whether any particular ecosystem will endure or whether the whole life support system of the biosphere is safe from dangerous perturbation. When seeking to

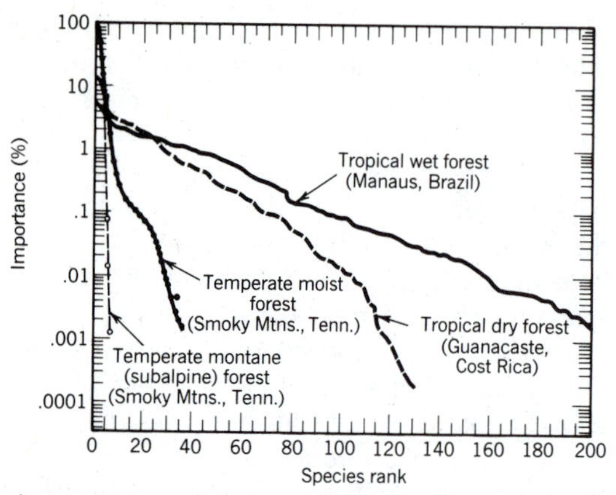

Figure 26.28 Abundance–diversity curves of tropical rain forest and temperate forests compared.
The Amazon forest apparently is without dominants. (From Hubbell, 1979.)

answer these legitimate questions about ecosystem stability, however, we face difficulties in defining exactly what we mean by stability.

In some physical senses, stability can mean immutability or the power to continue unchanged in the presence of a would-be upsetting force. This is the stability of a heavy rock or the foundations of a castle, both of which resist pressure. Very great force applied to either will, of course, knock them over, after which they adopt new stable states.

Those who have worried about the possibility of biospheric collapse probably have this thought of simple stability in mind. They ask if the main forces holding the biosphere in its present state can be overturned. The answer seems clearly to be "no." The state of the biosphere is a function of solar energy received, the spin of the earth, and the mass of gases in the atmosphere, as well as of rates of exchange within the biosphere itself (Chapters 18 and 19). Without cosmic calamity, forces sufficient to overcome this system do not exist.

But we also mean something different and more subtle when we talk of ecosystem stability. We ask if ecosystems are subject to changes of large amplitude, tacitly assuming that *constancy* is *stability*. If constancy is what we desire, then we have a problem of time scale. Over very short intervals all systems appear constant; over geological time all systems change. Furthermore, all ecosystems show rhythmic changes tracking seasons of the year. Fluctuating ecosystems of the arctic repeat themselves every year and thus in one sense are constant. Tropical rain forests have more muted annual cycles, though these are always present. If stability as constancy is our interest, then much of that stability certainly is fixed by the solar system and is outside the control of the biota.

Ecosystem constancy can, however, be a biological property if the ecosystem depends on a successional community. All seres before climax are unstable in the sense that change is certain. If grazing, fire, or human manipulation as in agriculture maintains a seral stage as a *proclimax* a prolonged but vulnerable constancy results. These proclimax systems are in a sense unstable. They can be likened to a rock poised on its edge so that a push from either side will knock it over to a fresh and more stable state.

The stability of agricultural ecosystems is the stability of a proclimax. Social critics are correct to say that agricultural yields depend on continued input of fossil fuel energy or other human interference (H. T. Odum, 1971), but the stability of these agricultural systems is, by this argument, the stability of the human social systems in charge of them. Ecosystems can even be forgiving of human social change, like the agricultural proclimax fields of parts of Europe that have been farmed successfully for 5000 years as empires rose and fell.

The Hypothesis That High Species Diversity Leads to Population Stability

A meaning of stability that is more important in ecology refers to the *resilience* of ecosystems. This is *the power of a system under stress to return to its original state when the stress is removed.* In the sense of *resilience* a rubber string is stable when stretched below the yield point because it will return to its original state as soon as it is let go. A rock perched on edge lacks resilience because any deflection causes it to fall into a different stable state.

The resilience of ecosystems is of practical interest. The concept is invoked when we argue the dangers of clear-cutting, or of polluting lakes with detergents. Lakes have long-term resilience in that they will return to the original state set by the watershed when the input of pollutants is ended (Chapters 21 and 22). Temperate forest ecosystems are resilient to minor perturbations, since they regenerate rapidly if the soil is not disturbed. But, as the Hubbard Brook study shows (Chapter 17), if the disturbance continues until the soil is lost the ecosystem falls into another state as does a toppled rock.

Of more general interest is the resilience of species populations within ecosystems. What are the ecosystem properties that prevent large changes in its populations? More important still is the power of ecosystems to regulate population fluctuations so that species do not go extinct. It is in this context that ecological theory has been concerned with looking at a possible connection between species diversity and stability.

A STABLE ECOSYSTEM IS ONE WHERE THE CHANCE OF EXTINCTION IS LOW. This is an elaboration of Preston's (1969) definition of A SPECIES IS TEMPORALLY STABLE IF ITS FLUCTUATIONS OVER A LARGE NUMBER OF YEARS (OR WHATEVER IS THE APPROPRIATE UNIT OF TIME) CONFORM TO A LOG-NORMAL DISTRIBUTION. This language recognizes that the relative abundance of a species across its range is distributed log-normally, and concludes that a similar log-normal distribution over time represents safety from extinction. The definition then has the merit of allowing very wide fluctuations in number within the definition of stability.

These definitions treat *stability* as *resilience* and they use the chance of extinction as a test of resilience. One of the more difficult tasks of theoretical ecology has been to explore the possibility of a direct link between species diversity and resilience. Do more species promote resilience, thus reducing extinction and granting the "stability" of a prolonged existence? As ecologists worked to answer this question their thoughts were taken up by environmental debates with the claim, now known not to be true, that high species diversity should represent population or ecosystem stability. It is on this claim, for instance, that arctic ecosystems are often called "fragile," though in fact their populations may be some of the most resilient on earth.

Very simple predator–prey systems can be shown to be unstable if the predator is efficient and the prey has few chances to escape. This is the condition in simple laboratory systems like those of Gause described in Chapter 10. These systems can be made stable if the predator is sufficiently incompetent or the prey has avenues of escape. The azuki bean weevil–wasp system, for instance, persists indefinitely (Chapter 10), as does the wolf–moose system on Isle Royale. Natural selection works to promote stability in these simple systems because the traits leading to instability are fatal to their carriers.

A different theoretical background to stability is provided by the conception of more complex communities in which each animal relates to many others. Too successful predators might be spared the consequences of over-eating their prey if they had alternative food. Prey that had been harried to near extinction might be saved if their predators found it more profitable to switch their attack to other more abundant victims. This is the essence of the hypothesis relating increased stability to increased ecosystem complexity. Population fluctuations are postulated to be damped when food webs are complex so that many species interact.

Well-known facts of natural history can be marshaled in support of the hypothesis. The best known of simple ecosystems with low diversity are those of the arctic, which are notorious for extreme population fluctuations like those of lemmings (Chapter 13). The most complex and species-rich ecosystems known are those of the tropical rain forests where, until recently, less was known about population fluctuations. Best known of all are agricultural systems that are both extremely simple and subject to population outbreaks of organisms called weeds or pests. The hypothesis that diversity is related to stability was directly interesting because it suggested that replacing complex natural vegetation with monoculture led to a somewhat frightening instability.

An apparently strong theoretical base for complexity–stability theory arose when information theory was first applied to biological systems. The information theory measure H' was used as a diversity measure combining species richness with equitability (page 651). This theory was originally developed to measure probabilities of information transfer in an information network (Shannon and Weaver, 1949). When used for its original purpose the measure also describes the capacity, and hence the stability, of the information network, since capacity and stability of that message-flow system are increased in proportion to the extra alternate channels provided. In an information system, therefore, high diversity in the network can be shown to cause high stability of information flow.

MacArthur (1955) and Margalef (1957) used the logic of information theory first to use H' as a measure of species diversity, and then to suggest that information theory mathematics made it reasonable to argue that high species diversity led to population stability. This was the important development that gave wide public prominence to the hypothesis that ecosystem complexity caused stability.

Essential to the hypothesis is the analogy that compares an information network to a food web. The analogy depends on the facts that both systems transfer energy, either as units of information or as food particles. Both food and information can be treated as forms of negentropy, thus making the mathematical part of the analogy acceptable, but otherwise the analogy is very imperfect. Linkages in an information network are designed to promote the transfer of information, but every link in a food web is more likely to impede the passage of food than to promote it. The food-energy to be passed is actually the body of an animal preserved by natural selection for the maximum avoidance of being eaten. Thus the main analogy behind complexity–stability theory in its information theory guise is unsound. The information theory measure H' is not a measure of ecosystem stability. Lately, of course, ecologists have also come to doubt the value of H' as a measure of diversity as well.

The analogy between an information network and a food web allows restatement of the hypothesis in a form that could, if not actually be tested, then at least be subjected to critical scrutiny. *The hypothesis is that species populations in complex food webs fluctuate less than similar populations in simple food webs.* Computer simulation of food webs of various complexity suggests that there is no such correlation between food web complexity and population fluctuation (Goodman, 1974).

But more important to a general loss of confidence in complexity–stability theory was mathematical analysis that investigated interactions in food webs by families of differential equations, or difference equations, with some family relationship to Lotka–Volterra systems (Leigh, 1965; Chapter 6). May (1972, 1973) has shown that complex systems of these equations do not lead to stability. Instead, his analysis suggests that the more complex the system of these equations, the greater the amplitude in species number. Piling more species into one of these systems is more likely to make original instability resonate than to dampen it. Certainly May's (1973) monograph shows that previous belief in an established mathematical relationship between complexity and increased stability was completely unfounded.

Environmental Stability: The Arbiter of Both Diversity and Ecosystem Stability

Analysis of diversity along latitudinal clines shows an unmistakable causal relationship between environmental factors and diversity. Strongly seasonal climates in unproductive arctic systems lead to low diversity, probably working through high extinction rates. Constant climates of moderate deserts lead to high diversity, notably working by permitting rapid invasion. Our latest ideas about the tropical rain forests suggest that high

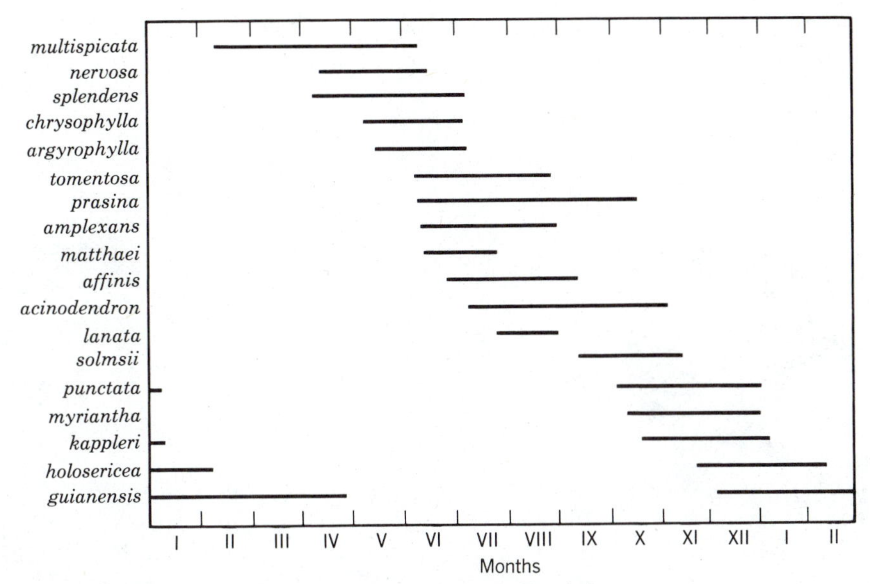

Figure 26.29 Seasonal *fruiting* of *Miconia* species in Trinidad. This staggered pattern of fruiting appears to be an adaptation to using birds as seed dispersers. If the miconias fruited synchronously, the birds would be satiated and many *Miconia* fruits would not be eaten. (From Snow, 1964.)

diversity results because the system is almost constantly perturbed. In each case the ultimate determinant of ecosystem complexity is the state of the physical environment. Environmental stability determines species complexity, not the other way around.

Under the broad umbrella of environmental control the relationships of species in food webs certainly have consequences for population histories, but these consequences may not be simple and they may be different in different ecosystem types. One of the possibilities suggested by modern theory is that the particularly rich species complexity of tropical rain forests may in fact reduce the resilience of their populations to perturbing forces.

Figure 26.29 shows how species of *Miconia* (a tropical shrub) have adapted their flowering times to their several needs to use fruit-eating birds as seed dispersal agents. It is necessary to flower at a time when the birds are not satiated with other *Miconia* seeds. The result is a tightly ordered society in which every species is dependent on predicted events like the times of flowering of all the rest. The way in which these species of a tropical system have adapted to their numerous contemporaries makes them actually vulnerable to any alteration of the system.

Ecologists are now more worried about interference with tropical rain forests, once thought to be so stable, than they are with interference with arctic systems, the organisms of which tend to be generalists well equipped by natural selection to cope with environmental change. We are also aware that whole tropical rain forest ecosystems are particularly vulnerable to large-scale perturbations of the habitat. Tropical climates have resulted in soils and nutrient retrieval systems such that removal of the forest takes with it much of the forest support system, thus letting the ecosystem fall into another and less productive state.

The overriding fact of existence is that 98 to 99% of solar energy incident on all ecosystems is dissipated in physical processes. Action is where the energy is. The animals and plants of a place adapt to physical necessity. The numbers of species adapted to any local reality depend on the time and space available.

REFERENCES

Ab'Saber, A. N., 1982, The paleoclimate and paleoecology of Brazilian Amazonia, p. 41–59 *in* Prance, G. T., *ed., Biological Diversification in the Tropics,* New York, Columbia Univ. Press.

Ager, T. A., 1982, Vegetational history of western Alaska during the Wisconsin glacial interval and the Holocene, p. 75–93 *in* Hopkins, D. M., J. V. Matthews, Jr., C. E. Schweger, and S. B. Young, *eds., Paleoecology of Beringia,* New York/London, Academic Press.

Alcaraz, M., G. A. Paffenhofer, and J. R. Strickler, 1980, Catching the algae: A first account of visual observations on filter-feeding calanoids, p. 241–248 *in* Kerfoot, W. C., *ed., Evolution and Ecology of Zooplankton Communities.* Hanover, N.H. Univ. Press of New England.

Allee, W. C., A. E. Emerson, O. Park, T. Park, and K. P. Schmidt, 1949, *Principles of Animal Ecology.* Philadelphia, Saunders.

Allen, J. A., 1871, Mammals and winter birds of east Florida, and a sketch of the bird fauna of eastern North America. *Bull. Museum of Comp. Zool.,* v. 2, p. 161–450, pt. V, p. 375–450.

Allen, L. H., E. R. Lemon, and L. Muller, 1972, Environment of a Costa Rican forest, *Ecol.* v. 53, p. 102–111.

Alyea, F. N., D. M. Cunnold, and R. G. Prinn, 1975, Stratospheric ozone destruction by aircraft-induced nitrogen oxides, *Science,* v. 188, p. 117–121.

Andrewartha, H. G., and L. C. Birch, 1954, *The Distribution and Abundance of Animals,* Chicago, Univ. of Chicago Press.

Andrews, R. V., 1968, Daily and seasonal variation in adrenal metabolism of the brown lemming, *Physiol. Zool.,* v. 41, p. 86–94.

Armitage, K. B., 1962, Social behavior of a colony of the yellow-bellied marmot (*Marmota flaviventris*). *Anim. Behav.,* v. 10, p. 319–331.

Armstrong, E. A., 1955, *The Wren.* London, Collins.

Armstrong, J., 1960, The Dynamics of *Dugesia tigrina* populations and of *Daphnia pulex* populations as modified by immigration, Ph.D. dissertation, Univ. of Michigan.

Ashton, P. S., 1969, Speciation among tropical forest trees: Some deductions in the light of recent evidence. *Biol. Jour. of Linneaean Soc. of London,* v. 1, p. 155–196.

Ayala, F. J., 1968, Genotype, environment, and population numbers. *Science,* v. 162, p. 1453–1459.

Bacastow, R., and C. D. Keeling, 1973, Atmospheric carbon dioxide and radiocarbon in the natural carbon cycle. II. Changes from A.D. 1700 to 2070 as deduced from a geochemical model, p. 86–135 *in* Woodwell, G. M., and E. V. Pecan, *eds., Carbon and the Biosphere.* Springfield, Va., U.S. AEC.

Baker, H. 1937, Alluvial meadows: A comparative study of grazed and mown meadows. *Ecol.* v. 25, p. 408–420.

Baker, H. G., and I. Baker, 1975, Studies of nectar constitution and pollinator-plant coevolution, p. 100–140 *in* Gilbert, L. E., and P. R. Raven, *eds.,* Austin/London, Univ. of Texas Press.

Baskerville, G. L., 1965, Estimation of dry weight of tree components and total standing crop in conifer stands. *Ecology,* v. 46, p. 867–869.

Baumgartner, A., 1969, Meteorological approach to the exchange of CO_2 between the atmosphere and vegetation, particularly forest stands. *Photosynthetica,* v. 3, p. 127–149.

Beard, J. S., 1955, Tropical American vegetation types. *Ecol. Michigan Botanist,* v. 5, p. 109–114.

Beddington, J. R., and D. B. Taylor, 1973, Optimal

age specific harvesting of a population. *Biometrics,* v. 29, p. 801–809.

Benecke, V., 1972, Wachstum, CO_2-Gaswechsel und Pigmentgehalt einiger Baumarten nach Ausbringung in verochiedene Hohenlagen. *Angew. Bot.,* v. 46, p. 117–135.

Benemann, J. R., 1973, Nitrogen fixation in termites. *Science,* v. 18, p. 164–166.

Bennet, C. L., 1979, Radiocarbon dating with accelerators. *Amer. Scientist,* v. 67, p. 450–457.

Bennet, F. B., and J. A. Ruben, 1979, Endothermy and activity in vertebrates. *Science,* v. 206, p. 649–654.

Benninghoff, W. S., 1966, The Relevé method of describing vegetation. *Michigan Botanist,* v. 5, p. 109–114.

Bertram, B. C. R., 1978, Living in groups: Predators and prey, p. 64–96 *in* Krebs, J. R., and N. B. Davies, *eds., Behavioural Ecology,* Sunderland, Mass., Sinauer Associates.

Bidgood, B. F., 1974, Reproductive potential of two lake whitefish (*Coregonus clupeaformis*) populations. *J. of Fisheries Res. Board of Canada,* v. 31, p. 1631–1639.

Bigarella, J. J., and Andrade–Lima, D., 1982, Paleoenvironmental changes in Brazil, p. 27–40 *in* Prance, G. T., *ed., Biological Diversification in the Tropics.* New York, Columbia Univ. Press.

Billings, W. D., 1938, The structure and development of old field short-leaf pine stands and certain associated physical properties of the soil. *Ecol. Monog.,* v. 8, p. 437–499.

Billings, W. D., 1978, *Plants and the Ecosystem,* 3rd ed. Belmont, Calif., Wadsworth.

Birch, L. C., and H. G. Andrewartha, 1941, The influence of weather on grasshopper plagues in South Australia. *J. of Dept. of Agriculture So. Australia,* v. 49, p. 95–100.

Birks, H. H., M. C. Whiteside, D. M. Stark, and R. C. Bright, 1976, Recent paleolimnology of three lakes in northwestern Minnesota. *Quat. Res.,* v. 6, p. 249–272.

Birks, H. J. B., and H. H. Birks, 1980, *Quaternary Paleoecology.* Baltimore, Univ. Park Press.

Bischoff, J. L., and R. J. Rosenbauer, 1981, Uranium series dating of human skeletal remains from the Del Mar and Sunnyvale sites, California. *Science,* v. 213, p. 1003–1005.

Bjorkman, O., and J. Berry, 1973, High efficiency photosynthesis. *Scientific Am.,* v. 229(4), p. 80–93.

Blum, U., and E. L. Rice, 1969, Inhibition of symbiotic nitrogen-fixation by gallic and tannic acid, and possible roles in old field succession. *Bull. Torrey Bot. Club,* v. 96, p. 531–544.

Blumenstock, D. I., and S. W. Thornthwaite, 1941, *Climate and the world pattern: Climate and man (USDA Yearbook),* Washington, D.C., U. S. Government Printing Office, p. 98–127.

Bode, H. R., 1958, Beitrage zur Kenntnis alleloysathischer Erscheinungen bei einigen Juglandaceen. *Planta,* v. 51, p. 440–480.

Bodenheimer, F. S., 1938, *Problems of animal ecology.* London, Oxford Univ. Press.

Boersma, P. D., 1977, An ecological and behavioral study of the Galapagos penguin. *The Living Bird,* v. 15, p. 43–93.

Bohner, R. N., 1981, Experimental studies of diel vertical migration, p. 111–121 *in* W. C. Kerfoot, *ed., Evolution and Ecology of Zooplankton Communities.* Hanover, N.H., Univ. Press of New England.

Bolin, B., 1970, The carbon cycle. *Scientific Am.,* v. 223, p. 124–132.

Bonner, J., 1962, The upper limit of crop yield. *Science,* v. 137, p. 11–15.

Bonner, J., and J. E. Varner, *eds.,* 1976, *Plant Biochemistry,* 3rd ed. New York, Academic Press.

Bormann, F. H., 1976, An inseparable linkage: Conservation of natural ecosystems and the conservation of fossil energy. *BioSci.,* v. 26, p. 754–760.

Bormann, F. H., and G. E. Likens, 1967, Nutrient cycling. *Science,* v. 155, p. 424–429.

Bormann, F. H., G. E. Likens, and J. S. Eaton, 1969, Biotic regulation of particulate and solution losses from a forest ecosystem. *BioSci.,* v. 19, p. 600–610.

Bormann, F. H., G. E. Likens, and J. M. Melillo, 1977, Nitrogen budget for an aggrading northern hardwood forest ecosystem. *Science,* v. 196, p. 981–983.

Bormann, F. H., G. E. Likens, T. G. Siccama, R. S. Pierce, and J. S. Eaton, 1974, The export of nutrients and recovery of stable conditions following deforestation at Hubbard Brook. *Ecol. Monog.,* v. 44, p. 255–277.

Bormann, F. H., T. G. Siccama, G. E. Likens, and R. H. Whittaker, 1970, The Hubbard Brook ecosystem study: Composition and dynamics of the tree stratum. *Ecol. Monog.,* v. 40, p. 373–388.

Botkin, D. B., J. F. Janak, and J. R. Wallis, 1972, Some ecological consequences of a computer model of forest growth. *Ecol.,* v. 60, p. 849–872.

Boyce, J. B., 1946, The influence of fecundity and

egg mortality on the population growth of *Tribolium confusum* duval. *Ecol.*, v. 27, p. 290–302.

Bradbury, J. P., 1971, Paleolimnology of Lake Texcoco, Mexico. Evidence from diatoms. *Limnology and Oceanography*, v. 16, p. 180–200.

Bradbury, J. P., B. Leyden, M. Salgado–Labouriau, W. M. Lewis, Jr., C. Schubert, M. W. Bindford, D. G. Frey, D. R. Whitehead, F. H. Weibezahn, 1981, Late quaternary environmental history of Lake Valencia, Venezuela. *Science*, v. 214, p. 1299–1444.

Braun, E. L., 1950, *Deciduous Forests of Eastern North America*. New York, McGraw–Hill (Blakiston).

Braun, E. L., 1955, The phytogeography of unglaciated Eastern United States and its interpretation. *Botan. Rev.*, v. 21, p. 297–375.

Braun–Blanquet, J., 1932, Plant Sociology: *The Study of Plant Communities* (trans., rev., and ed. by Fuller, G. D., and H. S. Conard). New York, McGraw–Hill.

Breznak, J. A., W. J. Brill, J. W. Mertins, and H. C. Coppel, 1973, Nitrogen fixation in termites. *Nature*, v. 244, p. 577–580.

Bridges, E. M., 1978, *World Soils*, 2nd ed. Cambridge, England, Cambridge Univ. Press.

Broecker, W. S., 1970, Man's oxygen reserves, *Science*, v. 168, p. 1537–1538.

Broecker, W. S., 1971, A kinetic model for the chemical composition of sea water. *Quat. Res.*, v. 1, p. 188–207.

Broecker, W. S., 1973, Factors controlling CO_2 content in the oceans and atmosphere, p. 32–50 *in* Woodwell, G. M., and E. V. Pecan, *eds.*, *Carbon and the Biosphere*. Springfield, Va., U.S. AEC.

Broecker, W. S., and D. L. Thurber, 1965, Uranium-series dating of corals and oolites from Bahaman and Florida limestones. *Science*, v. 149, p. 58–60.

Brooks, J. L., 1946, Cyclomorphosis in *Daphnia*. *Ecol. Monog.*, v. 16, p. 409–447.

Brooks, J. L., 1965, Predation and relative helmet size in cyclomorphic *Daphnia*. *Proc. Nat. Acad. Sci.*, v. 53, p. 119–126.

Brooks, J. L., and S. I. Dodson, 1965, Predation, body size, and composition of plankton. *Science*, v. 150, p. 28–35.

Brower, L. P., and J. V. Z. Brower, 1964, Birds, butterflies and plant poisons: A study in ecological chemistry. *Zoologica*, v. 49, p. 137–159.

Brown, J. H., 1971, Mammals on Mountaintops: Nonequilibrium insular biogeography. *Am. Natur.*, v. 105, p. 467–478.

Brown, J. H., and R. C. Lasiewski, 1972, Metabolism of weasels: The cost of being long and thin. *Ecol.*, v. 53, p. 939–943.

Brown, J. L., 1969, Territorial behavior and population regulation on birds. *Willson Bull.*, v. 81, p. 293–329.

Brown, L. L., and E. O. Wilson, 1956, Character displacement. *Systematic Zool.*, v. 5, p. 49–64.

Brown, R. T., and J. T. Curtis, 1952, The upland conifer-hardwood forests of northern Wisconsin. *Ecol. Monog.*, v. 22, p. 217–234.

Brugam, R. B., 1978, Pollen indicators of land-use change in southern Connecticut. *Quat. Res.*, v. 9, p. 349–362.

Bryson, R. A., 1966, Air masses, streamlines, and the boreal forest. *Geog. Bull.*, v. 8, p. 228–269.

Buchanan, C., and J. F. Haney, 1981, Vertical migration of zooplankton in the Arctic: A test of the environmental controls, p. 65–79 *in* Kerfoot, W. C., *ed.*, *Evolution and Ecology of Zooplankton Communities*. Hanover, N.H., Univ. Press of New England.

Buckman, H. O., and N. C. Brady, 1969, *The Nature and Properties of Soils*, 7th ed. Toronto, The Macmillan Co./Collier–Macmillan Canada, Ltd., p. 653.

Buechner, H. K., and H. D. Roth, 1974, The lek system in Uganda Kob antelope. *Amer. Zool.*, v. 14, p. 145–162.

Bunt, J. S., 1975, Primary productivity of Marine Ecosystems, p. 169–215 *in* Leith, H., and R. H. Whittaker, *eds.*, *Primary Productivity of the Biosphere*. New York, Springer–Verlag.

Burnett, T., 1958, Dispersal of an insect parasite over a small plot., *Can. Entomol.*, v. 90, p. 279–283.

Burris, R. H., and C. C. Black, *eds.*, 1976, *CO_2 Metabolism and Plant Productivity*. Baltimore, Univ. Park Press.

Burton, T. H., and G. E. Likens, 1973, The effect of strip-cutting on stream temperatures in the Hubbard Brook experimental forest, New Hampshire. *BioSci.*, v. 23, p. 433–435.

Calhoun, J. B., 1962, Population density and social pathology. *Scientific Am.*, v. 206, p. 139–148.

Calvin, M., 1976, Photosynthesis as a resource for energy and materials. *Amer. Scientist*, v. 64, p. 270–278.

Caughley, G., 1979, Eruption of ungulate populations, with emphasis on Himalayan thar in New Zealand. *Ecol.*, v. 51, p. 53–72.

Chapman, R. N., 1928, The quantitative analysis of environmental factors. *Ecol.*, v. 9, p. 111–122.

Chapman, R. N., 1931, *Animal Ecology with Especial Reference to Insects.* New York, McGraw–Hill.

Chesson, P. L., and R. R. Warner, 1981, Environmental variability promotes coexistence in lottery competitive systems. *Am. Natur.,* v. 117, p. 923–943.

Chew, R. M., 1974, Consumers as regulators of ecosystems: An alternative to energetics. *Ohio J. of Sci.,* v. 74, p. 359–370.

Chew, R. M., and A. E. Chew, 1970, Energy relationships of the mammals of a desert shrub (*Larrea tridentata*) community. *Ecol. Monog.,* v. 40, p. 1–21.

Child, J. J., 1976, New developments in nitrogen fixation research. *BioSci.,* v. 26, p. 614–617.

Chitty, D., 1958, Self regulation of numbers through changes in viability. *Cold Spring Harbor Symp. of Quantitative Biol.,* v. 22, p. 277–280.

Christian, J. J., 1950, The adreno-pituitary system and population cycles in mammals. *J. Mammol.,* v. 31, p. 248–259.

Christian, J. J., 1970, Social subordination, population density, and mammalian evolution. *Science,* v. 168, p. 84–90.

Christian, J. J., V. Flyger, and D. E. Davis, 1960, Factors in the mass mortality of a herd of Sika deer, *Cervus nippon. Chesapeake Science,* v. 1, p. 79–95.

Church, D. C., and W. G. Pond, 1974, *Basic animal nutrition and feeding.* Corvallis, Oreg., O. & B. Books.

Clements, F. E., 1916, Plant Succession: An Analysis of the Development of Vegetation. Carnegie Inst. of Wash. Publ. 242. Facsimile reprint by Haffner, New York.

Clements, F. E., 1936, Nature and structure of the climax. *Ecol.,* v. 24, p. 252–284.

Clements, F. E., and V. E. Shelford, 1939, *Bioecology.* New York, Wiley.

CLIMAP, 1976, The surface of the ice-age earth. *Science,* v. 191, p. 1131–1144.

Cloud, P., 1972, A working model of the primitive earth. *Am. J. Sci.,* v. 272, p. 537–548.

Cloud, P., 1973, Paleoecological significance of the banded iron formation. *Econ. Geol.,* v. 68, p. 1135–1143.

Cloud, P., 1976, Beginnings of biospheric evolution and their biochemical consequences. *Paleobiology,* v. 2, p. 351–387.

Cloud, P., and A. Gibor, 1970, The oxygen cycle. *Scientific Am.,* v. 223(3), p. 110–123.

Cody, M. L., 1966*a,* A general theory of clutch size. *Evol.,* v. 20, p. 174–184.

Cody, M. L., 1966*b,* The consistency of inter- and intra-specific continental bird species counts. *Amer. Natur.,* v. 102, p. 107–147.

Cody, M. L., 1974, *Competition of the Bird Communities.* Princeton, Princeton Univ. Press.

Cody, M. L., 1975, Towards a theory of continental species diversity, p. 214–257 *in* Cody, M. L., and J. M. Diamond, *eds., Ecology and Evolution of Communities.* Cambridge, Harvard Univ. Press.

Coetzee, J. A., 1967, Pollen analytical studies in East and Southern Africa. *Paleoecology of Africa and of the Surrounding Islands and Antarctica,* v. 3, p. 1–146, Capetown, A. A. Balkema.

Cohen, J., 1966, *A Model of Simple Competition.* Cambridge, Harvard Univ. Press.

Cohen, J., 1968, Alternative derivations of a species-abundance relation. *Am. Natur.,* v. 102, p. 165–172.

Cole, L. C., 1954*a,* The population consequences of life history phenomena. *Quart. Rev. of Biol.,* v. 29, p. 103–137.

Cole, L. C., 1954*b,* Some features of random population cycles. *J. Wildlife Management,* v. 18, p. 1–24.

Colinvaux, P. A., 1964*a,* Environment of the Bering land bridge. *Ecol. Monog.,* v. 34, p. 297–329.

Colinvaux, P. A., 1964*b,* Origin of ice ages: Pollen evidence from arctic Alaska. *Science,* v. 145, p. 707–708.

Colinvaux, P. A., 1965, Pollen from Alaska and the origin of ice ages. *Science,* v. 147, p. 633.

Colinvaux, P. A., 1968, Reconnaissance and chemistry of the lakes and bogs of the Galapagos Islands. *Nature,* v. 219, p. 590–594.

Colinvaux, P. A., 1972, Climate and the Galapagos Islands. *Nature,* v. 240, p. 17–20.

Colinvaux, P. A., 1973, *Introduction to Ecology.* New York, Wiley.

Colinvaux, P. A., 1978*a,* On the use of the word "absolute" in pollen statistics. *Quat. Res.,* v. 9, p. 132–133.

Colinvaux, P. A., 1978*b, Why Big Fierce Animals Are Rare: An Ecologist's Perspective.* Princeton, Princeton Univ. Press.

Colinvaux, P. A., 1979, The ice age amazon. *Nature,* v. 278, p. 399–400.

Colinvaux, P. A., 1981, Historical ecology in Beringia: The south land bridge coast at St. Paul Island. *Quat. Res.,* v. 16, p. 18–36.

Colinvaux, P. A., 1982, Towards a theory of history: Fitness, niche and clutch of *Homo sapiens. Ecol.,* v. 70, p. 393–412.

Colinvaux, P. A., and B. D. Barnett, 1979, Lindeman and the ecological efficiency of wolves. *Am. Nat.,* v. 114(5), p. 707–718.

Colinvaux, P. A., and M. Steinitz–Kannan, 1981, Species richness and area in Galapagos and Andean lakes: Equilibrium phytoplankton communities and a paradox of the zooplankton communities, p. 697–712 *in* Kerfoot, W. C., *ed., Evolution and Ecology of Zooplankton.* Hanover, N.H., Univ. Press of New England.

Collier, B., G. Cox, A. Johnson, and P. Miller, 1973, *Dynamic Ecology.* Englewood Cliffs, N.J., Prentice–Hall.

Connell, J. H., 1961, The influence of interspecific competition and other factors on the distribution of the barnacle *Chthamalus stellatus. Ecol.,* v. 42, p. 710–723.

Connell, J. H., 1978, Diversity in tropical rain forests and coral reefs. *Science,* v. 199, p. 1302–1310.

Connell, J. H., and E. Orias, 1964, The ecological regulation of species diversity. *Am. Natur.,* v. 98, p. 387–414.

Connell, J. H., and R. O. Slatyer, 1977, Mechanisms of succession in natural communities and their role in community stability and organization. *Am. Natur.,* v. 111, p. 1119–1144.

Connor, E. F., and D. Simberlof, 1978, Species number and compositional similarity of the Galapagos flora and avifauna. *Ecol. Monog.,* v. 48, p. 219–248.

Cook, R. M., and B. J. Cockrell, 1978, Predator ingestion rate and its bearing on feeding time and the theory of optimal diets. *J. of Animal Ecol.,* v. 47, p. 529–547.

Cooke, G. D., 1967, The pattern of autotrophic succession in laboratory microecosystems. *BioSci.,* v. 17, p. 717–721.

Coope, G. R., 1959, A Late Pleistocene insect fauna from Chelford, Cheshire. *Proc. R. Soc. Lond. Ser. B.,* v. 151, p. 70–86.

Coope, G. R., 1977, Fossil coleopteran assemblages as sensitive indicators of climatic changes during the Devensian (last) cold stage. *Philos. Trans. of R. Soc. Lond. Ser. B.* v. 280, p. 313–340.

Coope, G. R., 1979, Late Cenozoic fossil Coleoptera: Evolution, biogeography, and ecology: *An. Rev. Ecol. and Systematics,* v. 10, p. 247–267.

Cooper, W. S., 1939, A fourth expedition to Glacier Bay, Alaska. *Ecol.,* v. 20, p. 130–155.

Cottam, G., and J. T. Curtis, 1949, A method for making rapid surveys of woodlands by means of pairs of randomly selected trees. *Ecol.,* v. 30, p. 101–104.

Cottam, G., and J. T. Curtis, 1956, The use of distance measures in phytosociological sampling. *Ecol.,* v. 37, p. 451–460.

Cowgill, V. M., G. E. Hutchinson, A. A. Racek, C. E. Goulden, R. Patrick, and M. Tsukada, 1966, The history of Laguna de Petenxil. *Memoirs of Conn. Acad. Arts and Sci.,* v. 17, 1–126.

Cowles, H. C., 1899, The ecological relations of the vegetation on a structural basis. *Ecol.,* v. 32, p. 172–229.

Cox, A., 1973, *Plate Tectonics and Geomagnetic Reversals.* San Francisco, Freeman.

Cox, C. P., and P. D. Moore, 1980, *Biogeography: An Ecological and Evolutionary Approach.* Oxford, Blackwell.

Crocker, R. L., and J. Major, 1955, Soil development in relation to vegetation and surface age at Glacier Bay, Alaska. *Ecol.,* v. 43, p. 427–428.

Crombie, A. C., 1946, Further experiments on insect competition. *Proc. R. Soc. of Lond. Ser. B.* v. 133, p. 76–109.

Crook, J. H., 1965, The adaptive significance of avian social organizations. *Symp. Zool. Soc. Lond.,* v. 14, p. 181–218.

Curtis, J. T., 1959, *The Vegetation of Wisconsin: An Ordination of Plant Communities.* Madison, Univ. Wisconsin Press.

Curtis, J. T., and R. P. McIntosh, 1950, The interrelations of certain analytic and synthetic phytosociological characters. *Ecol.,* v. 31, p. 434–455.

Curtis, J. T., and R. P. McIntosh, 1951, An upland forest continuum in the prairie–forest border region of Wisconsin. *Ecol.,* v. 32, p. 476–496.

Cutchis, P., 1974, Stratospheric ozone depletion and solar ultraviolet radiation on Earth. *Science,* v. 184, p. 13–19.

Cwyner, L. C., and J. C. Ritchie, 1980, Arctic steppe-tundra: A Yukon perspective. *Science,* v. 208, p. 1375–1377.

Daly, M., 1978, The cost of mating. *Am. Natur.,* v. 112(986), p. 771–774.

Dammermann, K. W., 1948, The fauna of Krakatau 1882–1933. *Verhandel. Koninkl. Ned. Akad. Wetenschap. Afdel. Natuurk.,* v. 44, p. 1–594.

Dansereau, P., 1951, Description and recording of vegetation on a structural basis. *Ecol.,* v. 32, p. 172–229.

Darlington, P. J., 1957, *Zoogeography: The Geographical Distribution of Animals.* New York, Wiley.

Darlington, P. J., 1965, *Biogeography of the Southern End of the World.* Cambridge, Harvard Univ. Press.

Davidson, J., and H. G. Andrewartha, 1948, The influence of rainfall, evaporation, and atmospheric temperature on fluctuations in the size of a natural population of *Thrips imaginis* (Thysanoptera). *J. of Animal Ecol.,* v. 17, p. 193–199.

Davies, N. B., 1977, Prey selection and social behaviour in wagtails (Aves: Motacillidae). *J. of Animal Ecol.,* v. 46, p. 37–57.

Davies, N. B., 1978, Ecological questions about territorial behavior, p. 317–350 *in* Krebs, J. R., and N. B. Davies, *eds., Behavioural Ecology: An Evolutionary Approach.* Sunderland, Mass., Sinauer Associates.

Davis, D. E., and F. B. Golley, 1963, *Principles of Mammology.* New York, Reinhold.

Davis, M. B., 1963, On the theory of pollen analysis. *Am. J. Sci.,* v. 261, p. 897–912.

Davis, M. B., 1965, Phytogeography and Palynology of northeastern United States *in* Wright, H. E., and D. G. Frey, *eds., Quat. of the U.S.* Princeton, Princeton Univ. Press.

Davis, M. B., 1969, Climatic changes in southern Connecticut recorded by pollen deposition at Rogers Lake. *Ecol.,* v. 50, p. 409–422.

Davis, M. B., 1982, Holocene vegetational history of the Eastern United States, p. 166–181 *in* Wright, H. E., Jr., *ed., Late-Quaternary Environments of the U.S.* Minneapolis, Univ. of Minn. Press.

Davis, R. B., 1967, Pollen studies on near-surface sediments in Maine lakes, p. 143–173 *in* Cushing, E. J., and H. E. Wright, *eds., Quarternary Paleoecol.* New Haven, Yale Univ. Press.

Dawkins, R., 1976, *The Selfish Gene.* New York, Oxford Univ. Press.

DeBach, P., *ed.,* 1964, *Biological Control of Insect Pests and Weeds.* New York, Reinhold.

deCandolle, A. P. A., 1874, Constiturion dans le regne vegetal de groupes physiologiques applicables a la geographie ancienne et moderne. *Archives des Sciences Physiques et Naturelles,* Geneva, Switzerland.

deCandolle, A. P. A., and C. deCandolle, 1824–1873, *Prodomus systematis naturalis regni vegetabilis,* 17 vols. Paris.

DeCosta, J., 1968, The history of the Chydorid (Cladocera) community of a small lake in the Wind River Mountains, Wyoming, U.S.A. *Archiv fur Hydrobiol.,* v. 64, p. 400–425.

Deevey, E. S., 1939, Studies on Connecticut lake sediments. I. A post-glacial climatic chronology for southern New England. *Am. J. Sci.,* v. 237, p. 691–724.

Deevey, E. S., 1947, Life tables for natural populations of animals. *Quart. Rev. of Biol.,* v. 22, p. 283–314.

Deevey, E. S., 1949, Biogeography of the Pleistocene. *Bull. Geol. Soc. Am.,* v. 60(9), p. 1315–1416.

Deevey, E. S., 1955, The obliteration of the hypolimnion. *Mem. 1st Ital. Idrobiol., Suppl.,* v. 8, p. 9–38.

Deevey, E. S., 1959, The hare and the haruspex: A cautionary tale. *Yale Rev.,* v. 49, p. 161–179.

Deevey, E. S., 1969, Coaxing history to conduct experiments. *BioSci.,* v. 19, p. 40–43.

Deevey, E. S., 1970, In defense of mud. *Bull. Ecol. Soc. Am.,* v. 51, p. 5–8.

Deevey, E. S., D. S. Price, P. M. Rice, H. H. Vaughan, M. Brenner, and M. S. Flannery, 1979, Mayan urbanism: Impact on a tropical karst environment. *Science,* v. 206, p. 298–306.

Delcourt, H. R., 1979, Late quaternary vegetation history of the eastern highland rim and adjacent Cumberland Plateau of Tennessee. *Ecol. Monog.,* v. 49(3), p. 255–280.

Delwiche, C. C., 1970, The nitrogen cycle. *Scientific Am.,* v. 223, p. 136–146.

Diamond, J. M., 1969, Avifaunal equilibria and species turnover rates on the Channel Islands of California. *Proc. Natl. Acad. Sci.,* v. 64, p. 57–63.

Diamond, J. M., 1973, Distributional ecology of New Guinea birds. *Science,* v. 179, p. 759–769.

Diamond, J. M., and R. M. May, 1976, Island biogeography and the design of natural reserves, p. 163–186, *in* May, R. M., *ed., Theoretical Ecology: Principles and Applications.* Philadelphia, Saunders.

Dirks, C. O., 1937, Biological studies of Maine moths by light trap methods. *Maine Agr. Exp. Station Bull. 389,* Orono, Maine.

Dodd, A. P., 1940, The Biological Campaign against Prickly Pear. Brisbane, Australia, Commonwealth Prickly Pear Board.

Dowding, P., F. S. Chapin, III, F. E. Wielgolaski, and P. Kilfeather, 1981, Nutrients in Tundra Ecosystems, p. 647–683 *in* Bliss, L. C., O. W. Heal, and J. J. Moore, *eds., Tundra Ecosystems: a Comparative Analysis.* Cambridge, Mass., Cambridge Univ. Press.

Downhower, J. F., 1971, Darwin's finches and the evolution of sexual dimorphism in body size. *Nature,* v. 263, p. 558–563.

Downhower, J. F., and K. B. Armitage, 1971, The yellow-bellied marmot and the evolution of polygamy. *Am. Natur.,* v. 105, p. 355–370.

Drury, W. H., and I. C. T. Nisbet, 1973, *Succession.* J. Arnold Arboretum, v. 54, p. 331–368.

Duffey, E., M. G. Morris, J. Sheail, L. K. Ward, D. A. Wells, and T. C. E. Wells, 1974, *Grassland Ecology and Wildlife Management.* London, Chapman & Hall.

DuRietz, G. E., 1929, The fundamental units of vegetation. *Proc Int. Cong. Plant Sci. Ithace,* v. 1, p. 623–627.

DuRietz, G. E., 1930, Classification and nomenclature of vegetation. *Svensk Botanisk Tidskrift,* v. 24, p. 489–503.

Duvigneaud, P., and S. Denaeyer–de Smet, 1970, Biological cycling of minerals in temperate deciduous forests, p. 199–225 *in* Reichle, D. E., *ed., Analysis of Temperate Forest Ecosystems.* Berlin/Heidelberg/New York, Springer–Verlag.

Edmondson, W. T., 1945, Ecological studies of sessile rotatoria, part II. Dynamics of populations and social structures. *Ecol. Monog.,* v. 15, p. 141–172.

Edmondson, W. T., 1970, Phosphorus, nitrogen, and algae in Lake Washington after diversion of sewage. *Science,* v. 169, p. 690–691.

Edmondson, W. T., and J. T. Lehman, 1981, The effect of changes in the nutrient income on the condition of Lake Washington. *Limnology and Oceanography,* v. 26, p. 1–29.

Edmunds, G. F., Jr., and D. N. Alstad, 1978, Coevolution in insect herbivores and conifers. *Science,* v. 199, p. 941–945.

Einarsen, A. S., 1942, Specific results from ring-necked pheasant studies in the Pacific Northwest. *Trans. No. Am. Wildlife Conf.,* v. 7, p. 130–145.

Eldredge, N., and S. J. Gould, 1976, Rates of evolution revisited. *Paleobiology,* v. 2, p. 174–179.

Elias, T. S., and H. Gelband, 1975, Nectar: Its production and functions in trumpet creeper. *Science,* v. 189, p. 289–291.

Elner, R. W., and R. N. Hughes, 1978, Energy maximization in the diet of the shore crab, *Carcinus naenas* (L.). *J. of Animal Ecol.,* v. 47, p. 103–116.

Elton, C. S., 1927, *Animal Ecol.* New York, Macmillan.

Elton, C. S., 1942, *Voles, Mice, and Lemmings.* New York, Oxford Univ. Press.

Elton, C. S., 1966, *The Pattern of Animal Communities.* London, Methuen.

Emiliani, C., 1966, Isotopic paleotemperatures. *Science,* v. 154, p. 851–856.

Emlen, S. T., and Oring, L. W., 1977, Ecology, sexual selection, and the evolution of mating systems. *Science,* v. 197, p. 215–223.

Erdtman, G., 1969, *Handbook of Palynology.* New York, Hafner.

Eriksson, E., 1959, The yearly circulation of chloride and sulfur in nature: Meteorological, geochemical and pedological implications, Part I. *Tellus,* v. 11, p. 375–403.

Eriksson, E., 1960, The yearly circulation of chloride and sulfur in nature: meteorological, geochemical and pedological implications, Part 2. *Tellus,* v. 12, p. 63–109.

Errington, P. L., 1946, Predation and vertebrate populations. *Quar. Rev. of Biol.,* v. 21, p. 144–177, v. 21, p. 221–245.

Errington, P. L., 1963, *Muskrat Populations.* Ames, Iowa State Univ. Press.

Evans, G. H., 1970, Pollen and diatom analysis of late-Quaternary deposits in the Blelham Basin, North Lancashire. *New Phytologist,* v. 69, p. 821–874.

Eyre, S. R., 1968, *Vegetation and Soils.* London, Arnold.

Faegri, K., and J. Iversen, 1964, *Textbook of Pollen Analysis,* 2nd ed. New York, Hafner.

Farmer, J. G., and M. S. Baxter, 1974, Atmospheric carbon dioxide levels as indicated by the stable isotope record in wood. *Nature,* v. 247, p. 273–275.

Farmworth, E. G., and F. B. Golley, *eds.,* 1973, *Fragile Ecosystems.* New York, Springer–Verlag.

Fedorov, V. D., and T. G. Gilmanov, 1980, *Ecology,* Moscow.

Feeny, P., 1975, Biochemical coevolution between plants and their insect herbivores, p. 3–19 *in* Gilbert, L. E., and P. H. Raven, *eds., Coevolution of Animals and Plants.* Austin, Austin Univ. Press.

Feller, W., 1951, *Probability Theory and its Applications.* New York, Wiley.

Ferguson, E. E., and W. F. Libby, 1971, Mechanism

for the fixation of nitrogen by lightning. *Nature,* v. 229, p. 37.

Finegan, B., 1984, Forest succession. *Nature,* v. 312, p. 109–114.

Finerty, J. P., 1980, *The Population Ecology of Cycles and Small Mammals.* New Haven, Yale Univ. Press.

Fischer, A. G., 1960, Latitudinal variations in organic diversity. *Evol.,* v. 14, p. 64–81.

Fitch, H. S., 1965, An ecological study of the garter snake, *Thamnophis sirtalis. Univ. of Kansas Mus. of Nat. Hist.,* v. 15, p. 493–564.

Fleischer, R. L., 1979, Where do nuclear tracks lead? *Am. J. Sci.,* v. 67, p. 194–203.

Flint, R. F., 1971, *Glacial and Quarternary Geology.* New York, Wiley.

Flohn, H., 1969, *Climate and Weather.* New York, McGraw–Hill.

Forbes, S. A., 1887, The lake as a microcosm. *Bull. of Peoria (Illinois) Sci. Assoc.,* Reprinted in *Ill. Nat. Hist. Surv.,* v. 15, p. 537–550, (1925).

Force, D. C., 1974, Ecology of insect host–parasitoid communities. *Science,* v. 184, p. 624–632.

Fretwell, S. D., 1972, *Populations in a seasonal environment.* Princeton, Princeton Univ. Press.

Frey, D. G., 1953, Regional aspects of the late-glacial and post-glacial pollen succession of southeastern North Carolina. *Ecol. Monog.,* v. 23, p. 289–313.

Frey, D. G., 1969, The ecological significance of Cladoceran remains in lake sediments. *Ecol.,* v. 41, p. 785–790.

Fridriksson, S., 1975, *Surtsey.* New York/Toronto, Wiley.

Friedman, I., and F. W. Trembour, 1978, Obsidian: The dating stone. *Am. J. Sci.,* v. 66, p. 44–51.

Fritts, H. C., 1976, *Tree Rings and Climate.* New York, Academic Press.

Gaarder, T., and H. H. Gran, 1927, Investigations of the production of plankton in the Oslo Fjord. *Rapp. et Proc. Verb., Cons. Int. Explor. Mer.,* v. 42, p. 1–48.

Gaastra, P., 1958, Light energy conversion in field crops in comparison with photosynthetic efficiency under laboratory conditions. *Medeleel, Landbouwhogeschool Wageningin,* v. 58(4), p. 1–12.

Gamms, H., 1918, Prinzipienfragen der vegetations forschung. Ein beitrag zur begriffsklarung und methodik der biocoenologie. *Naturf. Gesell. Zurich, Vierteljahrschr,* v. 63, p. 293–493.

Garbett, G. G., 1981, The elm decline: The depletion of a resource. *New Phytologist,* v. 88, p. 573–585.

Garrels, R. M., and F. T. Mackenzie, 1971, *Evolution of Sedimentary Rocks.* New York, Norton.

Garrells, R. M., F. T. Mackenzie, and C. Hunt, 1975, *Chemical Cycles and the Global Environment.* Los Altos, Ca., Kaufmann.

Gates, D. M., 1968*a,* Energy exchange between organisms and environment. *Australian J. Sci.,* v. 31, p. 67–74.

Gates, D. M., 1968*b,* Energy exchange between organisms and environment, p. 1–22 *in* Lowry, W. P., *ed., Biometeor. Proc. of 28th An. Biol. Colloquium 1967.* Corvallis, Oregon State Univ. Press.

Gauld, G. T., 1951, The grazing rate of planktonic copepods. *J. Marine Biol. Assoc. of U.K.,* v. 29, p. 695–706.

Gause, G. F., 1934, *The Struggle for Existence.* Baltimore, Williams & Wilkins.

Gause, G. F., 1936, The principles of biocenology. *Quart. Rev. of Biol.,* v. 11, p. 320–396.

Georgi, J., 1934, *Mid-Ice* (translated by Lyon, F. H.) London, Kegan Paul, Trench, Trübner & Co., Ltd.

Gerloff, G. C., and F. Skoog, 1954, Cell contents of nitrogen and phosphorus as a measure of their availability for growth of *Microcystis aeruginosa. Ecol.,* v. 35, p. 348–353.

Gieskes, J. M., 1974, The alkalinity–total carbon dioxide system in seawater, p. 123–151 *in* Goldberg, E. D., *ed., The Sea: Vol. 5.* New York, Wiley.

Gilbert, J. J., 1980, Feeding in the rotifer *Asplanchna:* Behavior, cannibalism, selectivity, prey defenses, and impact on rotifer communities, p. 158–172 *in* Kerfoot, W. C., *ed., Evolution and Ecology of Zooplankton Communities.* Hanover, N.H., Univ. Press of New England.

Gilbert, L. E., 1971, Butterfly-plant coevolution: Has *Passiflora adenopoda* won the selectional race with Heliconiine butterflies? *Science,* v. 172, p. 585–586.

Gilbert, L. E., and M. C. Singer, 1975, Butterfly ecology. *An. Rev. Ecol. and Systematics,* v. 6, p. 365–397.

Gilpin, M. E., 1973, Do hares eat lynx? *Am. Natur.,* v. 107, p. 727–730.

Gilpin, M. E., 1975, *Group Selection in Predator–Prey Communities.* Princeton, Princeton Univ. Press.

Gleason, H. A., 1917, The structure and development of the plant association. *Bull. Torrey Bot. Club,* v. 43, p. 463–481.

Gleason, H. A., 1926, The individualistic concept of the plant association. *Bull. Torrey Bot. Club,* v. 53, p. 7–26.

Glesener, R. R., and D. Tilman, 1978, Sexuality and the components of environmental uncertainty: Clues from geographic parthenogenesis in terrestrial animals. *Am. Natur.*, v. 112(968), p. 659–673.

Godwin, H., 1956, *The History of the British Flora, a Factual Basis for Phytogeography.* Cambridge, Cambridge Univ. Press.

Goldman, C. R., 1975, The role of minor nutrients in limiting the productivity of aquatic ecosystems. *Special Symp. Am. Soc. Limn. and Oceanog.*, v. 1, p. 21–33.

Goldsmith, E., and R. Allen, 1972, A blueprint for survival. *Ecol.*, v. 2, p. 1–25.

Goodman, D., 1974, Natural selection and a cost-ceiling on reproductive effort. *Am. Natur.*, v. 113, p. 735–748.

Goodman, D., 1975, The theory of diversity–stability relationships in Ecology. *Quar. Rev. of Biol.*, v. 50, p. 237–266.

Goodman, D., 1979, Management implications of the mathematical demography of long lived animals. National Technical Information Service (NTS No. PB-289 673), Springfield, VA, U.S. Dept. of Commerce.

Gorham, E., J. W. G. Lund, J. E. Sanger, and W. E. Dean, 1974, Some relationships between algal standing crop, water chemistry, and sediment chemistry in the English lakes. *Limnology and Oceanography*, v. 14, 317–326.

Gosz, J. R., R. T. Holmes, G. E. Likens, and F. H. Bormann, 1978, The flow of energy in a forest ecosystem. *Scientific Am.*, v. 238(3), p. 92–102.

Gosz, J. R., G. E. Likens, and F. H. Bormann, 1975, Organic matter and nutrient dynamics of the forest and forest floor in the Hubbard Brook Forest. *Oecologia*, v. 22, p. 305–320.

Goulden, C. E., 1966, La Aquada de Santa Ana Vieja: An interpretive study of the Cladoceran microfossils. *Archiv fur Hydrobiol.*, v. 62, p. 373–404.

Grant, P. R., 1975, The classical case of character displacement. *Evol. Biol.*, v. 8, p. 237–337.

Gray, J., and P. Thompson, 1977, Climatic information from $^{18}O/^{16}O$ analysis of cellulose, lignin and whole wood from tree rings. *Nature*, v. 270, p. 708–709.

Green, R. G., C. L. Larson, and J. F. Bell, 1939, Shock disease as the cause of the periodic decimation of snowshoe hares. *Am. J. Hygiene, Sect. b*, v. 30, p. 83–102.

Griffing, B., 1967, Selection in reference to biological groups, I. Individual and group selection applied to populations of unordered groups. *Aust. J. Bio. Sci.*, v. 20, p. 127–139.

Grigg, R. W., and J. E. Maragos, 1974, Recolonization of hermatypic corals on submerged lava flows in Hawaii. *Ecol.*, v. 55, p. 387–395.

Grime, J. P., 1979, *Plant Strategies and Vegetation Processes.* New York, John Wiley & Sons.

Grinnel, J., 1904, The origin and distribution of the chestnut-backed chicadee. *Auk*, v. 21, p. 364–382.

Grubb, T. C., Jr., 1975, Weather-dependent foraging behavior of some birds wintering in a deciduous woodland. *Condor*, v. 77, p. 175–182.

Grubb, T. C., Jr., 1978, Weather-dependent foraging rates of wintering birds. *Auk*, v. 95, p. 370–376.

Haeckel, E., 1866, *Generelle Morphologie der Organismen*, Berlin, Reimer, 2 vols.

Hainsworth, F. R., and L. L. Wolf, 1970, Regulation of oxygen consumption and body temperature during torpor in a hummingbird, *Eulampis jugularis. Science*, v. 168, p. 368–369.

Hairston, N. G., F. E. Smith, and L. B. Slobodkin, 1960, Community structure, population control, and competition. *Am. Natur.*, v. 94, p. 421, 425.

Hall, C. A. S. and J. W. Day, 1977, *Ecosystem Modelling in Theory and Practice.* New York, Wiley.

Hall, C. A. S., A. Eckdahl, and D. E. Wartenberg, 1975, A fifteen-year record of biotic metabolism in the northern hemisphere. *Nature*, v. 225, p. 136–138.

Hall, C. A. S., and R. M. Moll, 1975, Methods of assessing aquatic primary productivity, p. 19–53, *in* Lieth, H., and R. H. Whittaker, *eds.*, *Primary Productivity of the Biosphere.* New York, Springer–Verlag.

Hames, R. B. and W. T. Vickers, *eds.*, 1983, *Studies in Anthropology.* New York, Academic Press.

Hamman, O., 1979, On climatic conditions, vegetation types, and leaf size in the Galapagos Islands. *Biotropica*, v. 11, p. 101–122.

Hardin, G., 1960, The competitive exclusion principle. *Science*, v. 131, p. 1292–1297.

Hardy, A. C., 1924, The herring in relation to its animate environment: Part 1. *Min. Agri. and Fish, Fishery Investigations*, Series 2, v. 7, p. 1–57.

Hardy, A. C., 1956, *The Open Sea and the World of Plankton.* London, Collins.

Harper, J. L., 1969, The role of predation in vegetational diversity. *Brookhaven Symp. in Biol. No. 22, Diversity and Stability in Ecol. Syst.*, p. 48–62.

Harper, J. L., 1977, *Population Biology of Plants.* London, Academic Press.

Harris, W. F., R. A. Goldstein, and G. S. Henderson, 1973, Analysis of forest biomass pools, annual primary production and turnover of biomass for a mixed deciduous forest watershed, p. 41–64, *in* Young, H. E., *ed., IUFRO Biomass Studies: International Union of Forest Research Organization Papers.* Orono, Maine.

Harvey, H. W., 1950, On the production of living matter in the sea off Plymouth. *J. Mar. Biol. Assoc. of U.K.,* v. 29, p. 97–137.

Hatch, M. D. and C. R. Slack, 1979, Photosynthetic CO_2-fixation pathways. *An. Rev. Plant Physiol.,* v. 21, p. 141–162.

Hays, J. D., J. Imbrie, and N. J. Shackleton, 1976, Variations in the earth's orbit: Pacemaker of the ice ages. *Science,* v. 194, p. 1121–1132.

Heatewole, H., and R. Levins, 1972, Trophic structure, stability and faunal change during recolonization. *Ecol.,* v. 53, p. 531–534.

Hedges, R. E. M., 1979, Radioisotope clocks in archaeology. *Nature,* v. 281, p. 19–24.

Heinselman, M. L., 1963, Forest sites, bog processes, and peatland types in the glacial Lake Agassiz region, Minnesota. *Ecol. Monog.,* v. 33, p. 327–374.

Heinselman, M. L., 1975, Boreal peatlands in relation to environment, p. 93–103 *in* Hasler, A. D., *ed., Coupling of Land and Water Systems, Ecological Studies 10:* New York, Springer–Verlag.

Henderson, L. J., 1913, *The Fitness of the Environment.* New York, Macmillan.

Hensley, M. M., and J. R. Cope, 1951, Further data on removal and repopulation of the breeding birds in a spruce-fir forest community. *Auk,* v. 68, p. 483–493.

Hespenheide, H. A., 1979, Are there fewer parasitoids in the tropics? *Am. Natur.,* v. 113, p. 766–769.

Hibbard, C. W., C. E. Ray, D. E. Savage, D. W. Taylor, and J. E. Guilday, 1965, Quaternary mammals of North America, p. 509–525, *in* Wright, H. E., Jr., and D. G. Frey, *eds., The Quaternary of the United States.* Princeton, Princeton Univ. Press.

Hillis L., 1966, Distribution of marine algae in the Bay of Fundy, New Brunswick, Canada. *Proc. of 5th Int. Seaweed Symp., Halifax, N.S.,* p. 92–98.

Hillis–Colinvaux, L., 1980, The genus *Halimeda:* Primary producer in coral reefs. *Adv. in Mar. Biol.,* v. 17, p. 1–327.

Hobbie, J. E., 1975, Carbon and Eutrophication, *in* Likens, G. E., *ed.,* Nutrients and eutrophication: The limiting nutrient controversy. *Special Symp. Am. Soc. Limnol. and Oceanog.,* v. 1, p. 41–110.

Holland, H. D., 1984, *The Chemical Evolution of the Atmosphere and Oceans:* Princeton, Princeton Univ. Press.

Holling, C. S., 1959, The components of predation as revealed by a study of small mammal predation of the European pine sawfly. *Can. Entomol.,* v. 91, p. 293–320.

Holling, C. S., 1965, The functional response of predators to prey density and its role in mimicry and population regulation. *Mem. Entomol. Soc. of Can.,* v. 46, p. 60.

Holling, C. S., 1968, The Tactics of a Predator: Symp. of Royal Entomol. Soc. of London, Number 4, Insect Abundance, Southwood, T.R.E., *ed.*

Holloway, J. K., 1964, Projects in biological control of weeds, p. 650–670 *in* DeBach, P., *ed., Biological Control of Insect Pests and Weeds.* New York, Reinhold.

Hopkins, D. M., 1959, Some characteristics of the climate in forest and tundra regions in Alaska. *Arctic,* v. 12, p. 215–220.

Hopkins, D. M. *ed.,* 1967, *The Bering Land Bridge.* Stanford, Stanford Univ. Press.

Horie, S., 1972–1981, *Paleolimnology of Lake Biewa and the Japanese Pleistocene.* Otsu, Japan, Kyoto Univ. Press. Nine vol.

Horn, H. S., 1968, Adaptive significance of colonial nesting in the Brewer's blackbird (*Euphagus cyanocephalus*). *Ecol.,* v. 49, p. 682–694.

Horn, H. S., 1971, *The Adaptive Geometry of Trees.* Princeton, Princeton Univ. Press.

Horn, H. S., 1974, The Ecology of Secondary Succession. *An. Rev. Ecol. and Systematics.,* v. 5, p. 25–37.

Horn, H. S., 1975, Markovian Properties of Forest Succession, p. 196–211, *in* Cody, M. L., and J. M. Diamond, *eds., Ecology and Evolution of Communities.* Cambridge, Harvard Univ. Press.

Horn, H. S., 1976, Succession, p. 187–204, *in* May, R. M., *ed., Theoretical Ecology, Principles and Applications.* Philadelphia, Saunders.

Horne, R. A., 1969, *Marine Chemistry: The Structure of Water and the Chemistry of the Hydrosphere.* New York, Wiley.

Hornocker, M. G., 1969, Winter territoriality in mountain lions: *J. Wildlife Management,* v. 33, p. 457–464.

Howard, H. E., 1920, *Territory in Bird Life.* New York, Dutton.

Hrdy, S. B., 1981, *The Woman That Never Evolved.* Cambridge, Harvard Univ. Press.

Hubbel, S. P., 1979, Tree dispersion, abundance, and

diversity in a tropical dry forest. *Science,* v. 203, p. 1299–1308.

Huckel, W., 1951, Structural Chemistry of Inorganic Compounds (transl. by Long, L. H.) v. 11. New York, Elsevier.

Huffaker, C. B., 1958, Experimental studies on predation: Dispersion factors and predator-prey oscillations. *Hilgardia,* v. 27, p. 343–383.

Humphreys, W. F., 1979, Production and respiration in animal populations. *J. of Animal Ecol.,* v. 48, p. 427–453.

Hurlbert, S. H., 1971, The nonconcept of species diversity: A critique and alternative parameters. *Ecol.,* v. 52, p. 577–586.

Hutchinson, G. E., 1949, Circular causal systems in ecology. *An. of N.Y. Acad. of Sci.,* v. 51, p. 221–246.

Hutchinson, G. E., 1950, The biogeochemistry of vertebrate excretion. *Bull. Am. Mus. Nat. Hist.,* v. 96, p. 554.

Hutchinson, G. E., 1951, Copepodology for the ornithologist. *Ecol.,* v. 32, p. 571–577.

Hutchinson, G. E., 1957, Concluding remarks. *Cold Spring Harbor Symp. on Quantitative Biol., 22, Population Studies: Animal Ecology and Demography.* Cold Spring Harbor, Biol. Lab., p. 415–427.

Hutchinson, G. E., 1959, Homage to Santa Rosalia or why are there so many kinds of animals? *Am. Natur.,* v. 93, p. 145–159.

Hutchinson, G. E., 1961, The paradox of the plankton. *Am. Natur.,* v. 95, p. 137–145.

Hutchinson, G. E., 1965, *The Ecological Theater and the Evolutionary Play,* New Haven, Yale Univ. Press.

Hutchinson, G. E., 1967, *A Treatise on Limnology: Vol. 2, Introduction to Lake Biology and the Limnoplankton,* New York, Wiley–Interscience.

Hutchinson, G. E., 1970, Ianula: An account of the history and development of the Lago di Monterosi, Latium, Italy. *Trans. Am. Philo. Soc.,* v. 60, p. 1–170.

Hutchinson, G. E., 1978, *An Introduction to Population Ecology.* New Haven/London, Yale Univ. Press.

Hutchinson, G. E., and V. T. Bowen, 1950, Limnological studies in Connecticut. IX. A quantitative radiochemical study of the phosphorus cycle in Linsley Pond. *Ecol.,* v. 31, p. 194–203.

Huxley, J. S., 1932, *Problems of Relative Growth.* New York, Dial.

Imbrie, J., and N. G. Kipp, 1971, A new macropaleontological method for quantitative paleoclimatology: Application to a late Pleistocene Caribbean core, p. 71–181. New Haven/London, Yale Univ. Press *in* Turekian, K. K., *ed., Late Cenozoic Glacial Ages.*

Iversen, J., 1949, The influence of prehistoric man on vegetation. *Danmarks Geol. Undersogelse, IV,* 3(6), p. 1–25.

Ivlev, V. S., 1939, Transformation of energy consumption by *Tubifex tubifex* (Oligochaeta). *Int. Rev. Hydrobiol.,* v. 38, p. 449–458.

Janzen, D. H., 1968, Host plants as islands in evolutionary and contemporary time. *Am. Natur.,* v. 102, p. 592–594.

Janzen, D. H., 1970, Herbivores and the number of tree species in tropical forests: *Am. Natur.,* v. 104, p. 501–528.

Janzen, D. H., 1971, Escape of juvenile *Dioclea megacarp* (*Leguminosae*) vines from predators in a deciduous tropical forest. *Am. Natur.,* v. 105, p. 97–112.

Jenkins, D. G., and N. Shackleton, 1979, Parallel changes in species diversity and paleotemperature in the Lower Miocene. *Nature,* v. 278, p. 50–51.

Jerlov, N. G., 1951, Optical studies of ocean waters. *Rep. Swedish Deep Sea Expedition,* v. 3, p. 1–59.

Johnson, A. W., and J. G. Packer, 1965, Polyploidy and environment in arctic Alaska: *Science,* v. 148, p. 237–239.

Johnson, M. P., and P. H. Raven, 1973, Species number and endemism: The Galapagos archipelago revisited. *Science,* v. 179, p. 893–895.

Jones, M. G., 1933, Grassland management and its influence on the sward. *Empire J. Exper. Agri.,* v. 1, p. 43–57, 122–128, 224–234, 362–367.

Juday, C., 1940, The annual energy budget of an inland lake. *Ecol.,* v. 21, p. 438–450.

Kellog, C. E., 1949, Preliminary suggestion for the classification and nomenclature for great soil groups in tropical and equatorial regions. *Comm. Bur. Soil Sci. Tech. Commun.,* v. 46, p. 76–85.

Kellog, W. W., R. D. Cadle, E. R. Allen, A. L. Lazrus, and E. A. Martell, 1972, The sulfur cycle. *Science,* v. 175, p. 587–596.

Kemeny, J. G., and J. L. Snell, 1960, *Finite Markov Chains.* New York, Van Nostrand.

Kendall, R. L., 1969, An ecological history of the Lake Victoria basin: *Ecol. Monog.,* v. 39, p. 121–176.

Kendeigh, S. C., 1961, *Animal Ecology.* Englewood Cliffs, N.J., Prentice-Hall.

Kennedy, R. A., and W. M. Laetsch, 1974, Plant species intermediate for C3, C4 photosynthesis. *Science,* v. 184, p. 1087–1089.

Kenward, R. E., 1978, Hawks and doves: Attack success and selection in goshawk flights at woodpigeons. *J. of Animal Ecol.*, v. 47, p. 449–460.

Kerfoot, W. C., D. L. Kellogg, and J. R. Strickler, 1980, Visual observations of live zooplankters: Evasion, escape, and chemical defenses, p. 10–27, *in* Kerfoot, W. C., *ed.*, *Evolution and Ecology of Zooplankton Communities.* Hanover, N.H., Univ. Press of New England.

Keyfitz, N., 1968, *Introduction to the mathematics of population.* Reading, Mass., Addison Wesley.

Kibby, H. V., 1971, Energetics and population dynamics of *Diaptomus gracilis. Ecol. Monog.*, v. 41, p. 311–326.

King, C. E., 1964, Relative abundance of species and MacArthur's model. *Ecol.*, v. 45, p. 716–727.

Kira, T., and T. Shidei, 1967, Primary production and turnover of organic matter in different forest ecosystems of the western Pacific. *Japanese J. Ecol.*, v. 17, p. 70–87.

Kitchen, D. W., 1974, Social behavior and ecology of the pronghorn. *Wildlife Monog.*, v. 38, p. 1–96.

Kleiber, M., 1961, *The Fire of Life. An Introduction to Animal Energetics.* New York, Wiley.

Klopfer, P. H., 1969, *Habitats and Territories.* New York, Basic Books.

Kluijver, H. N., 1951, The Population Ecology of the great tit. *Parus m. major L. Ardea*, v. 39, p. 1–135.

Koblentz–Mishke, O. J., V. V. Volkovinsky, and J. G. Kabanova, 1970, Plankton primary production of the world ocean, *in* Wooster, W. S., *ed.*, *Scientific Exploration of the South Pacific.* Wash., D. C., Nat. Acad. Sci. p. 183–193.

Köppen, W., 1884, Die Warmezonen der Erde, nach der Dauer der Heissen, Gemassigten und Kalten Zeit, und nach der Wirkung der Warme auf die Organische Welt betrachter. *Meteorologische Zeitschrift*, v. 1, p. 215–226.

Köppen, W., 1900, Versuch einer Klassifikation der Klimate, Vorzugsweise nach ihren Beziehungen zur Pflanzenwelt. *Geographische Zeitschrift*, v. 6, p. 593–611.

Köppen, W., 1918, Klassifikation der Klimate nach Temperatur, Niederschlat, und Jahres lauf. *Petermann's Mitteilungen*, v. 64, p. 193–203, 243–248.

Krebs, C. J., 1964, The lemming cycle at Baker Lake, Northwest Territories, during 1959–1962. *Arctic Inst. of No. Am., Tech. Paper No. 15*, p. 104.

Krebs, C. J., 1970, *Microtus* population biology: behavioral changes associated with the population cycle in *M. ochrogaster* and *M. pennsylvanicus. Ecol.*, v. 51, p. 34–52.

Krebs, C. J., M. S. Gaines, B. L. Keller, J. H. Myers, and R. H. Tamarin, 1973, Population cycles in small rodents. *Science*, v. 179, p. 35–41.

Krebs, J. R., 1971, Territory and breeding density in the great tit, *Parus major L. Ecol.*, v. 52, p. 2–22.

Krebs, J. R., 1978, Optimal foraging: Decision rules for predator, p. 23–63, *in* Krebs, J. R., and N. B. Davies, *ed.*, *Behavioural Ecology.* Sunderland, Mass., Sinauer.

Krebs, J. R., 1979, Coevolution of bees and flowers. *Nature*, v. 278, p. 689.

Krebs, J. R., J. T. Erichsen, M. I. Webber, and E. L. Charnor, 1977, Optimal prey selection in the great tit (*Parus major*). *An. Behav.*, v. 25, p. 30–38.

Kruuk, H., 1972, *The Spotted Hyena.* Chicago/London, Univ. of Chicago Press.

Kubiena, W. L., 1953, *The Soils of Europe.* London, Murby.

Kuhn, T. S., 1962, *The Structure of Scientific Revolutions.* Chicago, Univ. of Chicago Press.

Kullenberg, B., 1955, Deep sea coring. *Rept. Swedish Deep Sea Expedition*, v. 4(2), p. 35–96.

Lack, D. L., 1944, Ecological aspects of species-formation in passerine birds: *Ibis*, v. 1944, p. 260–286.

Lack, D. L., 1945, Ecology of closely related species with special reference to cormorant (*Phalacrocorax carbo*) and shag (*P. aristolelis*). *J. of Animal Ecol.*, v. 14, p. 12–16.

Lack, D. L., 1947, *Darwin's Finches.* New York, Cambridge Univ. Press.

Lack, D. L., 1954, *The Natural Regulation of Animal Numbers.* New York, Oxford Univ. Press.

Lack, D. L., 1966, *Population Studies of Birds.* London, Oxford Univ. Press.

Lack, D. L., 1968, *Ecological Adaptations for Breeding in Birds.* London, Methuen.

LaMarche, V. C., 1974, Paleoclimatic influences from long tree-ring records. *Science*, v. 183, p. 1043–1048.

LaMarche, V. C., D. A. Graybille, H. C. Fritts, and M. R. Rose, 1984, Increasing atmospheric carbon dioxide: Tree-ring evidence for growth enhancement in natural vegetation. *Science*, v. 225, p. 1019–1021.

Lamb, H. H., 1972, *Climate: Present, Past and Future, v. 1.* London, Methuen.

Lamont, B. B., S. Downes, and J. E. D. Fox, 1977, Influence of temperature on cyanogenic polymorphisms. *Nature*, v. 265, p. 438–441.

Langmuir, I., 1938, Surface motion of water induced by wind. *Science*, v. 87, p. 119–123.

Lawrence, D. B., 1958, Glaciers and vegetation in southeastern Alaska. *Am. Scientist,* v. 46, p. 89–122.

Lawton, J. H. and D. Schroder, 1977, Effects of plant type, size of geographical range and taxonomic isolation on number of insect species associated with British plants. *Nature,* v. 265, p. 137–140.

Lee, D. W., and J. B. Lowry, 1975, Physical Basis and Ecological Significance of Iridescence in Blue Plants. *Nature,* v. 254, p. 50–51.

Leigh, E. G., 1965, On the relationship between productivity, biomass, diversity, and stability of a community. *Proc. Natl. Acad. Sci.,* v. 53, p. 777–783.

Lemon, E. R., 1969, Gaseous exchange in crop sands, p. 117–137, *in* Easterlin, J. D., *ed., Physiological Aspects of Crop Yield.* Madison, Am. Soc. Agronomists.

Lemon, E. R., L. H. Allen, and L. Muller, 1970, Carbon dioxide exchange of a tropical rain forest. II. *BioSci.,* v. 20, p. 1054–1059.

Leopold, A., 1936, *Game Management.* New York, Scribner's.

Leopold, A., 1943, Deer eruptions. Wis. Conserv. Bull., Originally in Wis. Conserv. Dept. Publ., v. 321, p. 1–11.

Leshniowski, W. O., P. R. Dugan, R. M. Pfister, J. I. Frea, and E. I. Randles, 1970, Aldrin: Removal from lake water by flocculent bacteria. *Science,* v. 169, p. 993–995.

Levins, R., 1968, *Evolution in Changing Environments.* Princeton, Princeton Univ. Press.

Levins, R., 1975, Evolution in communities near equilibrium, p. 16–50 *in* Cody, M. L., and J. M. Diamond, *eds., Ecology and Evolution of Communities.* Cambridge, Belknap Harvard.

Lewis, W. M., 1979, *Zooplankton Community Analysis.* New York, Springer–Verlag.

Libby, L. M., L. J. Pandolfi, P. H. Payton, J. Marshall, B. Becker, and V. Giertz–Sienbenlist, 1976, Isotopic tree thermometers. *Nature,* v. 261, p. 284–288.

Lieth, H., 1975*a,* Measurement of caloric values, p. 119–129 *in* Lieth, H., and R. H. Whittaker, *eds., Primary Productivity of the Biosphere.* New York, Springer–Verlag.

Lieth, H., 1975*b,* Primary productivity of the major vegetation units of the world, p. 203–216, *in* Lieth, H., and R. H. Whittaker, *eds., Primary Productivity of the Biosphere.* New York, Springer–Verlag.

Likens, G. E., *ed.,* 1975, Nutrients and eutrophication: The limiting-nutrient controversy. *Special Symp. Am. Soc. Limn. and Oceanog.*

Likens, G. E., and F. H. Bormann, 1972, Nutrient cycling in ecosystems, p. 25–67, *in* Wiens, J. A., *ed., Ecosystem Structure and Function.* Corvallis, Oregon State Univ. Press.

Likens, G. E., F. H. Bormann, N. M. Johnson, D. W. Fisher, and R. S. Pierce, 1970, Effects of forest cutting and herbicide treatment on nutrient budgets in the Hubbard Brook watershed-ecosystem. *Ecol. Monog.,* v. 40, p. 23–47.

Likens, G. E., F. H. Bormann, N. M. Johnson, and R. S. Pierce, 1967, The calcium, magnesium, potassium and sodium budgets for a small forested ecosystem. *Ecol.,* v. 48, p. 772–785.

Likens, G. E., F. H. Bormann, R. S. Pierce, J. S. Eaton, and N. M. Johnson, 1977, *The Biogeochemistry of a Forested Ecosystem.* New York, Springer–Verlag.

Likens, G. E., F. H. Bormann, R. S. Pierce, and W. A. Reiners, 1978, Recovery of a deforested ecosystem. *Science,* v. 199, p. 492–496.

Lindeman, R. L., 1941*a,* Seasonal food-cycle dynamics in a senescent lake. *Am. Midland Nat.,* v. 26, p. 636–673.

Lindeman, R. L., 1941*b,* The developmental history of Cedar Creek Bog, Minnesota. *Am. Midland Nat.,* v. 25, p. 101–112.

Lindeman, R. L., 1942, The trophic dynamic aspects of ecology. *Ecol.,* v. 23, p. 399–418.

Livingstone, D. A., 1955, A lightweight piston sampler for lake sediments. *Ecol.,* v. 36, p. 137–139.

Livingstone, D. A., 1957, On the sigmoid growth phase in the history of Linsley Pond. *Am. J. Sci.,* v. 255, p. 364–373.

Livingstone, D. A., 1963*a,* Data of geochemistry, chapter G. Chemical composition of rivers and lakes, p. 63 in Geol. Surv. Prof. Paper, 440-G, Wash., D.C.

Livingstone, D. A., 1963*b,* The sodium cycle and the age of the ocean. *Geoch. et Cosmoch. Acta,* v. 27, p. 1055–1069.

Livingstone, D. A., 1971, A 22,000-year pollen record from the plateau of Zambia. *Limnology and Oceanography,* v. 16, p. 349–356.

Livingstone, D. A., 1973, Summary and envoi, p. 366–367 *in* Woodwell, G. M., and E. V. Pecan, *eds., Carbon and the Biosphere.* Springfield, Va., U.S. AEC.

Livingstone, D. A., 1975, Late Quaternary climatic change in Africa. *An. Rev. Ecol. and Systematics,* v. 6, p. 249–280.

Livingstone, D. A., 1982, Quaternary geography of

Africa and the refuge theory, p. 523–536 *in* Prance, G. T., *ed., Biological Diversification in the Tropics.* New York, Columbia Univ. Press.

Livingstone, D. A., and W. D. Clayton, 1981, An altitudinal cline in tropical African grass floras and its paleoecological significance. *Quat. Res.,* v. 13, p. 392–402.

Livingstone, D. A., and T. Van der Hammen, 1978, Paleogeography and paleoclimatology. *Tropical Forest Ecosystems, UNESCO/UNEP/FAO,* Chapt. 3, p. 61–90.

Lloyd, M., and R. J. Ghelardi, 1964, A table for calculating the equitability component of species diversity. *J. of Animal Ecol.,* v. 33, p. 217–225.

Long, S. P., L. D. Incoll, H. W. Woolhouse, 1975, C4 photosynthesis in plants from cool temperate regions, with particular reference to *Spartina townsendii. Nature,* v. 257, p. 622–624.

Longwell, C. R., R. F. Flint, and J. E. Sanders, 1969, *Physical Geology.* New York, Wiley.

Loomis, R. S., W. A. Williams and A. E. Hall, 1971, Agricultural productivity. *An. Rev. Plant Physiol.,* v. 22, p. 431–468.

Lotka, A. J., 1925, *Elements of Physical Biology.* Baltimore, Williams & Wilkins. Reprinted as *Elements of Mathematical Biology.* New York, Dover Press, 1956.

Lund, J. W. G., 1965, The ecology of the freshwater phytoplankton. *Biol. Rev.,* v. 40, p. 231–293.

MacArthur, R. H., 1955, Fluctuations of animal populations, and a measure of community stability. *Ecol.,* v. 36, p. 533–536.

MacArthur, R. H., 1957, On the relative abundance of bird species. *Proc. of Natl. Acad. Sci.,* v. 43, p. 293–295.

MacArthur, R. H., 1958*a,* A note on stationary age distributions in single-species populations and stationary species populations in a community. *Ecol.,* v. 39, p. 146–147.

MacArthur, R. H., 1958*b,* Population ecology of some warblers of northeastern coniferous forests. *Ecol.,* v. 39, p. 599–619.

MacArthur, R. H., 1960, On the relative abundance of species. *Am. Natur.,* v. 94, p. 25–36.

MacArthur, R. H., 1961, Community, p. 262–264 *in* Gray, P., *ed., The Encyclopedia of the Biological Sciences.* New York, Reinhold.

MacArthur, R. H., 1964, Environmental factors affecting bird species diversity. *Am. Natur.,* v. 98, p. 387–397.

MacArthur, R. H., 1965, Patterns of species diversity. *Biol. Rev.,* v. 40, p. 510–533.

MacArthur, R. H., 1968, The theory of the niche, p. 159–176 *in* Lewontin, R. C., *ed., Population Biology and Evolution.* Syracuse, Syracuse Univ. Press.

MacArthur, R. H., 1972, *Geographical Ecology.* New York, Harper & Row.

MacArthur, R. H., and J. H. Connell, 1966, *The Biology of Populations.* New York, Wiley.

MacArthur, R. H., and J. MacArthur, 1961, On bird species diversity. *Ecol.,* v. 42, p. 594–598.

MacArthur, R. H., and E. R. Pianka, 1966, On optimal use of a patchy environment. *Am. Natur.,* v. 100, p. 603–609.

MacArthur, R. H., and E. O. Wilson, 1963, An equilibrium theory of insular zoogeography. *Evol.,* v. 17, p. 373–387.

MacArthur, R. H., and E. O. Wilson, 1967, *The Theory of Island Biogeography.* Princeton, Princeton Univ. Press.

Macdonnell, N. B., 1913, On the expectation of life in Ancient Rome, and in the province of Hispania, Lusitania, and Africa. *Biometrika,* v. 9, p. 366–380.

MacFayden, A., 1957, *Animal Ecology: Aims and Methods.* London, Pitman.

MacFayden, A., 1962, Energy flow in ecosystems and its exploitation by grazing, p. 3–20 *in* Crisp, D. J., *ed., Grazing in Terrestrial and Marine Environments.* Oxford, Blackwell.

Machta, L., 1973, Prediction of CO_2 in the atmosphere, p. 21–31 *in* Woodwell, G. M., and E. V. Pecan, *eds., Carbon and the Biosphere.* Springfield, Va., U.S. AEC.

MacIntyre, F., 1970, Why the Sea is Salt. *Scientific Am.,* v. 223(5), p. 104–115.

MacLulick, D. A., 1937, Fluctuations in the numbers of the varying hare (*Lepus americanus*). *Univ. Toronto Studies Bio. Ser.,* v. 43, p. 1–136.

Maguire, B., Jr., 1963, The passive dispersal of small aquatic organisms and their colonization of isolated bodies of water. *Ecol. Monog.,* v. 33, p. 161–185.

Maguire, B., Jr., 1971, Phytotelmata: Biota and community structure determination in plant-held waters. *An. Rev. Ecol. and Systematics,* v. 2, p. 439–464.

Maher, L. J., 1972, Absolute pollen diagram of Redrock Lake, Boulder County, Colorado. *Quat. Res.,* v. 2, p. 531–553.

Margalef, D. R., 1957, La teoria de la información en ecologia. *Memo. de la Real Acad. De Cienc. y Artes de Barcel.,* v. 23(13), p. 79.

Margalef, D. R., 1958, Information theory in ecology. *Gen. Syst.*, v. 111, p. 36–71.

Margalef, D. R., 1963, On certain unifying principles in ecology. *Am. Natur.*, v. 97, p. 374.

Margalef, D. R., 1968, *Perspectives in Ecological Theory.* Chicago, Univ. of Chicago Press, p. 111.

Margulis, L., J. C. G. Walker, and M. Rambler, 1976, Reassessment of roles of oxygen and ultraviolet light in precambrian evolution. *Nature,* v. 264, p. 620–624.

Marks, P. L. and F. H. Bormann, 1972, Revegetation following forest cutting: Mechanisms for return to steady-state nutrient cycling. *Science,* v. 176, p. 914–915.

Marshall, L. G., 1974, Why kangaroos hop. *Nature,* v. 248, p. 174–175.

Martin, P. S., 1958, Pleistocene ecology and biogeography of North America, p. 375–420 *in* Hubbs, C. L., *ed., Zoogeography: Publ. 51,* Amer. Assoc. Adv. Sci., Wash., D.C.

Martin, P. S., 1967, Prehistoric overkill: Pleistocene extinctions, Proc. of VII Cong. INQUA, v. 6, p. 75–120.

Matthews, J. V., 1976, Evolution of the sub-genus *Cyphelophorus* (Genus *Helophorus:* Hydrophilidae, Coleoptera): Description of two new fossil species and discussion of *Helophorus tuberculatus* Gyll. *Canadian J. Zool.*, v. 54, p. 652–673.

Maxwell, D. C., 1974, Marine primary productivity in the Galapagos Islands: Ph.D. dissertation, The Ohio State Univ.

May, R. M., 1972, Limit cycles in predator–prey communities. *Sci.*, v. 177, p. 900–902.

May, R. M., 1973, *Stability and Complexity in Model Ecosystems.* Princeton, Princeton Univ. Press.

May, R. M., 1975, Patterns of species abundance and diversity, p. 81–120, *in* Cody, M. L., and J. M. Diamond, *eds., Ecology and Evolution of Communities,* Cambridge, Belknap, Harvard Univ. Press.

May, R. M., 1976, Models for two interacting populations, p. 49–70, *in* May, R. M., *ed., Theoretical Ecology, Principles and Applications.* Philadelphia, Saunders.

McIntosh, R. P., 1980, The relationship between succession and the recovery process in ecosystems, p. 11–62, *in* Cairns, J., *ed., The Recovery Process in Damaged Ecosystems.* Ann Arbor, Ann Arbor Sci.

McLaren, I. A., 1963, Effects of temperature on growth of zooplankton and the adaptive value of vertical migration. *Jour. of Fisheries Res. Board of Canada,* v. 20, p. 685, 727.

McNab, B. K., 1963, Bioenergetics and the determination of home range size. *Am. Nat.,* v. 97, p. 133–140.

McNeill, S., and J. H. Lawton, 1970, Annual production and respiration in animal populations. *Nature,* v. 225, p. 472–474.

Mech, L. D., 1966, *The Wolves of Isle Royale: Fauna of Nat. Parks of U.S. Fauna* Series 7, Wash., D.C., U.S. Govt. Printing Office, p. 210.

Mech, L. D., 1970, *The Wolf.* New York, Natural History Press.

Merriam, C. H., 1890, The Geographic Distribution of Life in North America: Proc. of Biol. Soc. of Wash., 1892. Reprinted 1893 in Smith. Inst. An. Rep., p. 1–64, 365–415, v. 7.

Mertz, D. B., 1971, The mathematical demography of the California condor population. *Am. Nat.,* v. 105, p. 437–453.

Milliman, J. D., 1974, *Recent Sedimentary Carbonates Part I: Marine Carbonates.* New York, Springer–Verlag.

Mitchell, R., 1981, Insect behavior, resource exploitation, and fitness. *An. Rev. Entom.,* v. 26, p. 373–396.

Mohr, E. C. J., and F. A. VanBaren, 1957, *Tropical Soils.* New York, Interscience.

Monsi, M., and Y. Oshima, 1955, A theoretical analysis of the succession process of plant community, based upon the production of matter. *Jap. J. Bot.*, v. 15, p. 60–82.

Moore, H. B., 1934, The biology of *Balanus balanoides* I growth rate and its relation to size, season, and tide level. *Jour. Marine Biol. Assoc. of U.K., New Series,* v. 19, p. 851–868.

Moore, P. D., 1979, Next in succession. *Nature,* v. 282, p. 361–362.

Moore, P. D., and J. A. Webb, 1978, *Pollen Analysis.* New York, Wiley.

Morgan, A., 1973, Late Pleistocene environmental changes indicated by fossil insect faunas of the English midlands. *Boreas,* v. 2, p. 173–212.

Morowitz, H. J., 1968, *Energy Flow in Biology.* New York, Academic Press.

Morrow, P. A., and V. C. LaMarche, 1978, Tree ring evidence for chronic insect suppression of productivity in subalpine *Eucalyptus. Science,* v. 201, p. 1244–1246.

Mosby, H. S., 1969, The influence of hunting on the population dynamics of a woodlot gray squirrel population. *J. Wildlife Management,* v. 3, p. 59–73.

Mueller–Dombois, D., and H. Ellenberg, 1974, *Aims and Methods of Vegetation Ecology.* New York, Wiley.

Mullen, D. A., 1969, Reproduction in brown lemmings (*Lemmus trimucronatus*) and its relevance to their cycle of abundance. *Univ. of California Pub. Zool.,* v. 85, p. 1–24.

Muller, R. A., 1977, Radioisotope dating with a cyclotron. *Science,* v. 196, p. 489–494.

Murie, A., 1944, *The Wolves of Mount McKinley: Fauna of Nat. Parks of U.S. Fauna Series 5.* Wash., D.C. Govt. Printing Office, p. 238.

Nagy, K. A., 1972, Water and electrolyte budgets of a free-living desert lizard *Sauvomalus obesus.* J. Comp. Physiol., v. 79, p. 39–62.

Nelson, B., 1968, *Galapagos: Islands of Birds:* New York, Morrow.

Nicholson, A. J., 1954, An outline of the dynamics of animal populations. *Aust. J. Zool.,* v. 2, p. 9–65.

Nicholson, A. J., and V. A. Bailey, 1935, The balance of animal populations. *Proc. Zool. Soc. London,* v. 1935, p. 551–603.

Nutman, P. S., 1956, The influence of the legume in root-nodule symbiosis. *Biol. Rev.,* v. 31, p. 109–151.

Nye, R. H., and D. J. Greenland, 1960, The soil under shifting cultivation. *Comm. Bur. Soil Sci. (At. Brit.) Tech. Com., 51.*

O'Brien, W. J., 1979, The predator–prey interaction of planktivorous fish and zooplankton. *Am. Sci.,* v. 67, p. 572–581.

O'Brien, W. J., D. Kettle, and H. P. Riessen, 1979, Helmets and invisible armor: Structures reducing predation from tactile and visual planktivores. *Ecol.,* v. 60, p. 287–294.

Odum, E. P., 1969, The strategy of ecosystem development. *Science,* v. 164, p. 262–270.

Odum, E. P., 1971, *Fundamentals of Ecology,* 3rd ed. Philadelphia, Saunders.

Odum, H. T., 1956, Efficiencies, size of organisms, and community structure. *Ecol.,* v. 37, p. 592–597.

Odum, H. T., *ed.,* 1970, *A Tropical Rain Forest: A Study of Irradiation and Ecology at El Verde, Puerto Rico.* Wash., D.C.

Odum, H. T., 1971, *Environment, Power and Society.* New York, Wiley–Interscience.

Odum, H. T., and E. P. Odum, 1955, Trophic structure and productivity of a windward coral reef community on Eniwetok Atoll: *Ecol. Monog.,* v. 25, p. 291–320.

Olson, J.S., 1958, Rates of succession and soil changes on southern Lake Michigan sand dunes. *Bot. Gaz.,* v. 199, p. 125–170.

Oosting, H. J., 1956, *The Study of Plant Communities,* 2nd ed. San Francisco, Freeman.

Opler, P. A., 1974, Oaks as evolutionary islands for leaf-mining insects. *Amer. Scientist.,* v. 62, p. 67–73.

Orians, G., 1969, On the evolution of mating systems in birds and mammals. *Am. Natur.,* v. 103, p. 589–603.

Orians, G. H., 1980, *Some Adaptations of Marsh-nesting Blackbirds.* Princeton, Princeton Univ. Press.

Orr, R. T., 1971, *Vertebrate Biology,* 3rd ed.: Philadelphia, Saunders.

Osman, R. W., and R. B. Whitlach, 1978, Patterns of species diversity: fact or artifact? *Paleobiology,* v. 4, p. 41–54.

Oster, G. F., and E. O. Wilson, 1978, *Caste and Ecology in the Social Insects.* Princeton, Princeton Univ. Press.

Overton, W. S., 1971, Estimating the number of animals in wildlife populations, p. 403–453 *in* Giles, R. H., *ed., Management Techniques.* Wash., D.C., Wildlife Soc.

Ovington, J. D., 1956, The form, weights and productivity of tree species grown in close stands. *New Phytologist,* v. 55, p. 289–304.

Ovington, J. D., 1957, Dry-matter Production of *Pinus silvestris* L. *Ann. Bot. N.S.,* v. 21, p. 287–314.

Ovington, J. D., 1965, Organic Production, Turnover and Mineral Cycling in Woodlands. *Biol. Rev.,* v. 40, p. 295–336.

Owen, D. F., and J. Owen, 1974, Species diversity in temperate and tropical Ichneumonidae. *Nature,* v. 249, p. 583–584.

Owen–Smith, N., 1971, Territoriality in the white rhinocerus (*Ceratotherium simun*) Burchell. *Nature,* v. 231, p. 294–296.

Paine, R. T., 1966, Food web complexity and species diversity. *Am. Natur.,* v. 100, p. 65–75.

Paine, R. T., 1969, A note on trophic complexity and community stability. *Amer. Natur.,* v. 103, p. 91–93.

Park, T. D., B. Mertz, W. Grodzinski, and T. Prus, 1965, Cannibalistic predation in populations of flour beetles. *Physiol. Zool.,* v. 38, p. 289–321.

Pasteur, G., 1982, A classificatory review of mimicry

systems. *An. Rev. Ecol. Syst.*, v. 13, p. 169–199.

Patrick, R., 1954, A new method for determining the pattern of diatom flora. *Not. Nat. of Acad. Nat. Sci. Phil.*, v. 259, p. 1–12.

Pearl, R., 1928, *The Rate of Living.* New York, Knopf.

Pearl, R., 1932, The influence of density of population upon egg production in *Drosophila melanogaster. J. Exp. Zool.*, v. 63, p. 57–84.

Pearl, R., and S. L. Parker, 1922, Experimental studies on the duration of life, IV: Data on the influence of density of population on duration of life in *Drosophila: Am. Natur.*, v. 56, p. 312–322.

Peet, R. K., 1974, The measurement of species diversity. *Ann. Rev. Ecol. and Systematics,* v. 5, p. 285–307.

Pennak, R. W., 1953, *Freshwater Invertebrates of the United States.* New York, Ronals Press.

Pennington, W., R. S. Cambray, J. D. Eakins, and D. D. Harkness, 1976, Radionuclide dating of the recent sediments at Blelham Tarn. *Freshwater Biol.*, v. 6, p. 317–331.

Perkins, D. F., 1978, The distribution and transfer of energy and nutrients in the *Agrostis-Festuca* grassland ecosystem, p. 375–395, *in* Heal, W. O., and D. F. Perkins, *eds., Production Ecology of British Moors and Montane Grasslands.* New York, Springer–Verlag.

Petersen, R., 1975, The paradox of the plankton: An equilibrium hypothesis. *Am. Natur.*, v. 109, p. 35–49.

Phillips, J., 1934–1935, Succession, development, the climax, and the complex organism: An analysis of concepts. *Ecol.*, v. 22, p. 554–571; v. 23, p. 210–246, 488–508.

Pianka, E., 1969, Habitat specificity, speciation and species density in Australian desert lizards. *Ecol.*, v. 50, p. 498–502.

Pianka, E. R., 1970, On r and K selection. *Am. Natur.*, v. 104, p. 592–597.

Pielou, E. C., 1975, *Ecological Diversity.* New York, Wiley.

Pielou, E. C., 1979, *Biogeography.* New York, Wiley.

Pilcher, J. R., J. Hillam, M. G. Baillie, and G. W. Pearson, 1977, A long sub-fossil oak tree-ring chronology from the north of Ireland. *New Phytologist,* v. 79, p. 713–729.

Pitelka, F. A., 1957, Some characteristics of microtine cycles in the Arctic, p. 73–88 *in* Hansen, P. H., *ed., Arctic Biology, 18th Annual Colloquium.* Corvallis, Oregon State Univ.

Pitelka, F. A., P. Q. Tomich, and G. W. Treichel, 1955, Ecological relations of jaegers and owls as lemming predators near Barrow, Alaska. *Ecol. Monog.*, v. 25, p. 85–117.

Pough, F. H., 1980, The advantages of ectothermy for tetrapods. *Am. Natur.*, v. 115, p. 92–112.

Pough, F. H., 1983, Amphibians and reptiles as low-energy systems, p. 141–188 *in* Aspey, W. P., and S. Lustick, *eds., Behavioral Energetics: The Cost of Survival in Vertebrates.* Columbus, Ohio State Univ. Press.

Prance, G. T., *ed.*, 1982*a*, *Biological Diversification in the Tropics.* New York, Columbia Univ. Press.

Prance, G. T., 1982*b*, Forest refuges: Evidence from woody angiosperms, p. 137–158 *in* Prance, G. T., *ed., Biological Diversification in the Tropics.* New York, Columbia Univ. Press.

Prance, G. T., W. A. Rodrigues, and M. F. da Silva, 1976, Inventario florestal de uma hectare de mata de terra firme km. 30 estrada Manaus-Ilacoatiara. *Acta Amazonica,* v. 6, p. 9–35.

Preston, F. W., 1948, The commonness, and rarity, of species: *Ecol.*, v. 29, p. 254–283.

Preston, F. W., 1962, The canonical distribution of commonness and rarity. *Ecol.* v. 43, p. 185–215, 410–432.

Preston, F. W., 1969, Diversity and stability in ecological systems. *Brookhaven Symp. in Biol. 22,* New York, Upton.

Price, P. W., 1975, *Insect Ecology.* New York/London, Wiley.

Pyke, G. H., 1978, Optimal foraging in bumble bees in coevolution with their plants. *Oecologia,* v. 36, p. 281.

Pyke, G. H., H. R. Pulliam, and E. L. Charnov, 1977, Optimal foraging: A selective review of theory and tests. *Quart. Rev. of Biol.*, v. 52, p. 137–154.

Quinn, W. H., 1971, Late Quarternary meteorological and oceanographic developments in the equatorial Pacific. *Nature,* v. 229, p. 330–331.

Ralph, E. K., and J. Klein, 1979, Composite computer plots of ^{14}C dates for tree-ring-dated bristlecone pines and sequoias, p. 545–553 *in* Berger, R., and H. E. Suess, *eds., Radiocarbon Dating.* Berkeley, Univ. California Press.

Rankama, K., and T. G. Sahama, 1950, *Geochemistry.* Chicago, Univ. of Chicago Press.

Rasmussen, D. I., 1941, Biotic Communities of Kaibab Plateau, Arizona. *Ecol. Monog.*, v. II, p. 229–275.

Rathcke, B. J., and R. W. Pool, 1975, Coevolutionary

race continues: Butterfly larval adaptation to plant trichomes. *Science,* v. 187, p. 175–176.

Rathcke, B. J., and P. W. Price, 1976*a,* Anomalous diversity of tropical ichneumonid parasitoids: A predation hypothesis. *Am. Natur.,* v. 110, p. 889–893.

Rathcke, B. J., and P. W. Price, 1976*b,* Insect plant patterns and relationships in the stem-boring guild. *Amer. Midl. Nat.,* v. 96, p. 98–117.

Raunkiaer, C., 1934, *The Life Forms of Plants and Statistical Plant Geography; Being the Collected Papers of C. Raunkiaer.* London, Oxford Press.

Rechav, Y., R. A. I. Norval, J. Tannock, J. Colborne, 1978, Attraction of the tick *Ixodes neitzi* to twigs marked by the Klipspringer antelope. *Nature,* v. 275, p. 310–311.

Reiners, W. A., 1973, Terrestrial detritus and the carbon cycle, p. 303–327, *in* Woodwell, G. M., and E. V. Pecan, *eds., Carbon and the Biosphere.* New York, Upton, 24th Brookhaven Symp. in Biol.

Revelle, R., 1965, Atmospheric Carbon Dioxide, p. 111–133, Appendix Yr. in *Restoring the Quality of our Environment: Report of the Env. Pol. Panel,* Pres. Sci. Adv. Com. Wash., D.C.

Reynolds, R. C., 1965, The concentration of boron in Precambrian seas. *Geoch. et Cosmoch. Acta,* v. 29, p. 1–16.

Rhoades, D. F., 1979, Evolution of plant chemical defense against herbivores, p. 3–54 *in* Rosenthal, G. A., and D. H. Janzen, *eds., Herbivores. Their Interaction with Secondary Plant Metabolites.* New York/London, Academic Press.

Rhoades, D. F., and R. G. Cates, 1976, Toward a general theory of plant antiherbivore chemistry. *Recent Adv. Phytochem.,* v. 10, p. 168–213.

Rich, E. R., 1956, Egg cannibalism and fecundity in *Tribolium. Ecol.,* v. 37, p. 109–120.

Richards, F. A., 1968, Chemical and biological factors in the marine environment, p. 259–303 *in* Brahtz, J. F., *ed., Ocean Engineering.* New York, Wiley.

Richards, P. W., 1952, *The Tropical Rain Forest.* London, Cambridge Univ. Press.

Richards, P. W., 1973, Africa, the "odd man out," p. 21–26, *in* Meggers, B. J., E. S. Ayensu, and W. D. Duckworth, *eds., Tropical Forest Ecosystems in Africa and South America.* New York, Random House.

Richardson, J. L., and A. E. Richardson, 1972, History of an African rift lake and its climatic implications. *Ecol. Monog.,* v. 42, p. 499–534.

Richman, S., 1958, The transformation of energy by *Daphnia pulex. Ecol. Monog.,* v. 28, p. 273–291.

Ricklefs, R. E., 1973, *Ecology.* Newton, Mass., Chiron Press.

Ricklefs, R. E., and G. W. Cox, 1972, Taxon cycles in the West Indian avifauna. *Am. Natur.,* v. 106, p. 195–219.

Riessen, H. P., 1980, Diel vertical migration of pelagic water mites, p. 122–129 *in* Kerfoot, W. C., *ed., Evolution and Ecology of Zooplankton Communities.* Hanover, N.H., Univ. Press of New Eng.

Riley, G. A., 1963, Marine biology, I., p. 69–70 *in* Riley, G. A., *ed., Proceedings of the First International Interdisciplinary Conference.* Wash., D.C., Am. Inst. Bio. Sci.

Robertson, F. W., and J. H. Sang, 1944, The ecological determinants of population growth in a *Drosophila* culture, II: Circumstances affecting egg viability. *Proc. R. Soc. London Ser. B.* 132, p. 258–277.

Ronov, A. B., 1968, *Probable Changes in the Composition of Sea Water During the Course of Geologic Time.* Edinburgh, VI Int. Sed. Cong.

Root, J. B., 1967, The niche exploitation pattern of the blue-grey gnat catcher. *Ecol. Monog.,* v. 37, p. 317–350.

Rosenthal, G. A., and D. H. Janzen, *eds.,* 1979, *Herbivores. Their Interaction with Secondary Plant Metabolites.* N.Y./London, Academic Press.

Runge, M., 1973, Energievmsatze in den Biozonosen terrestrischer Okosysteme: Scripta Geobot. V. 4, Gottingen, Goltze Verlag.

Russell, J. E., 1953, *Soil Conditions and Plant Growth,* 8th ed. New York, Longmans, Green.

Ruttner, F., 1963, *Fundamentals of Limnology,* 3rd ed. (trans. by Frey, D. G., and F. E. J. Fry). Toronto, Univ. Toronto Press.

Ryther, J. H., 1959, Potential productivity of the sea. *Science,* v. 130, p. 602–608.

Ryther, J. H., 1969, Photosynthesis and fish production in the sea. *Science,* v. 166, p. 72–76.

Salgado–Labouriau, M. L., 1980, A pollen diagram of the Pleistocene–Holocene boundary of Lake Valencia, Venezuela. *Rev. Palaeobot. and Palynol.,* v. 30, p. 297–312.

Sanders, H. L., 1968, Marine benthic diversity: A comparative study. *Am. Natur.,* v. 102, p. 243–282.

Sanders, H. L., and R. R. Hessler, 1969, Ecology of the deep-sea benthos. *Science,* v. 163, p. 1419–1424.

Saunders, A. A., 1936, *Ecology of the Birds of Quaker Run Valley, Allegheny State Park.* New York State

Mus. Handbook, 16. Albany, New York.

Schaller, G. B., 1967, *The Deer and the Tiger.* Chicago, Chicago Univ. Press.

Schaller, G. B., 1972, *The Serengetti Lion.* Chicago/London, Chicago Univ. Press.

Schmidt–Nielsen, K., 1964, *Desert Animals. Physiological Problems of Heat and Water.* New York, Oxford Univ. Press.

Schmidt–Nielsen, K., 1972*a, How Animals Work.* London/New York, Cambridge Univ. Press.

Schmidt–Nielsen, K., 1972*b,* Locomotion: Energy costs of swimming, flying, and running. *Science,* v. 177, p. 222–228.

Schoener, T. W., 1971, Theory of feeding strategies. *An. Rev. of Ecol. and Systematics,* v. 11, p. 369–404.

Schoener, T. W., 1974, Resource partitioning in ecological communities. *Science,* v. 185, p. 27–39.

Schoener, T. W., and A. Schoener, 1970, Inverse relation of survival of lizards with island size and avifaunal richness. *Nature,* v. 274, p. 685–687.

Schofield, E., and V. Ahmadjian, 1973, Field observations and laboratory studies of some Antarctic cold desert cryptogams, p. 97–142 *in* Llano, G. A., *ed., Antarctic Terrestrial Biology.* Wash., D.C. Am. Geophy. Union.

Schofield, E. K., and P. A. Colinvaux, 1969, Fossil *Azolla* from the Galapagos Islands. *Bull. Torrey Bot. Club,* v. 96, p. 623–628.

Schopf, J. W., 1975, Precambrian paleobiology: Problems and perspectives: An. Rev. Earth and Planet, *Science,* v. 3, p. 213–249.

Schultz, A. M., 1969, A study of an ecosystem: The arctic tundra, p. 77–93 *in* VanDyne, G. M., *ed., The Ecosystem Concept in Natural Resource Management.* New York, Academic Press.

Sclater, P. L., 1858, On the general geographical distribution of the members of the class aves. *J. Linnaean Soc. London. (Zool.),* v. 2, p. 130–145.

Selye, H., 1950, *The Physiology and Pathology of Exposure to Stress; a Treatise Based on the Concepts of the General-Adaptation-Syndrome and the Diseases of Adaptation.* Montreal, Acta.

Sernander, R., 1908, On the evidences of postglacial changes of climate furnished by peat mosses of northern Europe. *Geol. Foreningens Forhandlingar,* v. 30, p. 465–473.

Shackleton, N., 1968, Depth of pelagic Foraminifera and isotopic changes in Pleistocene oceans. *Nature,* v. 218, p. 79–80.

Shannon, C. E., and W. Weaver, 1949, *The mathematical theory of communication.* Urbana, Univ. of Ill. Press.

Sheail, J., 1971, *Rabbits and Their History.* Newton Abbot, David and Charles.

Shelford, V. E., 1908, Life-histories and larval habits of the tiger beetles (Cicindelidae). *J. Linnean Soc. (Zool.),* v. 30, p. 157–184.

Shelford, V. E., 1911, Ecological succession II. Pond fishes. *Biol. Bull.,* v. 21, p. 127–151.

Shields, W. M., 1982, *Philopatry, Inbreeding, and the Evolution of Sex.* New York, SUNY Press.

Shimwell, D. W., 1971, *The Description and Classification of Vegetation.* Seattle, Univ. Washington Press.

Siccama, T. G., F. H. Bormann, and G. E. Likens, 1970, The Hubbard Brook Ecosystem Study: Productivity, nutrients, and phytosociology of the herbaceous layer. *Ecol. Monog.,* v. 40, p. 389–402.

Sih, A., 1980, Optimal behavior: Can foragers balance two conflicting demands? *Science,* v. 210, p. 1041–1043.

Sillen, L. G., 1961, The physical chemistry of sea water, p. 549–581 *in* Sears, M., *ed., Oceanography.* Wash., D.C., AAAS.

Simberloff, D., 1974, Equilibrium theory of biogeography and ecology. *An. Rev. Ecol. and Systematics,* v. 5, p. 161–182.

Simberloff, D. S., and E. O. Wilson, 1969, Experimental zoogeography of islands: The colonization of empty islands. *Ecol.,* v. 50, p. 278–296.

Simberloff, D. S., and E. O. Wilson, 1970, Experimental zoogeography of islands. A two-year record of colonization. *Ecol.,* v. 51, p. 934–937.

Simonson, R. W., 1962, Soil classification in the United States. *Science,* v. 137, p. 1027–1034.

Simpson, G. G., 1949, Measurement of diversity. *Nature,* v. 136, p. 688.

Skutch, A. F., 1949, Do tropical birds rear as many young as they can nourish? *Ibis,* v. 91, p. 430–455.

Slobodkin, L. B., 1962, Energy and animal ecology. *Adv. Ecol.,* v. 4, p. 69–101.

Slobodkin, L. B., and H. L. Sanders, 1969, On the contribution of environmental predictability to species diversity, p. 82–95 *in Brookhaven Symp. in Biol. No. 22, Diversity and Stability in Ecological Systems.*

Smiley, J., 1978, Plant chemistry and the evolution of host specificity: New evidence from *Heliconius* and *Passiflora. Science,* v. 201, p. 745–747.

Smith, M. W., 1948, Preliminary observations upon

the fertilization of Crecy Lake, New Brunswick. *Trans. Am. Fisheries Soc.*, v. 75, p. 165–174.

Smuts, J. C., 1926, *Holism and Evolution*. New York, Macmillan.

Snow, D., 1963, The evolution of manakin displays. Proc. XIII Int. Ornith. Cong. New York, Ithaca.

Soholt, L. F., 1973, Consumption of primary production by a population of kangaroo rats (*Dipodomys merriami*) in the Mojave Desert. *Ecol. Monog.*, v. 43, p. 357–397.

Sollbrig, O. T., and B. B. Simpson, 1974, Components of regulation of a population of dandelions in Michigan. *Ecol.*, v. 62, p. 473–486.

Solomon, M. E., 1949, The natural control of animal populations. *J. of Animal Ecol.*, v. 18, p. 1–35.

Spedding, C. R. W., 1971, *Grassland Ecology*. London, Oxford Univ. Press.

Stahl, J. B., 1959, The developmental history of the chironomid and *Chaoborus* faunas of Myers Lake. *Invest. Indiana Lakes and Streams*, v. 2, p. 47–102.

Steeman–Nielsen, E., 1952, The use of radioactive carbon (^{14}C) for measuring organic production in the sea. *J. Cons. Perm. Int. Explor. Mer.*, v. 18, p. 117–140.

Stehli, F. R., R. G. Douglas, and N. D. Newell, 1969, Generation and maintenance of gradients in taxanomic diversity. *Science*, v. 164, p. 947–949.

Steinitz–Kannan, M., P. A. Colinvaux, and R. Kannan, 1983, Limnological studies in Ecuador: I. A survey of chemical and physical properties of Ecuadorian lakes. *Archiv Fur Hydrobiol.*, v. 1, p. 61–105.

Stewart, R. E., and J. W. Aldrich, 1951, Removal and population of breeding birds in a spruce-fir forest community. *Auk*, v. 68, p. 471–482.

Stiles, F. G., 1977, Coadapted competitors: The flowering seasons of hummingbird-pollinated plants in a tropical forest. *Science*, v. 198, p. 1177–1178.

Stockner, J. G., 1972, Paleolimnology as a means of assessing eutrophication. *Verh. Int. Verein. Limnol.*, v. 18, p. 1018–1030.

Strahler, A. N., 1969, *Physical Geography*, 3rd ed. New York, Wiley.

Strickland, J. D. H., and T. R. Parsons, 1972, A practical handbook of seawater analysis. *J. of Fisheries Res. Board of Canada*, v. 167, p. 1–311.

Strickler, J. R., 1977, Observation of swimming performances of planktonic copepods. *Limnology and Oceanography*, v. 22, p. 165–170.

Strong, D. R., Jr., 1974, Rapid asymptotic species accumulation in phytophagous insect communities: the pests of cacao. *Science*, v. 185, p. 1064–1065.

Strong, D. R., Jr., 1984, Exorcising the ghost of competition past: phytophagous insects, p. 28–41, *in* Strong, D. R., Jr., D. Simberloff, L. G. Abele, and A. B. Thistle eds., *Ecological communities. Conceptual issues and the evidence*. Princeton, Princeton Univ. Press.

Strong, D. R., E. D. McCoy, and J. R. Rey, 1977, Time and the number of herbivore species: The pests of sugarcane. *Ecol.*, v. 58, p. 167–175.

Struhsaker, T., 1977, Infanticide and social organization in the redtail monkey (*Cercopithecus ascanius schmidti*) in the Kibale Forest, Uganda. *Zeitschrift fur Tierpsych.*, v. 45, p. 75–84.

Stuiver, M., 1978, Atmospheric carbon dioxide and carbon reservoir changes. *Science*, v. 199, p. 253–258.

Stuiver, M., S. W. Robinson, and I. C. Yang, 1979, ^{14}C dating to 60,000 years B.P. with proportional counters. *Proc. Ninth Int. Radiocarbon Dating Conf.*, Berkeley, Univ. Ca. Press. p. 202–215.

Summerhayes, E. S., and C. S. Elton, 1923, Contributions to the ecology of Spitsbergen and Bear Island. *Ecol.*, v. 11, p. 214–286.

Sussman, R. W., and P. H. Raven, 1978, Pollination by lemurs and marsupials: An archaic coevolutionary system. *Science*, v. 200, p. 731–736.

Talbot, L. M., and M. H. Talbot, 1963, The wildebeest in Western Masailand, East Africa. *Wildlife Monog.*, v. 12, p. 88.

Tanner, J. T., 1978, *Guide to the Study of Animal Populations*. Knoxville, Univ. Tennessee.

Tansley, A. G., 1920, The classification of vegetation and the concept of development. *Ecol.*, v. 8, p. 118–149.

Tansley, A. G., 1935, The use and abuse of vegetational concepts and terms. *Ecol.*, v. 16, p. 284–307.

Tansley, A. G., and R. S. Adamson, 1925, Studies of the vegetation of the English chalk. III. The chalk grasslands of the Hampshire–Sussex border. *Ecol.*, v. 13, p. 177–223.

Taylor, C. R., and V. J. Rowntree, 1973, Running on two or on four legs: Which consumes more energy? *Science*, v. 179, p. 186–187.

Tedrow, J. C. F., 1965, *Arctic Soils*. Proc. First Permafrost Int. Conf., NAS-NRC, Wash., D.C. Publ. 1287, p. 50–55.

Tedrow, J. C. F., and D. E. Hill, 1955, Arctic brown soil. *Soil Sci.*, v. 80, p. 265–275.

Temple, S. A., 1977, Plant animal mutualism: Coevolution with dodo leads to near extinction of plant. *Science,* v. 197, p. 885–886.

Terborgh, J., 1973, Chance, habitat, and dispersal in the distribution of birds in the West Indies. *Evolution,* v. 27, p. 338–349.

Terborgh, J., 1974, Preservation of natural diversity: The problem of extinction prone species. *BioSci.,* v. 24, p. 715–722.

Thomas, M. F., 1974, *Tropical Geomorphology.* New York, Macmillan.

Thompson, D. W., 1942, *On Growth and Form.* London/New York, Cambridge Univ. Press.

Thompson, L. G., S. Hastenrath, and B. M. Arnao, 1978, Climatic ice core records from the tropical Quelccaya ice cap. *Science,* v. 203, p. 1240–1243.

Thompson, R., R. W. Battarbee, P. E. O'Sullivan, and F. Oldfield, 1975, Magnetic susceptibility of lake sediments. *Limnology and Oceanography,* v. 20, p. 687–698.

Till, A. R., 1979, Nutrient cycling, p. 277–285 *in* Coupland, R. T., *ed, Grassland Ecosystems of the World: Analysis of Grasslands and Their Uses.* London/New York, Cambridge Univ. Press.

Tilman, D., 1982, *Resource Competition and Community Structure.* Princeton, Princeton Univ. Press.

Tinbergen, N., 1953, *The Herring Gull's World.* London, Collins.

Towe, K. M., 1978, Early Precambrian oxygen: A case against photosynthesis. *Nature,* v. 274, p. 657–661.

Tranquillini, W., 1979, *Physiological Ecology of the Alpine Timberline.* New York, Springer–Verlag.

Transeau, E. N., 1926, The accumulation of energy by plants. *Ohio J. of Sci.,* v. 26, p. 1–10.

Troughton, J. H., P. V. Wells, and H. A. Mooney, 1974, Photosynthetic mechanisms and paleoecology from carbon isotope ratios in ancient specimens of C_4 and CAM plants. *Science,* v. 185, p. 610–612.

Tsukada, M., 1967, Fossil Cladocera in Lake Nojiri and ecological order. *Quat. Res.* (Japan), v. 6, p. 101–110.

Tucker, V. A., 1975, The energetic cost of moving about. *Amer. Scientist,* v. 63, p. 413–418.

Turekian, K. K., 1968, *Oceans:* Englewood Cliffs, N.J. Prentice–Hall.

Turekian, K. K., 1977, The fate of metals in the oceans. *Geoch. et Cosmoch. Acta,* v. 41, p. 1139–1144.

Urey, H. C., H. A. Lowenstam, S. Epstein, and C. R. McKinney, 1961, Measurement of paleotemperatures of the Upper Cretaceous of England, Denmark, and the southeastern United States. *Bull. Geol. Soc. of Am.,* v. 62, p. 399–416.

Utida, S., 1950, On the equilibrium state of the interacting population of an insect and its parasite. *Ecol.,* v. 31, p. 165–175.

Utida, S., 1957, Cyclic fluctuations of population density intrinsic to the host-parasite system. *Ecol.,* v. 38, p. 442–449.

Valentine, J. W., and C. A. Campbell, 1975, Genetic regulation and the fossil record. *Amer. Scientist,* v. 63, p. 673–680.

Vallentyne, J. R., 1956, Fossil pigments, p. 83–105 *in* Allen, M. B., *ed., Biochemistry of photoreactive systems.* New York, Academic Press.

Van Balgooy, M. M. J., 1971, Plant geography of the Pacific as based on a census of phanerogam genera. *Blumea Sup.* v. 6, p. 1–222.

Van der Hammen, T., and E. Gonzalez, 1960, Upper Pleistocene and Holocene climate and vegetation of the Sabana de Bogota (Colombia, South America). *Leidse Geol. Meded.,* v. 25, p. 261–315.

Van Valen, L., 1973, A new evolutionary law. *Evol. Theory,* v. 1, p. 1–30.

Varley, G. C., G. R. Gradwell, and M. P. Hassell, 1973, *Insect Population Ecology.* Oxford, Blackwell.

Vaurie, C., 1951, Adaptive differences between two sympatric species of nuthatches (*Sitta*): *Proc. Int. Ornithol. Cong.,* v. 19, p. 163–166.

Verner, J., 1964, Evolution of polygamy in the long-billed marsh wren. *Evol.,* v. 18, p. 252–261.

Waddington, K. D., and L. R. Holden, 1979, Optimum foraging: On flower selection by bees. *Am. Natur.,* v. 114, p. 179–196.

Waggoner, P. E., and G. R. Stephens, 1970, Transition probabilities for a forest. *Nature,* v. 255, p. 1160–1161.

Walker, D., 1970, Direction and rate in some British postglacial hydroseres, p. 117–139 *in* Walker, D., and R. West, *eds., The Vegetational History of the British Isles.* Cambridge, Cambridge Univ. Press.

Wallace, A. R., 1876, *The Geographic Distribution of Animals.* 2 vols. London, Macmillan.

Wallace, A. R., 1878, *Tropical Nature and Other Essays.* London, Macmillan.

Walter, H., 1973, *Vegetation of the Earth.* New York, Springer–Verlag.

Ward, H. B., and G. C. Whipple, 1918, *Fresh-Water Biology.* New York, Wiley.

Wardle, P., 1968, Englemann spruce (*Picea engel-*

mannii Engel) at its upper limits on the Front Range, Colorado. *Ecol.*, v. 49, p. 483–495.

Warming, E., 1909, *Oecology of Plants: An Introduction to the Study of Plant Communities.* English Translation by Groom, P., and I. B. Balfour. London, Oxford.

Wassink, E. D., 1959, Efficiency of light energy conversion in plant growth. *Plant Physiol.*, v. 34, p. 356–361.

Watts, W. A., 1970, The full-glacial vegetation of northwestern Georgia. *Ecol.*, v. 51 (1), p. 17–33.

Watts, W. A., 1980, The late Quaternary vegetation history of the southeastern United States. *An. Rev. Ecol. Syst.*, v. 11, p. 387–409.

Watts, W. A., and T. C. Winter, 1976, Plant macrofossils from Kirchner Marsh, Minnesota—A paleoecological study. *Bull. Geol. Soc. Am.*, v. 77, p. 1339–1360.

Webb, T., and R. A. Bryson, 1971, Late and postglacial climatic change in the northern Midwest, USA: Quantitative estimates derived from fossil pollen spectra by multivariate statistical analysis. *Quat. Res.*, v. 2, p. 70–115.

Wegener, A., 1966, *The Origin of Continents and Oceans* (trans. by Biram, J.). New York, Dover.

Werner, E. E., and D. J. Hall, 1974, Optimal foraging and the size selection of prey by the bluegill sunfish (*Lepomis macrochirus*). *Ecol.*, v. 55, p. 1216–1232.

West, R. G., 1968, *Pleistocene, Geology and Biology.* New York, Wiley.

Westlake, D. F., 1963, Comparisons of plant productivity. *Biol. Rev.*, v. 38, p. 385–425.

Wetzel, R. G., 1975, *Limnology.* Philadelphia, Saunders.

White, T. C. R., 1969, An index to measure weather-induced stress of trees associated with outbreaks of Psyllids in Australia. *Ecol.*, v. 50, p. 905–909.

White, T. C. R., 1978, The importance of a relative shortage of food in animal ecology. *Oecologia*, v. 33, p. 71–86.

Whitehead, D. R., 1964, Fossil pine pollen and full glacial vegetation in southeastern North Carolina. *Ecol.*, v. 44, p. 403–406.

Whitehead, D. R., 1965, Palynology and Pleistocene phytogeography of unglaciated eastern North America, p. 417–432 *in* Wright, H. E., and D. C. Frey, *eds., The quaternary of the United States.* Princeton, Princeton Univ. Press.

Whitehead, D. R., 1967, Studies of full-glacial vegetation and climate in southeastern United States, p. 237–248 *in* Cushing, E. J., and H. E. Wright, *eds., Quarternary Paleoecology,* New Haven, Yale Univ. Press.

Whitham, T. G., 1977, Coevolution of foraging in *Bombus* and nectar dispensing in *Chilopsis:* A last dreg theory. *Science,* v. 197, p. 593–596.

Whitmore, T. C., 1984, *Tropical Rain Forests of the Far East.* London, Oxford Univ. Press.

Whittaker, R. H., 1953, A consideration of climax theory: The climax as a population and pattern. *Ecol. Monog.*, v. 23, p. 41–78.

Whittaker, R. H., 1956, Vegetation of the Great Smoky Mountains. *Ecol. Monog.*, v. 26, p. 1–80.

Whittaker, R. H., 1960, Vegetation of the Siskiyou Mountains, Oregon and California. *Ecol. Monog.*, v. 30, p. 279–338.

Whittaker, R. H., 1962*a*, Classification of natural communities. *Botan. Rev.*, v. 28, p. 1–239.

Whittaker, R. H., 1962*b*, Net production relations of shrubs in the Great Smoky Mountains. *Ecol.*, v. 44, p. 176–182.

Whittaker, R. H., 1965, Dominance and diversity in land plant communities. *Science,* v. 147, p. 250–260.

Whittaker, R. H., 1966, Forest dimensions and production in the Great Smoky Mountains. *Ecol.*, v. 47, p. 103–121.

Whittaker, R. H., 1975, *Communities and Ecosystems,* 2nd ed. New York, Macmillan.

Whittaker, R. H., 1977, Evolution of species diversity in land communities. *Evol. Biol.*, v. 10, p. 1–67.

Whittaker, R. H., F. H. Bormann, G. E. Likens, and T. G. Siccama, 1974, The Hubbard Brook Ecosystem Study: Forest biomass and production. *Ecol. Monog.*, v. 44, p. 233–254.

Whittaker, R. H., and P. P. Feeny, 1971, Allelochemics: Chemical interactions between species. *Science,* v. 171, p. 757–770.

Whittaker, R. H., and G. E. Likens, 1973, Carbon in the biota, p. 281–302 *in* Woodwell, G. M., and E. V. Pecan, *eds., Carbon and the Biosphere.* Wash., D.C., U.S. AEC.

Whittaker, R. H., and P. L. Marks, 1975, Methods of assessing terrestrial productivity, p. 55–118 *in* Lieth, H., and R. H. Whittaker, *eds., Primary Productivity of the Biosphere.* New York, Springer–Verlag.

Whittaker, R. H., and G. M. Woodwell, 1968, Dimension and production relations of trees and shrubs in the Brookhaven forest, New York. *Ecol.*, v. 56, p. 1–25.

Whittaker, R. H., and G. M. Woodwell, 1971,

Measurement of net primary production of forests, p. 159–175 *in* Duvigneaud, P., *ed., Ecology and Conservation 5*. Paris, UNESCO.

Wicklow, D. T., and G. C. Carroll, 1981, *The Fungal Community*. New York, Dekker.

Wiegert, R. G., and R. Mitchell, 1973, Ecology of Yellowstone thermal effluent systems: Intersect of blue-green algae, grazing flies (*Paracoenia, Ephydridae*) and water mites (*Partnuniella,* Hydrachnellae). *Hydrobiol.*, v. 41, p. 251–271.

Wiens, J. A., 1966, On group selection and Wynne–Edward's hypothesis. *Am. Sci.*, v. 54, p. 273–287.

Wiley, R. H., 1973, Territoriality and non-random mating in sage grouse, *Centrocercus urophasianus. An. Behav. Monog.*, v. 6, p. 85–169.

Williams, G. C., 1975, *Sex and Evolution*. Princeton, Princeton Univ. Press.

Wilson, D. S., 1980, *The Natural Selection of Populations and Communities*. Benjamin/Cummings, Menlo Park, Calif.

Wilson, E. O., 1959, Adaptive shift and dispersal in a tropical ant fauna. *Evol.*, v. 13(1), p. 122–144.

Wilson, E. O., 1969, The species equilibrium, p. 38–47 *in* Woodwell, G. E., and H. H. Smith, *eds., Diversity and Stability in Ecological Systems*. Brookhaven Symp. in Biol., v. 22, p. 38–47.

Wilson, E. O., 1971, *The Insect Societies*. Cambridge, Belknap/Harvard Univ. Press.

Wilson, E. O., 1975, *Sociobiology*. Cambridge, Harvard Univ. Press.

Wilson, E. O., and D. S. Simberloff, 1969, Experimental zoogeography of islands. Defaunation and monitoring techniques. *Ecol.*, v. 50, p. 267–278.

Wilson, E. O., and R. W. Taylor, 1967*a,* An estimate of the potential evolutionary increase in species density in the Polynesian ant fauna. *Evol.*, v. 21, p. 1–10.

Wilson, E. O., and R. W. Taylor, 1967*b,* The ants of Polynesia (Hymenoptera: Formicidae). *Pacific Insects Monog.*, v. 14. p 1–109.

Winterhalder, B. P., 1980, Canadian fur bearer cycles and Cree–Ojibwa hunting and trapping practices. *Am. Natur.*, v. 115(6), p. 870–879.

Wolda, H., 1978, Fluctuations in abundance of tropical insects. *Am. Natur.*, v. 112, p. 1017–1045.

Wolf, L. L., and F. R. Hainsworth, 1982, Economics of foraging strategies in sunbirds and hummingbirds, p. 223–259 *in* Aspey, W. P., and S. Lustick, *eds., Behavioral Energetics: Vertebrate Costs of Survival*. Columbus, Ohio State Univ. Press.

Wood, B. J., 1971, Development of integrated control programmes for pests of tropical perennial crops in Malaysia, p. 422–457 *in* Huffaker, C. B., *ed., AAAS Symposium on Biological Control, Boston*. New York, Plenum Press.

Woodruff, L. L., 1912, Observations on the origin and sequence of the protozoan fauna of hay infusions. *Jour. Exp. Zool.*, v. 12, p. 205–264.

Woodwell, G. M., and D. B. Botkin, 1970, Metabolism of terrestrial ecosystems by gas exchange techniques: The Brookhaven approach, p. 73–86 *in* Reichle, D. E., *ed., Analysis of Temperate Forest Ecosystems:* New York, Springer–Verlag.

Woodwell, G. M., and W. R. Dykeman, 1966, Respiration of a forest measured by CO_2 accumulation during temperature inversions. *Science,* v. 154, p. 1031–1034.

Woodwell, G. M., R. H. Whittaker, W. A. Reiners, G. E. Likens, C. C. Delwiche, and D. B. Botkin, 1978, The biota and the world carbon budget. *Science,* v. 199, p. 141–146.

Wynne–Edwards, V. C., 1962, *Animal Dispersion in Relation to Social Behavior*. New York, Hafner.

Yadara, P. S., and J. S. Singh, 1977, *Grassland Vegetation, its Structure, Function, Utilization and Management*. New Delhi, India (Indo-Am., Globe, Arizona), Today and Tomorrow's Printers.

Yoda, K., 1967, A preliminary survey of the forest vegetation of eastern Nepal. II. General description, structure and floristic composition of the sample plots chosen from different vegetation zones. *J Col. Arts and Sci., Chiba Univ., Nat. Sci. Ser.*, v. 5, p. 99–140.

Zajic, J. E., 1969, *Microbial Biogeochemistry*. New York/London, Academic Press.

Zaret, R. E., 1980, The Animal and Its Viscous Environment, p. 3–9 *in* Kerfoot, W. C., *ed., Evolution and Ecology of Zooplankton Communities*. Hanover, N. H., Univ. Press of New Eng.

Zaret, T. M., 1980, *Predation and Freshwater Communities*. New Haven, Yale Univ. Press.

Zaret, T. M., and R. T. Paine, 1973, Species introduction in a tropical lake. *Science,* v. 182, p. 449–455.

Zaret, T. M., and J. S. Suffern, 1976, Vertical migration in zooplankton as a predator avoidance mechanism. *Limnology and Oceanography,* v. 21, p. 804–813.

GLOSSARY

adiabatic lapse rate: cooling of rising air without external source of energy for expansion; 10°C (dry), 6°C (wet) per kilometer
albedo: reflectivity of the earth surface to sunlight
alfisol: gray-brown podzolic soil of temperate latitudes
allelochemics: study of use of chemical agents by organisms to influence other organisms, most particularly as defenses or as lures
allelopathy: use of secondary compounds by plants for herbivore defense
allometric growth: principle that growth in total size involves disproportionate growth in different body parts
allometry: relative changes of parts as a structure is enlarged
allopatric: living in regions that are clearly separated
alluvium: water-borne deposit
alpha diversity: within-habitat
altricial: has naked or helpless offspring
anemophily: wind pollinated
annual: plant completing life-history from germination to death in one year
ant-guard: symbiosis of plants providing resources for ants and ants providing defense for plants
apomixis: asexual reproduction, especially when resembling sexual reproduction but in which the egg and sperm do not fuse
arena: alternative term for lek
aridisol: desert soil; A-C soil
artiodactyl: 2-toed ungulate (e.g., cow)
assimilation: for animals, absorption of food energy from the gut; for plants the uptake of carbon dioxide
assimilation efficiency: ratio of food absorbed to food ingested
association: plant community unit; variously, often subjectively, defined
autecology: ecology of single species populations
benthos: organisms living on the bottom of sea or lake (adjective: benthic)
beta diversity: between-habitat
biennial: plant completing life-history in two years, typically with the main reproductive episode in the second year
biocoenose: (biocoenosis) animal and plant communities combined to one unit, roughly equivalent to "ecosystem," though a version more closely comparable to ecosystem is biogeocoenosis, as widely used in Soviet literature
biological accommodation: fitting life-styles more to the presence of other organisms than to physical factors
biomass: mass of living organisms, originally expressed as a mass-density (e.g., grams per square meter) but now sometimes expressed as calories per unit area
biome: ecosystem of a large geographic area in which plants are of one formation and for which climate sets the limits
biotype: genotype reproduced asexually
Blytt-Sernander: sequence of Holocene climatic events in northern Europe
BOD: biological oxygen demand
boundary layer: thin layer alongside leaf or other object in which air or other fluid is motionless
brown earth: soil of temperate forest; alfisol

caliche: carbonate deposit at the line of maximum water penetration in mollisol or chernozem, or term may refer to the carbonate layered soil itself
CAM: crassulacean acid metabolism, use of C4 photosynthesis and storage of carbon dioxide collected at night as C4 acids for photosynthesis by day behind closed stomates

catena: related set of soil series differentiated by drainage or topography
character displacement: selection for characters that avoid interspecific competition
character species: indicator of qualitatively defined community type
chernozem: black earth of prairies through which percolation is incomplete; mollisol
climax: plant community resulting after an ecological succession and in which further change is slow
clone: asexually produced offspring of a common ancestor
cohort: standard-size population used in construction of life tables
colluvium: deposit of material slowly worked down a slope
competitive exclusion: principle that strongly competing species cannot coexist indefinitely
constant species: one present in at least 80% of communities sampled (Uppsala school of phytosociology)
consumer: herbivores and carnivores that consume energy originally transformed into producer biomass
continuum: community unit describing changing species composition along a gradient
croppers: browsers or grazers but not animals that eat seeds or fruits that must be sought out in a manner analogous to hunting
cropping principle: predation that keeps prey populations low (works to allow increased diversity)
cyclomorphosis: changing of shape with succeeding generations

dark reaction: synthesis of reduced carbon compounds following light energy transformation in photosynthesis
decomposer: organism that receives energy by oxidizing dead organic matter (most are bacteria or fungi)
deep scattering layer: plankton at depths in the oceans by day as recorded by sonar
deme: panmictic population with general and random mating
dendrochronology: dating by tree rings
density: number of individuals per unit area
detritivore: consumer of detritus
detritus: fragmented dead plant matter
diapause: process of surviving hostile season with resting stage of reduced metabolic rate
dominant: (ecological dominant) more abundant than allowed for in a random assortment, or competitively superior; sometimes also used to refer merely to the most common species

ecdysis: molting of exoskeleton during growth of arthropods
ecocline: gradient of changing species composition
ecological time: duration of interactive process among existing species
ecological (Lindeman) efficiency: ratio of gross production between trophic levels
ecological equivalents: organisms of comparable function or niche
ecological succession: sequential appearance of species or communities
ecotone: region of rapidly changing species composition at an environmental disjunction
ecotope: term intended to define niche or niche space as a function of habitat
ectocrine: chemical secreted into the environment that influences other organisms
ectothermy: use of environmental heat (principally the sun) to regulate body temperature
Eltonian pyramid: the pyramid of numbers
eluvial horizon: soil layer from which matter has been removed
endorheic: interior drainage (used of closed basin lakes)
endothermy: use of metabolic heat to regulate body temperature
energy flow: concept that food transfer between prey and predator is calorie transfer with energy release, hence a unidirectional flow
enthalpy: state of order in a thermodynamic system
entomophilous: insect pollinated
entropy: physicist's term describing state of maximum disorder or randomness toward which natural systems spontaneously move
epideictic display: sexual display postulated to serve the purpose of a density measuring device (a view not generally held by ecologists)
epilimnion: warm surface water of a lake
equilibrium species: species adapted to persist; *K*-strategist
equitability: component of relative abundance in measures of diversity
eukaryote: has nucleus in cells
euphotic: zone of surface water (about 100 m in sea) where light sufficient for photosynthesis penetrates
eury- : broadly tolerant, as in eurythermal or euryhaline for organisms tolerating wide ranges in temperature or salt

eusociality: insect social systems with divisions of labor among morphs many of which do not reproduce themselves
eustacy: changing sea level resulting from changes in ice or ocean volumes
eutrophic: fertile
exclusion: see competitive exclusion
exploitation efficiency: ratio of food ingested to net production of food species

fecundity: rate at which female produces offspring
fertility: (demography studies) population reproductive rate
floristic: refers to species composition of vegetation
food chain: predation series linking animals to ultimate plant food
food cycle: old concept now replaced with concepts of nutrient cycles (nutrients being the "food") and energy flow
food web: concept of intersecting food chains
formation: vegetation of a large climatic region recognized by a characteristic shape or life-form
frass: solid insect excrement
frequency: proportion (%) of samples in which a species occurs
frugivore: seed or fruit eater
fugitive species: opportunist or *r*-strategist adapted to disperse away from competitors
functional response: high prey density leads to changes in predator behavior
fundamental niche: niche in the absence of interspecific competition

gamma diversity: regional over geographic clines
GAS: general adaptation syndrome—hormonal response to stress in vertebrates
graben: basin formed by geological faulting
gross production: used in ecological energetics, term means energy input, but used differently in related disciplines
guild: a set of coexisting species that share a common resource

hemimetabolous: insect life history with no distinctive larval stage
hemolymph: body fluid of arthropods
heterogonic: life cycle in which sexual and asexual reproductive epidodes alternate
Holocene: time since end of last ice age (about 10,000 years)
holometabolous: insect life histories with separate larval and pupal stages
homeostasis: maintenance of constancy or near-uniformity in organism, community or other entity
homeotherm: animal that maintains a core temperature by release of metabolic heat
home range: wandering or feeding area of an animal
hydrarch: succession on wetlands
hydrarch succession: sequential colonization of marshy habitats
hypolimnion: cold bottom water of a lake

illuvial horizon: soil layer to which matter has been added from above
inclusive fitness: includes genes carried by offspring and genes carried by relatives into the next generation
ingestion: taking of food into the gut
instar: insect form between molts
interspecific: between members of different species
intraspecific: among members of a species population
intrinsic rate of increase: exponential growth rate of a population with a stable age distribution
inversion: stable system of warm air floating over cold air
isohyet: line on a map connecting places of equal precipitation
isopleth: line of constant precipitation–evaporation ratio
isostacy: changing sea level due to crustal movements under ice masses
iteroparity: breeding repeatedly

karst: lake occupying a solution basin
kin selection: selection favoring individuals by preserving genes carried in relatives
kranze syndrome: leaf morphology required for C4 photosynthesis in which chloroplasts are in cells below the epidermis
***K*-strategy:** life-style in which fecundity is reduced to divert resources to persistence

landnam: hypothetical land clearing by early European agriculturalists and recorded by elm pollen decline
langley: (ly) flux of solar energy measured as gram-calories per unit area in unit time
large young strategy: concentrates resources into a few young, either by making young large or through parental care
laterite: tropical red earth, oxisol or latosol

lek: communal breeding ground where males display and females come to be mated
light reaction: stage of photosynthesis in which light energy is transformed
limiting factors: concept that distribution or abundance may be set by a critically limiting resource
Lincoln index: mark and recapture index
loess: wind-blown deposits of characteristic grain size
luxury consumption: uptake of nutrients by algae in excess of immediate needs

maar: volcanic explosion crater
margallitic soil: tropical mollisol important to tropical agriculture
meromictic lake: permanently stratified lake
mesic: environment midway between extremes; especially habitats supporting dense, diverse plant communities
mesotrophic: moderately fertile
metalimnion: volume of lake water across which a thermocline runs
microcosm: once used as synonymous with "ecosystem," now widely taken to mean a small experimental ecosystem
minimum area: sum of area of quadrats to be sampled to include all but chance species of a habitat
mixolimnion: upper water of a chemically stratified lake
mollisol: black and brown soils of prairies through which percolation is incomplete
monimolimnion: bottom water of a chemically stratified lake
monolayer: postulated leaf arrangement for shade trees in which leaf shape and pattern achieves maximum light occlusion in a narrow vertical distance
mor: acid humus under podzols
mortality: rate of death
mull: humus type collecting in temperate soils
multilayer: postulated leaf arrangement for trees of bright sunlight that allows deep diffusion of light
mutualism: living together of species in which both partners gain and both incur cost
mycorrhizae: fungal root symbionts

nannoplankton: the very smallest plankton that passes through most collecting nets, includes protozoa as well as small phytoplankters
natality: rate of offspring production
net production: energy input less respiration (but has other meanings in related disciplines)
niche: role, function, or place of organism
niche-packing: conceptual process of community building with final species number set by maximum possible interspecific competition
niche space: environmental parameters defining a niche, or the resource flux required for an individual to survive and reproduce
nidicolous: young fed in nest
nidifugous: young leave nest as soon as hatched
numerical response: high prey density results in high reproduction of predators

oligotrophic: infertile
opportunist: colonist or *r*-strategist
optimal defense: theory that secondary compounds are allocated to defense in a way that minimizes cost and maximizes inclusive fitness
optimal foraging: theory that animals should behave so as to maximize energy intake for the time spent foraging
orogeny: mountain building
oxisol: tropical red soil depleted of silicate minerals; lateritic soil

paleoecology: use of fossils to test ecological hypotheses (but see alternative meanings in text)
palynology: study of pollen and spores
paradox: statement apparently conflicting with common sense but actually well founded or reasonable
parapatric: living in adjacent regions
parthenogenesis: reproduction with unfertilized eggs
pedalfer: soil through which water percolates to a drainage system
pedocal: soil through which water does not percolate completely
pedon: smallest describable patch of a soil series
peneplain: flat dissected surface from erosion by an ancient river system
perennial: plant living many years
periphyton: organisms attached to aquatic plants
perissodactyl: 1-toed ungulate (e.g., horse)
permafrost: permanently frozen ground, only a thin surface layer of which thaws in summer
phenology: properly the study of phenomena with constrained timing but now loosely used in ecology to refer to the process of constraining biological process like flowering into particular times of year or day as part of a system of coevolution
pheromone: chemical used as a social cue
phoresy: dispersal of communal group by host organism

photorespiration: respiration of carbon dioxide in light resulting from properties of the enzyme RuBP carboxyllase
phycobilisomes: colored pigments of algae that are adaptations to increase light absorption in dim light
physiognomic: refers to shape, structure, or form
phytolith: silica concretion within a plant cell
phytoplankter: individual plant of the plankton
phytoplankton: plant plankton
phytosociology: study of plant communities
plankter: organism of the plankton
plankton: drifting organisms of open water, mostly small to microscopic
pleuston: organisms suspended from water surface
pluvial: time of increased rainfall
podzol: soil with bleached A horizon, typical of boreal forest; spodosol
poikilotherm: animal that does not rely on metabolic heat to maintain a core temperature
point diversity: in very small samples
pollen influx: pollen grains deposited per unit area in unit time
pollen spectrum: histogram of pollen percentages
precocial: has fully formed young able to feed themselves
primary succession: ecological succession on ground that has not hitherto supported vegetation
proclimax: community maintained by repeated disturbance
producer: primary energy transformer of an ecosystem, most are green plants but chemosynthetic bacteria are producers in a few ecosystems
production: used variously in ecology as energy input, energy stored as biomass, or biomass sequestered for reproduction
production efficiency: ratio of growth and reproduction to assimilation
profligate reproduction: relies on large numbers of very small eggs, seeds, etc.
prokaryote: no nucleus in cells; principally bacteria and some algae
propagule: unit of propagation, whether live young, egg, seed, or asexually cloned individual or dispersal unit
protandrous flower: flower that changes from male to female with age
prudential reproduction: relies on large young or parental care
pubescent: hairy
pyramid of numbers: observation that food chains run in parallel with animals at any level having comparable size and relative abundance, the animals high on the food chains being both large and rare

quadrat: quantitative sample; particularly number per unit area
qualitative defense: allelopathy employing specific toxins (e.g., alkaloids)
quantitative defense: allelopathy employing universal toxins (e.g., tannin)

***r* and *K* selection:** selection within a population for traits tending towards *r* or *K* strategies
R-C-S continuum: postulated replacement series of plants of ruderal, competitive, and stress-tolerant strategies
reactive distance: effective targetting range of a predator
realized niche: niche in the presence of interspecific competition
realm: (biogeographic) latitudinal expanse in which organisms have comparable adaptations
red queen: hypothesis that evolution results as selection tracks persistent environmental change
region: (biogeographic) continental expanse the organisms of which are taxonomically related
regolith: top layer of unconsolidated material covering crustal rocks
relaxation: in island biogeography means loss of species following decreased access or reduction in area
respiration: energy represented by carbon dioxide given off from oxidation of reduced carbon compounds to do work
richness: total species number
***r*-strategy:** colonist life-style of high fecundity, low persistence
ruderal: plant with colonizer strategy

saprobe: synonym for decomposer—hence "saprobe chain"
saprophage: synonym for decomposer
secondary compounds: plant physiological term for chemicals with no obvious physiological function and which ecologists generally consider to have coercive functions on other organisms
secondary plant metabolites: toxic compounds manufactured by plants, frequently tending to be autotoxic
secondary succession: ecological succession on ground from which previous vegetation was removed

seed bank: accumulation in soil of seeds with delayed germination
seedling bank: accumulation of seedlings maintaining themselves in dim light on a forest floor
semelparity: breeding only once in a lifetime
seral stage: community identifiable as part of a successional sequence
sere: a particular successional sequence
sesquioxides: generic term for a mixture of iron and aluminum oxides
sexual dimorphism: physical differences between sexes
sigmoid: growth or response curve that is s-shaped
signal receiver: term for dupe in a mimicry system
small egg strategy: relies on many small young
sociation: plant community unit defined by measured minimum area
sociobiology: the study of how selection for individual fitness leads to group or social phenomena
soil horizon: soil layer resulting from weathering processes
soil series: local soil type dependent on parent material and drainage
solar constant: flux of solar energy reaching the upper atmosphere (about 2 langleys)
spodosol: podzol
standing crop: biomass present at time of sampling
steno- : narrowly restricted, as in stenothermal or stenohaline for organisms with narrow tolerances to temperature or salt
stomate: gas-exchange aperture through plant epidermis that can be controlled by turgor in surrounding cells
stress-tolerator: plant adapted to allocate resources between competition and physical stress resulting from competition
stromatolite: carbonate structure made by marine algae
structured deme: system of isolated populations without general and continuous mixing of genes throughout the population
subduction: descent of leading edge of a continental plate at line of collision
succession: allogenic = responding to habitat changes imposed by outside physical forces; autogenic = from biological interactions within the ecosystem; primary = on land never before vegetated; secondary = on old habitat that has been devegetated
Suess effect: dilution of atmospheric ^{14}C by ^{12}C from burning fossil fuels
survivorship: cohort survivors of stipulated age
symbiosis: living together of species in which both partners benefit
sympatric: living in the same region
synecology: ecology of communities, study of community properties

taxon cycle: selection series on islands as *K*-strategists evolve from colonists, followed by *r*-selection among island endemics
teleology: the doctrine that developments are due to the purpose that is served by them
tetrapod: 4-legged animal
thermocline: middle depths of a stratified lake where temperature changes rapidly
thermokarst: lake occupying a melted depression in permafrost
till: unsorted mineral deposit left by a retreating glacier (also called boulder clay)
time-stability: hypothesis that places with stable and permanent physical environments collect many species because extinction is unlikely
tramp species: extreme opportunists, particularly species dispersed by human commerce
transpiration: transfer of water from soil to atmosphere by plants, the motive force of which is evaporation from the leaves
trichome: fine barbed hair, a mechanical defense
trophic level: common feeding at the same link on food chains—level in a pyramid of numbers
trophogenic: upper zone of lake or ocean where light allows photosynthesis

VAM: vesicular arbuscular mycorrhizae, fungal symbionts important to nutrient uptake
varve: couplet of bands in sediment deposited in a single year

xerarch succession: sequential colonization of dry or desert habitats

yield: may mean net production or any harvestable portion of net production according to context

zooplankter: an animal of the zooplankton
zooplankton: animal plankton
zooxanthellae: filamentous algae symbiotic with reef-building corals

PHOTO CREDITS

Chapter 1
Opener: Ylla/Photo Researchers. Fig. 1.1: National Audubon Society/Photo Researchers. Fig. 1.2: NOAA. Fig. 1.4: Paul A. Colinvaux. Fig. 1.3: Photos by Jack Dermid.

Chapter 2
Opener: D. Bancroft/National Film Board of Canada, Phototeque. Fig. 2.3: G. Gely/National Film Board of Canada, Phototeque. Fig. 2.4: Fred Breunner. Fig. 2.5: Fred Breunner. Fig. 2.13: Richard Ellis/Photo Researchers.

Chapter 3
Opener: Lawrence Pringle/Photo Researchers. Fig. 3.11: Jack Dermid. Fig. 3.12: Ho-Yih Liu, Ohio State University. Fig. 3.16: Paul A. Colinvaux. Fig. 3.18: J.S. Horn, Department of Biology, Princeton University. Fig. 3.21: Paul A. Colinvaux. Fig. 3.22: Paul A. Colinvaux. Fig. 3.26: Robert C. Hermes/Photo Researchers. Fig. 3.27: Paul A. Colinvaux.

Chapter 4
Opener: Jim Yoakum. Fig. 4.2: Runk Schoenberger/Grant Heilman. Fig. 4.3: M. Cognac/National Film Board of Canada, Phototeque. Fig. 4.4: Egon Bork/National Film Board of Canada, Phototeque. Fig. 4.11: Jack Dermid.

Chapter 5
Opener: Bureau of Sport Fisheries & Wildlife. Fig. 5.6: Arthur Ambler/National Audubon Society/Photo Researchers. Fig. 5.7: (a) Verna R. Johnston/Photo Researchers, (b) Photo Researchers. Fig. 5.9: (a) Verna R. Johnston/Photo Researchers, (b) Robert Lamb/Photo Researchers, (c) Jim Yoakum. Fig. 5.10: Paul A. Colinvaux. Fig. 5.11: Peter Zwerger/Forest Research Institute. Fig. 5.16: Jen & Des Bartlett/Photo Researchers. Fig. 5.17: Ron Austing/Photo Researchers. Fig. 5.18: (left) Len Rue, Jr./Photo Researchers, (right) A.W. Ambler/National Audubon Society/Photo Researchers. Fig. 5.19: Leonard Lee Rue/Photo Researchers. Fig. 5.20: Paul A. Colinvaux. Fig. 5.21: Tom McHugh/Photo Researchers. Fig. 5.22: Stephen Dalton/Photo Researchers. Fig. 5.25: Julie Bartlett/Photo Researchers. Fig. 5.28: Leonard Lee Rue/Photo Researchers.

Chapter 6
Opener: Walter Dawn/Photo Researchers.

Chapter 7
Opener: Whittlesey Birnie/Leo De Wys. Fig. 7.5: Dr. Eric R. Pianka, Department of Zoology, University of Texas, Austin. Fig. 7.10: Dr. Peter Grant, Department of Zoology, University of Michigan. Fig. 7.14: Paul A. Colinvaux.

Chapter 8
Opener: F.A.O. Fig. 8.4: Haverschmidt/National Audubon Society/Photo Researchers. Fig. 8.7: Eric Hosking/National Audubon Society/Photo Researchers.

Chapter 9
Opener: Pietro Francesco Mele/Photo Researchers. Fig. 9.1: Steve McCutcheon/Alaska Pictoral Service. Fig. 9.2: Jack Dermid. Fig. 9.8: Robert Orr/Photophile. Fig. 9.13: L. David Mech.

Chapter 10
Opener: Hubertus Kanus/Photo Research. Fig. 10.10: F.E. Skinner. Fig. 10.22: Jack Clark, Cooperative Extension, University of California. Fig. 10.23: Department of Lands, Queensland, Australia. Fig. 10.24: Paul A. Colinvaux. Fig. 10.25: Serge Jauvin/National Film Board of Canada, Phototeque. Fig. 10.28: Stephen Dalton/National Audubon Society/Photo Researchers. Fig. 10.29: Nancy Tucker/Photo Researchers.

Chapter 11
Opener: Gordon Smith/Photo Researchers. Fig. 11.2: Paul A. Colinvaux. Fig. 11.8: George Harrison/Grant Heilman. Fig. 11.10: Paul A. Colinvaux. Fig. 11.12: D. Overcash/Bruce Coleman. Fig. 11.13: Paul A. Colinvaux.

Chapter 12
Opener: Leonard Lee Rue/Photo Researchers. Fig. 12.5: Paul A. Colinvaux. Fig. 12.6: Bob & Ira Spring. Fig. 12.11: Toni Angermayer/Photo Researchers. Fig. 12.12: Leonard Lee Rue/Photo Researchers. Fig. 12.13: Tom McHugh/Photo Researchers. Fig. 12.14: S. Nagendra/Photo Researchers. Fig. 12.15: Department of Photography, Ohio State University. Fig. 12.21: National Marine Fisheries Service, Honolulu Laboratory. Fig. 12.22: Fred Breunner.

Chapter 13
Opener: Tom McHugh/Photo Researchers.

Chapter 14
Opener: NOAA. Fig. 14.5: Steve McCutcheon/Alaska Pictoral Service. Fig. 14.6: NASA. Fig. 14.12: Paul A. Colinvaux. Fig. 14.19: Paul A. Colinvaux. Fig. 14.20: U.S. Forest Service. Fig. 14.21: Jack Dermid. Fig. 14.22: R.H. Noailles. Fig. 14.23: Paul A. Colinvaux. Fig. 14.24: Tom McHugh/Photo Researchers. Fig. 14.25: L. & D. Klein/Photo Researchers. Fig. 14.26: Mark N. Boultan/National Audubon Society/Photo Researchers. Fig. 14.27: Howard Sochurgk/Woodfin Camp. Fig. 14.28: Russ Kinne/Photo Researchers. Fig. 14.29: French Government Tourist Office. Fig. 14.32: Paul A. Colinvaux.

Chapter 15
Opener: Grant M. Haist/Photo Researchers.

Chapter 16
Opener: Andy Bernhaut/Photo Researchers. Fig. 16.8: From *A Tropical Rain Forest* by Howard T. Odum, Book 3, 1970, Courtesy AEC. Fig. 16.12: Paul A. Colinvaux. Fig. 16.19: Robert Hermes/National Audubon Society/Photo Researchers.

Chapter 17
Opener: David E. Reichle/Oak Ridge National Laboratory. Fig. 17.6: G.E. Likens, Cornell University.

Chapter 18
Opener: George Daniell/Photo Researchers.

Chapter 19
Opener: Fritz Henle/Photo Researchers. Fig. 19.1: Josephus Daniels/Photo Researchers. Fig. 19.9: Paul A. Colinvaux. Fig. 19.16: Tracy Borland.

Chapter 20
Opener: Nedra Westwater/FPG. 20.8: Paul A. Colinvaux.

Chapter 21
Opener: Keith Gunnar/Photo Researchers. Fig. 21.2: Paul A. Colinvaux. Fig. 21.3: Paul A. Colinvaux. Fig. 21.4: Department of Photography, Ohio State University. Fig. 21.5: Department of Photography, Ohio State University. Fig. 21.6: K.K. Everett, Ohio State University, Institute of Polar Studies. Fig. 21.7: Paul A. Colinvaux. Fig. 21.21: Rudi Strickler.

Chapter 22
Opener: Courtesy Scott W. Rogers. Fig. 22.2: Paul A. Colinvaux. Fig. 22.4: Paul A. Colinvaux. Fig. 22.17: Paul A. Colinvaux. Fig. 22.26: Dr. G.R. Cope, Department of Geological Sciences, University of Birmingham, England.

Chapter 23
Opener: Russ Kinne/Photo Researchers. Fig. 23.8: Paul A. Colinvaux. Fig. 23.9: Paul A. Colinvaux. Fig. 23.16: Paul A. Colinvaux. Fig. 23.17: Frank J. Staub/The Picture Cube. Fig. 23.18: Bob & Ira Spring. Fig. 23.19: Paul A. Colinvaux.

Chapter 24
Opener: Georg Gerster/Photo Researchers. Fig. 24.3: George Harrison/Grant Heilman.

Chapter 25
Opener: Karl Maslowski/Photo Researchers. Fig. 25.1: Toni Angermayer/Photo Researchers. Fig. 25.2: Len Lee Rue/Photo Researchers. Fig. 25.3: Lawrence E. Gilbert, Department of Zoology, University of Texas, Austin. Fig. 25.4: Grant Heilman. Fig. 25.6: Thomas S. Elias/Rancho Santa Ana Botanic Garden. Fig. 25.7: Courtesy Carl W. Rettenmeyer, University of Connecticut. Fig. 25.8: B. Miller/Biological Photo Service. Fig. 25.9: Z. Leszczybki/Animals, Animals.

Chapter 26
Opener: H.W. Kitchen/Photo Researchers. Fig. 26.1: Bob & Ira Spring. Fig. 26.7: Russ Kinne/Photo Researchers.

INDEX